5th International Conference on Transportation and Traffic Engineering (ICTTE 2016)

MATEC Web of Conferences Volume 81 (2016)

Lucerne, Switzerland
6 - 10 July 2016

Editors:

Monteiro Figueia
Zhihua Guo

ISBN: 978-1-5108-3168-1

Printed from e-media with permission by:

Curran Associates, Inc.
57 Morehouse Lane
Red Hook, NY 12571

Some format issues inherent in the e-media version may also appear in this print version.

Printed by Curran Associates, Inc. (2016)

For additional information, please contact EDP Sciences – Web of Conferences at the address below.

EDP Sciences – Web of Conferences
17, Avenue du Hoggar
Parc d'Activité de Courtabœuf
BP 112
F-91944 Les Ulis Cedex A
France

Phone: +33 (0) 1 69 18 75 75
Fax: +33 (0) 1 69 28 84 91

contact-edps@webofconferences.org

Additional copies of this publication are available from:

Curran Associates, Inc.
57 Morehouse Lane
Red Hook, NY 12571 USA
Phone: 845-758-0400
Fax: 845-758-2633
Email: curran@proceedings.com
Web: www.proceedings.com

TABLE OF CONTENTS

PREFACE

It is a pleasure to welcome all of you to the 2016 5th International Conference on Transportation and Traffic Engineering (ICTTE 2016). ICTTE was founded in 2012, and has been successfully held in Hong Kong, Maldives, Portugal, and Spain during the past four years. This year, ICTTE 2016 will be held in University of Lucerne, Lucerne, Switzerland. It is a conference dedicated to issues related to transportation planning and system optimization, transportation security, transportation planning, controlling methods for supply chains, quantitative and qualitative performance measures, Managing logistics service quality, etc.

The major goal and feature of the conference is to bring academic scientists, engineers, industry researchers together to exchange and share their experiences and research findings and results, and discuss the practical challenges encountered and the solutions adopted. We hope you will have a delightful and rewarding experience, and use this occasion to meet old friends and make many new ones in related research fields.

The proceedings of ICTTE 2016 contain 71 papers selected from the conference which are presented orally at the conference or via poster sessions, and provide up-to-date, comprehensive and worldwide state-of-the art knowledge in this field. Each contributed paper was rigorously peer-reviewed by international reviewers who were drawn from the organizing and advisory committee members as well as other external reviewers invited from all over the world. The proceedings are divided into 8 chapters based on the different topics of the papers, which include: transportation planning and system optimization, transportation security, traffic control, transportation system modeling and forecasting, transport infrastructure, supply chains, construction project modeling and prediction, analysis and improvement of vehicle

We would like to thank all the organization staff, the members of the program committees and reviewers. They have worked very hard in reviewing papers and making valuable suggestions for the authors to improve their work. We also would like to express our gratitude to the external reviewers, for providing extra help in the review process, and the authors for contributing their research results to the conference.

Hopefully, we'll be seeing all of you again in ICTTE next year.

Prof. Roberto Montemanni
Dalle Molle Institue for Artificaial Intelligence, Switzerland
Prof. Monteiro Figueira
University Lusofona, Portugal

June 28, 2016.

Congestion Characteristics of Interrupted Flow for Urban Roads with Heterogeneous Traffic Structure

Hemant Kumar Sharma[1,2] and B. L. Swami[1]

[1] *Malviya National Institute of Technology, Jaipur- 302 017, India*
[2] *Rajasthan Urban Infrastructure Finance Development Corporation, Jaipur- 302 017, India*

Abstract. Traffic congestion is a severe trepidation to transportation engineers for its unrestrained growth and consequential losses. This paper presents congestion models and speed-flow analysis for urban roads with interrupted flow comprising of heterogeneous structure of traffic. Models have been developed for heterogeneous traffic under constraints of roadway geometry, vehicle characteristics, driving behaviour and traffic controls. The growth of congestion with flow in unsaturated and oversaturated states of flow has been analysed and quantified. The congestion model developed in this paper shows that there exist different regions of congestion- flow behaviour that can be characterized by different rate of change of congestion and the severity of congestion becomes tremendous for oversaturated flows. Different levels of service have been proposed to define operating conditions using more realistic parameter 'congestion'.

1 Introduction

Traffic congestion is one of the main liabilities of present time [1]. It occurs when the cost of travel is increased by the presence of other vehicles, either because speed falls or because greater attention is required to drive safely [2]. It is a price that people pay for the different benefits derived from agglomeration of population and economic activity. Since the land is scarce and road capacity is costly to construct, it would be uneconomical to spend heavily to increase capacity to the extent that travel become congestion-free. Since demand for travel depends on the cost, improvements in travel conditions induce people to take more trips, and it would probably be impossible to eliminate congestion [3]. Traffic engineers often compare traffic with fluid, assuming that certain volume must flow through the road system. But urban traffic may be more comparable to a gas that expands to fill available space [4]. Congestion can be defined as traffic condition caused by a downstream bottleneck, or increase in travel time compared to low traffic conditions. The selection of congestion performance indicators and data collection techniques must be carefully coordinated with the definition adopted to capture the true nature of congestion [5].

Transportation researchers have long struggled to find satisfactory ways to describe and analyse congestion, as evident from the large number of often competing approaches and models that have been developed. Early researchers developed models based on fluid dynamics, but congestion, unlike fluid flow, is not a purely physical phenomenon but rather the result of people's trip-making decisions and minute-by-minute driving behaviour. One should therefore expect the quantitative and qualitative, characteristics of congestion to vary with automobile and road design, rules of the road, pace of life, and other factors. Efforts have been made by many researchers to understand and quantify congestion, however, it is found that most of the works have been concentrated on the area wise quantification of congestion i.e., Congestion Severity Index [6], Roadway Congestion Index [1], Percentage of Congested Freeway [1], K-Factor [1], Lane Mile Duration Index, Freeway Congestion Index [7].

Models calibrated and validated in a developed country may not fit well a developing country for the difference in traffic conditions [2].

1.1 Quantification of congestion

The level of congestion varies with roadway, traffic control conditions, and operating traffic volume. While majority of studies regarding congestion are based on highways, quantification of congestion on urban roads has been attempted by several researchers [8-11]. Since, the loss in freedom of movement can be measured as the area under the speed-flow envelope (between the free-flow operation and the actual operating condition), the congestion can be quantified as the percentage loss in freedom of movement under prevailing roadway, traffic, and control conditions [12].

Fig. 1 (a) and 1 (b) represents quantification of congestion proposed by [12]. Although, the worst operating condition represented by point D (*Figure 1a*) normally be defined as 100% congestion; the behaviour of midblock traffic is seldom observed in the range BD (the lower portion of speed-flow envelope). Therefore, point B (i.e., the capacity) should be taken as traffic condition representing 100% congestion. With the capacity as 100% congested operation, the generality of the philosophy of quantification of congestion remains valid, and when the lower envelope or a part of it is observed, it will be representing forced flow with the quantified congestion more than 100%. Therefore, in that case, a point near the capacity in the stable flow zone may be taken to represent 100% congestion. In *Figure 1(b)* Point A_1 represents the operation with free-flow speed v_f, whereas point B_1 represents a 100% congested operation in the stable flow zone with a realized speed v_n and volume level Q_1. Therefore, 100% congested operation represented by point B_1 indicates an amount of loss in freedom of movement equivalent to an area $A_1B_1G_1$ under the envelope. Similarly, the amount of loss in freedom of movement for an operating point P_1, represented by a flow level q_a', and realized speed v_u' under the same operating conditions, can be expressed as the area $A_1P_1D_1$. Consequently, congestion at operating point P_1 can be expressed as (Area $A_1P_1D_1$/ Area $A_1B_1G_1$)*100 (Maitra *et al.*, 1999).

1.2 Unsaturated, saturated and oversaturated flows

In Fig. 1 (c) region A represents under- saturated conditions with arrival flows (q_a) below capacity ($q_a \le Q$). Travel speed at a given flow rate (v_u) is between v_f and v_n ($v_f \ge v_u \ge v_n$) where v_f is the free-flow speed and v_n is the speed at capacity. With increasing flow rate in Region A, speeds are reduced below the free-flow speed due to traffic delays resulting from interactions between vehicles. Region B represents the forced (congested) flow conditions with flow rates reduced below capacity ($q < Q$) which are associated with further reduced speeds ($v < v_n$) as observed at a reference point along the road. In this region, flow rates (q) are reduced flow rates due to forced flow conditions, not demand flow rates (q_a). Region C represents oversaturated conditions, i.e. arrival (demand) flow rates above capacity ($q_a > Q$) which are associated with reduced travel speeds ($v < v_n$) observed by travel through the total section, e.g. by an instrumented car. In this case, the flow represents the demand flow rate which can exceed the capacity value as measured at a point upstream of the queuing section [13-14].

1.2.1 Interrupted traffic flow

Flows may be further classified as uninterrupted flows and interrupted flows. In uninterrupted flow, traffic flow condition results from interactions among vehicles in traffic stream between vehicles and the geometric and environmental characteristics of the roadway. These flows do not have external elements such as traffic signals which might interrupt the traffic flow. However, interrupted-flows have controlled and uncontrolled access points that can interrupt the traffic flow. These access points include traffic signals, stop signs, yield signs and other types of control that stop traffic periodically or slow it significantly, irrespective of amount of traffic [15].

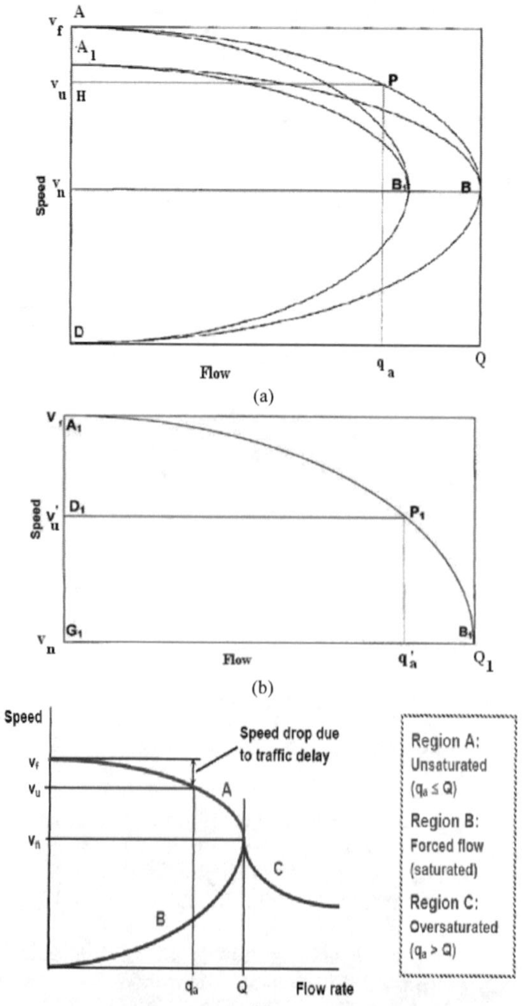

Figure 1. Representation of quantification of congestion (a) Unsaturated and saturated flow (b) Unsaturated flow (c) Unsaturated, Saturated and Oversaturated Flows.

1.3 Level of Service (LOS) based on congestion

LOS has also been defined by several researchers based on quantified congestion. Since quantified congestion is a measure of loss in freedom of movement that accounts for the variation of speed level with increase in traffic volume (along with its character) under prevailing roadway, traffic, and control conditions, the congestion level is a logical and better measure of effectiveness (MOE) to define LOS in a quantitative manner.

The degradation of quality of operation and increase in congestion level due to traffic volume is not uniform at all flow levels. To represent the variation of level of service through congestion in a complete manner, 10 different levels of service have been proposed by researchers with congestion levels of 5, 10, 20, 30, 40, 50, 60, 80, and 100%; distinguishing nine LOS (A – I) within the stable flow zone, and one LOS (J) with congestion more than 100%, indicating the unstable (forced) flow. The limiting values of congestion for different LOS are stated to be consistent with the usual shape of the speed-flow curve with a relatively lesser speed drop near free-flow operation and sharp drop near the capacity flow. The service volumes are governed by congestion levels, and therefore, any change in prevailing roadway, traffic, and control conditions would normally be reflected by the changes in service volumes corresponding to various levels of service [12].

2 Simulation model

The simulation tool used in this paper is VISSIM 5.3 (official license available). VISSIM uses the psycho-physical driver behaviour model developed by Wiedemann [16]. The basic concept of this model is that the driver of a faster moving vehicle starts to decelerate as he reaches his individual perception threshold to a slower moving vehicle. Since he cannot exactly determine the speed of that vehicle, his speed will fall below that vehicle's speed until he starts to slightly accelerate again after reaching another perception threshold. This results in an iterative process of acceleration and deceleration. Stochastic distributions of speed and spacing thresholds replicate individual driver behaviour characteristics. VISSIM's traffic simulator not only allows drivers on multiple lane roadways to react to preceding vehicles, but also neighbouring vehicles on the adjacent travel lanes, are taken into account. Moreover, approaching a traffic signal results in a higher alertness for drivers at a distance of 100 meters in front of the stop line. VISSIM simulates the traffic flow by moving "driver-vehicle-units" through a network. Every driver with his specific behaviour characteristics is assigned to a specific vehicle. As a consequence, the driving behaviour corresponds to the technical capabilities of his vehicle. Attributes characterizing each driver-vehicle-unit are (1) Technical specifications of the vehicle, e.g. length, maximum speed, potential acceleration, actual position in the network, actual speed and acceleration (2) behaviour of driver-vehicle-unit, e.g., psycho-physical sensitivity thresholds of the driver (ability to estimate, aggressiveness), memory of driver, acceleration based on current speed and driver's desired speed (3) interdependence of driver-vehicle-units, e.g. reference to leading and following vehicles on own and adjacent travel lanes, reference to current link and next intersection, reference to next traffic signal [16].

3 Data collection, model calibration and validation

Jaipur city in India is chosen to investigate interrupted heterogeneous flow as the city size, its roads, type of vehicles, mixed traffic and driver's aptitude represent most of the developing countries where interrupted, heterogeneous and oversaturated flow is a daily phenomenon. A representative network of urban road network, comprising of two signalized intersections (label 1&5 in *Fig 2*), one un-signalized rotary intersection in between (label 4 in *Fig 2*) and two minor T-junctions (label 2&3 in *Fig2*) was decided to explore speed-flow and congestion characteristics of urban roads. The total length of the representative network was 4 km; it was neither too long nor too short and is a reasonable distance to have preciseness of results incorporating elements of bigger network. The study is conducted for this network as a whole so that curves for this network will provide a realistic estimate of traffic variables in urban road scenario.

The simulation model construction procedure consists of (i) identification of important geometric features (ii) collection and processing of traffic data (iii) analysis of mainline data to identify recurring bottlenecks (iv) VISSIM coding (v) calibration based on observations from (iii). Calibration is the process by which individual components of simulation model are refined and adjusted so that simulation model accurately represents field measured or observed traffic conditions. With regard to calibration, traffic simulation model contain numerous variables to define and replicate traffic control operations, traffic flow characteristics and driver behaviour. VISSIM simulation model contains default values for each variable, but also allows a range of user applied values for each variable. These variables are changed as per field measurements and observed conditions [16].

The geometry of existing network from Danik Bhaskar intersection to JDA intersection on Jawahar Lal Nehru road (4-lane divided with lane width 3.5m) was created using links and connectors which are the building blocks of VISSIM network. The number of lanes per road and width of each lane, left turning lanes on each approach road, central median, traffic islands and other road features were specified as per existing. After creation of network, the vehicle input for various links was given. The traffic composition is heterogeneous with Car/Jeep/Taxi/SUV :37%; Two wheeler : 41%; Three wheeler :11%; Bus : 5%; HGV :1%; Slow moving vehicle : 5% (traffic data for time : 08:00-20:00 on February 28, 2012). This is followed by specifying the various routes in which vehicles travelled and the volume of these vehicles in each route is specified. The other features viz. positioning of speed limits, conflict zones, stop signs, signal heads are specified as per existing. The data collection points, travel time sections, queue counters and nodes are placed. The Indian driving behaviour is calibrated for the following parameters: standstill longitudinal distance between the stopped vehicles, headway time in seconds, following variation which restricts the longitudinal oscillation and indicates how much more distance than desired distance a driver allows before he intentionally moves closer to vehicle in front, threshold for entering 'following' controlling the start of deceleration process, following threshold which controls the speed differences during the 'following' state, speed dependency of oscillation, oscillation acceleration, standstill acceleration, minimum headway, maximum deceleration of vehicle and trailing vehicle for lane change, overtaking characteristics, minimum lateral distance at different speeds, waiting time for diffusion. The vehicles are calibrated for desired speed distribution, weight distribution, power distribution and model distribution. The links are assigned behaviour according to driving behaviour. On roads of most of the developing countries including India, because of heterogeneity of traffic, it is difficult to enforce lane discipline. Hence, vehicle occupies lateral positions on any part of road based on space availability; overtake within lane from both the sides. The validation of the model was carried out by comparing maximum queue length simulated by model for existing intersections on each approach road with field observed values. The simulation model was given multi-run with 20 random seed numbers and average of 20 simulation runs was taken as final output of the model. The value of t-statistic, calculated based on observed data (to) for all the four approach road on both the signalized intersections is below 2.00. The critical value of t-statics for level of significance of 0.05, at 19 degrees of freedom is 2.093. Thus, value of t-statistic, calculated on the basis of observed data, is less than the corresponding table value. This shows that there is no significant difference between the simulated and observed queue lengths.

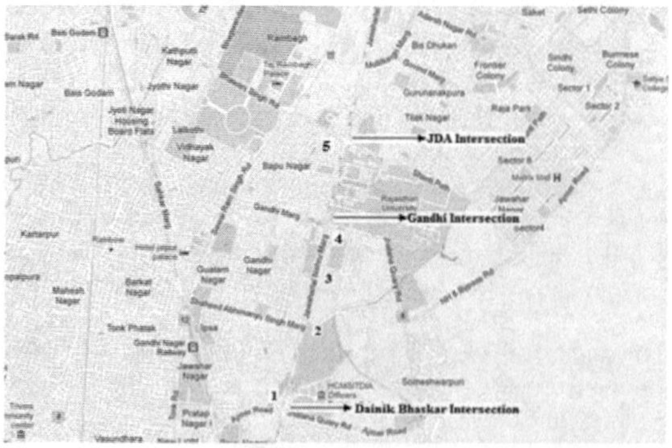

MNIT

Figure 2. The representative network for study from dainik bhaskar intersection to JDA intersection

4 Estimation of congestion

The model validated as above is used to investigate the shape of speed-flow curve for varying volumes and estimating congestion. The values shown are the average of 20 simulation runs with different random numbers so as to have reasonable results to conclude.

4.1 Speed–flow curves and determination of capacity

Fig. 3(a), 3 (b) and 3 (c) shows the shape of Speed-Flow curve. The oversaturated portion of curve (Fig. 3 a) follows the equation (1)

$$\text{Speed} = -3E - 13\,(\text{Flow})^4 + 5E - 09\,(\text{Flow})^3 - 2E - 05(\text{Flow})^2 + 0.0404\,(\text{Flow}) - 0.2396 \quad (1)$$

The R^2 value is 0.9977.

In order to do detailing of Speed-Flow curve for unsaturated part of flow, investigations are carried out to explore it with more flows values up to capacity. Fig. 3 (b) shows the shape of Speed-Flow curve for unsaturated condition. The unsaturated portion of curve follows the equation (2).

$$\text{Speed} = -2E\text{-}12(\text{Flow})^4 + 1E\text{-}08(\text{Flow})^3 - 4E\text{-}05(\text{Flow})^2 + 0.039(\text{Flow}) + 6.162 \quad (2)$$

$R^2 = 0.999$

Since the oversaturated flow starts where unsaturated flow ends, the capacity may be defined as the point of intersection of these two curves i.e., curves defined by (1) and (2) respectively. In this case the capacity is 2200 veh/hr.

Fig. 3 (c) shows the consolidated Speed-(v/c) curve for unsaturated as well as oversaturated states of interrupted flow.

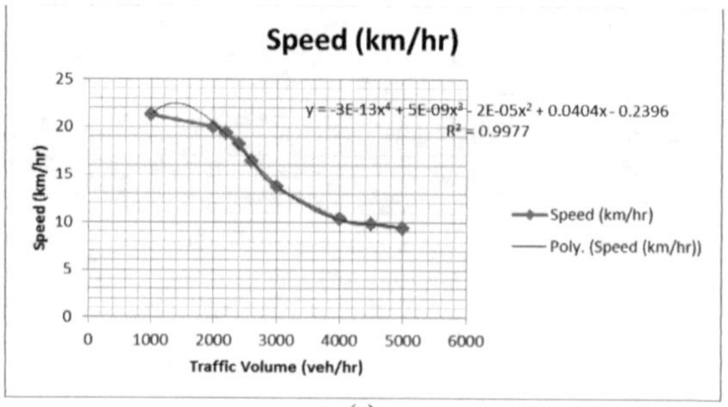

(a)

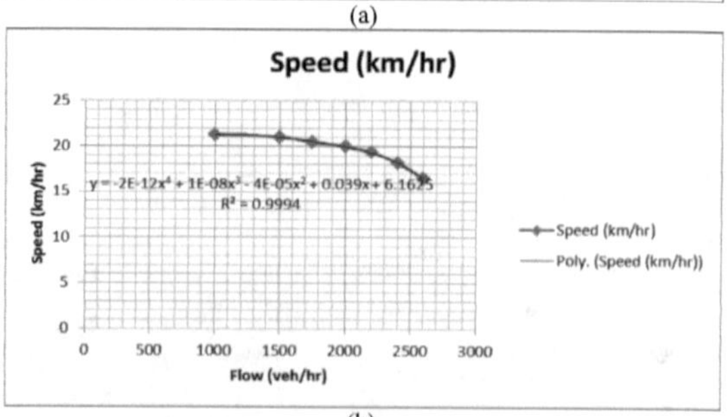

(b)

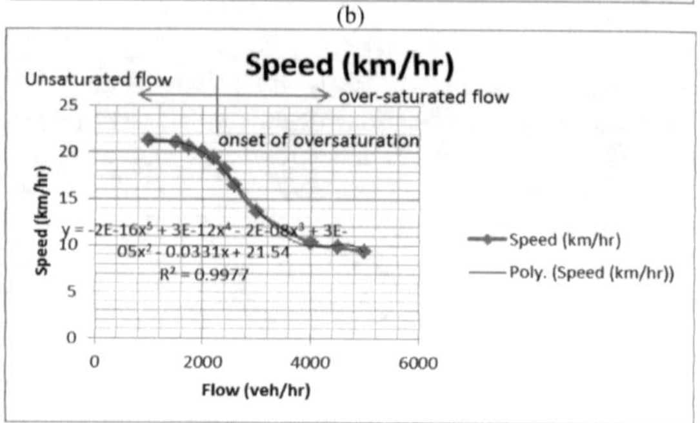

(c)

Figure 3. Speed-Flow Curve (a) oversaturated details (b) unsaturated details (c) consolidated curve.

4.2 Model for congestion

The area under the Speed-Flow envelope between any two operating points on it represents loss in freedom of movement between these two traffic conditions, hence congestion. Similarly congestion for oversaturated flow can be defined as loss in freedom of movement with respect to traffic condition prevailing at capacity to account for oversaturated state of flow.

4.2.1 Congestion for unsaturated flow conditions

For unsaturated flow condition congestion curve is shown in Fig. 4. The equation of curve is given by *equation (3)*. Here the congestion value at capacity is taken as 100%.

$$\text{Congestion (\%)} = 415.54(v/c)^5 - 696.94(v/c)^4 + 600.07(v/c)^3 - 277.49(v/c)^2 + 64.623(v/c) - 5.8154 \quad (3)$$

$R^2 = 1$

The curve follows fifth order polynomial as given by equation (3).

The Fig. 4 Shows that within under-saturated condition of flow the congestion starts visibly growing up at v/c approximately equal to 0.5. The rate of congestion becomes tremendous after v/c ratio 0.8 and achieves full congestion at capacity. Congestion which is merely 25% at v/c ratio 0.8 increases four times at capacity and becomes 100%.

4.2.2 Congestion in over-saturated condition of flow

Fig. 5 shows the congestion curve for oversaturated state. The congestion is quantified with respect to traffic condition at capacity. Here, in order to demonstrate a clear depiction of congestion of oversaturated flow with respect to traffic condition at capacity, congestion at a point near capacity is taken as zero and a new term congestion ratio is defined similar to v/c i.e., (congestion at v/ congestion at c). The curve may be given by the following *equation (4)*:

$$\text{Ratio of Congestion at a flow to congestion at capacity} = 247.7\,(v/c)^2 - 461.1\,(v/c) + 212.9 \quad (4)$$

Here R^2 value is 0.999.

Here, ordinate represents ratio of congestion at an operating point and congestion at capacity while abscissa represents volume to capacity ratio.

4.2.3 Overall congestion scenario

If congestion is plotted for both unsaturated and oversaturated flows, taking congestion at capacity as 100%, the congestion characteristics shows that rate of built up congestion up to capacity is with slower rate which increases significantly after capacity. The consolidated congestion curve for both unsaturated and over-saturated condition is shown in Fig. 6.

$$\text{Congestion (\%)} = -4884.2(v/c)^4 + 31093(v/c)^3 - 46704(v/c)^2 + 24182(v/c) - 3659.7 \quad (5)$$

$R^2 = 0.9998$

Equation (5) represents the total congestion curve. Equation (3) and (4) can be used for precise calculation in unsaturated and over-saturated condition respectively, while equation (5) represents overall scenario. It can be seen that at v/c ratio 1.25, congestion is 20 times that of capacity; at v/c ratio 1.5, congestion grows by 80 times of value at capacity. Beyond v/c ratio 1.5 the traffic condition deteriorates severely and at v/c ratio 2.0, congestion becomes 300 times while at v/c ratio 2.4 it becomes 520 times that of value at capacity. Since, any change in prevailing roadway, traffic, and control conditions gets reflected by the changes in service volumes and the service volumes corresponding to various levels of service are governed by congestion levels, the level of service (LOS) may be redefined based on congestion levels. LOS may be defined for a network with interrupted heterogeneous flow, separately for under-saturated flow and oversaturated flow. For under-saturated flow proposed LOS (A-E) may correspond to congestion level (0.05, 0.10, 0.30, 0.50, and 1.00) and for oversaturated flow LOS (I-L) may correspond to (20, 60, 300 and beyond). These levels correspond to congestion curve.

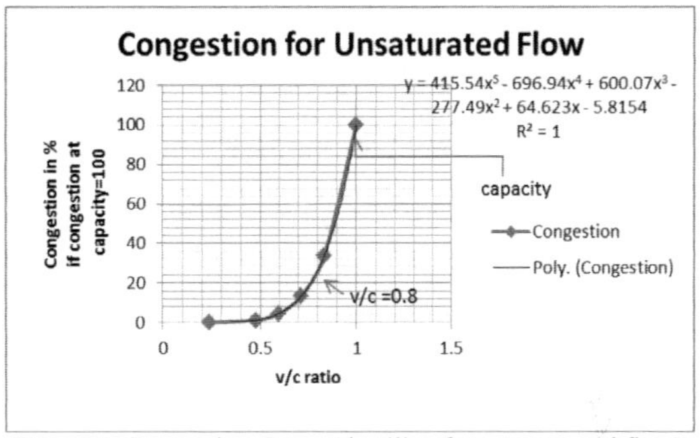

Figure 4. Congestion Curve (detailing for un-saturated flow).

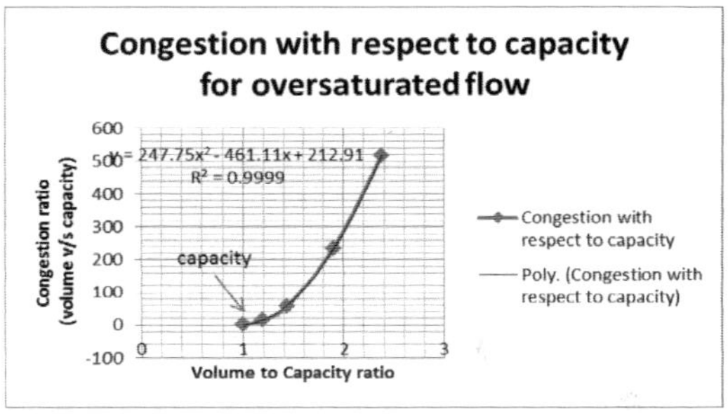

Figure 5. Congestion curve for oversaturated flow.

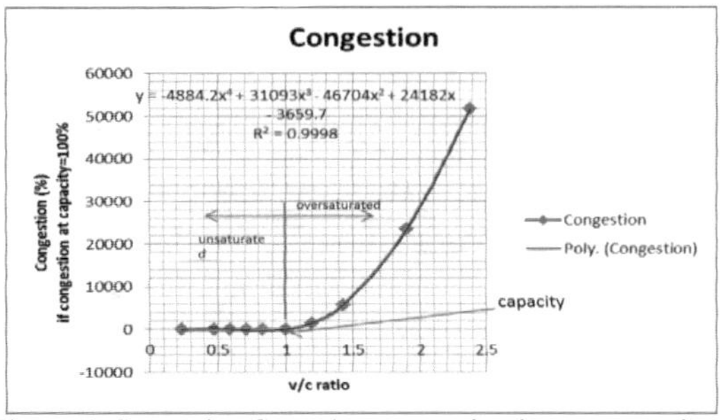

Figure 6. Congestion Curve (un-saturated and oversaturated flow combined).

5 Conclusion

This paper presents speed-flow and congestion-flow curves for urban roads observing interrupted flow with heterogeneous structure of traffic. The models are developed for speed-flow in unsaturated and oversaturated flow conditions under constraints of roadway geometry, vehicle characteristics, driving behaviour and traffic controls. The capacity of road is determined as the point of intersection of Speed-Flow curve for oversaturated flow and Speed-Flow curve for unsaturated flow. The speed-flow curves are used to quantify congestion and models are developed for congestion in unsaturated state and oversaturated state. The onset of congestion, its growth with flow and its behaviour in unsaturated flow as well as oversaturated flow is analysed. It has been found that within under-saturated condition of flow the congestion starts visibly growing up at v/c approximately equal to 0.5; rate of congestion becomes tremendous after v/c ratio 0.8 and achieves full congestion at capacity. Congestion which is merely 25% at v/c ratio 0.8 increases four times at capacity and becomes 100%. In oversaturated condition of flow, at v/c ratio 1.25, congestion is 20 times that of capacity; at v/c ratio 1.5, congestion grows by 80 times of value at capacity. Beyond v/c ratio 1.5 the traffic condition deteriorates severely and at v/c ratio 2.0, congestion becomes 300 times while at v/c ratio 2.4 it becomes 520 times that of value at capacity. The investigation and analysis presented in this paper gives more realistic values of speed, capacity and congestion as the investigation is based on total performance of network and the model captures both operational characteristics as well as volume characteristics of interrupted heterogeneous flow through simulations.

Acknowledgement

Authors wish to acknowledge the support provided by Dr. Ritu Sharma, MNIT Jaipur to realize this research work.

References

1. T. J. Lomax, Methodology for estimating urban roadway system congestion. *Transp. Res. Rec. 1181*, Transportation Research Board, National Research Council, Washington, D.C, 38-49 (1988).
2. C. R. Lindsey and E. T. Verhoef, Congestion Modelling. Tinbergen Institute Discussion Papers 99-091/3, Tinbergen Institute (1999).
3. K. Hashimoto, Monitoring road traffic congestion in Japan. Transp. Rev., **10**(2), 171–186(1990)
4. T. Litman, Generated Traffic and Induced Travel: Implications for Transport Planning. VTPI Discussion Papers. Victoria Transport Policy Institute, (2011).
5. C. A. Moran and K. L. Bang, Congestion Performance Indicators for Urban Road Network Case Study: Stockholm Congestion Charging Trial. Paper #08-0483. Transportation Research Board Annual Meeting, (2008).
6. J. A. Lindley, Urban freeway congestion problem and solutions: An update. ITE J., **59**(12), 21-23. (1989).
7. M. V. L. R. Anjaneyulu and B. N. Nagaraj, Modeling Congestion on Urban Roads Using Speed Profile Data. IRC Journal, **70**, 549, pp 65-73, (2009).
8. D. K. Witheford, Traffic jam-relieving congestion of the national highways. TR News No. 139, Transportation Research Board, National Research Council, Washington, D.C., 2–4, (1988).
9. K. M. Lakshmana Rao and B. Sridhar, Sectorial view on traffic congestion in a transportation system. Proc., Int. Conf. on Rd. and Rd. Transport (ICORT), University of Roorkee, India, 1037–1044, (1995).
10. D. K. Parbat, Quantification of congestion index on major corridors. Proc., Nat. Conf. on Transp. Sys. Studies (NCOTSS), University of Mumbai, India, 103–110, (1996).
11. S. M. Mateen, Estimating and relieving urban congestion. Proj. Rep., R.E.C. Warangal, India, (1990).
12. B. Maitra, P. K. Sikdar and S. L. Dhingra, Modeling Congestion on Urban Roads and Assessing Level of Service. Journal of Transportation Engineering, vol. **125**, no. 6, pp. 508-514, (1999).
13. R. Akçelik, Speed-Flow Models for Uninterrupted Traffic Facilities. Technical Report, Akcelik & Associates Pty Ltd. (2003).
14. R. Dowling, Speed-Flow Curves for Arterials. Dowling Associates, (2004).
15. Transportation Research Board, Highway Capacity Manual 2000. 500 Fifth St. NW, Washington, D.C. (2001).
16. PTV Vision, VISSIM tutorial. PTV AG, Karlruhe, (2010)

MATEC Web of Conferences **81**, 01002 (2016) DOI: 10.1051/matecconf/20168101002
ICTTE 2016

Evaluation of Operational Improvement Strategies to Reduce Congestion on Urban Corridors

Kaniz Fatima and Sara Moridpour

School of Civil Chemical & Environmental Engineering, RMIT University, Melbourne, Australia

Abstract. In recent years, traffic congestion has become a major problem in transport networks particularly in peak periods. The congestion problem has become worse in large cities, particularly in the CBDs over the last few decades and the problem intensifies every year. Traffic congestion results in longer travel times, larger delays, more fuel consumptions and more emission. Australian Bureau of infrastructure, transport and regional economics estimates that the avoidable social costs of congestion will be more than doubled in 2020. This research intends to propose a solution to reduce traffic congestion and consequently traffic delay for a congested corridor in Melbourne metropolitan area. To achieve this, a number of strategies are introduced and their influence in traffic congestion is evaluated. Dedicated bus /tram lane, Bus/tram stop location change, Parking restriction, Bicycle lane restriction, Passenger car movement restriction, Lane configuration Advanced Green Signal and Extended Green signal will be applied to reduce congestion strategies. The site selected for this research is a corridor from inner suburbs of Melbourne.

1 Introduction

Traffic congestion problem in Australia may be the consequence of many factors. Congestion is generally worse in inner suburbs than outer area. A large number of commuters travel to/from work and school on same time intervals representing the most significant contributors of congestion. Rapid increase in population and cars are the key reasons of traffic congestion. Other key causes relating to both supply and demand side include under-pricing of road use, lack of proper road infrastructure design/operation, travel demand growth and inadequate public transport alternatives. Traffic congestion is consequence of these factors. Therefore, effective demand management can reduce traffic congestions in urban areas. Congestion results in longer travel times, larger delays, more fuel consumptions and more emission. Time wasted in congestion has a huge cost for the society. The Texas Transport Institute (TTI) 2012 Urban Mobility Report reveals that U.S.A commuters wasted $5.5 billion hours of extra time and fuel due to traffic congestion in 2011. This is equal to the amount of time that business and individual taxpayers spend lodging their tax applications. In Australia, the Australian bureau of infrastructure, transport and regional economics (BTRE) preliminary study estimates wastage of total $9.4 billion as social cost of congestion for the year 2005 over the major Australian cities [1].

2 Aims and objectives

The aim of this research is to propose strategies to reduce traffic congestion in a particular corridor of Melbourne by modal management strategies. Corridor is a network of intersections. Melbourne is operating the largest tram network in the world. The tram system is a popular and convenient mode of transport in Melbourne. The case study corridor operates almost every mode of transport tram, bus, cars, pedestrian and bicycle. The research objectives are as follows:

a. Identify different strategies (i.e time and space priority for buses/trams) can be used to reduce traffic congestion and delay for different modes of transport.

b. Apply the strategies to a corridor in Melbourne as case study.

c. Analyse and compare the influence of different strategies on traffic congestion and delay of different modes of transport.

d. Propose better strategies to reduce congestion and travel delay.

3 Background of research

From literature review congestion management strategies can be defined as infrastructure expansion, congestion pricing management, modal congestion management. The study is focused on modal congestion management as it is the more economical solution than adding infrastructure and pricing congestion strategies [2].

4 Research methodology

The study intends to propose better strategies to control congestion for a selected corridor through SIDRA analysis. The steps are as follows:

4.1 Modelling of the corridor in SIDRA

Simulate the traffic conditions on four intersections with same number of lanes, traffic volume, bus/tram frequency and signal cycle time. In the case of any differences of actual network condition and SIDRA simulated intersection model, input parameters need to be calibrated. If the observed intersection condition and SIDRA simulated intersection model is similar, the SIDRA model can identified as a validated model.

4.2 Identifying of different strategies

Identify different modal congestion management strategies to reduce traffic congestion. Space priority for trams/buses at intersection/corridor is one of the proposed and applied strategies in different cities. Modifying signal time (i.e extended green or early green signal) during peak periods of the day is another applied strategy to control traffic congestion. Parking restrictions and removal of bicycle lanes are also utilized to reduce congestion around different intersections.

4.3 Simulation of different strategies

Provide priority to only bus/tram or any other strategy is not effective. The research will simulate different identified strategies to see the effect on intersections and corridor/network. The study will apply each strategy on intersection to see the effect of congestion.

4.4 Comparison of applied strategies

The study will compare simulated models of network/corridor with present conditions and propose better methods to control congestion. Three main parameters travel speed, queue length and average delay will be used to compare strategies.

Table 1. Summary of intersection grouping

Intersection Name	Site Grouping according to VicRoads.
Princess Street/Studley Street/High Street South/ Denmark Street/ High Street	3662
Denmark Street/Wellington Street	3663
Denmark Street /Stevenson Street	1190
Denmark Street/Power Street/Barkers Road	3002

5 Study area

The study corridor starts from Manningtree Road and Power Street intersection to the intersection of Princes Street and Beatrice Street (M21, Melways Ref: 45, C11-Street Directory) in Melbourne, Australia. The corridor is about 2.1 km long which approaches various lanes per direction. The daily traffic volume of the corridor is (annual average) 7, 91, 013 per day. The corridor includes fifteen different intersections.

Table 1 represents the summary of intersection grouping name provided by VicRoads. The numbers are used althrough the research to identify the intersections.

6 Datasets for corridor simulation

Datasets have been collected from five different sources for one hour morning peak period. Table 2 represents the summary of collected data.

Table 2. Summaries of collected datasets.

Data Sets (From 7.30 AM to 8.30 AM)			
VicRoads	**Site Survey**	**Yarra trams/Bus system Information**	**Google Maps**
Intersection Grouping: Identify intersection with a site number. Controller Time Setting: Green, Red and yellow signal time. Detector details: Installed position of pedestrian push button detector, bicycle push button detector, radio controlled vehicle detector and video detector). Traffic Volume: Traffic volume from May 2012 till May 2013 (15 mins interval). Phasing Diagram: Order of traffic signals, priority and extended green signal upon demand.	Intersection Geometric details: Lane no, length, width, approaches for vehicle and bicycle. Movement: Straight/left/ right, restriction. Speed: Speed limit for all approaches. Average Speed: Average speed, travel time and back of queue have been measured for all approaches to validate of the corridor model.	Time table and frequency for buses. Time table and frequency for trams.	Distances between intersection

7 Model development

Micro simulation models contain a number of independent parameters to describe traffic control operation, traffic flow characteristics and drivers' behaviour. These models contain default values for each variable but they also allow users to input values.

Changes of these parameters during calibration should be based on field-measured conditions and should be justified by the users [3][4]. SIDRA Intersection version 6 has a number of input and output parameters to model and validate model. To configure the corridor each intersection has to model separately according to the datasets [5][6]. Output results will produce on table and layout format. The output results can form for individual approaches, intersections and networks.

7.1 Calibrated parameter

Saturation flow rate: The flow rate of vehicle per lane at which vehicles can pass through a signalized intersection if the signal remains green all the time.

Critical gap: The travel time for following vehicle that can reach to immediate front vehicle on signal.

Follow up headway: The travel time for following vehicle that can reach to immediate front vehicle on move.

SIDRA presents a number of parameters for output results. Three main parameters, travel speed, delay average and queue distances are calibrated for each intersection. Origin-destination (OD) trip matrices volume is applied for this network volume along the corridor [7][8][9][10][11]. Main three parameters travel speed, delay average and queue length has been calibrated with actual datasets for this network/corridor. The network is validated and ready for the application of strategies.

7.2 Level of Service (LOS)

The level of service (LOS) of the simulated intersections is the main target. Strategies can be applied on worst LOS lane to reduce congestion. Intersection 3662 is the most congested intersection of this corridor. Table 3 represents the summary of LOS.

Table 3. Summary of LOS.

Summary of LOS for Intersection 3662					
Denmark St	High St	Princess St	Studley St	High St south	Intersection
D	E	D	C	E	D

Summary of LOS for Intersection 3663			
Denmark St, South	Wellington St	Denmark St, North	Intersection
B	C	C	C

Summary of LOS for Intersection 1190			
Denmark St, South	Denmark St, North	Stevenson St	Intersection
B	B	B	B

Summary of LOS for Intersection 3002				
Power St	Barker Rd, East	Denamrk St	Barker Rd, West	Intersection
D	C	D	D	D

8 Applied strategies on corridor

From literature review modal management strategies can be apply on this corridor as follows:

- Priority for buses and trams
- Remove bicycle lanes
- Change lane configuration
- Restrict movements
- Change locations of tram stops
- Signal Cycle Change

The strategies will be applied on intersections separately. Better individual resultant strategies will apply on network/corridor and simulate to determine the best strategies. Table 4 represents the summary of applied and efficient strategies for each intersection.

Table 4. Summary of applied and efficient strategies

Applied Strategies on Site			
3662	**3663**	**1190**	**3002**
• Movement Restriction • Change of Phase time and phase order • Tram Stop relocation from High St to High St south • Bicycle lane restriction • Lane Configuration	• Lane Configuration: Configure parking lane to an exit lane at Wellington St and changed signal phase order. • Parking Restriction: Configure parking lane to exit lane at Wellington St. • Changed signal phase order.	Changed signal phase order.	Changed signal phase order.
Efficient Strategies for Sites			
3662	**3663**	**1190**	**3002**
• Change of phase time and phase order. • Movement restriction: 1. Restrict movement from Denmark St towards High St South. 2. Restrict Movement from High St south towards Princess St and from Denmark St towards High St South. • Lane Configuration: 1. Restrict High St South towards Princess St and from Princess St towards High St South. 2. Restrict High St lane1 towards Studley and from Denmark St towards High St South.	• Parking Restriction: Configure parking lane to exit lane at Wellington St. • Changed signal phase order.	Changed signal phase order.	Site 3002 Intersection has less congestion during morning peak period. Applied strategies for signal phase order change do not affect the intersection.

9 Efficient strategies for case study

1. Modify signal phase order for intersection 3662 and other three intersections can operate according to present condition.

2. Movement restriction from High Street South to Princess Street and from Princess Street towards High Street South at intersection 3662 and other three intersections kept same as present.

3. Configure parking lane to exit lane at Wellington Street.

Fig. 1 represents travel speed comparison of efficient strategy 1 & 2. Traffic condition improves for first through efficient strategies application. Arrows on layout shows vehicle movement direction. Numbers in black is the simulate value for travel speed. Percentage number in blue indicates the improvement of each direction.

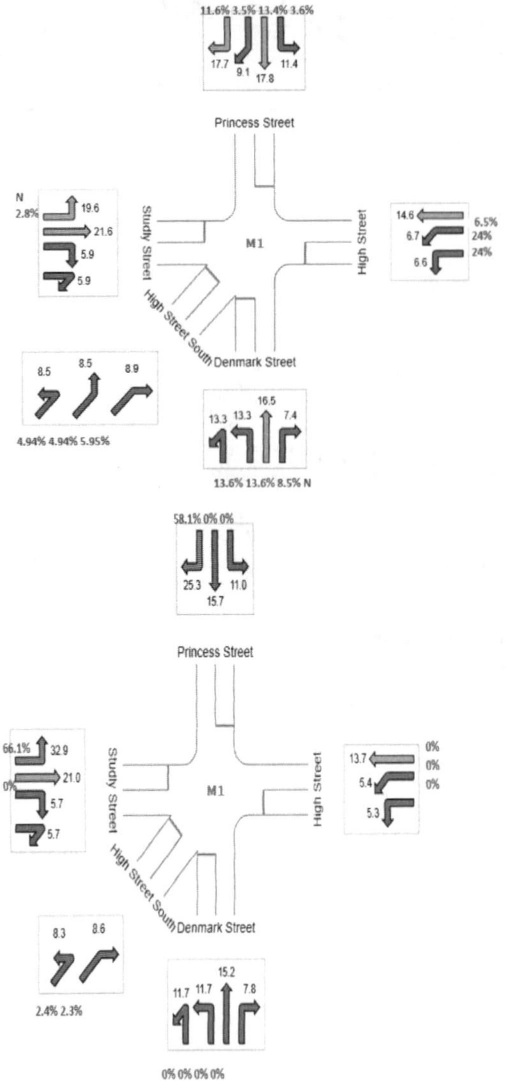

Figure 1. Travel speed comparison strategy 1& strategy 2.

Fig. 2 represents queue distance comparison of efficient strategy 1 & 2. Traffic condition improves for first through efficient strategies application. Arrows on layout shows vehicle movement direction. Numbers in black is the simulate value for travel speed. Percentage number in blue indicates the improvement of each direction.

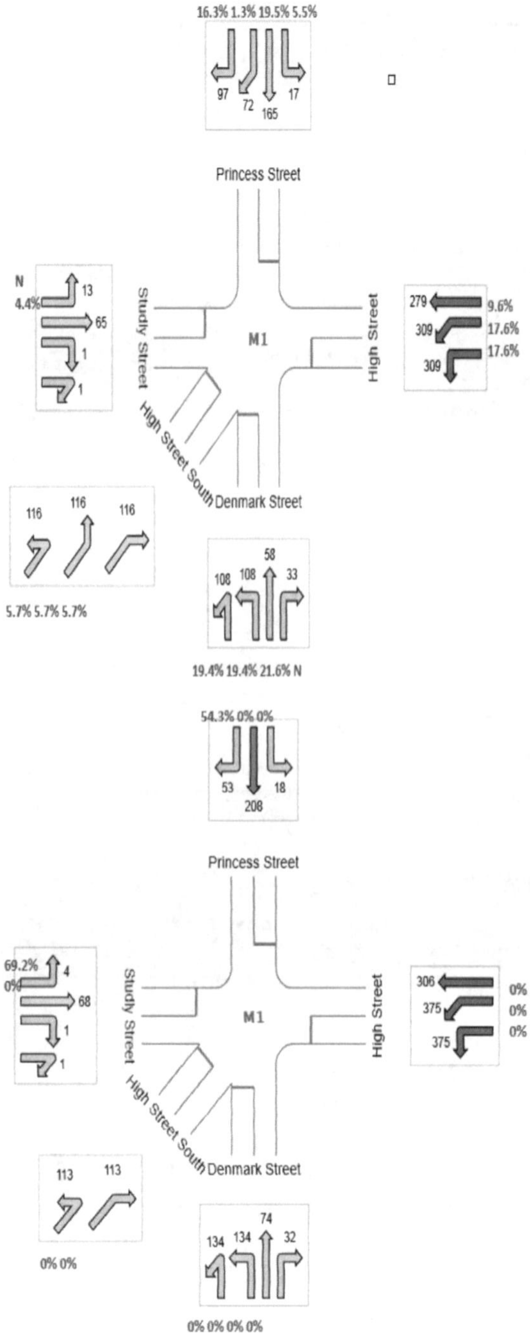

Figure 2. Queue distance comparison strategy 1& strategy 2.

Again Fig. 3 represents delays comparison of efficient strategy 1 & 2. Traffic condition improves for first through efficient strategies application. Arrows on layout shows vehicle movement direction. Numbers in black is the simulate value for travel speed. Percentage number in blue indicates the improvement of direction.

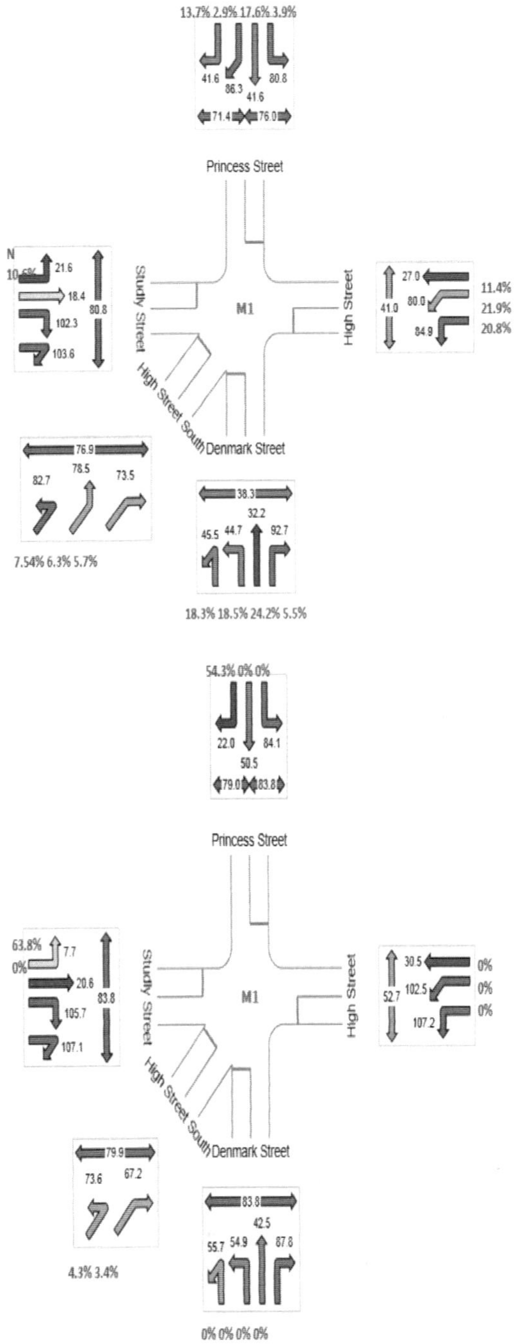

Figure 3. Delays average comparison strategy 1& strategy 2.

Table 5 shows comparison of output parameters after simulation as a network or corridor. Observed values have been collected from site survey. Applying the strategies improved network operation. Travel speed has been increased. Travel time, delay and cost decreased. Modification of signal phase order and cycle is more effective than movement restriction.

Table 5. Summary of network output parameter comparison.

Output Parameter	Observed Network	Strategy 1 Network	Strategy 2 Network
Travel Speed, km/h	19.8	20.7(4.5%)	20.3(2.5%)
Travel Time (Total), veh-h/h	219.8	210.0(-4.4%)	214.1(-2.5%)
Travel Delay, sec	42.0	39.3(-6.4%)	40.4(-3.8%)
Control Delay (Total), veh-h/h	128.26	119.31(-6.9%)	123.14(-3.9%)
Congestion Co-efficient	3.03	2.90(-4.3%)	2.95(-2.6%)
Cost (Total), $/I	7622.78	7273.90(-4.8%)	7447.17(-2.3%)

10 Conclusion

Two best strategies have been identified to control congestion of this corridor. Time priority and movement restriction is the two best strategies for this corridor. Parking restriction for morning peak period on Wellington street intersection also increase the LOS of this corridor. Parking restriction on Wellington street increase the corridor traffic condition but it also has a minimal bad effect on Wellington Street and Stevenson Street. As these two streets are not congested and highly traffic volumed. Parking restriction on Wellington street during morning peak period can consider one of the proposed strategies. The study proposes several individual strategies which increased travel speed but decreases travel time. For T-intersection rearrange of signal phase can be efficient. Again for busy intersection with more than four approaches movement restriction and lane configuration improves the intersection as well as corridor performance.

In future the research can continue applying strategies on different intersections and roundabout. To control congestion with modal management strategies more intersections can be simulated. The research can direct as follows:

- More intersections can be added to simulate larger corridor.
- Adjoin simulated corridor as a network and apply strategies to see the effect.
- Adjoin signalised, un-signalised and roundabout and simulate strategies.

References

1. Transport, A. *wp_071.pdf.* (2007).
2. J. Leonardo, C. Angelo G. Antonio Gschwendera, M.Fusterb Basso, L. J., et al. Congestion pricing, transit subsidies and dedicated bus lanes: Efficient and practical solutions to congestion. Transport Policy **18**(5): 676-684.4(2011).
3. B. Park. Microscopic simulation model calibration and validation.pdf. Transportation Research Record: Journal of the Transportation Research Board, **1856/2003**, 185-192 (2003).

4. B.P.H. Qi. Microscopic simulation model calibration and validation for freeway work zone network.pdf. Intelligent Transportation Systems Conference, 2006. ITSC '06. IEEE, 1471 – 1476, (2006).
5. R. Akcelik. Fundametal traffic varriables in adaptive controland the SCATS DS parameter.pdf. 3rd International conference ITS Australia, (1997).
6. R. Akcelik. An Improved Method for Estimating Sign-Controlled Intersection Capacity.pdf. New Zealand Modelling User Group NZMUGS 2012 Conference. Auckland, (Sep 2012).
7. G.-L. Chang & J. Wu. Recursive estimation of time-varying origin-destination flows from traffic counts in freeway corridors. Transportation Research Part B: Methodological, **28**, 141-160, (1994).
8. M. Cremer. A new class of dynamic methods for the identification of origin-destination flows.pdf. Transport Research Board, **21B**, 117-132, (1986).
9. H. Yang, T. Sasaki, Y. Iida, & Y. Asakura. Estimation of origin-destination matrices from link traffic counts on congested networks. Transportation Research Part B: Methodological, **26**, 417-434, (1992).
10. A. Medina, N. Taft, K. Salamatian , S. Bhattacharyya, C. Diot. Traffic Matrix Estimation: Existing Techniques and New Directions, (2002).
11. M. Papageorgiou. Dynamic modeling, assignment, and route guidance in traffic networks. Transportation Research Part B: Methodological, **24**, 471-495, (1990).

The Homogeneity Research of Urban Rail Transit Network Performance

Wang Fu-jian and Ruan Shu-bin

College of Civil Engineering and Architecture, Zhejiang University, Hangzhou 310058, China

Abstract. Urban Rail Transit is an important part of the public transit, it is necessary to carry out the corresponding network function analysis. Previous studies mainly about network performance analysis of a single city rail transit, lacking of horizontal comparison between the multi-city, it is difficult to find inner unity of different Urban Rail Transit network functions. Taking into account the Urban Rail Transit network is a typical complex networks, so this paper proposes the application of complex network theory to research the homogeneity of Urban Rail Transit network performance. This paper selects rail networks of Beijing, Shanghai, Guangzhou as calculation case, gave them a complex network mapping through the L, P Space method and had a static topological analysis using complex network theory, Network characteristics in three cities were calculated and analyzed form node degree distribution and node connection preference. Finally, this paper studied the network efficiency changes of Urban Rail Transit system under different attack mode. The results showed that, although rail transport network size, model construction and construction planning of the three cities are different, but their network performance in many aspects showed high homogeneity.

1 Introduction

Urban Rail Transit does not occupy ground road resources, large capacity, fast and convenient features, ease urban traffic congestion to a great extent, so the major cities in China have been increased investment in rail transportation. The total length of rail transit lines in operation by the 621 km (2006) soared to 2, 408 km (2013), the respective rail vehicle operators also increased to 14,366 units from the 2, 764 units [1]. Beijing, Shanghai, Guangzhou has basically formed a stable operation of the rail transit network, Hangzhou, Nanjing, Wuhan, Xi'an, Qingdao also in actively construction of rail transportation, Urban Rail Transit will become an important part of the public transit. In order to ensure efficient and safe operation of the urban rail system, it is necessary to make the corresponding network performance analysis for Urban Rail Transit system.

Urban Rail Transit network is a typical complex network of Nodes and Links composition [2]. With the maturation of complex network theory, more and more domestic and foreign scholars with their research to analyze the performance of Urban Rail Transit network, mainly including the prevention of topological analysis, node dynamics, network attacks and subsequent failure of the network Chaos and synchronization of complex networks analysis. As Latora *et al* [3] found a small-world characteristics of Boston subway system, and the corresponding mathematical modeling; Angeloudis et al [4] studied the overall efficiency of the rail network changes situation under different network nodes attack mode; Derrible *et al* [5] conducted a thorough analysis of the Urban Rail Transit scale-free and small world, and studied the relationship between transfer hub station and network robustness; Rui *et al* [6] conducted a simulation of complex network evolution Research for Kuala Lumpur Metro. Yang *et al* [7] studied the dynamic characteristics of urban rail network performance under cyber attack using complex network theory; Liu et al[8] conducted a complex topology modeling and the appropriate network parameters analysis for Wuhan subway network using the L, P Space method; Zhang et al[9] studied abroad metro reliability and robustness on the basis of the system analysis for complex network parameters.

Previous studies focused on analysis of network performance parameters of Urban Rail Transit, less studied the homogeneity of Urban Rail Transit network performance in different cities. The so-called homogeneous, referring to the inherent unity of the network topology properties. That is different network topologies have the same or similar network performance. Through horizontal comparison analysis of Urban Rail Transit network performance in Beijing, Shanghai and Guangzhou, it was found to have a high degree of homogeneity in the topology and network performance, provide an effective reference to expand research the urban rail transit in other cities.

2 Network construction of Urban Rail Transit system

Urban Rail Transit system is a complex network having a defined topological relations, The Urban Rail Transit system network methods include the R Space method and the L, P Space method. The R Space method is to track the rail-line as a complex network node, and the site of line intersects as a complex network link, the number of the site what two lines intersecting as a complex network connection weights, to construct the R spatial network model. In contrast, the L, P Space method is to track the site as a complex network node, and the rail-line as the complex network link, so this method is a direct mapping to track network. the configuration differences of the L,P Space method as follows:

(1). The L space network model. The network model is a direct mapping of the rail network diagram that if two adjacent sites in a track circuit, then connect it with the link.

(2). The P space network model. When two sites are connected by at least one rail line, then connect it with the link. i.e. $Line=(p_1, p_2, p_3 \ldots p_n)$, the site p_1 and the sites $p_2, p_3 \ldots p_n$ are to establish a connection.

Fig. 1 shows the display method of the L, P Spcce.

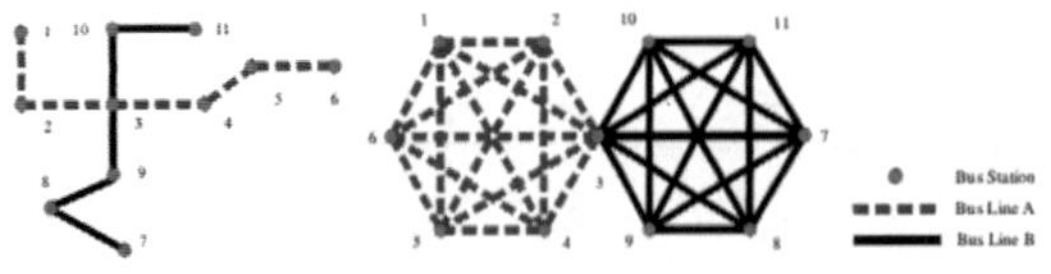

Figure 1. The display method of the L, P Spcce.

This paper selected the L, P spatial network method for urban rail network mapping. The L space network model is applied to study the degree distribution and network efficiency of the track network. The P space network model is suitable for the study of transfer parameters. The above two methods have their limitations respectively, so this paper will combine the use of these two methods.

Taking into account the Urban Rail Transit system in Beijing, Shanghai, Guangzhou developed more perfect and mature, so this paper make the Urban Rail Transit networks of above three cities as the study object. Urban Rail Transit network data including line network operators table, name of the site, ID number of the site, name of the line, station location (latitude and longitude) and other items, All data items got through the Baidu maps API interface. Convert the data file for the site ID retrieval matrix, site latitude and longitude matrix, adjacency matrix and a series of data basis for complex network analysis based on the above L, P spatial network method.

3 Topology homogeneity of urban rail transport network

3.1 Static topological features

Static topological characteristics of complex networks including degree and degree distribution, the network diameter and the average path length, betweenness and clustering coefficient.

(1). Degree and degree distribution. A degree of the node i is the number of edges associated with the node i. The average degree value of each node is the degree of the entire network, i.e.

$$<k>=\overline{k_i}=(\sum_{i\in N} k_i)/N \quad (1)$$

$p(k)$ represents the probability of each node degree distribution.

(2). The network diameter and the average path length. In the complex network theory, define the distance between nodes n_i, n_j is the number of edges in the shortest path connecting nodes n_i, n_j . The network diameter D is the maximum distance between any two nodes in the network, i.e.

$$D=\max_{i,j}(d_{ij}) \quad (2)$$

The average path length of the network is the average distance between all nodes.

(3). Betweenness. Betweenness divided the edges betweenness and the node betweenness. The so-called edge betweenness, refers to the network between all nodes on the shortest path through which the ratio of the edge, the node betweenness and the edge betweenness have the similar meaning.

(4). Clustering coefficient. clustering coefficient reflects the degree of aggregation of nodes in the network, the degree of node n_i is k_i , i.e. adjacent to the surrounding k_i nodes. If the interconnectivity between these k_i nodes, even up to the $k_i(k_i-1)/2$ links, Suppose even number of links it actually exists is n_{k_i} , then the node clustering coefficient C_i is:

$$C_i=n_{k_i}/(k_i\times(k_i-1)) \quad (3)$$

Network clustering coefficient C is the arithmetic mean of each node clustering coefficient:

$$C=\frac{1}{N}\sum_{i=1}^{N}C_i \quad (4)$$

3.2 Homogeneous topology

Through computer programming to Beijing, Shanghai and Guangzhou Urban Rail Transit network topology packet characteristic calculation, the specific results are shown in Table 1.

By the Table 1, it can be seen, Urban Rail Transit network scale of the three cities, but the average degree of nodes, the network diameter, the average path length, the average clustering coefficient roughly equivalent, It illustrates the complex network topology homogeneity of the three cities.

DOI: 10.1051/matecconf/20168101003

Table 1. Topology eigenvalues of the Urban Rail Transit network

Space	City	Nodes	Average degree	Network diameter	Average path length	Average Betweenness	Average clustering coefficient
Space L	Beijing	275	2.24	45	15.0662	2309.03	0.0024
	Shanghai	291	2.24	41	15.1194	2445.5	0.0029
	Guangzhou	144	2.17	44	12.8969	962.2	0
Space P	Beijing	275	26.61	5	2.5349	543.9236	0.9231
	Shanghai	291	29.52	4	2.2184	469.2096	0.9127
	Guangzhou	144	19.47	4	2.3344	168.9444	0.928

L spatial clustering coefficient of the three cities substantially 0, which is determined by the properties of the construction of urban rail transit systems, expensive urban rail construction costs and determine the sparsity of oligomeric class line network, Most sites track just two sites adjacent to and substantially on the same line, so the clustering coefficient is 0. In contrast spatial clustering coefficient of the P Space, then it was higher than 0.91, indicating clustering between different lines are relatively strong, Urban Rail Transit network transfer functions are strongly reflected and the average path length of the P Space network also corroborated this judgment. It should be noted here. The transfer here refers to the transfer of different urban rail transit lines, and the transfer between urban rail transit sites and other means of transport (such as cars, buses, etc.) is not included. that can see, the average path length of the Space P network to Beijing, Shanghai, Guangzhou, respectively 2.5349, 2.2184, 2.3344. Adjacency matrix DP of the Space P network reveals transfer relationship about urban rail complex network, processing to obtain the distance matrix of the Space P network using the shortest path algorithm, i.e. PDist. $PDist(i,j)-1$ represents the number of transfers site i to the site j, and if we can get the site of the travel demand distribution $S_{i,j}$, that can calculate the average number of transfers to the entire urban rail complex networks, i.e. Aver_P:

$$Aver\_P=\frac{\left(\sum_{i}^{m}\sum_{j}^{n}\left(S_{i,j}*\left(PDist(i,j)-1\right)\right)\right)}{\sum_{i}^{m}\sum_{j}^{n}S_{i,j}} \tag{5}$$

Aver_P reflects the level of transfers to the entire urban rail complex networks, the smaller the value indicates that the better direct accessibility of the system and the higher the efficiency of the system. When the demand for travel distribution rail network between sites is uniform distribution function, The average transfer times of rail transit network of the three cities, i.e. Aver_P, is 1.5349, 1.2184, 1.3344, It showed that the number of rail lines network, topology, the number of transfer hubs of the three cities are different, but have a high homogeneity in the transfer performance.

3.3 Uniformity of distribution

Respectively, performed the function fitting to the site degree probability distribution of Beijing, Shanghai, Guangzhou's Urban Rail Transit network, the results shown in Fig. 2.

Fitting functions are:

Beijing: $$P(K)=0.04000+\frac{1.02329\times e^{-2\times\left(\frac{K-1.69619}{0.45999}\right)^2}}{0.45999\times\sqrt{\pi}} \tag{6}$$

Shanghai: $$P(K)=0.01947+\frac{0.75825\times e^{-2\times\left(\frac{K-1.91034}{0.77438}\right)^2}}{0.77438\times\sqrt{\pi}} \tag{7}$$

Guangzhou: $$P(K)=0.03241+\frac{0.73099\times e^{-2\times\left(\frac{K-1.89232}{0.74341}\right)^2}}{0.74341\times\sqrt{\pi}} \tag{8}$$

where: K is the site degree, and $p(k)$ represents the probability of each node degree distribution. Specific topology eigenvalues of the Urban Rail Transit network Can refer to Table 1.

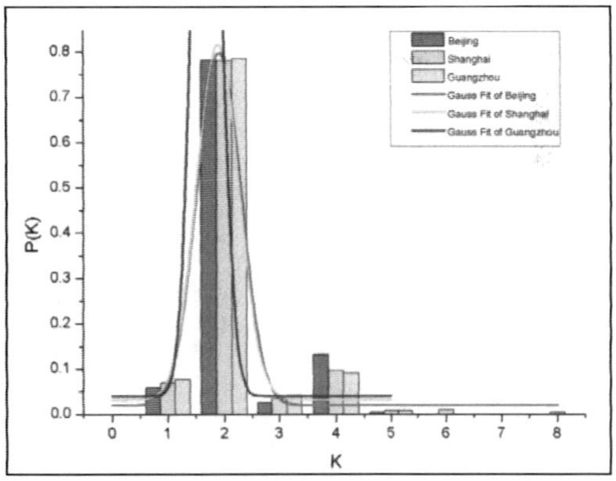

Figure 2. Degree distribution fitting (Space L).

The rail transit network degree distribution fitting function of the three cities was Gaussian distribution, between poisson distribution (ER random network and the small-world networks) and Power laws distribution (BA scale-free network) between, While the average degree of the Space L,P are small,, indicating good small-world characteristics, it was found that three urban rail transport network are random scale-free networks. Although, the construction planning and the construction model of different cities' rail transit network are different, with a high degree of unity which generates the type of network topology, indicating the internal homogeneity of the urban rail network.

3.4 Assortative coefficient of the Urban Rail Transit network

Assortative coefficient of the complex network describes the connection between nodes preferences, indirectly reflects the evolution of the network organization strategy. If relatively large degree nodes tend to connect relatively large degree nodes, Then the network is the degree - the degree of positive correlation, it is also called Assortativeness. Newman [10] use the Pearson correlation coefficient (r) of edge nodes at both ends to describe the degree - the degree of correlation.

Pearson correlation coefficient of node degrees, also known as the Assortativity coefficient that defined as:

$$r=\frac{M^{-1}\sum_{e_{ij}\in E}k_ik_j-\left[M^{-1}\sum_{e_{ij}\in E}\frac{1}{2}\left(k_i+k_j\right)\right]^2}{M^{-1}\sum_{e_{ij}\in E}\frac{1}{2}\left(k_i^2+k_j^2\right)-\left[M^{-1}\sum_{e_{ij}\in E}\frac{1}{2}\left(k_i+k_j\right)\right]^2} \tag{9}$$

where: respectively, k_i, k_j is the degrees of the two nodes v_i, v_j to link e_{ij} ; M is the total number of network links; E is the set of links of the network. The range for the Pearson correlation coefficient(r) is $0 \le |r| \le 1$. When r<0, the network is negative, i.e. No-Assortativeness; when r> 0, the network is positively, i.e. Assortativeness; When r = 0, the network is irrelevant.

Processing the urban rail network adjacency matrix of Beijing, Shanghai, Guangzhou, to give the Pearson correlation coefficient were 0.0198, 0.0190, 0.0827. The results illustrated the rail network of the three cities are assortative network, i.e. relatively large degree nodes tend to connect relatively large degree nodes. Since the rail network is sparse network and the high cost of construction, the new lines will always want to be able to connect the hub sites which have larger degree together, and to radiate more areas. With well-developed network structure, the new lines will be more emphasis on the connection between the suburb sites of the network which have smaller degree, the pearson correlation coefficient of Beijing, Shanghai was about 0.02, while Guangzhou was 0.08, had higher assortative coefficient than Beijing and Shanghai. Shows that Urban Rail Transit network of Beijing and Shanghai has been the better development, the Urban Rail Transit network of Guangzhou is still in the growth evolution period.

4 Functionality homogeneity of rail network

With people's increasing reliance on the urban rail network, Analysis of functionality and reliability for urban rail networks are also becoming more important. Urban rail network failures include internal faults and external faults, Internal fault system mainly refers to the aging of internal disturbance fault itself generated, External fault mainly refers to the network fault external factors such as natural disasters, terrorist attacks, traffic congestion. External fault including the random attack, the maximum degree attack, the maximum betweenness attack from the attack mode[11]. Since the terrorist attacks and other malicious attacks against the subway station mainly, so fragile and reliability studied in this paper is about the node attack of the urban rail network.

4.1 Attack mode

Specific steps of the three kinds of network nodes attack mode are as follows:

(1). The random attack. Initially, randomly select a node in the network to delete, update the network adjacency matrix, update network topology parameters. Repeated the operation, until all the nodes are removed from the network.

(2). The maximum degree attack. Initially, compute node degree of the network, randomly select a node form a set of nodes which have the same maximum node to delete, update the network topology parameters. constantly repeating the above operations until all nodes are removed from the network.

(3). The maximum betweenness attack. Similar to the maximum degree attack, but the attack node selection rule becomes the maximum betweenness nodes.

4.2 Network efficiency analysis

To measure changes in urban rail network performance under attack, network efficiency herein chosen to measure the performance of the network topology.

Defined the node efficiency between nodes n_i, n_j as τ_{ij}, i.e.

$$\tau_{ij}=1/d_{ij} \tag{10}$$

d_{ij} is the shortest path between network nodes. According to the complex network theory, the average efficiency of all nodes on the network is the global efficiency of the network, i.e. E_{glob}:

$$E_{glob}=(2\sum_{i<j\in N}\frac{1}{d_{ij}})/(N(N-1)) \tag{11}$$

where: d_{ij} is the shortest path between network nodes; N is the number of nodes of entir network.

Fig. 3-Fig. 5 are schematic network efficiency changes of Beijing, Shanghai, Guangzhou rail transit network under the random attack, the maximum degree attack and the maximum betweenness attack. As can be seen, variation rail network efficiency of the three cities are common under the above three attack modes--showing "waterfall" down. Among them, the maximum degree attack and the maximum betweenness attack have

DOI: 10.1051/matecconf/20168101003

the quite same destroying effect on the Urban Rail Transit network, they were significantly higher than the random attacks.

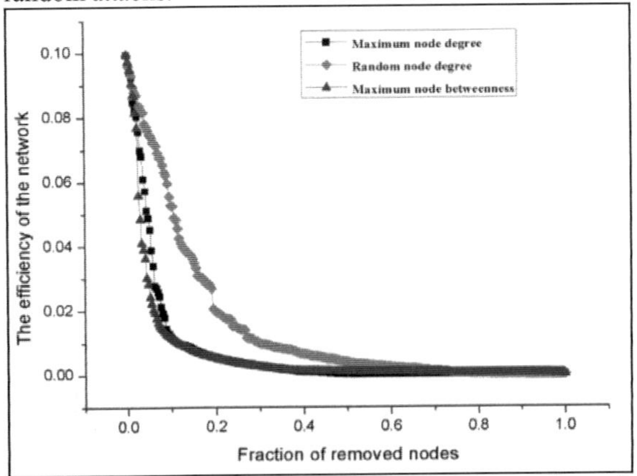

Figure 3. Network efficiency changes of Beijing urban rail network.

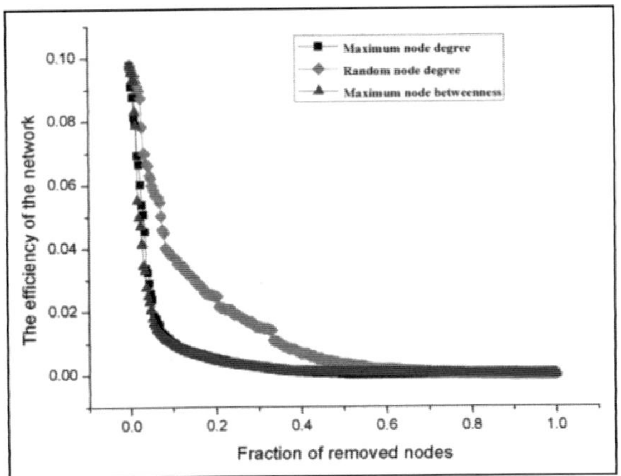

Figure 4. Network efficiency changes of Shanghai urban rail network.

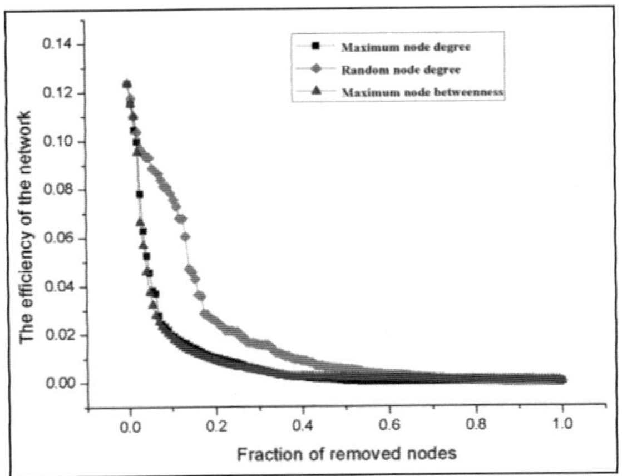

Figure 5. Network efficiency changes of Guangzhou urban rail network.

Under the maximum degree attack and the maximum betweenness attack, removed only 8% of the network nodes, it makes Beijing rail network efficiency decreased by 80% and 85%, Shanghai and Guangzhou are similar. It shows that in the early attacks, a single node removal can cause a lot of network failure. And under the random attack mode, "waterfall" down still showed more obvious, just a little gentle slope only, This is due to the degree and degree distribution of the Urban Rail Transit network. The urban rail network are random scale-free networks, only a few high degree nodes, the degree most nodes are 2, so that performed homogeneity under the three attack modes, i.e. the random attack is robust , but the network performed extremely fragile under the maximum degree attack and the maximum betweenness attack.

4.3 Analysis of the functional homogeneity

This paper make the horizontal comparison on the efficiency of the rail network in the three cities, Fig.6 shows the changes in network efficiency of the Urban Rail Transit in the three cities under the maximum degree attack. Note that because the network nodes and the initial value of efficiency are different, in order to perform horizontal comparison, this paper was carried out normalization process-- The abscissa is the ratio of a node is removed, the vertical axis is the percentage of network efficiency. As can be seen, although the Guangzhou rail transit network nodes far less than Beijing and Shanghai, but the robustness of its network is strongest, Guangzhou has the best description of the network topology in the three cities structure.

At the same time it can be seen from the Fig.6, when the ratio of removed nodes reaches a "threshold", Urban Rail Transit network efficiency will drop to zero. By accessing the relevant parameters matrix of the program, specific values can be obtained as shown in Table 2. From Table 2, when fraction of removed nodes is in the range of 0.51273 to 0.51636, The efficiency of the Beijing Urban Rail Transit network will drop to zero, the fraction range of Shanghai is 0.51203 to 0.51546, and Guangzhou is 0.53472 to 0.54167, so the threshold can be weighed as 0.52, Further illustrates the homogeneity of the Urban Rail Transit network topology in the three cities, also provide an effective reference to expand research the urban rail transit in other cities.

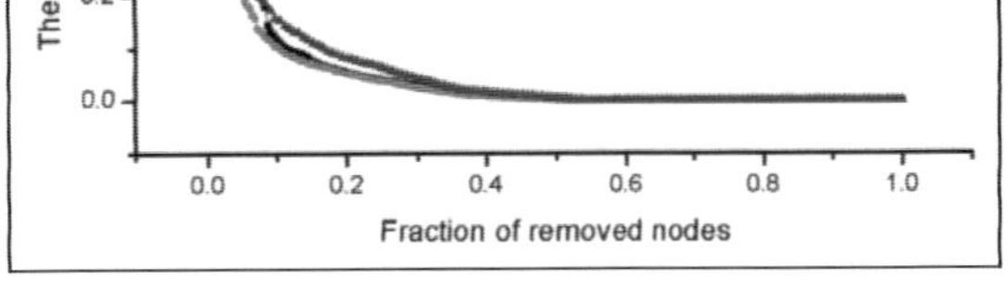

Figure 6. The comparison of network efficiency changes.

Table 2. The threshold of network efficiency

Beijing		Shanghai		Guangzhou	
Fraction of removed nodes	**The efficiency of the Network**	**Fraction of removed nodes**	**The efficiency of the Network**	**Fraction of removed nodes**	**The efficiency of the Network**
0.50182	0.00107	0.50172	0.00097	0.51389	0.00314
0.50545	0.00080	0.50515	0.00073	0.52083	0.00236
0.50909	0.00053	0.50859	0.00049	0.52778	0.00157
0.51273	0.00027	0.51203	0.00024	0.53472	0.00079
0.51636	0.00000	0.51546	0.00000	0.54167	0.00000
0.52000	0.00000	0.51890	0.00000	0.54861	0.00000

5 Conclusion

This paper selects urban rail networks of Beijing, Shanghai, Guangzhou as calculation case, gave them a complex network mapping through the L, P Space method and had a static topological analysis using complex network theory, The main conclusions are as follows:

(1). Urban Rail Transit network of Beijing, Shanghai, Guangzhou have the similar network topology, and clustering between sites is extremely small, high clustering between the lines, belong to the random scale-free networks.

(2). The rail network of the three cities are assortative network, i.e. relatively large degree nodes tend to connect relatively large degree nodes. the pearson correlation coefficient of Guangzhou was higher than Beijing and Shanghai. Shows that Urban Rail Transit network of Beijing and Shanghai has been the better development, the Urban Rail Transit network of Guangzhou is still in the growth evolution period.

(3). Network efficiency of urban rail network under the random attack, the maximum degree attack and the maximum betweenness attack, all showed a "waterfall" down, and the "zero threshold" of network efficiency in the three cities is substantially the same, It shows the homogeneity of Urban Rail Transit network topology under different attack patterns.

However, the research object of this paper is mainly for the physical topology of Urban Rail Transit network, and did not take into account the effect of actual line network passengers and railway operation management for the performance of Urban Rail Transit network, whether the performance of Urban Rail Transit network is still exhibit homogeneity under the influence of many factors, this requires further research.

References

1. National Bureau of Statistics of the People's Republic of China. http://www.stats.gov.cn/
2. S. Boccaletti, V. Latora, Y. Moreno, M. Chavez, & D.U. Hwang. Complex networks: Structure and dynamics[J]. Physics Reports, **424**(4-5):175–308(2006).
3. V. Latora, M. Marchiori. Is the Boston subway a small-world network ?[J]. General Information, **314**(1):109-113, (2002).
4. P. Angeloudis, D. Fisk. Large subway systems as complex networks[J]. Physica A Statistical Mechanics & Its Applications, **367**:553–558(2006).
5. S. Derrible, C. Kennedy. The complexity and robustness of metro networks[J]. Physica A Statistical Mechanics & Its Applications, **389**(17):3678-3691, (2010).
6. R. Ding, N. Ujang, H.B.Hamid, & J. Wu. Complex Network Theory Applied to the Growth of Kuala Lumpur's Public Urban Rail Transit Network[J]. Plos One, **10,** (2015).
7. Y. Yang, Y. Liu, M. Zhou, F. Li, & C. Sun. Robustness assessment of Urban Rail Transit based on complex network theory: A case study of the Beijing Subway[J]. Safety Science, **79**:149-162, (2015).
8. Y. Liu, Y. Tan. Complexity Modeling and Stability Analysis of Urban Subway Network: Wuhan City Case Study[J]. Procedia - Social and Behavioral Sciences, **96**:1611-1621, (2013).
9. J. Zhang, M. Zhao, H. Liu, & X. Xu. Networked characteristics of the Urban Rail Transit networks[J]. Physica A Statistical Mechanics & Its Applications, **392(6)**:1538-1546, (2013).
10. M.E.J. Newman. 2003b, 'Mixing patterns in networks[J]. Physical Review E, **67**(2):241-251, (2003).
11. J. Zhang, X. Xu, L. Hong, S. Wang, & Q. Fei. Networked analysis of the Shanghai subway network, in China[J]. Physica A Statistical Mechanics & Its Applications, **390**(23):4562-4570, (2010).

Study on Multi-Depot Collaborative Transportation Problem of Milk-Run Pattern

Zhenkai Lou, Zhenzhen Li, Lieying Luo and Xiaozhen Dai

Wenzhou Business College, Wenzhou, China

Abstract. Analyze the relevance between Milk Run mode and collaborative transportation problem, put forward collaborative transportation problem of multiple-depot on Milk Run mode under the supply and demand separate nodes, consider the value of transport and transport costs, introduce the concept of node - arc flow, by comparing the size of traffic flow determine nodes collection, and then constructed multi-transport model of the problem. Considering one-way pickup and delivery closed, construct two-stage algorithm model, use dynamic programming recursive solution to determine the best route to pick up, and then solving delivery routing problem with different start and return point based on geometric method of Cosine. Finally use a numerical example illustrates the effectiveness of the algorithm and reasonable model.

1 Introduction

Transportation is the essential means to ensure the orderly conduct of production and life. There is a proverb "to get rich first build roads", which is fully demonstrates the promotion of transport on the economy. Transportation costs accounting for up to half in total logistics costs. A large part of the reason is the lack of cooperation among enterprises, resulting in insufficient capacity or load phenomenon occurs frequently. In this context, it is very important strengthen the cooperation between enterprises, through collaborative transportation combined transport tasks, integrate transport resources, thus reducing transport costs. In collaborative transportation, cost allocation, delivery order, route planning, etc are the focus of the enterprise. Collaborative transportation is the development and the important content of collaborative logistics operation mode in recent years. Collaborative transportation demands relatively independent enterprise work together, share logistics resources and logistics information and customer requirements to reduce the traffic light, and reduce overall logistics costs, which improving performance and the service level of transportation purposes. This paper research the suppliers and demanders distribution in different regions with collaborative transportation problem of less-than-one carload logistics.

Literature [1] summarize the situation of vehicle mixed load, limit load and full coverage etc. about collaborative transportation problem, by constructing a two-stage greedy algorithm to solve the problem of large-scale multi-coordinated transport depot fully loaded. Algorithm consider the radius of customer distribution and route design, and it can fit actual situations well. But it only consider the total transport mileage without considering the level of logistics services, the value of goods, etc., so the research is one-sided.

Literature [2] analyze the problem of collaborative transportation route optimization in straight-through transportation, a single point and two points of transit models, built integer programming model of transportation costs and achieved good results with using Lagrangian relaxation algorithm. Literature [4], [5] research collaborative transportation cost control from the perspective of select transit point and route coverage. Literature [7] take the efficiency of the distribution into key account, by choosing a reasonable set of nodes to establish with Hybrid Synergy distribution model to improve the client base's distribution levels. The disadvantage is not considered cost factor, and also lack scientific evidence of the number of set point. A single distribution point is often unable to meet the set requirements in reality of large-scale cooperative distribution problems.

2 Question and models

Coordinated transport refers to multiple transportation enterprises that according to certain organization integrated transportation tasks and resources, reasonable schedule the transportation resources, collaborative optimization is carried out on the transport task allocation so as to reduce the total cost and increase the enterprise profit and competitiveness. As for collaborative transportation now are very seriously, and has made remarkable benefit in practice.

First, compare traditional transportation with collaborative transportation. Traditional transportation refers to each enterprise only consider their own transport tasks, namely do the job in the form of single node. The collaborative transportation refers to a number of enterprises need to complete a number of tasks transportation, vehicles set out from a node, shipping and delivery tasks for a plurality of transport tasks and finally return to the yard.

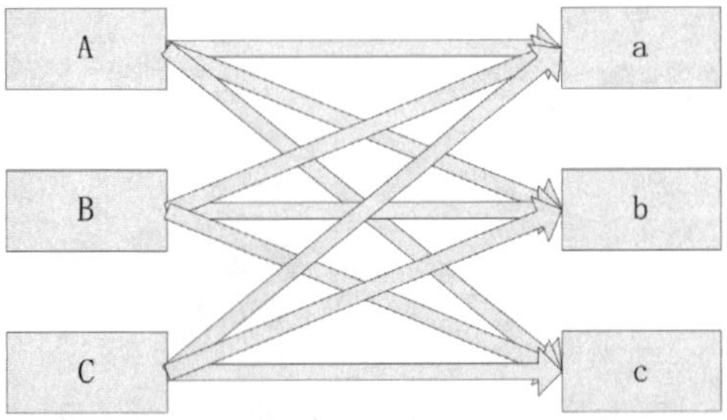

Figure 1. Traditional transportation model.

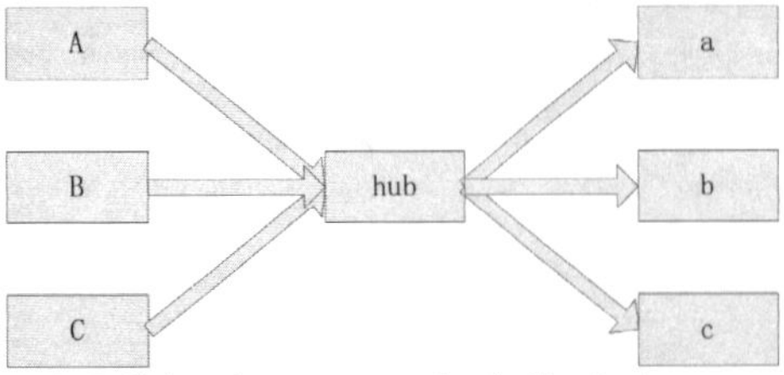

Figure 2. Collaborative transportation in distribution center.

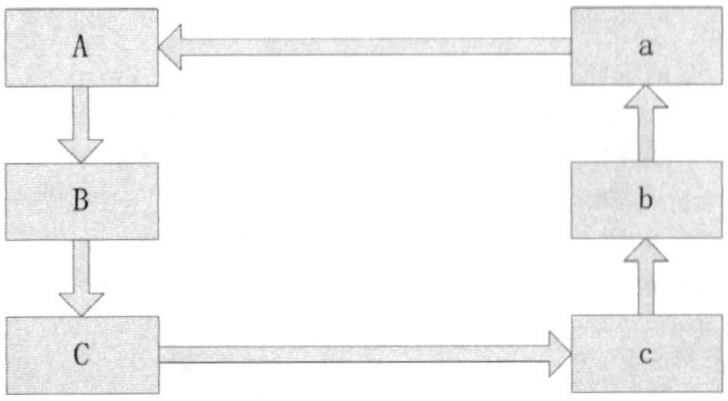

Figure 3. Milk run mode.

This paper adopts the model of the cooperative transport in Fig. 3. In collaborative transportation of different the shipper, the carrier will transport business integration in the process of optimization, and the participant enterprises must carry out individual decisions, collaborative alliance also must carry out the overall decision. Individual decision making and the overall decision-making process of the alliance, need to arrange a schedule based on the path of transport vehicles. At the same time, the cooperative transport in a number of important issues, such as participating in the mission of the organization combinatorial auction bid, collaborative design problems of transportation costs and benefits of allocation strategies, must be oriented to collaborative transportation vehicle routing problem based on optimization results.

To build the model convenient, make the following assumptions with the problems:

(1) The geographical location of each node is known, node's mileage expressed by linear distance;

(2) Each node is taken delivery and only once;

(3) All the tasks have been confirmed before shipping, the supply and demand situation is no longer change;

(4) The source of supply and demand are scattered in different areas;

(5) As our emphasis is sporadic freight transportation, every car's transport tasks are less than capacity;

(6) As a result of the gas is convenient, there are no biggest form distance constraints of the vehicle.

In the collaborative transportation process which by one or more business enterprises to take a common transport task, the smooth conduct of the premise is to adopt a uniform standard transport costs. The shipping costs only depends on the volume, weight and pick up of yard goods, and goods independent of the actual value, defined as the value of transport. The transportation costs only associated with transport mileage. Optimization goal of this paper is to make the maximum revenue of synergies transport, namely the difference costs between transport and transport of the value. Model as follows:

$$\max Z=\sum_{i\in M} p_i - q\sum_{l\in L}\sum_{k\in K}\sum_{i\in O}\sum_{j\in O} c_{ij}x_{ij}^{lk} \tag{1}$$

$$s.t.\sum_{i\in L}\sum_{j\in M} x_{ij}^{lk}=1, x_{ij}^{lk}\in\{0,1\} \tag{2}$$

$$\sum_{i\in L}\sum_{j\in N} x_{ij}^{lk}=1, x_{ij}^{lk}\in\{0,1\} \tag{3}$$

$$\sum_{i\in O\cup L} x_{ih}^{k} - \sum_{j\in O\cup L} x_{hj}^{k}=0, h\in O\cup L \tag{4}$$

$$p_i - q\sum_{l\in L}\sum_{k\in K}\sum_{j\in O} c_{ij}x_{ij}^{lk} > 0, i\in M \tag{5}$$

$$\sum_{l\in L}\sum_{k\in K}\sum_{i\in M} d_i^{lk} = \sum_{l\in L}\sum_{k\in K}\sum_{j\in N} d_j^{lk} \tag{6}$$

In the formula: M are nodes of a set of the supply, N are nodes of a set demand, O = M∪N; L is a collection of parking lots; K is the set of available transport vehicles; pi is the value of transport; q is units mileage costs; cij is the distance between i to j; xijlk is the decision variables, if the vehicle k starting from yard l and drive from i towards j, take 1, then take 0 otherwise; di is the traffic of node i.

In the Model: the formula (1) is the revenue - cost function of synergies transportation; the formula (2)means each supply point pickup and only once; the formula (3) refers to each point needs to be and only be shipping once; the formula (4) represents in each node the inflow and outflow of service vehicle is counterpoise; formula(5) represents the value of each supply point transport is more than transportation costs; formula (6) represents the total amount of transport supply and demand set point set point of balance .

For solving vehicle routing problem model, there are a variety of methods, including traditional algorithm and heuristic algorithm. In view of the separation of supply and demand of collaborative distribution problem, paper designs two-stage algorithm to solve. The purpose to design the algorithm is, from local optimal approximate global optimal, and obtain a satisfactory solution.

DOI: 10.1051/matecconf/20168101004

3 Algorithm design

Generally speaking, the vehicle routing problem can be divided into point path problem and arc routing problem. This paper combine the concept of point and arc , introduces the concept of node - arc discharge.

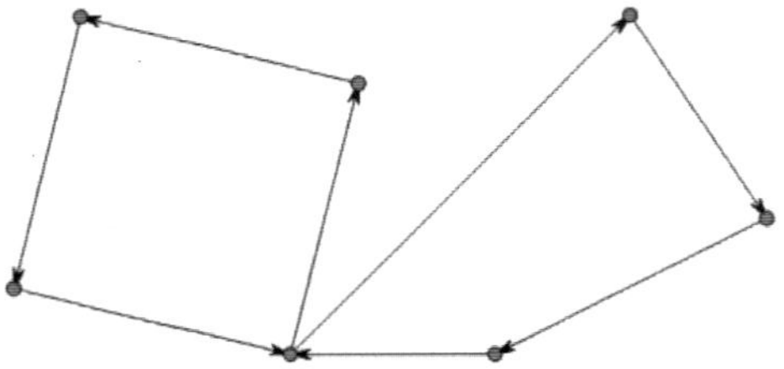

Figure 4. Node routing problem.

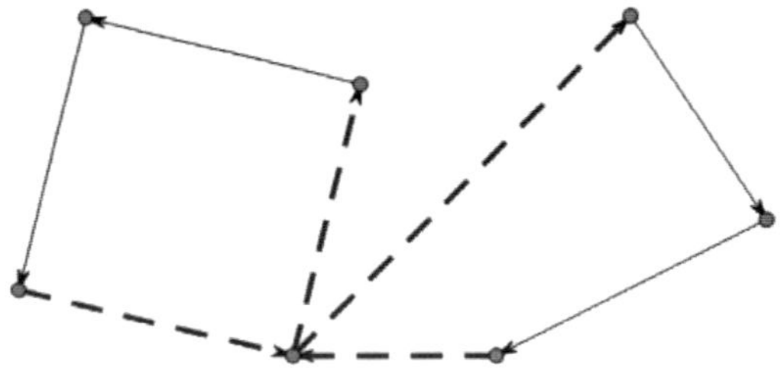

Figure 5. Arc routing problem.

Milk run mode of collaborative transportation problem is belong to the yard vehicle scheduling problem, the article considers the perspective of transportation benefits, establishes the income - cost model, and put forward the thinking of solving the two-phase algorithm on the basis of this. The first stage it introduces the concept of node - arc discharge and split all supply center to each depot, then set the stage and the state, using the dynamic programming algorithm for optimal route of pick up; The second phase uses geometric method by covering arc will delivery routes connect circuits.

The rationality of two phase algorithm is that the supply notes are always far away from the demand notes, the finally pick up location of vehicle changes have less effect on the delivery path length, just as the train passed through a little way, and the distance of the train between the reference has not changed much, this can be explained from the Angle of the differential.

3.1 The first phase of dynamic programming

First of all, illustrates the concept of node - arc discharge, This concept is a comprehensive supplier of transport and yard to transport mileage, and the supplier. To transfer the traffic of the node and arc mileage united into a quantity of transportation, need a unified dimensional. The supplier of goods on the basis of size, weight, turning traffic into transport value, The cost of transportation is equal to the yard to supply level between arc length multiplied by the unit transportation cost, Transport value and the difference between the cost of transportation, is the number of arc of the node - arc discharge.

The next major is to determine a collection of major yard service node. A yard l1,l2,……,li, Supply nodes m1,m2,……,mj, Supply node behavior, as a yard, the node - arc flow matrix of the:

$$\begin{bmatrix} a_{11}\ a_{12} \cdots\cdots a_{1j} \\ a_{21}\ a_{22} \cdots\cdots a_{2j} \\ \cdots\cdots\cdots\cdots\cdots\cdots \\ a_{i1}\ a_{i2} \cdots\cdots a_{ij} \end{bmatrix}$$

Determine the set of rules is as follows: For each line in the matrix of the node – arc discharge, take one of the biggest, and divided into corresponding yard, it can be divided all supply nodes in turn.

To determine the significance of the collection is that each station is responsible for the pickup service outlets are transportation value larger nodes, which will be a big problem is decomposed into several small problems.

Do not break general, a yard l1 service node set {m1,……,mk}, if expressed in position state, this problem is a typical multistage decision problem, its initial state is l1, according to the model shows that the k status to the arrangement of m1~mk, which is similar to the calculation of the determinant rule.

3.2 The second stage geometry method

After the completion of the first stage of the unstuffing need to arrange delivery of each demand point. A yard li finally pickup node for mj is Michael Jackson, mj is the starting point of delivery, delivery after the return to li, pickup route and route of delivery constitute a loop.

According to the geometric method principle, circuit obtuse Angle in the polygon, the more the greater the Angle, then this polygon Zhou Chang small, easy to understand its rationality from the Angle of mathematics. Setting up a yard is responsible for the delivery of the demand points for n, combining with the cosine theorem, improved geometric method steps are given:

(1) For each station, responsible for pickup outlets, and responsible for the delivery of the demand points, lists its geographical location, for convenience, expressed in plane rectangular coordinates;

(2) Based on cosine function is monotone decreasing in the 0~π features, To calculate the cosine of every node in (1) based on any two nodes, and the smallest to the An+2n cosine value, more nodes in the circumstances, there can be realized by programming;

(3) According to cosine list select nodes and the corresponding edges, since the rules are as follows: select a node and two nodes, in the list of cosine cross to the rest of the cosine value associated with this node, then choose not to scratch the smallest cosine values and the corresponding node, until form a loop.

4 Analysis of example

The process of algorithm design fully considers the background of the problem and the applicable scope, and targeted to the filtered solution, make it meet the requirements of the model. Test of practical examples for the rationality of the model and algorithm is indispensable.

DOI: 10.1051/matecconf/20168101004

To analyze the feasibility and validity of the model, and constructed two yard, supply level and requirements on the separation of small example, yard location L1(8，10)，L2(16，16), unit mileage charge set to 1, the fewer number of yard, assuming the same goods to different transport value equal to the yard, each supply level is set to 50.

Table 1. Supplier point information

A	B	C	D	E	F	G	H
4,4	8,5	6,12	18,12	12,12	16,10	12,18	18,20

Table 2. Demand point information

A	B	C	D	E	F	G	H
4,4	8,5	6,12	18,12	12,12	16,10	12,18	18,20

So L1 is responsible for the collection of pickup outlets for {A, B, C, E}, L2 is responsible for the collection of pickup outlets for {D, F, G, H}. This is one-way traveling salesman problem; By traveling salesman problem of dynamic programming method, we know that smaller case can use the discrete problem in dynamic programming method for the optimal solution. It is easily obtained that in the supplier concentration area, the first optimal path for L1→B→A→C→E, the second optimal path for the L2→G→H→D→F.

Next is the route of delivery. Respectively by the end of each unstuffy as a starting point, to deliver goods is regard as the starting point for unstuffy as the end of the delivery operation so as to forming a loop.

Improvement principle of geometric method is established on the basis of Vogel method and geometric method. The establishment of the cosine table make delivery of the nodes selected in turn becomes simple.

Every time when choose a node, we can make the next step of alternative nodes greatly reduced. Delivery can be obtained by applying the method of geometric path, and take the order form a loop.

Loop1 : L1→B→A→C→E→e→c→b→a→L1， The total mileage is 92.94;

Loop2: L2→G→H→D→F→f→g→h→d→L2，The total mileage is 83.26.So The cooperative transport income is 223.8.

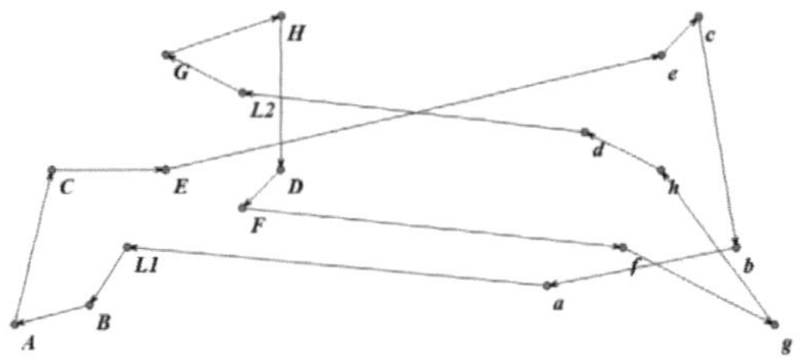

Figure 6. Coordinated transport vehicle trajectory.

5 Conclusion

The object of this study is supply level and demand point separation gathered collaborative transportation problem, considering the subject need to provide transportation service to ensure profitability, Therefore, from the perspective of income - cost model is established. According to the complexity of multiple yard to introduce the concept of node - arc discharge and supply level corresponding to yard can be divided into several collections, the essence of this idea is a NP problem into several small problems under certain conditions, to approach the global optimal solution by local optimum.

The design of the two-phase algorithm improves the dynamic programming method and geometry method. Further research can be from the time window, different models such as the angle of the multi-objective optimization.

Collaborative transportation is a new mode of transportation, which is a very complicated system; there are many factors related to it, the problems that need to be solved are very complex and difficult. With the unceasing change of environment, it will also appear new features and problems in this field. We can also coordinate transport problems in the research of the subsequent study combined with other problems. Such as collaborative transportation problem and the profit distribution problems in combination with in-depth study, or combination coordinated transport and cooperative inventory problem, research collaborative logistics oriented collaborative management of inventory and transportation problem is also a very meaningful study direction.

References

1. R. Liu, Z. Jiang, F. Chen, Journal of Shanghai jiaotong University **3**, 43, 455-459, (2009)
2. K. Wen, Z. Xu, Industrial Engineering and Management **12**, 19, 51-56, (2014)
3. K. Wen, K. Zhu, G. Liu, Journal of Systems & Management **1**, 24, 38-42, (2015)
4. S. A. Alumur, B. Y. Kara, O. E. Karasan, Omega **40**, 6, 927-939, (2012)
5. J. Puerto, A. B. Ramos, Computers & Operations Research **38**, 2, 559-570, (2011)
6. K. Wen, K. Zhu, G. Liu, Chinese Journal of Management Science **1**, 23, 135-140, (2015)
7. M. Liu, L. Zhao, Control and Decision **1**, 26, 96-100, (2011)
8. W. Tan, Q. Wen, Mathematics in Practice and Theory **12**, 45, 235-242, (2015)

 DOI: 10.1051/mateccconf/20168101005

Simulation Framework for Rebalancing of Autonomous Mobility on Demand Systems

Katarzyna A. Marczuk[1,4], Harold S.H. Soh[2], Carlos M.L. Azevedo[3], Der-Horng Lee[1,4], and Emilio Frazzoli[3,4]

[1] *National University of Singapore*
[2] *University of Toronto*
[3] *Massachusetts Institute of Technology*
[4] *Singapore-MIT Alliance for Research and Technology*

Abstract. We are observing a disruption in the urban transportation worldwide. The number of cities offering shared-use on-demand mobility services is increasing rapidly. They promise sustainable and affordable personal mobility without a burden of owning a vehicle. Despite growing popularity, on-demand services, such as carsharing, remain niche products due to small scale and rebalancing issues. We are proposing an extension to the traditional carsharing, which is Autonomous Mobility on Demand (AMOD). AMOD provides a one-way carsharing with self- driving electric vehicles. Autonomous vehicles can make the carsharing more attractive to customers as they (i) reduce the operating cost, which is incurred when a manually driven system is unbalanced, and (ii) release people from the burden of driving.
This study is built upon our previous work on Autonomous Mobility on Demand (AMOD) systems. Our methodology is simulation-based and we make use of SimMobility, an agent-based microscopic simulation platform. In the current work we focus on the framework for testing different rebalancing policies for the AMOD systems. We compare three different rebalancing methods: (i) no rebalancing, (ii) offline rebalancing, and (iii) online rebalancing. Simulation results indicate that rebalancing reduces the required fleet size and shortens the customers' wait time.

1 Introduction

During the post-World War II period automobiles became more widespread, especially in the US. With large-scale suburban areas, the commuting trend accelerated and people became car-dependent [1]. The mass use of motor vehicles led to some unforeseen consequences in terms of congestion, pollution, safety and climate change. This has spurred a growing interest in shared-use mobility—particularly a one-way vehicle sharing—as a sustainable alternative to privately owned vehicles. The challenge is to ensure flexibility of private vehicles while removing the need of car ownership. It is commonly known that most of the vehicles used in urban areas are heavily *underutilized*, i.e., private vehicles are parked around 22 hours daily and their driving speed is usually five to ten times slower than their design speed [2]. One of the hopes for shared-use mobility is that it will reduce congestion and costly parking requirements. Despite these prominent advantages, an inadequate and unbalanced fleet of shared vehicles can result in service unavailability problems, particularly during periods of high demand. One potential solution for this issue is to leverage on recent developments in robotics technology and use vehicles with self-driving capabilities. Through automated rebalancing, Autonomous Mobility on Demand (AMOD) can redistribute cars to better meet demand. Autonomous vehicles can make the carsharing more attractive to customers as they can (i) reduce the operating cost, which is incurred when a manually driven system is unbalanced, and (ii) release people from the burden of driving.

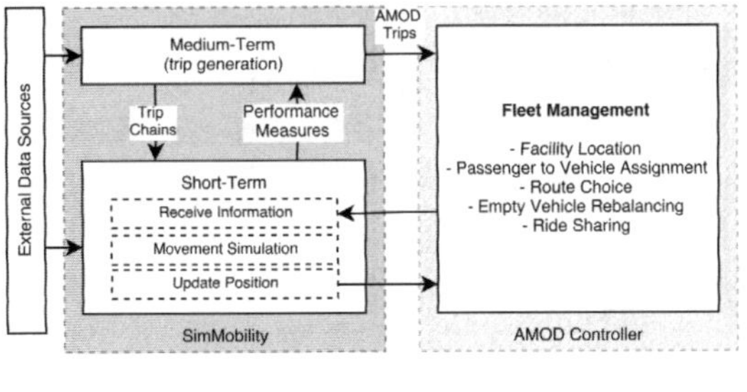

Figure 1. The AMOD Controller handles fleet management of autonomous vehicles and consists of five main components responsible for: (i) facility location, (ii) passenger to vehicle assignment, (iii) routing, (iv) empty vehicle rebalancing and (v) ridesharing. AMOD Controller dispatches orders to SimMobilityST, which performs a 0.1 second scale simulation of the vehicles and returns vehicular information (e.g., speed and location) back to the fleet management module. The AMOD trips are generated in SimMobilityMD.

A main motivation for the development of AMOD systems is sustainable transportation, yet no standard methodology has been established to accurately and consistently design and evaluate this new service. Existing methods for operating as well as modeling urban transport require extensions to credibly incorporate AMOD systems as benefits analysis requires controlled experiments that compare transportation behavior with and without the new mode. Breakthroughs in real-time management of the entire transportation system can lead to transit models that better optimize resources and improve efficiency.

This work is built upon our work presented in [3, 4] and sets out to highlight importance of rebalancing for autonomous mobility on demand system. In our previous studies, we presented the AMOD Controller, which handles fleet management of autonomous vehicles, specifically facility location, passenger to vehicle assignment, vehicle routing, empty vehicle rebalancing and ridesharing. The AMOD Controller is developed as an extension of SimMobility—an agent-based microscopic simulation platform—to model and evaluate different scenarios of AMOD systems (Fig. 1). In the current work we focus on the framework for evaluating the impact of different rebalancing policies on the performance of the AMOD system. The system performance is measured in terms of the level of service, i.e., the customers' waiting time and travel time (travel time is understood as an indicator of road congestion).

The rest of this paper is structured as follows. Section 2 provides a literature review of fleet management of mobility on demand systems. Section 3 discusses our methodology to test different rebalancing policies. Results are presented in Section 4. Conclusion and future directions are highlighted in Section 5.

2 Background and literature review

Automated mobility on demand systems attempt to provide a one-way carsharing with self-driving electric vehicles. AMOD vehicles hold great promise for mobility on demand systems because they can cooperate with each other and rebalance themselves. Through the system-level coordination, autonomous vehicles can use existing roads more efficiently i.e. by reducing the distance headway to the autonomous vehicle in front or by routing vehicles via not heavily congested roads [5]. To the consumer, AMOD offers an alternative transportation mode to the private vehicles. AMOD vehicles are demand-responsive, which means that they do not operate on a regular schedule like buses or trains, but rather only run when there is a request for the service. This allows for a long- term environmental sustainability and potential cost savings for customers, while relieving people from the burden of driving. Autonomy could also potentially increase safety as the road accidents are mainly caused by the human errors.

The system of autonomous shared vehicles combines benefits of both, standard carsharing systems and taxis. In general, carsharing system might be seen as not as attractive as taxis because the customers might not be able to find a free vehicle close to their position. On the contrary, taxi can pick them up from any location, but with the cost of hiring people to drive the vehicles. AMOD reduces the need of hiring drivers while ensuring vehicles' availability not only within a walking distance to parking lots. Despite these advantages, an unbalanced and inappropriately sized fleet of AMOD vehicles can result in service unavailability problems, particularly during periods of high demand. To address these problems, a few directions in the literature have been established.

To estimate the minimum required fleet size of shared-use vehicles, many researchers have focused on rebalancing strategies for both station-based and free-floating carsharing system [6–8]. Some studies attempt to estimate the fleet size of autonomous shared-use vehicles [3–5, 9–11].

[9] shows a theoretical solution to fleet sizing by introducing rebalancing assignments that minimize the number of empty vehicles traveling in the network and the number of rebalancing drivers needed, while ensuring stability. The introduced rebalancing policy (based on a fluid model) was tested in a simulation framework developed in Matlab. Following on this work, [5] discusses a thought experiment of replacing all vehicles in Singapore with a fleet of autonomous and on demand cars to serve for the personal mobility of the entire population. The authors investigate the potential benefits of the AMOD system. Their work shows that the AMOD system requires a number of robotic vehicles equal about 1/3 of the current number of passenger vehicles in Singapore. [11] proposes a queuing-theoretical model for autonomous mobility on demand systems, which minimizes the number of rebalancing vehicles by solving a linear program. The model is applied to the New York City. Their case study shows that the current taxi demand in Manhattan can be met with about 60% of the size of the current taxi fleet.

[12] presents algorithms for dynamic vehicle routing (DVR) for autonomous transportation-on-demand. In each specific DVR scenario the author adopts the methods from queuing theory, combinatorial optimization, and stochastic geometry for the automatic planning of optimal multi-vehicle routes to provide service to demands. Similarly, [13] and [14] consider routing of shared-use autonomous vehicles. The authors design a routing policy that minimizes the average steady-state time delay between the generation of an origin-destination pair and the time the trip is completed. [2] reviews modeling, control and evaluation of AMOD systems presented in [5, 11–14].

An overview of the modeling, control and evaluation of AMOD systems is presented in [2]. The review is built upon work discussed in [5, 11–14].

[3, 4] is published in continuance of [2, 5, 9, 11]. These two studies make use of SimMobility, a microscopic simulation tool, which allows mode choice and congestion to be taken into account while performing evaluation of the AMOD system. One of their findings for a specific policy indicates that introducing the AMOD impacts the mode choice of commuters as some of them switch to the new service. The authors also show that the rebalancing has a positive effect on system performance (when com-

pared to the system without rebalancing model), however, the empty vehicle trips increased vehicle miles traveled (VMT) and road congestion. In this paper we extend the work on rebalancing by introducing the model which directly estimates the fleet size of AMOD system.

3 Methodology

This work is built upon the studies presented in [3, 4]. A long-term goal of our analysis is to establish a complete and comprehensive research that compares transportation efficiency with and without autonomous mobility system. The research seeks to answer the following questions: (i) What fleet sizes and vehicles distribution are necessary to serve a given demand? (ii) What vehicles assignment and route choice policies are likely to improve system performance? (iii) How to perform rebalancing of the vehicles? (iv) How the ride-sharing should be incorporated into the system?

To better capture the behavior and dynamics of travel patterns, we use a multi-agent modeling approach in a microscopic simulation framework (Fig. 1). We extended SimMobility with a dedicated controller for managing autonomous vehicles. As shown in Figure 1, SimMobility is made up of three pillars: (i) the Short-Term (ST) simulator, which simulates movement of agents at a microscopic granularity, (ii) the Mid-Term (MT) simulator, which simulates agents' behavior in terms of their activities and travel patterns, and (iii) the Long-Term (LT) simulator, which captures land use and economic activity on a year-to-year scale. SimMobility's framework is fully modular in a way that each level can run independently and only interact with the other level when necessary. Every agent exists and at all levels and in this way agent's behavior and characteristics are consistent across the three pillars. More details on SimMobility can be found in [15].

As presented in our previous papers, the fleet management problem is divided into five sub-problems: (i) facility location, which attempts to find the best locations of the stations to park and recharge the vehicles, (ii) passenger to vehicle assignment, which is to find the best vehicle for each customer, (iii) routing, which aims to find an optimal assignment of a vehicle to the route between the origin and destination, (iv) empty vehicles rebalancing, which is to redistribute vehicles to better meet the demand, and (v) ride-sharing, which attempts to match more than one passenger to one vehicle.

The current work sets out to highlight the importance of rebalancing for autonomous mobility on demand system. In our methodology we distinguish three rebalancing methods: (i) no rebalancing, (ii) offline, and (iii) online rebalancing. In (i) vehicles are only moved when assigned to customers and parked at the destination of the trip. This is our baseline scenario for the analysis. (ii) is run based on the historical data and the rebalancing counts are decided before starting the simulation, while (iii) is run during the simulation time and the rebalancing counts are optimized based on the predicted requests at the time of invoking the rebalancing function. From the perspective of system manager, the offline model solves problem of strategic planning and long-term operation, specifically what is the number of vehicles needed and how to distribute them between stations. The online model solves for the short-term operating decisions, specificalljmy how to move vehicles based on the current situation on the network. This study aims to analyze and compare the three models.

3.1 Offline rebalancing

The offline rebalancing model developed in this study extends the linear program introduced in [9]. The objective of our model is to find the minimum number of vehicles at each station at the beginning of the day (Fig. 2). As proven in [9] this is equivalent to minimizing the rebalancing effort. This policy requires a priori knowledge of the demand $d_{ij}(t)$. We assume that this knowledge is available through the historical data and we obtained it from the simulation in SimMobility MT.

The model is formulated as follows. Let $N_i(t)$ and $d_i^w(t)$ be the number of vehicles owned by station i at time t and anticipated number of customers waiting at station i at time t such that $d_i^w(t) = \sum_j [d_{ij}(t) - d_{ji}(t - \tau_{ji})]$. Two decision variables are the number of empty (rebalancing) vehicles to send from station i to station j at each rebalancing period, $r_{ij}(t)$, and the number of vehicles idling at station i at time t, $v_i(t)$. Note, that $N_i(t)$ is the sum of vehicles departed for i plus vehicles idling at i. Therefore, at any time t, $N_i(t) = v_i(t) + \sum_j \left[d_{ji}(t-\tau_{ji}) - d_{ij}(t)\right] + \sum_j \left[r_{ji}(t-\tau_{ji}) - r_{ij}(t)\right]$ for all i,t. The number of vehicles departed for station i, but not available for assignment yet, is $dep_i(t)$. Then, the total number of vehicles $N = \sum_i N_i(t) + \sum_i dep_i(t)$; N is constant over time. We define $v_i^{reb}(t) = \sum_j \left[r_{ji}(t - \tau_{ji}) - r_{ij}(t)\right]$ and assume that the travel time from station i to station j is known and given as τ_{ij}; τ_{ij} is constant over time for any i and j, $i \neq j$. Length of the rebalancing interval is Δt; τ_{ij} is aggregated based on Δt.

The objective of this problem is to minimize the number of rebalancing trips at all times t.

$$
\begin{aligned}
\min \quad & \sum_i r_{ij}(t) & \\
\text{s.t.} \quad & N(t) = N(t - \Delta t) & \forall t \\
& v_i(t) + v_i^{reb}(t) \geq d_i^w(t) & \forall i, t \\
& v_i(t + \Delta t) = v_i(t) + v_i^{reb}(t) - d_i^w(t) & \forall i, t \\
& v_i(t = 0) = v_i(t = T_p) & \forall i \\
& r_{ij}(t = 0) = r_{ij}(t = T_p) & \forall i, j \\
& r_{ij}(t), v_i(t) \geq 0 & \forall i, j, t
\end{aligned}
\tag{1}
$$

The first constraint is set to ensure that the number of vehicles in the system is constant over time. The second constraint tells us that the number of available vehicles at station i has to be sufficient to serve the demand. The third constraint describes a flow conservation at each node. The fourth and fifth are the periodicity constraints. The sixth are the non-negativity constraints.

The model solves for the number of vehicles required to service given booking requests and the rebalancing

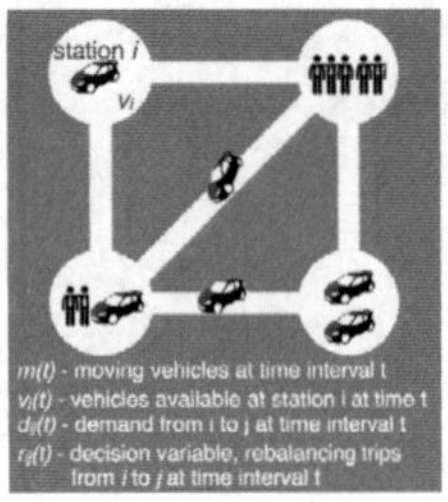

Figure 2. A simple example of an unbalanced AMOD system, i.e., we can observe that the top right station is heavily underserved, while there are vehicles available at the bottom right station. The rebalancing aims to align the demand with the supply.

counts between different intervals. The number of vehicles at the beginning of the simulation and the rebalancing counts resulted from the model were fed to the simulator and later compared against the solutions of the online model.

3.2 Online rebalancing

The online model used in this study is based on the fluid model first introduced in [5, 9]. The model repeatedly solves the optimization formulated in [4] and it finds the rebalancing counts to match the anticipated demand at all stations.

In the online model, the number of excess demand at station i, $\hat{d}_i$, is the number of customers that cannot be served using only the vehicles available at station i, i.e., $\hat{d}_i = v_i - d_i$. A negative $\hat{d}_i$ indicates that there are vehicles available to send. We assume that the cost of sending one vehicle from station i to station j is equivalent to the travel time between the i and j and given as c_{ij}, which is constantly being updated by SimMobility (travel time is time-varying). Our decision variable r_{ij} is the number of empty (rebalancing) vehicles to send from station i to station j. Note, that for simplification, t is omitted in this formulation (the model can be understood as a time invariant model). The objective function minimizes the rebalancing effort.

$$
\begin{aligned}
\min \quad & \sum_{ij} c_{ij} r_{ij} \\
\text{s.t.} \quad & \sum_{j} (r_{ji} - r_{ij}) \geq \hat{d}_i \qquad \forall i, j \\
& \sum_{j} r_{ij} \leq v_i \qquad \forall i \\
& r_{ij} \geq 0 \qquad \forall i, j
\end{aligned} \tag{2}
$$

The first constraint ensures that the number of rebalancing counts is greater or equal to the excess demand. The second constraint prevents us from sending more vehicles than we have available in at station i. The third constraint is the non-negativity constraint.

The online model is implemented such that the AMOD controller executes it at every time $t = n\Delta t$, where n is the number of intervals and Δt is the length of the interval, i.e., in this study we focus on online rebalancing every one hour, which is explained in Section 4.1.

4 Results

In this section we study the relation between the rebalancing models, number of vehicles in the simulation and customers' waiting times (Fig. 3). The waiting time is defined as the time difference between the trip request time and the pick-up time.

4.1 Simulation setup

To evaluate performance of the three rebalancing policies, we use a $56km^2$ network of the Central Business District in Singapore. The network consists on 1229 nodes, 14948 lanes and 313 traffic signals.

The demand generation process for AMOD is based on integration of SimMobility MT with SimMobility ST. The simulated population in SimMobility MT was estimated based on the Household Interview Travel Survey for 2012 (for more details please refer to [4]). Given the output from SimMobility MT, we replaced all transport modes except of the Mass Rapid Transit (subway) and public buses with the AMOD service. If only a part of the trip was inside CBD, then we cut the trip and simulated it from or to the border of the network. We run the simulation for the period of 3am-12pm. The total number of AMOD trips for this period was 363,859. Customers do not drop the booking requests and leave the system only when they finish their trips.

Locations of the stations are optimized based on the maximum coverage model (detailed description in [4]). In this study, we assumed that the coverage radius of a station is 1000 meters, which is equivalent to 2-3 minute ride at the average speed of 30 km/h (based on the Land Transport Authority's data, the average speed for arterial roads during peak hours in Singapore is 28.9 km/h [16]). Solution of the model is 34 stations within the analyzed zone. In the online model, we initialized stations with equal number of vehicles across the stations. We simulated fleet sizes of 10,000 to 40,000 vehicles. In the offline version the stations were initialized with the number of vehicles based on the optimization output (to recall, the offline model gives a unique solution for the fleet size, vehicle distribution between the stations and the number of rebalancing trips, while the online model is solved during the simulation for the rebalancing counts).

Both models are sensitive to the rebalancing interval. Therefore, the interval for the offline model is 15 minutes, while for the online model is 1 hour. The reason of this difference is the following: too long interval in the offline model results in overestimation of the fleet size (as the travel time is discretized based on the interval size); too short interval in the online model results in rebalancing during the peak period (especially if there is a booking queue).

All simulations were run in SimMobility ST, which simulates the individual decisions and the transportation

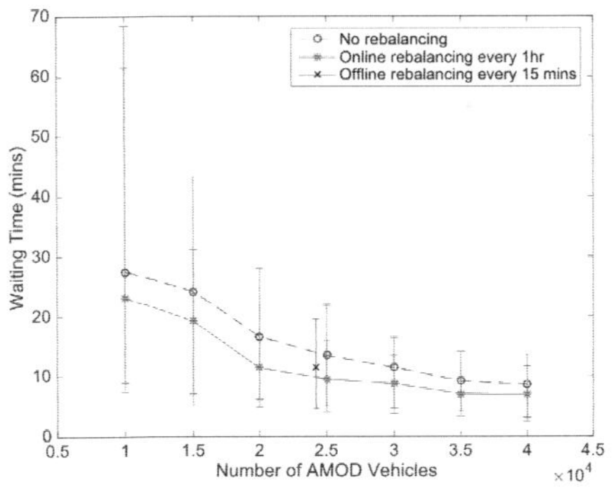

Figure 3. The performance of the three different rebalancing policies expressed in terms of the waiting time across different fleet sizes. The no-rebalancing policy (the blue line) performs quite poorly compared to the online and offline rebalancing methods. Both rebalancing methods help in reducing the fleet size, i.e., we need 35,000 vehicles without rebalancing for the average waiting time to be below 10 minutes and only 25,000 if we perform rebalancing. This finding translates to significant savings in the number of required parking lots.

network at the sub-second level (microscopic level). Our AMOD Controller is an integrated, but detachable, component that imbues SimMobilityST with the capability to simulate an AMOD system (Figure 1).

4.2 Policy comparison

Figure 3 summarizes the performance of the three rebalancing methods introduced in Section 3. It shows the average customer waiting time (with upper and lower quartiles) for different rebalancing policies as a function of the fleet size. We observe a decrease in waiting time with the increase in the fleet size. The trend lines for the no rebalancing model (the blue line) and the online rebalancing model (the red line) are nonlinear. The no rebalancing policy performs quite poorly in terms of the waiting time. Given the level of service we want to achieve, it requires the biggest fleet size for the AMOD operation, e.g., performance curve for the online model indicates that for the fleet size of 25,000 vehicles and more the average waiting time falls below 10 minutes. If we do not perform rebalancing we need as many as 35,000 vehicles, which translates to 28% increase in the required fleet size. Our finding that rebalancing reduces the fleet size is consistent with [2, 4, 5, 11].

The offline model (the grey cross) optimizes the fleet size and rebalancing effort by providing 24,216 vehicles. For this fleet size the average waiting time is 11.62 minutes, which is almost 2 minutes longer than for a similar fleet size under online rebalancing model. With the fleet size of 24,216 (offline model) and 25,000 (online model) we could serve 93% and 89% of the trips, respectively. Note that not all the generated trips were served because a proportion of the passengers had not yet arrived by the end of the simulation.

The differences in the results can be explained as follows: (i) Time-invariant travel time in the offline model results in underestimation of the fleet size. The online model performs assignment and rebalancing based on the current travel time on the network (taken from Simmobility), while the offline model does it based on the average travel-time, which does not account for congestion. (ii) In the offline model all booking requests are served immediately causing overestimation of the fleet. In the online model the customers are waiting for service a queue. The queue is always taken into account during rebalancing. (iii) The simulation of the offline model accounts for the optimized initial vehicle distribution, while in the online model vehicles are distributed evenly across the stations.

5 Conclusion and future work

The work described in this paper has been concerned with the development of algorithms for rebalancing of mobility on demand systems in Singapore. Three rebalancing models were proposed: (i) no rebalancing, (ii) offline rebalancing, and (iii) online rebalancing; (i) is our baseline model. In this configuration, vehicles are only moved when assigned to customers and parked at the destination of the trip. The results show that this model requires the biggest fleet size for the average waiting time to fall below 10 minutes. In (ii) and (iii) for the same level of service, we need about 28% and 23% less vehicles, respectively. We observed that the offline model gives a reasonable estimation of the fleet size, however further analysis has to be done to reduce the gap between online and offline model. For the future work we are planning to enrich rebalancing by combining the benefits of both models into one robust rebalancing policy. We have also an interest in incorporating the demand management policies in our AMOD Controller, e.g., dynamic pricing.

Acknowledgement

This research was supported by the National Research Foundation Singapore through the Singapore-MIT Alliance for Research and Technology, Future Urban Mobility IRG research program. We would also like to thank the researchers and developers from SMART for their contributions in the development of SimMobility.

References

[1] *Automobile dependency* (2016), accessed: 23-03-2016, `http://www.vtpi.org/tdm/tdm100.htm`

[2] M. Pavone, *Autonomes Fahren* (Springer Berlin Heidelberg, 2015), chap. Autonomous Mobility-on-Demand Systems for Future Urban Mobility, pp. 399–416

[3] K.A. Marczuk, H.S.H. Soh, C.M.L. Azevedo, M. Adnan, S.D. Pendleton, E. Frazzoli, D.H. Lee, IEEE 7th International Conference on Cybernetics and Intelligent Systems and Robotics, Automation and Mechatronics pp. 167–172 (2015)

[4] C.L. Azevedo, K. Marczuk, S. Raveau, H. Soh, M. Adnan, K. Basak, H. Loganathan, N. Deshmunkh, D.H. Lee, E. Frazzoli et al., Transportation Research Record: Journal of the Transportation Research Board (2016)

[5] K. Spieser, K. Treleaven, R. Zhang, E. Frazzoli, D. Morton, M. Pavone, in *Road Vehicle Automation* (Springer International Publishing, 2014), pp. 229–245

[6] D. Jorge, G. Correia, European Journal of Transport and Infrastructure Research **13**, 201 (2013)

[7] J.A. Barrios, J.D. Godier, Transportation Research Record: Journal of the Transportation Research Board pp. 1–9 (2014)

[8] F. Ciari, B. Bock, M. Balmer, Transportation Research Record: Journal of the Transportation Research Board **2416**, 37 (2014)

[9] M. Pavone, S. Smith, E. Frazzoli, D. Rus, *Load Balancing for Mobility-on-Demand Systems*, in *Proceedings of Robotics: Science and Systems* (Los Angeles, CA, USA, 2011)

[10] C. Brownell, A. Kornhauser, Transportation Research Record: Journal of the Transportation Research Board **2416**, 73 (2014)

[11] R. Zhang, M. Pavone, Robotics: Science and Systems Conference (2014)

[12] M. Pavone, Ph.D. thesis, Massachusetts Institute of Technology (2010)

[13] K. Treleaven, M. Pavone, E. Frazzoli, IEEE Conference on Decision and Control pp. 5691—-5698 (2013)

[14] K. Treleaven, M. Pavone, E. Frazzoli, *Asymptotically optimal algorithms for one-to-one pickup and delivery problems with applications to transportation systems*, in *IEEE Transactions on Automatic Control* (2012), Vol. 58, pp. 2261—-2276

[15] Y. Lu, M. Adnan, K. Basak, F.C. Pereira, C. Carrion, V.H. Saber, H. Loganathan, M.E. Ben-Akiva, TRB 93rd Annual Meeting. Washington DC. (2015)

[16] *Singapore land transport statistics in brief* (2015), singapore Land Transport Authority. Accessed: 2016-02-16, `http://www.lta.gov.sg/content/ltaweb/en/publications-and-research.html`

The Application of Policy Leverage in Promoting the Sustainable Development of Intermodal Freight Transport

Yu Yuan

China Academy of Transportation Sciences, Research Centre of Comprehensive transportation, 100029 Beijing, China

Abstract. With the rapid growth of transportation, environmental issues emerges. In order to achieve the society healthy development objective, transportation needs to develop sustainably. This paper proposes that intermodalism is crucial and efficient to transportation sustainability. The traditional intermodal transportation only considers about cost, time and reliability in choosing transportation mode and routes. First, this paper proposes the methodology of sustainable intermodal freight transport, which consider not only temporal and distance attributes, but also cost, energy, and emissions attributes, including emissions of carbon dioxide(CO_2), particulate matter(PM10), sulfur oxides(SO_x), volatile organic compounds(VOC_s), and oxides of nitrogen(NO_x). Then, this paper indicates that public policies is very important for sustainable transportation, because of the market failure in some cases. At last, the paper propose some policies that will help us a lot to achieve sustainable freight transport. The main policies include mode subsidies, infrastructure investment, economic incentives and technology improvement.

1 Introduction

In recent years, pollution and environmental issues become the crucial factors when people make decisions. Now they play the paramount important role. This phenomenon appeared in all ranges of our daily life, as well as in all the industries in the world, including the freight transport industry.

Truck and air sectors are traditionally the most energy-intensive and carbon-intensive modes of freight transport. Because truck transport has a higher energy- and carbon-intensity compared to rail or ship, these trends do not bode well for developing sustainable freight transportation. One method of reducing the energy use and emissions from freight transportation is intermodalism and mode-shifting.

Previously, when people choose transportation modes or routes, they always consider three factors, which are total travel time, travelling cost as well as reliability. Truck used to be the most popular transport mode in the world. Based on the existing shortest path algorithm, the route with lowest total travel cost will be chosen, if people consider the cost element as the most important. We can also translate total travel time and reliability factors into travel cost, in order to implement the three factors together to help us make decision.

Now, people also began to care the environmental impact of emissions by freight transport. People want to find a method that can be used on freight transport in order to make improvements on the impact on environment. In other words, people are working hard to change the existing freight transport into sustainable freight transport. This shows our responsibility to protect the environment.

We are going to find a way to choose the optimal route based on variety of factors. The method should not only consider traditional factors such as total cost, delivery of time and reliability, but also take energy and environmental factors into consideration.

2 Methodology

First, we should construct a hub-and-spoke system to present our real world transport network (Seen as in Fig. 1) [1].

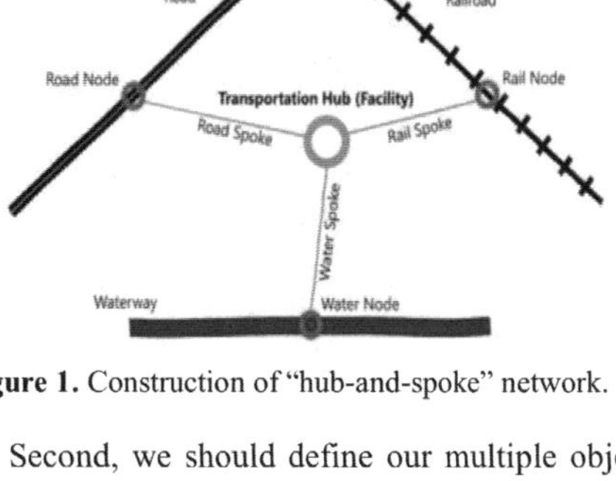

Figure 1. Construction of "hub-and-spoke" network.

Second, we should define our multiple objectives of the freight transport.

Third, translate those non-monetary objectives into monetary objective like the total cost of the route. In this step, we should find and define some criteria to translate them.

For transport time and transfer time, we can use "value of time" (VOT) to translate them into cost. Specifically, we can adopt Delphi method to determine VOT.

For reliability, we can use a rate to represent the goods' relative importance. The higher value the goods has, the higher rate we'll assign to it. In general, we assign lower cost to the goods that have the higher rates. In this case, there is more to high value than the merely high monetary value of the goods, but also include the need for JIT delivery.

To represent the energy use and emission of a mode, we use an emissions calculator (Seen as in Fig. 2). The inputs of the emission calculator are some general information of the mode we are researching, such as horsepower, fuel economy, and cargo capacity. Then, it can output the cost-factor. An example of cost-factor is the amount of CO_2 emitted by a truck carrying a twenty-foot equivalent unit (TEU) container of freight for one mile (grams of CO_2 per TEU-mile). In addition, the result of the cost-factors can be modified by general information of specific mode, such as the travel speed, horsepower, fuel types and so on. Having obtain the cost-factors, we can use the "price for carbon" to implement the CO_2 emissions results into the total cost of the transport route [2].

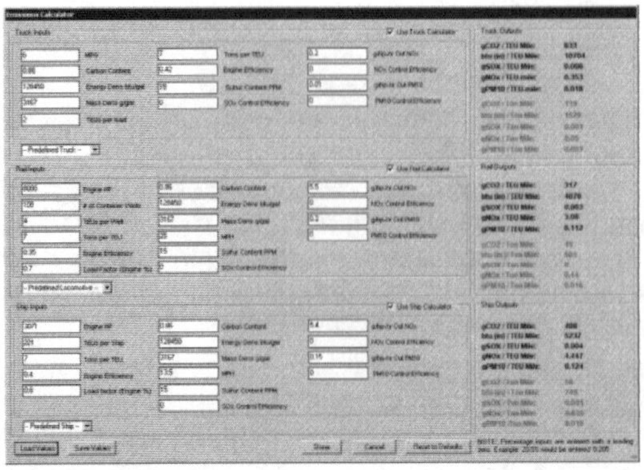

Figure 2. The GIFT Emissions calculator.

At last, the multi-objective problem has been translated into single-objective problem, so we adopt the shortest path algorithm to determine the optimal route we'll choose.

3 The importance of policies in reducing energy use and emissions in fright transport

The policies have the power to adjust and control the freight transport industry in a macro scope. It can make up the limitations of market economics, and can powerfully change something in the freight transport industry that the industry can't change to the extent by itself. The policies just give a push to the freight transport industry to reform.

Public policies are necessary to reduce freight transport energy use and emissions in order to: (1) address a market failure; (2) provide economic security.

3.1 Construct a framework of the policy levers to decrease the energy use and CO_2 emissions by freight transport

First, we construct a framework of policy levers. It includes 6 policy levers that can affect the energy use and emissions of CO_2 [3]. They are:
- Mode-shift: use efficient modes
- Fuel: use low-carbon fuel
- Technology: use advanced technology during
- Intermodal freight transport
- Operations: do the best and efficient operations.
- Logistics: improve supply-chain management in
- Order to enhance efficiency

3.2 Infrastructure investment.

Infrastructure investment includes 2 kinds of cases. The first case is that we need to build new transfer infrastructure along the routes. Another case is that we need to make extensions on the existing transfer infrastructure, in order to make it become more efficient.

In order to evaluate our infrastructure investment, we need to go through the following 4 steps:

First, we need to know the freight volumes along specific corridors;

Second, we need to measure the costs and benefits of investment;

Third, we need to consider this infrastructure investment from stakeholder's views;

Fourth, we need to know and determine where funding would come from.

A thorough cost-benefit analysis is very important; it helps portray the tradeoffs associated with a proposed public policy.

Stakeholders would include policymakers; the public; freight transport businesses and organizations; and other special interest groups (perhaps environmental groups). If the policy can be shown to result in greater benefits than costs, it is more likely that stakeholders will support the proposal.

Stakeholders will be most interested in where funding would come from. Would the government encourage this investment and provide subsidies or grants for construction? If so, what agencies would disburse these funds and administer such a program? The answers to these questions will help determine the economic and political feasibility of the policy.

Firstly, we should know where major freight flows using containerization exist. This will create an origin and destination pair for the route. Then we may have an idea of potential sites for new intermodal transfer facilities or locations of existing facilities that could be upgraded along those freight corridors.

We use the hub-and-spoke approach. In the network, the transfer facility is the "hub" and the "spokes" are created by a connecting the facility to available road, rail, and water networks.

At last, we can solve an objective function (least-time, least-cost, least-CO_2, etc.) from origin and destination before and after adding the new facility and determine if the addition affects the results.

But for the upgrade of the existing transfer infrastructure, we need to measure how much transfer time can decrease after upgrading the existing transfer infrastructure. So we need to get and know the data of current container handling capacity and container volume through the facility. In this way, we can get the transfer time of the current system. Second, we need to get the data of container handling capacity after upgrading and container volume through the facility now. So we have got the original transfer time and the transfer time after improving the facility. At last, we can get total time reduction achieved by improving the container handling infrastructure at the facility.

Therefore, to reduce the intermodal transfer time for a container, infrastructure investments should affect the rate of incoming container offloading, the on-site capacity for container storage, or the rate of outgoing container loading (or a combination of these variables). (Seen as in Fig. 3)

However, we can't increase just one among the three factors. It's not meaningful to just increase the rate of incoming container offloading if they do not have the ability to store those containers. Also, if the rate of incoming container offloading is greater than the rate of outgoing container loading, there will not be a transfer time reduction.

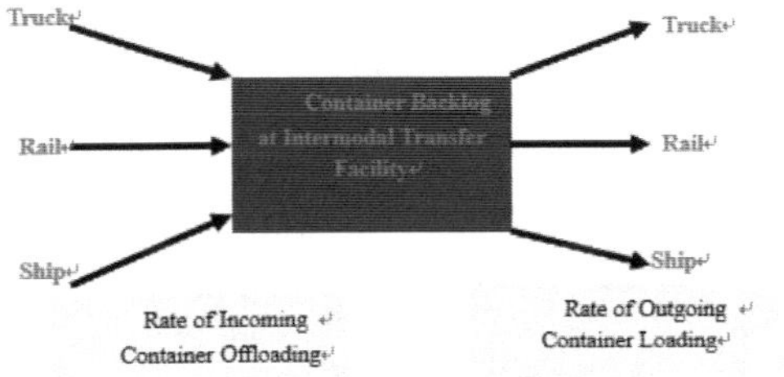

Figure 3. Variables that affect the intermodal transfer time.

3.2.1 Cost-benefit analysis of infrastructure investment

The costs of implementing such a policy might include the cost of new container handling equipment, fuel costs for new equipment, construction materials, and labor. In addition, economic loss due to closure or limited use of the facility during the upgrade should be considered.

The benefits of implementing such a policy can be the increased profit because of larger capacity of the new facility; reduced emissions due to more efficient equipment; and the decreasing of transfer times.

3.2.2 Stakeholders related to this policy

The stakeholders of facility investment include government (policy makers), public, freight transport companies, and environmental protectors and so on. If it seems to the stakeholders that the benefit of it exceeds the cost, stakeholders will support this proposed policy.

3.2.3 The result this policy can make

This policy may result in reduction of transfer time. As the transfer time reduced, the total time of intermodal transport will decrease. Intermodal transport offers an important opportunity to lower shipping costs in medium-to long-distance transportation [4]. So many people will turn to use intermodal transport to replace the single truck mode. However, we need to define how many percent of single truck transport will change into intermodal transport.

3.3 Economic incentives and penalties.

3.3.1 Modal subsidies

1) Definition

Knowing which modes would be subsidized is very important. Truck is the most carbon-intensive and fuel consumption mode. Air transport is also relatively carbon-intensive. Compared with truck and air transport, rail and ship transport are low-carbon transport modes. So government encourages people to use low-carbon transport modes through giving economic incentives to people to make them shift transport modes from truck to rail or ship transport. This means that if you use the government encouraged transportation modes, you will get subsidies, convenience and better service from it. For example, Government may reduce the per container-mile cost for the rail and ship segments of the network.

We need to find appropriate subsidy level. Freight modes like rail and ship are already less expensive than truck, so the point of the subsidy is not to make the per container-mile cost of transporting goods by rail and ship less expensive than truck; rather than that, we should offset the other "costs", such as longer time-of-delivery. We also need to determine what price per container-mile for rail and ship will offset the additional time-of-delivery for these modes [5].

2) Cost and benefit analysis of this policy

The costs of the policy might include the amount of money budgeted for the policy; the wages of the employees that would devise, implement, and evaluate the policy; economic loss to the trucking industry; and potential environmental and social costs from the increased use of rail and ship.

The benefits of the policy might include the social and environmental benefits of reduced CO_2 emissions and economic benefits to the rail and ship industries.

3) Stakeholders

The stakeholders related to this policy are government (policy makers), truck companied industry, rail and ship industries, and public.

4) The results this policy can make

The policy would likely result in an increased usage of less carbon-intense modes of freight transport such as rail and ship and a decrease in the use of heavy-duty trucks.

In order to measure the impact of a modal subsidy or grant, we can calculate per container-mile cost for each mode affected by the policy.

3.3.2 Fuel taxes

1) Definition
Fuel taxes based on carbon content encourages people to use low-carbon fuels. So the demand of low-carbon fuels will increase; on the other hand, the demand and consumption of high-carbon fuels will decrease. In addition, the "technology" lever will be affected, which push companies and industries to innovate and install fuel-saving technologies.

In order to know the feasibility of this policy, we need to get the data of the cost and benefit as well as the stakeholders' attitude toward the policy.

2) Cost and benefit of this policy
The costs of the policy might include increased economic cost to the truck, rail, and ship industries and increased cost to those industries' customers (note: the additional operating cost would likely be passed on). In addition, the oil demand will decrease, so the oil companies will have economic loss, which should also be considered in the cost analysis.

When doing analysis of the benefits of the policy, we should consider the reduced air pollution, savings for fuel and increased profit to alternative fuel suppliers and developers.

3) Stakeholders
Stakeholders in this case should include government (policy makers), oil companies, freight transport companies (including truck, rail, air and ship companies), and publics.

4) The results this policy can make
The policy is likely result in an increased use of more fuel-efficient modes, such as rail and ship, instead of truck. In addition, it'll push toward technologies that improve fuel economy.

In order to measure the impact of a fuel tax, we would first increase the per-container-mile cost of each mode since operating costs would increase due to the fuel tax. We also want to determine how much each mode's per container-mile cost should be increased in order to make people shift their modes, so data on trucks' fuel economy and energy use, horsepower, and speed of rail and ship would be necessary for us.

In order to encourage mode-shifting, the fuel tax should be determined high enough for shippers to be unwilling to use unimodal truck routes.

3.3.3 Carbon taxes

1) Definition
Carbon taxes are based on the amount of CO_2 emitted by vehicles. It encourages people to use the less carbon intensive transport modes, like rail and ship, and low-carbon fuels. In addition, it encourages people to implement fuel efficient technologies. The carbon taxes are to decrease the demand of carbon-intensive transport modes and high-carbon fuels. It is similar to fuel taxes.

2) Cost and benefit of this policy
The cost of this policy include the more money that companies and people will pay for the freight transport due to the carbon taxes; the cost to implement this policy, including some labor cost; the decreased sales of some oil companies; and the decreased amount of taking truck mode.

The benefit of this policy might include the alleviated air pollution, which can be translated into cost by the "price of CO_2"; increased revenue of government from rail and ship companies; and increased profit of some fuel suppliers.

3) Stakeholders
The stakeholders related to this policy may include government (policy maker); freight transportation companies, including truck, rail, air and ship companies; oil companies; publics.

4) The results this policy can make
The policy would likely result in an increased use of more fuel-efficient modes (like rail and ship versus truck) and a push toward technologies that improve fuel economy.

The amount of the increase for each mode depends on the level of taxation for each unit of CO_2 emitted and the efficiency of each mode. We should find the point at which the shipper will find that it is not economic efficient to transport goods by unimodal truck routes, and they'll be pushed to shift to use other modes such as rail-ship intermodal routes.

Because the cost of rail and ship are lower than truck initially, what we can do to push people to not use truck is not decreasing the cost of rail and ship. Instead, we should do some improvement on the delivery of time on the two modes, like improving the facility infrastructure to let it be more efficient and largely decrease the transfer time. People have the VOT (value of time) in their mind, so the tradeoff of time of delivery in of paramount importance.

3.4 Technology implementation.

1) Definition
Technology implementation refers to that we implemented advanced technology in order to decrease the CO_2 emissions and improve fuel efficiency by the intermodal freight transport industry. For example, low-rolling resistance tires and aerodynamic improvements for heavy duty diesel trucks can improve the fuel efficiency of intermodal freight transport [6].

We need to analyze its cost and benefits, as well as the stakeholders' viewpoint, so as to evaluate whether this policy is attractive or not, and we can estimate whether the stakeholders will support or not.

2) Cost and benefit of this policy
The cost of this policy might be the purchase cost of the technology; the labor cost, including the cost for them to evaluate and implement this policy; and the decreased profit for some fuel companies due to possible decreased fuel consumption.

The benefit of this policy might include revenue for companies that produce the technologies, reduced pollution, and cost savings due to greater fuel economy.

3) Stakeholders
Stakeholders in this case include policy makers (government), companies that produce the technologies, fuel companies, freight transport companies, publics.

4) The results this policy can make

A technology implementation policy might result in lower CO_2 emissions for trucks, due to reduced fuel consumption. This counts to the reduction of the total fuel consumption and CO_2 emissions in the intermodal transport process.

3.5 Legislation.

Government can act laws related to the freight transport to give enforcement to the environmental. In that way, not only the environmental-concerned people, but all the people will take environmental issues into consideration. For example, government can determine a point (CO_2 grams/TEU/km) at which if people arrived they will be fined at a predefined rate.

4 The benefits of intermodalism and mode-shifting

All of our policies above are to encourage people to make shift to intermodalism. The benefits of intermodalism and mode-shifting include:
- Lower CO_2 emissions;
- Lower economic costs;
- Increased efficiency.

4.1 Lower CO_2 emissions

Switching to intermodal routes that make use of less carbon-intense modes of freight transport (rail and ship) or switching from unimodal truck routes to unimodal rail and ship routes has the effect of reducing CO_2 emissions. Reduction of CO_2 emissions results in lower social and environmental costs.

4.2 Lower economic costs

Lower CO_2 emissions can also lead to economic cost savings if a "price for carbon" is implemented, such as a carbon tax or a cap-and-trade program for CO_2 emissions; this results in a lower economic cost. Also, switching to intermodal or unimodal routes that use rail and ship are inherently less expensive due to increased efficiency, that is, more goods (by volume or weight) can be moved by rail and ship compared to truck at a lower price.

4.3 Increased efficiency

Switching to intermodal or unimodal routes that make use of rail and ship leads to greater efficiency. Increased fuel efficiency leads to lower fuel consumption.

5 Limitations

In order to get the results, we must make some assumptions, such as the fuel economy, horsepower, container capacity, engine load factor, engine efficiency, and speed of the vehicles. However, these assumptions are also limitations. Also, emissions are represented on a per mile basis. No consideration is given to changes in emissions as a function of speed, grade, or load.

6 Conclusion

This paper focuses on the impacts and tradeoffs of multi factors associated with intermodal freight transport, including energy, environmental, and economic attributes. In order to let the intermodal transport become sustainably, the untraditional factors such as energy use, CO_2 emissions and the efficiency of fuel consumptions should take into consideration. The methods include using VOT and emission calculator to translate the multiple non-monetary factors into cost. In that way, we can make a decision using shortest path model.

In order to motivate people to change their behavior to care about the environment, policies are necessary for us. Policies has it's a series of advantages to encourage people to make mode shift and turn to less polluted way to choose routes. The proposed policies include four aspects, such as infrastructure investment, economic incentives, technology implementation and legislation. Public policies are necessary, because it can address market failures such as negative social and environmental externalities, promote economic growth, and improve national security by reducing our dependence on foreign oil. However, there are both costs and benefits to any policy. We need to determine a point at which shippers will turn to use intermodal transport.

References

1. M. P. Boile (2000). Intermodal transportation network analysis - A GIS application. Paper presented at the 10th Mediterranean Electro technical Conference
2. J. J. Winebrake; J. J. Corbett, A. Falzarano, J. S. Hawker, K. Korfmacher, S. Ketha, S. Zilora (2008). Journal of the Air & Waste Management Association, **58**(8)
3. E. Taniguchi, R.G. Thompson (2004). Logistics Systems for Sustainable Cities. Proceedings of the 3rd International Conference on City Logistics
4. K. Trusty, E. Malstrom (1998). A Feasible assessment of Truck-barge Intermodal Freight Transportation. Journal of Costs
5. B. Comer (2009). Sustainable intermodal freight transportation: applying the geospatial intermodal freight transport model
6. Baasankhuu manduul (2007). Preliminary discussion paper on guidance for the developing countries to build a sustainable freight transport system. World Road Association (PIARC)

MATEC Web of Conferences **81**, 01007 (2016) DOI: 10.1051/matecconf/20168101007
ICTTE 2016

Intelligent Safety Transport Framework for Schools: A Review of Route Planning and Tracking Systems

Manoharan Ramachandran , Reza Sahandi, Simant Prakoonwit and Wajid Khan

Creative Technology Department, Faculty of Science and Technology, Bournemouth University, Bournemouth, United Kingdom

Abstract. This work presents a review of recent literature in intelligent school transportation frameworks, particularly focusing on route planning, real time vehicle and children tracking. The focus on route planning and tracking is to identify the hidden practical problems and threats present in school transportation, bearing in mind safety. Different methods and technologies used for route planning and vehicle as well as children tracking are reviewed. A discussion is provided on the current frameworks along with the challenges and future research direction.

1 Introduction

School Transportation System (STS) is safety critical which involves school children (students) who are the most vulnerable users of roads [1]. Recent events and accidents highlighted the need for improving the existing school transportation and bus monitoring systems [52], [53]. Statistics demonstrate that in England alone 85,814 children were wounded near the schools during the period 2007 to 2013, which is equivalent to 1,190 a month [2], [3]. Moreover, during the period 2006 to 2011 there were 557,200 vehicle collisions around schools [2], [3]. As a result, proper safety route planning and monitoring strategies for school vehicles and children are in high demand [4], [5]. Efforts to reduce school transport injuries have begun through constraint-based school bus routing [12] and peer to peer children monitoring [13] while they are on route. Recent innovations [6]-[8] in STS, gave birth to Intelligent School Transportation Systems (ISTSs), which addressed some of the issues [9], [10], [11] in STS such as optimal routing of school vehicles, continuous monitoring of school buses and children. An ISTS incorporates the innovation and adoption of recent technologies to create applications for the benefit of school transportation. Major technologies involved in ISTS are school bus route planning systems, school vehicle-children tracking and monitoring systems. School bus route planning systems rely on the history of School Bus Routing Problems (SBRPSs) which have been studied since the first published work by Newton and Thomas in 1969 [14]. There are only limited publications available for reviewing school bus routing problems [15]-[17]. Junhyuk and Byung-in Kim [17] describe school bus routing problems for a fleet of school buses as an efficient schedule planning, where children are picked up by each bus from various geographical locations and delivered to their respective schools, while satisfying a set of constraints. The main reasons behind the school bus routing problems were frequent changes in the number of children per stop, rerouting during temporary roadworks and traffic jams. According to Desrosiers et al. [15], there are five steps to reduce school bus routing problems, namely, data preparation, bus stop selection, bus route generation, school bell time adjustment and route scheduling. Parents spend more time on the streets and making phone calls while waiting for school buses due to the unpredictable nature of the traffic, particularly during the winter months. For this very reason, vehicle-children tracking systems were made. School vehicle-children tracking is a process of tracking the school bus and children inside it using tracking devices such as the Global Positioning System (GPS) [38] and Radio Frequency Identification (RFID) systems tags [39], which are commonly used in tracking technologies. The tracked data may be utilized by the school transportation department and shared with parents to let them recognize the location of their children. Most of the published work so far on ISTS have focused on route planning systems and only limited work has been done to address the whole school transport system [18], [19]. In this paper, a review of recent studies in intelligent safety school transportation systems with a focus on route planning and tracking is presented. A fresh perspective and functions for an ISTS framework are provided along with the challenges and further research direction.

2 Intelligent school transportation systems

It was indicated earlier that ISTSs attempt to provide safe and secure transportation for school children through design, development and integration using appropriate technologies. However, limited works have been done so far in addressing this for school transportation and child safety. A recent Europe funded project conducted in Sweden "SAFEWAYTOSCHOOL" provided a holistic and safety framework for children [1], [11], [18], [19], [38]. This project addresses most of the difficulties faced during school transportation through a door to door approach (between home and school). Figure 1 shows a holistic approach to this project. At first, safest routes are considered for children to reach the bus stop. When a child reaches the bus stop, an alert light is automatically turned on at the bus stop. When the bus reaches the bus stop to pick up the child, a warning sign in the bus is turned on. During the journey towards the school, the children are notified to put the seat belts on. Also, audio-visual information is provided about the destination before reaching the school. Once the bus reaches the school stop, a warning notification light is turned on in the bus and the bus stop is also turned on. Finally, on arrival at the school, a notification message is sent to parents.

Atsushi et al., proposed a safety support system through ad-hoc network formed using mobile phones to increase the safety of children when walking between bus stops and home [44]. Volunteers who maintain safety in the local community assist the children when walking between home and bus stop. Both the volunteers and children are provided with mobile phones which facilitate the Bluetooth function. A safety support mobile application is installed on these mobile phones. Bluetooth poles are placed at even distance intervals between the home and bus stop. When children's and volunteer's mobile phones come into contact with these poles, the mobile application automatically connect them. The volunteer's mobile application obtains information about the exact location of the child and shares it with the parents and caretakers. A warning message is sent to parents and volunteers if any abnormalities like deviating from the normal path are found.

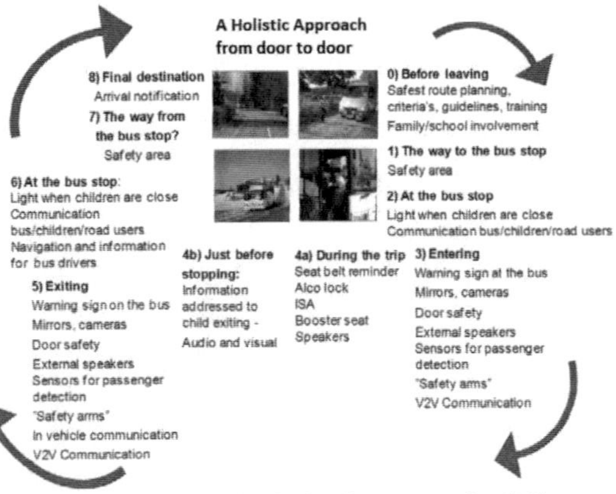

Figure 1. Holistic approach of school transportation [18].

3 Functions of ISTSs

School transportation systems have been evolving in the past decades. ISTSs were developed due to recent advancements in route planning systems [6], [12], [23], [29]-[33], [35]-[37], school vehicle-student tracking and monitoring systems [8], [18], [38]-[43]. Figure 2 shows the functions of a typical ISTS. Each ISTS function applies its own methods to find the best solutions [6], [7], [8], [12], [17], [21] for the challenges in each function. For instance, GPS [39] and RFID [18] tags are used for vehicle and student tracking. However, providing an accurate and a non-intrusive service at all time is challenging [7].

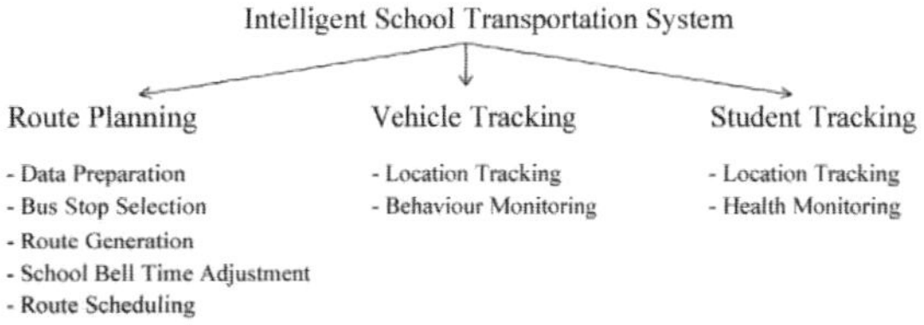

Figure 2. Functions of a typical intelligent school transportation system.

3.1 Route planning systems

In schools, manual route planning is an intensive task and it consumes a considerable amount of time for selecting appropriate safety routes, as well as the number of busses required for a route. Typically, school transportation departments perform manual route planning at the beginning of each term due to the changes in the number of students using the service. To make this process more efficient, automated route planning systems are used. To avoid school bus routing problems, automated route planning systems utilize the following five steps sequentially: preparation of data, selection of bus stops, generation of school bus routes, adjustment of school bell time and scheduling of school bus routes [15].

3.1.1 Data Preparation (DP)

Data for the next four sub problems (Bus Stop Selection (BSS), Bus Route Generation (BRG), School Bell Time Adjustment (SBTA) and Route Scheduling (RS) of school bus routing problem are created through data preparation. In this step, a dataset containing a single road network with student residence coordinates, the school constraints (start-end time of schools, maximum ride time, etc.), number of vehicles with seating plan and an Origin-Destination (OD) matrix are created. In an OD matrix, the shortest distance between the school location, student locations and the bus origin location are stored. Various shortest path algorithms and a geographic information system (GIS) are applied to calculate the OD matrix [17]. Performances of the shortest path algorithms are compared by Kim and Jeong and an approximation approach for the OD matrix generation is developed [20]. Thus, the dataset (students, schools, vehicles and OD matrix) created in the data preparation is used to prevent school bus routing problems.

3.1.2 Bus Stop Selection (BSS)

It is the process of selecting a number of school bus stops and assigning students to them. Bus stop selection can be done through heuristic approaches [29], [36], [37] or by exact methods based on mathematical programming [12], [21], [22], [23], [37]. The heuristic approach follows either the Location-Allocation-Routing (LAR) [15], [24], [25], [26] or the Allocation-Routing-Location (ARL) [27], [28], [29] strategies. In LAR strategy, a set of bus stops for a school are selected first and the students are allocated to these bus stops based on the distance between the bus stop and their home. Once the children are assigned to these bus stops, the routes are generated. LAR strategy tends to create more routes since bus stop selection and children assignments are done before the route generation. ARL strategy overcomes the drawback of LAR by performing it in the opposite order. At first, children are grouped into clusters with respect to vehicle capacity constraints. Subsequently, bus stops are selected and routes between each cluster are generated. Finally, the clusters are assigned to the bus stop by satisfying constraints such as maximum walking distance between home and bus stop, maximum number of children for a bus stop and minimum distance between each bus stop [17].

3.1.3 Bus Route Generation (BRG)

School bus routes are generated by analysing the data gathered from the previous two processes. The route generation algorithms classified into either "route-cluster" or "cluster-route" approaches [24]. In the "route-cluster" approach, all the stops in a large route are built using the traveling salesman algorithm from which small routes are partitioned based on the constraints [14], [24]. In the "cluster-route" approach, students are grouped into clusters and each cluster is served as a route satisfying the constraints [25], [27], [28]. The constraints that are categorized to minimise school bus routing problem are represented in Table 1. They are, Vehicle Capacity (VC) – number of children allowed for a vehicle, maximum Ride Time (RT) – travel time of each child, school Time Window (TW) – arrival time of vehicle at school, maximum Walking Distance (WD) – distance between student home and bus stop, minimum number of Students for a Route (SR), Passenger Demand (PD) – route request by student during travel, maximum Bus stops Visited (BV), maximum Waiting Time (WT) – allowed waiting time for student at the bus stop and Terrain Type (TT) – type of road selected.

3.1.4 School Bell Time Adjustment (SBTA)

It has performed for schools where the start and end times are not considered as constraints but as decision variables. This is to identify the optimal start and end time to reduce the number of buses used [17].

3.1.5 Route Scheduling (RS)

The start and end times for each route are scheduled through route scheduling using a series of routes that can be visited successfully by the same bus. SBTA and RS are implemented when buses transport students from more than one school.

Minimisation of school bus routing problems is performed based on the nature of the underlying problem characteristics as follows; the number of schools: single (vehicle serves only one school) [35] or multiple (vehicle serves more than one school) [37]; fleet type: homogeneous (vehicles seating capacity remains the same) [23] or heterogeneous (vehicles seating capacity varies) [37]; service type: rural [31] or urban [30]; [17]. Table I also represents a comparison of methods applied by various researchers since 2010 for reducing school bus routing problems. For the work done before 2010, Park and Kim in [17] provide a comprehensive review of school bus routing problem.

3.2 School vehicle tracking and monitoring systems

School vehicle tracking is similar to common vehicle tracking which involves the use of GPS devices to track the vehicle. A GPS device in the school vehicle automatically provides updates of its location coordinates to a cloud server, which process and plots these coordinates on a virtual map. This map can be accessed by the stakeholders (transport department and parents) for information about the location of the school vehicle. School vehicle tracking systems can be utilized for two purposes: location tracking and vehicle behaviour monitoring

3.2.1 Vehicle Location Tracking (VLT)

The most commonly applied technology for position tracking systems is GPS. Grounded along the tracking nature, location tracking can be classified as real-time tracking [39] and lag-time tracking [38]. In real-time tracking, the location sensor in the vehicle (GPS) frequently updates a server in the real-time [39], [40]. In lag-time tracking, an active RFID unit in the vehicle communicates with the bus stop when it reaches near it and through that the vehicle is tracked [38].

Table 1. Representation of recent works in school bus routing problem (2010-2016).

Study	Considered school bus routing problem sub Problems	Constraints	Method	Goal
Jalel Euchi et al., 2012 [30]	BSS,BRG,RS	VC,RT,PD	Artificial ant colony with variable neighborhood local search algorithm	Minimizing the total number of buses required

Riera et al., 2012 [23]	BSS,RG	VC,WD	Branch and cut approach based Exact algorithm	Number of routes and route length minimization
Pacheco et al., 2012 [31]	RG,RS	RT	Multi-objective adaptive memory programming	Minimizing the duration of the longest routes and total distance traveled
Byung In et al., 2012 [32]	RS	TW	Branch and bound approach based on assignment problem	Optimization of bus schedules to serve all the given trips within the given TW
Junhyuk et al., 2012[33]	RG,RS	VC,RT,TW	Mixed load improvement algorithm	Minimizing the total number of buses required
Patrick et al., 2013 [29]	BSS,RG	VC,TW,SR,WD	GRASP+VND metaheuristic approach	To integrate BSS and RG through meta-heuristic approach with simplified implementation
Riera et al., 2013 [12]	BSS,RG	VC,RT,SR,BV,WD	Set partitioning formulation based branch and price algorithm	Minimizing the number of routes, route length and walking distance
Kinable et al., 2014 [35]	BSS,RG	VC,SR,RT	Exact branch and price framework	Minimizing the routing costs
Michael Bogl al., 2015 [36]	BSS,RG,RS	VC,WD,WT	Heuristic solution framework allowing transfers	Minimizing the operational costs
Xiaopan et al., 2015 [37]	RG,RS	TW,RT	Exact mixed integer programming and two-stage metaheuristic method	Minimizing the number of routes and total number of buses
Cristiano et al., 2015 [6]	RG,RS	VC,TT,SR,WD	Mixed load approach	Minimizing the total traveled distance of a heterogeneous fleet

Table 2. Representation of various research in school vehicle tracking and monitoring.

Study	Sensor type	Type of tracking considered	Nature of tracking	Goal
Anund et al., 2011 [38]	Active RFID	VLT	Lag time	To monitor bus location and providing bus time information to parents
Khaled et al., 2013 [39]	GPS	VLT	Real time	To smartly track the school bus and share the location info with parents
Zambada et al., 2015 [40]	GPS,accelerometer	VLT, VDBM	Real time	Increasing the safety of school bus monitoring through InterneT of things (IoT)

Table 3. Representation of various research in school student tracking and monitoring.

Study	Sensor/Technology used	Type of tracking considered	Nature of tracking	Goal
Anund et al., 2010 [18]	RFID	SLT, IVA	Active	To track and monitor the school students inside the school bus
Saranya et al., 2012 [41]	GPS	SLT	Active	To track the location and emotional status of the student
Khaled et al., 2013 [39]	RFID	SLT	Passive	To track and monitor the student during their trip to and from school on school bus
Huang et al., 2014 [8]	GPS/Bluetooth	SLT	Active	To develop a mobile based child security monitoring system in school travel
Anwaar et al., 2015 [42]	RFID	SLT, IVA	Passive	To track the student location and monitoring bus boarding times
Peter et al., 2015 [43]	GPS/Heart rate	SLT, HM	Passive	To monitor the student physical activities from school to home travel

3.2.2 Vehicle-Driver Behaviour Monitoring (VDBM)

It is used to detect the vehicle and driver behaviour to find the vehicle Estimated Time of Arrival (ETA) and rash driving. This is accomplished through various speed sensors on the vehicle like accelerometer along with GPS [40]. Table II shows characteristics of school vehicle tracking and monitoring systems. The main focus in the table is the different types of vehicle tracking systems used in ISTS.

3.3 Student tracking and monitoring systems

Similar to bus tracking, student tracking also employs similar technologies which are used to traverse the accurate placement of students. RFID is commonly employed for student tracking [42]. The student tracking system can be classified into two types: location tracking and health monitoring.

3.3.1 Student Location Tracking (SLT)

The most common technique used for student location tracking is the utilization of RFID tags. RFID tags can show the nature of tracking. There are two types: active and passive RFID [39]. An active RFID tag has its own power source. Passive RFID have a chip embedded, which withdraws power from the RFID scanner that emits electromagnetic waves. Students are also tracked using GPS embedded devices [8], [41] which share their location with all the stakeholders (i.e) school authority, bus drivers and parents. In-Vehicle Attendance (IVA) is part of the student tracking system in which the presence of students is determined through RFID tags. The school bus with a RFID reader is used to read student' tag [42]. When a student enters the vehicle, it is automatically recorded and it is considered as active registration. If a student manually registers, it is considered as passive registration.

3.3.2 Health Monitoring (HM)

Health is an important factor for student monitoring. It not only helps the parents to understand their children's attentiveness but also helpful during emergencies. Health monitoring is performed through a heart rate sensor which is worn by the student [43]. Table III represents the characteristics of school student tracking and monitoring systems. The table shows different types of student tracking systems used in ISTS.

4 ISTS challenges

The implementation of various technologies (route planners, trackers, monitors) in the school transportation systems comes with many challenges which have to be resolved for the welfare of school children. A real-time school bus routing system has to be efficient and safe enough to transport the children. But considering all the safety aspects, selecting the school bus route is a challenging task. The routing problems get more complex when the nature of the problem changes (e.g. single school to multiple schools). Moreover, the safety aspects considered have to be easy to embed into the existing routing system without compromising performance. For an example, in [45] the energy consumption of vehicles is considered as one of the constraint for the route and it is directly added to the routing software without reducing the performance of the existing system. Dynamic traffic navigation (re-routing) will be of progressive importance in future years [46]. Implementing an efficient and dynamic traffic safety routing in urban school transportation is a challenging task. Tracking the school vehicle in real-time without the GPS network lag is important. The vehicle tracker lets the parents and children to comfortably plan their journey to the bus stop. A recent study on improving GPS-based vehicle positioning [47] stated that the GPS technology based conventional localization techniques are not able to provide reliable and accurate positioning in all situations. Also, providing real-time access to the stakeholders without network latency is a challenging task. Smartphone based vehicle-driver monitoring [50], [51] may improve the existing school bus tracking system because of the embedded sensors they use and the easy implementation. However, customizing it for school bus transportation by considering all safety characteristics will be challenging. Many student tracking systems have been developed in which mostly employ radio frequency based devices. Recent studies in radiation exposure [48], [49] reported that children are more likely to be vulnerable than healthy adults to radio frequency electromagnetic radiation. So the development of children tracking system with less or no radiation based methods will reduce the risk of a child's vulnerability towards radiations which is challenging.

5 Conclusion and future directions

Millions of people use school transportation in their daily lives. Children are seen as vulnerable users of the road and there is demand for safe school transportation organizations. Several technologies have been evolved to increase safety and to avoid fatal injuries to children to and from school. This paper presented a review of major technologies and approaches used in school transportation, including route planning systems, vehicle – student monitoring and tracking systems. Each category is reviewed and a comparison was made between different systems.

The existing systems suffer from many limitations. Safety levels of the existing systems can be improved in several ways. Proper safety frameworks are required to avoid unexpected and fatal child injuries. The comparison made between routing problems in this study reveals that, most of the existing literatures focus on the economy and the shortest routes to carry students, but they do not regard the safety aspects of the routes as precedence. Identification of vulnerable routes and safety of the routes should be considered in future systems. GPS is commonly utilized in most of the school bus tracking systems. Tracking the school bus in real-time without network latency is important. Improvement in GPS accuracy and reduction in network latency can enhance the efficiency and reduce safety risks in school transportation. Tracking the driving behaviour of school bus drivers can help transportation departments to take the necessary action, thereby enhancing the quality of transportation.

Development of radiation free devices for school children tracking systems will reduce children expose to radiation. We aim to propose a framework for safe route planning and tracking of school transportation with improved safety and a reduced risk of accidents.

References

1. Kostas Kalogirou, Eleni Chalkia, Evangelos Bekiaris and Frederik Diederichs, "An application for the information of children according their school transportation", in procedia - Social and behavioral sciences, Elsevier, (2012), pp. 363-372

2. UK Road safety data, https://data.gov.uk/dataset/road-accidents-safety-data, Accessed on (Dec. 2015)
3. Angela Harrison, "Road safety: Insurers show accidents near schools", BBC news, (Aug. 2013)
4. Vasumita s adarsh, ET bureau, "Startups like Ospox and North star offer safety solutions for school kids parents", Economic times of India, (Aug. 29, 2014)
5. Teresa Yuan, "New GPS tracking devices on Kent school buses", King 5 News, (Jan. 4, 2016)
6. C. M. Silva, J. F. M. Sarubbi, D. F. Silva, M. F. Porto, and N. T. R. Nunes, "A Mixed Load Solution for the Rural School Bus Routing Problem," IEEE 18th International Conference on Intelligent transportation systems, pp. 1940-1945, (2015)
7. Faraj, M. F., Sarubbi, M., Silva, C. M., Porto, M. F., Tadeu, N., & Nunes, R, "A Real Geographical Application for the School Bus Routing Problem". IEEE 17th International Conference on Intelligent Transportation Systems, Qingdao, China, (2014), pp. 2762–2767
8. Huang, Z., Gao, Z., Lu, H., Zhang, J., Feng, Z., & Xia, H. "A Mobile Safety Monitoring System for Children". 10th IEEE International Conference on Mobile Ad-Hoc and Sensor Networks, (2014), pp. 323–328.
9. Mammen, G., Stone, M. R., Buliung, R., & Faulkner, G, "Putting school travel on the map: Facilitators and barriers to implementing school travel planning in Canada". Journal of Transport & Health, (2015), pp. 318-326
10. Harrison, F., Burgoine, T., Corder, K., van Sluijs, E. M. F., & Jones, A, "How well do modelled routes to school record the environments children are exposed to?: A cross-sectional comparison of GIS-modelled and GPS-measured routes to school", International Journal of Health Geographics, (2014)
11. Falkmer, T., Horlin, C., Dahlman, J., Dukic, T., Barnett, T., & Anund, A, Usability of the SAFEWAY2SCHOOL system in children with cognitive disabilities", European Transport Research Review, pp. 127–137, (2014)
12. Riera-Ledesma, J., & Salazar-González, J. J., "A column generation approach for a school bus routing problem with resource constraints". Computers and Operations Research, pp. 566–583, (2013)
13. Yang, S.-H., & Hsiung, P.-A, "Innovative Application of RFID Systems to Special Education Schools", IEEE Fifth International Conference on Networking, Architecture, and Storage, pp. 299–304, (2010)
14. Rita M. Newton, Warren H. Thomas," Design of school bus routes by computer", Socio-Economic Planning Sciences, Elsevier, pp. 75-85, (1969)
15. Desrosiers, J., Ferland, J.A., Rousseau, J.-M., Lapalme, G., Chapleau, L., "An overview of a school busing system", Scientific Management of Transport Systems, pp. 235–243, (1981)
16. Braca, J., Bramel, J., Posner, B., Simchi-Levi, D., "A computerized approach to the New York City school bus routing problem", IIE Transactions, pp. 693–702, (1997)
17. Junhyuk Park, Byung-In Kim, "The school bus routing problem: A review", European Journal of Operational Research, Elsevier, pp. 311-319, (2010)
18. Anund, A., Dukic, T., Börsbo, B., & Falkmer, T., "Piloting smart safe school bus: exploration of security gains from implementation of a driver support system, additional technical equipment and intelligent bus stops", European Transport Research Review, pp. 157–163, (2010)
19. Anund, A., Dukic, T., Thornthwaite, S., & Falkmer, T., "Is European school transport safe?—The need for a "door-to-door" perspective", European Transport Research Review, pp. 75–83, (2011)
20. Byung-In Kim and Sangwon Jeong, "A comparison of algorithms for origin-destination matrix generation on real road networks and an approximation approach," Computers and Industrial Engineering, vol. **56**, no. 1, pp. 70–76, (2009)
21. Riera-Ledesma J, Salazar-Gonza´lez JJ., "Solving a school bus routing problem", Odysseus 2006, third international workshop on freight transportation and logistics, Altea, Spain. pp. 285–8, (2006)
22. Schittekat P, Sevaux M, S¨ orensen K "A mathematical formulation for a school bus routing problem", IEEE 2006 international conference on service systems and service management, Troyes, France, pp. 1552–7, (2006)
23. Riera-Ledesma J, Salazar-Gonza´lez JJ, "Solving school bus routing using the multiple vehicle traveling purchaser problem: a branch-and-cut approach", Computers & Operations Research, pp. 391–404, (2012)
24. Bodin L, Berman L, "Routing and scheduling of school buses by computer", Transportation Science, pp. 113–129, 1979.
25. Dulac G, Ferland JA, Forgues PA, "School bus routes generator in urban surroundings", Computers & Operations Research, pp. 199–213, (1980)
26. Desrosiers J, Ferland JA, Rousseau JM, Lapalme G, Chapleau L, "Transcol: a multi-period school bus routing and scheduling system", TIMS Studies in the Management Sciences, pp. 47–71, (1986)
27. Chapleau L, Ferland JA, Rousseau JM, "Clustering for routing in densely populated areas", European Journal of Operational Research, pp. 48–57, (1985)
28. Bowerman R, Hall B, Calamai P, "A multi-objective optimization approach to urban school bus routing: formulation and solution method", Transportation Research Part A, pp. 107–23, (1995)
29. Schittekat, P., Kinable, J., Sörensen, K., Sevaux, M., Spieksma, F., & Springael, J, "A metaheuristic for the school bus routing problem with bus stop selection", European Journal of Operational Research, pp. 518–528, (2013)
30. J. Euchi and R. Mraihi, "The urban bus routing problem in the Tunisian case by the hybrid artificial ant colony algorithm," Swarm Evolutionary Computation, vol. **2**, pp. 15–24, (2012)
31. J. Pacheco, R. Caballero, M. Laguna, and J. Molina, "Bi-Objective Bus Routing: An Application to School Buses in Rural Areas," Transportation Science, vol. **1655**, pp. 1–15, (2012)

32. B.-I. Kim, S. Kim, and J. Park, "A school bus scheduling problem," European Journal of Operational Research, vol. **218**, no. 2, pp. 577–585, (2012)
33. J. Park, H. Tae, and B. I. Kim, "A post-improvement procedure for the mixed load school bus routing problem," European Journal of Operational Research, vol. **217**, no. 1, pp. 204–213, (2012)
34. S. M. Song and T. Kim, "Customer-oriented school bus operations for childcare centers in Korea," Computers and Industrial Engineering, vol. **66**, no. 1, pp. 116–124, (2013)
35. J. Kinable, F. C. R. Spieksma, and G. Vanden Berghe, "School bus routing-a column generation approach," International Transactions in Operational Research, vol. **21**, no. 3, pp. 453–478, (2014)
36. M. Bogl, K. F. Doerner, and N. P. Sophie, "The School Bus Routing and Scheduling Problem with Transfers," Networks, vol. **47**, no. 1, pp. 26–36, (2015)
37. X. Chen, Y. Kong, L. Dang, Y. Hou, and X. Ye, "Exact and Metaheuristic Approaches for a Bi-Objective School Bus Scheduling Problem," PLoS One, vol. **10**, no. 7, (2015)
38. A. Anund and T. Dukic, "From user wishes to system architecture safety for children going to and from school by bus," IEEE 11th International Conference on ITS Telecommunications, ITST 2011, pp. 327–331, (2011)
39. K. Shaaban, A. Bekkali, E. Ben Hamida, and A. Kadri, "Smart Tracking System for School Buses Using Passive RFID Technology to Enhance Child Safety," Journal of Traffic and Logistics Engineering, vol. **1**, no. 2, pp. 191–196, (2013)
40. J. Zambada, R. Quintero, R. Isijara, R. Galeana, and L. Santillan, "An IoT based scholar bus monitoring system," IEEE First International Smart Cities Conference, (2015)
41. J. Saranya and J. Selvakumar, "Implementation of children tracking system on android mobile terminals," IEEE International Conference on Communication and Signal Processing, ICCSP 2013 - Proc., pp. 961–965, (2013)
42. A. Al-lawati, S. Al-jahdhami, A. Al-belushi, D. Al-adawi, M. Awadalla, and D. Al-abri, "RFID-based System for School Children Transportation Safety Enhancement," 8th IEEE GCC Conference and Exhibition, Muscat, Oman, (1-4 February, 2015)
43. P. Collins, Y. Al-Nakeeb, and M. Lyons, "Tracking the Commute Home From School Utilising GPS and Heart Rate Monitoring: Establishing the Contribution to Free-Living Physical Activity", Journal of Physical Activity & Health, vol. **12**, pp. 155–162, (2015)
44. A. Ito, Y. Kakuda, T. Ohta, and S. Inoue, "Safety Support System on School Routes Based on Grouping of Children in Mobile Ad Hoc Networks," IEEE Tenth International Symposium on Autonomous Decentralized Systems, pp. 533–538, (2011)
45. K. Kraschl-Hirschmann and M. Fellendorf, "Estimating energy consumption for routing algorithms," 2012 IEEE Intelligent Vehicles Symposium, Proceedings, pp. 258–263, (2012)
46. M. Richter and H. Kabza, "Strategic dynamic traffic routing algorithm for globally optimized urban traffic," IEEE Intelligent Vehicles Symposium, Proceedings, pp. 756–762, (2013)
47. A. Amini, R. M. Vaghefi, J. M. De Garza, and R. M. Buehrer, "Improving GPS-Based Vehicle Positioning for Intelligent Transportation Systems," Intelligent Vehicles Symposium Proceedings, 2014 IEEE, pp. 1023–1029, (2014)
48. S. Fiocchi, I. a. Markakis, P. Ravazzani, and T. Samaras, "SAR exposure from UHF RFID reader in adult, child, pregnant woman, and fetus anatomical models," Bio electromagnetics, vol. **34**, no. April, pp. 443–452, (2013)
49. M. Redmayne and O. Johansson, "Radiofrequency exposure in young and old: different sensitivities in light of age-relevant natural differences," Rev. Environ. Health, vol. **30**, no. 4, pp. 323–335, (2015)
50. H. Eren, S. Makinist, E. Akin, and A. Yilmaz, "Estimating driving behavior by a smartphone," 2012 IEEE Intelligent Vehicles Symposium, Proceedings, no. **254**, pp. 234–239, (2012)
51. R. Araújo, Â. Igreja, R. De Castro, and R. E. Araújo, "Driving coach: A smartphone application to evaluate driving efficient patterns," IEEE Intelligent Vehicle Symposium, Proceedings, vol. **1**, no. 1, pp. 1005–1010, (2012)
52. Amanda McElfresh, "Official: School bus tracking needs improvement", theadvertiser part of the USA today network, (Jan. 5 2016)
53. The associated press, "Official: Lafayette school bus tracking needs improvement", kansascity news, (Jan. 6 2016)

Cooperative Intelligence in Roundabout Intersections Using Hierarchical Fuzzy Behavior Calculation of Vehicle Speed Profile

Ivan Bosankić[1] and Lejla Banjanović Mehmedović[2]

[1]*IVL, 75000 Tuzla, Bosnia and Herzegovina*
[2]*University of Tuzla, Faculty of Electrical Engineering, 75000 Tuzla, Bosnia and Herzegovina*

Abstract. In this paper, a new fuzzy-behavior-based algorithm for roundabout intersection management is presented. The algorithm employs cooperative intelligence and includes intelligent vehicles and infrastructure to calculate speed profiles for different vehicles, in order to achieve more comfortable driving profiles, as well to reduce congestion and CO_2 emissions. The algorithm uses adaptive spatio-temporal reservation technique and was tested in MATLAB/Simulink environment. The algorithm is designed to function in different scenarios with both cooperative and non-cooperative vehicles, as well as optional intersection infrastructure. Results have show that using the proposed algorithm different vehicle communication types can be successfully combined in order to increase traffic flow through roundabout intersections.

1 Introduction

Roundabout intersections have recently become very popular, since they reduce number of conflict points which is characteristic for classic intersections, reduce driving speeds and increase driver attention [1]. The traffic flow can be as high as 70.000 vehicles per day. In such roundabouts, waiting times are high, and collisions happen often [2]. When traffic is heavy, waiting time is a significant problem. Normal approach would be building elevated road lanes, but often free space is an issue. Possible solutions include installation of traffic lights at roundabout entry, which can decrease waiting times during increased traffic flow times [3], especially if it is optimized [4].

Modern approaches, such as flower and turbo roundabouts present recent solutions that improve road safety and reduce number of collisions, but waiting time when traffic flow is above 1500 vehicles per hour increases more than five times when compared to 1000 vehicles per hour [5]. Increased capacity also increases pollutant emissions [6].

1.1 State of the art

With introduction of the smart vehicles, alternative methods for roundabout management have emerged. Although it has been considered for a long time that traffic lights are only way to achieve efficient traffic control actuator function [7], with increased development of self driving vehicles, it becomes clear that a new approach to enforce traffic control on intersections can be expected in the future.

V2V and V2I communications represent an important role in modern Intelligent Transportation System (ITS), and they can be used to exchange inter-vehicular signals, therefore avoiding classic traffic lights. New models implement Virtual Traffic Lights (VTL) [8] as method that allows each smart car to receive its individual traffic signal, therefore acquiring better control and throughput. Using virtual traffic lights it is possible to increase average speed up to 26% in low and up to 41% in high density traffic, while reducing CO_2 emissions up to 18% [9].

This system can further be improved by advanced algorithms for intersection management, which also calculates speed profiles for each vehicle that passes through roundabout, spatio-temporal intersection management etc.

There are many recent works that address intersection, but they rarely address roundabout intersections. Basically, there are approaches that employ centralized control, by implementing intersection management unit which collects data from all participating vehicles and then calculates best speed profiles and assigns priorities: [10] implements supervisor which receives state information from the cars and then estimates possible collisions and sends overriding signal if necessary, [11] presents reservation-based intersection control algorithm. Others approaches include distributed intersection management, e.g. [12] assigns priorities to vehicles while introducing various V2V protocols.

Evolutionary algorithms (GA and SFLA) via V2V communication are used in [13], while [14] employs fuzzy logic for intersection detection and crossing. In [15] mutual exclusion algorithm is introduced, which is organized via V2V (distributed, without need of intersection controller). Some papers use linear programming algorithms, e.g. [16] uses job-shop scheduling. Centralized V2I coordination that employs Active set method and Interior point method for optimization is used in [17]. In [18] and [19] genetic optimization of fuzzy controller are used to achieve better results. Agent based approaches are often used, e.g. zoning for trajectory adjustment in [20]. In [21] a hybrid automaton for designing intersection management algorithm is used.

Since vehicles without communication capabilities cannot be excluded, [19], [22] and [23] suggest several algorithms both in roundabouts and classical intersections, based on spatio-temporal reservation technique.

In this paper, a new fuzzy-behavior-based algorithm for roundabout intersection management is presented. The advantage of our Cooperative Intelligence approach is that various different vehicle communication types – combinations of cooperative and non-cooperative vehicles as well as possibility of faulty or missing infrastructure controller can be modeled and examined. Our algorithms increase the traffic flow and reduce the travel time through a roundabout intersection. In Section 2, Roundabout intersection model used for testing the proposed algorithm is explained, as well as concept of spatio-temporal reservation and hierarchical fuzzy behavior algorithm. In Section 3, cooperative intelligence principle is explained and several different cases are analyzed, while Section 4 contains research discussion. The fifth section is the conclusion and future work.

2 Roundabout intersection model

Roundabout intersection (RI), analyzed in this work consists of 5 of branches, marked with letters A-E, as shown in Fig. 1. This model can be extended to any finite number of branches. Each branch has entry and exit lane (indexed 1 and 2), and each part is divided in equal segments which are numbered and used for spatio-temporal reservation.

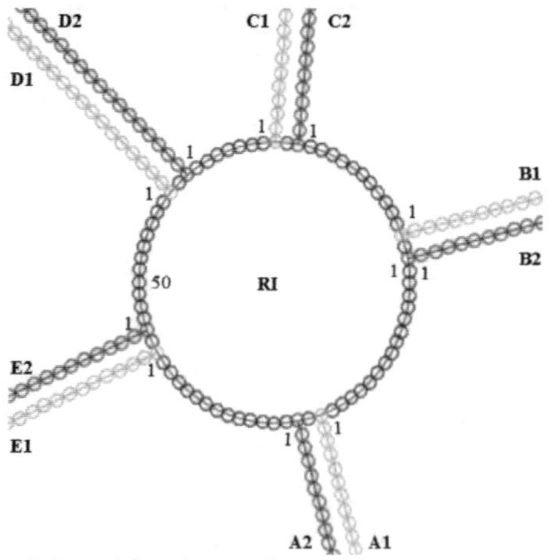

Figure 1. Roundabout intersection model.

Intersection may have intersection management computer (IMC), which collects data about vehicles inside roundabout and coordinates vehicles via V2I communication.

There are two types of vehicles in this model: cooperative (intelligent) and non-cooperative (ordinary) vehicles. Vehicles can exchange information between each other (V2V) or with IMC (V2I). It is supposed that IMC, if exists and functions correctly, can obtain location and speed data for every non-cooperative vehicle in the intersection.

Proposed algorithm includes fuzzy hierarchical behavior coordination of speed profiles for each vehicle.

2.1 Reservation array

Each cooperative vehicle forms and updates spatio-temporal reservation array. Roundabout is divided in finite number of discrete segments. In order to provide intersection management, this array must be filled for each vehicle. Cooperative autonomous vehicles maintain their own reservation array, since they have complete information about entrance and desired exit from the roundabout. This array is then exchanged with other vehicles via V2V and with IMC, via V2I. For other vehicles, IMC must use its sensors to store information about the vehicles. IMC forms reservation array based on measured speed, but here all exits must be considered, since real trajectory of the vehicle is unknown. IMC has no information on exit branch of non-cooperative vehicle, so it has to fill all possible exiting branches for this vehicle. Example of spatial reservation arrays for two vehicles is shown in Fig. 2. As we can see here, segments 4-16 of Roundabout intersection ring are both occupied by both vehicles, but it does not mean that vehicles occupy them at the same time. Spatio-temporal reservation is analyzed in Section III.

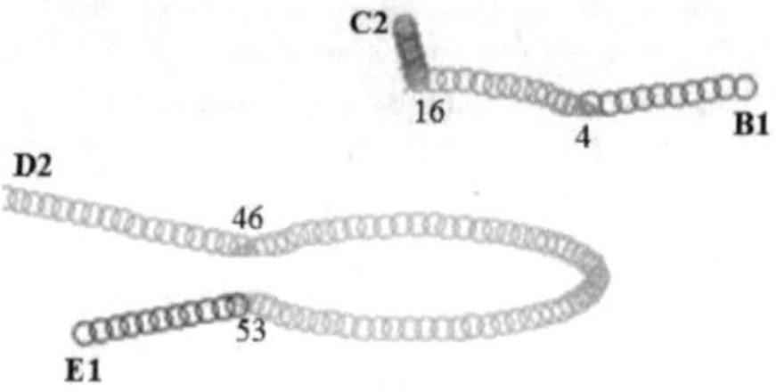

Figure 2. Example of spatial reservation arrays for two vehicles.

2.2 Fuzzy spatial speed profile

Speed profile on an intersection depends on various parameters: speed limit, roundabout radius, entry and exit road angles, vehicle limits and dynamics, lane widths, lane speeds etc. In order to ease calculation of reservation array, Fuzzy spatial speed profile block is proposed. It provides intuitive and easy way to determine vehicle speed, based on available parameters. Example of one speed profile is shown in Fig. 3.

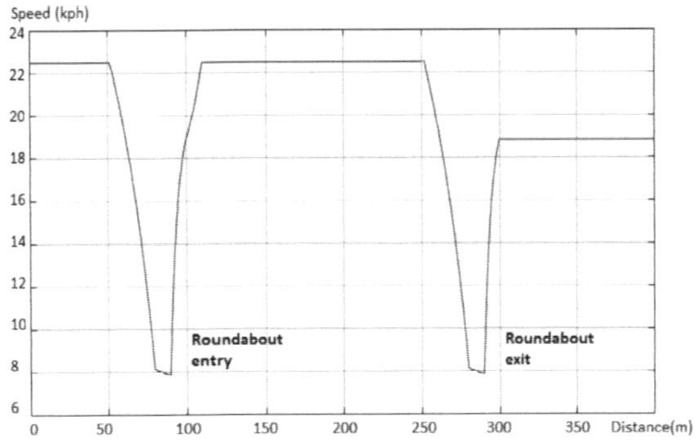

Figure 3. Example of fuzzy speed profile in roundabout intersection

2.3 Fuzzy hierarchical speed behavior profile

To implement proposed speed calculation algorithm, a hierarchical fuzzy-based speed calculation is used. The vehicle speed is influenced by three behaviors:

- ***Normal speed profile behavior*** is used to calculate normal speeds based on vehicle entry and exit point, as well as speed limits inside the roundabout. Spatial speed profile from Section 2.1 is converted in time domain so it can be used in hierarchical fuzzy
- ***Collision avoidance behavior*** is used to avoid collisions in case vehicles unexpectedly change its course, mostly as interaction with non-cooperative vehicles. It is implemented as simple reactive algorithm, which utilizes sensor inputs and IMC warnings to make corrections to Normal speed profile. Output of this block is speed change and can only be negative.
- ***Cooperative intelligence behavior*** is used to coordinate vehicles, in order to avoid congestions. Due to its nature, it can provide both positive and negative speed change commands as output. Cooperative algorithm is explained in detail in Section 3.

All proposed behaviors are fused in Hierarchical fuzzy speed profile block, as shown in Fig. 4.

Hierarchical fuzzy speed profile block forms corrected speed profile for an autonomous vehicle, based on sensor and cooperative corrections to Normal speed profile, Fig. 5. It is used to calculate updated reservation array, and if it overlaps with reservation arrays from other vehicles, additional cooperative actions need to be employed.

2.4 MATLAB/Simulink model

Since reservation arrays were needed to be updated in regular discrete intervals, combination of MATLAB script and Simulink models was used in simulation. In order to convert Normal speed profile from spatial to time domain, speed profile calculation was formed in separate model. Hierarchical fuzzy behavior speed profile is obtained by new simulation in regular intervals of 1s, Fig. 6. Reservation array was then recalculated based on speed profile and compared with other profiles.

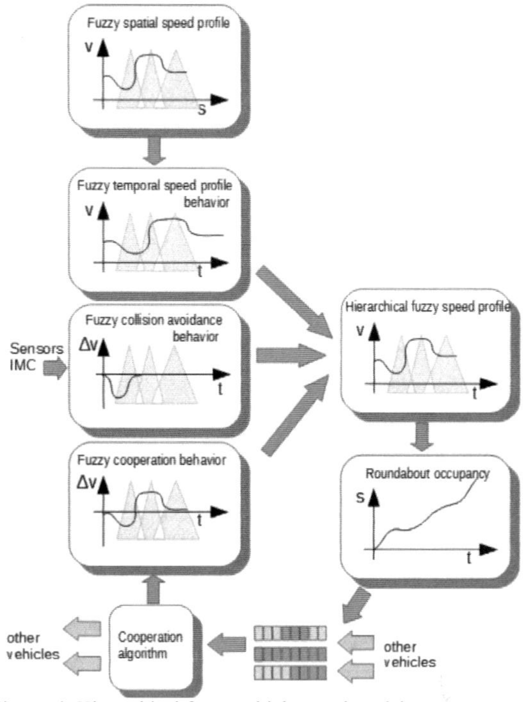

Figure 4. Hierarchical fuzzy vehicle speed model.

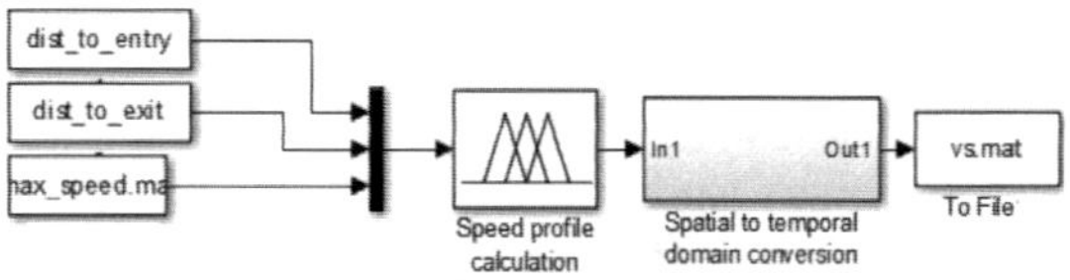

Figure 5. Simulink model for normal speed profile.

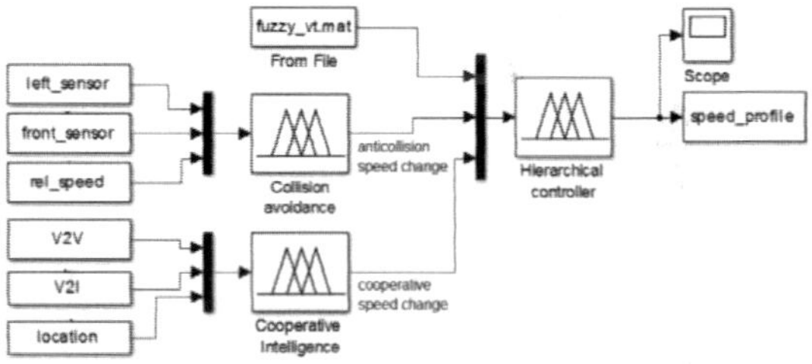

Figure 6. Hierarchical fuzzy vehicle speed model.

3 Cooperative intelligence

A proposed coordination algorithm aims to cover different cases of cooperative and non-cooperative vehicles, with or without intersection management controller (e.g. if IMC is broken).

Base for coordination algorithm are reservation arrays. Obtained reservation arrays for all vehicles are compared in order to detect overlapping. If overlapping occurs, normal action would be for vehicle that is already inside the roundabout to continue driving at normal speed, and other vehicle to slow down. If there are intelligent agents, roundabout priorities can be rearranged based on type of overlapping, if it can be proven that total waiting times can be reduced.

The advantage of this approach is that various different combinations can be modeled and examined. In this article, four different cases are identified, for different combinations of cooperative and non-cooperative agents in roundabout intersection segment:

- ***Case 1*** is used if there is an IMC and two independent cooperative vehicles. Obviously, here exists largest possibility for improvement, since IMC can provide additional data that vehicles themselves can not, providing even larger speeds above speed limit. In this scenario, if two vehicles have overlapped reservation arrays, IMC can provide speed-up signal for one vehicle, and slowdown signal for another.
- ***Case 2*** is applied if there is an IMC with one cooperative and one non-cooperative vehicle. IMC calculates best scenario based on received reservation matrices and sends an adjustment to cooperative intelligence block. Since it can not affect the non-cooperative vehicle, it checks if speedup signal can avoid reservation array overlapping. Otherwise, it sends slowdown signal.
- ***Case 3*** is provided if there are two cooperative vehicles in intersection without IMC. Vehicles need to exchange their reservation matrices and adjust their speed profiles, simultaneously providing collision avoidance for all non-cooperative vehicles that can only be detected with vehicles' sensors.
- ***Case 4*** is used as backup in all other cases (no IMC, none or one cooperative vehicle) and here ordinary roundabout rules apply.

Different cases are depicted in Fig. 7.

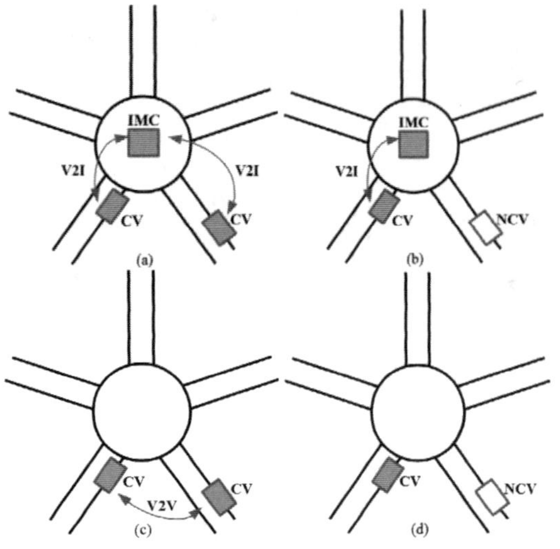

Figure 7. Possible combinations of agents in a roundabout intersection (a) case 1 (b) case 2 (c) case 3 (d) case 4.

Based on these data, intersection manager employs different algorithms in case two vehicles attempt to enter the same roundabout intersection segment at the same time.

If we have both cooperative vehicles and an intersection (Case 1), IMC coordinates with vehicles to make minimal waiting time, as follows:

- Intersection controller compares reservation arrays
- If at current speed limit one vehicle can enter the roundabout before the other vehicle, a zero "speed change" signal is issued to both vehicles
- Controller attempts to correct both vehicles by calculating which vehicle is expected to leave conflicted segment first, so it gets positive speed change signal, while the other gets small negative speed change signal if necessary
- Else larger negative speed change signal is issued to one of the vehicles.

Similar results will be achieved in Case 2, not all options are possible, so slightly worse performance is expected, since one of the vehicles cannot be controlled by IMC (therefore, all scenarios where both vehicle adjust speeds are not possible)

If there are two cooperative vehicles with no IMC (Case 3), two vehicles coordinate to make minimal waiting time, as they calculate which vehicle enters the conflict area first and it has advantage (other vehicle gets negative "speed change" signal.

In Case 4 congestion cannot be improved, since there are not enough intelligent agents to employ coordination. However, fuzzy controller can calculate smooth speed profile based on inter-vehicular distance and relative speed. In this case, normal roundabout traffic rules conditions are used.

4 Results and discussion

After implementation of the MATLAB/Simulink model, all four cases were examined in various combinations of entry and exit location, as well as speed limit.

Fig. 8 shows initial reservation arrays for 3 vehicles. As it can be seen, two vehicles have overlapping.

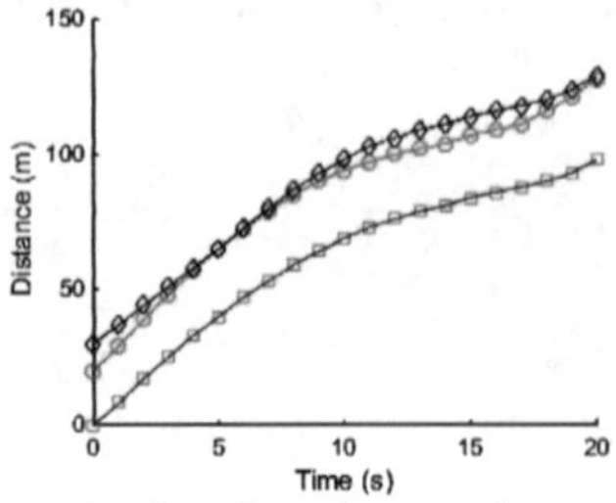

Figure 8. Example of overlapped reservation array.

Fig. 9 shows regular roundabout travel rules – one vehicle slows down while other continues normally. Travel time to 100m is 10.45s and 14.23s for each vehicle, which brings total travel time 24.68s for these two vehicles.

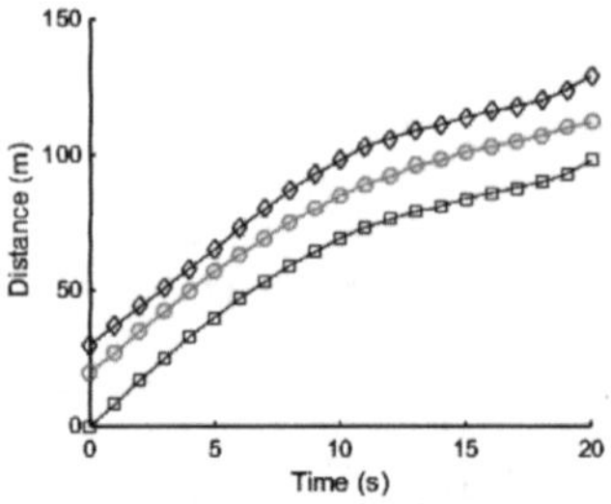

Figure 9. Example of reduced speed for one vehicle (middle line).

At Fig. 10 we see the best case, where one vehicle speeds up and the other slows down. Travel time to 100m is 8.29s and 12.37s for each vehicle, which brings total travel time 20.66s, therefore saving more than 4 seconds.

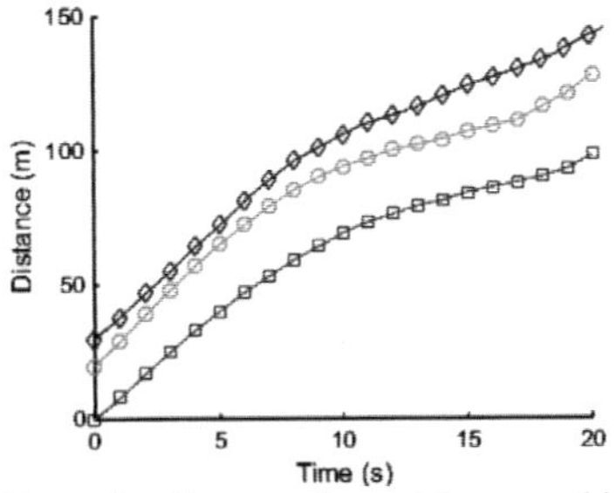

Figure 10. Example of increased speed for one vehicle (top line).

5 Conclusion and future work

Simulation has shown that travel time through a roundabout intersection can be reduced. It is shown that system brings improvements in every case – even if there is not intersection controller present.

In future work, we plan to include experiments this case on real-time robotic platform, as well as improving the model (slowing down while entering intersection, more lanes in roundabout, traffic lights, pedestrians, platooning) and improving controlling algorithm. Also, plan is to extend algorithm to coordination with multiple vehicles inside roundabout, so different congestion scenarios can be analyzed.

Acknowledgment

This paper is supported by the Federal Ministry of Education and Science, Government of Federation of Bosnia and Herzegovina.

References

1. P. Lipar, J. Kostjansek, "Traffic Calming in Slovenia", *3rd Urban Street Symposium, Seattle, Washington* (2007)
2. I. Legac, H. Pilko, N. Subic, "Analysis of traffic safety on roundabout Jadranska Avenue - Avenue Dubrovnik in Zagreb" *Conference proceedings Transport, Maritime and Logistics Science, 14th ICTS 2011, Slovenia* (2014)
3. J. Ben-Edigbe, A. Abdelgalil, I. Abbaszadehfallah, "Extent of Delay and Level of Service at Signalised Roundabout", International Journal of Engineering and Technology, **2**(3), 419 (2012)
4. Y.-J. Gong, J. Zhang, "Real-Time Traffic Signal Control for Modern Roundabouts by Using PSO-Based Fuzzy Controller", *2012 IEEE Congress on Evolutionary Computation* (2012)
5. T. Tollazzi, M. Rencelj, "Comparative analyse of the two new alternative types of roundabout - turbo and flower roundabout" The Baltic Journal of Road and Bridge Engineering, **9**(3), 164 (2014)
6. F. Corriere, M. Guerrieri, D. Ticali, A. Messineo, "Estimation of air pollutant emissions in flower roundabouts and in conventional roundabouts archives of civil engineering", Archives of Civil Engineering, LIX, **2** (2013)
7. S. Jianjun, W. Xu, G. Jizhen, C. Yangzhou, "The analysis of traffic control cyber-physical systems", *13th COTA International Conference of Transportation Professionals (CICTP 2013)*, (2013)
8. R. Sinha, P.S. Roop, P. Ranjitkar, "Virtual Traffic Lights+: A Robust, Practical, and Functionally Safe Intelligent Transportation System", Transportation Research Record Journal of the Transportation Research Board, (2013)
9. M. Ferreira, P.M. d'Orey, "On the Impact of Virtual Traffic Lights on Carbon Emissions Mitigation", *IEEE Transactions on Intelligent Transportation Systems*, (2012)
10. H. Ahn, A. Rizzi, A. Colombo, D. Del Vecchio, "Experimental Testing of Semi-autonomous Multi-vehicle Control for Collision Avoidance at Intersections", *IEEE/RSJ International Conference on Intelligent Robots and Systems*, (2015)
11. Z. Li, M.V. Chitturi, L. Yu, A.R. Bill, D.A. Noyce, "Sustainability Effects of Next-generation Intersection Control for Autonomous Vehicles", TRANSPORT - Special Issue on Smart and Sustainable Transport, **30**(3), 342 (2015)
12. R. Azimi, G. Bhatia, R. Rajkumar, P. Mudalige, "Intersection Management using Vehicular Networks", SAE Technical Paper 2012-01-0292 (2012)
13. Y. Zheng, L. Jin, L. Gao, K. Li, Y. Wang, F. Wang, "Development of a Distributed Cooperative Vehicles Control Algorithm Based on V2V Communication", *Procedia Engineering* **137**, 649 (2016)
14. V. Milanés, J. Pérez, E. Onieva, C. González, "Controller for Urban Intersections Based on Wireless Communications and Fuzzy Logic", *IEEE Transactions on Intelligent Transportation Systems*, **11**(1) (2010)
15. W. Wu, J. Zhang, A. Luo, J. Cao, "Distributed Mutual Exclusion Algorithms for Intersection Traffic Control", *IEEE Transactions on Parallel and Distributed Systems,* **26**(1), 65 (2015)
16. H. Ahn, D. Del Vecchio, "Semi-autonomous Intersection Collision Avoidance through Job-shop Scheduling", arXiv:1510.07026, (2015)
17. H. Ahn, A. Rizzi, A. Colombo, D. Del Vecchio "Experimental Testing of Semi-autonomous Multi-vehicle Control for Collision Avoidance at Intersections", *IEEE/RSJ International Conference on Intelligent Robots and Systems (IROS)* (2015)
18. E. Onieva, V. Milanés, J. Villagrá, J. Pérez, J. Godoy, "Genetic optimization of a vehicle fuzzy decision system for intersections", Expert Systems with Applications **39**, 13148 (2012)
19. E. Onieva, U. Hernandez-Jayo, E. Osaba, A. Perallos, X. Zhang, "A Multi-Objective Evolutionary Algorithm for the Tuning of Fuzzy Rule Bases for Uncoordinated Intersections in Autonomous Driving", Information Sciences (2015)

20. M. N. Mladenovic, M. Abbas, "Priority-based Intersection Control Framework for Self-Driving Vehicles: Agent-based Model Development and Evaluation", *2014 International Conference on Connected Vehicles and Expo (ICCVE),* **377** (2014)
21. S. H. HosseinNia, B.M. Vinagre, V. Milanés, C. González, "Controller for Urban Intersections Based on Hybrid Automaton", *Intelligent Transportation Systems (ITSC), 2010 13th International IEEE Conference*, **261** (2010)
22. L.C. Bento, R. Parafita, S. Santos, U. Nunes, "Intelligent Traffic Management at Intersections: Legacy Mode for Vehicles not Equipped with V2V and V2I Communications", *IEEE Conference on Intelligent Transport Systems, ITSC2013* (2013)
23. L. C. Bento, R.Parafita, U. Nunes "Intelligent traffic management at intersections supported by V2V and V2I communications", IEEE Conference on Intelligent Transportation Systems (2012)

A Rear-End Collision Avoidance Scheme for Intelligent Transportation System

Chen Chen[1] , Hongyun Liu[1], Hongyu Xiang[1], Meilian Li[1], Qingqi Pei[1] and Shengda Wang[2]

[1]*State Key Laboratory of Integrated Service Networks Xidian University, Xi'an 710071, China*
[2]*Jilin Electric Power Company Limited Communication Branch Transport Department, Changchun 130021, China*

Abstract. In this paper, a rear-end collision control model is proposed using the fuzzy logic control scheme for the autonomous or cruising vehicles in Intelligent Transportation Systems (ITSs). Through detailed analysis of the car-following cases, our controller is established on some reasonable control rules. In addition, to refine the initialized fuzzy rules considering characteristics of the rear-end collisions, the genetic algorithm is introduced to reduce the computational complexity while maintaining accuracy. Numerical results indicate that our Genetic algorithm-optimized Fuzzy Logic Controller (GFLC) outperforms the traditional fuzzy logic controller in terms of better safety guarantee and higher traffic efficiency.

1 Introduction

According to the investigations and statistics, of all the traffic accidents that ever took place, about 60 to 70 percent were caused by vehicle collisions, particularly rear-end collisions [1]. To alleviate the growing risks of rear-end collisions, the auxiliary or automatic control methods proposed for autonomous or cruising vehicles in ITSs look promising. Actually, there have been many classic automatic control schemes proposed, such as PID control [2], sliding mode control [3], and the linear quadratic optimal control [4]. Although these methods enable accurate control to be exercised, they are highly dependent on precise mathematical models. In practice, an accurate control model for vehicle collision avoidance is almost unavailable.

Instead of precise control, the fuzzy logic-based controllers (FLC) are also well studied to reduce the possible vehicle collisions in the last few years. Due to its superiority in solving problems in multi-parameter, non-linear systems as well as capturing the driving characteristics in a vehicular environment, it is feasible and preferable to apply FLC-related models to vehicular active control issues. However, since the effect of a FLC relies on the number of fuzzy rules, an excessive number of such will directly impair its effectiveness. Moreover, the number of fuzzy rules increases exponentially with the number of fuzzy subsets, resulting in greater computational complexity of the controller. Among various schemes that have been proposed to address the above-mentioned problems and have become an active research topic recently is combining neural network (BP) with fuzzy control [5, 6]. However, the neural network converges slowly and is prone to being stuck in local optimum. Unlike the neural network, the genetic algorithm (GA) is capable of searching for the global optimal solution and is almost application-independent. To efficiently reduce the risk in car-following scenario, this paper proposes a rear-end collision avoidance system that uses the fuzzy logic control model with its fuzzy rules optimized by the generic algorithm. This GA-based FLC should be able to allow drivers to promptly avoid approaching risks and accurately make a brake, as well as accommodating driving behaviours of drivers. In addition, the model is also applicable to the popular autonomous vehicles or vehicles in cruising state to keep a safe inter-vehicle space and at the same time improve the driving efficiency by controlling the headway.

2 GA-based FLC

In practical cases, the factors that influence driving safety are very complicated and usually include weather conditions, road surface situations, response time of drivers, vehicle dynamics, etc. Since most of these factors are non-linear and time-varying, obtaining an accurate mathematical model of the vehicle control system is difficult. In this context, some of the traditional methods (e.g., PID control, linear quadric-form optimal control) are unable to exercise accurate control. On the other hand, fuzzy control has advanced considerably in control and reason applications in recent years mainly due to its independence from accurate mathematical models. Its performance relies on the number of fuzzy rule bases, the number of linguistic variables as well as the types and ranges of membership functions. An excessive number of fuzzy control rule bases and linguistic variables results in longer FLC search time and higher computational complexity. Therefore, the key to FLC design is to determine the fuzzy control rule base and input/output membership function, which are usually acquired by expert summarization. However, in practical applications, an FLC generally cannot always guarantee its effectiveness, highlighting the need for optimization. By reviewing previous studies, there are many optimization schemes that optimize fuzzy control rules [7-9] or the input/output quantified scale factors [10]. In this paper, GA is used to optimize the control rules by finding the optimal input/output variables combination. The flowchart of our proposed GA-based FLC to avoid the rear-end collisions in car-following scenarios is shown in Fig. 1.

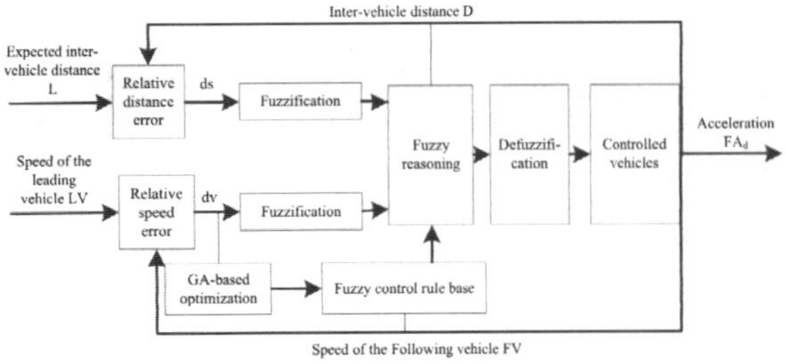

Figure 1. GA-based FLC.

To exercise effective control over vehicles, a dual-input single-output fuzzy controller is designed as Fig. 1 illustrates. The relative distance error ds and the relative speed error dv between the leading and trailing vehicles are defined as the input variables. The output acceleration FA_d is defined as the output variable. The input variables are given as

$$ds = D - S, \tag{1}$$

$$dv = LV - FV, \tag{2}$$

where D denotes the inter-vehicle distance, S denotes the expected inter-vehicle space, LV and FV are the speeds of the leading and trailing vehicles, respectively. Note that D here denotes the current longitudinal inter-vehicle distance measured using such as line laser [11] or wireless signal [12] while S indicates the inter-vehicle space by which the driving safety could be guaranteed considering the vehicular dynamics and braking process. The speed of the trailing vehicle FV should be known since the collision warning system has been installed on it, while the speed of the leading vehicle LV could be obtained through GPS sharing [13] or wireless communication systems.

2.1 Fuzzification of input/output variables

In our work, the ranges of the input/output variables are properly set based on practical considerations. According to the braking distance algorithm from Mazda[14], we have:

$$S = \frac{1}{2}\left(\frac{FV^2}{a_{-\max}} - \frac{LV^2}{a_{-\max}}\right) + FV \cdot \tau_1 + dv \cdot \tau_2 + d_0, \tag{3}$$

where τ_1 and τ_2 denote the delay. d_0 denotes the distance between the two vehicles after they stop. Generally, $\tau_1 = 0.1s$, $\tau_2 = 0.6s$, and $d_0 = 1.5m$, where τ_1 , τ_2 and d_0 are the system delay, driver reaction time and minimum distance needed to prevent a collision if both vehicles begin braking with their respective maximum decelerations, respectively. In the case of the leading vehicle braking immediately at a speed of 120km/h (i.e., 33.33m/s), the expected inter-vehicle space is 67.5m. So the relative distance error is set from -67.5m to 67.5m. According to the Chinese-laws on the speedway[15], vehicles drive at a speed varying from 60km/h to120km/h, so the range of dv in our work is [-60km/h, 60km/h]. Taking into account the comfort of drivers and passengers, the acceleration of the trailing vehicle is set to [-8m/s^2, 8m/s^2].

The relative distance error ds , relative speed error dv and acceleration FA_d are first fuzzified into seven fuzzy subsets: negative large (NL), negative medium (NM), negative small (NS), zero (Z), positive small (PS), positive medium (PM), and positive large (PL). Before the fuzzification, scaling transformations need to be made to the actual input variables such that they can be transformed into the specified domains. Therefore, we have

$$x_0 = \frac{x_{\min} + x_{\max}}{2} + k\left(x_0^* - \frac{x_{\min}^* + x_{\max}^*}{2}\right), \tag{4}$$

$$k = \frac{x_{\max} - x_{\min}}{x_{\max}^* - x_{\min}^*}, \tag{5}$$

where x_0^* denotes the actual input variable, $\left[x_{\min}^*, x_{\max}^*\right]$ denotes the range of the variable, $\left[x_{\min}, x_{\max}\right]$ denotes the specified domain of the variable and k denotes the scaling factor. For the relative distance error ds , the scaling factor $k_1 = 0.05$. Using Equation (4), it can be determined that the fuzzy domain of the relative distance error is [-6, 6]. Since the range of the relative speed error dv is [-16.67m/s, 16.67m/s] and the scale factor $k_1 = 0.36$, the fuzzy domain of the relative speed error is [-6, 6] by (4). Similarly, since acceleration FA_d is limited to [-8m/s^2, 8m/s^2] and the scale factor $k_1 = 0.75$, its fuzzy domain is [-6, 6].

2.2 Determination of membership functions

In our work, the triangle function is used as the membership functions for the relative distance error ds , the relative speed error dv and the acceleration FA_d . The domain of each linguistic variable is partitioned into seven parts for ds , dv and FA_d as shown in Fig. 2

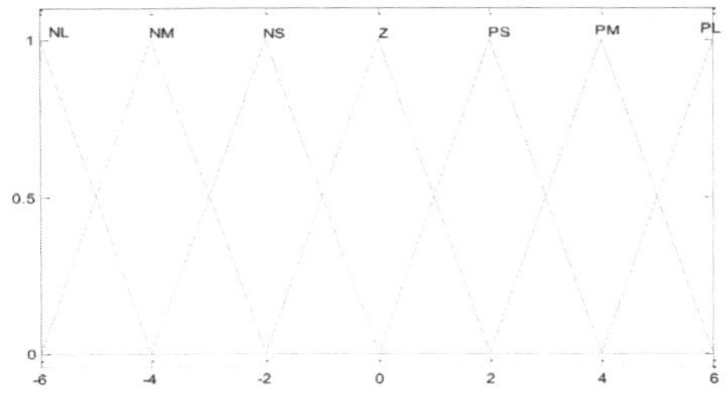

Figure 2. Membership function of ds, dv and FA_d.

2.3 Establishment of the fuzzy rules

To construct the fuzzy rules used to control the risky vehicles when a specific condition is met, the experience of drivers should be fully considered. According to the common response for a normal person, the following typical driving scenarios are taken into account:

(1) If the inter-vehicle distance is small and the speed of the trailing vehicle is substantially larger than that of the leading vehicle, then rear-end collision has a high probability of occurrence. In this case, the trailing vehicle should brake immediately to decelerate for collision avoidance.

(2) If the inter-vehicle distance is still large and the speed of the trailing vehicle is close to that of the leading vehicle, then rear-end collision has a small probability of occurrence. In this case, the speed of the trailing vehicle can remain or increase properly.

(3) If the inter-vehicle distance is small and the speed of the trailing vehicle is close to that of the leading vehicle, then the driver of the trailing vehicle should brake slightly in order to prevent rear-end collision and ensure driving safety.

Based on above driving experiences as well as trial-and-error, a set of 49 rear-end avoidance fuzzy control rules representing the actual driving experience is devised and shown in Table 1. The gth rule R^g is:

$$if\ x_{ds}\ is\ A^g\ and\ x_{dv}\ is\ B^g\ then\ y_{FA_d}\ is\ C^g, \quad (6)$$

where x_{ds} is the input value of ds, x_{dv} is the input value of dv, y_{FA_d} is the output value, A^g and B^g denote the fuzzy subsets of ds and dv, respectively. C^g denotes the fuzzy subset of FA_d. The fuzzy rule R^g can be regarded as the fuzzy implication $A^g \times B^g \rightarrow C^g$ over the product space $U_{fz} \times V_{fz}$, where $U_{fz} = A^g \times B^g$.

Table 1. Fuzzy control rules for the real-end avoidance system.

Control variable		Relative distance error						
		NL	NM	NS	Z	PS	PM	PL
relative speed error	NL	NL	NL	NL	NM	NM	NS	NS
	NM	NL	NL	NM	NM	NS	NS	Z
	NS	NL	NM	NS	NS	NS	Z	Z
	Z	NM	NS	Z	NS	Z	PS	PS
	PS	NM	Z	Z	Z	Z	PM	PL
	PM	NS	Z	Z	PS	PM	PL	PL
	PL	NS	Z	Z	PS	PL	PL	PL

2.4 Optimizing to the fuzzy rules using genetic algorithm

The fuzzy rules determined in the previous subsection will be optimized with the genetic algorithm to reduce the problem scale of our collision control strategy. The detailed steps are given as follows:

2.4.1 Coding of the fuzzy control rule

To optimize the fuzzy rules, we first encode it using the binary coding scheme [15]. Each rule is represented by four binary numbers, the first of which is the control bit and the other three are the rule representation bits. The tandem coding method is employed by connecting these 49 rules to form a chromosome. According to our predefined fuzzy rules, these 49 rules will be the parameters to be optimized. The antecedents of the rules are ds and dv, and the consequent is FA_d. Because the control variable has seven fuzzy sets, the consequent of the rules can be represented via three-digit binary coding, i.e., 000, 001, 010, 011, 100, 101 and 110. If a certain fuzzy rule does not exist, then let X denote the consequent of the rule and its code is 111. Thus, the 49 control rules in Table 1 can be numbered as rule1 to rule49, sequentially. Next, the above 49 rules can be connected in a tandem manner to provide the chromosome of the rules as shown in Fig.3. The chromosome of the rules is the unit of the population, i.e., the operand of GA. Each bit of the control gene controls the corresponding rule. As a result, the GA-based optimization yields the fuzzy rules consisting of fewer but more effective rules.

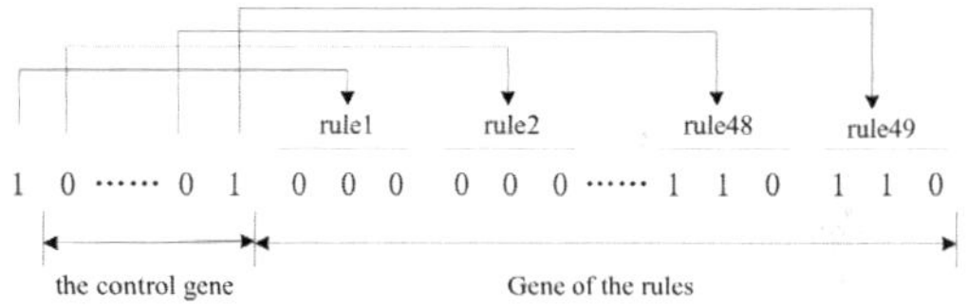

Figure 3. Chromosome of the rules

2.4.2 Determination of the fitness function

Since our FLC is designed to avoid rear-end collisions, the key to achieving this objective is to ensure both the relative distance error ds and the relative speed error dv are small. Therefore, the objective function J is defined as:

$$J = |ds| + |dv|. \quad (7)$$

It is worth noting from (8) that the smaller the value of J, the better the performance of our FLC. However, generally, the GA prefers individuals with high fitness. So, our objective function needs to be properly transformed into a fitness function, which is given by:

$$f = \frac{1}{1+J}. \quad (8)$$

2.4.3 Determination of the genetic parameters

The effectiveness of the GA relies on the related genetic parameters, including the population size n, the crossover probability P_c and the mutation probability P_m. Usually, the large size of the initial population means a high space and time complexity. The size of the initial population is usually within the range [20,100] [16]. In our work, the initial population is obtained in the following way:

a) Randomly extract 80 matrixes with 7 rows and 7 columns as the matrixes of the control variables. Elements of all matrixes are integers ranging from 1 to 7.

b) Compute the fitness f of each individual, and sort these n individuals in descending order of f.

c) Eliminate m=30 individuals with small fitness, define the remaining individuals as the initial population S1, thereby obtaining an initial population whose size $n_{s1} = 50$.

Note that each individual represents a chromosome of the rules. In this paper, the crossover probability P_c is 0.7. Because a small mutation probability P_m can effectively prevent damage to the good individuals, P_m is set to 0.001. The number of iterations is set to G=30.

2.4.4 Determination of selection, crossover and mutation operations

Biological evolution is achieved via genetic operations. Therefore, genetic operators are essential for optimization of fuzzy rules.

a) Selection operator. The fitness proportional selection operator is used. The fitness is first calculated using the fitness function and then the replication probability of each individual is obtained. The number of replications of this individual in the next generation is equal to the product of the replication probability and the population size. The probability P_{si} that each individual is selected is denoted by:

$$P_{si} = \frac{f_i}{\sum_{j=1}^{n_{s1}} f_j}, \tag{9}$$

where n_{si} is the population size, and f_i is the fitness of the ith individual in the population.

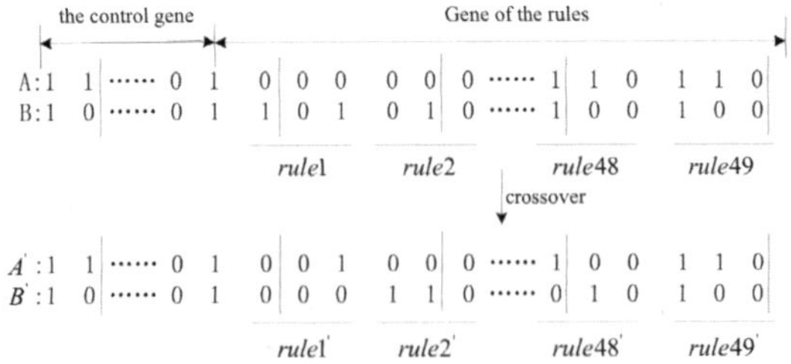

Figure 4. Illustration of the crossover operation.

b) Crossover operator. During the crossover operation, the chromosomes before the cross point undergo the mutation operation, while those after the cross point undergo crossover and mutation operations. The crossover probability $P_c = 0.7$, and the crossover operation is illustrated in Fig. 4.

c) Mutation operator. The basic-bit mutation method is used in our work. Because binary coding is carried out on the fuzzy control rules, the mutation operator may involve reversing genes of some gene bits (e.g., $0 \to 1$ or $1 \to 0$). The mutation probability $P_m = 0.001$. Note that the mutation operations improve the local search ability of GA and avoid premature convergence.

Finally, through the GA optimization, the number of the fuzzy rules has been reduced from 49 to 28.

2.5 Defuzzification of the output variables

The defuzzification process takes the fuzzy sets and produces a single-value output. In our paper, the control variable is defuzzified via the gravity method as:

$$Y_{out} = \frac{\sum_{g=1}^{7} l \cdot \eta_{C^g}(l)}{\sum_{g=1}^{7} \eta_{C^g}(l)}, \tag{10}$$

where l, as we have stated before, is the value of the output in the domain, $\eta_{C^g}(l)$ is the membership of l with respect to the fuzzy set C^g.

3 Simulations

To demonstrate the superiority of our GFLC controller, both FLC and GFLC are simulated under typical car-following scenarios. The simulation setup is as follows: the sampling time is 0.1s, total simulation time is 80s, initial distance between two vehicles is 20m, the initial speeds of the leading and trailing vehicles are 20m/s and 30m/s, respectively. Note that in our simulation, dv is initially set to -10m/s to increase the possibility of rear-end collision. The leading vehicle keeps a constant speed before 20s and then accelerates at $1 m/s^2$ during 20-30s. During 31-33s, the leading vehicle begins to brake at $4 m/s^2$. Note that the risk of rear-end collision with this configuration is very high between the leading and trailing vehicles, so that the trailing vehicle ought to brake immediately to avoid the collision. To obtain the statistical average, each test is repeated 30 times.

From Fig. 5, it can be observed that both controllers instruct the trailing vehicle to brake during the early period of the simulation for rear-end collision avoidance. However, it is worth noting that there is actually a slight difference between two controllers. GFLC executes the decelerate action from approximately seconds 30 while FLC at about seconds 32. This difference verifies that our controller has a better response speed to the possible dangers than FLC. In addition, during 20-30s, when the leading vehicle accelerates at $1 m/s^2$, GFLC instructs the trailing vehicle to gradually increase the acceleration from approximately $0.1 m/s^2$ at 20s to almost $1 m/s^2$ at 30s. On the contrary, the FLC shows a fluctuant process

DOI: 10.1051/matecconf/20168102001

which first makes the acceleration increase from 0.28m/s^2 at 20s to 0.9m/s^2 at about 23s, and then suddenly decrease to almost 0.5m/s^2 at 30s. This dramatic variation implies that the rules used in the traditional FLC are unreasonable to some extent while our rules, optimized by GA, are well suited for the car-following scenario. During 31-33s, when the leading vehicle begins to decelerate with 4m/s^2, GFLC immediately outputs a deceleration for the trailing vehicle of about 4m/s^2 (equal to the deceleration of the leading vehicle), thus reducing the collision risk. However, FLC outputs a smaller deceleration which is actually insufficient to promptly avoid the possible rear-end collision. During 34-80s, since the rear-end collision risk is low, our GFLC almost keeps a zero acceleration for the trailing vehicle. However, during this period, FLC still yields non-zero accelerations which vibrate around 0m/s^2.

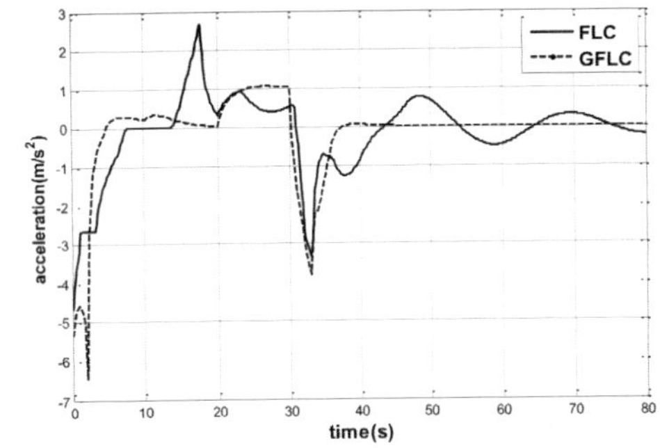

Figure 5. Comparison for accelerations of the trailing vehicle.

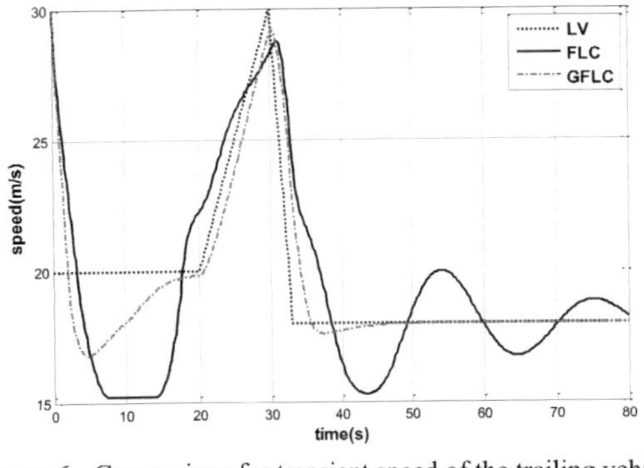

Figure 6. Comparison for transient speed of the trailing vehicle.

The transient speed comparisons between two models are depicted in Fig. 6. It can be observed that the leading vehicle keeps a steady speed before 20s while the trailing vehicle adapts itself according to the instructions of the corresponding controllers to avoid collision. It is worth noting that approximately before 8s, FLC outputs continuous deceleration for the trailing vehicle while GFLC outputs a rapid decrease before about 5s and then an increase during approximately 5-8s. Actually, since the dv is initially set to -10m/s, the trailing vehicle will reduce the gap to zero between two vehicles after 2 seconds. Under such circumstance, our GFLC first outputs a deceleration to avoid the possible collision in advance. After a continuous deceleration before about 5s, the inter-vehicle distance returns to a safe level and then a persistent acceleration has been given out by our GFLC till about 20s. In comparison, FLC instructs the trailing vehicle to drive at a very low speed, i.e., about 16m/s, during approximately 8-14s, which seems to overestimate the risk and result in a lower driving efficiency. During 20-30s, due to the acceleration of the leading vehicle, both controllers suggest a fast speed-up for the trailing vehicles. However, by checking the points carefully, it has been determined that our GFLC outputs more precise controlling instructions which are almost strictly consistent with our simulation configurations. On the contrary, FLC instructs the trailing vehicle to accelerate too fast for too long, say about 14-32s. During 31-33s of simulation, GFLC instructs the trailing vehicle to decelerate as soon as the leading vehicle decelerates, effectively avoiding the possible rear-end collision. With FLC, the trailing vehicle's speed changes slowly, thus increasing the rear-end collision risk. During 34-80s, GFLC enables the trailing vehicle to drive at a stable rate of 18m/s, reducing the risk and at the same time improving the driving efficiency. With FLC, it shows great fluctuation indicating the instability of its adopted fuzzy rules.

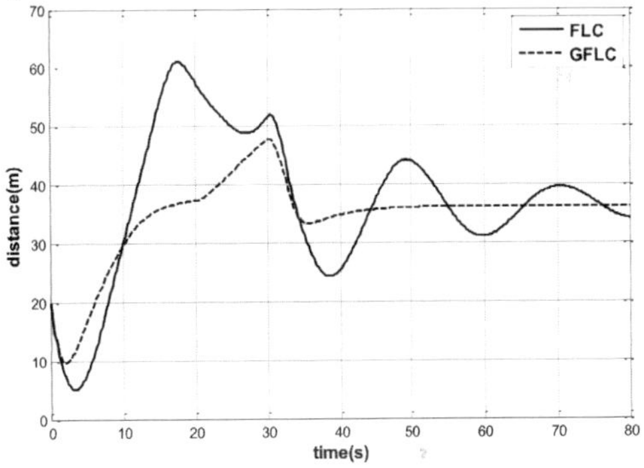

Figure 7. Comparison of inter-vehicle distance.

From Fig. 7, it can be observed that the inter-vehicle distance actually changes with the variation of acceleration. With GFLC, the inter-vehicle distance changes slightly during the entire simulation procedure and almost remains the same during 40-80s. In contrast, FLC causes continuous fluctuations. At about 4s, FLC even outputs an inter-vehicle distance of no more than 6 meters, which greatly increases the rear-end collision risk. Similarly, FLC also gives out a very large value, i.e., over 60 meters around 18s, which is really unnecessary considering the previous continuous low speed of the trailing vehicle. It is also worth noting that the lasting duration of inter-vehicle distance over 40 meters exceeds approximately 27s for FLC while it is only about 9s for our GFLC. This large difference confirms once again that GFLC is more effective than FLC in terms of driving efficiency.

4 Conclusions

In this paper, a novel fuzzy controller is proposed, which leverages GA to optimize the fuzzy control rules, to improve the performance on rear-end collision avoidance of the vehicular control system, and enhances the emergency response ability. Simulation results successfully indicate that the proposed controller outperforms the traditional one and is capable of avoiding the possible rear-end collision while increasing the traffic efficiency.

Our future work will attempt to incorporate the advanced sensor system into our vehicular dynamics control framework and propose an effective active-control system. This could further reduce collision risks by introducing precise prediction algorithm depending on the online data collection module.

Acknowledgement

This work was supported by the National Natural Science Foundation of China (61201133, 61571338), the National Science and Technology Major Project of the Ministry of Science and Technology of China(2015zx03002006-003), the Natural Science Foundation of Shaanxi Province(2014JM2-6089), the National High-tech R&D Program of China (863 Program-2015AA015701), the Research collaboration innovation program of Xi'an(BD31015010001), the Hong Kong, Macao and Taiwan Science and Technology Cooperation Special Project (2014DFT10320, 2015DFT10160) and the "111 Project" of China (B08038).

References

1. S. H. Kim, P. S. Moon, W. S. Jang, K. S. Kim, S. C. Lee, M. Woo, *et al.*, "An experimental investigation of a CW/CA system for automobiles," SAE Technical Paper, (1999)
2. S. Moon, I. Moon, and K. Yi, "Design, tuning, and evaluation of a full-range adaptive cruise control system with collision avoidance," *Control Engineering Practice,* **17**, pp. 442-455, (2009)
3. A. Ferrara and C. Vecchio, "Second order sliding mode control of vehicles with distributed collision avoidance capabilities," *Mechatronics,* vol. 19, pp. 471-477, (2009)
4. J. van den Berg, D. Wilkie, S. J. Guy, M. Niethammer, and D. Manocha, "LQG-obstacles: Feedback control with collision avoidance for mobile robots with motion and sensing uncertainty," in Robotics and Automation (ICRA), 2012 IEEE International Conference on, pp. 346-353, (2012)
5. Y. J. Mon and C. M. Lin, "Supervisory recurrent fuzzy neural network control for vehicle collision avoidance system design," Neural Computing and Applications, **21**, pp. 2163-2169, (2012)
6. N. B. Hui, V. Mahendar, and D. K. Pratihar, "Time-optimal, collision-free navigation of a car-like mobile robot using neuro-fuzzy approaches," Fuzzy Sets and Systems, **157**, pp. 2171-2204, (2006)
7. H. S. Hwang, Y. H. Joo, H. K. Kim, and K. B. Woo, "Identification of fuzzy control rules utilizing genetic algorithms and its application to mobile robots," Algorithms and Architectures for Real-Time Control (Korea, 1992), pp. 249-254, (2014)
8. A. Jalali, F. Piltan, A. Gavahian, and M. Jalali, "Model-free adaptive fuzzy sliding mode controller optimized by particle swarm for robot manipulator," International Journal of Information Engineering and Electronic Business (IJIEEB), **5**, p. 68, (2013)
9. D. K. Pratihar, K. Deb, and A. Ghosh, "A genetic-fuzzy approach for mobile robot navigation among moving obstacles," International Journal of Approximate Reasoning, **20**, pp. 145-172, (1999)
10. F. Betin, A. Sivert, A. Yazidi, and G. A. Capolino, "Determination of scaling factors for fuzzy logic control using the sliding-mode approach: Application to control of a DC machine drive," Industrial Electronics, IEEE Transactions on, **54**, pp. 296-309, (2007)
11. S. Kang, I. Yoo, M. Shin, and S. Seo, "Accurate inter-vehicle distance measurement based on monocular camera and line laser," IEICE Electronics Express, **11**, pp. 20130932-20130932, (2014)
12. L. Cheng, B. E. Henty, D. D. Stancil, F. Bai, and P. Mudalige, "Mobile vehicle-to-vehicle narrow-band channel measurement and characterization of the 5.9 GHz dedicated short range communication (DSRC) frequency band," Selected Areas in Communications, IEEE Journal on, **25**, pp. 1501-1516, (2007)
13. M. Rohani, D. Gingras, V. Vigneron, and D. Gruyer, "A new decentralized Bayesian approach for cooperative vehicle localization based on fusion of GPS and inter-vehicle distance measurements," in Connected Vehicles and Expo (ICCVE), 2013 International Conference on, pp. 473-479, (2013)
14. P. Seiler, B. Song, and J. K. Hedrick, "Development of a collision avoidance system," Development, **4**, pp. 17-22, 1998
15. W. Zhang, O. Tsimhoni, M. Sivak, and M. J. Flannagan, "Road safety in China: analysis of current challenges," Journal of safety research, **41**, pp. 25-30, (2010)
16. L. Fan and E. M. Joo, "Design for auto-tuning PID controller based on genetic algorithms," in Industrial Electronics and Applications, 2009. ICIEA 2009. 4th IEEE Conference on, pp. 1924-1928, (2009)

Basic Framework and Significance on the Economics of Port Safety

Zhang Shijie, Liu Yan, Zhuang Rong and Wang Xuting

Tianjin Research Institute of Water Transport Engineering of Ministry of Transport, Tianjin, China

Abstract. This article is based on drawing lessons from the basic theory of safety economics, such as the basic concepts, principles, theories, research methods, etc.; and is combined with the actual on the safety production of port enterprises. It has expounded the basic theory on the economics of port safety, such as basic connotation, subject investigated and research method; and has pointed out the basic features. The significance and value of research on the economics of port safety have been analysed in article. The basic framework on the economics of port safety has sketched in this article.

1 Introduction

With the continuous development of national economy and society, the guidelines, policies, and theories of safety management have being continuous innovation, and the models on port safety management have significantly changed. It is transforming from paying attention to after the accident treatment, to attaching great importance to accident prevention before accident. Port enterprises gradually establish the safety concept such as safety first, prevention first, safety development, etc.; increase safety investments; introduce the methods of risk management; and have achieved good results.

However, port enterprises generally encountered some common problems in practical work. The first, how to define and measure safety investment (cost) and to know and evaluate safety benefit (output). The second, in the case of that safety investments are limited, how to optimize the distribution the resources of human, material and financial, in order to obtain the biggest safety benefits. The third, in the premise that fulfilling the requirements for the safety regulations and standards and ensuring the due level of safety, how to reduce the safety investments or to reduce unnecessary investments. The fourth, how to scientifically evaluate the risk of the safety in production, how to reasonably determine the indexes of risk control, and how to properly handle the relationship between safety investments and safety benefits. These problems are so confused not only port enterprises, and including with the correlative government departments on port safety production, scientific research and consulting institutions.

The above problems are the basic problems of safety economics [1,2]. Over the years, according to the research on safety economics carried out less in China, and research on the economics of port safety at home and abroad is in the initial stage of starting, so the research results cannot see much.

The economics of port safety is a new subject. Because of the lack of theoretical guidance and scientific evaluation method, evaluation standard, for the research of this subject, there are some unreasonable, undesirable method of work, for safety investments, in some port enterprises.

For example, some port enterprises one-sided focus on the safety equipment and facilities investments (hardware) and ignore the safety management investments (software, including organization, personnel, education, training and so on), or on the contrary. Some port enterprises mistakenly believe that there are only investments and no benefits in safety production, or large investments and small benefits, therefore they are not willing to do investments, or reduce investments. And there are a small number of enterprises which constitute too high safety objectives that divorced from reality, causing that port aspire after zero accidents or zero risk, and blindly increase safety investments, too much to increase the economic burden of port enterprises.

With the importance of port safety has become increasingly prominent, the research of the actual effect and the economic relationship problems for port safety is especially urgent and important. From this point of view, there are very realistic basic functions that be as soon as possible to carry out the research of the economics of port safety, to research on the basic characteristics, content, research methods for the economics of port safety, and to define some basic concepts as the initial phase of research for the economics of port safety.

2 Subject investigated

From the angle of subject property and task, the definition for the economics of port safety can be given

[1,3,4]. The economics of port safety is a science that studying, which study economy (the interests, investment, benefit) forms and conditions for port safety, through the reasonable organization, control and adjustment of port safety activities, in order to reach the biggest benefits of safety for human, technology, and environment [3]. It is thus clear that the economics of port safety is a science which is to research the relationships between the safety activities of ports with the economic activities of ports. This definition has the following meanings: the subject investigated for the economics of port safety is the economic form and condition of port safety, that is, through theoretical research and analysis, in order to reveal and clarify the form of expression and the realization condition of the port safety behalves, safety investments, and safety benefits. The objectives for the economics of port safety are to achieve optimum safety benefit [5] the three which are human, technology, and environment.

The goal for the economics of port safety is achieved through the control and adjustment of the port safety activity. Port safety activities refer to the activities carried out in order to ensure the safety for port enterprises, which is including that: setting regulation and police for port safety, safety education and management, the implementation of safety engineering and technology, etc. Safety activities require a fair amount of human resources, material resources and financial resources, these resources are called port safety investments [6,7]. It will hope to obtain the corresponding rewards when to invest. The investment return for port safety is to ensure safety, and reduce accidents and reduce the loss of accident in port transportation. This also indirectly promote the added value for port economy production. Therefore, the safety activities of port is not only a kind of "consumption activity", or "investment activities" and "benefit activities".

The economics of port safety also has itself subject investigated, is to study and solve the economic problems of port safety. It is not only a special economics, and is a specific application science which be in order to the specific application area for the management activities of port safety. The subject investigated for the economics of port safety, is the economic relations in the field of port safety, that is, the economic relations of port safety. The so-called economic relations of port safety contains the following four layers meanings: division of labor and cooperation relationship for port safety, economic interests relationship for port safety, economic quantity relationship among the various factors related to the port safety, the economic benefit relationship for port safety. These 4 aspects constitute the system of economic relationship for port safety.

The research objects for the economics of port safety, specifically, are in the operation process of the ports, according to the unity of opposites between port safety reality and economic effect, to research on how to combine port safety activities with the best way with port transportation production, to explore the best input and output ratio in production of port safety, to make the benefits maximization of port safety investments. Specifically, it will study the following questions: ① The macro theory for the economics of port safety, including that the influence on port safety with social development, national economic conditions and port industry development, and research on the influence of port safety which impact on the whole social economy, national security, and even the country's foreign policy. ② Established the objective for the economics of port safety. ③ From theoretically to discuss the proportion relationship between the growth rate of port safety investments and the speed of development of port transport, to grasp and control the development direction and speed of the scale for the economy of port safety. ④ Research on the range of port safety investment. ⑤ To define the scope of the direct economic loss and the indirect economic loss, and to do some discussions to estimate the direct economic benefits and the indirect economic benefits. ⑥ On the basis of the above research, combining with the development trend port development goals, to analyze and forecast the basic trend for that adapt to port safety investment, based on fast development of the ports.

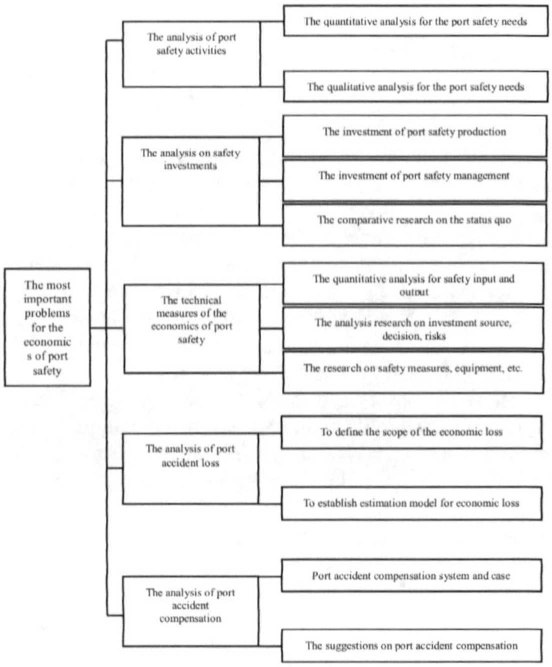

Figure 1. The most important problems for the economics of port safety.

There are many problems which worth studying in the economics of port safety. In view of the uniqueness and complexity of the economic problems for port safety, it cannot include all the contents of all. In the initial stage of research, it can choose the most important problems for the economics of port safety to analyze (Figure 1.).

First of all, it should be from the analysis of port safety activities, to the scope and type of, to illustrate the role of the safety for port transportation. On this basis, to analyze the influencing factors of port safety, to induce and summarize that the principle which the activities of port safety should adhere to. How to analyze the needs of the port safety? It needs to start from the two aspects of quantitative and qualitative, such as that: ① the quantitative analysis for the port safety needs, which be including that the historical investigation of port business and safety signs, the analysis of growth trend for port traffic, and the forecasting for port safety needs; ② the qualitative analysis for the port safety needs, which be including that the changes in the international relations, the changes of political situation and social security, the changes in the social security mechanism, etc.

The analysis on safety investments is one of the important research contents on the economics of port safety. It mainly includes the following aspects: ① the investments of port safety production, (the types of safety production activities, the investments of the resources on human, financial and material, investment amount ratio); ② the investments of port safety management, (the types of safety management activities, the investments of the resources on human, financial and material, investment amount ratio). ③ the comparative research on the status quo of Chinese port safety investment with foreign developed countries.

In the technical measures, based on the input and output theory of economics being applied to the research process, the related quantitative analysis is studying on the input and output of port safety. Using the investment theory and technology to the research on the analysis research for investment source, safety decision and investment risks of port safety and other issues. And being used the basic theory of economic management to study that such as the raising and management of funds for safety measures, the depreciation of safety equipment and facility, and the safety and economic management of port enterprises, and other issues.

The analysis of port accident loss is based on the necessary limits analysis and investigation of port accident and accident loss, to define the scope of the direct economic loss and the indirect economic loss, to establish the estimation model for direct economic loss and the indirect economic loss.

The analysis of port accident compensation should first introduce Chinese accident compensation system and international compensation of port accident, and to compare the different. This analysis is taking port compensation case as the analysis object, to study and put forward the suggestions on port accident compensation.

The safety benefit is the destination and goal of the research on port safety economics. The economic benefit for port safety should be including two aspects that the economic benefit for port safety and social benefit, to study on the basic content and evaluation method of the evaluation on the economic benefit for port safety, and to make prediction for the trend of port safety investments, based on the fast growing of port transportation demand. On the basis of this study, to put forward views and suggestions for the decision-making of the country's port safety.

3 Basic characteristics

The economics of port safety is a new special cross subject, which is being across the natural sciences and social. The subject investigated and task for the economics of port safety determine the subject characteristics (Figure 2.). From the research method for the economics of port safety, which is systemic, advance and superiority [1-4]. From the discipline essence for the economics of port safety, which is marginal and practicability.

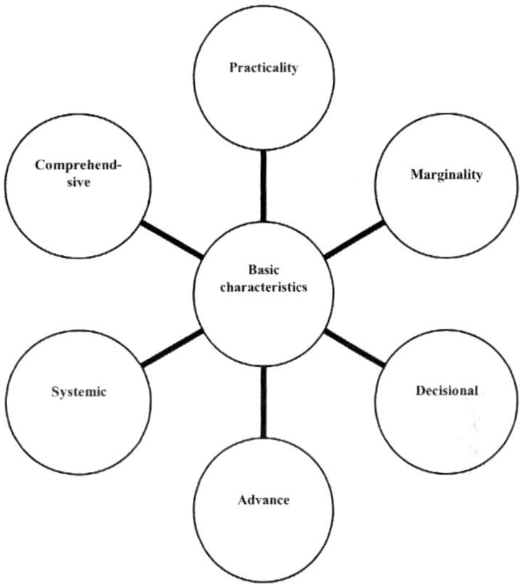

Figure 2. The basic characteristic for the economics of port safety.

① Comprehensive. From research content, the economics of port safety related to economics theory, and related to the basic method of safety management and safety technology, also include the basic content of port transportation. From research method, the economics of port safety is to use the basic method of economics, also make the use of safety management, safety engineering theory, but also consider the basic rules and characteristics of port. Therefore, the economics of port safety needs very strong comprehensive to carry out research, and many of the necessary factors are indispensable.

② Systemic. The problems for the economics of port safety are often complicated problems of multi-objective, multi variables. The safety factors and economics factors are considered in solving the problems for the economics of port safety. Also it should analyze its own factors of the subject investigated, and analyze the variety factors which associated with subject investigated. This constitutes the systematic for the research process and scope.

③ Advance. The output for the economics of port safety is often delay and lag behind, and the essence of port safety activities has the characteristics which are foresight and preventive. Therefore, the economic activities of port safety should have to be advance which adapt to the requirements of economic production activities

④ Decisional. The economics activities of port safety should be based on scientific decision-making. In the port transportation, the decision-making of safety aspects is always the priority among priorities of port management. But to ensure safety need invest. What kind of method used in the process of port safety investment is the most reasonable and guaranteed for port safety, and what way can make the maximum benefit for safety investment. So the economics of port safety will provide optimization techniques and methods for the economic decision of safety.

⑤ Marginality. The economic problems of port safety like other economic problems, not only restricted by the laws of nature (the restriction of safety objective law), but also controlled by the economic laws. That is, the economics of port safety is necessary to study some natural laws of safety, also to study the economic laws for port safety. Therefore, the economics of port safety is a borderline science which is across the natural science and social science of safety.

⑥ Practicality. The economic problems of safety, which being researching on the economics of port safety, are all with strong technical and practical. The research on the economics of port safety is to service and protect the safety and health development of port production transportation, and port safety is as a required service in the operation of port enterprises. The economics of port safety provide the technical methods and guidance for how to ensure the safety of port transportation production, and how to achieve the rationalization, maximization, scientific for safety input and output. The economics of port safety is directly applied to the specific safety operation in port, and is really solve the practical problems of the port, which is be of great practical value.

4 Basic method

The basic method used in the research on the economics of port safety is the method of dialectical materialism [1, 3,8]. The basic method should pay specifically attention to reverse thinking method, comparative analysis method, investigation and study method, and the quantitative and qualitative analysis method, also should pay attention to the combination method of macroscopic and microscopic [9,10]. And some scholars have also proposed the opportunity cost method. At the same time, it should absorb and apply the existing relevant disciplines, with the system method of multidisciplinary comprehensive research.

① Reverse thinking method. In order to achieve safety, it should be to necessary research on the condition and cause of the accident. It should start from the observation and analysis of safety accident, reversely trace the reason and condition for the occurrence of safety accident, thus to eliminate the cause and condition for the occurrence of safety accident.

② Comparative analysis method. Port safety system is a multi-variable, multi object system which is involving a wide range and complex contact factor. Therefore, the research way and method should be scientific and reasonable, and should be in line with the objective requirement. The comparison analysis method is one of the basic method to grasp the characteristics and law of system. It can get accurate understanding only through analysis and comparison.

③ Investigation and study method. In order to understand the law for the economics of port safety, it should carried out according to the existing experience and data, and to get the truth from practice. Therefore, the investigation and study method is an important method to know the rules of port safety. The rules of accident loss can be prompted and reflected only through lots of investigation and study.

④ The quantitative and qualitative analysis method. The number relationship analysis and the scientific quantitative analysis for the economic problems of port safety, is the inevitable requirement of the development for the economics of port safety. Because of the constraint on the objective factors and the basic theory, some propositions for the economic field of port safety are not absolute quantification. It is necessary to use the method of quantitative and qualitative analysis, in order to obtain the reasonable and correct conclusion.

The concept of opportunity cost can lead into the process of research on the economics of port safety. In the real world, "no free lunch" is the true saying. Everything in the world has a price to pay. The economics of port safety also can be analyzed using the related theory of opportunity cost, in order to obtain greater benefits for input and output. In addition, it should also pay attention to set up the system and the dynamic point of view, and use the method of macro and micro.

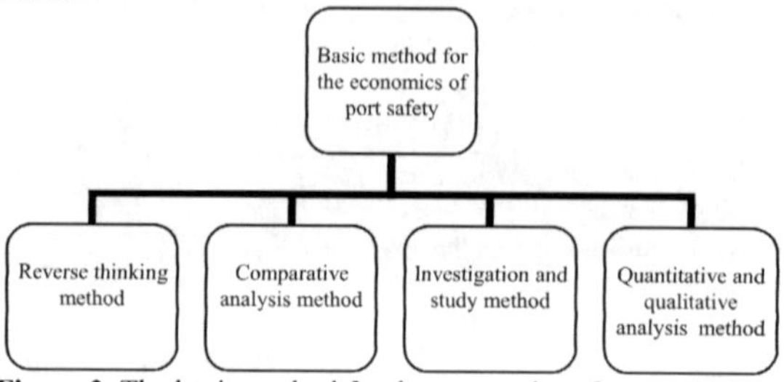

Figure 3. The basic method for the economics of port safety.

5 Research significance

The task for the economics of port safety is to reveal the objective law on the occurrence, development and movement for the economic relations of port safety. And the task is to discuss the realization of port safety production, safety of life, and that the way, method and measures for safety of survival. The economics of port safety not only provides the basis theoretical and reference for the scientific safety guidelines and policies formulated by country and government, but also provide technical support and basic working method for the port enterprise to organize the safety resource allocation optimization and the safety technical management. The economics of port safety can not only provide theoretical guidance for the port safety work, but also provides the implementation options for the actual work of port safety management. Through research on the economics of port safety, thus it can provide the maximum protection for the health and development of port transportation production, so as to promote the harmonious development of the society and economy thriving and prosperous.

To discuss the relationship between port safety and port economy, and to study economic problems of port safety, it is an important topic which constantly sought by many experts and scholars, and the actual workers. The real development of the port also continuously puts forward new requirements to the theory workers who study port problems.

Safety economics [1-4,11,12] itself is in the growth stage, and the development of the discipline is imperfection. The economics of port safety as a part of the safety economics, also is imperfection, and there are very few existing research achievements. In this background, the economic significance for research on the economics of port safety is particularly important. Specifically, embodied in the following aspects:

① the research has great significance for port safety in production. Port safety is the basis guarantee for the development of economic benefits. There is the dialectical relationship such as correlation dependence and mutually reinforcing, between port safety and economic benefits.

② the research provides a theoretical basis for government decision-making in the aspect of port safety, especially to reverse the wrong understandings on port safety for some of the government departments and the port enterprises. In the port industry, it also exists the fuzzy recognition that the safety investment is not to create economic benefits, and safety investment is a completely free return. The research purpose on the economics of port safety is to make the government, corporate decision makers and safety workers who can correct understanding, on the role of safety and the benefit created by safety. So that the decision-making of port safety in the relevant departments is more scientific and practical. It must not only see the political significance, but also pay attention to the economic benefit.

③ the research provides a theoretical tool for safety activities of port enterprises. The realization of port safety needs to invest, and the reasonable investment should be how much, how to achieve greater safety benefit and economic benefit based on limited investment. These must be the guiding theory and method of scientific selection. The related research results can be used to guide port enterprises for the reasonable selection of safety investment and safety means, in order to improve the level for the economic decision-making of safety and management.

④ the research has practical value for establishing the system of safety management system for the throughout port industry, and for further improving the level of port safety in China. The correlation research on port safety in China is overall relatively weak. Undoubtedly, the research on the economics of port safety can greatly enrich the research content for port safety in China, and promote scientific research for port safety in China.

⑤ the research has important significance to promote the study on safety economics. Safety economics is a macro research for the safety economic problems. The safety economics for port is from a field to study on the economic problems. And it is based on the related theories of safety economics to research on safety of port industry. The safety economic problems of this field can enrich the research content of safety economics from the perspective of scientific research. It has positive role in promoting the research on safety economics. In a sense, the study on the economics of port safety is groundbreaking.

6 Conclusion

Based on the research on the basic characteristics, basic contents and research methods for the economics of port safety, this article tries to make a little contribution in the research framework aspect for establishing the economics of port safety. The research on the economics of port safety can from different angles to study, and it is a matter of opinion. The article is from start out the general system of economics, based on the basic theory of safety economics, along the idea of the input channels, cost accounting and benefit analysis. Without a doubt, the point of view in this article is only the statements of a school, only carries on the discussion to the economics of port safety from one aspect.

The complexity of the economics of port safety put forward the challenge of milepost significance to every interested person. It needs to give more attention and research, in order to continuously expand and improve the research on the economics of port safety, for the port service, the water transport industry service and the social service.

References

1. L. Yun, Safety Economics, 2st ed. Beijing, China: Chemical Industry Press, pp. 121-123, (2010).
2. S. Dacheng, Enterprise Safety Economics, Beijing, China: Meteorological Press, (2000).
3. L. Quan, "Basic Framework and Significance on Aviation Safety Economics," Journal of Civil Aviation University of China, **27**, no. 3, pp. 37-41, June 2009.
4. T. Shuicheng, Xiandai Anquan Jingji Lilun Yu Shiwu, Xuzhou, China: China University of Mining and Technology Press, (2004).
5. M. Cheng, "Port safety input and safety cost," Technological Development of Enterprise, **31**, no. 17, pp. 30-31, June (2012).
6. Z. Yuhai, "The theory of enterprise's safety investment", Safety, China, no. 1, pp. 5-7, (2008).
7. W. Shengxia, Research on Enterprise Safety Investment and the System Development, Central South University, Changsha, China, May, (2011).
8. Q. Zhanyu, "Using Safety Economics in Rail Delivery Operation on the Guiding role of Safety

Investment Decision-making," Shanghai Construction Science & Technology, China, no. 3, pp. 64-66, (2013).
9. L. Yun, "Prospects of Safety Management Science in the 21st Century", China Safety Science Journal, **10**, no. 10, pp. 27-32, February (2000).
10. Ian Savage, The Economics of Railroad Safety, Kluwer Academic Publishers, the United States, (1998).
11. L. Calvin, B. Avendaño, and R. Schwentesius, "The Economics of Food Safety: The Case of Green Onions and Hepatitis A Outbreaks," Electronic Outlook Report from the Economic Research Service, December (2004).
12. P. Dorman, "The Economics of Safety, Health, and Well-Being at Work: An Overview," May (2000).

Impact Velocity Prediction in a Traffic Accident

Ali Can Yilmaz and Kadir Aydin

Cukurova University, Automotive Engineering Department, 01330 Adana, Turkey

Abstract. Reconstruction of traffic accidents has been so crucial scientific process in order to make impartial and judicious decisions. This study focuses on impact speed prediction of accident sufferers just before the collision in a comprehensive scientific way by using an accident reconstruction software called "vCrash" and Function Fitting Neural Network (FITNET) artificial intelligence method (predictor) in case of absence of skid marks or other clues about calculating impact speeds. A sample real world accident was simulated on the software several times by changing collision speeds to form different deformation on the collision regions of the vehicles in every simulation. Every single deformation amount corresponding to each impact velocity was recorded and used as teaching data for FITNET prediction model. Using 10-fold cross validation, mean squared error (MSE) and multiple correlation coefficients (R) were observed to exhibit performance of the prediction model. The model performed high R (close to 1) and acceptable MSE values. This method aims that, in a probable similar accident scene in future, it will be possible to analyze the impact speeds just by entering average deformation amounts into an application on a portable device at the accident scene without requirement of expensive reconstruction tools and it will be a guide for analysis of other accident types.

1 Introduction

Traffic engineering focuses on safety and efficiency in transportation. Public agencies strive to reduce traffic accidents on roadways in most countries all over the world. Financial burden of traffic accidents places another problem on society. Driver and roadway design are the major factors in a traffic accident. Safety strongly depends on the gender and the age of the driver [1]. Death rates related to diseases are prone to fall in developed countries in which deaths due to traffic accidents represent a notable exception. As a country flourishes, growing economy brings a substantial increment in motor vehicles which causes substantial probable accidents. Traffic accidents are the first cause of injury-related deaths all around the world. April 7 was devoted to prevention of traffic injuries by World Health Organization (WHO) in 2004. As in many countries, in Turkey, fatalities and injuries due to traffic accidents are major problems for public [2]. Reckless driving with excessive speed gives rise to most of the traffic accidents [3]. According to WHO, traffic fatalities will be the sixth leading cause of death worldwide by the year 2020 [4].

According to 2002 annual reports, traffic accidents are the 11th reason for fatalities in Turkey. Traffic accidents have the 2nd priority for fatalities at the age interval of 5-29 and the 3rd for the age interval of 30-44. In 1999, while the number of registered traffic accidents was 466000, it was 501000 in year 2000. General Directorate of Highway reports indicate 2954 fatalities and 94497 injuries involving 409407 accidents in year 2001 and 570419 accidents in year 2005 [5].

Transport policy substantially corresponds to accessibility, safety and environmental issues. Safety policies are of great importance almost for all countries worldwide [6]. There are some methods in evaluating plans based on infrastructure and Cost–Benefit Analysis (CBA) is now deemed the standard method in most countries [7].

Almost all of the passenger and freight transportation are carried out by highway in Turkey. In Western countries, this load on highways is shifted to railway and seaway, thus lowered highway accidents can be provided. In Turkey, highway transportation constitutes the biggest place within transportation types which unfortunately results increased number of highway vehicles [8].

Rapid increment in motor vehicles is inevitable especially during the last decades which leads to increased traffic density, several traffic accidents. Therefore, accident reconstruction is crucial in terms of solving the question of liability for these accidents. Accident experts investigate accident scenes and generally conclude about causes and procession of the accident referring to police reports. However, growing vehicle numbers and accidents make challenges for police officers to report the scene accurately. Furthermore, anti-lock braking system (ABS) in modern vehicles, when it is active, avoids locking wheels by controlling wheel slip which entails missing brake traces left on the road surface. However, brake traces are one of the most important clues for accident investigators to reconstruct the vehicle motion leading to an accident. More complex situations occur for vehicles equipped with vehicle dynamic control (VDC). Moreover, there are systems interfering the latest systems into the steering system to manipulate the steering system automatically [9]. There are many uncertain factors in road traffic accidents. It is necessary to consider the uncertainty influence of these factors by random theory to improve the analysis veracity. In order to improve the precision and reliability of traffic accident analysis and reconstruction, random theory and the probabilistic perturbation method are introduced for the uncertainty analysis of traffic accident. Precision and uncertainty analyses of accidents yield expedient results within expected confidence levels which also aid accident reconstruction [9].

There have been many studies about traffic accident analysis and reconstruction. Kallberg [10] conducted an analysis in Finland and made a report related to his investigations. The report is mainly for the use of the Finnish road accident investigation teams. Formulas for calculating pre-crash speeds from skid marks are presented as well as other methods for estimating speeds and locations of vehicles during different phases of the accident process. Instructions for graphic description of the pre-crash movements of vehicles are also presented. A method for determining the causes of the accident on the basis of accident reconstruction is suggested. Buck et al. [11] showed the benefits of the 3D documentation and computer-assisted, drawn-to-scale 3D comparisons of the relevant injuries with the damages to the vehicle in the analysis of the course of accidents, especially with regard to the impact situation on two examined cases. Davis also [12] studied on attempts to explain the probable reasons of accidents with physical properties and Bayesian networks statistical models together with the aid of artificial intelligence. Xu *et al.* [13] studied on effects of vehicle impact speed in pedestrian–vehicle accident by comprising reconstruction model and they verified their analysis by ten real-world accident cases to validate their results and to yield an approach for investigators. Raedt and Ponjaert [14] made investigations on predicting at-fault car accidents of older drivers and they concluded that although accident prediction is difficult, the predictability of car accidents by neurocognitive measurements and a road test increases when the kind of accident is specified. Yannis et.al. conducted [15] an analysis on driver age and vehicle engine size effects on fault and severity in young motorcyclists accidents. They concluded that accident severity modeling revealed a significant second-order interaction between severity, driver age and two-wheeler engine size. On the contrary, no second-order effects were identified in fault risk modeling. Moreover, a significant effect of driver age on accident fault risk was identified. The effect of engine size was not significant.

2 Examinations and analysis

2.1 Analysis of a sample accident

A passenger car (Aston Martin DB9) and a bus (Volvo B10B) were crashed into each other on an intersection within simulation software. A sample snapshot and instant data related to vehicles are shown in Fig. 1 and Fig. 2 (specifications that are not legible in the figure are depicted under figure captions in italic).

In case of absence of skid marks (mostly anti-lock braking system equipped vehicles), the biggest clue about the speeds was the damage formed on the vehicles. More damage on the vehicle(s), more energy transformed into deformation energy which is defined as crush depth (ε) in terms of meters. For the scenario, vehicles were crashed into each other at angle of 90° with estimated initial speeds and deformations were recorded. Coefficient of friction (μ) on the road surface was assumed as 1. Vehicles with features similar to those in real-world accident were exposed to different deformations (minimum of 500 data for each scenario) that also corresponded to varied speed values. While one of speeds was kept constant, the other's was increased by 5 km/h. Currently in Turkey, determination of impact velocities is conducted by examining skid marks "if applicable". Considering most of the vehicles are equipped with anti-lock braking system, skid marks may not form on the road surface. In this case, speed analysis cannot be achieved and the situation avoids having contact with scientific analysis. Thus, in simulation, for every speed value, the simulation was repeated and the crush depths were recorded from the simulation report. One of the speed values was kept constant and the other was changed every 20 trials until 500 data were obtained for the scenario to obtain more accurate prediction values. The data were randomly divided into two disjoint subsets using 10-fold cross. The training set had 450 observations (90% of data) and testing set had 50 (10% of data). Consequently, the data of each deformation datum corresponding to each speed value were achieved to be used as teaching data for FITNET model.

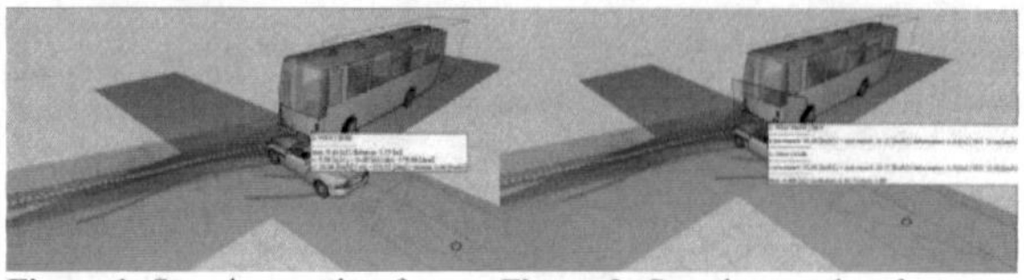

Figure 1. Sample snapshot from the first contact moments of the vehicle (Volvo)
Pre-impact velocity: 25 km/h
Post-impact velocity: 21.3 km/h
Deformation: 0.20 m (average on collision region)
Friction: 1.00

Figure 2. Sample snapshot from the first contact moments of the vehicles (DB9)
Pre-impact velocity: 50 km/h
Post-impact velocity: 26.2 km/h
Deformation: 0.25 m (average on collision region)
Friction: 1.00

Two phases can be distinguished during the crash of a vehicle: there is a compression phase and a restitution phase. The compression phase lasts from the contact of the vehicle with an obstacle (another vehicle or anything else) to the point of maximum compression. During this phase, the energy is stocked until the maximum deformation. The restitution phase begins when deformation is maximum and ends when the vehicle separates from the obstacle. During this phase, the deformation energy is released [16].

3 Overview of FITNET model

The FITNET model consists of 3 layers (input, hidden and output). The input, hidden and output layers have 4, 10 and 2 neurons respectively (Fig. 4). The tan-sigmoid activation function is used in FITNET model in the hidden layers. A pure-linear activation function is used in the output layer and Levenberg–Marquardt (LM) algorithm is utilized for training the networks. FITNET is also a modified form of Multilayer Feed Forward (MFFNN) with the default tan-sigmoid transfer function (Fig. 3a) in the hidden layer and linear transfer function (Fig. 3b) in the output layer.

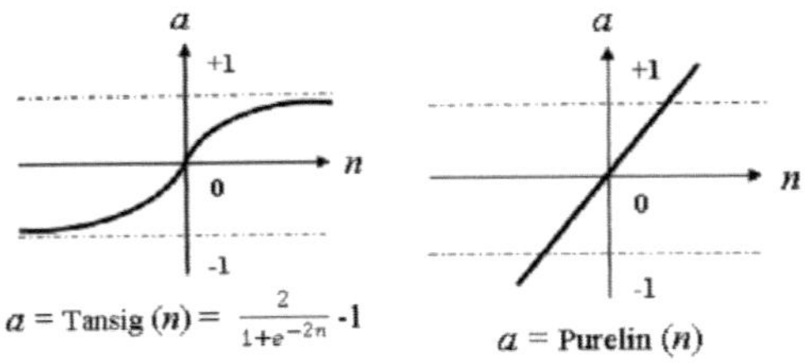

Figure 3. Transfer functions for FITNET [17]

The other important parameters of the FITNET model are the number of epochs, the learning rate and momentum which were selected as 1000, 0.02 and 0.5, respectively.

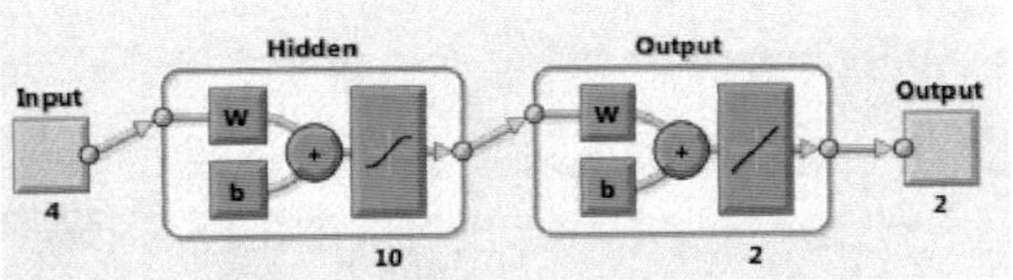

Figure 4. Architecture of FITNET model

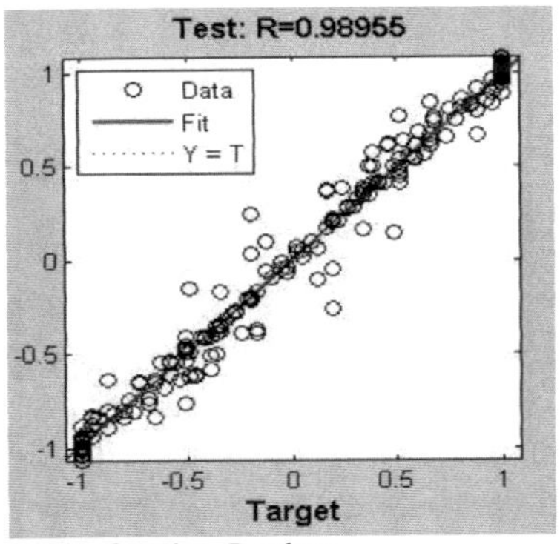

Figure 5. trainlm function, R values

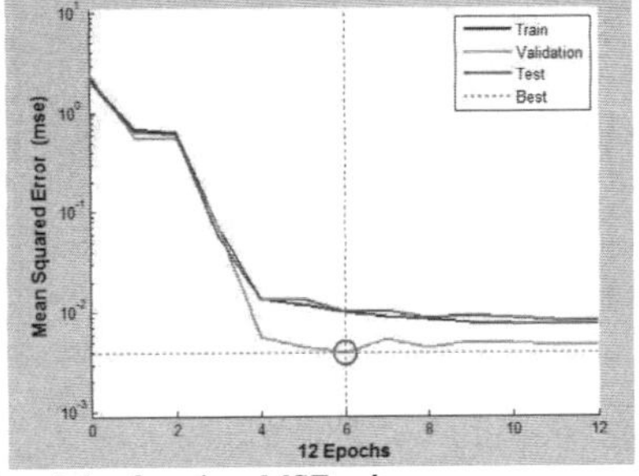

Figure 6. trainlm function, MSE values

Two different training algorithms (trainlm, trainrp) within FITNET were used in order to compare the R and MSE values. "trainlm" is a network training function that updates weight and bias values according to Levenberg-Marquardt optimization. "trainlm" is often the fastest backpropagation algorithm in the toolbox, and is highly recommended as a first-choice supervised algorithm, although it does require more memory than other algorithms. "trainrp" is a network training function that updates weight and bias values according to the resilient backpropagation algorithm. Multilayer networks typically use sigmoid transfer functions in the hidden layers. These functions are often called "squashing" functions, because they compress an infinite input range into a finite output range. Sigmoid functions are characterized by the fact that their slopes must approach zero as the input gets large. This causes a problem when you use steepest descent to train a multilayer network with sigmoid functions, because the gradient can have a very small magnitude and, therefore, cause small changes in the weights and biases, even though the weights and biases are far from their optimal values. The purpose of the resilient backpropagation (Rprop) training algorithm is to eliminate these harmful effects of the magnitudes of the partial derivatives. Only the sign of the derivative can determine the direction of the weight update; the magnitude of the derivative has no effect on the weight update. The size of the weight change is determined by a separate update value. The update value for each weight and bias is increased by a factor "delt_inc" whenever the derivative of the performance function with respect to that weight has the same sign for two successive iterations [18]. It was observed that best validation performance and R values were achieved within trainlm training algorithm compared to trainrp function (Fig. 5, Fig. 6, Fig. 7 and Fig. 8).

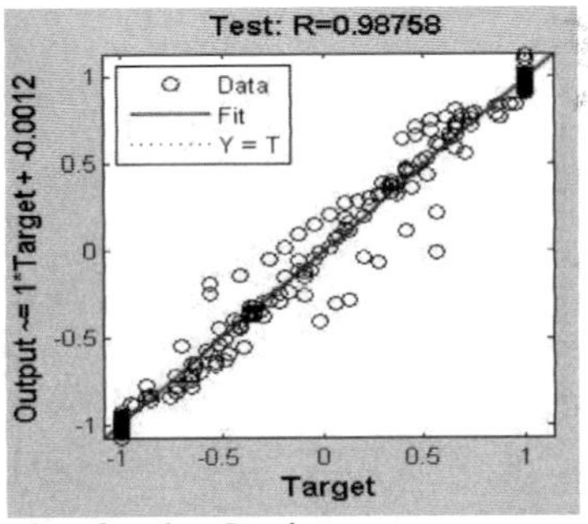

Figure 7. trainrp function, R values

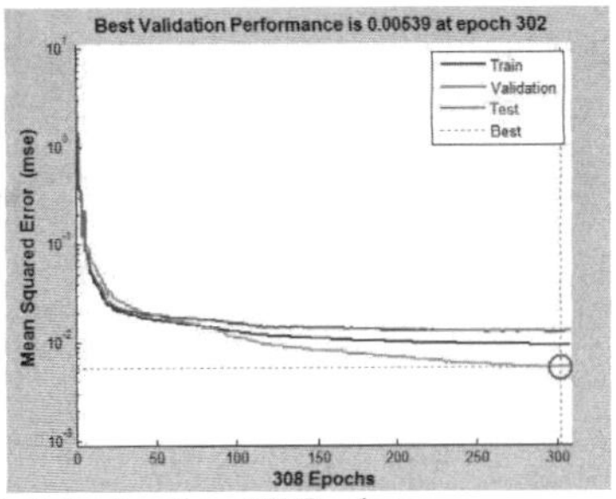

Figure 8. trainlm function, MSE values

MSE in Equation 1 and R in Equation 2 were utilized in order to observe the prediction performance of the developed model by using MATLAB software [18].

$$MSE = \left[\sum_{1}^{N} \left(\frac{Q_{exp} - Q_{cal}}{n} \right)^2 \right] \quad (1)$$

$$R = \sqrt{1 - \frac{\sum_{i=1}^{N} (Q_{exp} - Q_{cal})^2}{\sum_{i=1}^{N} (Q_{exp} - \bar{Q}_{exp})^2}} \quad (2)$$

where;

Q_{exp} : observed value; Q_{cal} : predicted value

$\bar{Q}_{exp}$: mean predicted value; N: number of data points

4 Results and discussion

Deformation is the key parameter to estimate impact velocities. In simulation, 3 parameters related to crash analysis were also examined with deformation (ε) which were deformation energy (E), pre-impact velocity (v_1) and post-impact velocity (v_2). Descriptive statistics for the dataset are given in Table 1.

Table 1. Descriptive statistics of the dataset

Data	Parameter			
	Mean	**Max.**	**Min.**	**Std. Deviation**
E (J)	63583.51	1013511.43	652.12	215873.74
ε (m)	**0.24**	**0.84**	**0.03**	**0.15**
v_1 (km/h)	28.32	98.87	0	26.46
v_2 (km/h)	21.81	64.32	0.06	20.42

However, for predicting the impact speeds using FITNET models, deformations formed on vehicles (ε1, ε2) were used as input variables (training data) due to its bigger effect on speed analysis than the others and impact speeds (learning data) were the output variables. It was aimed that, in case of absence of clues about impact speed analysis such as skid marks, police reports, witnesses etc., deformation could be used as the main parameter for speed prediction. It may be possible that an application that can run on portable devices may be coded in order to estimate impact speeds by entering average deformations on the collision regions of vehicles.

Table 2. Descriptive statistics of the dataset

Fold Number	MSE	R
1	0.409939	0.999843
2	1.241129	0.999393
3	0.223066	0.999924
4	1.353711	0.999473
5	1.451669	0.999366
6	1.046015	0.999596
7	1.359004	0.999425
8	1.528730	0.999434
9	0.251186	0.999890
10	0.935292	0.999574
Average	**0.989554**	**0.999592**

Within performance analysis of FITNET model, it was observed that average of MSE values was in acceptable range and average of R values was close to 1 (Table 2). The results showed that FITNET model with trainlm training function is an appropriate impact velocity predictor.

5 Conclusions

This study focuses on exhibiting new aspect and scientific approach for determining impact speeds of involvements in a sample most frequent accident occurring in Turkey without regarding to initiative and/or experience of experts. As subject matter, referring to direct proportion between speed and deformation energy, a new systematic approach for determining impact velocities was carried out. vCrash software was used to simulate the accident scene several times. The obtained dataset consisted of 4 parameters related to accident scene where deformations were used as main input parameter for FITNET model. The model was developed to predict impact speeds in which R and MSE values were also calculated. For the scenario, the FITNET model gave acceptable MSE and R values. Based on these approaches, various speed studies may be conducted within different prediction methods in a probable similar accident scene. It will be possible to analyze the impact speeds just by entering average deformation amounts into an application on a portable device at the accident scene without requirement of expensive reconstruction tools and it will be a guide for analyses of other accident types.

Acknowledgment

We would like to thank Cukurova University, Faculty of Engineering-Architecture, Automotive Engineering Laboratories and Scientific Research Unit for their resources that we were benefited from during conducting this study.

References

1. S. Miaou, H. Lum, Accident Anal. & Prev., **25**, 689–709 (1993)
2. F. Temel, H. Ozcebe, Turkish Medic. Assoc., **15,** 192-198 (2006)
3. E. Cangul, *Development of an expert system for the quantification of fault rates in traffic accidents* (Middle East Technical University, Engineering Faculty. MSc Thesis, 129 pages, 2010)
4. C. Murray, A. Lopez, K. Elissa. The Lancet, **349**, 1498-1504 (1996)
5. Rep. of Turkey Ministry of Internal Affairs, Traffic Statistical Bulletin (2006)

6. R. Wee, P. Rietveld, Transportation, **40,** 295-314 (2013)
7. Y. Hayashi, H. Morisugi, Transport. Policy, **7,** 73–88 (2000)
8. E. Karadayi, *Uluslararası 1. Trafik ve Yol Güvenliği Kongresi*, May, 8-12, Ankara, Turkey.
9. P. Koehn, A. Pauly, R. Fleck, M. Pischinger, T. Richter, M. Schnabel, R. Bartz, M. Wachinger, S. Schott, Automobiltechnische Zeitschrift Extra, **105**, 96-105 (2003)
10. V.P. Kallberg. *Motorcycle traffic accidents* (Transport Res. Board, 49, 1986)
11. U. Buck, S. Naether, M. Braun, S. Bolliger, H. Friederich, C. Jackowski, E. Aghayev, P. Vock, R. Dirnhofer, M.J. Thali, Forensic Science Int., **170**, 20–28 (2007)
12. G.A. Davis. Law, Prob. and Risk, **2,** 69-89 (2003)
13. J. Xu, L. Yibing, L. Guangquan, W. Zhou, Accident Anal. & Prev., **36,** 783–788 (2009)
14. R.D. Raedt, I. Ponjaert-Kristoffersen, Accident Anal. & Prev., **33,** 809-819 (2001)
15. G. Yannis, J. Golias, E. Papadimitriou, Accident Anal. & Prev., **37,** 327-333 (2005)
16. E. Tomasch, *Accident reconstruction guidelines* (Graz University of Technology, 2004)
17. M.H. Beale, M.T. Hagan, H.B. Demuth, *Neural network toolbox: for use with MatLab* (7th Ed., The MathWorks Inc. Michigan, USA, 2015)
18. MathWorks, MatLab. 2014. Available at: http://www.mathworks.com/products/matlab/. (Accessed May 4, 2015).

Thailand Ranks Second in the World for Number of Road Accidents under Thailand's Codes of Geometrical Design and Traffic Engineering Concept When Compared with AASHTO

Weeradej Cheewapattananuwong

Bureau of Road Design, Department of Rural Road, 9 Paholyothin Rd, Bangkhen Bangkok Thailand
Faculty of Engineering, Thummasat University (Rangsit Campus), Bangkok, Thailand

Abstract. Traffic problems in Bangkok have an influence on road users during peak hours. Especially, the traffic bottleneck on curves under the saturation flow situation must be remedied in order to increase the roadway capacity and speed. However, the appropriate speed for heavy vehicles is taken into consideration during off peak after the increasing lanes. This leads to the Rollover of heavy truck and rear-end collisions which are the main causes of vehicles accidents on curves. In addition, road accidents on curves account for the majority of all accidents in Thailand. According to the road accidents data collected in Thailand, 44 road deaths per 100,000 people, the country ranks second in the world for road accidents. When Thailand's Code of Geometrical Design is compared with AASHTO (The American Association of State Highway and Transportation Officials), the super elevation length of Thailand's Code is more than AASHTO. As a result, drivers are not made aware of the appropriate speed and the stooping sight distances (SSD) on curves. Therefore, the Design of Traffic Signage under the Perception and Reaction Times (PRT) for Thai Drivers will be taken into account.

1 Introduction

The Development of ITS (Intelligent Transportation System) has been developed since 2009. Image Sensing Technique by VLD (Virtual Loop Detector) for Checking Speed of Vehicles and LPR (License Plate Recognition) for Traffic Classified Counts in Thailand are taken into account. The traffic volumes on road curve are over saturation situation or F - Level of Service (L.O.S.). In case of Roadway Capacity, the increasing lanes on curve are taken into consideration. Although the traffic problem can be solved by this method, the over speed on curve must be controlled and the appropriate speed needs to be informed to drivers by the warning signage. In addition, the researcher is concerned with the geometrical design factors such as, super-elevation length on curves and Stopping Sight Distances (SSD) of PRT for Thai drivers. The Traffic Signage will also be calculated and installed on curved roads under the design factors as mentioned above. However, the number of traffic accidents on curves rises dramatically when Super Elevation Length of Thailand's Codes is more than AASHTO's Length. This is owing to the fact that sizes of letters on the traffic signage have an effect on road users.

2 Traffic collected data by ITS

There are two cases of applied ITS for collected data. One is a measurement of vehicle speed and the other is vehicle classification count by LPR off-line method as described in the following paragraph.

2.1. Image processing concept

FG/BG Detection module performs foreground/ background segmentation for each pixel. Blob Entering Detection module uses the result (FG/BG mask) of FG/BG Detection module to detect new blob object entered to a scene on each frame. Blob Tracking Module initialized by Blob Entering Detection results and tracks each new entered blob. Trajectory Generation module performs a saving function. It collects all blobs positions and saves each whole blob trajectory to hard disk when finished. Trajectory post processing module performs a blob trajectory smoothing function. The application of Blob Entering Detection Techniques is shown in Fig. 1.

Figure 1. Results of the Application of Blob Entering Detection Techniques.

2.2 Measurement of speed

The positions of Lane Changing, which are 3 points, are located on Fig. 2. The first curve of Lane Changing and Speed of circular curve is composed of coordinates among points (x1,y1), (x2,y2), and (x3,y3) respectively. The calculation of angle, length of Lane Changing and speed used the concepts of geometrical design under the specific times as shown in Fig. 2 and Fig. 3.

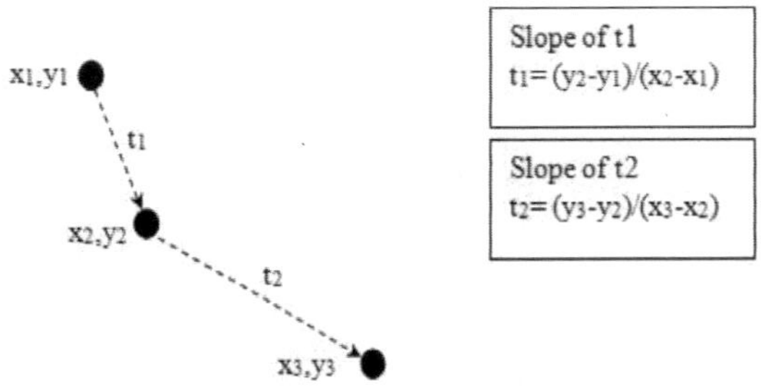

Figure 2. Three points of lane changing and speed concept.

Figure 3. Speed of each vehicle.

2.3 LPR

There are 22 types of classified vehicles [4]. In case of the accuracies of LPR Classified Counting at night time, it is not clear to read the number of license plate, although the researchers choose both high resolution of optical lens (50 mm) and the zero LUX of CCTV. The Traffic Volumes on Curve during peak hour is 684 vehicles per hour and the average speed is 13 KPH (L.O.S - F). This is due to the fact that the saturation flow from next intersection (Bromrajchonne Road) has an influence on this road as presented in the Fig. 4.

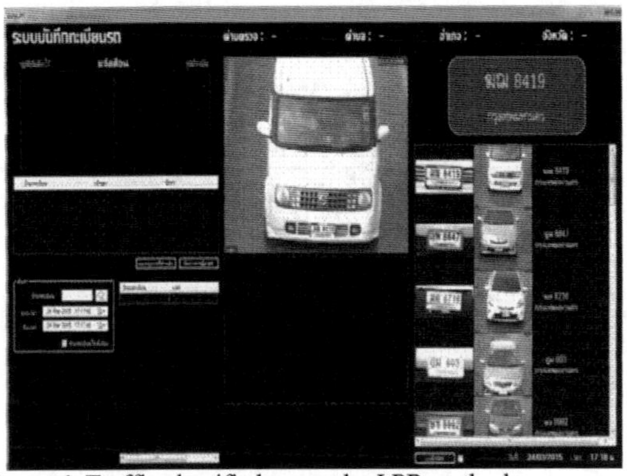

Figure 4. Traffic classified survey by LPR method.

2.4 Traffic collected data and evaluated existing situation

The Existing Traffic Volumes from LPR Method are input in the Trafficware Software [7]. As a result, the traffic situation is a stop and go model as presented in Fig. 5.

SimTraffic Performance Report
Baseline 20/7/2015

Total Network Performance

Total Delay (hr)	169.2
Delay / Veh (s)	374.3
Total Stops	1669
Travel Dist (km)	1308.1
Travel Time (hr)	198.0
Avg Speed (kph)	13
Fuel Used (l)	557.2
HC Emissions (g)	352
CO Emissions (g)	7188
NOx Emissions (g)	679
Vehicles Entered	1769
Vehicles Exited	1485
Hourly Exit Rate	8910
Denied Entry Before	165
Denied Entry After	1151

Figure 5. Traffic factors on curve (Exit Ramp) before situation.

Figure 6. Increasing of lane capacity on curve.

SimTraffic Performance Report
Baseline 20/7/2015

Total Network Performance

Total Delay (hr)	157.8
Delay / Veh (s)	336.9
Total Stops	2674
Travel Dist (km)	1538.2
Travel Time (hr)	191.4
Avg Speed (kph)	19
Fuel Used (l)	567.6
HC Emissions (g)	358
CO Emissions (g)	7526
NOx Emissions (g)	722
Vehicles Entered	1809
Vehicles Exited	1563
Hourly Exit Rate	9378
Denied Entry Before	194
Denied Entry After	1140

Figure 7. The raised speed after mitigation.

3 Roadway capacity and number of lanes on curve (exit lane)

Roadway Capacity is very important for Highway Planners when considering traffic lanes on curve [6]. The solution of this case study is to increase the number of lanes from 2 lanes to 3 lanes as can be seen in the Figure

3. However, the road width is lower than the standard lane (3,5 m). For example, the comparison of lane capacity between 3.5 m and 3.0 m of lane width are 1,288 and 1,261 vehicles/hour/lane or 2,575 vehicles/hour/2Lanes (3.5 m) and 3,783 vehicles/hour/3Lanes (3.0 m) respectively. Therefore, the increasing of roadway capacity is tested on curve to improve the traffic flow. As a result, it can also raise the average speed of vehicles when driving on curve (see Fig. 6, Fig. 7).

4 The geometrical design factors on curve

There are two Geometrical Design Factors as compared between Thailand's Codes and AASHTO. The Super Elevation Length on Curve and Stopping Sight Distances lead to the traffic accidents and these factors under Thailand's Codes are of higher values than AASHTO. Furthermore, Fatalities in Thailand rank second out of all the road accidents in the world, which is the rate of 44 road deaths per 100,000 people in Thailand. The first Ranking of 45 road deaths per 100,000 occurs in Namibia [1]. The traffic factors will be described in the following parts.

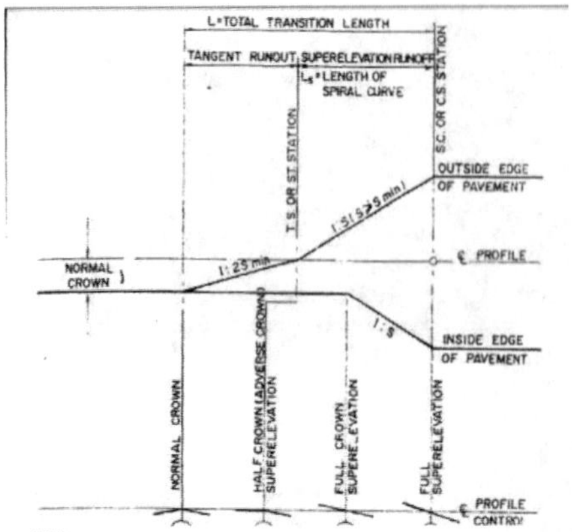

Source : [3]

Figure 8. Super elevation length of AASHTO.

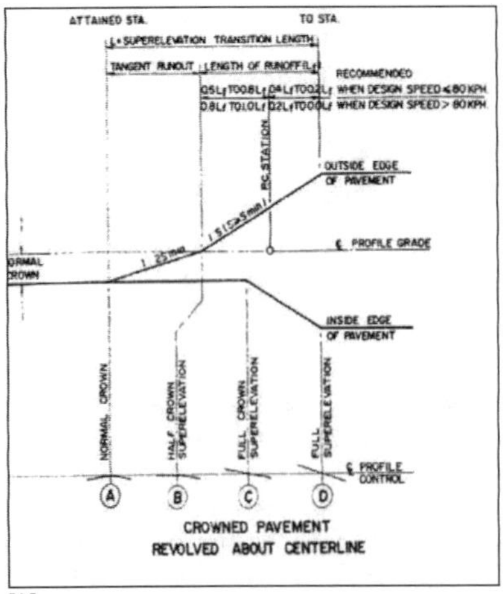

Source : [2]

Figure 9. Super Elevation Length of Thailand's Code.

4.1 Super elevation curve length

The comparison of Super Elevation between AASHTO and DRR (Thailand's Codes) will be described in the Fig. 8, Fig. 9 and Table 1, Table 2 and Table 3.

Table 1. Comparison between super elevation length of Thailand's code and AASHTO.

Thailand's Code	AASHTO
Lf=HC*s+(e*W/2)-HC)*s	Lf= (((wn1)*ed/δ)*bw
Lt= HC*2*s	Lt= (enc/ed)*Lf

Where;
Lf = Length of Run Off
Lt = Tangent Run Out
HC = enc = (Crown Slope%/100)* Lane Width/2
e= ed = Super Elevation
S = d = 75+1.5V
V = Speed KPH
W = Lane Width
n_1 = Number of Super Elevation – Lane
bw = (1+0.5*(n1-1))/n1

Table 2. Results of super elevation length of Thailand's Code.

	DRR Method					
	Speed <= 80			Speed > 80		
V	0.7*Ts = Lf= Y+Z	0.3*Ts = Lf	Lt = X	0.9*Ts = Lf	0.1*Ts = Lf	Lt
30	4.5	1.9	10.8			
40	9.4	4.0	12.2			
50	10.4	4.5	13.5			
60	11.4	4.9	14.9			
70	12.5	5.3	16.2			
80	13.5	5.8	17.6			
90				18.7	2.1	18.9
100				20.0	2.2	20.3

Table 3. Results of super elevation length of AASHTO.

	AASHTO Method					
	Speed <= 80			Speed > 80		
V	0.7*Ts = Lf	0.3*Ts = Lf	Lt	0.9*Ts = Lf	0.1*Ts = Lf	Lt
30	4.5	1.9	10.8			
40	5.1	2.2	12.2			
50	5.7	2.4	13.5			

60	6.2	2.7	14.9			
70	6.8	2.9	16.2			
80	7.4	3.2	17.6			
90				7.9	3.4	18.9
10 0				8.5	3.6	20.3

As can be seen from the above Tables, the Length of Curve – Y+Z in the case of Thailand's Codes is dramatically more than AASHTO. However, the length of X in case of Thailand's Codes is equal to AASHTO. In case of C.G. of Vehicle, the Running Speed during testing with C.G. Equipment is illustrated in Fig. 10 and Fig. 11.

Source : [5]
Figure 10. mSuper Speeding of a Test Vehicle under the Standard Radius.

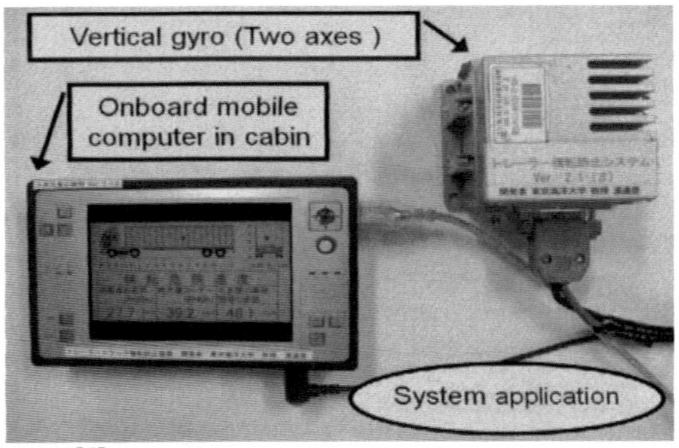

Source : [5]
Figure 11. C.G. Equipment.

During the testing, a trailer was tested with the C.G. equipment and the critical speed or Safety Speed was lower than the design speed. The Safety Equation is presented in the next equation.

$$\text{Safety Speed} = 44.6153846153846 + 14.1862550031749 * \sin(0.805821839123194 / \operatorname{asinh}(\operatorname{mod}((\text{Design Speed, km/hr}), 0.987455336652175)) + \operatorname{delay}((\text{Design Speed, km/hr}), 5) + \operatorname{erf}(\sin(1.78663889085124 - \operatorname{delay}((\text{Design Speed, km/hr}), 3)))) - 9.15707764551099 * \cosh(\operatorname{mod}(\operatorname{delay}((\text{Design Speed, km/hr}), 1), 1.2422838109887)) * \operatorname{mod}(1.17493434602731, \operatorname{mod}((\text{Design Speed, km/hr}), 0.793435970498463)) * \cosh(\operatorname{mod}(\operatorname{delay}((\text{Design Speed, km/hr}), 1), 1.2422838109887)^{1.95733458837149} \quad (1)$$

R^2 Goodness of Fit = 0.89883218 and Correlation Coefficient = 0.95710217

4.2 Stopping sight distances (SSD)

According to Running Speed, the researcher will give the Safety Speed of Heavy Vehicle as based on Equation (2, 3 and 4). For example, the Design Speed is about 60 KPH and the Safety Speed will be 44 KPH. Therefore, the installation of traffic signs and sizes of letters will be located before the point of curve (PC) as described in the following parts.

*4.2.1 Formatting the title concept of PIEV (Perception, Interpretation, Emotion and Volition and SSD (See Figure 12). L = 0.278*V*T*

$$A = 6.30442 + (0.08586*V) + (7.44363e\text{-}10)* V5 \quad (2)$$

$$K = 0.278 * \text{PIEV Time} \quad (3)$$

$$SSD = 0.896671 * Vf2 \quad (4)$$

where;
L = Reading at the Meaning (0.88 seconds) of Traffic Sign – Distance and Reading Time (2 Syllable per second)
A = Acute Angle of Driver's Vision - Distance
K = PIEV (2.5 Seconds) Distance
SSD = Stopping Sight Distance
V = Running Speed
V_f= Final Running Speed

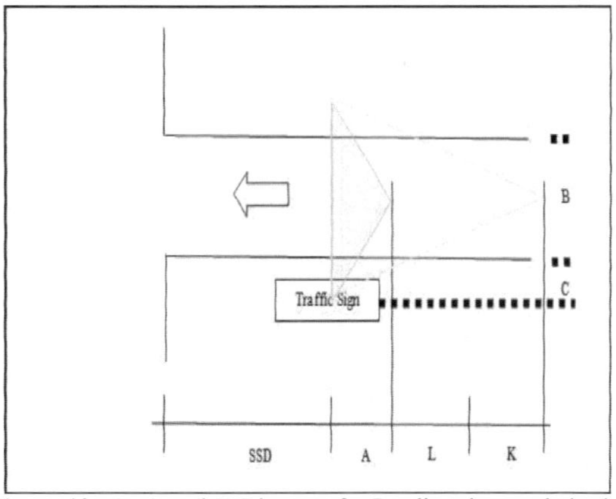

Figure 12. L, A and K Distance for Reading the words in the Traffic Sign.

4.2.2 Concepts of sizes of letters in traffic sign

$$H = 1.3 + 0.195*[(L+A+K) + 0.139*V*N] \quad (5)$$

$$T = H/6 \quad (6)$$

$$N = [(H - 1.3)/0.195 - (L+A+K)] / (0.139*V) \quad (7)$$

where;
H = Height of Letters (cm.)

DOI: 10.1051/matecconf/20168102004

T = Thick of Letters (cm.)
N = Syllable

The Table 4 presents 30 cm. of Height, Thickness and Syllable of Letters in a Traffic Sign at various Running or Safety Speed. For example, speed at urban area (60 KPH) will be proposed 30 cm. of letters, two rows of Letters with 5 Syllables per row at a traffic sign under 101.77 meters of the L+A+K Distances.

Table 4. 30 CM, Height, thickness and syllable of letters on traffic sign based on speed.

Height = 30 cm

Speed	L+A+K	N	Level
30	54.00	22	7
40	70.00	14	5
50	86.00	9	3
60	102.00	5	2
70	119.00	3	1
80	136.00	1	0
90	153.00	0	0
100	172.00	-2	-1
110	209.00	-4	-1

5 Mitigation of traffic accidents

Thai drivers use the SSD which is lower than the SSD of ASSHTO as based on the Equation (2, 3 and 4). Therefore, the Traffic Signage by ITS can warn Thai's Drivers to increase the length of SSD in order to operate their vehicles safely by using VMS (Variable Message Sign) as described in Fig. 13 and Table 5.

Figure 13. Controlled Speed by the Size of Letter in Variable Message Sign (VMS).

Table 5. 7 Collected Traffic Changing Lane and Speed Data during Installed VMS.

Mitigation Scenarios	Number of Lane Changing Vehicles	Number of Over Speed Vehicles	Decreasing Percentage (%)
W/O VMS	47	192	
Wt VMS « No Fine »	29	139	-38.30, 27.60
Wt VMS « Fine »	20	118	-62.45,-38.54

6 Conclusion

The increasing of traffic volumes by the decreasing of normal lane width and the increasing of lanes from 2 lanes to 3 lanes are taken into consideration. In addition, the application of Thailand's Code for Road – Curve Design Length is longer than AASHTO. However, the number of Traffic Accidents has substantially increased. The important mitigation is to install traffic signage and to focus on various sizes of letters when compared with speed. This is pointed out in the message on the traffic sign to warn drivers to drive on curve safely. Strict and Severe Law Enforcement is the best solution for Thai people who disobey the traffic laws. The researcher wishes to decrease the number of accidents as soon as possible.

References

1. Asian Correspondent, Study: Thailand's roads 2nd most dangerous in the world, https://asiancorrespondent.com/2014/02/study-thailand-roads-2nd-most-dangerous-in-the-world/ ,(2014).
2. CALTRANS, Highway Design Manual, California Department of Transport, California, (1995)
3. Department of Rural Roads, Standard of Super Elevation, Ministry of Transport, Bangkok, (2014)
4. Land Transport Department. Vehicle License Plate ACT. (1979).
5. Y. Watanabe, Detection of three Dimension Centre of Gravity (D3DCG), Tokyo University of Marine Science and Technology, Managed by Japan Association of D3DCG. (2014).
6. Transport Road Board (TRB), Highway Capacity Manual, fifth edition, (2010).
7. Trafficware Synchro Version 6, Traffic Engineering and Simulation Software (1993-2005), (2005).

Maneuvering Route Safety Planning for Military Vehicles during Wartime

Tan Zhao , Jincai Huang, Guangquan Cheng, Chao Chen and Jianmai Shi

Science and Technology on Information Systems Engineering Laboratory, National University of Defense Technology, Changsha 410073, P. R. China.

Abstract. Maneuvering route planning for the military vehicles during wartime is a noteworthy problem in the field of military transportation. Existing research paid little attention to security of vehicles, which we consider as a non-ignorable aspect with the military background. Based on scenarios, we build a k-th shortest paths planning model aiming at vehicles safety improving. A genetic algorithm with specific encoding rules is proposed. And two groups of computational experiments based on the real road network of Fujian, China are performed. The results of the experiments show this is an effective algorithm to improve the vehicles' safety during maneuvering.

1 Introduction

Vehicle route planning is a traditional problem in communication and transportation area. With regard to the military application, it is indeed an intractable problem for commanders to plan the military vehicles' maneuvering routes during wartime, on account of the effects of weathers, road statuses, operations of enemies, etc. The operations of enemies, especially, turn this problem completely different from a normal route planning problem, considering the enemies' capability to destroy the traveling vehicles, which interrupts the course of action (COA) significantly. We can present the Military Vehicles Route Planning during Wartime (MVRPW) problem as follows: design a set of routes for a fleet of i identical military vehicles required to transfer to j preliminary positions in preparation for the forthcoming mission. On the premise that all vehicles arrive on schedule, our scheme shall ensure vehicles' security as possible, in all considered scenarios.

With respect to route planning for military vehicles, Guan, L. [1] put forward a two-stage algorithm to get the optimal vehicles distribution for multitask using types of military vehicles with the highest efficiency. Zhong, Liu, et al. [2] presented a multiple destinations route planning algorithm based on dynamic programming considering the target values. Their focus is the efficiency of vehicles' allocation. YiePinedo, Ruben D. [3] studied a problem of routing several vehicles from multiple origins to multiple destinations in an unsafe environment, and presented solution to allocate zone-limited escorts to minimize the average threat level. The same as what they did, vehicles' security was stressed in our paper. To ensure the course of action not breaking, avoiding the enemies' attack and ensuring the maneuvering action as safe as possible is our main purpose.

In view of our military background, we are supposed to provide a feasible scheme at the very beginning of a mission, rather than determine the next road in every intersection for vehicles. Thus, a model having each vehicle's path specified is required. An effective solution is the k-th shortest path technique, which is able to give the shortest k paths from sources to destinations intactly for each vehicle. The applications of this technique can be seen in the emergency evacuation field. It is first seen in the dissertation of Campos and Netto [4]. They applied the k-th shortest path to their model, aiming at a larger traffic capacity and shorter clearance time. Alexander and Smith [5] combined the k-th shortest path with the M/G/c/c queuing theory in their model to deal with congestion and time delay. He, Yunyue, *et al.* [6] proposed a k-th-shortest-path-based technique that uses explicit congestion control to optimize evacuation routing and police resource allocation. Their objection is to minimize the overall clearance time in city emergency evacuation. Noticing that few researchers apply this technique in military field, we attempt to use the k-th shortest path tech in our military vehicles route planning problem.

With an eye to the complexity in battlefield, kinds of situations exist and we may not know what situation comes next. It is a great challenge that all situations are supposed to be considered in one scheme. Inspired by work of Zhang, Ying, *et al.* [7], we will consider situations as discrete scenarios in this paper. Scenario describes the future situation may occur, and also presents the uncertainty of reality [8]. A scenario, with specific chance of happening, specifies states of all factors in a situation. Our purpose is to search for an eclectic scheme, not optimal in a certain scenario though, performing robustly in scenarios.

In the rest of this paper, we propose a scenario-based integer programming model with application of k-th shortest path technique for MVRPW problem in section 2; to solve this model, a genetic algorithm with a novel genetic encoding rule is introduced in section 3; and in section 4, we give two computational examples based on a real road network; section 5 concludes the discussion.

2 K-shortest-path-based model for MVRPW

In this section, we introduce the problem description, underlying assumptions and mathematic formulation.

This problem is a type of optimal routing problem (ORP) with consideration of multiple scenarios. The objective of the model is to maximize the overall security, indicated by expectation of the safe arrival of all vehicles in all scenarios. We present a transportation network consisting of origins (e.g. bases or campsites), destinations (i.e. the positions) and related road links, and intersections in this problem. The military vehicles are supposed to maneuver from origins to destinations through this network, with consideration of safe passing probabilities (SPPs, considering enemies' disruption) of each road link respectively in each scenario. Meanwhile, a mission deadline is considered, before which the vehicles shall accomplish the maneuver mission.

It is a tough job to describe a scenario exactly, as so many dimensions or factors should be considered. Recognizing it is not our main work, we can present the scenario in a brief way. While saying a scenario, we suppose that all statuses in all dimensions are set. In other word, a scenario is an intersection of statuses in each dimension. For example, 2 dimensions are considered to describe a scenario. Fig. 1 shows the scenarios we will consider in this example. Besides, the probability of a scenario occurring is assumed to be known. In this way, considering all dimensions, we can present scenarios S as S = {s1, s2, s3, …} with probabilities P = {P1, P2, P3, …}, one to one correspondence.

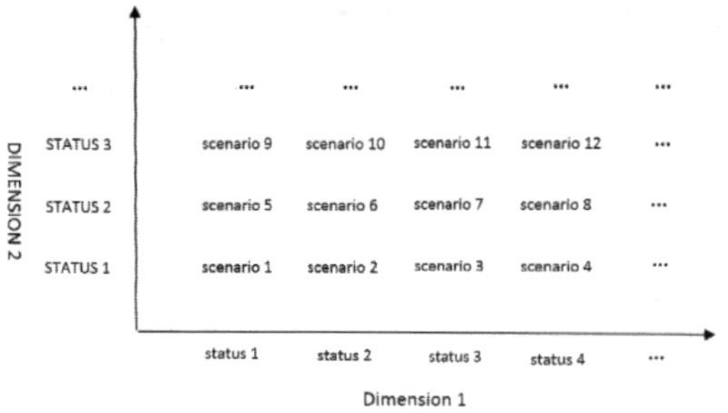

Figure 1. An example to explain the scenario considered, 2 dimensions.

2.1 Assumptions

To highlight the nature of the problem and clarify the boundaries of the problem concerned, some rational assumptions are proposed.

1) The SPPs vary based on scenarios.

Without describing the scenarios comprehensively though, it is supposed that we can still get the safe passing probability (i.e. SPP) of each road link in each scenario via simulation technique and the analysis of practice data. As scenarios are different, the SPP of each road link varies.

2) The travelling time and capacity of each road link are set constant.

Taking no account of road destruction, the travelling time shows little different when scenario changes. What is noteworthy, the capacity of each road link is a value set by the commander rather than an objective value. Considering the practical bearing capacity of a road is meaningless in the case of confrontation. We suppose that the commander will integrate all factors into account and set clear numbers as road capacity values. Yang, Ping et al [9] figured that uniform traffic flow in each road brings a rise of elusiveness of the whole road network. Thus the value of each road capacity shall not be set large.

3) Each destination can accommodate 1 vehicle at most.

Military vehicles, such as missile launchers and movable radars, are usually able to execute assignments independently. To ensure the safety of vehicles and to reduce losses after arrival, sending vehicles to separate terminals is a bright choice. In this paper, we suppose that each destination can accommodate 1 vehicle at most.

2.2 Notation

We propose integer programming to describe the MVRPW problem. The notation used in the problem formulation is introduced below:

Indices. i an index for military vehicles, $i = 1,2,...,I$, I is the total number of vehicles;

j an index for destinations, $j = 1,2,...,J$, J is the total number of destinations ;

k an index for k-th-shortest paths, $k = 1,2,...,K_{ij}$, K_{ij} is the total number of shortest paths vehicle i can travel to destination j, set or get via limitation;

s an index for scenarios, $s = 1,2, ...,S$, S is the total number of scenarios considered;

l an index for road links, $l = 1,2,...,L$, L is the total number of road links;

n an index for road nodes, $n = 1,2,...,N$, N is the total number of nodes in the transportation network;

Parameters. λ_l the capacity of road link l;

t_l the travelling time of road link l;

R_{ijk} the set of road links, which the k-th shortest path vehicle i travelling to destination j covers;

P_s the probability scenario s occurs, $\sum_{s=1}^{S} P_s = 1$;

p_{sl} the SPP of road link l in scenario s;

P_{sijk} the SPP of the k-th shortest path vehicle i travelling to destination j in scenario s;

T_0 time allowed for maneuver(i.e. deadline);

Decision variables. x_{ijk} 0–1 variable. 1 for vehicle i travels to destination j in the k-th shortest path, 0 for otherwise.

2.3 Formulation

We formulate the problem as follows:

$$\text{(MVRPW)} \max Z=\sum_s P_s \sum_i \sum_j \sum_k x_{ijk} P_{sijk} \tag{1}$$

Subject to:

$$x_{ijk} \sum_{l \in R_{ijk}} t_l \le T_0 \text{ , } \forall i.j,k \tag{2}$$

$$\sum_j \sum_k x_{ijk} = 1 \text{ , } \forall i \in I \tag{3}$$

$$\sum_i \sum_k x_{ijk} \le 1 \text{ , } \forall j \in J \tag{4}$$

$$\sum_{l \in R_{ijk}} x_{ijk} \le \lambda_l \text{ , } \forall i,j,k,l \tag{5}$$

$$P_{sijk} = \prod_{l \in R_{ijk}} p_{sl} \tag{6}$$

$$x_{ijk} \in \{0,1\} \text{ , } \forall i,j,k \tag{7}$$

The objective function of the MVRPW formulation is the sum of expectation of the safe arrival of all vehicles in all scenarios, indicating the robustness of the scheme. Constraints (2) ensure that all vehicles arrive on schedule. Constraints (3) represent that each vehicle arrive at a destination. Constraints (4) ensure that each destination can accommodate 1 vehicle at most, as noted in assumption 3). Constraints (5) show that the traffic flow on road link l shall not exceed the set capacity, as assumption 2) implies. Constraints (6) are based on the definition of P_{sijk} , representing the safe passing probability of the k-th shortest path in which vehicle i travels to destination j in scenario s. Constraints (7) indicate that the decision variables x_{ijk} are 0–1 variables. This is a typical Integer Programming Problem (IPP). We try to solve it with a genetic algorithm.

3 Genetic algorithm for MVRPW

In this section, genetic operators which are used in GA for MVRPW will be explained.

The first application of GA to transportation/distribution problem is the study of Michalewicz, Z [10] in 1991. They used matrix-based representation to represent transportation tree. They used $|K|$ and $|J|$ to represent the number of sources and depots respectively, the dimension of matrix is$|K|\cdot|J|$. While Gen, M. and Cheng, R [11] applied *Prüfer* number to represent transportation tree in 2000, it successfully reduced the digits to represent a transportation tree with $|K|$ sources and $|J|$ depots to $|K|+|J|$-2. Altiparmak, Fulya, et al. [12] developed a priority-based encoding from research of While Gen, M. and Cheng, R in 2006. In priority-based encoding, the position of a gene is used to represent a node (source/depot in transportation network), and the value is used to represent the priority of corresponding node for constructing a tree among candidates. A similar encoding rule is applied in our paper.

3.1 Encoding algorithm

In our encoding rule, I identical vehicles are requested to maneuver to J destinations ($J \ge I$). Here, we set J as D. If we set the sum of origins as O ($O \ge 1$, and for one origin can accommodate more than one vehicle, $I \ge O$), the length of our chromosome is set $O + D$. The front O genes represent the origin nodes, while the rear D genes the destination nodes. The gene values represent priorities of origins and destinations to obtain transportation tree. The transportation tree corresponding with a given chromosome is generated by sequential arc appending between origins and destinations. At each step, only one arc is added to tree, selecting an origin/destination with the highest priority and connecting it to a destination/origin through a k-th shortest path considering the biggest *SPP*, with the capacity of road link and deadline time in account. A length-equal operator is given to each chromosome, called *surplus*. The *surplus* shows the remaining capability to send/accommodate vehicles of an origin/destination. For each time a military vehicle is sent/accommodated, the corresponding value of *surplus* shall be reduced by 1. When a value of *surplus* reduces to 0, showing this origin/destination has no ability to send/accommodate a vehicle more, we set the corresponding gene as 0.

An example showing how the selecting procedure works is given in Fig. 2. In this example, 2 origins and 5 destinations are considered. There are 1 and 2 military vehicles waiting to maneuver from the two origins separately. With time and road capacity constraints considered, Table. 1 shows the decoding algorithm for the priority-based encoding in detail.

Sample	No.	1	2	3	4	5	6	7
	Chrom	3	7	4	6	2	5	1
	Surplus	1	2	1	1	1	1	1

Step 1	No.	1	2	3	4	5	6	7
	Chrom	3	7	4	6	2	0	1
	Surplus	1	1	1	1	1	0	1

Step 2	No.	1	2	3	4	5	6	7
	Chrom	3	0	4	6	0	0	1
	Surplus	1	0	1	1	0	0	1

Step 3	No.	1	2	3	4	5	6	7
	Chrom	0	0	4	0	0	0	1
	Surplus	0	0	1	0	0	0	1

Final Result: 2→6; 2→5; 1→4

Figure 2. An example to show the selecting procedure.

Table. 1. Pseudo-code of the decoding algorithm.

Procedure: Decoding of the chromosome for transportation tree

Input:	O	:	set of origins;
	D	:	set of destinations;
	a_o	:	capacity of origin o, $\forall o \in O$;
	b_d	:	accommodation of destination j, $b_d=1$, $\forall d \in D$;
	λ	:	an integer matrix, showing the road capacity between two adjacent nodes;
	c	:	chromosome;
	Time	:	deadline time;

T_{odk} : time cost from origin o to destination d through the k-shortest path, $\forall o,d,k$;

R_{odk} : the route from origin o to destination d through the k-shortest path, $\forall o,d,k$;

SPP : the set of safe passing probabilities of all k-shortest paths from origins to destinations, with scenarios already considered in.

P_s : probability scenario s will happen, $\forall s \in S$

Output: M : an O-by-J matrix. //If $M_{od} = k$, a vehicle is arranged through the k-shortest path from origin o to destination d; if $M_{od}=0$, no vehicle is arranged from o to d.

Index : *Index* shows the advantage of the chromosome. The bigger *Index* is, the more robust the scheme is.

Step 1. $M_{od} \leftarrow 0$, $\forall o \in O, \forall d \in D$, $surplus \leftarrow [a, b]$, $Index \leftarrow 0$;//initialize

Step 2.$mark \leftarrow \arg\max\{c(i) | i \in O+D\}$;//select a node

Step 3.**if** $mark \in O$,**then** $o^* \leftarrow mark$;//select a origin

Select an available path to a destination with the highest *SPP*.

$d^*,k^* \leftarrow \arg\max\{\sum_s P_s \cdot SPP_{o^*dk} \mid T_{o^*dk} \le Time$,

$\prod_{(w,v)\in R_{o^*dk}} \lambda_{wv} > 0$,$c(d)\neq 0$, $\forall d,k\}$;

else $d^* \leftarrow mark$;//select a destination

Select an available path from an origin with the highest *SPP*.

$o^*,k^* \leftarrow \arg\max\{\sum_s P_s \cdot SPP_{od^*k} \mid T_{od^*k} \le Time$,

$\prod_{(w,v)\in R_{od^*k}} \lambda_{wv} > 0$,$c(o)\neq 0$, $\forall o,k\}$;

Step 4.**if** d^*=**none**, **then return**; //If a vehicle from origin o^* cannot finds an available destination, this chromosome is infeasible.

else $M_{o^*d^*} \leftarrow k^*$; //assign available path

Update capacity of origins and destinations, and λ .

$surplus(o^*) \leftarrow surplus(o^*)$ -1,

$surplus(d^*) \leftarrow surplus(d^*)$ -1;

$\lambda_{wv} \leftarrow \lambda_{wv}$ -1, $\forall (w,v) \in R_{o^*d^*k^*}$;

Step 5.**if** $surplus(i) = 0$, **then** $c(i) \leftarrow 0$, $\forall i \in O+D$

Step 6.**if** $surplus(o) = 0$, $\forall o \in O$,

then $Index \leftarrow$sum(SPP_{odk}, $\forall \{o,d,k \mid M_{od}=k\}$) and **return**,

else goto Step 2.

3.2 Genetic operators

3.2.1 Selection

In this GA, initial population is randomly generated. To get non-worse results for every generation, we adopt an elitist preservation strategy. In this strategy, we evaluate the SPPs of all chromosomes in population, and set them as the fitness of chromosomes. A fixed proportion, named *gap*, is given as the selected portion. The worst *gap* of the population, measured by fitness, will be selected. Thus an amount of *gap**population chromosomes are selected to take part in the following genetic operating, while an amount of (1-*gap*)*population of advantageous chromosomes are preserved.

9	5	1	3	7	4	2	10	8	6
10	5	4	6	3	8	7	2	1	9

After cross:

9	5	1	6	3	8	7	10	*	*
10	5	*	3	7	4	2	*	1	9

Partial Mapping:

9	5	1	6	3	8	7	10	4	2
10	5	8	3	7	4	2	6	1	9

Figure 3. A partial mapping crossover example.

3.2.2 Crossover

The crossover is done to explore new solution space. We employ a partial mapping crossover (PMX) operator. In this operator, we have two parents crossover. Two random positions are selected, and the segments between the two positions are interchanged. Then if repetitive values exist in the same chromosome, we keep the exchanged parts and non-redundant segments, and use the mapping crossover to eliminate the conflicts. An example is shown in Figure 3, assuming length of chromosome as 10, and the two random numbers as 4 and 7. In this figure, conflicted positions are marked by stars.

3.2.3 Mutation

Mutation is used to prevent the premature convergence and explore new solution space. The mutation operator is modifying the gene within a chromosome with a mutation chance. While mutating, two random positions of the chosen chromosome are selected, and the values of the two positions will be exchanged.

4 Computational examples

In this part, we apply the GA for MVRPW in the road network of Fujian Province, China, to evaluate how this algorithm performs. A road network of 200 nodes are selected, i.e. nodes 1-200 in Figure 4. The dataset is from an open resource [13]. The k-th shortest paths data are calculated by a commercial GIS software, ArcGIS, with application of a related algorithm [14]. Other related parameters, including S, P_s, p_{sl} and λ, are either set or generated with probabilities. We coded the algorithm in MATLAB R2013a, and executed the programs on a 32-bit Windows XP computer with 2.93GHz Dual-Core CPU and 3.0 GB of physical RAM. Two groups of experiments are performed as followed.

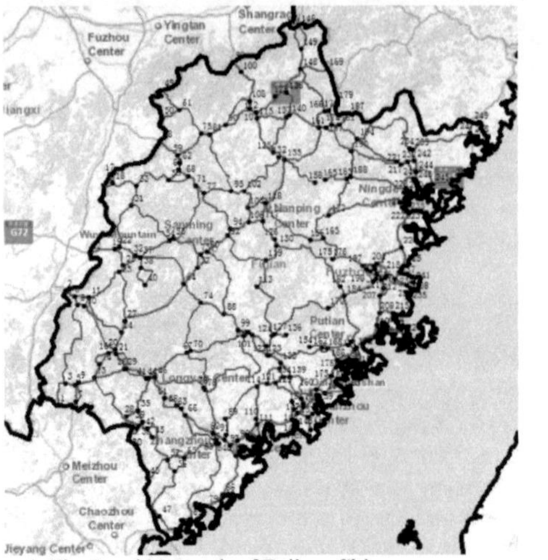

Figure 4. The Road Network of Fujian, China.

The speed of vehicles are all set 40km/h. Robustness of a scheme is indicated by the average SPP of vehicles. In experiment group I, 7 vehicles are arranged to maneuver from 3 origins to 26 destinations, with 10 shortest paths of each origin-destination pair considered; in experiment group II, 12 vehicles are arranged to maneuver from 6 origins to 30 destinations, with 20 shortest paths considered. 20 random scenarios are considered in both experiments. Thus, experiment in group II is around 4 times the k-th shortest paths scale of in group I. In the GA, the value of population is set 200, and the stopping criterion is if no better result appears in a continuous 15 generations. To evaluate the performances of our algorithm, 10 repetitions are carried out respectively in each experiment. Other experimental data and performances of the GA are listed in Table 2.

Table 2. Performances of the Algorithm.

Experiment	No.	Capacity of Road Links	Time permitted (h)	Proportion of Unavailable Paths (%)	Optimization Results			Computing Time(s)		
					max	min	average	max	min	average
I	1	[2,3;0.6,0.4] *	10	34.87	0.1392	0.1379	0.1390	437.3	264.4	342.4
	2	2	12	5.64	0.1592	0.1465	0.1567	571.0	373.7	514.5
	3	[2,3;0.6,0.4]	12	5.64	0.1592	0.1435	0.1555	948.8	374.2	676.7
II	1	2	8	43.11	0.0939	0.0930	0.0933	663.7	345.5	386.4
	2	2	10	26.47	0.0980	0.0936	0.0957	867.3	403.1	695.2
	3	[2,3;0.6,0.4]	10	26.47	0.0999	0.0938	0.0986	732.1	329.3	586.4
	4	[2,3;0.6,0.4]	12	5.08	0.1003	0.0935	0.0981	1096.2	621.4	915.9
	5	3	12	5.08	0.1008	0.0911	0.0962	1307.4	717.1	965.8

*[2,3;0.6,0.4] means element values are 2 with 60% probability or 3 with 40% probability.

According to Table 2, the algorithm can handle both the groups of experiments. The gap between the minimum and the maximum of optimization results varies. The biggest gap is 10.94%, appearing in experiment I-3. By this token, it is a relatively stable algorithm. As the constraints loosening, the maximum of optimization results increases steadily, indicating that better results are searched out. However, looser constraints brings a larger solution space, which results in a greater probability of local optimal solution. This can be observed from the enlarging gap between the minimum and the maximum. On account of that initial solutions are randomly generated, and the randomness in the genetic process as well, time cost in computing differs greatly. Time consumed is between 4.5 minutes to 22 minutes, which is an acceptable time cost in practical route planning.

Table 3. Changes of K-th shortest paths selected.

Part 1:Permitted Time=12h

Capacity	K-th Shortest Path Selected											
[1,2;0.1,0.9]	8	20	2	5	1	10	3	4	2	12	1	1
2	17	20	2	5	1	1	3	4	2	12	1	1
[2,3;0.6,0.4]	17	20	2	5	1	1	3	4	2	1	1	1
3	3	20	2	5	5	1	1	4	2	1	1	1
inf *	5	13	2	1	1	1	3	4	2	1	1	1

Part 2:Road Capacity=2

Permitted Time	K-th Shortest Path Selected											
8h	3	20	2	5	5	1	3	4	2	2	1	1
9h	17	20	2	5	1	1	1	4	2	2	1	1
10h	17	20	2	5	5	1	3	2	2	2	1	1
11h	17	20	11	5	1	1	3	4	2	2	1	1
12h	17	20	2	5	1	1	3	4	2	12	1	1

* *inf* means no constraint is set here.

To observe the changes of *k*-th shortest paths selected of each vehicle, we expand experiment group II. Listed in Table 3, we set the permitted travelling time and the road capacities constant respectively, to see how the shortest paths selected change. In part 1, permitted travelling time is set 12 hours. With the road capacity grows, an obvious trend is that vehicles tend to select the shortest path to travel. As in each scenario, SPP of each road link is set randomly, path with fewer road link is safer in common. And because no ultra-long road link exists in this road network, path with fewer road link is usually shorter as well. This explains the trend of selecting shortest paths with loosening road capacity constraint. While setting road capacity constant and varying the permitted time in Part 2, we get an opposite trend. More vehicles maneuver through a longer path. In that longer time is permitted, safer paths with longer distances get unlocked. This is why more vehicles are arranged to maneuver in a path with a long distance.

5 Conclusion

In this paper, we present a MVRPW model to describe the safety planning problem of military vehicles maneuvering during wartime. This integration optimization model allows decision makers to get a robust scheme to improve the security of vehicles with the consideration of scenarios. A GA algorithm with a novel encoding rule dealing with the selection problem of *k*-th shortest paths is presented. Two groups of computational examples based on a real road network are conducted to verify this algorithm. These experiments indicate that this algorithm is effective and relatively stable.

In the future, in order to search for better and more stable results, we will test other heuristic algorithms in the MVRPW model. Moreover, to improve the security of maneuvering military vehicles further, optimal allocation of protective resources will be considered in this problem.

Acknowledgment

This project was supported partially by National Natural Science Foundation of China (71471174).

References

1. L. Guan. "The Research of the Optimal Distribution for the Multitask using Types of Vehicles in the Military Road Transportation." International Conference on Transportation Engineering 2007 ASCE:539-544, (2012).
2. L. Zhong, et al. "A Task Assignment Algorithm for Multiple Aerial Vehicles to Attack Targets With Dynamic Values." IEEE Transactions on Intelligent Transportation Systems **14**.1:236-248, (2013).
3. R. D. YiePinedo. "Route optimization while improving safety using escort vehicles." Dissertations & Theses – Gradworks, (2013).
4. V.B.G. Campos, P.A. da Silva, P.O.B. Netto. "Evacuation transportation planning: A method of identify optimal independent routes." Urban Transport V, (2000):555-564.
5. A. Stepanov and M. G. Smith. "Multi-objective evacuation routing in transportation networks." European Journal of Operational Research, **198**.2:435-446, (2009).
6. Y. He, et al. "K-Shortest-Path-Based Evacuation Routing with Police Resource Allocation in City Transportation Networks." Plos One **10**.7, (2015).
7. Y. Zhang, et al. "A metaheuristic approach to the reliable location routing problem under disruptions." Transportation Research Part E Logistics & Transportation Review **83**:90-110, (2015).
8. S. T. Zhang, Y. J. Dou, and Q. S. Zhao. "Evaluation of Capability of Weapon System of Systems Based on Multi-Scenario." Advanced Materials Research 926-930:3806-3811, (2014).
9. P. Yang, W. D. Liu, and L. I. Ming yu. "An optimization model of transportation scheme in preoperational preparations of conventional missile." Fire Control & Command Control, (2007).
10. Z. Michalewicz. "A Non-Standard Genetic Algorithm for the Nonlinear Transportation Problem." Orsa Journal on Computing **3**:307-316, (1991).
11. M. Gen, R. Cheng. "Genetic algorithms and Engineering Optimization." New York: Wiley, (2000)
12. F. Altiparmak, et al. "A genetic algorithm approach for multi-objective optimization of supply chain networks." Computers & Industrial Engineering **51**.1:196-215, (2006).
13. Openstreetmap. Available from: http://www.openstreetmap.org.
14. H. Aljazzar, and S. Leue. "K ⋆: A heuristic search algorithm for finding the k shortest paths." Artificial Intelligence **175**.18(2011):2129-2154.

MATEC Web of Conferences **81**, 02006 (2016) DOI: 10.1051/matecconf/20168102006
ICTTE 2016

The Types Division and Severity Threshold of Traffic Conflict on Freeway Interchange

Shen Li, Qiaojun Xiang, Xin Gu, Junchen Dai, Chao Tang and Shipeng Zhao

School of Transportation, Southeast University, Si Pai Lou #2, Nanjing 210096, China

Abstract. This paper selects time to collision (TTC) as the traffic conflict index. Shooting high-definition video from high altitude above freeway interchange by unmanned aircraft, the TTC was calculated by processing software. The mechanism of traffic conflict on Freeway Interchange area was researched by combining subjective and objective method, and then defined severity. Firstly, conflict observers were trained. Twelve observers were trained repeatedly on the judgment of traffic conflict type and severity cognitive until a consensus was reached. Secondly, the subjective judgment on traffic conflict types and severity was made. The video clips containing traffic conflict were selected, and then all observers together determined the type of each conflict and severity. Thirdly, TTC samples were analyzed. The obviously unsuitable TTC samples were excluded, and the rest would be tested whether meet the sample size requirement of each type of traffic conflict. Finally, the TTC threshold value was defined. According to the cumulative frequency distribution of each type TTC, 85% value was selected as severity of traffic conflict threshold.

1 Introduction

Freeway interchange, including diverge, merge, and weaving area, is the traffic facilities providing continuous traffic flow. However, interference is still existed in traffic flow, which often leads to sideswipe and rear-end accidents. The traditional traffic accident statistics record has the defect of lacking in standardization and systematisms, and the small sample size. Traffic Conflict could be observed with high efficiency, and could be given quantitative analyzed. Itis a Traffic safety Evaluation Index of non-accident [1]. Since the 1980s, the technology has been extensive researched and applied in Intersection Traffic Safety evaluation in China [2], [3], W. Cheng, improved the traffic safety of urban road intersection with traffic conflict [4], [5], Q. Xiang, improved the highway intersection [6], [7], F. Pan, analyzed the unsignalized intersection[8], W. Guo renovated the traffic conflict discrimination model[9]. However, it lacks research on application in the Interchange. And before applying traffic conflict, the criteria traffic conflict must be determined primarily, including conflict type division and define the severity rating.

In order to ensure the traffic flow video quality, high-altitude unmanned aircraft hanging shot was used to overcome two problems. One was that traffic conflict points occurred were not fixed in interchange area. The other was that the digital video could not be placed because there was no high-rise building around interchange.

Tracker is a free video analysis and modeling tool built on the Open Source Physics (OSP) Java framework. Features include object tracking with position, velocity and acceleration overlays and graphs, special effect filters, multiple reference frames, calibration points, line profiles for analysis of spectra and interference patterns, and dynamic particle models. It is designed to be used in introductory college physics labs and lectures.

In this paper, determining the type division and definition severity of traffic conflict of freeway interchange was dealt with the method of combining objective and subjective analysis. TTC was selected as traffic conflict's evaluation index. Firstly, the severity and type of traffic conflict was subjectively judged by 12 trained observers. And conflict threshold was defined objectively by statistical analysis, then the safety evaluation method of interchange was formed which based on traffic conflict technology.

2 Lecture review

2.1 Traffic safety evaluation method

There were two evaluation traffic safety methods of freeway, the speed consistency and comprehensive evaluation.

2.1.1 The speed consistency

T. Guo, safety performance of 37 exit ramps are appraised using three different measures namely Δ V85, 85(Δ V), and 85(Δ V)R. The result show that 85(Δ V) is 1.42 to 2.02 times of Δ V85, and that the evaluation result of 85(Δ V)R was between the other two [10]. Y. Zhang, built specific route design process based on the operating speed consistency, according to the service level of freeway safety, constructed alignment design consistency evaluation and evaluation criteria, and finally the use of research results for a proposed highway route design consistency analysis [11]. W. Wu, according to the interchange ramp speed variation, the ramp was divided into three sections: the deceleration section, uniform section, and the acceleration section. Analyze the relationship between speed change and influencing factors between the various segments, determined interchange ramp speed prediction correlation coefficient factors to construct interchange ramp operating speed prediction model for each segment. And prediction models for the relevant test to verify the applicability of the model [12].

2.1.2 Comprehensive evaluation

S. Fan, by analyzing factors affecting the safety of the interchange, presented a mathematical evaluation interchange security model based on the fuzzy credibility [13]. X. Jiang, according to North Third Ring Road highway interchange design in Guangzhou City, proposed the problems interchange security design [14]. K. Zhang, analyzed the main factors affecting the safety impact of the interchange area, established the safety evaluation system of highway interchange area based on road conditions [15].

2.2 Traffic conflict technique application

In addition, traffic conflict analysis was as a method for non-traffic safety accidents, increasingly used on the road and intersections.

2.2.1 Intersection

U. Shahidah, replaced the crashes by simulating the conflicts for different road and traffic conditions, researched on the correlation between the number of traffic conflict and crash modification factors (CMFs), integrated empirical bayesian before-and-after methods, changed left turn signal priority at Toronto signalized intersections from permissive to protected-permissive [16]. A. Stevanovic, in order to optimize design in signalized intersection balancing safety and efficiency based on VISSIM multi-objective genetic algorithm to, used traffic conflict instead of traffic accident data, analysis the 12-intersection corridor on Glades Road in Boca Raton and two smaller synthetic network served, then optimized the signal timing[17]. L. Yuan, based on correlation analysis of road traffic conditions and traffic conflicts by gray relational theory in signalized intersections, established a relationship model between road traffic conditions and traffic conflicts rates [18].

2.2.2 Basic segment

J. Zhou, chosen the ratio between the number of traffic conflicts and the traffic volume in certain road section length as the index, simulating traffic conflicts by software VISSIM, selected fuzzy evaluation method to evaluate the freeway security situation, the examples verified effect [19]. Q. Shen, constructed the correlation relationship model between the traffic conflicts and interchange spacing, divided the level of risk by gray clustering method, and ultimately determined the new minimum interchange spacing [20]. S. Luo, focused on the criterion and grading standards, improved the traffic conflict theory, and provide a new means for the road traffic safety analysis and evaluation [21].

Traffic Conflict Technique was applied in intersection and road. In this paper, the traffic conflict would be used in researching on the interchange traffic safety, as a supplement of the interchange traffic safety research methods.

3 Interchange traffic conflict type analysis

Due to changes of structure or operations of road users in the interchange area, the road users would be close to each other at a certain time and space, leading to the rear vehicle must take evasive measures to change the driving state excluding the danger of collision. Interchange is the traffic facility keeping continuous flow; with the same direction of traffic, impact angle is less than 90^0. According to the conflict angle partitioning method in SSAM analysis software, the interchange conflict was divided into two types: sideswipe conflict, rear-end the conflict. Sideswipe conflict belongs to $[15^0,85^0]$, rear-end conflicts $[0^0,15^0)$.

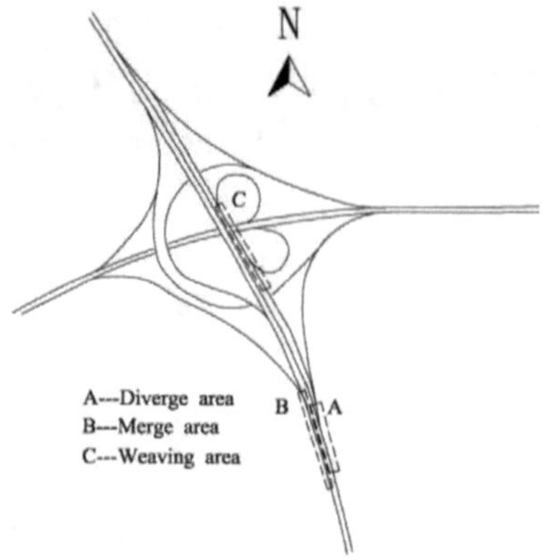

Figure 1. Maqun interchange investigation site

4 Data collection and analysis

4.1 Data collection

The investigation site was selected at the northeast of Nanjing Maqun interchange. It is the semi-clovers interchange type interchange, with a typical diverge, merge and weaving area. Unmanned aircraft was allowed to shoot video in the surroundings, including the wind force below grade 4, sunny weather, good light, no electromagnetic interference. Finally four hours high definition video traffic was obtained. Fig. 1 shown the case of observation points.

4.2 Traffic conflict calculation method

4.2.1 The rear-end conflict

Set the Tracker player steps as twenty frames per one second. According to the specifications, freeway lane-dividing line, solid line six meters, dotted line nine meters, determined the video scale. Set the junction of interchange as the origin pixel of the coordinate system. The new mass A was selected the front vehicle of the existence conflict group, set the template tracking, set the mass A center at the rearmost end, started tracking, until traffic conflict disappears. A particle point coordinates of travel locus was (X_A, Y_A). The trajectory point set was $L_A = \{(x_{A1}, y_{A1}), (x_{A2}, y_{A2}), (x_{A3}, y_{A3}), \ldots, (x_{An}, y_{An})\}$.

Figure 2. The Tracker demonstration of rear-end conflict TTC calculation.

Rewinding the Tracker video to the time of the conflict occurred, A was set as the origin of coordinates now. Meanwhile, the new mass B was selected the rear vehicle of the existence conflict group, and the mass B centre was set at the vehicle forefront. The track method was same to A. B mass point coordinates of travel locus was (X_B, Y_B). The trajectory point set was $L_B=\{(x_{B1}, y_{B1}), (x_{B2}, y_{B2}), (x_{B3},y_{B3}),\ldots, (x_{Bn}, y_{Bn})\}$.

ΔL_B = the changes of distance of B.

$$\Delta L_B = \sqrt{(x_{B2} - x_{B1})^2 + (y_{B2} - y_{B1})^2} \quad (1)$$

Δv= one frame time traveling the changes distance.

The Tracker created the plot of relative distance changes and time.

T_m = the horizontal axis value when the maximum slope of relative distance slow down.

T_e = the maximum slope extended the horizontal axis point value.

$$\Delta T = |T_e - T_m| \quad (2)$$

Recorded the ΔT as TTC data. The rear-end conflict TTC calculation method was shown in Fig. 2 and Fig. 3 by Tracker. In Fig. 3, the vertical axis "r" was the distance of particle B away from the origin (particle A) each fame.

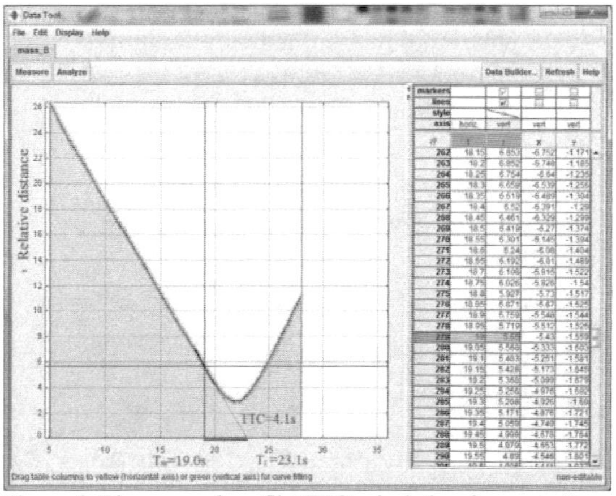

Figure 3. The rear-end conflict TTC calculation demonstration

4.2.2 The sideswipe conflict

For sideswipe conflict, since the two vehicles were in different lanes, it wasn't proved the conflict existence that the relative distance between two vehicles declined rapidly. Because even keeping this motion and doing nothing of evasive action, the intersection of the two vehicles might pass their track intersection independently. In rear-end collision, two vehicles had the same track. When the front vehicle speed declined, in the same time, the rear did not anything to change its speed, the crash would happen. But the sideswipe conflict tracks crossed a point (body length and width should be considered strictly speaking, a "block "). If one missed the only crossing point, no collision. So the sideswipe conflict calculation method was not same with rear-end.

Initial setups of the origin pixel of the coordinate system, the scale and the template were same with rear-end conflict. But, the origin pixel of the coordinate system was fixed. The potential conflict point was the trajectory of the intersection of particle A and B. Output the plot of the speed and time both mass A and B, and recode the speed suddenly slow down about VA and V_B, and then measured the distance about L_A, L_B,S_A and S_B, respectively calculation TTC_A, TTC_B.

V_A = the speed when mass A's speed slow down suddenly.

V_B = the speed when mass B's speed slow down suddenly.

L_A = the length of vehicle A.

L_B = the length of vehicle B.

S_A = the distance between mass A's speed slow down suddenly and the potential conflict point.

S_B = the distance between mass B's speed slow down suddenly and the potential conflict point.

$$TTC_A = S_A / V_A \quad (3)$$

$$TTC_B = S_B / V_B \quad (4)$$

 DOI: 10.1051/matecconf/20168102006

The results appeared in both cases.

a) $TTC_A \geqslant TTC_B$

If $TTC_A \leqslant TTC_B+L_B/V_B$, the conflict was effective, $TTC=TTC_A$. Otherwise, the conflict was not valid.

b) $TTC_A < TTC_B$

If $TTC_B \leqslant TTC_A+L_A/V_A$, the conflict was effective, $TTC = TTC_B$. Otherwise, the conflict was not valid.

And the sideswipe conflict TTC calculation method was shown in Fig. 4.

a) $TTC_A \geq TTC_B$ And, $TTC_A \leqslant TTC_B+L_B/V_B$ b) $TTC_A < TTC_B$ And, $TTC_B \leqslant TTC_A+L_A/V_A$

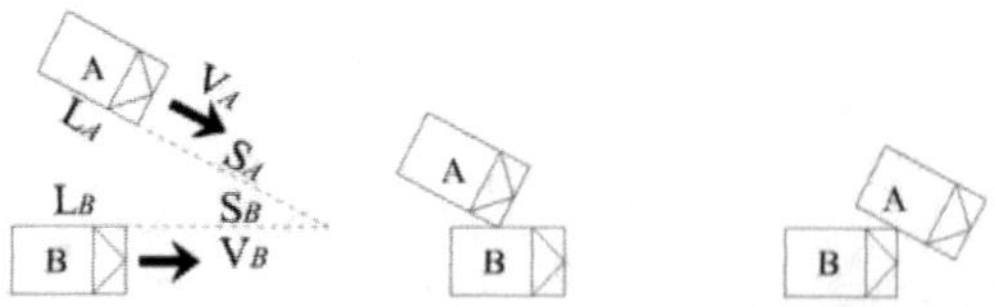

Figure 4. The sideswipe conflict TTC calculation method

4.3 Error Analysis

4.3.1 Unmanned aircraft altitude error

Unmanned aircraft, type INSPIRE 1, created by DJI-Innovations in China, can hover on altitude of four hundred meters. In order to the quality of video, it worked on two hundred miters. Removal of the unstable part at the beginning and end of the video, ensure constant shooting height.

4.3.2 Road surface gradient

A small slope is usually designed on interchange, represented by α , the design requirements $\alpha \leq 3^0$, according to extreme slope $\cos 3^0$ calculated, reduced projected length of the actual 0.14%. It can be neglected on the same slope gradient.

4.3.3 Tracer trajectory analysis

Software working speed setting was twenty frames per one second. Then the rest parameters were set, including the reference line, selecting the junction as the original pixel position, angle, proportion, the scale. The tracking model template should contain the whole vehicle selecting trajectory in every frame. In the same position of interchange, the vehicle mark travel time was calculated by artificial check. Compared with the software displays, an error was 1.26%, indicating that the result was statistically significant from software calculation.

5 Traffic conflict severity subjective judgment

Most observers have been trained in Nanjing at the School of Transportation, Southeast University. On request, the training crew also went to other places in or outside Nanjing. Training took one week, and about twelve people can be trained simultaneously as a group. The crew consists of three men whose task was teaching theory, background and practical skills to the video observers. These practical parts include appraisal of distance and speed, instruction from video tapes.

According highway accident statistics, with good road conditions and high speed, either a serious car crash accident or general scratch rear-end accidents could happened. So the traffic conflict subjective observer partitioned the severity of serious conflict and general conflict as the accidents varies in severity. Therefore, the interchange conflict was divided serious and general conflict by twelve observers. The next step, the samples of TTC should be test, and then the threshold would be defined

6 Division the severity of traffic conflict

6.1 Sample Test

Equation (5) was the minimum sample size required, according to the theory of error.

$$N = (\frac{SK}{E})^2 \tag{5}$$

Where: N is the smallest sample size;

E is the observed value of tolerance;

S is the estimated sample standard deviation;

K is determined by the confidence level of the coefficient.

6.1.1 Sample Calculation

After calculation, the sample standard deviations of various conflicts TTC are close to 1.0, so "S" is 1.0. Fixing the "S", with Equation (5), "E" was set interval of 0.2 and 0.1 second, the minimum sample size error situations was calculated for different confidence levels allowance.

6.1.2 Traffic conflict severity classification

According to the subjective judgment of learning, the severity of the conflict was divided into the serious conflict and general conflict. Observation of the classified data found that, serious conflict TTC value was in 2.5 second or less, the general conflict TTC value was between 2.5 second and 5 second.

The conflict sample distribution of different types and severity was shown in Table 1. In the allowable error E=0.2, 90% confidence level, the required minimum sample size was 68. The actual amount sample of the various conflicts met the minimum sample size requirements.

Table 1. The actual conflict sample

Severity	Conflict Type	
	Sideswipe	***Rear-end***
serious conflict	106	71
general conflict	153	84

6.2 TTC threshold calculation

According to cumulative frequency distribution of TTC, the 85% position value was defined the upper bound of TTC threshold, then getting all TTC thresholds. For example, the following was calculation of serious rear-end conflict threshold.

The TTC data of serious conflict between 0.6 second and 3.4 second were divided fourteen groups, setting interval 0.2 second, and calculated the frequency and cumulative frequency of each data. Fig. 5 shown the TTC cumulative frequency distribution histogram. The 85% position value was between 2.8 second and 3.0 second, thus 2.8 second was defined as the serious conflict threshold.

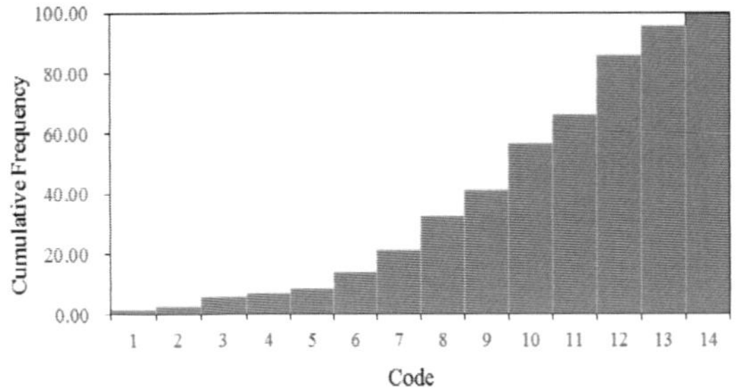

Figure 5. Serious rear-end conflict TTC cumulative frequency distribution histogram.

Using the same method, the other interchange conflict thresholds definition were shown in Table 2.

Table 2. The TTC threshold of interchange

Conflict Type	Rear-end		Sideswipe	
	serious conflict	***general conflict***	***serious conflict***	***general conflict***
threshold （TTC/s）	0~2.8	2.8~4.7	0~2.3	2.3~4.2

7 Conclusion

Based on the characteristics of freeway interchange continuous flow and collision angle less than right angle, the interchange conflicts were divided into sideswipe conflict and rear conflict. Twelve conflict observers were unified training with factors from traffic accident analysis, until cognitive consistency was reached. Then jointly determination was made on the type and severity of the conflict. With the help of unmanned aircraft and trajectory tracking software, the traffic conflict indicators TTC data were calculated. In the Plot of relative distance and time, one time was horizontal axis value when the maximum slope of relative distance changes slow down, and the other time was the maximum slope extended the horizontal axis point value; the absolute distance of two time was TTC value. Sideswipe conflicts, inappropriate using relative time, using minimum time of driving to the trajectory intersection point when the conflict occurred. The large error TTC data were removed. Based on statistics, the 85% position data of cumulative frequency was defined as the severity threshold. In sideswipe conflict, the serious and general threshold value was 2.3 second and 4.2 second; the rear-end conflict, threshold value was2.8 second and 4.7 second. This paper have described the type of traffic conflict on interchange which is divided by traffic collision angle. In the future, further refined conflict severity may be defined by the specific environment.

References

1. C. Hydén, BLIT, **70**, (1987)
2. S. Zhang, China traffic conflict technique (SCUP, Chengdu, 1998)
3. X. Liu, and H. Duan, HT. J. **14**, 29 (1997)
4. W. Cheng, and J. Li, TSEIT. J. **4**, 48 (2004)
5. W. Ceng, and G. Wang, KUST. J. ST. **30**, 106 (2005)
6. Q. Xiang, J. Lu, Q. Wu, J. Lu, and Y. Ma, HTRD. J. **26**, 132 (2009)
7. Q. Xiang, J. Lu, C. Lu, and X. Ge, Analysis and application of road traffic conflict technique, (Science Press, Beijing, 2008)
8. F. Pan, L. Zhang, and J. Lu, SJTU. J. **47**, 259 (2013)
9. W. Guo, Z. Qu, and D. Wang, JLU. J. ET. **41**, 35 (2011)
10. T. Guo, W. Deng, and J. Lu, TSEIT. J. **10**, 76 (2010)
11. Y. Zhang, NT. J. **4**, 3 (2010)
12. W. Wu, D. BJUT, (2009)
13. S. Fan, S. Zhao, Y. Yan, and H. Gao, HPC. J. **12**, 42 (2009)
14. X. Jiang, and H. Yuan, Hig. J. **4**, 167 (2006)
15. K. Zhang, D. CSUST. (2011)
16. U. Shahdah, F. Saccomanno, and B. Persaud, AAP. J. **71**, 228 (2014)
17. Stevanovic, J. Stevanovic, and C. Kergaye, TRP C. J. **32**, 159 (2013).
18. L. Yuan, H. Yuan, Q. Xiang, and J. Lu, TIS. J. **28**, 117 (2010)
19. J. Zhou, Y. Chang, M. Guo, and G. Wang, CQJTU. J. NS. **30**, 974 (2011)
20. Q. Shen, S. Yang, Y. Zhao, and H. Cao, SETP. **35**, 160 (2015)
21. W. Zhou, and S. Luo, CJHT. J. **13**, 82 (2000)

Potential Safety Benefit of the Blind Spot Detection System for Large Trucks on the Vulnerable Road Users in Taiwan

Ming-Hang Wang[1] and Chien-Hung Wei[2]

[1]*Assistant Professor, Department of Transportation Technology and Management, Kainan University, Taoyuan City, Taiwan*
[2]*Professor, Department of Transportation and Communication Management Science, National Cheng Kung University, Taiwan City, Taiwan*

Abstract. Considering motorcyclists, pedestrians and bicyclists as vulnerable road users (VRUs), more than 75 percent of the victims of fatal crashes involving large trucks in Taiwan are VRUs. Most crashes occurred at or were due to the blind spots of large trucks because of the size and traveling locations of the VRUs. This study applies typology and statistical methods to estimate the potential safety benefit of blind spot detection (BSD) systems for large trucks on VRUs. The pre-crash scenarios associated with the blind spots of large trucks were derived by counting the maneuvers of large trucks and VRUs, prior to crashes, the truck drivers' improper behaviors (cause of crashes), and the crash types. The number of crashes and fatalities were counted for the pre-crash scenario relevant to the BSD systems. A value of 0.8 of human machine interface factor (HMIF) based on a previous study was applied to estimate the potential safety benefits of the BSD system. The results show that the implementation of BSD systems on all large trucks could help avoid about 24, 10, and 11 percent of large truck-involved crashes with pedestrians, bicycles, and motorcycles, respectively. The BSD systems could also save 5 pedestrians, 3 bicyclists, and 15 motorcyclists per year from crashes involving large trucks.

1 Introduction

Advanced vehicle safety systems (AVSs) are designed to help avoid vehicle crashes or reduce the severity of crashes by providing warning signals to drivers or automatic actions for vehicles when vehicles are facing potential collision circumstances. As innovative wireless and image detection technologies are developed, various types of AVSs are embedded into new cars. However, the popularity of AVSs is still limited due to a lack of laws and awareness of AVS technologies. To enhance traffic safety, several countries have been emphasizing the standardization of AVSs on new vehicles. Since 2007, the United States [1], and Europe [2] and Australia have been making the electronic stability control (ESC) mandatory on new vehicles.

Several researches have been conducted to estimate the potential safety benefit of AVSs using various methodologies. Blower et al.[3] reviewed the current state of knowledge regarding the effectiveness of several advanced collision-avoidance technologies (ACATs) and presented an assessment of the effect of each technology on traffic safety for light-duty vehicles. Overall, the systems reviewed were estimated to be substantially effective in reducing their target crash types.

In Taiwan, due to the high number of motorcycles, more blind spot detection functions that can detect motorcycles traveling beside trucks along the street, pedestrians crossing in front of the large trucks are needed. Considering motorcyclists, pedestrians and bicyclists as vulnerable road users (VRUs), more than 75 percent of the victims of fatal crashes involving large trucks are vulnerable road users according to the national police-reported crash database from 2009 to 2014. Among crashes involving large trucks with pedestrians and bicyclists, about 40 percent of the fatal crashes occurred when the large trucks were starting off and making right or left turns. It is believed that most crashes occurred at or are due to the blind spots of the large trucks. Thus, the implementation of blind spot detection (BSD) system in large trucks is an essential and urgent task to avoid large trucks crashing into VRUs.

In order to increase the public's awareness of the function of AVSs on road safety, particularly for large trucks on VRUs, this study applies typology and statistical methods to estimate the potential safety benefit of BSD systems for large trucks on VURs based on the national police-reported crash database in Taiwan. The pre-crash scenarios associated with the blind spot of the large truck were derived by combining the truck drivers' improper behaviors (cause of crashes), the truck's movement prior to the crash, and the collision part of the large trucks corresponding to the crash types. Next, the statistical method is used to calculate the potential safety benefits in terms of the frequency and fatalities of the crashes relevant to the BSD systems. Finally, the results and the limitations of this study, as well as further studies for extending the method to other AVS technologies, are discussed and recommended.

2 Blind spot detection systems

The blind spot is the area not covered by the driver's line of sight and mirrors, alongside and off-set to the rear of the moving car on both sides. The BSD technology uses radar, cameras or ultrasonic technologies to monitor the blind spot area of the vehicle. If a moving object is detected within the specified zone, a warning signal is issued. Warning signals vary from one version of the system to another and include visual, audio or haptic signals. Essentially, most BSD systems were designed to alert the driver that another vehicle may be present and to use caution if planning a lane change [4]. Some BSD systems were also developed particularly for detecting vulnerable road users in a truck's blind spot zone [5][6][7] using camera detection technology.

The potential benefits of the BSD systems were estimated by several researches. In Europe, the BSD could save approximately 975 lives each year and avoid 2,100 injuries, through the crash causation analysis [8]. In the U.S., it can avoid 457,000 injuries and 428 fatalities per year [9], and prevent 39,000 large truck-involved crashes including 79 fatal crashes per year [10] based on the pre-crashes scenario analysis. Other research used depth clinical reviews for each case [11] and found that the BSD can avoid 5.9 percent of crashes involving large trucks. In Germany, it could avoid about 24.7 percent of crashes [12].

However, the benefits of the BSD systems were assessed for passenger cars only. Most BSD systems may not always detect all vehicles located in a car's blind spot. In particular, narrow and often fast moving vehicles such as motorcycles are easy to miss [13]. The potential of the technology to assist Heavy Goods Vehicles (HGVs) drivers to detect VRUs in their blind spot should also be investigated. Users, such as cyclists, are particularly at risk when they find themselves in the blind spot of an HGV. The development and deployment of reliable technologies enabling the recognition of pedestrians and cyclists while they are in the blind spot of HGVs should be a priority.

In Taiwan, due to specific traveling characteristics of a high number of motorcycles, additional blind spot detection functions that can detect the motorcycles traveling beside the trucks along the streets are needed. A blind spot detection system was developed by the Automotive Research & Testing Center (ARTC), a benchmark automotive R&D institute in Taiwan [14]. The Taiwanese BSD system was developed by using image-based object recognition, image-feature analysis, object/space measurement, and signal integration. The BSD system is adopted to detect the status of approaching objects (including all types of vehicles, bicycles and pedestrians) in both left and right side blind areas by active image recognition processes of machine vision. This BSD system has been verified that the accuracy rate is 91 percent and the frame rate is more than 20 frames per sec (fps), in day and night and all weather conditions [14]. The blind spot areas which the Taiwanese BSD can detect include the front side, rear side, front lateral sides and rear lateral sides. Based on the detected areas, the systems provide the approaching object detection and door opening warning function.

The Taiwanese BSD system was extended to large trucks by mounting cameras on several parts of a semi-trailer to detect any object in front of the heavy vehicle. As the object with defined characteristics in the blind area is detected, the system issues the pre-warning (lighting) or warning (beep and lighting) messages to the driver.

Although the BSD technologies are developed, the potential safety benefit on traffic crashes and fatality reduction have not been estimated. This study will estimate the potential safety benefit if all large trucks are equipped with truck BSD systems.

3 Crash database

The crash data for this study is provided by the National Police Administration (NPA) of Taiwan. Data on crashes involving large trucks and VRUs, such as pedestrians, bicycles, and motorcycles for the years 2009–2014, were extracted from the nationwide traffic crash database. Large trucks are defined as cargo vehicles with a gross vehicle weight (GVW) over 3,500 kg. The NPA crash database only includes crashes in which at least one road user is killed or injured. The crash is considered fatal if at least one person died within 30 days of the collision.

The NPA crash dataset contains the information on the characteristics of the crash, the road environment, people (e.g., drivers, bicyclists, and pedestrians) and units. The characteristics of the crash include the month, date, time of day, and crash location. The road environment data contain the information on road types, control devices, and geometry deployment (e.g. the direction or lane divisions, and sight distance condition). The characteristics of truck drivers and VRUs including the personal information (e.g., age, gender), impaired condition, and the injuries sustained (e.g., fatal, injury, or no injury) are also provided. The unit data recorded the movement manners of the vehicles prior to the crashes, damage, first collision part of the vehicle, and crash type, as well as the major/minor causes of the crashes. This data can then be used to identify the possible scenarios prior to the crashes.

4 Methods

4.1 Pre-crash Scenario technologies

Due to low penetration rates of vehicles with the AVSs, most studies relied on simulations or limited field operational tests to evaluate the effectiveness of AVSs [4]. In the U.S., a number of crash typologies have been developed and widely been used to provide a common foundation to develop and estimate potential safety benefits of effective crash countermeasure systems [15][16][17][18]. Yanagisawa et al., [19] analyzed the national crash databases from National Highway Traffic

 DOI: 10.1051/matecconf/20168102007

Safety Administration (NHTSA) to identify the pre-crash scenarios for the Pedestrian Crash Avoidance/Mitigation (PCAM) systems. Safety benefits, in terms of pedestrian injuries avoided and pedestrian injuries mitigated, are expressed in terms of reductions in annual harm measures. Toma et al. [20] describes pre-crash scenarios involving at least one large truck, which might be avoided with crash-imminent warning systems based on short range vehicle-to-vehicle (V2V) communications.

In Europe, Papps et al., [21] estimated the potential proportion of injury accidents that could be avoided and/or the severity of the accidents that could be reduced, for future and existing safety functions, by means of the Traffic Accident Caution in Europe (TRACE) projects. They assessed the effectiveness for 19 advanced safety devices including the BSD system. They counted the main maneuver of vehicles and crash types, which is relevant to the BSD. The results indicate that the BSD could lead to about a 2 percent reduction in injury accidents and related casualties. However, they only considered the effective-ness between passenger cars.

In this study, the possible crash scenario due to or at the blind spots of large trucks are categorized by using the "if the drivers were warned, would the crash be avoided?" method. Thus, five major attributes available in the existing national police administration (NPA) crash database were investigated. They are: (1) maneuver of the large truck prior to the crashes; (2) maneuver of the VRUs prior to the crashes; (3) crash contributions due to the large trucks; (4) crash types; and (5) first collision spot on the trucks. These attributes were used to generate the pre-crash scenarios which could possibly be avoided by the implementation of the BSD systems.

4.1.1 Maneuvers of large trucks prior to the crashes

Truck's Maneuvers is counted to identify whether the objects (VRUs) appeared within the blind spots, and determine whether the crash could be avoided or not. It also can be used to identify if the crashes resulted from the off tracking of the trucks. In general, crashes occur due to or at the blind spots when the trucks are:

- Starting off (TSO) from the road side or stop line;
- Turning Right (TTR);
- Turning Left (TTL)/U-turn (TUT);
- Changing Lane (TCL);
- Parking or backing operation (TPB); or
- Parked (TPK).

A parked truck could result in crashes when the driver opens the door carelessly (ODC) without checking for approaching bicycles/ motorcycles.

4.1.2 Maneuvers of the VRUs prior to the crashes

The VRU's maneuver was checked to further confirm the direction and relative position between trucks and VRUs, and also to determine if the VRUs appear within the blind area. Because pedestrians are in slow moving situations comparing to the trucks, any maneuvers could possibly be detected and avoid the crashes in time. For bicycles and motorcycles, changing lane or turning movements may be detected by the BSD system, but may not prevent the crashes in time. Thus, the maneuvers of the bicycles or motorcycles relevant to the BSD system are:

- Going Straight (VGS): same direction with truck;
- Parked (including temporary parked) (VPD): which is in still condition.
- Turning Right (VTR) ; or
- Turning Left" (VTL): same direction with truck.

4.1.3 Cause of crash by trucks

Truck's contribution factor was inspected to identify whether the drivers were driving carefully or already noticed the appearance of the VRUs or not. Theoretically, under careless or improper driving scenarios, the crashes could be avoided if the drivers were warned of the appearance of the VRUs. Particularly when the objects appear within the blind spots, these crashes could be avoided. The major causes due to the truck drivers are:

- Failing to look properly (FLP);
- Turning Improperly (TIP);
- Backing Carelessly (BCL);
- Failing to Keep the Safe Side Gap (FKSG); or
- Changing Lane Improperly (CLIP).

4.1.4 Crash types and first collision spot on trucks

Crash types and collision spots on trucks should be checked simultaneously to identify the relative position between trucks and VRUs. For pedestrian-related crashes, the first collision spot on the truck body should be taken into account to check which spot of the truck hit the pedestrian. For instance, pedestrians hit by the front head of trucks may be due to the blind spot in front of the trucks.

For crashes with bicycles or motorcycles, the crash types occurring at the blind spots of trucks could be side effect (SE), same direction side swipes (SSW). When trucks are starting off from the stop line or road sides, the trucks may collide into stopped bicycles or motorcycles due to the blind spot in front of the trucks. The collision spots are recorded in the NPA crash database as the first collision spot (FCS), including:

- Front head (FH);
- Right Front(RF)/Right side (RS)/Right Rear(RR);
- Rear end (RE);
- Left Front(LF)/Left side (LS)/Left Rear (LR)

Table 1. Pre-crash Scenario for large truck crashes with pedestrian relevant to the BSD system.

Pre-crash Scenario	Maneuvers of large trucks	Maneuvers of VRUs	Cause by LT	Crash Type	First collision spot on truck
P1	TSO	Any	FLP/FKSG	Crash with Pedestrian	Any expect RE
P2	TPBO		FLP/BCL /FKSG		Any spots
P3	TTR		FLP/TIP /FKSG		FH/RF/RS/RR
P4	TTL		FLP/TIP/FKSG		LF/LS/LR
P5	TCL		FKSG/CLIP		RF/RS/RR

4.2 Pre-crash Scenario

Based on the principles described above for determining the scenarios which could be avoided by BSD systems, the pre-crash scenarios were proposed in Table 1 and Table 2 for crashes with pedestrians, bicycles and motorcycles, respectively. The maneuvers and characteristics of bicycles and motorcycles are similar expect for the traveling speeds. The crash scenario for bicycles and motorcycles should be similar.

Table 2. Pre-crash Scenario for large truck crashes with bicycles/motorcycles relevant to the BSD systems.

Pre-crash Scenario	Maneuvers of large trucks	Maneuvers of VRUs	Cause by truck	Crash type	First collision spot on truck
B1/MB1	TSO	VPD/VSP /VSO/VGS	FLP/FKSG	any	Any but rear end
B2/MB2	TPBO	VPD/VSP /VSO/VGS	FLP/BCL /FKSG	any	Any spots
B3/MB3	TTR	VGS/VPD /VTR	FLP/TIP /FKSG	SE/SSW	FH/RF /RS/RR
B4/MB4	TTL/TUT	VPD/VGS /VTL	FLP/TIP/FK SG	SSW	LS/LF/FR
B5/MB5	TCL	VGS	FKSG/CLIP	SE/SSW	FR/RS/RR
B6/MB6	TPD	VGS	ODC	ODC	*LS

*: NPA database does not specify the door as collision spot.

4.3 Potential safety benefits estimation

The descriptive statistical method is used to calculate the potential safety benefits in terms of the frequency and fatalities of the crashes for each scenario relevant to the truck's BSD system. Theoretically, the crashes could be avoided by the drivers' reaction when the BSD sends the warning signal (or sound) to the driver. However, in reality, some drivers may not react in time or may not act the proper way to prevent the crashes. The human machine interface factor (HMIF) will affect the appropriate ratio of drivers' reactions. The HMIF is a value between 0 and 1. The theoretical BSD should be multiplied by the HMIF in order to determine the achievable safety potential of BSD systems.

A value of HMIF=0 indicates the system is not achievable in reality because of a poor interface design. A value of HMIF=1 means that the theoretical BSD can be achieved in reality. For the BSD systems, the HMIF can be assumed to be 0.8 in accordance to the research conducted in Germany [22] referred in [12]. However, an additional filed test of the BSD system is necessary to estimate a precise value of HMIF.

5 Results analysis

5.1 Facts of crashes involving large trucks and VRUs

From 2009 to 2014 (6 years), a total of 34,945 large trucks were involved in 33,580 crashes, leading to 2,141 deaths (6.4 percent) and 38,261 injuries. Among those large truck involving crashes, more than 70 percent were VRU related (64.8 percent of motorcycles, 3.6 percent bicycles, and 2.7 percent pedestrians). In terms of deaths, 1,339 are motorcyclists (62.5 percent), 147 are cyclists (6.9 percent), and 144 are pedestrians (6.7 percent). The fatality rate of a motorcyclist crashing with a large truck is 5.3 percent. For pedestrians and bicyclists, the fatality rate is 15.2 percent and 11.9 percent, respectively, when they collide with a large truck.

To identify the major causes of the crashes, crashes with other non-VRUs were excluded for the analysis. Thus, crashes with only one large truck crashing with VRUs were used for the analysis. A total of 21,820 crashes (19,879 crashes with motorcycles, 742 crashes with pedestrians, and 1,199 crashes with bicycles) were analyzed. Among these crashes, 123 pedestrians, 145 bicyclists, and 1,120 motorcyclists were killed.

5.2 Crashes and fatalities relevant to BSD systems

Table 3 summarizes the numbers of crashes and fatalities relevant to the BSD systems based on the pre-crash scenarios.

Table 3. Numbers of crashes and fatalities relevant to the BSD systems.

vs. pedestrians			vs. Bicycles			vs. Motorcycles			vs. Al VRUs	
Pre-crash scenario	No. of Fatalities	No. of Crashes	Pre-crash scenario	No. of Fatalities	No. of Crashes	Pre-crash scenario	No. of Fatalities	No. of Crashes	No. of Fatalities	No. of Crashes
P1	29	86	B1	5	20	MB1	13	491	47	597
P2	4	58	B2	1	18	MB2	32	783	37	859
P3	2	30	B3	15	70	MB3	38	1004	55	1104
P4	3	45	B4	0	1	MB4	1	59	4	105
P5	0	5	B5	1	18	MB5	26	303	27	326
			B6	0	15	MB6	0	116	0	131
Sum	38	224		22	142		110	2756	170	3122
% of Total	30.9%	30.2%		15.2%	11.8%		9.8%	13.9%	12.2%	14.3%
Total	123	742		145	1199		1120	19879	1388	21820

Theoretically, if all large trucks were equipped with BSD systems, they could prevent 14.3 percent of large truck-involved crashes with VRUs, and save 170 lives of VRUs in 6 years (28.3 per year).

For the large truck-involved crashes with pedestrians, about 27 percent of crashes, and 24 percent of pedestrian fatalities were relevant to the blind spots of the trucks. Among these crashes during the "starting off" stage (86 crashes), half of all fatal crashes, pedestrians were hit by the front head of the trucks.

For the crashes with bicycles, the BSD system can possibly prevent about 12 percent of crashes, and 15 percent of bicyclist fatalities. Most crashes and fatalities could be avoided, by about 5 percent and 10 percent, respectively, by the BSD systems when trucks were turning right at the intersections (Scenario B3).

For the crashes with motorcycles, the BSD system could possibly decrease the crashes by 14 percent and fatalities by 10 percent. Most crashes (about 5 percent) could be avoided by BSD systems during the "turning right" stage of the trucks (Scenario MB3). That is because most motorcycles are traveling on the right side of the trucks. Potential traffic conflicts occur between trucks and motorcycles at trucks' blind spots when trucks turn right. The BSD systems are more effective for preventing the crashes when the trucks are in slow movements, such as 'starting off", "parking or backing operation". During these processes, the BSD could possibly decrease more than 6 percent of crashes and 4 percent of fatalities of motorcyclists.

5.3 Potential safety benefits of BSD systems

If the 0.8 HMIF was applied, the BSD system could avoid about 24 percent of large truck-involved crashes with pedestrians, 10 percent of crashes with bicycles, and 11 percent of crashes with motorcycles. The BSD system could also reduce 19 percent of fatalities for pedestrians, 12 percent for bicyclists, and 8 percent for motorcyclists among the large truck-involved crashes. These are about, on average, saving 5 pedestrians, 3 bicyclist, and 15 motorcyclists per year.

6 Conclusions and recommendations

The benefits of the AVSs based on the data from other countries are not applicable during the process of the standardization of AVSs. Prior AVS systems are also different from other countries' systems. Due to the high number of motorcycles in Taiwan, additional blind spot detection functions that can detect motorcycles traveling beside the trucks are needed.

To increase the public's awareness of the functions of advanced vehicle safety systems (AVSs) on road safety, particularly for large trucks on VRUs, this study applies typology and statistical methods to estimate the potential safety benefit of BSD systems for large trucks on the VRUs. The results show that theoretically the BSD system could prevent 14.3 percent of large truck-involved crashes with VRUs, and save 170 lives of VRUs in 6 years (28 per year). When the HMIF (0.8) was considered, the BSD system could avoid about 24 percent of crashes with pedestrians, 10 percent of crashes with bicycles, and 11 percent of crashes with motorcycles. The BSD system could also save about 5 pedestrians, 3 bicyclists, and 15 motorcyclists per year from the crashes involving large trucks. However, since the HMIF may vary from country to country, further field tests are needed to obtain the suitable HMIF for the Taiwanese BSD. The combinations of other AVSs, such as automatic emergency braking systems, will increase the safety benefit of the BSD systems.

Acknowledgement

This research is sponsored by Ministry of Science and Technology of Taiwan (No.MOST-104-221-E-424-001).

References

1. US Department Of Transportation, 49 CFR Parts 571 and 585- Federal Motor Vehicle Safety Standards, Docket No. NHTSA 2005-20586
2. The European Parliament and Council, REGULATION (EC) No 661/2009, (13 July 2009)
3. D. Blower, Final Report, The University of Michigan Transportation Research Institute, (2014)
4. G. Forkenbrock, R. L. Hoover, E. Gerdus, T. R. Van Buskirk, and M. Heitz, Report No. DOT HS 812 045, Washington, DC: NHTSA, (2014)
5. K. Van Beeck, T. Goedem´e, and T. Tuytelaars, Proc. MVA, Nara, Japan, (2011)
6. K. Van Beeck, T. Tuytelaars, and T. Goedem´e, Proc. ICINCO, (2012)
7. K. Van Beeck, and T. Goedemé, Proc. the 3rd International Conference on Pattern Recognition Applications and Methods, 412-420, Angers, France, (March 2014)
8. eImpact, Assessing the Impacts of Intelligent Vehicle Safety Systems, Socio-economic impact assessment of stand-alone and co-operative Intelligent Vehicle Safety Systems (IVSS) in Europe, (2008)
9. Insurance Institute for Highway Safety (IIHS), Researchers Estimate Potential Benefits of Crash Avoidance Features, Status Report **43**, 3 (April 17, 2008) | SPECIAL ISSUE: CRASH AVOIDANCE FEATURES |
10. Insurance Institute for Highway Safety (IIHS), Large Trucks to Benefit from Technology Designed to Help Prevent Crashes, Status Report **45**, 5 (May 20, 2010)
11. K. J. Kingsley, NHTSA, Paper No. 09-0460, (2009)
12. M. Kuehn, T. Hummel, and J. Bende, No. 09-0317. 21st International Technical Conference on the Enhanced Safety of Vehicles, Stuttgart, Germany
13. Fédération Internationale de l'Automobile (FIA), policy position on the general safety regulation, (2014)
14. C. T. Chen and Y. S. Chen, ITSC '09. 12th International IEEE Conference on Intelligent Transportation Systems, (2009). St. Louis, MO
15. NAO Engineering, Safety & Restraints Center, Crash Avoidance Department, "44 Crashes". General Motors Corporation, Version 3.0, (January 1997)
16. Crash Avoidance Metrics Partnership, "Enhanced Digital Mapping Project – Final Report". U.S. Department of Transportation, National Highway Traffic Safety Administration, (November 2004)
17. W. G. Najm, B. Sen, J. D. Smith, and B. N. Campbell, DOT-VNTSC-NLTSA-02-04, DOT HS 809 573, (February 2003)
18. W. G. Najm, J. D. Smith, and M. Yanagisawa, Report No. DOT HS 810 767. Washington, DC: NHTSA, (2007)
19. M. Yanagisawa, E. Swanson, and W. G. Najm, Report No. DOT HS 811 998. Washington, DC: NHTSA, (2014)
20. S. Toma, D. Swanson, and W. G. Naim, Report No. DOT HS 812 023. Washington, DC: NHTSA, (2014)
21. Page, Y et al. D 4.2.1 - A posteriori evaluation of safety functions effectiveness –Methodologies, TRACE Report, (2007)

22. B. Färber, F. Freyberger, and B. Färber, Nutzenpotenzial von Fahrerassistenzsystemen. Literaturstudie, Expertenbefragung, Bewertung, On behalf of the German Insurers Accident Research. Unpublished , (2008)

DOI: 10.1051/matecconf/20168102008

Load Response Analysis of Asphalt Pavement

Xiao Zhijun[1,2] and Luan Liqiang[3,4]

[1]*Tongji University, China*
[2]*Key Laboratory of Road and traffic Engineering of the Ministry of Education, Shanghai, China*
[3]*Guangxi Key Laboratory of Geomechanics and Geotchniacal Engineering Guilin China;*
[4]*Guilin University of Technology, Guilin, China*

Abstract. Based on the filed data of Fuzhou test road, load response regulation of nonlinear pavement structure was analyzed in this paper; Mechanics software was adopted to analyze linear elastic and nonlinear of the pavement structure, analyzed results and road filed data were compared, and determine the correlation between them and the field data.

1 Introduction

In recently years, loading reaction on Asphalt pavement gradually become a new research direction, with the purpose of studying the behaviors reaction upon different conditions on asphalt pavement radically. However due to limited financial and technical means, current research is mainly conducted by the mechanical analysis and theoretical calculation, which was lack of support of field data, obtained very limited loading response data through Experimental road paving. This article relies on the field data inspected from Fuyang high speed test road, NO.2 test section, carrying the in-depth analysis of load response of asphalt pavement.

2 Introduction of the test road

The structural of the Fuzhou test road is a typical long-life pavement composition, as shown in Fig. 1, adopting SMA asphalt mixture as the surface layer, other asphalt layers are made of SBS modified asphalt mixture, using local limestone and basalt in Fuyang as aggregate.

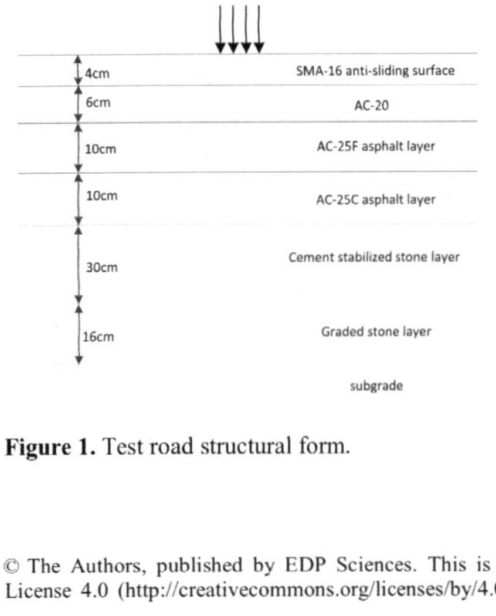

Figure 1. Test road structural form.

A lot of stress and strain sensors were laid under the test road, to measure the load response when vehicles pass by. Sensors buried in each structure layer of the wheel path and the apparatus are shown in Fig. 2 to Fig. 3.

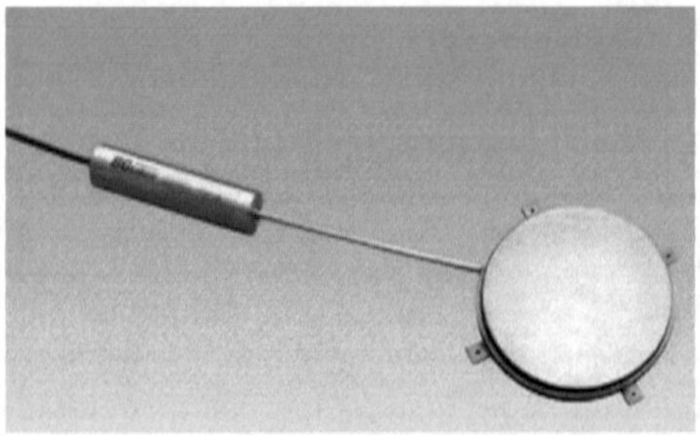

Figure 2. Earth pressure cell.

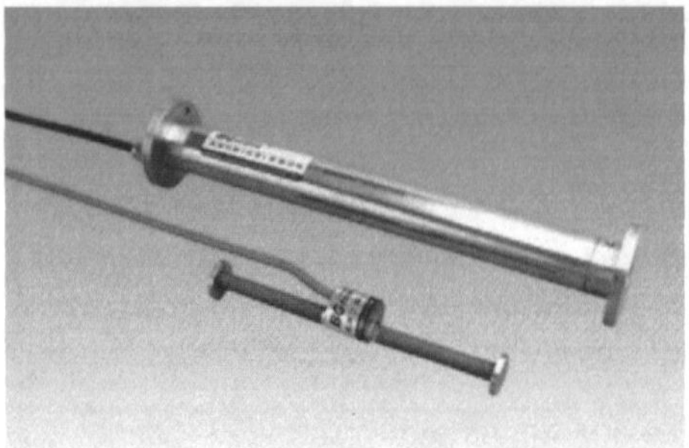

Figure 3. Embeded strain gage.

3 Test data on experimental road

Trucks with rear axle of 100KN were used as standard vehicle, from a distance of 20m to the instrument Embedment position, marching forward at 10km/h and continuing the driving for another 20m after reaching the embedment position. Data were recorded every 0.01m,

DOI: 10.1051/matecconf/20168102008

after wavelength and temperature correction treatment, the result are shown as Fig. 4~Fig. 9.

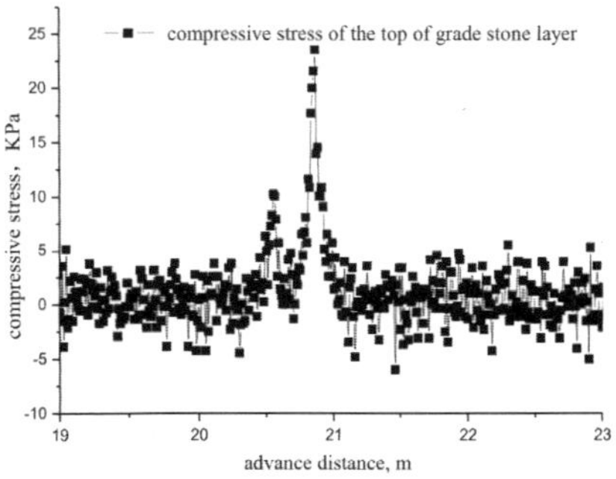

Figure 4. Compressive stress of the top of graded stone.

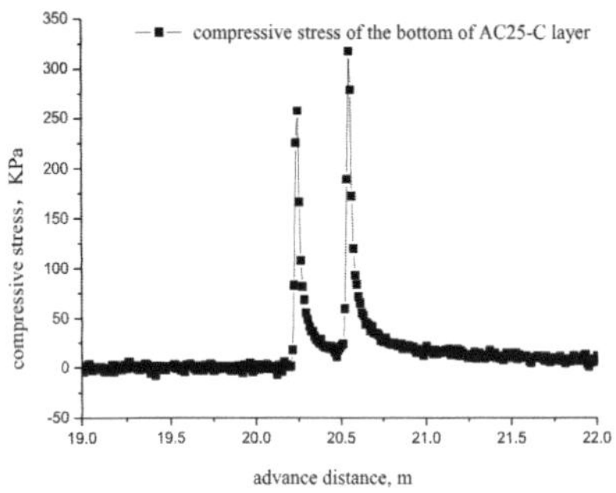

Figure 5. Compressive stress of the bottom of AC25-C.

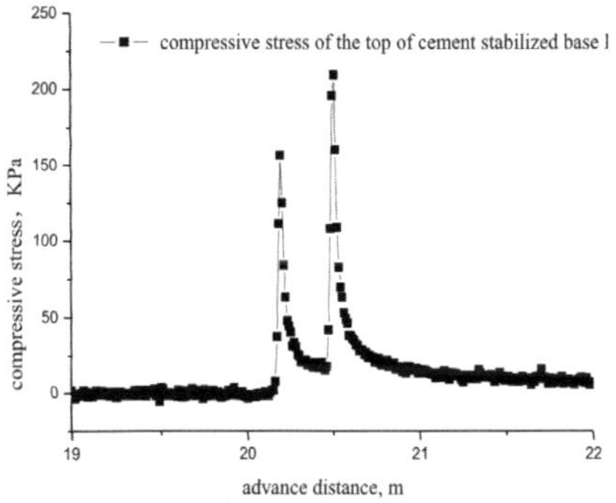

Figure 6. Compressive stress of the top of cement stabilized base.

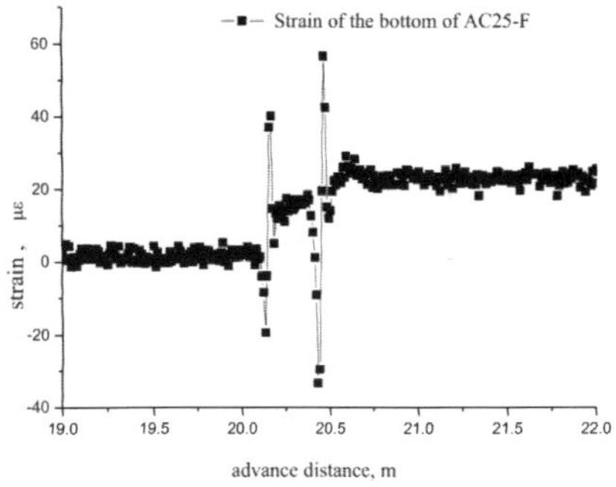

Figure 7. Strain of the bottom of AC25-F.

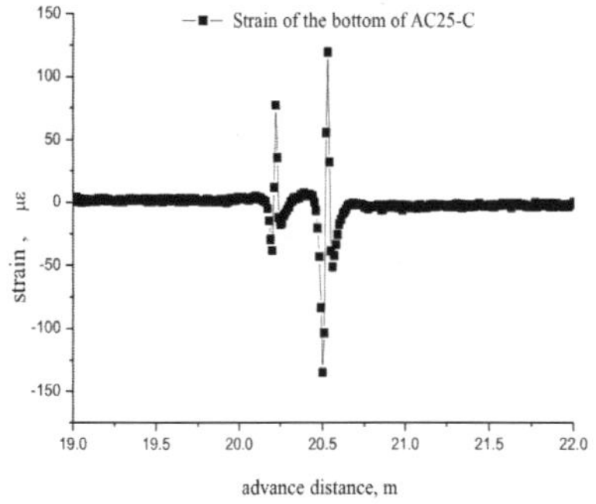

Figure 8. Strain of the bottom of AC25-C.

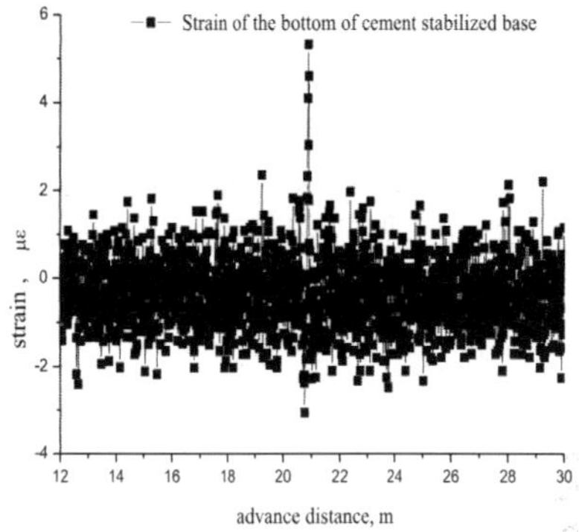

Figure 9. Strain of the bottom of cement stabilized base.

The front and rear wheels of the standard vehicle pass through the instrument embedment position, in the direction of the driving lane, the stress and strain can produce twice wave crest, as the rear wheel is heavier than the front wheel, the second wave significantly higher than the first time. The stress and strain in different structural layers upon the rear are shown as in table 1.

Table 1. Loading response when the rear wheel passed by.

Type of field data	Test Data
Strain in bottom of AC-25F along the lane, με	56.47
Strain in bottom of AC-25C along the lane, με	124.22
Strain in bottom of cement stabilized base along the Lane, με	5.32
vertical stress in bottom of AC25-C	317.45
vertical stress top of cement stabilized base	209.30
vertical stress top of graded gravel layer	23.50

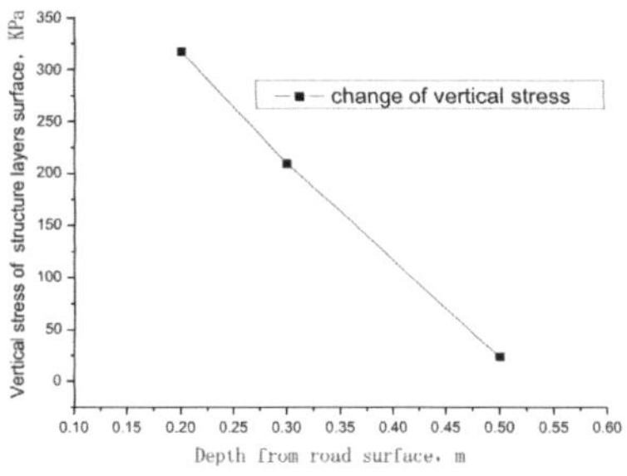

Figure 10. The vertical compressive stress of structural layers surface under loading.

4 Test data analysis

Analyzing the test data, it can be found that relationship between vertical compression and road surface depth is shown as in Fig. 10.

Can be seen from Figure 10, the vertical stress produced by road surface loading is compressive stress, and with depth increasing, stress values gradually decrease, that is stress gradually decreases with loads passing from the top to foundation of the road.

Can be seen from Figure 4 to Figure 9, After the rear wheel of standard truck passed the instruments laying position, strain gradually approach a constant value, Because the viscous and elastic recovery of strain can be completed within a very short period of time, the process of strain reaching a constant value from a maximum value can be considered as a process of standard truck still have an impact on the position of instruments laying. After the standard truck leaving instruments laying position, it can still have an impact on the position of instruments laying within a certain distance. Assuming this distance is "loading effect radius". Loading effect radius of various structural layers is shown in Table 2.

Table 2. Radius of loading influence of the structural layers under loading.

Layer	Depth, m	loading effect radius, m
AC25-F	0.18	0.17
AC25-C	0.26	0.12
cement stabilized base	0.60	0.10

Can be seen from Table 2, as the depth increases, the standard truck had a diminishing lading effect radius on different structural layers, This is primarily because vertical loading transmission in the pavement structure is diminishing with depth increases, when the loading passes each layer structure, it will be reduced by dispersing when handed over from upper level, while larger vertical stress in the same structural layers cause wider range of transverse wave.

Instrumental sensor tells strain from extreme gradually returns to a non-zero value before and after loading. Loading makes residual strain happens on structural layers, this fact illustrate that pavement structure as a whole is not a simple linear elastic body, the inherent viscosity and plastic characteristics of structural layer material contribute strain not fully recover after the pavement loading. Specific strain change is shown in Table 3.

Table 3. Strain changes of the structural layers before and after loading.

	、	Extreme strain i n bottom, με	Residual strain in bottom, με	Modulus, MPa
AC25-F	0	56.47	22.2	1335
AC25-C	0	-134.78	-5.0	1685
Cement stabilized base	0	5.32	0.1	2850

Can be seen from Table 3, cement stabilized macadam layer is more stable compare with AC25-C layer and AC25-F layer. After the Standard truck leaving the instruments laying position, tension strain can be fully recovered in cement stabilized macadam layer, while AC25-F layer and AC25-C layer has markedly residual tensile strain. This is mainly due to stiffness of cement stabilized macadam is relatively large, and does not have the visco-elastic-plastic properties of asphalt concrete, presents mainly the flexibility before and after the load performance. Visco-plastic property is significant in AC25-F and AC25-C layers, tensile strain at the end layer cannot be fully recovered after the load cancellation. AC25-C is larger in module and higher stiffness than AC25-F, resulting in residual strain smaller than AC25-F.

5 Theoretical analysis

In China, current pavement design of asphalt concrete pavement for highway taken the pavement as a continuous static layer system, while pavement structure is not a simple elastic layer system, Asphalt concrete is a visco-elastic-plastic material, modulus and load response of grade stone depending on the loading on it, with elastic-plastic characteristics, subgrade is a typical nonlinear materials. All these lead to load response is not a simple elastic response of pavement.

This article takes our country's pavement design specification on the basis, using Abaqus calculation software to analyze the transient load response when the rear wheel reach the instrument laying position, and calculating linear and nonlinear loading response of the road .Model width takes 6m, depth takes 6.76m (assuming the subgrade depth to be 6m), 0.7MPa surface loads are taken on. Establishing model is shown in Fig. 11.

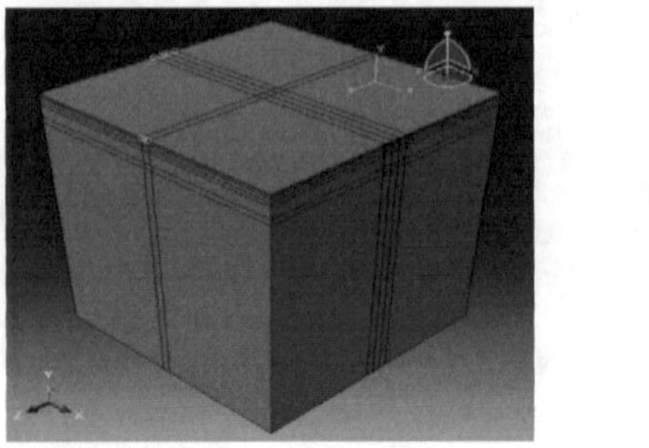

Figure 11. mechanical model.

5.1 Linear-elastic mechanical models

Traditional mechanical analysis take the pavement structure as a linear-elastic layer system, various material parameters are shown in Table 4.

Table 4. Material parameters.

	Modulus，MPa	Poisson's ratio
SMA-16	1644	0.3
AC-20	1426	0.3
AC-25F	1335	0.3
AC-25C	1685	0.3
cement stabilized base	2850	0.25
Graded gravel	325	0.35
Subgrade	50	0.35

Results of Linear elastic analysis are shown in Table 5.

Table 5. Results of Linear elastic analysis.

Type of field data	Calculating Result
Strain in bottom of AC-25F along the lane, με	23.08
Strain in bottom of AC-25C along the Lane, με	6.89
Strain in bottom of cement stabilized base along the lane, με	40.20
vertical stress in bottom of AC25-C	211.61
vertical stress top of cement stabilized base	122.22
vertical stress top of graded gravel layer	11.15

5.2 Non-linear layered system model

5.2.1 Mechanical model considering with the material nonlinearity

Mechanical characteristics for Asphalt concrete, graded gravel and subgrade material do not conform to Hooke law, should not be seen as a simple elastic material. Considering that too much material nonlinearity will lead to model non-convergence, this article only considers subgrade non-linear factor. Mohr Coulomb model, in Abaqus software's materials library, with fewer parameters, can be a good simulation of the soil and mechanical properties of porous materials, So it is taken Mohr Coulomb as subgrade material model, the cohesive forces take 2000Pa, internal friction angle and dilatancy angle respectively taking 20 degrees.

5.2.2 Considering contacting layers mechanics model

Assuming that contact surface will never vertically break away while stressing, and pass shear stress through friction in the horizontal direction. In order to guarantee the precision and ease of calculation, this paper processed below treatment to the model in modeling procedure.

(1): consider two surfaces in contact with the same grid, nodes one to one correspondence as far as possible.

(2): when specify mutual contact surfaces, big modulus as the primary surface, small modulus as from surface.

(3): slide mode takes as small slip

Results of nonlinear analysis are shown in Table 6.

Table 6. Calculated values of nonlinear models.

Type of field data	Subgrade non linear data	Contacting n onlinear Data
Strain in bottom of AC-25F along the lane, με	26.02	33.22
Strain in bottom of AC-25C along the lane, με	12.92	13.89
Strain in bottom of cement stabili zed base along the lane, με	35.03	37.20
vertical stress in bottom of AC25 -C	245.65	233.45
vertical stress top of cement stabilized base	134.94	144.70
vertical stress top of graded grave l layer	6.54	8.55

5.3 Results comparison

Correlation coefficient is a statistical indicator, used to reflect the close relationship between variable variables, can determine the linear dependence of two sets of variables. Results of the analysis model as shown in Table 7.

Table 7. Correlation coefficient between test data and calculated values.

	Testing Data	Linear elastic model	Subgrade nonlinear model	Contacting nonlinear model
Testing Data	1	0.898677	0.920612	0.919697

Can be seen from Table 7, three models are able to react loading response of pavement very good, Calculation results of subgrade non-linear model & non-linear contacting model is more similar to test result compare with linear elastic models, which simulate better actual loading response of pavement conditions, which is also consistent with actual working condition of the road.

6 Conclusions

(1) When the standard truck pass through instruments laying position, vertical stress produced by loads within the pavement structure is declining from top layer to bottom layer.

(2) Due to the decline characteristics when loads passed from top to bottom layer in pavement structure, as the depth increases, the standard truck "load effect radius" in different structural layers is diminishing

(3) Before and after loading, tensile strain at the end of the cement stabilized macadam layer can be fully recorded, AC25-C and AC25-F layers with greater residual tensile strain; Due to the low modulus and more significant visco-plastic than AC25-C, AC25-F gains more residual strain.

(4) Both linear elastic model and nonlinear model can well simulate the load response of road structure; Calculation results of subgrade non-linear model & non-linear contacting model is more similar to test result compare with linear elastic models, which is also consistent with actual working condition of road.

References

1. D. Zhang, *Asphalt and Asphalt Mixture*, Beijing: People's Communications Publishing House,:42–54, (1993)
2. X. Zhang, *Experimental Viscoelastic Theory* (Heilongjiang: Harbin Shipbuilding Engineering Institute Press, 1990: 20-60)
3. J. Yuan, X. Huang, Journal of Chang'an University (Natural Science Edition) **6**, (2007)
4. J. Mallela, K. P. George, Transportation Research Record **l448**, 92-99, (1995)
5. H. Luo, H. Zhu, X. Hao, Journal of China University of Geosciences **17**, 4, 349-354, (2006)
6. X. Deng, C. Li, Chinese Journal Of Geotechnical Engineering **4**, (2002)

Effect of Supplemental Signal on Start-Up Lost Time at Signalized Intersection

Yongfeng Ma and Jian Lu

Jiangsu Key Laboratory of Urban ITS, Southeast University, Si Pai Lou No.2, Nanjing 210096, China

Abstract. Supplemental signals have been widely used at intersections to reduce the impact of heavy vehicles. This study aims to measure the effects of supplemental signals on Start-Up Lost Time (SULT). A corrected method of calculating SULT considering heavy vehicles are presented. The method also introduce into basic saturation headways, locations of the heavy vehicles and the four types of headways. Field studies were implemented at 18 approaches of 6 intersections in Nanjing, China. The results show that the existence of heavy vehicles has significant influence on both of headways and SULT at signalized intersection. Supplemental signal is effective countermeasure to reduce the influence of heavy vehicles. The study will be helpful when measuring SULT at intersection with supplemental signal.

1 Introduction

As a critical node, signalized intersection plays an important role in the whole road network. There are much more vehicles driving into and crowded around signalized intersections, especially during the peak period, which lead to further poor performance of the whole road network. More and more heavy vehicles, containing buses or single unit trucks, also contribute to the phenomena from two aspects: (a) slow-start factor. Compared with cars, heavy vehicles start more slowly during green time. A heavy vehicle among the first few queuing vehicles will significantly increase the Start-Up Lost Time (SULT), which results in low traffic capacity; (b) block factor. Heavy vehicles will block views of the subsequent queuing vehicles (especially for cars) due to different height of vehicles, which will prevent related drivers from seeing and reacting to the change of the signals timely at the intersections. Then this slow reaction will also increase the SULT and lessen the traffic capacity. According to our survey, the proportion of heavy vehicles around signalized intersections has already reached 15% ~ 30%, which is 5 to 10 times more than the default proportion 3% assumed in the Highway Capacity Manual 2010 [1]. This high level of heavy vehicles makes an important impact on SULT and the traffic capacity. Therefore, how to reduce the influences of heavy vehicles at signalized intersection and the measurement of effectiveness of related countermeasures are very important and meaningful.

Many researchers are focusing on developing different countermeasures to improve the performance of signalized intersections. For the former factor, it is difficult to be eliminated, due to the heavy vehicles' design. But for the latter factor, much more countermeasures are put forward and one of which is to install the supplemental signals. In Manual on Uniform Traffic Control Devices for Streets and Highways (MUTCD, Edition 2009) [2], it is indicated that "Supplemental signal face is not a primary signal face but which is provided for a given approach or separate turning movement to enhance visibility or conspicuity". In recent years, to reduce the block factor at signalized intersections, supplemental signals have been increasingly deployed in urban areas in China. However, there are still very little publications or data to demonstrate the effectiveness of supplemental signals.

The calculation of SULT under different traffic environment is very important and also attracts many related researchers. Generally, the SULT can be derived by measuring the elapsed time for the first four vehicles, including the start-up response time (SRT) of the first vehicle, and adjust it with the average saturation headway [3]. The Highway Capacity Manual (HCM) is an important publication and keeps update of SULT and saturation headway in last twenty years. In edition 1994, the saturation flow rate is recommended as 1,800 pc/hr/ln, which corresponds to a saturation headway of 2 sec. In edition 1997 and 2000, the saturation flow rate is increased to 1,900 pc/hr/ln, which corresponds to a saturation headway of 1.895 sec [4]. And also, in edition 2000, SULT ranges from 1.0 sec to 2.0 sec based on typical observation. But the result is limited to the through movement. In the latest edition (edition 2010), the SULT is recommended as 2.0 sec [1]. It also presents that the elapsed times of the first four vehicles in a queue are usually to estimate SULT and the average headway of the latter vehicles in a queue are usually recommended as saturation headway. Some other researchers also measure SULT and saturation flow rate with none or less heavy vehicles through field study [5]-[8].

With the development of economy, more and more heavy vehicles are on urban arterials [9]. A survey about motor vehicle use between 1972 and 2003 also reports that heavy vehicles make up more than 10% and are still increasing [10]. Many publications also pay their attentions on the disadvantage and propose some countermeasures. In HCM (edition 2010), a coefficient of equivalent is introduced to differ the passenger car and heavy vehicles. More and more studies [11]-[15] are focusing on effects of heavy vehicles. Rafai Alyami et al. examined the effects of truck percentages on SULT at a signalized intersection [9]. A new model is proposed to calculate SULT under mixed traffic. According to the field observations, the SULT is generally higher when trucks are in a queue and reach the highest when trucks are among the first four vehicles of a queue.

The primary objective of this study is to evaluate the effects of supplemental signals on SULT when heavy vehicles exist in a queue. More specifically, the study includes the following tasks: (1) Measuring the impacts of heavy vehicles on SULT at signalized intersections; (2) Determining whether supplemental signals can reduce the block effects of heavy vehicles at signalized intersections.

2 Methodology

Traditional calculation of SULT involves saturation headways. However, if we use saturation headways to calculate the SULT, it is often negative when heavy vehicles exist in the queue. In HCM (edition 2010), any headway affected by a heavy vehicle is not ideal, in other words, only headways that occur before the first vehicle's arrival can be treated as "ideal". However, under real traffic and existing method of data collection, many other factors affect the headways and their measurement, such as drivers, vehicles and traffic environment. Thus, in this study, the observed saturation headway is defined as basic saturation headways.

Generally, the first four headways of each cycle are discarded, which are taken as unstable. Thus, for each cycle, headways from the 5th vehicle to the one followed by a heavy vehicle are included to calculate the basic saturation headway. The basic saturation headway can be calculated using the following equation:

$$h = \frac{\sum_{i=1}^{n}\sum_{j=5}^{k_i} h_j'}{\sum_{i=1}^{n}(k_i - 4)} \tag{1}$$

where,

h – the basic saturation headway;

h_j' –the headway of the jth vehicle before the heavy one;

k_i–the number of vehicles in each cycle which are involved in this calculation; and

n – the total number of cycles observed.

$$l_s = \frac{\sum_{i=1}^{n'}\sum_{j=1}^{4}(h_j - h)}{n'} \tag{2}$$

where,

l_s - SULT;

h_j– the headway of the jth vehicle in the queue of ach cycle; and

n – the number of cycles.

In order to distinguish the influences brought by the different locations of heavy vehicles in a queue, all observed signal cycles are divided into four categories before the calculation of SULT. The classification is shown as below:

Case 1: one or more heavy vehicles are among the first four vehicles at the red light, but none among the latter vehicles;

Case 2: one or more heavy vehicles are among the 5th vehicle and followed vehicles at the red light, but none among the first four vehicles;

Case 3: one or more heavy vehicles are both among the first four vehicles and the latter vehicles; and

Case 4: no heavy vehicles were among a queue at the red light.

SULT for each case is calculated separately and then multiplied by their corresponding percentage of each case to obtain an overall SULT.

$$l_s' = \sum_{i=1}^{4} l_i \times p_i \tag{3}$$

where,

l_s'– the corrected SULT;

l_i – the SULT of each case; and

p_i – the percent of each case in a period of time.

3 Data collection

There are two basic kinds of deployment for supplemental signals. Scenario 1: supplemental signals for left-turning vehicles are installed at the left side of the approach, and those for through and right-turning vehicles are installed above the approach. Scenario 2: supplemental signals for left-turning vehicles are installed at the left side of the approach and those for right-turning and through vehicles are installed at the right side of the approach.

Field data collection was conducted at 18 approaches at 6 signalized intersections. The selected sites are all located on arterials in Nanjing, China. All of the intersections are under count-down traffic signal control. The traffic environments of these intersections are similar. For each intersection, at least two approaches (one for left-turning and one for through) are equipped with supplemental signals, while only primary signals for other approaches. Since vehicles can turn right during the red phase at the intersections, the study only focused on left-turning and through vehicles. The following principles were used in selecting sites and field study:

- The selected sites should have enough left-turning and through vehicles to queue up on their lanes;
- There are no parking and bus stops adjacent to the intersection (about 250 ft. upstream or downstream of the stop line);
- Each approach of the intersection has exclusive lanes for left-turning and through vehicles;

- Traffic conditions are steady during field study;
- No major pedestrian and non-motor activity; and
- Cycles with very slowly moving vehicles for some unusual condition (such as traffic congestion and traffic incident) were not included.

A typical layout of selected signalized intersections is illustrated in Fig. 1. Two video cameras were deployed on the top of a building near the intersection to record traffic. Data were collected during the AM peak hours (7:00~9:30a.m.).

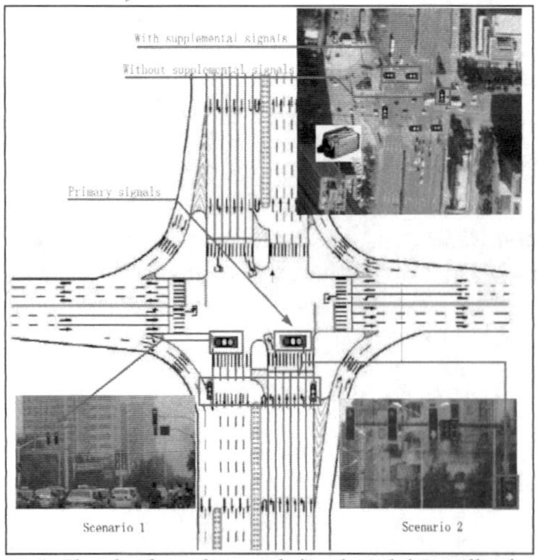

Figure 1. Sketch of supplemental signals and data collection.

Headways are generally measured at STOP line of approaches, and the reference point is generally the crossing of the rear axle of a vehicle over a STOP line. The first headway is recommended as the time from the initiation of the green indication to the crossing of the rear axle of the first vehicle over the STOP line. Only those vehicles that have come to a complete stop before the initiation of green indication are included. To better differ the heavy vehicles and passenger cars, four different types of headways were defined in this study [16], as follows:

- Type 1: (H-NH) the headways between a heavy vehicle and a car;
- Type 2: (H-H) the headways between two heavy vehicles;
- Type 3: (NH-NH) the headways between two cars;
- Type 4: (NH-H) the headways between a car and a heavy vehicle.

In total, 419 cycles for through movement and 285 cycles for left-turning movement were obtained. When the condition is "with supplemental signals", the queuing vehicles are affected only by the slow-start factor of heavy vehicles. Otherwise, the queuing vehicles are influenced by the slow-start factor and block factor of heavy vehicles.

4 Data analysis

4.1 Saturation headways

Since the first four headways in each cycle are affected by SULT, only the headways after the fourth vehicle were measured. A total of 2775 headways were obtained, including 970 left-turning headways and 1805 through headways. The summary for headway data is presented in Table 2 and Fig. 2. The mean saturation headways of left-turning vehicles vary from 1.30s to 2.24s with supplemental signals, and from 1.42s to 2.42s without supplemental signals. The mean saturation headways of through vehicles vary from 1.41s to 2.13s with supplemental signals, and from 1.44s to 2.53s without supplemental signals.

T-test was applied to identify whether saturation headways with supplemental signals were significantly different from that without supplemental signals. Most results, as shown in Table 2 were found to be significantly different statistically. However, there was no significantly difference for left turning and through vehicles in type 3, and through vehicles in type 4. It may be caused by the following two reasons: (a) cars do not have block effects on their following vehicles. So there is little difference in saturation headways for vehicles in type 3 and type 4, even after installing supplemental signals; (b) insufficient data were obtained when the approaches were controlled only by primary signals. The huge difference in number of samples may influent on the results of T-test. From Table 2, the mean saturation headways were decreased after installing supplemental signals, except for left turning vehicles in type 4. Furthermore, the mean saturation headways for vehicles in type 1 and type 3 are generally less than that for vehicles in type 2 and type 4, which may be affected by the slow-start factor of heavy vehicles.

Table 1. Summary statistics of saturation headways (seconds).

Type	Direction	Signal	N	Mean	Min.	Max.	Std. Dev.	p-value
1	L	with	55	1.30	0.84	2.24	0.32	0.007
		N/A	66	1.52	0.80	3.48	0.55	
	T	with	126	1.41	0.74	2.32	0.38	0.044
		N/A	37	1.55	1.15	2.48	0.32	
2	L	with	25	2.00	1.01	2.79	0.47	0.026
		N/A	25	2.42	1.57	3.40	0.50	
	T	with	64	2.13	0.80	3.28	0.55	0.040
		N/A	30	2.53	1.41	3.89	0.76	
3	L	with	196	1.36	0.66	2.78	0.34	0.062
		N/A	484	1.42	0.73	2.93	0.36	
	T	with	1142	1.42	0.58	3.20	0.39	0.452
		N/A	281	1.44	0.59	3.17	0.41	
4	L	with	52	2.24	0.93	3.31	0.50	0.001
		N/A	67	1.89	0.89	3.15	0.60	
	T	with	93	2.13	0.66	3.43	0.59	0.811
		N/A	32	2.10	1.02	3.30	0.52	

4.2 Start-up lost time (SULT)

The cycles only with cars in type 3 are selected to calculate basic saturation headways. The results of basic saturation headways and saturation headways are presented in Fig. 3. As shown in Fig. 3, the basic saturation headways are 1.36s and 1.42s for left turning and through vehicles with supplemental signals respectively. However, for left turning and through vehicles without supplemental signals, the basic saturation headways, 1.42s and 1.44s respectively, are little bigger than that with supplemental signals. The saturation headways with supplemental signals are 1.34s for left turning vehicles and 1.39s for through vehicles, while without supplemental signals are 1.42s for left turning vehicles and 1.46s for through vehicles. After installing supplemental signals, saturation headways and basic saturation headways both decreased whatever for left turning or through vehicles. Generally, the basic saturation headways are little bigger than saturation headways.

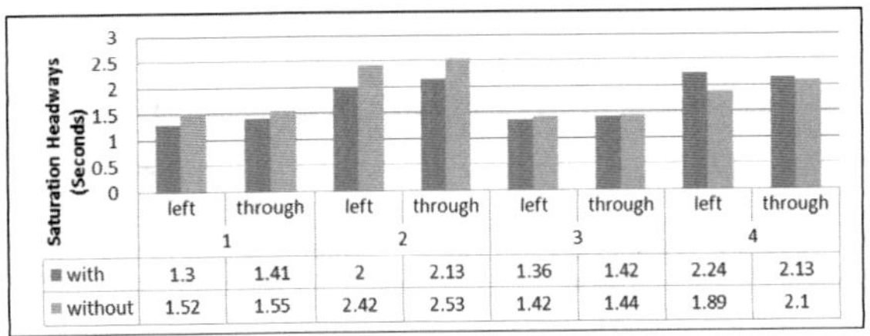

Figure 2. Comparison of the saturation headways.

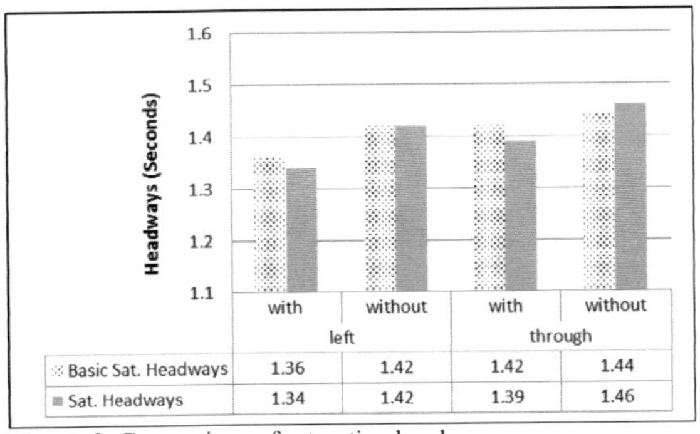

Figure 3. Comparison of saturation headways.

Summary of cycle numbers and means of SULT are presented in Table 3. The SULTs in case 1 and case 3 are larger than that in case 2 and case 4 for both movements. Meanwhile, the SULTs in all cases with supplemental signals are smaller than that without supplemental signals. T-test was used to identify whether the mean values of SULT with supplemental signals were significantly different from that without supplemental signals. With a 95% confidence level, the results showed that SULTs with supplemental signals were significantly different from those without supplemental signals statistically only in case 1 and case 3. In case 2 and case 4, the means of SULTs for left turning and through vehicles also decreased after installing supplemental signals, even though the results of t-test were not significant. In general, the supplemental signals were effective on decreasing SULT, especially when heavy vehicles existed in the first four vehicles.

Frequencies of the four cases were presented in Table 3. With the corrected method of SULT (Equation (3)) and collected data, SULTs at intersections with and without supplemental signals were calculated. For left turning vehicles, SULT with and without supplemental signals were 1.45 seconds and 1.60 seconds respectively. And for through vehicles, SULT with and without supplemental signals were 1.69 seconds and 2.10 seconds respectively.

Table 2. Summary of cycle number and mean values of SULT

No.	left turning						
	with			without			p-value
	N.	Mean	Std.	N	Mean	Std.	
case1	21	1.94	0.671	19	2.50	0.963	0.044
case2	31	1.08	1.057	36	1.16	0.746	0.700
case3	26	1.82	0.971	32	2.44	1.122	0.032
case4	36	1.23	0.883	84	1.27	1.142	0.056
sum	114	—	—	171	—	—	—
No.	through						
	with			without			p-value
	N	Mean	Std.	N	Mean	Std.	
case1	47	2.19	1.572	32	2.95	1.267	0.025
case2	53	1.08	1.090	24	1.58	1.181	0.069
case3	35	2.19	0.945	19	2.77	1.092	0.047
case4	146	1.64	1.190	63	1.67	1.184	0.879
sum	281	—	—	138	—	—	—

5 Conclusions

The study measures the impacts of heavy vehicles and supplemental signals on SULT for both left-turning and through vehicles at signalized intersections. Saturation headways and SULT with supplemental signals are analysed. Based on data analysis, the following conclusions can be drawn:

1) The most saturation headways with supplemental signals are smaller than that without supplemental signals for both movements. The results of T-test also show that saturation headways with supplemental signals are significantly different from that without supplemental signals. The supplemental signals are effective to reduce the block of heavy vehicles. Meanwhile, it is not reasonable to ignore the block of heavy vehicles in traditional calculation of SULT.

2) At intersections with supplemental signals, SULTs are less than that at intersections only with primary signals, especially when heavy vehicles exist in the first four vehicles. The results of T-test also show that SULT with supplemental signals are significantly different from that without supplemental signals.

The study is only implemented in Nanjing, China. More samples from other areas should be included to test the conclusions. And also, more data should be collected to measure correction coefficients. In the study, only locations of heavy vehicles are included. If the number or proportion of heavy vehicles is also introduced, a better understand on effects of heavy vehicles may be obtained.

Acknowledgement

This study is funded by National Key Technology Support Program (No. 2014BAG01B06). The authors also would like to thank the graduate research assistants from the School of Transportation, Southeast University for their assistance in field study.

References

1. Highway Capacity Manual. TRB, National Research Council, Washington, D.C. (2010)
2. Manual on Uniform Traffic Control Devices for Streets and Highways. U.S.Department of Transportation Federal Highway Administration. (2009)
3. H. Li, P.D. Prevedouros. Detailed observations of saturation headways and start-up lost times. Transportation Research Record (TRR)". TRB, National Research Council, Washington, D.C., (2002)
4. Highway Capacity Manual. TRB, National Research Council, Washington, D.C., (2000)
5. J. Bonneson, B. Nevers, J. Zegeer, T. Nguyen and T. Fong. Guidelines for quantifying the influence of area type and other factors on saturation flow rate. Technical Report. Texas Transportation Institute, College Station, Texas. (2005)
6. J.W. McMahon, J.P. Krane and A.P. Federico. Saturation flow rates by facility type. ITE Journal, **67**, 46-50 (1997)
7. T. Limanond, P. Prabjabok, Kr. Tippayawong. Exploring impacts of countdown timers on traffic operations and driver behavior at a signalized intersection in Bangkok. Transport Policy. **17**, 420–427. (2010)
8. J. Joseph, G.L. Chang. Saturation flow rates and maximum critical lane volumes for planning applications in maryland. Journal of Transportation Engineering. **131**(12), 2066-70. (2005)
9. R. Alyami, B.N. Janson. Effects of trucks on start-up lost time at signalized intersections. Transportation Research Board Annual Meeting. Washington, D.C. (2010)
10. Survey of Motor Vehicle Use ABS (1972-2003). Series 9208.0, Australian Bureau of Statistics, Canberra, ACT. (2004)
11. X. Zhao, G. Ren. Analysis of vehicle headway characters at a signalized intersection. Journal of Transportation Engineering and Information. **8**, 103-108 (2010)
12. X. Yang, B. Zhuang, K. Li. Analysis of saturation flow rate and delay at signalized intersection. Journal of Tongji University (Natural Science). **34**, 738-743 (2006)
13. S. Tsao, S. Chu. Adjustment factors for heavy vehicles at signalized intersections. Journal of Transportation Engineering, ASCE. **121**(2), 150-157, (1995)
14. K. Kockelman, R. Shabih. Effect of vehicle type on capacity of signalized intersections. Journal of Transportation Engineering, ASCE. Vol. **126**(6), 506-512 (2000)
15. E. Ramsay, J. Bunker, R. Troutbeck. Signalized intersection capacity reduction of trucks, In the Proceedings of the Fourth International Conference on Traffic and Transportation Studies. Science Press: Beijing, China. (2004)
16. P. Liu, J. Wan, W. Wang, Z. Li. Evaluating the impacts of an unconventional outside left-turn lane design on traffic operations at signalized intersections. Transportation Research Record (TRR). TRB, National Research Council, Washington, D.C. (2011)

Intrinsic Safety Evaluation for Urban Road Traffic Based on Information Entropy and Triangular Fuzzy Number

Gao Tingting[1, 2] Wang Wuhong[1] and Xing Zhongyi[1]

[1]*Beijing Institute of Technology, School of Mechanical Engineering, 100081, Beijing, China*
[2]*Tian Jin University of Technology and education, School of Automobile and Transportation line, 300222, Tianjin, China*

Abstract. Road traffic safety is a major problem, especially in low-and middle-income countries. This study attempted to evaluate traffic safety based on the concept of intrinsic safety, which can provide theoretical support for the road traffic management departments. The evaluation weight calculation about the pre-weight of the second level evaluation index based on the information entropy, and then calculates the final weight by combing qualitative and quantitative methods. The weight of the third level evaluation index using the triangular fuzzy number method. In order to verify the effectiveness and feasibility of the method via a case study. by computing the proportions of evaluation indexes based on the model, we found that the ranking is intrinsic human>intrinsic road and environment>intrinsic vehicle. However, the gap about the proportion of the three is small. Furthermore, the sub factors ranking influencing factors as follows: the human's alertness = road and the environment's completeness>vehicle's technical>human's self-consciousness >natural conditions >vehicle's maintenance. Given the objectivity of the model in the paper, especially using the entropy weight method to calculate the weight of the index factors. The result of the evaluation for the model is similar with current situation of road traffic safety, thus, it can be taken as a valuable reference to promoting the traffic safety. So it can provide theoretical basis for urban road transport departments to formulate relevant policies.

1 Introduction

As the degree of motorization increase, improving traffic safety has become a main goal of transportation researchers across the world. Traditionally, traffic safety evaluations have relied mostly on crashes statistics as the main data source, and it is an important method to evaluate the traffic safety. To date, there are a large number of domestic and foreign experts or scholars who have been studying deeply on traffic safety, but they worked in most cases on their own specific road safety problems, to a large extent because the socioeconomic conditions, the motorization levels, and the road safety experiences are different from country to country and from region to region. By contrasting China and Japan in the Asian countries in 2012, it can be found that the death toll from traffic accidents was respectively 59997 and 4411, the mortality rate per ten thousand cars was 2.5% and 0.58%, and the mortality rate per 100000 people was 4.45% and 3.45%. Therefore, it is necessary to evaluate the road traffic safety. To date, there are a large number of domestic and foreign experts or scholars who have been studying extensively on traffic safety: R • J • Smeed in University of London established a Smeed model about ownership of motor vehicles based on population and the number of road deaths in 1949; Solomon used vehicle speed to evaluate road traffic safety in 1964; Grazawa established an urban Traffic Safety Evaluation System with a relatively wide layered structure. Jake Kononov and Bryan Allery proposed SPFs (Safety Performance Functions), and subdivided the road traffic safety level into Loss Ⅰ, Loss Ⅱ, Loss Ⅲ, Loss Ⅳ four levels. Road traffic safety evaluation of China was relatively new, Liu Shiqi built a highway safety evaluation model Lu in 2003 and Zhang in 2004 evaluated the study transportation system from the perspective of its security; Shao established a road traffic safety evaluation system more comprehensively in 2004. The paper combined with information entropy to determine the index weight evaluation method based on triangular fuzzy number, and applies it to the nature of traffic safety evaluation, reduce the interference of subjective factors too much, Different from the traditional the third indexes, which focus on the characteristics of the second indexes, the number of the third indexes in the study is a few but its scope is generally fairly comprehensive. the essence of improving road safety evaluation index that affect the ranking on the level of road traffic safety and provide technical support and theoretical support for the positive [1]-[6].

2 Road intrinsic safety evaluation index

Evaluation indexes are the key of the evaluation model, and the premise of obtaining the real objective results. To choose the evaluation indexes correctly, the urban traffic system has to be analyzed comprehensively. The selected indexes should be able to reflect the impaction of various factors maximal degree, and the selected indexes should be able to determine the safety of transportation system maximal degree intrinsic.

Road traffic system is composed of human, vehicles, roads and the environment. Any one factor's disorder of transportation system can lead to accidents. Therefore, the indexes should meet the following principles: comprehensiveness and level, initiative and passive, and static and dynamic. Evaluation based on intrinsic safety by questionnaires shown in Fig. 2. The evaluation system of road traffic safety as the target layer, the human, vehicle and road-layer model as the rule layer, the consciousness and alertness of human, the technical and maintenance of vehicles, the completeness and natural condition of roads and environment as decision layer.

2.1 Human

Human is an active factor that has a great impact on road traffic safety. Whether drivers or ride personnel comply with traffic regulations and, have high safety consciousness constitute consciousness evaluation index; whether the driver's operation is skilled, and whether there are errors constitute alertness evaluation index.

2.2 Vehicle

Vehicle is a passive factor that also affects road traffic safety. Human - vehicle - environment is a closed loop control system that has a feedback loop. The vehicle plays a major role in the feedback loop. Whether the vehicle's technical performance is in good condition, and the maintenance is proper directly determine whether the transportation system is safe or not, and therefore vehicle design based on intrinsic safety is an important aspect regarding road safety.

2.3 Road & environment

Roads & environment is a subsystem of the transportation system as about the people - vehicle - environment, which is a place to accommodate people and vehicles. It is a necessary condition for the protection of people and cars to work. Whether road marking and labeling are clear and whether signal control facilities are complete constitute the completeness index. In addition, the bad natural conditions and the difference between day and night have a direct impact on road traffic safety.

The evaluation system compared with the previous evaluation system, are based on road transport system of humane, vehicles and roads, but there are several advantages: Use either SI (MKS) or CGS as primary units. (SI units are encouraged.) English units may be used as secondary units (in parentheses). An exception would be the use of English units as identifiers in trade, such as "3.5-inch disk drive".

1) Though the three-grade index has a small number, but covers the whole;
2) The three-grade indexes are easy to calculate standardized evaluation matrix;
3) The three-grade indexes are based on intrinsic safety.

3 Model

The model is a method based on fuzzy mathematics theory, and it is a more practical approach, and it can evaluate for the influence factors on multivariate fuzzy evaluation, especially for harder to quantify, and have strong fuzzy problem.

3.1 Standard matrix

General the dimension and magnitude of the model will be divided into two categories, one is the bigger the more excellent, and the other is the smaller the more excellent. Corresponding to these two types of matrix are shown as formula (1) and formula (2): [7]-[9]

The bigger the more excellent:

$$r_{ij} = \frac{x_{ij} - \min(x_j)}{\max(x_j) - \min(x_j)} \tag{1}$$

The smaller the more excellent:

$$r_{ij} = \frac{\max(x_j) - x_{ij}}{\max(x_j) - \min(x_j)} \tag{2}$$

where: x_{ij}——score of the j index about i index;

$\max(x_j)$——the max of j column about score matrix;

$\min(x_j)$——the min of j column about score matrix.

3.2 The weight of the index w_i

Determining the weight of the corresponding index is the key process of the evaluation, compared AHP method has a strong subjective to entropy method, it can better reflect the evaluation index weights. Weight refers to the standard is as follow formula (3) and (4): [10]-[11].

$$H_j = -\frac{\sum_{i=1}^{m} f_{ij} \ln f_{ij}}{\ln m} \quad (i = 1,2,\ldots,m;\ j = 1,2,\ldots n) \tag{3}$$

$$f_{ij} = r_{ij} / \sum_{i=1}^{m} r_{ij} \qquad w_i = \frac{(1 - H_j)}{(n - \sum_{i=1}^{n} H_j)} \tag{4}$$

where: r_{ij}——judgment matrix after specification;

m——the number of rows of the matrix;

n——the number of columns of the matrix

3.3 Triangular fuzzy number model

Triangular fuzzy number is a special form of trapezoid fuzzy number, Fig. 1 is a function of the triangular fuzzy number, membership function is as shown in formula (5).

$$f=\begin{cases}0 & x\in(-\infty,l] \\ \dfrac{x}{m-l}-\dfrac{l}{m-l} & x\in(l,m] \\ \dfrac{x}{m-n}-\dfrac{n}{m-n} & x\in(m,n] \\ 0 & x\in(n,+\infty)\end{cases} \quad (5)$$

With respect to the addition, the multiplication and the scalar-multiplication of triangular fuzzy number in accordance with classical mathematical calculation rule, evaluation factors of triangular fuzzy number described more objective than traditional scoring assignment method. Triangular fuzzy number and scoring assignment respectively express road traffic safety evaluation indexes showed in Table 1 [12], [13].

Table 1. road traffic safety evaluation indexes based on triangular fuzzy number

road traffic safety evaluation level	**very poor**	**Poor**	**ordinary**	**good**	**very good**
fuzzy number (l', m', n')	$(0,0,\frac{1}{4})$	$(0,\frac{1}{4},\frac{1}{2})$	$(\frac{1}{4},\frac{1}{2},\frac{3}{4})$	$(\frac{1}{2},\frac{3}{4},1)$	$(\frac{3}{4},1,1)$
score (total point)	$[0\sim5.5)$	$[5.5\sim6.5)$	$[6.5\sim7.5)$	$[7.5\sim8.5)$	$[8.5\sim10]$

Table 2. Tianjin city road traffic safety quantization indexes.

		score (tota l10point)	r_{ij}	W_i	**triangular fuzzy number**	p_i	$I(p_i)$	w_i
human	Consciousness	6.5	0	0.2589	$(\frac{1}{4},\frac{1}{2},\frac{3}{4})$	(0.0647,0.1295,0.1942)	0.1295	0.16
	alertness	7.8333	1		$(\frac{1}{2},\frac{3}{4},1)$	(0.1295,0.1942,0.2589)	0.1942	0.24
vehicle	technical	7.5	0.8571	0.2248	$(\frac{1}{2},\frac{3}{4},1)$	(0.1124,0.1686,0.2248)	0.1686	0.21
	Maintenance	5.8333	0.1428		$(0,\frac{1}{4},\frac{1}{2})$	(0,0.0562,0.1124)	0.0562	0.07
road & environment	Completeness	7.6667	1	0.2589	$(\frac{1}{2},\frac{3}{4},1)$	(0.1295,0.1942,0.2589)	0.1942	0.24
	natural condition	5.5	0		$(0,\frac{1}{4},\frac{1}{2})$	(0,0.0647,0.1295)	0.0647	0.08

Index weight W_i were calculated by the entropy weight, and multiplied by the fuzzy number for each index, obtained the fuzzy number of the pre-sorted by the expected formula

$$I(p_i)=\frac{W_i\cdot(l'+2m'+n')}{4} \quad (6)$$

Final weight of each evaluation factors w_i, can be sorted

$$w_i=\frac{I(p)}{\sum I(p)} \quad (7)$$

4 Case study

Take Tianjin as an example, we study on the impact of road traffic safety level of the essence of man, nature and he nature of road vehicles and the environment by the use of a questionnaire, we distributed 100 questionnaires were returned 95 valid questionnaires, obtained the road intrinsic safety score matrix A as follows:

$$A=\begin{pmatrix}6.5 & 7.8333 \\ 7.5 & 5.8333 \\ 7.6667 & 1\end{pmatrix}$$

And obtained standard matrix by The bigger the more excellent:

$$R=\begin{pmatrix}0 & 1 \\ 0.8571 & 0.1428 \\ 1 & 0\end{pmatrix}$$

We obtained the data by using equation (3) - (6) as shown in Table 2.

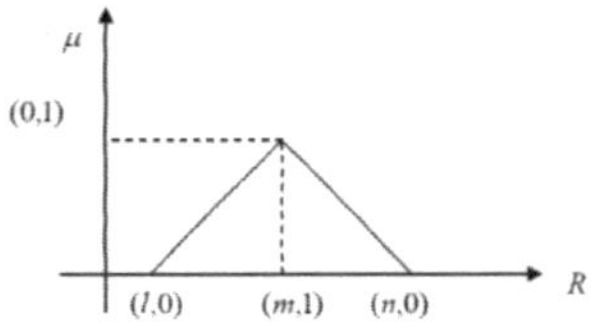

Figure 1. Function of the triangular fuzzy number.

The results show that, the intrinsic factors of urban road traffic safety, the human factor occupies 40%, the road and environment holds 32%, the vehicle holds 28%, the index order of factors: intrinsic human> intrinsic road and Environment> intrinsic vehicle, However, the difference of the proportion is little; and the proportion of sub-factors: human's alertness and completeness of the roads and environment are equally important, occupy 24%, the vehicle technical occupies 21%, the human consciousness occupies 16 %, natural conditions occupies 8%, vehicle maintenance occupies 7% .the order of sub-factors: human's alertness = road and environment completeness> vehicle technical> human consciousness> natural conditions> vehicle maintenance .

5 Conclusion

The main purpose of the paper is to analyze the evaluation indexes of impacting urban road traffic intrinsic safety, the paper calculated the index weight by using the information entropy, more objective than the conventional AHP method, to sort based on triangular fuzzy number method, more comprehensive than in the past, the calculated result may provide the basis for urban road transport sector. The main conclusions are as follows:

1) model was the impact of urban road traffic safety is the most essential factor in who holds 40% share, and the city that reason about 80% of road accidents are caused by humans is consistent;
2) The human's alertness and completeness of roads are the two most important factors that affect the intrinsic road traffic safety, occupy 24%, according to statistics, traffic accidents, especially in large-scale accidents, mainly due to the drivers' misuse, unresponsive and road transport facilities incomplete.

There are some limitations in the paper; the evaluation assignment was using the traditional questionnaire method, so questionnaire setting and its coverage problem will affect the value of the index, so this is a serious problem next paper.

Acknowledgment

This research has been financed by The National Natural Science Fund (NO.51408417), Tianjin University Science and Technology Development Fund (NO.20140917)

References

1. Andreessen D. Linking Deaths with Vehicles and Population [J]. Traffic Engineering and control, , **26**;547-549, Nov. (1985)
2. Nain, Rental. Road Safety Model Some Fundamental Ideas: Transportation Research Record, National Research Council, Washington,D.C,1963,48-57
3. Ghazwan, al-Haji. Road Safety Development Index(RSDI)[D]. Sweden: Linkoping University, (2003)
4. Jake Kononov, Bryan K.Alley. Explicit Consideration of safety in transportation planning and project scooping[C]. Transportation Research Board 2004 Annual Meeting, **10**:201-223, (2004)
5. Shao Zufeng. comprehensive evaluation of city road traffic safety[J]. Scholar Research, (2004): 22-24
6. Yu Rende, Shi Peng. The research on road traffic security evaluation based on the theory of fuzzy[J].Mathematics in Practice and Theory, **38** (7):109-116, (2008)
7. Chen Hong, Zhou Jibiao. Safety evaluation indexes and method for traffic environment of highway tunnels [J].Journal of Chang'an University, **33**(4):54-61, (2013).
8. Zheng Rui,Yang Zhenhong. Research on fuzzy comprehensive evaluation system of tailings based on entropy technology [J]. Journal of Science and Technology, **7**(6):107-111, (2011)
9. Zhou Wei, Li Xiaojing. A comprehensive evaluation method based on information entropy [J]. Science Technology and Engineering, **10**(23): 5839-5843, 2010
10. Qin Chen, Chen Xiaofang. Reaearch on safety Three-level evaluation of tailings reservoir based on FAHP-IE method [J]. Control Engineerin of China, **21**(6): 995-1000, (2014)
11. Liu Honghong, Wang Ni. Fuzzy assessment of city ecosystem health based on entropy weight method [J]. Engineering Journal of Wuhan University, **47**(6):755-759, (2014)
12. Metin Dadeviren. Developing a fuzzy analytic hierarchy process(AHP) model for behavior-based safety management[J]. Information Science, (178):1717-1733, (2008)
13. Si Shubin，Sun Shudong. A Weighted Analysis of an Integrated Logistics Support Based on Triangular Fuzzy Number [J]. Journal of Northwestern Polytechnical University, **22**(6):689-694, (2004)

DOI: 10.1051/matecconf/20168102011

Towards an Abnormal Bridge Location Identification Method Based on Novelty Detection Technique

Xiao-li Lu and Min-jie Wang

Wuhan University of Engineering Science, 432700 Wuhan, China

Abstract. The bridge structure abnormality recognition is one of the key steps of its health assessment. Using novelty detection technique based on BP neural network, the method to identify and locate abnormal bridge status was presented. It uses non-training-data in the original sample data to generate novelty indicator and determines threshold. If the difference between detection status indicator and normal value is larger than the threshold, the structure status is determined changed. The method adapts stepwise partition identification method. The method firstly determines damage position within a small range and then analyzes sensor data in detail, so as to locate specific position. The measured data on T beam model verifies the method can accurately carry out status identification and locate cracking position under cracking load conditions.

1 Introduction

Bridge health assessment is an effective method to assess health indicators as structure reliability and durability from processing data obtained from operational structure. The system can monitor operation of bridge scientifically, accurately and timely with information process and data fusion methods, so that the remaining life of key components can be estimated and its health can be objectively and quantitatively evaluated, which is the core of bridge monitoring system.

In order to ensure safe operation and provide technique basis for bridge design, load limit or demolition and reconstruction, it is necessary to carry out capacity test [1] and the fatigue test [2]. However, the existing assessment methods on bridge status abnormality diagnosis and health evaluation mostly rely on specific structure model, specific incentive and single evaluation indicator. Meanwhile, the researches on fatigue life of key components mostly use vehicle load simulating and fatigue stress spectrum estimating methods. In the actual bridge operation, the constantly changing bridge structure parameters lead to difficulty of establishing bridge model. It is difficult to achieve ideal result of abnormal diagnosis, endurance determination and health assessment because of unknown environmental load excitation. In the T beam cracking test, the determination of model crack is always made after comprehensive analysis of data obtained from test [3-4], while the judgment on cracking location commonly uses manual inspection method. Limited by test conditions, complicated sensor node distribution and small crack, it is difficult to determine and locate model crack with eye in the test. It is of practical importance to apply novelty detection technique based on BP neural network to T beam cracking test. The paper is organized as follows: Section 2 presents novelty detection technique based on BP network; Section 3 applies the proposed detection method to T beam cracking test; Section 4 concludes our work.

2 Novelty detection technique based on BP neural network

2.1 Novelty detection technique

The novelty detection is to identify new modes with obvious difference to existing known modes and to extract new, abnormal or unfamiliar features from input data [5]. The novelty detection techniques are mostly implemented with neural network. Using actually measured data under health status, establish mode of structural health status. If the new structural mode has significant difference compared with established mode, the structure is diagnosed as abnormal and deviated from health status. As the neural network has predominant advantages in the pattern recognition filed, the novelty detection can be achieved with neural network as shown in Fig. 1.

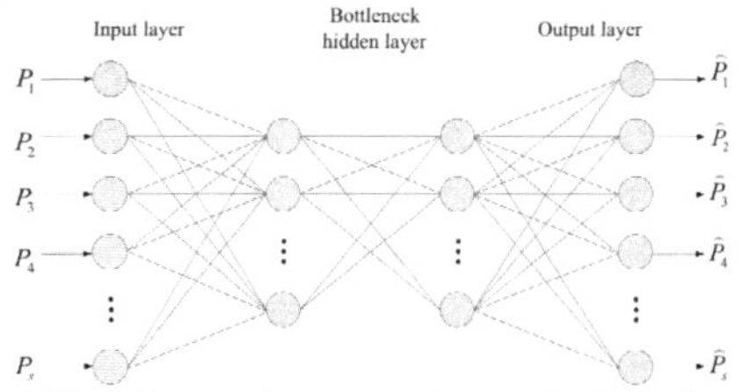

Figure 1. Neural network structure for novelty detection.

As seen from Fig. 1, the network is designed as a multi-layer network with bottleneck hidden layer. The number of input layer is equal to that of output layer. The unit number of bottleneck hidden layer is the smallest to force network learning dominant features of the input vector. The network can be completed with feed-forward BP neural network.

In case of network training, the actual measurement data under health status in the normal operation acts as the input vector and output target. With training, the neural network learns domain feature of input vector to establish mode under healthy structure status. In case of abnormal structure, as the structure mode has changed, the output vector will significantly deviate from input vector, thus status damage identification is achieved.

As the bridge model is complicated and prone to change, the general status identification result completely depends on model establishment. The novelty detection technique can effectively identify damage. The method only needs a certain amount of measure data but not completely depends on numerical model. It can also timely determine damage status to reduce alarm missing, which is of more practical value.

2.2 Detection indicator and threshold settings

In order to determine whether two modes have changed, it needs an indicator to determine the difference between input vector and output vector of BP neural network, which can be represented by a distance function. The paper regards the secondary norm of difference between input vector and output vector as the novelty indication [5].

The measured data under structural normal status is used as input vector for BP neural network training. After network training, the remaining groups of normal data is input into prepared neural network once time as input vector to produce corresponding output vectors. Based on (1) [5], the novelty indicator λ of structural normal status can be arrived.

$$\lambda = \|\hat{P} - P\| = \left(\sum_{i=1}^{N}(\hat{P}_i - P_i)^2\right)^{\frac{1}{2}} \tag{1}$$

where, P is the input vector of network; $\hat{P}$ is the output vector produced by vector P; N is the dimension of input vector; $\hat{P}_i$ and P_i are the i-the element of vector P and $\hat{P}$ respectively. in order to definitely determine abnormal phenomenon, select threshold as the determination criteria as shown in (2):

$$\delta_\lambda = \bar{\lambda} + 4\sigma_\lambda \tag{2}$$

where, δ_λ is the threshold to determine abnormal data. The $\bar{\lambda}$ and σ_λ are mean and standard difference of multiple normal input vector novelty indicators after neural network training.

The former definition of novelty indicator and threshold all use the indicators in the training phase [6], when the computation of novelty indicator uses input and output vector of training. The flow for novelty indicator and threshold setting is shown in Fig. 2(a). After neural network training, the input vector may infinitely approximate to output vector. In such circumstance, the obtained novelty indicator and threshold may be relatively small, which may lead to false alarm.

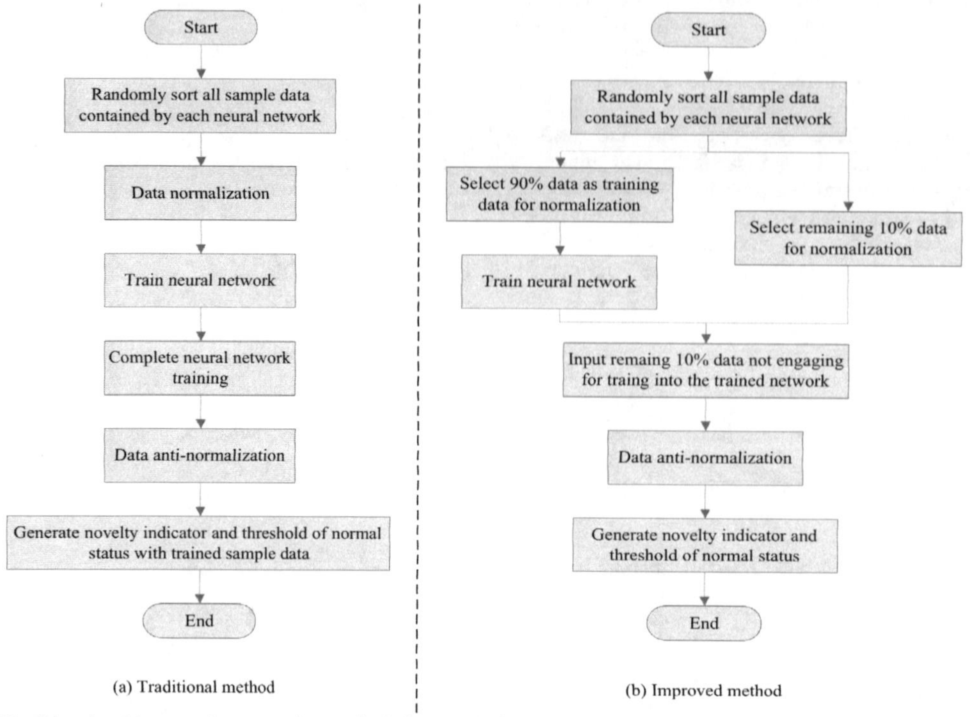

Figure 2. Traditional and improved computation methods for novelty indicator and threshold.

The paper uses non-training-data in the original sample data to generate novelty indicator as the normal indicator $\bar{\lambda}$ and determines threshold δ_{λ}. The improved flow is shown in Fig. 2(b). The obtained novelty indicator and threshold can accurately determine structural status.

In the detection phase, the collected measure data is input into trained neural network as the input vector Pt to arrive the output vector $\widehat{Pt}$. In accordance with (3), the novelty indicator in the detection phase can be arrived.

$$\lambda t = \left\| \widehat{Pt} - Pt \right\| \tag{3}$$

where, λt is the novelty indicator obtained in the abnormal detection phase. Compare the threshold δ_{λ} to λ_t to determine whether there is abnormal from the difference. If the structural status has changed, the novelty indicator under detection status will be quite different from that under normal structural status. When the difference is larger than determined threshold, we can determine the structural status has changed. In the same way, the degree of status can also be determined in accordance with value of novelty indicator.

3 T beam cracking test based on novelty detection

The paper firstly process collected data in the static load test to generate sample data needed for neural network training. After network training, the model of normal status for T beam based on novelty detection technique can be established to implement abnormal status identification and crack location identification of T beam.

3.1 Test conditions

The test model selected the bridge deck with maximum beam spacing and most unfavorable force for analysis. The load was determined following the equivalent stress principle. Before test, the lateral force of structure was analyzed. The equivalent model test was carried out according to the most unfavorable stress amplitude under designed load. The model selected flange 1:1 scale model. The steel structure is same with design. The model plate used simply supported structure, whose maximum width of flange plate span is 2.535m and plate width is 1m.

As the model test plate is general reinforced concrete, the cracking moment computation of middle section is as follows:

$$M_{cr} = \gamma f_{tk} W_0 = \frac{2S_0}{W_0} f_{tk} W_0 = 2S_0 f_{tk} \tag{4}$$

where,

$$S_0 = 1000 \times 100 \times 50 + 10 \times \left(\frac{20}{3.45} - 1 \right) \times \frac{\pi \times 12 \times 12}{4} \times 64 = 5347225mm^3.$$

So, $M_{cr} = 2S_0 f_{tk} = 2 \times 5347225 \times 2.65 = 28.34kN \cdot m$ and corresponding concentrated load is 39.0*kN*.

The crack test includes static load test before cracking, cracking static loading test and post-cracking static load test. In the static load test before cracking, as the theoretical value of cracking load is 39.0*kN*, we set 5*kN* as one level till 30*kN* and then repeat. As the stress load to 30*kN* not meet the cracking load, when the measurement data is used as training data of neural network. In case of cracking static load test, the stress is loaded till 35*kN* with 5*kN* as one level. In order to accurately capture cracking load, the load level was appropriately increase till 70*kN*. The measuring point layout is shown in Fig. 3.

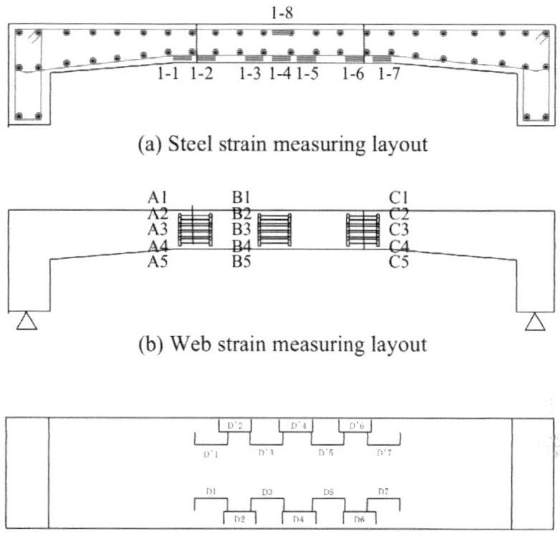

(a) Steel strain measuring layout

(b) Web strain measuring layout

(c) Floor strain measuring layout

Figure 3. Strain measuring point layout.

According to sensor distribution and finite element analysis model of T beam cracking test, the roles of different sensors are as follows:

Steel strain sensor. The sensor is embedded into T beam structure as shown in Fig. 3(a). It can reflect force status of internal structure in the load test, which is the basis to determine whether the whole structure has changed. The steel strain sensor includes 8 sensors with number from 1-1 to 1-8.

Web strain sensor. The sensors are located on two sides of model web as shown in Fig. 3(b). There are 5 sensors in the wet cross and middle section, total 6 groups, numbered by $A1$-$A5$, $B1$-$B5$, $C1$-$C5$, $A'1-A'5$, $B'1-B'5$, $C'1-C'5$. The aim is to compute height of central axis in the post data analysis. In the paper, each group is processed as a whole to determine status change in the section.

Floor strain sensor. The sensors are located on both outer sides of bridge floor as shown in Fig. 3(c). There are total 14 sensors, divided into 2 groups numbered by D1-D7 and $D'1-D'7$. The target is to capture whether there is crack in the model floor. In the strain loading process, the floor may be crack firstly in theory. Therefore, each group can be regarded as a while for status changing determination. Furthermore, test data of both groups is complementary, increasing accuracy for status determination.

3.2 Generate training sample

Before T beam structure crack test, it needs to carry out static load test before cracking. As the structural status does not change in the test, the test data before cracking can be used as training data for neural network. Under such circumstance, the training data model is neural network model not in cracking structure.

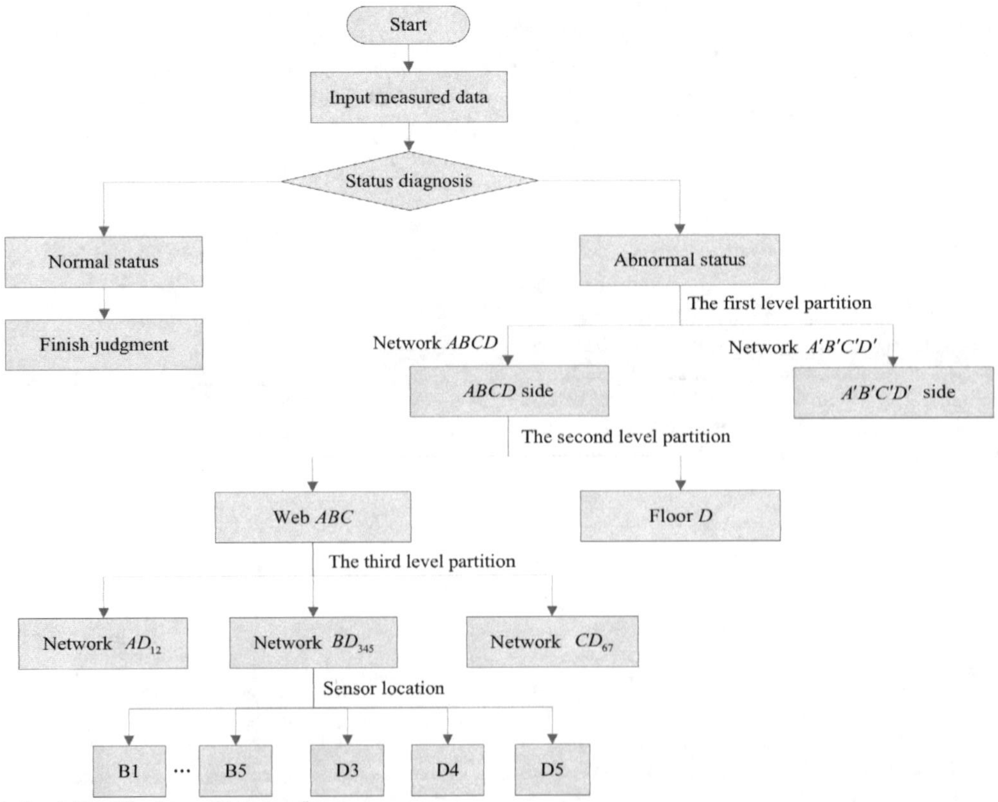

Figure 4. Crack identification and location flow.

As the problems of load process and test method, the sample number is limited. The original 30 groups test data is used as training sample for neural network. In accordance with training flow as shown in Fig. 2(b), the mean of novelty indicator can be arrived as $\bar{\lambda} = 49.9$, which is too large. We can modify this structural finite element model based on measured data to obtain more sample data so that the computation result is close to actual measure data from sensors.

Perform validity verification on neural network model trained by increased data with finite element model, taking 1-1 to 1-8 steel strain sensor as example for validation.

Firstly, use 110 groups of sample data, including 60 groups of uncrack steel strain data within 0-30kN simulated by finite element model and original 50 groups of data, as the training sample input into neural network. Input measured data into neural network produced by data from finite element model to obtain the mean of novelty indicator as:

$$\bar{\lambda} = 3.16,\ \lambda_{\max} = 5.59,\ \sigma_{\lambda} = 2.4.$$

The threshold can be computed with (2) that $\delta_{\lambda} = 12.7$. The result shows that the neural network model with expanded training sample has lower error and novelty indicator. It indicates that increasing sample amount in the manner of finite element model can also ensure correctness of the network model.

Secondly, in order to verify whether the neural network can identify abnormal structure, verify the steel strain data 0-30*kN* after cracking. The mean of novelty indicator is $\lambda t = 104.3$, which is much larger than that in uncrack status. It indicates that the neural network model can identify abnormal changes of the model.

Based on above verification, the cracking load of structure in theory is 40*kN*, so 80 groups of data of each sensor at load 0-40*kN* can be formed as the neural network training sample at normal status. The sample amount can be expanded using the method, thus increasing accuracy of neural network training.

3.3 Neural network level division

As the overall strain network model to measure structural health status, the data of 52 sensors within 0-40*kN* engages in training as a whole. The network structure is designed as 52-40-40-52 four-level feedforward BP network. The transfer functions of second and third layer select tan-sig function. The transfer function between first level and second level, third level and fourth level uses purelin function [7].

Along the loading process, input load data into trained network 1 to perform crack location identification in case of crack alarm. Combining with actual situation of T beam structure, the location identification uses three-level identification manner [7-9].

The first level partition is to determine the sensor side where crack located. The 22 strain values of sensors at *ABCD* and *A'B'C'D'* sides are used as input vector for neural network training to arrive at two trained networks, namely network *ABCD* and network *A'B'C'D'*. Based on these networks, the sensor side where crack located can be identified.

The second level partition aims at identify section of cracks. The structure is divided into 4 cross-sections, forming 4 groups neural networks with 15 and 7 sensor data, namely network web *ABC*, network web *A'B'C'*, network floor *D* and network floor *D'*.

The third level partition constitutes the section with 5 sensors at the web and sensors at the adjacent floor. The structure totally contains 6 sections to form 6 groups of neural networks, namely network AD_{12}, network BD_{345}, network CD_{67}, network $A'D'_{12}$, network $B'D'_{345}$ and network $C'D'_{67}$.

Based on above partition of network model, we can obtain 12 trained neural networks and corresponding novelty indicator and threshold under normal status. After the crack section was located by neural network, the sensor variable value within this section is analyzed to determine specific location. The training flow of ach BP neural network is shown in Fig. 2(b).

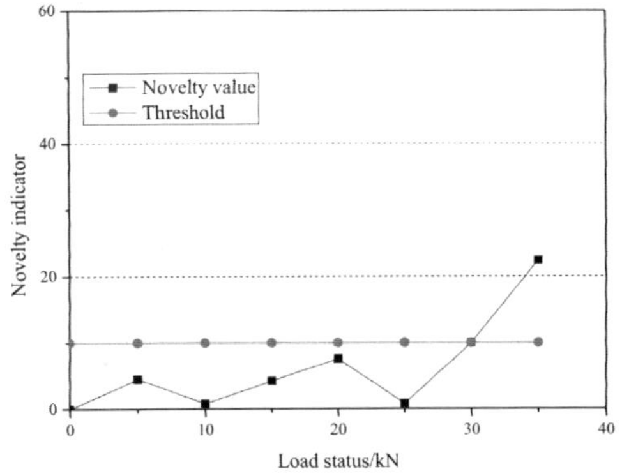

Figure 5. Crack load novelty indicator.

3.4 Crack location identification

The T beam status assessment can be divided into two phases of status identification and location identification. Firstly, the measurement data loaded by level is analyzed to input into trained status identification network, so as to arrive at status identification novelty indicator in the detection phase. If the indicator is normal, the identification ends. In case of abnormal situation, perform abnormal location identification. In the abnormal location identification phase, carry on level-by-level identification with former trained 12 networks to determine abnormal section. Finally, make comprehensive decision in accordance with sensor change rate within the section. The diagnosis flow is shown in Fig. 4.

3.5 Result analysis

When the crack loads 0-40*kN*, input data of 5*kN*, 10*kN*, 15*kN*, 20*kN*, 25*kN*, 30*kN*, 35*kN*, 37*kN*, 39*kN* and 40*kN* into network for determination respectively based on load status of each level. The obtained novelty indicator and threshold is shown in Fig. 5. As seen from the figure, the structure changes after loaded to 35*kN* determined from BP neural network, which is consistent with actual situation.

If the structure has changed, three-level partition manner is used to locate abnormal section. It is found that the novelty of network *ABCD* and network *A'B'C'D'* larger than threshold begin from 5*kN*. That is to say, two sections on side *ABCD* and side *A'B'C'D'* have abnormal phenomenon. In case of the second partition identification, the novelty indicator is larger than threshold. It is proven that there are abnormal crack in the web and floor sections, which is identified in the third partition. Part of network novelty indicators in the third partition is shown in Fig. 6 and Fig. 7.

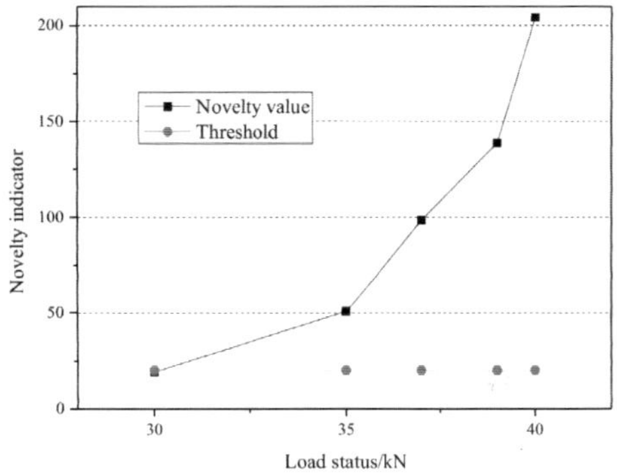

Figure 6. Novelty indicator of network BD_{345}.

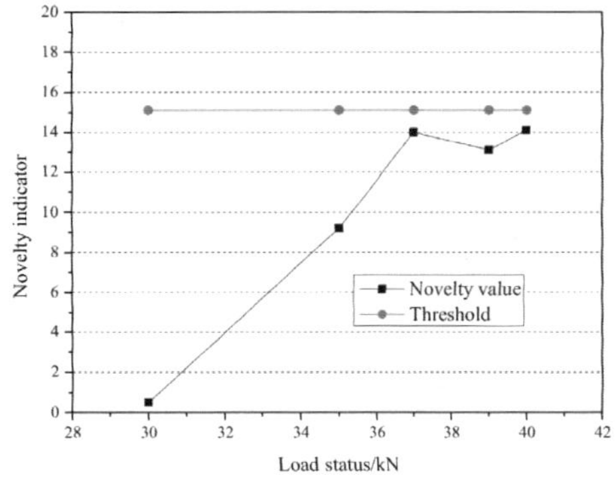

Figure 7. Novelty indicator of network AD_{12}.

We can know from the final section identification that there are abnormal data in the network BD_{345} and network $B'D'_{345}$. Furthermore, the specific position can be located by sensor change rate within the section. Table 1 shows the change rate φ of network BD_{345}.

Table 1. Sensor load change rate φ within BD_{345}.

	B1	B2	B3	B4	B5	D3	D4	D5
30	-2.88	-3.4	0.54	1.1	2.92	3.26	1.9	4.16
35	-3.3	-3.1	-0.22	2.28	4.36	3.58	1.08	5.32
37	-4.6	-0.7	-1.4	3.65	19.2	-0.5	-2.45	25
39	-7.2	-0.25	0.85	11.15	31.5	6.05	-8.05	39.25
40	-1.5	-3.3	10.5	18.7	44.1	5.5	-15.1	53.7

As seen from Table 1, the strain sensor change rate is relatively little compared to 30*kN* in case of 35*kN*. It means the manner using single sensor may not determine whether the whole structure has changed. The method can reduce missing alarm to diagnose abnormal structure. Along with load increasing and changing rate, it can determine that the possibility of crack at sensor B5, D4 and D5 is large. Finally, the determine result proves that the measurement values at B5, D4, D5, $B'4$, $B'5$, $D'3$, $D'4$, $D'5$ are abnormal. The determination result is same as that of manual inspection, which proves the validity of the crack identification and location method in the T beam static load test.

4 Conclusion

It is very important of bridge structure health monitoring and damage detection to timely locate structural damage and assess status. The paper discussed method to identify abnormal status of bridge structure and location crack with novelty detection technique based on BP neural network. In order to verify the feasibility of the method in abnormal stratus and location identification, the method was applied in T beam cracking location. In the near future, the data model should be studied so that the method can be applied to overall bridge abnormal status identification more conveniently.

References

1. E. Aranha, P. Borba, J. Lima, Model Simulation for Test Execution Capacity Estimation, *Proceedings of 17th International Symposium on Software Reliability Engineering*, pp. 231-236 (2006)
2. Zhang Chan-yin, Ji Wen-yu, An Ming-zhe, Experimental study on early autogenous shrinkage of 20m-span low-height reactive powder concrete T-beam, *Proceedings of 2010 International Conference on Mechanic Automation and Control Engineering (MACE)*, pp. 1068-1071 (2010)
3. Xu Lihua, Zeng Hao, Xu Feng, An experimental study on fatigue performance of T-shape RC beams strengthened with externally prestressed CFRP tendons, *Proceedings of 2011 Second International Conference on Mechanic Automation and Control Engineering (MACE)*, pp. 7162-7166 (2011)
4. H. Xiang, Z. X. Zhou, Y. Xu, The Finite Element Analysis of the Large Span Steel Truss Arch Bridge's Node Fatigue Test, *Proceedings of 2010 International Conference on E-Product E-Service and E-Entertainment (ICEEE)*, pp. 1-6 (2010)
5. Sun Zong-guang, Ni Yi-qing, Gao Zan-ming, Damage alarming of bridge structures based on novelty detection technique, Journal of Highway and Transportation Research and Development, **24**, no. 10, pp. 72-75 (2007)
6. Wang Xiu-yong, Wen Qing, Yang Qi, Yu Jin-xiu, He Xiong-wei, Condiction assessment for stay cables of cable-stayed bridge based on novelty detection technique, Journal of Highway and Transportation Research and Development, **28**, no. 9, 53-59 (2011)
7. Y. Liu, X. Wang, The Application of BP Neural Network in Cable-Stayed Bridge Construction Monitoring, *Proceedings of 2010 International Conference on Computational and Information Sciences (ICCIS)*, pp. 1190-1193 (2010)
8. Xie Xiaoyao, Yan Xinping, Cable tension prediction of Hongfeng Lake cable-stayed bridge using BP neural network, *Proceedings of 2nd International Conference on Anti-counterfeiting, Security and Identification*, pp. 18-21 (2008)
9. Q. Sun, X. Guo ; D. Zhang ; X. Guan, Research on the application of BP neural network in construction control for cable replacement of the cable-stayed bridge, *Proceedings of 2010 Sixth International Conference on Natural Computation (ICNC)*, **2**, pp. 914-918 (2010)

MATEC Web of Conferences **81**, 02012 (2016) DOI: 10.1051/matecconf/20168102012
ICTTE 2016

Study the Relationship between Pavement Surface Distress and Roughness Data

Muhammad Mubaraki

Department of Civil Engineering, Jazan University, Saudi Arabia

Abstract. In this paper, pavement sections from the highway connected Jeddah to Jazan were selected and analyzed to investigate the relationship between International Roughness Index (IRI) and pavement damage including; cracking, rutting, and raveling. The Ministry of Transport (MOT) of Saudi Arabia has been collecting pavement condition data using the Road Surface Tester (RST) vehicle. The MOT measures Roughness, Rutting (RUT), Cracking (CRA), raveling (RAV). Roughness measurements are calculated in terms of the International Roughness Index (IRI). The IRI is calculated over equally spaced intervals along the road profile. Roughness measurements are performed at speed between at 80 kilometers per hour. Thus RST vehicle has been used to evaluate highways across the country. The paper shows three relationships including; cracking (CRA) verses roughness (IRI), rutting (RUT) verses IRI, and raveling (RAV) verses IRI. Also, the paper developed two models namely; model relates IRI to the three distress under study, and model relates IRI to ride quality. The results of the analysis claim at 95% confidence that a significant relationship exist between IRI and cracking, and raveling. It's also shown that rutting did not show significant relationship to IRI values. That's leads to conclude that the distresses types: cracking and raveling may possibly be described as ride quality distresses at different level of significant. Rutting distress described as non-ride quality type's distresses.

1 Introduction

Pavement evaluation is a process by which field surveys and testing are carried out to characterize the condition of an existing pavement structure, both structurally and functionally. The structural condition of a pavement refers to its ability to support the current and future traffic loadings, whereas the functional condition refers to its ability to provide a safe, smooth, and quiet riding surface for the traveling public (Hass *et al* 1994).

Network-level evaluations are conducted on pavement sections within the network of pavements for which the agency is responsible, with the general purpose to document current conditions, to identify projects for maintenance, preservation and rehabilitation, to help prioritize projects and allocate budgets, and to help determine funding needs. In addition, the collection of performance data on a pavement network over time provides a valuable tool for tracking pavement performance as well as a mechanism for developing performance models that can be used to predict future conditions (both with and without the application of treatments).

Visual condition surveys are generally included determining type of distress, its severity, its extent and location. There are

Many methods for conducting pavement condition surveys, as there are agencies involved. Pavement condition index (PCI) ranges from zero to 100, where 100 represents excellent pavement condition. The PCI values are calculated based on pavement distress type, severity and quantity collected through visual inspection. To collect the appropriate distress data an inspector is sent to a particular pavement section to record the existing distress types, severities and quantities. The information is collected by actually walking through the section. The procedure provides detailed information on pavement section condition. However, considering the size of the city pavement network the procedure is very time consuming and very costly.

Roughness is primary a measure of riding quality of the road pavement surface. Pavement roughness is very much related to pavement serviceability, which is a measure of physical characteristics of pavement surface. Roughness measurement systems that currently used can be grouped into profilometric, vehicle response and subjective evaluation. Profilometric methods are the most accurate and best suited for detailed analysis.

Roughness measurements are calculated in terms of the International Roughness Index (IRI). The IRI is calculated over equally spaced intervals along the road profile, the IRI computation method converts the longitudinal and vertical profile data into a vehicle motion response using mathematical model. The IRI value is expressed as the units of displacement over units of length. Roughness measurements are performed at speed between

(40 to 50) kilometers per hour. Thus, roughness measurements collected in relatively very short time without excessive cost (Mubaraki, 2014).

2 Research significant

Application of pavement evaluation of pavements should be on a well-planned basis and be an integral part of the overall pavement management system. Pavement roughness data fulfills a wide variety of needs in the management function. These needs occur at network and project levels, for rural and urban road pavements. For example, at the network level, roughness data provides a functional evaluation of the pavement surface which can be used in project selection and programming. The roughness data evaluation helps in construction quality control and evaluation of rehabilitation options (Hass et al 1994).

Application of pavement distress data has recognized by pavement engineers as an important parameter for quantifying the quality of pavement surface. It is important at both the network and projects levels of pavement management system. For example, at both levels, the pavement distress information is useful in selecting appropriate treatments. At the network level, the concern is with determining what class of treatment is required for example, continued routine maintenance or resurfacing. At the project level, the concern is with estimating the extent and the methods of pavement repair such as to patch a certain area of alligator cracking (Hass et al 1994).

Roughness is primary a measure of riding quality of the road pavement surface. Pavement roughness is very much related to pavement serviceability, which is a measure of physical characteristics of pavement surface. The applicability of pavement roughness as a predictor variable of pavement condition index is a vital in pavement evaluation and maintenance to set criteria or trigger levels for different types of maintenance. On other hand, the interaction between pavement roughness and pavement distress types is very important to distinguish between ride quality distress type and non-ride quality type.

Therefore, the expected achievement of the research will greatly assist in determining maintenance needs by defining the types of distress that most probability will be accrued, determining the effective timing of applying the maintenance needs by defining the time after which the rate of deterioration will increase drastically, and help the municipalities in setting priority index for the budget and the maintenance program, and hopefully the results can help in local research in the areas of pavement maintenance.

3 Objectives

The main objective of this study is to study the relationship between pavement condition evaluation and roughness measurements and examine the effect of pavement distress types on pavement roughness for the Jazan entire road network.

4 Methodology

4.1 Correlation of IRI and type of distress

The Pearson correlation analysis was conducted to estimate the correlation factor and correlation hypothesis and to find out whether a significant statistical relation between IRI and types of distress exist. The statistical hypothesis test displays a p- value to determine the significant effect of a distress on IRI values (Park et al, 2004).

4.2 IRI and distress model

The modeling procedure used in the analysis is based on statistical regression. The basic form of regression equation is as follows:
$y = b0 + b1X$; Where:
Y: Dependent Variable (response)
X: Independent Variable
b0, b1: regression coefficients

To attain the analysis procedure, a spreadsheet is used to examine and study the data. The spread sheet program was used to tabulate the data, draw the charts (graphs) that describe the trend of the collected data, building the models and find out the correlations and the relationship between pavement roughness (IRI) and types of distress.

The IRI and distress model is constructed in order to predict types of distress in terms of a distress behavior as a function of pavement roughness (IRI). The procedure involved using sufficient regression techniques to construct models linking IRI and a distress values. The most suitable model was then selected based on proper statistical tests. The process of constructing the regression model was based on three types of regression functions (Myers and Keying, 2011):

1-Linear regression approach, assuming the relation between the two variables is linear. The regression equation is in the form $Distress = b0 + b1 * IRI$. In this equation the regression coefficients are b0 and b1.

2-Logarithm nonlinear relation approach, assuming the relation between IRI and a distress is nonlinear. The distress was considered as a function of logarithm of IRI. The nonlinear relation is then transformed to linear regression by calculating new parameter called Log (IRI). The regression equation is in the form: $Distress = b0 + b1$ Log (IRI). In this equation the regression coefficients are b0 and b1.

3-Quadratic nonlinear relation approach, assuming the relation between IRI and a distress is nonlinear. The distress was considered as a function of quadratic of IRI.

4-Distress =f IRIa. The nonlinear relation is transformed to linear regression by calculating new parameter called IRIa; where: a is constant. The regression equation is in the form: $Distress = b0 + b1$ IRIa. In this equation the regression coefficients are b0 and b1.

After the regression equations are constructed, a process of testing was used. The aim of the testing process is to find out the best suitable model that represents the given data. The analysis procedure was based on three main statistical regression tests, which are:

1-Test of hypothesis (T-test).

2-Coefficient of multiple determination (R2)

3-Analyze of Variance Test (ANOVA – Test)

The T-test of hypothesis is incorporated to study the relationship between each of regression coefficients (b0, b1) and the response variable at certain degree of confidence (assumed 95% in our study).Hence the test was conducted for all of the regression coefficients in the regression model. The coefficient of multiple determinations (R2) was used to illustrate the adequacy of a fitted regression model. The R2is a percentage indicates the proportion of the response variable variations explained by the regression model. The coefficient takes values from 0 (no correlation exists between variables) to 100 (perfect correlation between variables). The Analysis Of Variance test (ANOVA) is powerful tool used to investigate the relationship between the response variable (a distress in this study) and the independent variables (IRI). The ANOVA test extends the t-test for more general null hypothesis to estimate the quality of the regression line. That is whether the variation in distress readings is dependent on the IRI readings. This proves the significant evidence to accept or reject the regression model. (Myers and Keying, 2011).

4.3 Site selection

The study involves the analysis of roughness and distress data from a highway connects Jeddah city with Jazan city. Many readings from the surveys have been deployed to this study.

4.4 Data collection

For the selected highway, cracking, rutting, raveling, and road surface profiles were obtained from Ministry of Transport (MOT), Pavement Management System (PMS) database. Pavement distress values and roughness values were available for 2009, 2011, and 2013. The pavement distress and roughness data was available for 400 km for each km intervals. Detailed distress data in the form of distress type, distress density, and as well as measurements of roughness.

5 Data analysis

5.1 Cracking vs.IRI

Fig. 1 presents graph relation between IRI and cracking density. It could be noticed that the IRI never exceeded above 5 in most data points, and that is due to the fact that data collecting van drivers are instructed to avoid driving heavily deformed segments of the road. Also, for a cracking distress between 4% & 6%, the IRI values were clustered around "1". This irregularity most probably represents segments of the road with a very good IRI but had longitudinal cracks between the road and its paved shoulders. The graph verifies the positive relation between the cracks and IRI values. However, the graph does not show a clear cut decision about the linearity of the relation. The correlation coefficient of IRI and patching density is 0.500 and the P value is equal to.019.The positive sign of the coefficient indicates a positive relation between cracking density and IRI. The value of correlation factor indicates that about 50 % of the IRI and cracks density observations were distinguished by the linear relation. The best representation for the cracking data was by logarithmic formula and it's as $Cracking = 8.6 + 2.7 \log IRI$.

In the hypothesis test, the $P - value = 0.019$, therefore there is sufficient evidence at 95% confidence level of the relationship between IRI and patching density. The result of the statistical test doesn't comply with previews experiences that point toward cracks distress is on ride quality type of distress at all severity levels.

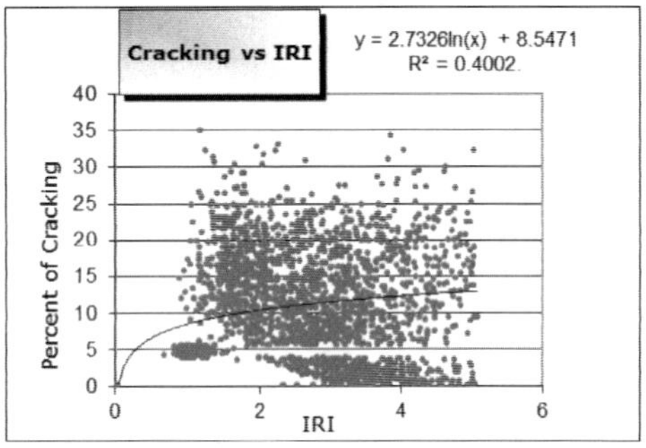

Figure 1. IRI and cracking distress density relationship.

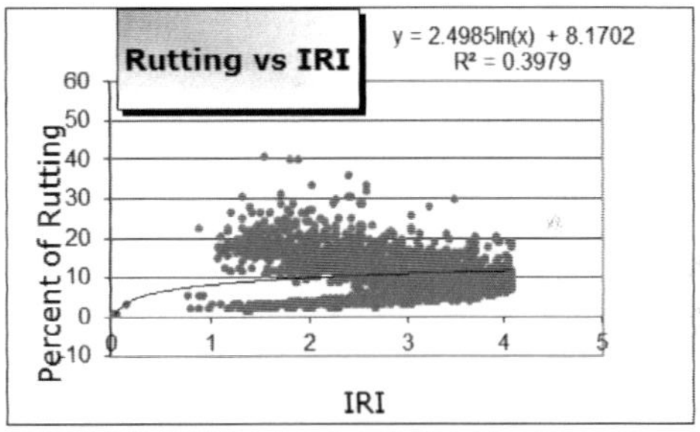

Figure 2. IRI and rutting density relationship.

5.2 Rutting vs. IRI

The relationship between IRI and rutting density is presented in Fig. 2. The Graph Indicates A Positive relation between the rutting and IRI values. There is no clear cut decision about the linearity of the relation. The Correlation coefficient of IRI and rutting density is 0.464 and P-value is 0.151.

The value of the correlation coefficient equals0.464 indicates that only about 46.4% of the IRI and running density observation were not able by the linear relation. For this reason, logarithm nonlinear relation approach was used. The regression model was as; Rutting = 8.2 + 2.5 log IRI. The P-value is 0.15 which is higher than the 0.05 level of significant; therefore there is no sufficient evidence of the relationship between IRI and rutting density.

5.3 Ravelling vs. IRI

Fig. 3 represents a graph relation between IRI and raveling distress density. The raveling data appears in 50 pavement sections. The graph indicates a positive relation between raveling and IRI. The value of correlation factor indicates that about 45 % of the IRI and raveling density observation were distinguished by the linear relation. The best model formula is as Raveling = 28.2 + 9.00 log IRI.

In the hypothesis test the P-value equals 0.006. Therefore there is sufficient evidence at95% confidence level of the relationship between IRI and raveling density.

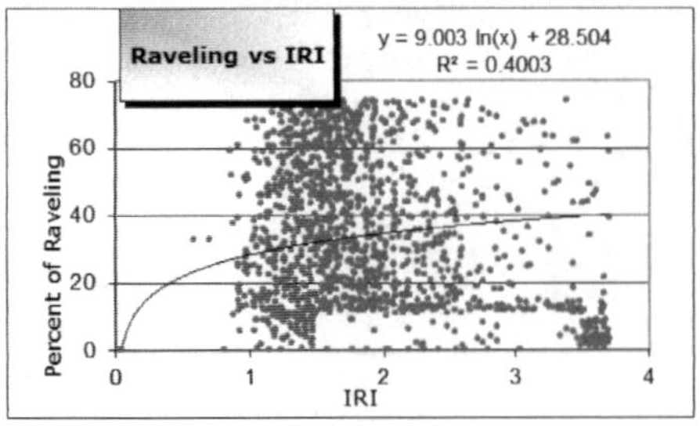

Figure 3. IRI and raveling density relationship.

6 Discussion

The effect of all the distress types that appears in the study area on the IRI reading was investigated. The study indicated the riding quality related types of distress. The relationship between IRI and distress types has been investigated through woke statistical tests:

- The individual Correlation hypothesis test between IRI and each distress types.
- The regression coefficients test for overall regression model relating IRI and distresses types.

The-correlation-hypothesis- test for cracking, rutting and raveling, demonstrates that P is zero except rutting. Therefore cracking and raveling and I have significant-relationship. Hence the result of the analysis climate 95% confidence that a significant relationship exist between IRI and cracking and raveling. Furthermore, the low correlation factors for all the three types of distress showed that there is no clear cut decision about the linearity of the relation. Therefore, the best representation is a nonlinear equation. In this study, it was found the logarithm shows reasonable fitting for the data.

Those results were proved through the statistical tests on the overall model between IRI and types of distress. The regression coefficients test compute p values for each regression coefficient. The regression coefficients for cracking and raveling are equal zero. This indicates a significant linear relationship between IRI and each regression coefficients. Obviously, different distress types related to IRI differently. It was found that from the values of the correlation factors, that patching and depression have the highest correlation to IRI. While the raveling has less correlation, cracking has the least relation to IRI. It was also found that the correlation factors for all the distressed against IRI were not exceed 50%, which mean that less than 50 percent of these relationships are definable by the relationship.

This relationship is not strong enough for pavement condition to be used as measure for IRI. Hence, the result of both correlation and regression coefficients tests indicates at 95% confidence that cracking, and raveling are rid equality type distressed. However, it's not realistic to conclude that IRI may completely reflect pavement distress conditions. The rutting, pointed toward no significant relationship to IRI values. The P Values (which larger than 0.05level of significance) of rutting demonstrate that this types of distress has no strong evidence to significant relationship with IRI.

Those results were proved through the statistical tests on the overall model between IRI and type of distress. The regression coefficients test computed p-values for each regression coefficient. The regression coefficient for rutting is greater than 0.05 level of significance, indicating no significant linear relationship between IRI and rutting. Hence, the result of both correlation and regression coefficients tests indicates at 95% confidence that rutting is not ride quality type of distress, and that IRI could not reflect rutting. There as on for these results may be the fact that running is in the longitudinal profile of pavement not detected roughness machine.

7 Conclusion

Pavement sections from the street network of Jazan city have been selected to evaluate correlation of three types of distress including cracking, rutting, and raveling distress densities values to IRI values. Based on the results, correlations, and statistical models that relate IRI values to the distress density values were established.

A statistical correlation analysis performed to analyze the interaction between IRI in pavement distresses. The aim of this analysis was to determine the riding quality related type's distresses. The correlation statistical test was us to determine the correlation between distress density and IRI values.

The model related IRI and distresses types were estimated in the pavement sections. The relationship between IRI and distresses types has been investigated through correlation hypothesis test and regression coefficients test for overall model relating IRI and distresses types.

The results of the analysis claim at 95% confidence that a significant relationship exist between IRI and cracking, and raveling. It's also shown that rutting did not show significant relationship to IRI values. That's leads to conclude that the distresses types: cracking, and raveling may possibly be describe d as ride quality distresses at different level of significant. Rutting distress described as non-ride quality type's distresses.

References

1. R. Hass, W. Hudson, and J. Zaniewnski, *Modern Pavement Management*, Krieger publishing, (1994)
2. M. Mubaraki, Identification of Pavement Distress Types and Pavement Condition Evaluation Based on Network Level Inspection for Jazan City Road

Network, The Journal of Engineering Research (TJER), **11**, No. 1, pp 44-54, (2014)

3. M. Mubaraki, *A methodology for project-level maintenance for urban road*, Journal of Transport-ICE, **168**, 3, pp 239 –255, (2015)
4. T. El-Korchi and W. Norman, Visual Appearance of surface Distress in Pavement, TRB No. 1260 Washington, D. C., (2004)
5. L. Aultman and E. Jackson, *Model Relating Pavement Quality Measures*. Journal of Transportation Research Board, Washington, D. C., (2004)
6. Myers and Keying E. Ye 2011. *Probability and Statistics for Engineers and Scientists*. (2011)
7. M. Shahin and S. Kohn, *Pavement Maintenance Management for Road And Parking Lots,* (2002)
8. K. Park, K. Lee, and T. Natacha, *Applicability of the International Roughness Index as a Predictor of Asphalt Pavement Condition*. Journal of Transportation Research Board, Washington, D. C., (2004)
9. S. Dewan and R. Smith, *Estimating IRI from Pavement Distresses to Calculate Vehicle Operating Costs for the Cities and Counties of San Francisco Bay Area*. Transportation Research Record 1816, Transportation Research Board; Washington D. C., (2002)
10. J. Lin, J. Yau, and L. Hsiao, *Correlation analysis Between International Roughness Index (IRI) and pavement Distress by Neural Network*, Journal of Transportation Research Board, Washington, D. C., (2003)

MATEC Web of Conferences **81**, 02013 (2016) DOI: 10.1051/matecconf/20168102013
ICTTE 2016

The Analysis of Serious Fatal Road Traffic Accidents in China

Miaomiao Liu[1], Yongsheng Chen[1] and Guangquan Lu[2]

[1]*Research Institute of Highway, Ministry of Transport, 8 Xitucheng Road, Haidian District, Beijing 100088, China*
[2]*School of Transportation Science and Engineering, Beihang University, 37 Xueyuan Road, Haidian District, Beijing 100191, China*

Abstract. A large number of road traffic accidents occur each year in China, which results in huge losses to the safety of people's lives and property. Among all accidents, the accident with over 10 fatalities in one accident is extraordinarily serious, which has immeasurable impact on the security. In this paper, we defined these accidents as serious fatal road traffic accidents (SF-RTAs). The aim of this study was to analyze SF-RTAs in China and its provinces and investigate the causes of SF-RTAs. Firstly, we described and analyzed the annual trend of SF-RTAs based on the data from 1990 to 2014. Then we analyzed region, time and accident type distribution characteristics of SF-RTAs during 2004-2014. Then the main factors influencing SF-RTAs were analyzed by linear regression and decision tree methods. Finally, we proposed corresponding control strategies for preventing SF-RTAs according to the analysis results. It is expected that the findings of this study are useful in indentifying effective prevention measures for serious fatal accidents in China, and protecting human lives and properties.

1 Introduction

Global Status Report on Road Safety 2015, reflecting information from 180 countries, indicates that worldwide the total number of road traffic deaths remains unacceptably high at 1.25 million per year, and non-fatal injuries as a result of road traffic crashes are between 20 to 50 million. These deaths and injuries have an immeasurable impact on people's lives and property safety. Currently, road traffic accidents are estimated to be the 8th position among the 100 important causes of death. However, it is reported that road traffic injuries will become the 5th leading cause of death by the year 2030 [1]. According to previous documents, traffic accidents on roads depend on many factors, such as changes in seasons, weather conditions, road traffic volume, traffic intensity, driving experience, driving culture and technical state [2]. All these factors are related to traffic safety, human lives and health. Hence, it is essential to analyze the various aspects of road traffic accidents through detailed research.

According to the statistics [3], 6.76 million road traffic accidents occur in China in the year 2014. Among them, 196812 accidents involve fatality or injury, which make 58523 persons killed and 211882 injured. From the year 1990 to 2014, more than 8.88 million accidents involved fatality or injury occur in China, which kill more than 1.90 million people and wound more than 7.29million people. And these accidents cause huge losses to the safety of people's lives and property. Among all accidents, the accident with over 10 fatalities in one accident is extraordinarily serious, which has immeasurable impact on the safety. We define these accidents as serious fatal road traffic accidents (SF-RTAs) in this paper. 1084 SF-RTAs occur in China from 1990 to 2014, which result in 17317 deaths and 18106 injuries (as shown in Fig. 1).

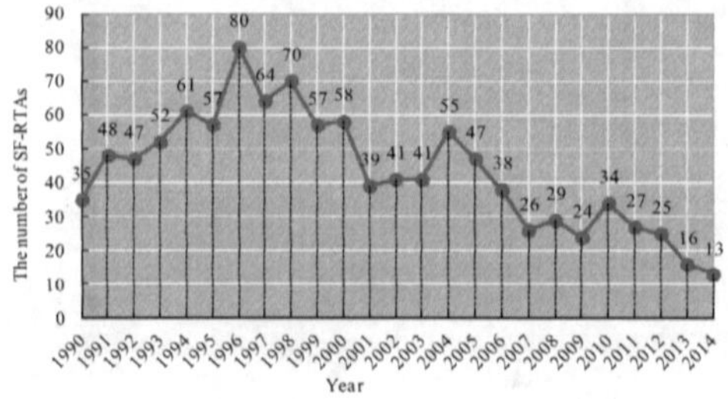

Figure 1. The number of SF-RTAs occurred in China during 1990 and 2014.

These SF-RTAs usually cause a lot of casualties, bring in directly or indirectly property losses, and have serious social impact. Although some traffic safety administration departments of China have taken some measures to prevent SF-RTAs, SF-RTAs still occur from time to time. In addition, the now existing prevention methods lack in-depth analysis for the characteristics of SF-RTAs. Thus, in this study, we mainly focused on the characteristics of SF-RTAs occurred in China, investigated the main factors influencing SF-RTAs, and proposed effective measures for preventing SF-RTAs.

2 Materials

All SF-RTAs information can be collected by Road Accident Statistics Annual Report of the People's

Republic of China, but it lacks detailed description in the Annual Report for each of SF-RTA before the year 2003. Detailed information, including date, time, province, address, the number of deaths and injuries, road grade, accident type, vehicle type and vehicle owner of SF-RTA is recorded in Annual Report from 2004. Thus, in this study, we mainly collected and analyzed the SF-RTAs occurred in China during January 2004 and December 2014 [3]. To investigate the main factors influencing SF-RTAs, we also collected much data information involved human, vehicle, road and society. Then we analyzed the correlation between the SF-RTAs and influence factors using the statistical software SPSS and data analysis methods.

3 Methodology and results

3.1 Data statistics

It was known from Fig. 1 that, since 1990s, the number of SF-RTAs every year decreased after increasing with two obvious peaks, 80 SF-RTAs occurred in 1996 and 55 SF-RTAs in 2004. From 2004 to 2007, the number of SF-RTAs decreased greatly with the improvement of overall road safety in China. Then the number of SF-RTAs increased slightly in 2008 and 29 SF-RTAs happened, killing 476 and injuring 504(as shown in Table 1). In 2009, the number of SF-RTAs decreased to 24. But in 2010, that number surged to 34 with 461 deaths and 432 injuries. Since 2010, the number of SF-RTAs starts to decrease again. In 2014, 13 SF-RTAs happened, killing 229 and injuring 137. Number of SF-RTAs and injuries reached to the historical low in 2014.

Table 1. SF-RTAs occurred in China during 2004 and 2013

Year	The number of SF-RTAs	The number of deaths	The number of injuries
2004	55	852	877
2005	47	807	705
2006	38	558	463
2007	26	389	449
2008	29	476	504
2009	24	329	345
2010	34	461	432
2011	27	455	403
2012	25	361	367
2013	16	208	259
2014	13	229	137
Total	334	5125	4941

The accidents data in Fig. 1 and Table 1 show that China is standing at a high frequent stage of SF-RTAs. Thus, it is important to analyze the distribution characteristics of SF-RTAs and then make more efforts to reduce the number of SF-RTAs.

3.1.1 Regional distribution of SF-RTAs

The statistics data indicated that SF-RTAs occurred in different regions of China. The number of SF-RTAs in different regions and different provinces from 2004 to 2014 is shown in Fig. 2 and Fig. 3, respectively.

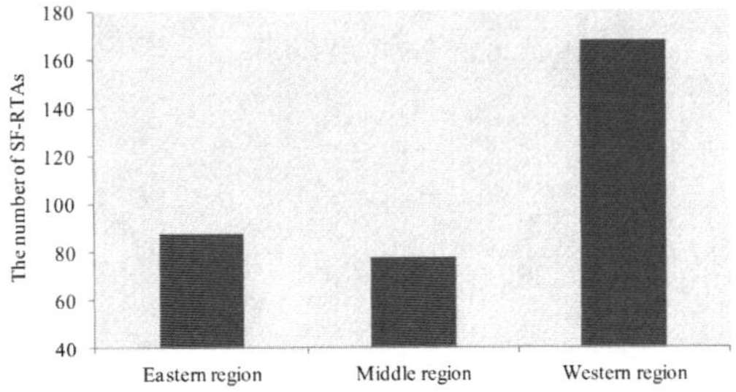

Figure 2. The number of SF-RTAs in different regions from 2004 to 2014

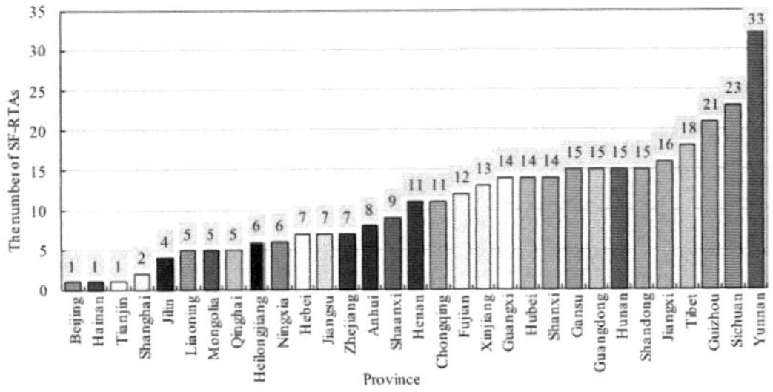

Figure 3. The number of SF-RTAs in different provinces from 2004 to 2014

It was known from Fig. 2 and Fig. 3 that the distribution of SF-RTAs had obvious regional features. Number of SF-RTAs in western region was more than that of eastern and middle regions. From 2004 to 2014, SF-RTAs occurred in Yunnan province was the most, followed by Sichuan and Guizhou provinces. All of these provinces are in south-west region of China. The total number of SF-RTAs happened in these three provinces above was 77, accounting for 23.1% of the total number of SF-RTAs. In addition, the number of SF-RTAs occurred in Beijing, Hainan, and Tianjin was the least (was all 1), and these provinces are all in eastern region.

3.1.2 Time distribution of SF-RTAs

In order to know the time distribution of SF-RTAs, 334 SF-RTAs occurred during the year 2004 and 2014 were analyzed (Fig. 4-Fig. 6). Fig. 4 shows the hour distribution of SF-RTAs.

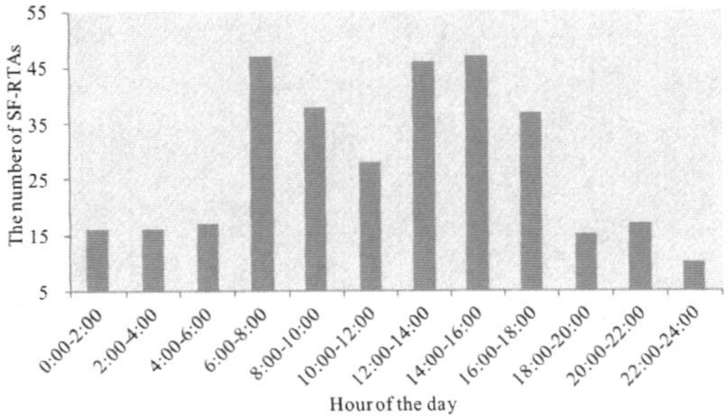

Figure 4. The hour distribution of SF-RTAs

It was known from Fig.4 that the number of SF-RTAs happened between 6:00 to 8:00 and between 12:00 to 16:00 was more than other time periods. During 22:00

and 24:00, SF-RTAs were the least. Fig. 5 shows the weekday distribution of SF-RTAs.

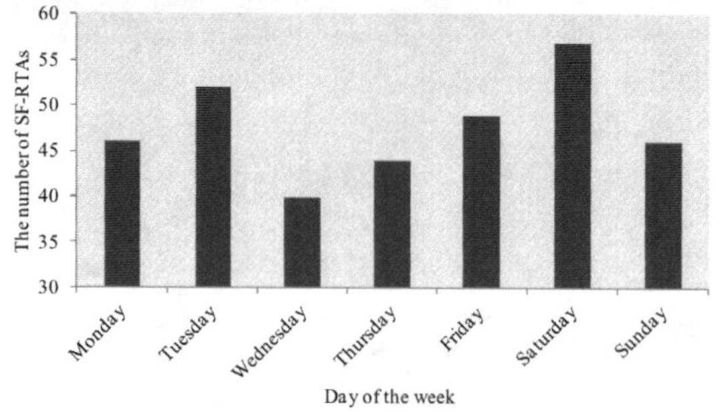

Figure 5. The weekday distribution of SF-RTAs

According to Fig.5, the number of SF-RTAs occurred on Saturday was the most, followed by Tuesday and Friday. On Wednesday, the number of SF-RTAs was the least. Fig. 6 shows the month distribution of SF-RTAs.

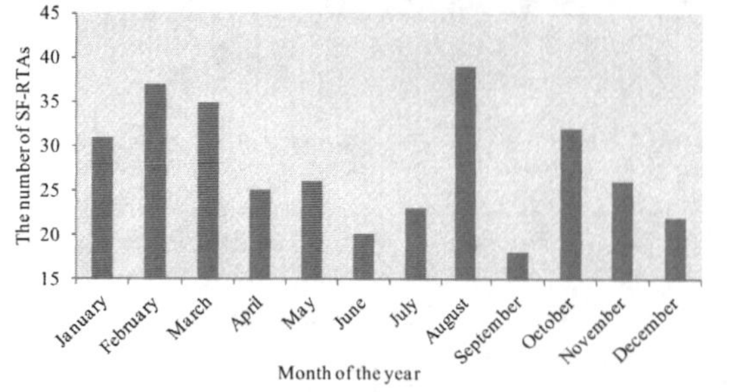

Figure 6. The month distribution of SF-RTAs

It was seen from Fig. 6 that SF-RTAs were more likely to occur in August, followed by February and March. The number of SF-RTAs occurred in September was the least.

3.1.3 Accident type distribution of SF-RTAs

SF-RTAs can be divided into many types, including collision, facing scrape, turnover, hit on fixed objects, fire, run-off-road, rolling compaction and explosion, etc. Of which, the collision accidents include head-on collision, offset collision and rear-end collision. And the run-off-road accidents include falling to cliffs, valleys, rivers and bridges. The accident type distribution of SF-RTAs during 2004 and 2014 is shown in Fig. 7.

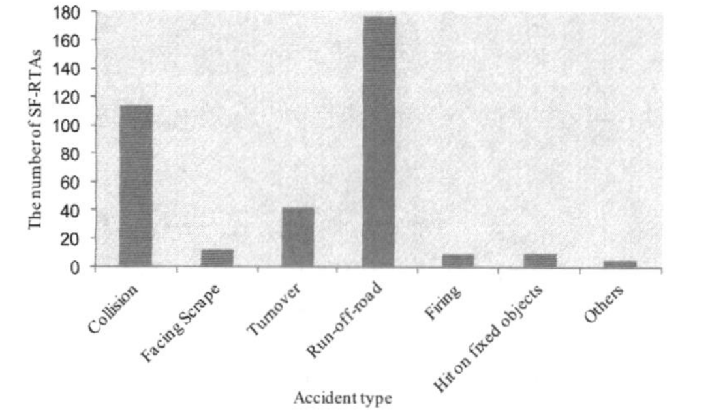

Figure 7. The accident type distribution of SF-RTAs

Fig. 7 indicated the SF-RTAs related to run-off-road were the most, accounting for about 48.0% of the total number of SF-RTAs. And the number of SF-RTAs related to collision was the second most.

3.1.4 Other distribution characteristic of SF-RTA

Vehicle type The distribution of vehicle type related to SF-RTAs is shown in Fig. 8.

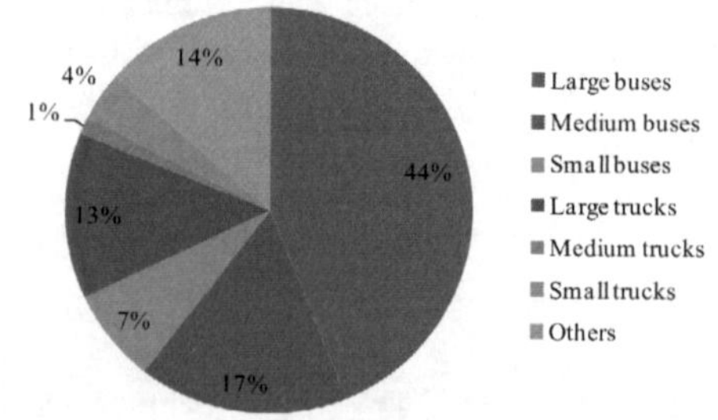

Figure 8. The distribution of vehicle types causing SF-RTAs

Fig.8 showed that 44% of the SF-RTAs related to large buses, and 17% related to medium buses. That is, the SF-RTAs related to large buses and medium buses accounted for more than 60% of the total. It could be concluded that large buses and medium buses were main vehicle types which easily resulted in SF-RTAs. Road type The distribution of road types related to SF-RTAs is shown in Fig. 9.

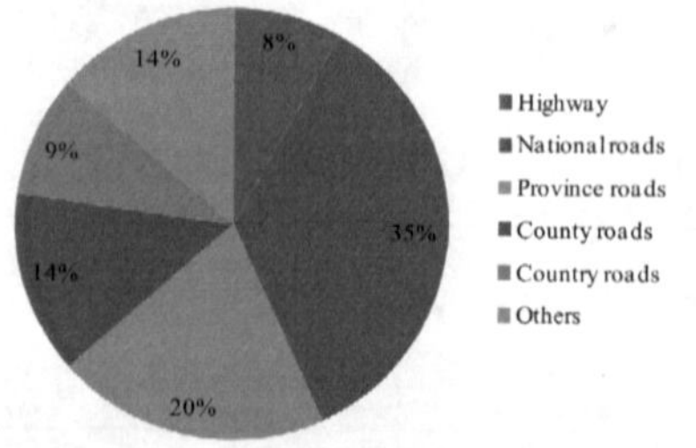

Figure 9. The distribution of road types related to SF-RTAs

It could be seen from Fig.9 that, for the 334 SF-RTAs from 2004 to 2014, the accidents related to national roads, province roads and county roads accounted for 35%, 20% and 14%, respectively. That is, SF-RTAs were more likely to happen on national roads and province roads.

Table 2. The factors we chose to analyze SF-RTAs

Different types of factors	Variables	Definition
Human	H1	The total population
	H2	The number of motor vehicle drivers
	H3	The number of car drivers
Vehicle	V1	The number of commercial vehicles
	V2	The number of passengers cars
	V3	The number of trucks
Road	R1	Highway mileage
	R2	Expressway mileage
	R3	Grade highway mileage
	R4	Substandard highway mileage
Society	S1	Gross domestic product (GDP)

3.2 Main Factors Influencing SF-RTAs

As we all known, many factors relate to human, vehicle, road and society may have effect on SF-RTAs. Based on the data available, these factors we can obtain are shown in Table 2.

To better understand the main factors influencing SF-RTAs, we chose two analysis methods to analyze the relationship between the number of SF-RTAs and different factors in this paper, which included regression analysis and classification tree analysis.

3.2.1 Regression analysis

In the traffic field, the application of regression analysis is extensive [4], [5]. In this paper, we chose linear regression module of SPSS to analyze the main factors that affected SF-RTAs in different provinces. The number of SF-RTAs occurred in different province in each year was considered as dependent variable. The factors in Table 2 were independent variables. Then, for each province, the correlation between the number of SF-RTAs and various factors (including human, vehicle, road and society) could be identified. Table 3 shows the variables entered the established model for each province.

Table 3. Variables entered the model for each province

Province		Province	
Name	*Variables Entered*	*Name*	*Variables Entered*
Beijing	S1, H1, H2, V2, V3, R1, R2, R4	Shanxi	S1, H1, H2, H3, V1, V2, R2, R3, R4
Tianjin	S1, H1, H2, H3, V2, V3, R1, R2, R4	Mongolia	S1, H1, H2, H3, V1, V2, R1, R2, R4
Hebei	S1, H1, H2, V1, V2, R1, R2, R4	Liaoning	S1, H1, H2, H3, V1, V2, R2, R3, R4
……	……	……	……

The results showed that, S1, H1 and R4 (denoted gross domestic product, the total population and substandard highway mileage, respectively) almost entered all regression models. This indicated that these 3 variables had larger impact on the number of SF-RTAs occurred in China than other factors. In addition, there existed significantly positive relationships between the number of SF-RTAs and H1 as well as R4 for all linear regression models.

3.2.2 Decision tree analysis

Decision tree analysis is a popular statistical method which emerges from the field of machine learning and data mining [6]-[8]. In the traffic field, the application of decision trees is also extensive [10], [11]. However, no study has been conducted to date on the analysis of SF-RTAs in China. Thus, in this study, we also chose decision tree method to analyze the influence of different factors on SF-RTAs. The main factors influencing the number of SF-RTAs can be obtained by the analysis for root node and leaf nodes in the decision tree results. Similarly, the number of SF-RTAs occurred in different province in each year was the dependent variable of the decision tree model. The factors shown in Table 2 were the independent variables. Fig. 10 shows the result of the established decision tree model.

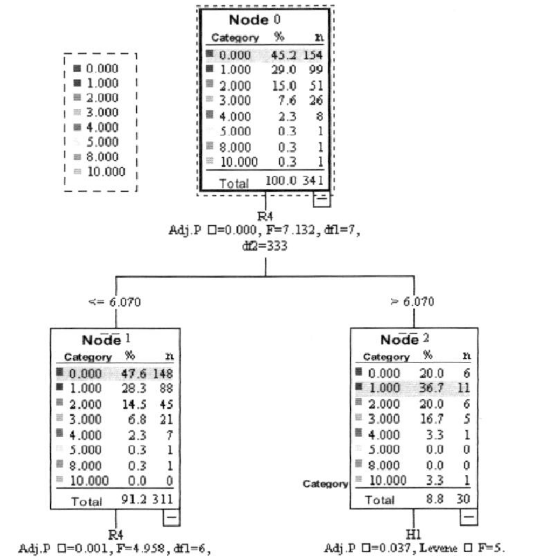

Figure 10. The result of the established decision tree model

According to Fig. 10, the main factors that affected the number of SF-RTAs were R4 (the variable occurred in Node 0 and Node 1) and H1 (the variable occurred in Node 3). This result was similar to the result of regression analysis above. It showed that if we wanted to control the number of SF-RTAs, we should focus on R4 (substandard highway mileage) and H1 (the total population) in each province. Additional detailed descriptions of these decision tree algorithms were beyond the scope of the present study. For further discussions of the tree methodology, please refer to the study of Breiman *et al.* [9].

4 Discussion and conclusion

According to the analysis, SF-RTAs in China have special statistic characteristics. Data distribution in Fig. 2 and Fig. 3 shows the number SF-RTAs occurred in western region of China is the most, especially in southwest region. According to the statistics of China's Ministry of Public Security (Ministry of Public Security Traffic Management Bureau, 2015), the provinces in south-west region have more hilly land, where topography, meteorology, hydrology, and geological structure conditions are different with other places. These reasons may create dangerous driving conditions and result in more SF-RTAs. Thus, it is needed to deeply analyze accident mechanism for western provinces, and then put forward targeted strategies for preventing SF-RTAs.

The analysis for time distribution of SF-RTAs (as shown in Fig. 4) shows that SF-RTAs occurred between 6:00 to 8:00 and between 12:00 to 16:00 are more than in other time periods. This is because 6:00 to 8:00 belongs to morning rush hour of China, and traffic volume usually increases in this period. During 12:00 and 16:00, drivers are more likely to feel drowsy, and maybe cannot focus on driving. These may make SF-RTAs occurred in these periods more than in other time periods. Fig. 5 indicates that SF-RTAs occurred on Saturday are more than on other weekdays. As we all known, Saturday is the first day of weekend in China, thus, more people choose this day to go out for travel. On this day, the traffic volume of the roads increases and traffic problems, such as traffic congestion, traffic delay and road rage of drivers are more serious. These can result in the number of SF-RTAs occurred on Saturday increases. In addition, Fig.6 demonstrates SF-RTAs occurred in August and February is more than in other months. This is because, in China, August is the hottest month, when drivers are easily distracted in driving. And the spring festival of China (i.e. the Chinese New Year) is usually in February, which is a traditional day for family reunions. Many people choose to go home or go out to travel. These may make the number of SF-RTAs occurred in above two months more than in other months. Based on the analysis above, we may be able to reduce SF-RTAs by strengthening traffic control measures at accident prone periods and increasing traffic controllers in key and dangerous road sections.

Fig. 7 shows accidents related to run-off-road are the most. Thus, we'd better deeply analyze accident causes of run-off-road accidents, and take appropriate methods to contain devastating accidents with group deaths and accidents. The analysis of Fig. 8 shows that large buses and medium buses more easily result in SF-RTAs. That is, if we want to reduce SF-RTAs, it is important to enhance supervision of large buses and medium buses and strengthen the safety education and training to these vehicle drivers. Besides, Fig. 9 indicates SF-RTAs are more likely to happen on national roads and province roads. It suggests that, to control and decrease SF-RTAs, an effective way is to increase safety control equipments and put forward effective strategies for national and province roads.

The main factors influencing SF-RTAs are analyzed by linear regression and decision tree methods in this study. Table 3 and Fig.10 indicate that the total population (H1) and substandard highway mileage (R4) have the strongest influence on the number of SF-RTAs in each province. The positive coefficients of H1 and R4 imply that the number of SF-RTAs increases with increase in H1 and R4. That is, if we want to reduce the number of SF-RTAs, another method is to strengthen the safety education to the drivers in the provinces with larger population and decrease the mileage of substandard highway in these areas.

We hope this analysis can attract attention of traffic administrative department and be helpful to reduce SF-RTAs, ensure traffic safety, and protect human lives and properties in China. In addition, one of the main limitations of this study is the unavailability of a long enough time series in the period analyzed. More regional data (such as weather information, motorization, use of seat belts, quality of the road network, etc.) will also help improve the analysis of serious fatal road traffic accidents (SF-RTAs) in China. Furthermore, we will choose more variables and analyze the relationship between the SF-RTAs and various variables, and then investigate the mechanism of SF-RTAs for each province and propose more effective measures to prevent SF-RTAs.

Acknowledgment

This work was supported by the Opening Project of Key Laboratory of Road Safety Technologies, Ministry of Transport, P.R.China (2015RST06) and the Fundamental Research Funds for the National-level Research Institutes (Z2060302150009038).

References

1. WHO. *Global status report on road safety*, (2015).
2. M. Stanislav, S. Henrikas, Transport **28**, 108-115, (2013).
3. The Ministry of Public Security Traffic Management Bureau. *Road accident statistics annual report of the people's republic of China* (2004-2014).
4. D. E. Clark, B. M. Cushing, Accident Analysis Prevention **36**, 967-972, (2004).
5. R. Fu, Y. Guo, W. Yuan, H. Feng, Y. Ma, Safety Science **49**, 416-420, (2011).
6. M. G. Washington, M. G. Karlaftis, F. L, *Mannering, statistical and econometric methods for transportation data analysis* (Chapman & Hall, London, 2003).
7. P. Kitsantas, M. Hollander, L. Li, Artificial Intelligence in Medicine **38**, 275-289, (2006).
8. T. S. Lee, C. C. Chiu, Y. C. Chou, C. J. Lu, Computational Statistics and Data Analysis **50**, 1113-1130, (2006).
9. L. Breiman, J. Friedman, C. J. Stone and R. A. Olshen, *Classification and regression trees* (Chapman & Hall, London, 1984).
10. L. Y. Chang, J. T. Chen, Safety Science **51**, 17-22, (2012) .
11. O. de Juan, O. de Rocio and F. J. Calvo, Expert Systems with Applications **39**, 11164-11171, (2012).

DOI: 10.1051/matecconf/20168102014

Modeling Crossing Behavior of Drivers at Unsignalized Intersections with Consideration of Risk Perception

Miaomiao Liu[1], Yongsheng Chen[1] and Guangquan Lu[2]

[1]*Research Institute of Highway, Ministry of Transport, 8 Xitucheng Road, Haidian District, Beijing 100088, China*
[2]*School of Transportation Science and Engineering, Beihang University, 37 Xueyuan Road, Haidian District, Beijing 100191, China*

Abstract. Drivers' risk perception is vital to driving behavior and traffic safety. In the dynamic interaction of a driver-vehicle-environment system, drivers' risk perception changes dynamically. This study focused on drivers' risk perception at unsignalized intersections in China and analyzed drivers' crossing behavior. Based on cognitive psychology theory and an adaptive neuro-fuzzy inference system, quantitative models of drivers' risk perception were established for the crossing processes between two straight-moving vehicles from the orthogonal direction. The acceptable risk perception levels of drivers were identified using a self-developed data analysis method. Based on game theory, the relationship among the quantitative value of drivers' risk perception, acceptable risk perception level, and vehicle motion state was analyzed. The models of drivers' crossing behavior were then established. Finally, the behavior models were validated using data collected from real-world vehicle movements and driver decisions. The results showed that the developed behavior models had both high accuracy and good applicability. This study would provide theoretical and algorithmic references for the microscopic simulation and active safety control system of vehicles.

1 Introduction

Traffic safety in intersections has attracted increasing attention. According to the National Highway Traffic Safety Administration (NHTSA) [1], approximately 47% of the total 10,064,000 crashes in the United States in 2013 occurred at intersections and nearby areas. The American Institute of Transportation Engineers have declared that road intersection safety was an important subject that required careful solutions [2]. Moreover, previous researches showed that drivers' behavior at intersections could significantly affect the capacity and safety of intersections [3, 4]. Thus, conducting research on drivers' behavior at intersections is necessary.

Unlike signalized intersections, unsignalized intersections have no positive indication to inform drivers when it is appropriate to enter the intersection. Drivers' behavior at unsignalized intersections is more complex than that at signalized intersections. Each driver makes decisions about when, where and how to complete a required maneuver based on the elements of a decision context, including his/her perceptions of distance, velocity, and the performance of his/her vehicle [5]. Accordingly, the number of vehicular conflicts and accidents at unsignalized intersections is higher than that at other intersections. According to NHTSA [1] statistical data, approximately 46% of intersection crashes in the United States in 2013 occurred at unsignalized intersections. In China, crashes also occurred at unsignalized intersections in 2013, accounting for approximately 60% of total intersection crashes [6]. Therefore, we mainly focused on drivers' behavior at unsignalized intersections and established drivers' behavior models in this study.

The information analysis of drivers, particularly risk perception, is notably a key problem in understanding driving behavior and improving traffic safety [7]. The level of drivers' risk perception changes dynamically in the dynamic interaction of a driver-vehicle-environment system. Once a perceived risk is out of an accepted region (i.e., an acceptable risk perception level), a driver will accordingly speed up or slow down to adjust his/her perceived risk. That is, drivers' risk perception significantly influences driving behavior. Thus, establishing drivers' risk perception models and determining acceptable risk perception levels are important to analyze drivers' behavior at unsignalized intersections.

In addition, the lack of stop signs and roundabouts at unsignalized intersections in China causes difficulty in controlling or guiding traffic. In most cases, when one vehicle encounters another vehicle at an unsignalized intersection in China, neither of the drivers will completely stop their vehicle. Instead, one driver will gradually approach the intersection and adjust his/her driving behavior by gaming with the other vehicle. That is, the encounter between the two vehicles is actually a game process between the two drivers, and thus, can be analyzed using game theory.

In this study, we selected typical crossing cases at unsignalized intersections in China (in each case, one straight-moving vehicle encountered another straight-moving vehicle from the orthogonal direction) and developed drivers' behavior models based on risk perception and game theory. The models were expected to possibly reflect drivers' psychological characteristics and describe drivers' behavior at unsignalized intersections with improved accuracy.

2 Quantifying risk perception of drivers

Drivers' risk perception exhibits nonlinear characteristics because of the effects of various factors. Thus, describing the risk perception of drivers using one particular formula is difficult. According to recent literature [8], an adaptive neuro-fuzzy inference system (ANFIS) is a combination of neural network and fuzzy logic approaches; hence, it inherently has the advantages of both, such as having a good learning mechanism and reasoning capability. Accordingly, we adopted ANFIS to model drivers' risk perception at unsignalized intersections in China in this study.

A typical unsignalized intersection, shown in Fig. 1(a), where pedestrians as well as non-motor and motor vehicles traveled together, was selected for this study.

(a) The observed unsignalized intersection

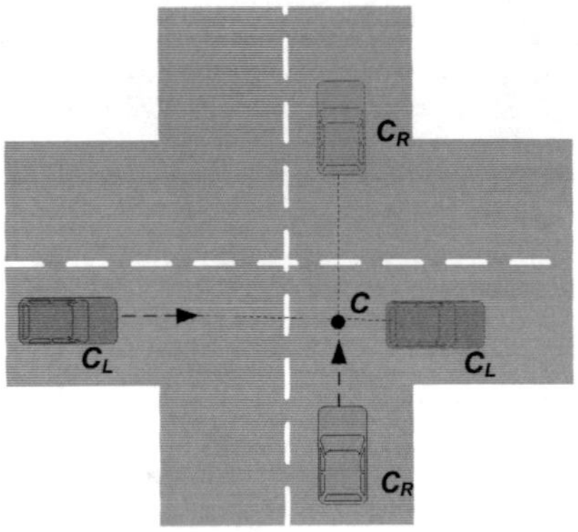

(b) The Sketch of the two vehicles' trajectories

Figure 1. The observation site

We chose a tall building on the southwest corner of the intersection and used a video-based system to record traffic conditions. Then, we selected 150 typical crossing cases, including two straight-moving vehicles from the orthogonal direction, as shown in Fig. 1(b). To avoid interference from other vehicles to drivers' crossing behavior, we selected only the simplest crossing cases. In each case, one straight-moving vehicle encountered another straight-moving vehicle and no other object. Furthermore, the vehicle from the right side was labeled as "right vehicle," that is, C_R in Fig. 1(b) in this study, whereas the vehicle from the left side was labeled as "left vehicle," that is, C_L in Fig. 1(b). Point C in Fig. 1(b) indicates the crossing point of the trajectories of the two vehicles.

2.1 Risk perception parameters

In our previous study [9], we identified the main factors that influenced the decision of straight-moving drivers when they encountered another vehicle moving at unsignalized intersections in China. For right vehicle drivers, SPEED_DIFFERENCE (i.e., the relative speed between the right vehicle and the left vehicle), DISTANCE_DIFFERENCE (i.e., the relative distance between the right vehicle to the crossing point and the left vehicle to the crossing point), and SPEED_L (i.e., the speed of the left vehicle) were main factors that affected drivers' risk perception and their yielding/preemptive decisions. For left vehicle drivers, the main factors that influenced their risk perception and decisions were SPEED_DIFFERENCE, CROSS_TIME_R (i.e., the travel time of the right vehicle to the crossing point), CROSS_DISTANCE_R (i.e., the distance between the right vehicle and the crossing point). In this study, we used these parameters to model drivers' risk perception in driving processes.

2.2 Quantitative method

In general, ANFIS has a six-layer feedforward neural network structure. For a detailed structure of ANFIS, see Jang et al. [10]. According to our previous study [9], both right and left vehicle drivers completed their yielding/preemptive decisions at 0.9 s before reaching the crossing point. To understand and model drivers' risk perception conveniently, we used the values of the input and output variables at the drivers' decision moment to train ANFIS of drivers' risk perception in this study. All the input variables were fuzzy variables, which should be described and measured using linguistic rather than precise numerical values. The membership functions of all the input variables were initially supplied exogenously. The output variable was the quantitative value of drivers' risk perception. For two drivers in a crossing process, if one driver decelerates or stops the vehicle before reaching the crossing point and makes a yielding decision, then this condition generally represents a more serious perceived risk of this driver than the other driver. Under this condition, we considered the risk perception value of this driver as "1." By contrast, if one driver makes a preemptive decision, then the risk perception value of this driver is smaller than that of the other driver. We considered the risk perception value of this driver as "0."

The fuzzy logic toolbox in MATLAB was used to develop and test the drivers' risk perception models. The specific steps were described as follows.

Step 1: Generating and inputting training and testing data. We chose 100 crossing cases as training samples and extracted the training data for this study. The parameter data obtained from the other 50 cases were used as testing data.

Step 2: Identifying the type of membership functions. In this study, a Gaussian membership function was selected to fuzzify all the input variables.

Step 3: Using the "genfis1" function to generate the original fuzzy inference structure (FIS).

Step 4: Setting the training parameters of ANFIS.

Step 5: Using the "anfis" function to train ANFIS.

Step 6: Using the "evalfis" function to test the obtained performance of FIS.

2.3 Model results

Drivers' risk perception models can be obtained using the quantitative method described earlier. Fig. 2 shows the input–output risk perception values of drivers.

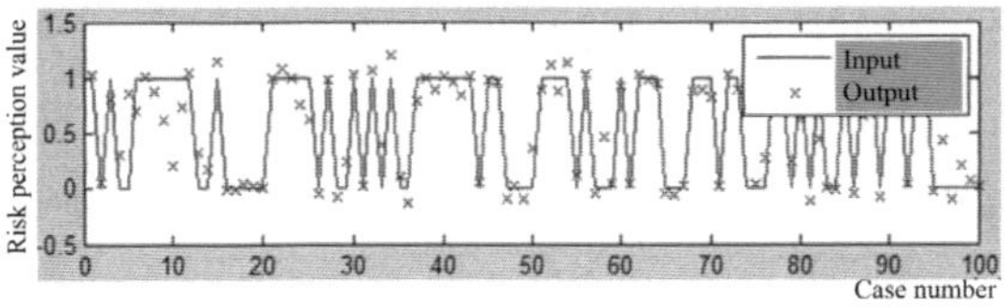

(a) For right vehicle drivers

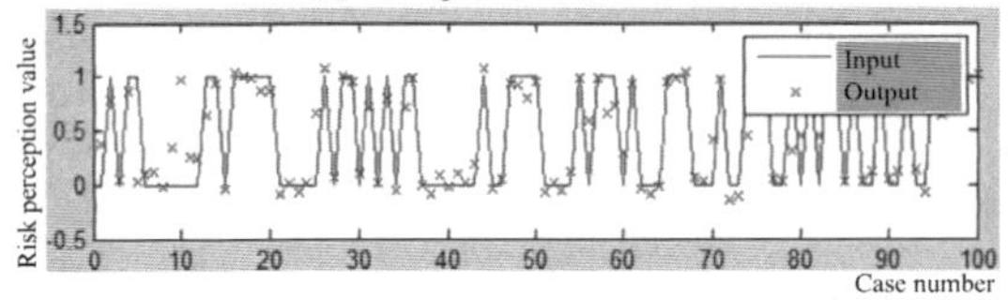

(b) For left vehicle drivers

Figure 2. Input-output risk perception values of drivers

From Fig. 2, it was known that the model result could possibly reflect the changes in drivers' risk perception within a small model error, which indicated that the model could satisfy practical requirements. The comparison between the observed and predicted values for drivers' risk perception also showed that the structure and algorithm of the established models were reasonable.

3 Indentifying acceptable risk perception level

Previous studies have indicated that the risk perception abilities of drivers varied under different characteristics of drivers during the actual driving process. However, each driver has a fixed level of acceptable risk perception that can guide his/her behavior. According to risk compensation theory (RCT), individuals tend to adjust their behavior in response to perceived changes in risk. After training, drivers should have a similar acceptable risk level in similar driving situations. In this study, we considered this level as a "range," which was assumed as [α, β]. According to risk homeostasis theory (RHT), drivers would adjust their driving procedures based on this range to maintain a stable and constant condition. When the perceived risk level of drivers was less than α, they would speed up to achieve their driving purpose. During this time, these drivers generally made preemptive decisions in the crossing process at unsignalized intersections. If the risk perception value was larger than β, then drivers would slow down and make yielding decisions to ensure safety. If the risk perception value of drivers was within the range of [α, β], then they would make either yielding or preemptive decisions when encountering another vehicle.

The acceptable risk perception level is related to the risk perception abilities of drivers and their driving operation. Thus, identifying the acceptable risk perception level is critical in this study. According to the established quantitative method, we can obtain the ranges of drivers' risk perception under preemptive and yielding conditions. Then, the acceptable risk level can be determined according to these two ranges. As shown in Fig. 3, assuming the overlapping region of both ranges is [A, B]. When the perceived risk value is within [A, B], drivers may make either yielding or preemptive decisions. Thus, [A, B] can be considered as the acceptable risk level [α, β] in this study.

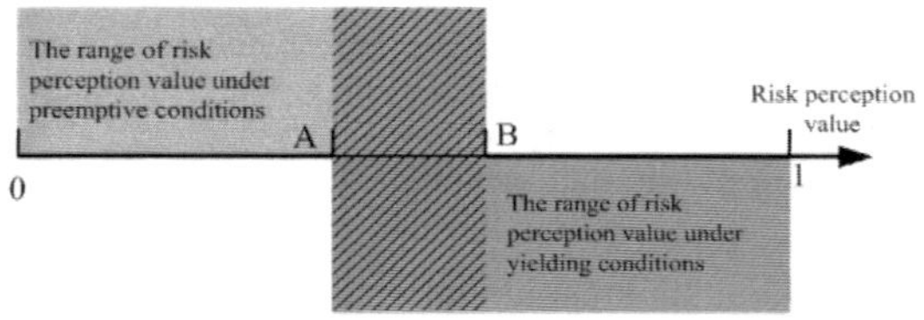

Figure 3. The sketch of drivers' acceptable risk perception level

By using the quantitative method, we determined that the range of risk perception of right vehicle drivers under preemptive conditions was [-0.31, 0.64]. However, the range of risk perception under yielding conditions was [0.59, 1.18]. Thus, the overlapping range was [0.59, 0.64]. That is, the acceptable risk perception level of right vehicle drivers was [0.59, 0.64]. For left vehicle drivers, the ranges of risk perception under preemptive and yielding conditions were [-0.26, 0.61] and [0.44, 1.17], respectively. Thus, the overlapping range [0.44, 0.61] was the acceptable risk perception level of left vehicle drivers.

4 Modeling drivers' crossing behavior

4.1 Modeling method

The lack of control measures for right-of-way at unsignalized intersections in China makes gap-forcing behavior common [9]. The previous studies and models may not be applicable in China. Moreover, they cannot explicitly consider the dynamic interactions among drivers and cognitive decision features. To address such shortcomings, an approach based on game theory inspired by the early works of Kita et al. [11] was adopted to model drivers' behavior in the present study.

In general, various factors influence drivers' crossing behavior. Among these, the velocity change of vehicles can effectively reflect the change in drivers' risk perception and their decision behavior. To simplify the model, in this study, we analyzed the game process of drivers and established game behavior models based on the following factors: vehicle speed and drivers' risk perception. In addition, a previous study [11] showed that the interaction process between right and left drivers could be considered a zero-sum non-cooperative game under complete information. This type of game indicated that crossing decisions were made when drivers were certain about the decisions of other drivers. Moreover, the payoffs of the drivers was obtained by choosing different strategies and depended on the preferences and characteristics of other drivers.

According to game theory, the player set for the crossing process in this study was $C = \{C_R, C_L\}$, as shown in Fig. 1(c). C_R and C_L might take different measures, such as acceleration, deceleration, or uniform motion, according to perceived risk. Thus, the strategy set for drivers could be considered $S = \{S_1, S_2, S_3\}$, where S_1, S_2 and S_3 denoted acceleration, uniform motion, and deceleration, respectively.

To study the game process between C_R and C_L, we assumed that the payoffs of C_R and C_L were $B_R(S_1, S_2, S_3)$ and $B_L(S_1, S_2, S_3)$, respectively. Assuming that C_R and C_L drivers began their game decision at moment T_A and completed it at moment T_E; then, we divided period $[T_A, T_E]$ into N time steps.

Assuming that the velocity of C_R and C_L at the beginning of time step i were $V_R{}^i$ and $V_L{}^i$, their acceleration (or deceleration) were $a_R{}^i$ and $a_L{}^i$, and the distance from both vehicles to the crossing point were $L_R{}^i$ and $L_L{}^i$, respectively. Assuming that the risk perception values of C_R and C_L drivers at time step i were $P_R{}^i$ and $P_L{}^i$, respectively. As mentioned earlier, SPEED_DIFFERENCE (i.e., $V_R{}^i - V_L{}^i$), DISTANCE_DIFFERENCE (i.e., $L_R{}^i - L_L{}^i$), and SPEED_L (i.e., $V_L{}^i$) were the main factors that affected the risk perception of right vehicle drivers. Meanwhile, the main factors that affected the risk perception of left vehicle drivers were SPEED_DIFFERENCE, CROSS_TIME_R (i.e.,$T_R{}^i$), and CROSS_DISTANCE_R (i.e., $L_R{}^i$). We assumed that all drivers would continue driving with the same acceleration at time step i to reach the crossing point. Thus, $T_R{}^i$ can be calculated using Equation 1:

$$T_R{}^i = \sqrt{({V_R{}^i}/{a_R{}^i})^2 + {2L_R{}^i}/{a_R{}^i}} - ({V_R{}^i}/{a_R{}^i}) \tag{1}$$

Then, $P_R{}^i$ and $P_L{}^i$ could be obtained according to the established risk perception models.

For time step i+1, the velocity of C_R and C_L at the beginning ($V_R{}^{i+1}$ and $V_L{}^{i+1}$), the distance from both vehicles to the crossing point ($L_R{}^{i+1}$ and $L_L{}^{i+1}$), and the velocity change of both vehicles ($\Delta V_R{}^{i+1}$ and $\Delta V_L{}^{i+1}$) could be estimated as follows:

$$\begin{cases} V_j{}^{i+1} = V_j{}^i + a_j{}^i \Delta t^i \\ L_j{}^{i+1} = L_j{}^i - V_j{}^i \Delta t^i - {a_j{}^i (\Delta t^i)^2}/{2} \\ \Delta V_j{}^{i+1} = V_j{}^{i+1} - V_j{}^i = a_j{}^i \Delta t^i \end{cases} \tag{2}$$

where $j = R, L$; $i = 1, 2, \ldots, N-1$. Δt^i was the time length at time step i.

The risk perception value of C_R driver at time step i+1 ($P_R{}^{i+1}$) could be analyzed by considering the parameters ($V_R{}^{i+1} - V_L{}^{i+1}$, $L_R{}^{i+1} - L_L{}^{i+1}$, and $V_L{}^{i+1}$) in the established risk perception model of right vehicle drivers. Where,

$$\begin{cases} V_R{}^{i+1} - V_L{}^{i+1} = V_R{}^i + a_R{}^i \Delta t^i - V_L{}^i - a_L{}^i \Delta t^i \\ L_R{}^{i+1} - L_L{}^{i+1} = L_R{}^i - V_R{}^i \Delta t^i - {a_R{}^i (\Delta t^i)^2}/{2} \\ \qquad -L_L{}^{i+1} + V_L{}^i \Delta t^i + {a_L{}^i (\Delta t^i)^2}/{2} \\ V_L{}^{i+1} = V_L{}^i + a_L{}^i \Delta t^i \end{cases} \tag{3}$$

At time step i+1, the risk perception value of C_L driver ($P_L{}^{i+1}$) could be determined by considering parameters ($V_R{}^{i+1} - V_L{}^{i+1}$, $T_R{}^{i+1}$, and $L_R{}^{i+1}$) in the risk perception model of left vehicle drivers. Where,

$$\begin{cases} V_R{}^{i+1} - V_L{}^{i+1} = V_R{}^i + a_R{}^i \Delta t^i - V_L{}^i - a_L{}^i \Delta t^i \\ T_R{}^{i+1} = T_R{}^i - \Delta t^i \\ L_R{}^{i+1} = L_R{}^i - V_R{}^i \Delta t^i - {a_R{}^i (\Delta t^i)^2}/{2} \end{cases} \tag{4}$$

Similarly, the risk perception values of both drivers at time steps $i+2$, $i+3$ to N could be calculated.

For both drivers, when the increase in velocity change was high and the risk perception value was low, the payoff of the drivers was high. Thus, we defined the payoff function of the drivers as follows:

$$F_j{}^i = m_j{}^i F(\Delta V_j{}^i) + n_j{}^i G(1 - P_j{}^i) \tag{5}$$

where $j = R, L$; $i = 1, 2, \ldots, N$. $F(\Delta V_j{}^i)$ and $G(1 - P_j{}^i)$ denoted the normalization for $\Delta V_j{}^i$ and $(1 - P_j{}^i)$, respectively. $m_j{}^i$ and $n_j{}^i$ represented the expectation coefficients of the drivers to vehicle speed and risk perception, and $m_j{}^i \geq 0$, $n_j{}^i \geq 0$, $m_j{}^i + n_j{}^i = 1$.

In actual driving processes, the speed and risk perception factors had different influences on drivers' payoff. If the risk perception value of a driver was lower than his/her acceptable risk perception level, then he/she would focus on improving vehicle speed. If the perceived risk value of a driver was higher than his/her acceptable risk perception level, then he/she would focus on changing risk perception. However, if the risk perception value of a driver was within the range of the acceptable risk level, then we could conclude that the speed and risk perception factors were equally important to the driver's payoff. Therefore, based on the acceptable risk perception level of drivers $[\alpha_j, \beta_j]$, the speed coefficient $m_j{}^i$ can be defined as follows:

$$m_j{}^i = \begin{cases} 1 - {0.5P_j{}^i}/{\alpha_j}, & P_j{}^i < \alpha_j \\ 0.5, & \alpha_j \leq P_j{}^i \leq \beta_j \\ {0.5(1 - P_j{}^i)}/{(1 - \beta_j)}, & P_j{}^i > \beta_j \end{cases} \tag{6}$$

To analyze the game behavior between C_R and C_L drivers, solving the game model at each time step was necessary. The method for this process was as follows.

Step 1: According to the established motion state and risk perception model, the velocity changes of the two drivers and their risk perception values at the first time step (i.e., $\Delta V_R{}^1$, $\Delta V_L{}^1$, $P_R{}^1$, and $P_L{}^1$) under different strategy combinations, such as $\{S_1, S_1\}$, $\{S_1, S_2\}$, $\{S_1, S_3\}$, $\{S_2, S_1\}$, $\{S_2, S_2\}$, $\{S_2, S_3\}$, $\{S_3, S_1\}$, $\{S_3, S_2\}$, and $\{S_3, S_3\}$, were calculated.

Step 2: The payoff values of the two drivers under different strategy combinations were evaluated based on Equation 5.

Step 3: The Nash equilibrium strategy at the first time step was obtained according to the payoff value, and consequently, the behavior strategies of the two drivers were determined. Then, the motion information of the two vehicles at the beginning of the second time step was calculated based on the determined behavior strategies.

Step 4: According to the motion state and the established risk perception models, the velocity changes of the two drivers and their risk perception values at the second time step under different strategy combinations were calculated. By repeating steps 2 and 3, the behavior strategies of the two drivers at the second time step and the motion state of the two vehicles at the beginning of the third time step were obtained.

Step 5: Similarly, the behavior strategies of the two drivers at time steps 3 to N were calculated.

4.2 Model results

To facilitate understanding, we analyzed the game behavior of the drivers within a certain period as follows: 2 s before the preemptive driver reached the crossing point. Then, we divided this period into 10 time steps. Each time step was 0.2 s. According to related literature, we assumed that if C_R and C_L drivers chose acceleration strategy (S_1), then acceleration was considered $1.0\ m/s^2$ (i.e., $a_R = 1.0\ m/s^2$, $a_L = 1.0\ m/s^2$). If the two drivers adopted deceleration strategy (S_3), then deceleration was determined as $1.5\ m/s^2$ (i.e., $a_R = -1.5 m/s^2$, $a_L = -1.5\ m/s^2$). If the drivers chose uniform motion (S_2), then $a_R = a_L = 0$.

For a crossing case (C_R passed the crossing point first), the motion state at the beginning of the first time step was $L_R{}^1 = 9.0$ m , $L_L{}^1 = 12.3$ m , $V_R{}^1 = 4.8$ m/s , $V_L{}^1 = 5.0$ m/s , $a_R{}^1 = 0$, $a_L{}^1 = 0$. As described earlier, the acceptable risk perception level of the right vehicle driver was [0.59, 0.64]; thus, $\alpha_R = 0.59$, $\beta_R = 0.64$. The acceptable risk perception level of the left vehicle driver was [0.44, 0.61]; that is, $\alpha_L = 0.44$, $\beta_L = 0.61$.

Based on Equation 5, the payoff values of the two drivers under different strategy combinations at the first time step were calculated. The result showed that at the first time step, the Nash equilibrium strategy was both C_R and C_L decelerated when the payoffs of C_R and C_L drivers were the largest, that is, 0.47 and 0.54, respectively. Then, we could determine the motion information of the two vehicles at the beginning of the second time step according to the estimated strategies: $L_R{}^2 = 11.27$ m , $L_L{}^2 = 7.97$ m , $V_R{}^2 = 4.7$ m/s , and $V_L{}^2 = 4.5$ m/s . Similarly, the Nash equilibrium at the second time step could be inferred, when both C_R and C_L also decelerated. Based on this analogy, the Nash equilibrium strategy at each time step could be determined. Table 1 shows the estimated drivers' strategies when reach each Nash equilibrium.

Table 1. The estimated strategies when reach each Nash equilibrium point

Parameters		Time step									
		1	2	3	4	5	6	7	8	9	10
Game strategy	C_R	S_3	S_3	S_3	S_1	S_1	S_1	S_1	S_1	S_1	S_1
	C_L	S_3	S_3	S_3	S_3	S_3	S_3	S_3	S_3	S_3	S_3

4.3 Model test

As shown in Table 1, both C_R and C_L drivers decelerated at the beginning of the crossing game, and then C_L continued to decelerate; however, C_R chose to accelerate. According to the estimated strategies, vehicle speed and distance to the crossing point at the beginning of each time step could be calculated. Fig. 4 shows the comparison between the estimated and the actual results.

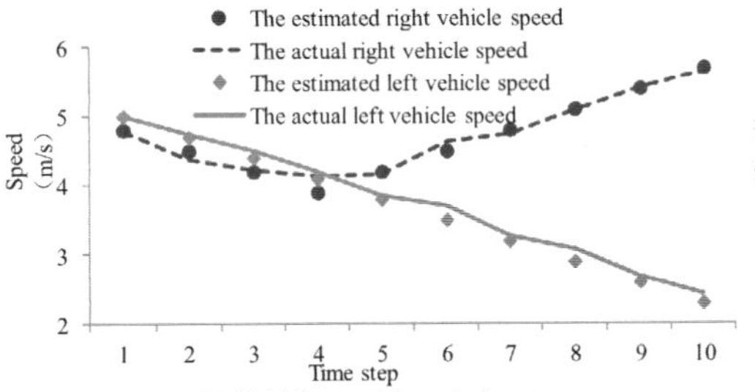

(a) Vehicle speed in each time step

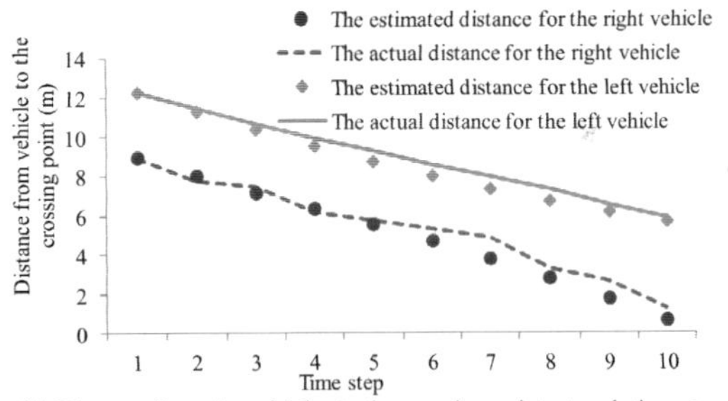

(b) Distance from the vehicles to the crossing point at each time step

Figure 4. The comparison between the estimated and the actual results

Fig. 4(a) showed the constant deceleration of C_L driver in actual driving. Meanwhile, C_R driver first decelerated and then accelerated. This indicated that the estimated result was consistent with the actual result. Based on the estimated strategies in Table 2 and Equation 2, the distance from both vehicles to the crossing point at the beginning of each time step could be estimated. Fig. 4(b) showed the estimated distance from the two vehicles to the crossing point could accurately reflect the actual distance. In addition, C_R was estimated to be closer to the crossing point than C_L at the last time step. That was, C_R driver would make a preemptive decision and pass the crossing point first. This result was also consistent with the observed data.

In this study, we also established the game models for another 50 crossing cases, and then calculated the game strategies of the drivers. We estimated vehicles' speed, distance from the vehicles to the crossing point, and final preemptive or yielding decisions of the drivers. Table 2 shows the error of the estimated speed and distance.

Table 2. The error of the estimated results

Parameters	Error	
	Mean error	***Standard error***
Right vehicle speed (m/s)	0.164	0.236
Left vehicle speed (m/s)	0.165	0.254
Distance from the right vehicle to the crossing point (m)	0.450	0.674
Distance from the left vehicle to the crossing point (m)	0.420	0.573

As shown in Table 2, the error between the estimated and the actual results was minimal. It indicated the established models could efficiently describe drivers' behavior. By comparing the estimated and the actual preemptive/yielding decision, 48 cases were estimated accurately. Accuracy was 96%.

5 Discussion and conclusions

In this study, we focused on typical crossing cases at unsignalized intersections. In each case, one straight-moving vehicle encountered another straight-moving vehicle. Based on one of our previous studies on typical crossing cases, the main parameters that affected the risk perception of right and left vehicle drivers were determined. Then, we used these parameters to establish the quantitative models of drivers' risk perception based on ANFIS. By comparing the actual data and the model prediction results, we verified the practicality of the proposed quantitative risk perception models. Then, according to a data analysis method that we developed, the acceptable risk perception levels for right and left vehicle drivers were identified.

Based on game theory, the relationship among the quantitative value of risk perception, the acceptable risk perception level, and vehicle motion state was analyzed. By analyzing the Nash equilibrium strategy, the acceleration/deceleration strategies of the drivers at each time step could be identified. Then, we could calculate vehicle speed and the distance from a vehicle to the crossing point at the beginning of each time step. It was shown that the estimated results were consistent with the fact. By analyzing a test intersection, the established game models were proven to have good applicability.

In summary, this study could promote the development of risk analysis theory from qualitative analysis to quantitative calculation, as well as expand the application of this theory to traffic safety. Moreover, we considered drivers' risk perception factor and used game theory to model their behavior. Such efforts made the model accurate and reasonable. Furthermore, the results of this study could provide a good theory and algorithm for the microscopic simulation and active safety control system of vehicles.

Acknowledgment

This work was supported by the Opening Project of Key Laboratory of Road Safety Technologies, Ministry of Transport, P.R.China (2015RST06) and the Fundamental Research Funds for the National-level Research Institutes (Z2060302150009038).

References

1. NHTSA. *Traffic safety facts 2013* (2015).
2. N. Elmitiny, X. Yan, E. Radwan, C. Russo, and D. Nashar, Accident Analysis and Prevention **42**, 101-111, (2010).
3. Y. Xiao, Q. Ran, J. Yang, and Z. Wang, Journal of Transportation Engineering **137**, 121-127, (2011).
4. A. Sharma, D. Bullock, S. Peeta, Transportation Research Part C **19**, 400-412 (2011).
5. TRB, Transportation Research Record **468**, 113-122, (1997).
6. National Bureau of Statistics of China, *Statistical yearbook of China* (2014).
7. F. P. McKenna, M. S. Horswill, J. L. Alexander, Journal of Experiment Psychological Application **12**, 1-10, (2006).
8. J. P. Sangole, G. R. Patil, Journal of Modern Transportation **22**, 235-243, (2014).
9. M. Liu, G. Lu, Y. Wang, Transportation Research Part F **24**, 244-255, (2014).
10. J. Jang, C. Sun, E. Mizutani. *Neuro-fuzzy and Soft Computing-A Computational Approach to Learning and Machine Intelligence* (Prentice Hall, NJ, 1997).
11. H. Kita, Transportation Research Part A **33**, 305-312, (1999).

Research on Heavy Truck Dynamic Load Coefficient and Influence Factors

Peng Hu, Xiao Ning Zhang and Sen Tian

Civil Engineering Department, Shan Dong Jiao Tong University, Ji'nan 250031, China

Abstract. In order to research the vehicle dynamic load coefficient(DLC) and influence factors, high class and low class pavement roughness is tested in fields, 2 degree of freedom vehicle model is built, DLC computing formula is given. Heavy truck is taken as an example, the relationship among DLC and traffic speed, vibrating frequency, loading mass are analyzed. The results show that resonation frequency of heavy truck is about 1HZ and 12HZ, the DLC increases with the increase of traffic speed while decreases with the increase of the loading. The relationship between DLC and travel speed(V) can be described as DLC=0.0047V0.5 for high class road and DLC=0.0102V0.5 for low class road

1 Introduction

Vehicle load is a kind of dynamic load which varies in size and frequency. When the pavement is applied with dynamic load, its response is different from static load. Some failure types are caused by dynamic load, such as fatigue cracking, rutting, etc. It is necessary to study the characters of dynamic load which is related with vehicle structure, loading and traffic speed.

In China the dynamic load mainly focus on theory of size, frequency. Huang[1](1993),Sun and Deng[2] from Southeast of China carried out the research of relationship between pavement roughness and dynamic load. From then on more scholars started to carry out the research of dynamic load characters, Zhang(2005), Song(2007), Yao(2008) and Zheng(2009) [3-6] have built multi degree of freedom vehicle model to research vehicle-pavement interaction force. In others countries the research started earlier, Dodds and Roson take pavement roughness as a function of vehicle departing off flat pavement and power spectral density (PSD) is employed to calculate dynamic load [7]. Todd（1987）[8]analyzed dynamic load by quarter truck model and half truck model. Hunt(1991)[9]gave calculating model of pavement vibrating PSD. With the development of research, Watts（1992）[10] started to calculate the vehicle vibrating characters when vehicle travel across pavement distress, such as pavement depression, rutting, pot holes. By means of a three-dimensional truck model, Javier(2013) [11] found that the vertical dynamic load applied by a tyre to the pavement depends not only on the profile under that wheel but also on the profile under the other wheels, mainly under the one on the same axle.Dae-Wook [12] formulated a continuous or distributed model using a two-dimensional (2D) half-truck finite element model. Using the 2D half-truck finite element model, numerical simulations were performed to obtain the dynamic loads using various parameters such as the road roughness, vehicle speed, suspension stiffness and damping in order to evaluate their individual effects on the dynamic axle load response.All these researches have supplied basic theories for dynamic load characters

In this research high class and low class pavement roughness is tested in field, relationship of pavement roughness indexes among variance(σ), international roughness index(IRI) and PSD is analyzed. Heavy truck is taken as example, DLC is calculated and influence factors are analyzed.

2 Pavement roughness and testing results in field

2.1 PSD of pavement roughness

According to the standard (GB/T 7031-2005)[13], PSD of pavement roughness can be described with formula(1) listed below:

$$G_x(n) = G_x(n_0)[\frac{n}{n_0}]^{-w} \quad (1)$$

where: n is space frequency, it is used to denote wave numbers within 1 meter equaling to reciprocal of wave length; n0 is referenced space frequency and equals to 0.1m-1 ; Gx(n0) is PSD of pavement roughness in condition that n0 is o.1m, which is determined by pavement class listed in Table 1. w is frequency exponent, which determines structure of PSD and equals to 2.

DOI: 10.1051/matecconf/20168102015

Table 1. PSD of pavement roughness Gx(n0)

Pavement class	$G_x(n_0)\times 10^{-6} m^2/m^{-1}, n_0=0.1m^{-1}$		
	Lower limit	Average	Upper limit
A	8	16	32
B	32	64	128
C	128	256	512
D	512	1024	2048

2.2 PSD formula expressed in time frequency

When the vehicle travels across the pavement with space frequency of n（m^{-1}） with the speed of v（m/s）,time frequency f（s-1）equals to v*n, then PSD formula expressed in time frequency:

$$G_x(f)=\frac{1}{v}G_x(n_0)(\frac{n}{n_0})^{-2}=G_x(n_0)n_0^{\ 2}\frac{v}{f^2} \tag{2}$$

2.3 Testing result for pavement roughness in field

2.3.1 Pavement roughness testing in field

Changxiao road is chosen as a low class testing road and Jinan belt expressway is chosen as a high class testing road. Changxiao road is a newly built low class road, asphalt pavement, single carriage with two lanes. Jinan belt expressway is a high class road, asphalt pavement, dual carriage with two lanes in each direction. Pavement roughness testing on Changxiao road is shown in figure 1.The testing results: variance of Chang-xiao road is 1.8, International Roughness Index (IRI) of Jinan belt expressway is 1.40m/km.

2.3.2 Relationship among international roughness indexes

Some scholars have studied the relationship among international roughness indexes of IRI, σ and PSD, the conclusions are almost the same. In this article we assume[14,15]:

$$IRI=0.6\sqrt{S_q(n_0)}$$

$$\sigma=0.6IRI$$

According to the two formulas and testing results in fields, Gx(n_0) of high class road equals to 5.4, low class road Gx(n_0) equals to 25.

3 Two degrees of freedom vehicle vibration model and DLC

3.1 Two degrees of freedom vehicle vibration model

Two degrees of freedom vehicle vibration model is employed here, it also named quarter vibration model shown in Fig. 2.

Figure 1. Pavement roughness testing in field

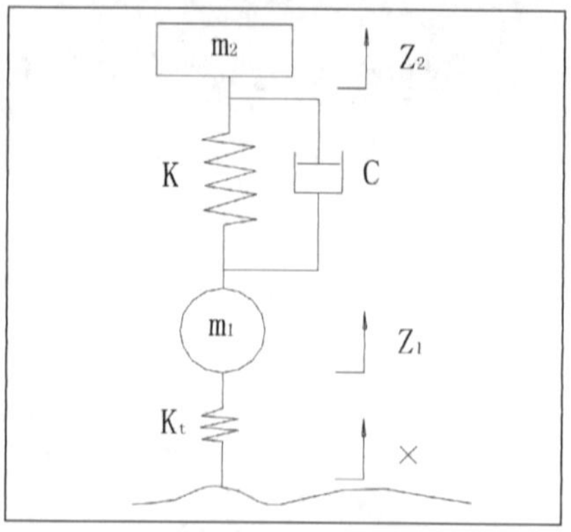

Figure 2. Quarter vehicle model

where: m1 is tire mass, m2 is mass on spring, K is suspension rigidity, Kt is tire rigidity, C is suspension damping, Z_2 is vertical displacement of m_2, Z1 is vertical displacement of m1, x is pavement roughness distance. According to Newton's laws of motion, formula(3) and (4) are obtained:

$$\mu_2\ddot{Z}_2+X(\dot{Z}_2-\dot{Z}_1)+K(Z_2-Z_1)=0 \tag{3}$$

$$\mu_1\ddot{Z}_1+X(\dot{Z}_1-\dot{Z}_2)+K(Z_1-Z_2)+K_r(Z_1-\xi)=0 \tag{4}$$

3.2 Amplitude-frequency solutions for dynamic load

According to stochastic process theory, for a linear time invariant system, if the input is stationary process, the output process is also stationary. When this system is applied to a stimulation $x_0e^{i\omega t}$ with amplitude x0 and frequency ω, after a transient lag, the output is produced which has the same form with the simulation. The output can be expressed as:

$$Z_1=Z_{10}e^{i(\omega t+\phi)}$$

$$Z_2=Z_{20}e^{i(\omega t+\phi)}$$

Z_{10} and Z_{20} are output amplitude.

DOI: 10.1051/matecconf/20168102015

Then:

$$\dot{Z}_1 = i\omega Z_1, \ \ddot{Z}_1 = -\omega^2 Z_1$$

$$\dot{Z}_2 = i\omega Z_2, \ \ddot{Z}_2 = -\omega^2 Z_2$$

Formula(3) and(4) can be transformed into (5) and (6):

$$Z_2(-\omega^2 m_2 + i\omega C + K) = Z_1(i\omega C + K) \quad (5)$$

$$Z_1(\omega^2 m_1 + i\omega C + K + K_t) = Z_2(i\omega C + K) + xK_t \quad (6)$$

From formula (5), we can get:

$$Z_2 = \frac{i\omega C + K}{-\omega^2 m_2 + K + i\omega C} Z_1 \quad (7)$$

Take formula (7) into formula(6) , frequency -response function of Z_1-x can be obtained:

$$H(\omega) = \frac{Z_1}{x} = \frac{A_2 k_t}{A_3 A_2 - A_1^2} \quad (8)$$

where: $A_1 = i\omega C + K$,

$A_2 = -\omega^2 m_2 + K + i\omega C$,

$A_3 = -\omega^2 m_1 + K + K_t + i\omega C$

3.3 Dynamic load coefficient computing

Vehicle static load $G = (m_1 + m_2)g$, dynamic load $F_d = K_t(Z_1 - x)$, dynamic load coefficient: DLC=Fd/G. The frequency-response function of DLC is:

$$H(\omega)_{F_d - x} = \frac{F_d}{Gx} = \frac{K_t(Z_1 - x)}{Gx}$$
$$= \left(\frac{A_2 k_t}{A_3 A_2 - A_1^2} - 1\right)\frac{K_t}{(m_1 + m_2)g} \quad (9)$$

PSD of dynamic load coefficient can be obtained [16]:

$$G_{F_d/G}(f) = \left|H(\omega)_{Fd-x}\right|^2 G_x(f) \quad (10)$$

As the probability of negative and positive dynamic load occurs, average dynamic load becomes zero, average DLC is also zero, so variance of DLC is the same with mean square value.

$$\sigma_{Fd/G}^{\ \ 2} = \int_0^\infty G_{F_d/G}(f) df$$
$$= \int_0^\infty \left|H(\omega)_{F_d-x}\right|^2 G_x(f) df = \mu^2_{F_d/G} \quad (11)$$

Take formula (9) into (11), we can get:

$$\mu^2_{F_d/G} = \int_0^\infty \left|H(\omega)_{F_d-x}\right|^2 G_x(f) df$$
$$= \int_0^\infty \left|\left(\frac{A_2 k_t}{A_3 A_2 - A_1^2} - 1\right)\frac{K_t}{(m_1 + m_2)g}\right|^2 G_x(n_0) n_0^2 \frac{v}{f^2} df \quad (12)$$

By formula (12), a program is developed in MATLAB which is used to compute DLC.

4 Example of heavy truck

The parameters of a heavy truck: tire mass m1 is 1000kg, mass on spring m2 is 9200kg, suspension stiffness K is 500000N/m, tire stiffness kt is 5000000N/m, suspension damping c2 is 15000Ns/m.

4.1 Relationship between PSD of DLC and vibration frequency

After computing by the program, the relationship between PSD of DLC and vibration frequency for the high class road is shown in Fig. 3 and Fig. 4. For the low class road it is shown is Fig. 5 and Fig. 6.

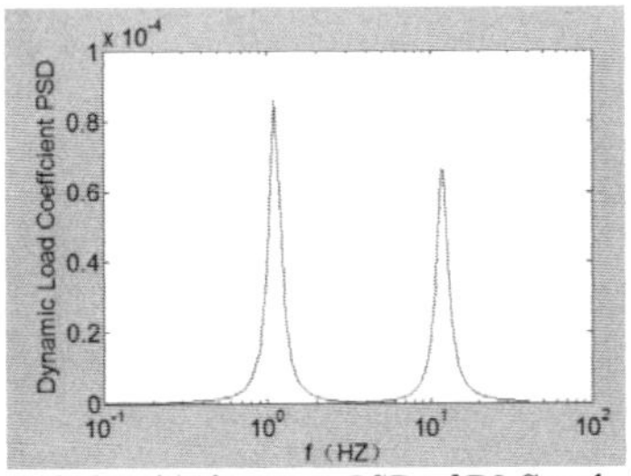

Figure 3. Relationship between PSD of DLC and vibration frequency（V=120km/h）

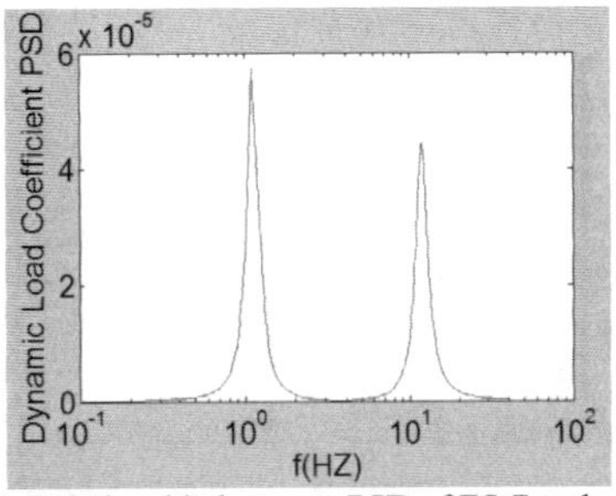

Figure 4. Relationship between PSD of DLC and vibration frequency (V=80km/h)

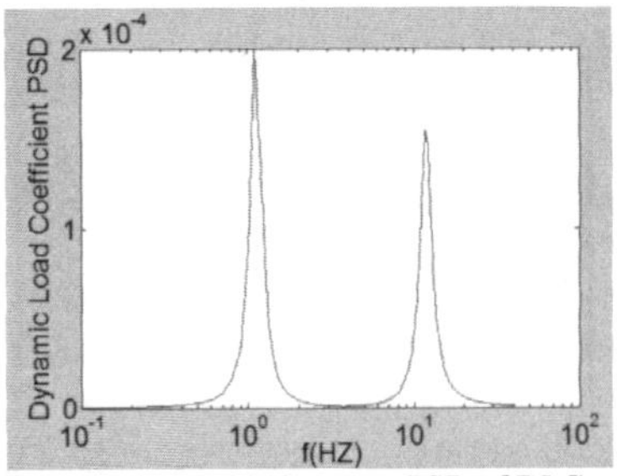

Figure 5. Relationship between PSD of DLC and vibration frequency（V=60km/h）

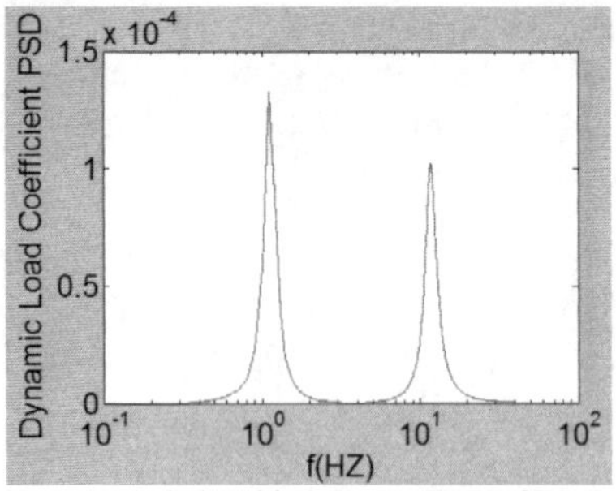

Figure 6. Relationship between PSD of DLC and vibration frequency（V=40km/h）

From the four figures, we can see that both in high class road and low class road vehicle tire will resonate at the frequency of 1HZ and 12HZ.

4.2 Relationship between DLC and travel speed

Dynamic coefficients are computed when the truck travels in different speeds and they are listed in Table 2.

Table 2. DLC in different travel speed

High class road		Low class road	
Speed(km/h)	DLC	Speed km/h	DLC
120	0.0522	80	0.0916
100	0.0467	60	0.0795
80	0.0426	40	0.0648
60	0.0369	20	0.0458

From the results in table 2, we can know that: (1)DLC is much lower on high class road than low class road; (2)DLC increases with the increase of travel speed, by regression analysis the relationship between DLC and travel speed(V) can be described as $DLC=0.0047V^{0.5}$ for high class road and $DLC=0.0102V^{0.5}$ for low class road .

4.3 Relationship between DLC and loading mass

As m2 is related with loading mass directly, it is taken as 12000kg, 8000kg and 4000kg individually, travel speed is 100km/h for high class road and 60km/h for low class road, and then DLC is calculated. The results are shown in table 3.

Table 3. Dynamic load and coefficients for different loading mass

Road class		Speed km/h	m2 /（kg）		
			12000	8000	4000
High class road	DLC	100	0.0372	0.0526	0.0921
	Dynamic load(N)	100	4464	4208	3684
Low class road	DLC	60	0.0801	0.1131	0.1981
	Dynamic load(N)	60	9612	9048	7924

From Table 3 we can know that DLC increases with the decrease of loading mass. But computing results also manifests that dynamic load decreases a little because the loading mass decreases.

5 Conclusions

PSD of pavement roughness is given according to testing results in high class and low class road, DLC computing program is developed. By analyzing DLC of heavy truck, conclusions are included:

(1) Resonation frequency of heavy truck is 1HZ and 12HZ;

(2) The relationship between DLC and travel speed (V) can be described as $DLC=0.0047V^{0.5}$ for high class road and $DLC=0.0102V^{0.5}$ for low class road .

(3)DLC decreases while dynamic load increases slightly with the increase of the loading.

Acknowledgements

This work was financially supported by the Basic Application Research Foundation of Ministry of Transport (2014319817250) and Shandong Scientific Rewarded Research Foundation from Outstanding Yong Scientists (BS2013SF007)

References

1. X. M. Huang. Relationship stochastic analysis between dynamic load and pavement roughnes. Journal of Southeast university, **23**, 1 (1993)
2. L. Sun and X. J. Deng. Dynamic Load Caused by Vehicle-Pavement Interactions. Journal of Southeast university, **26**, 5(1996).
3. S. Y. Yao and R. Y. Sun. Simulation Study on Handling and Stability Performance of Automobile Dynamic Model with 7 DOFs . Journal of Xihua university, **27**, 2(2008).
4. H.L.Zhang and C. S. Hu. A Study on The Allowable Differential Slope of The Approach Slab With Five Degree Fredom Vehicle Model. China Civil Engineering Journal, **38**, 6(2005).
5. Y. F. Song and R. F. Chen. Analysis method of vehicle vibration response caused by pavement roughness. Journal of Traffic and Transportation Engineering, **7**, 4(2007).
6. .Z. L. Zheng and P.M. Lv. Dynamic Load of Multi-axle Truck Acting on the Road, Journal of Zhengzhou University (Engineering Science), **7**, 4(2007).
7. C.J. Dodds and J.D. Robson The description of road surface roughness. Journal of sound and vibration, **31**, 2(1973).
8. K. B. Todd and B. T. Kulakowski. Simple computer models for predicting ride quality and pavement loading for heavy trucks.Transportation Research Record, 1987.
9. H E. Hunt. Modelling of road vehicles for calculation of traffic-induced ground vibration as a random process. Journal of Sound and Vibration, **44**, 1(1991).

10. G.R.Watts. The generation and propagation of vibration in various soils produced by dynamic loading of road pavement. Journal of Sound and Vibration, **156**, 2(1992).
11. O. Javier., José M.Goicolea Astiz, Á. Miguel and P. Antolín. Fully three-dimensional vehicle dynamics over rough pavement. Proceedings of the Institution of Civil Engineers: Transport, **166**, 3(2013).
12. Dae-Wook Park, A. T. Papagiannakis. Analysis of dynamic vehicle loads using vehicle pavement interaction model. KSCE Journal of Civil Engineering, **18**, 7(2014).
13. GB/T 7031-2005 Mechanical-Road surface profiles-Reporting of measured data. Beijing: Standards Press of China, 2005.
14. X. Q. Zhou and L .J. Sun. Relationship between international roughness index and speed of quarter car, Journal of Tongji University (Natural science), **33**, 10(2005).
15. Q.X. Wu, B. C. Chen and L. Z. Xi. Comparison of PSD method and IRI method for road roughness evaluation. Journal of Traffic and transportation Engineering, **8**, 1(2008).
16. Z.S. Yu. Vehicle theory (the fifth version). Beijing: China Machine Press, (2009).

MATEC Web of Conferences **81**, 02016 (2016) DOI: 10.1051/matecconf/20168102016
ICTTE 2016

Research on Secondary Impact Safety of Train Driver based on THUMS Dummy

Wenbin Wang[1], Jiawei Ji[1], Zhitao Yue[1], Dongdong Zhang[1], Honglei Tian[2] and Hongwei Zhao[2]

[1]*Institute of Rail Transit, Tongji University, Shanghai, China*
[2]*R&D Center, CRRC Qingdao Sifang Co., Ltd Qingdao, China*

Abstract. Based on biomechanical Total Human Model for Safety (THUMS) dummy and traditional rigid dummy, the train driver injury of secondary impact in railway crash events is analyzed and the results are compared. The results of THUMS dummy are more detail and comprehensive to describe the injury of train driver. It could be applied to the safety research of crashworthiness of train driver cab in future. The driver injury is serious and needed to be optimized the parameter of cab console to reduce injury risk for the driver in the next work.

1 Introduction

The injury of occupant is very serious in railway crash events, especially for the train driver. There are two main reasons. Firstly, the driver has got injured at the front zone of train. The large plastic deformation will be occurred at the train cab after the train collision and the survival space will be squeezed seriously. Secondly, owing to complicated interior structure of railway cab, the driver could be injury by interior decoration during secondary impact.

Research on railway vehicle crashworthy and occupant secondary impact safety has been developed for many years, and lots of achievements are made [1]-[7]. For example, British Rail developed a series of studies on secondary impact of occupant and passenger compartment structures, and drew up the crashworthiness for interior trim structures in vehicle and devised occupant protection standard AV/ST9001 [8]-[11]. The Volpe National Transportation Systems Center combined a number of tests with non-linear finite element simulations to research crashworthiness of railway vehicle and the occupant injury impacted with interior structure [12]-]=15]. The Japan Railway Technical Research Centre applied multi-body dynamics dummy to research secondary impact in subway [16]. However, the injury of train driver based on THUMS dummy is researched rarely at present.

In this paper, the train driver injury during secondary impact is researched based on THUMS dummy with the more detail and comprehensive injury index by THUMS.

2 Models and methods

2.1 Models

THUMS dummy model is developed by Japan's Toyota motor company (Fig. 1). All organs and skeletons are modeled to simulate body responses in crash. The injury index is more detail and reasonable than the traditional rigid dummy. So THUMS dummy is used in this research to improve some designs of structure and reduce injury of train driver in crash simulation.

The cab console is mainly made up of the equipment cabinet, the driver panel, the pedal and so on. Considering complicated structure, the model is simplified. A lot of equipments which have no effect on secondary impact are omitted (Fig. 2).

Figure 1. THUMS 50 percentile occupant dummy.

DOI: 10.1051/matecconf/20168102016

Figure 2. The finite element model of console.

2.2 Evaluation standard of biomechanical dummy

THUMS dummy model, which simulates human Biomechanical response, has a separate index evaluation system. A body or a part, an organ or a tissue under dynamic loads associated with the mechanical properties of the time is called biomechanical response. Not only to consider the stress and deformation of human organs but also physiological changes, such as pain, edema and arrhythmia. Recommended values of Human various organs and tissue damage are shown in Table 1.

Table1. Recommended values of human various organs and tissue damage.

Index / Organ	Strain	Stress
Brain tissue damage	0.21	±230kPa
Fracture	3%	*
Ligament	11%	*
Lung damage	*	±10kPa
Heart tissue damage	30%	*
Liver	30%	*
Kidney	30%	*
Spleen	30%	*
Small intestine	120%	*
Chest and blood vessels	100%	*

3 Working conditions and results

To make simulation results of the cab occupant more accurately reflecting human injury in secondary collision, THUMS dummy is applied to simulate cab occupant secondary impact damage, and compared the simulation results with the rigid dummy.

3.1 Working conditions

THUMS dummy and traditional rigid dummy sitting position are shown in figure 3. The same working conditions are set for both. Boundary condition is the triangular impact acceleration curve and the initial speed is 36km / h [14-15].

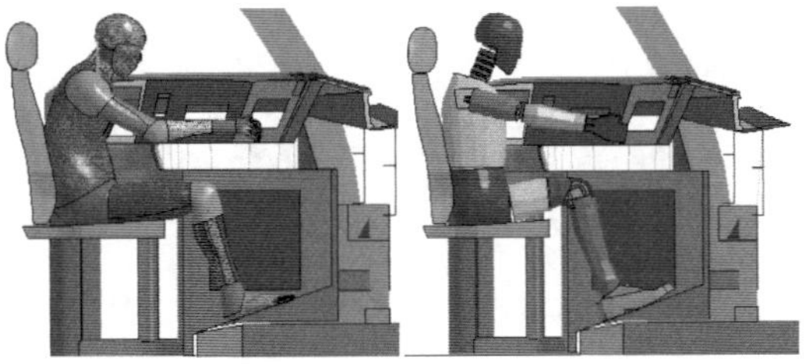

Figure 3. THUMS dummy and traditional rigid dummy.

3.2 Analysis of motion state

Dummy motions are shown in Fig. 4(a, b, c), when the collision time is 0.119s, THUMS dummy upper body including waist, hips and thigh section slides forward on the seat, and the abdomen and console bulkhead crashes. When the collision time is 0.161s, THUMS dummy's amount of compression of chest and abdomen reach the maximum and begin to recover slowly. The head started to move downward. The arm is supported on the console panel. Thigh and the lower surface of the console spacer is bonded tightly together. When the collision time is 0.225s, most abdominal compression of THUMS dummy has been restored, there is no contact with the console separator for head.

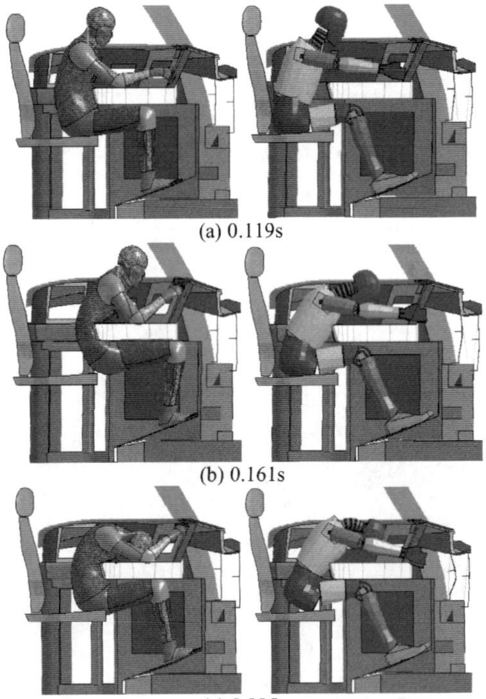

(a) 0.119s

(b) 0.161s

(c) 0.225s

Figure 4. The motion state of THUMS and rigid dummy.

As is shown in Fig. 5, Motion state of dummy is different. THUMS dummy thoracic and abdominal compression is larger than the rigid dummy. The THUMS dummy is softer and more accurate than rigid dummy. The contact force between the console and dummy is shown in Fig. 6. The contact force of THUMS dummy

 DOI: 10.1051/matecconf/20168102016

and the console is 14.9kN. However, the contact force of rigid dummy is only 4kN. Due to the different movement condition of THUMS dummy and rigid dummy, the two results are different because of the different contact part.

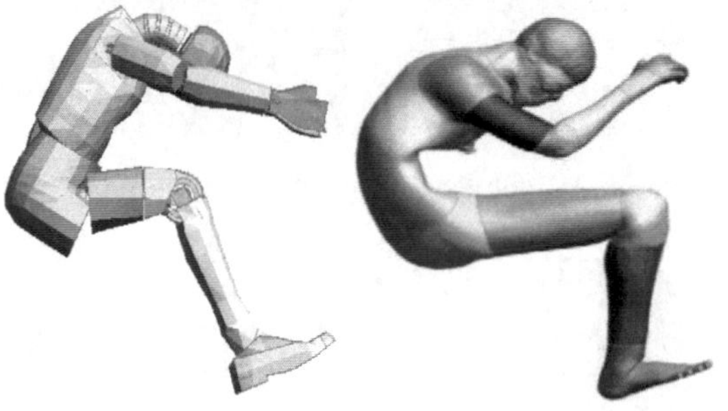

Figure 5. Deformation of dummy.

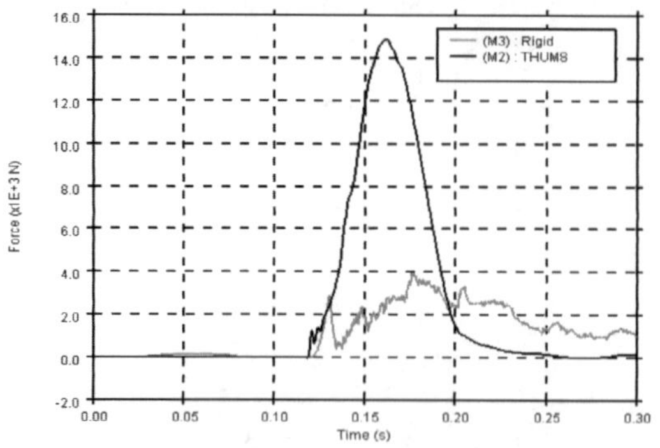

Figure 6. Curve of contact force.

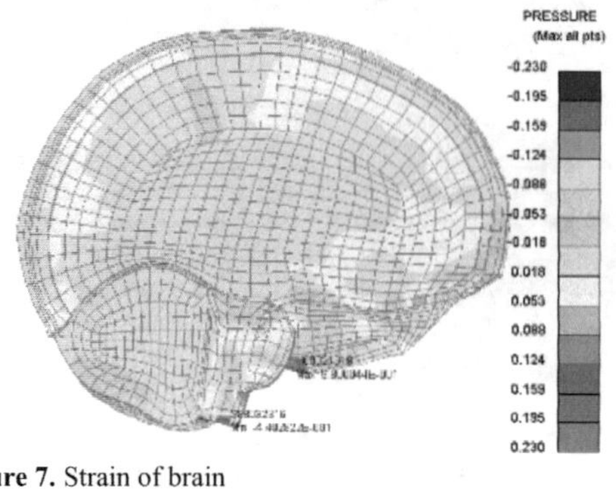

Figure 7. Strain of brain

3.3 Analysis of THUMS dummy damage

3.3.1 Head damage

The strain and stress of THUMS dummy brain injury are shown in Fig. 7 and Fig. 8. The maximum strain value of brain is 0.572, which exceeded the limited value 0.21. The maximum stress value of head is 232kPa, slightly larger than 230kPa.

The acceleration is shown in Fig. 9. HIC15 of THUMS dummy brain is 3045（center）and 251 (near the eye). There is huge difference at the different part of brain. So HIC index is not suitable for the THUMS brain damage. HIC15 of general rigid dummy brain is 195, almost same as the eye point of THUMS brain.

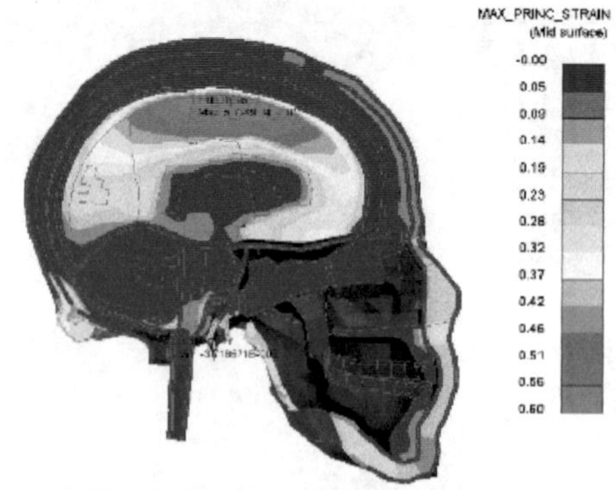

Figure 8. Stress of brain

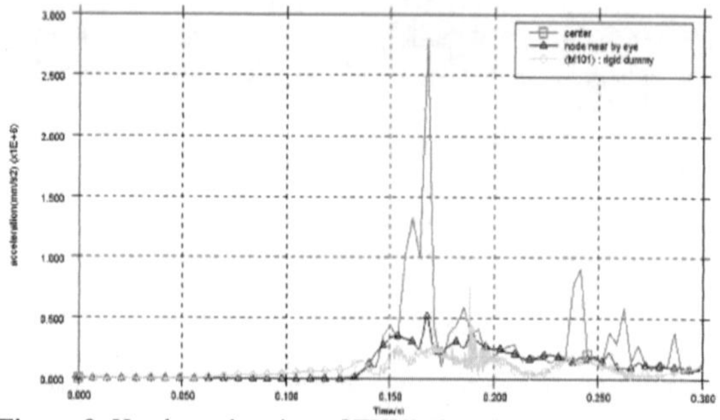

Figure 9. Head acceleration of THUMS and rigid dummy.

3.3.2 Chest and abdomen damage

THUMS dummy chest has three fractures (figure 10). Fractured ribs easily pierced soft tissue into the lungs, which resulted in haemothorax, pneumothorax and other serious injury. According to the result, most of ribs are very dangerous. The ligament bends seriously and is out of shape (Fig. 11). The maximum value of ligament strain is 1.7, which surpassed the limited value greatly. So, the possibility of the ligament injured is very large. The compression quantity of chest is shown in Fig. 12, the maximum of which is about 45 mm.

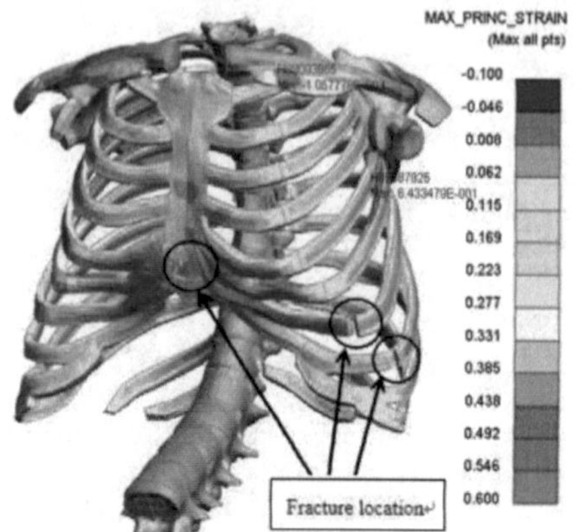

Figure 10. Strain of ribs

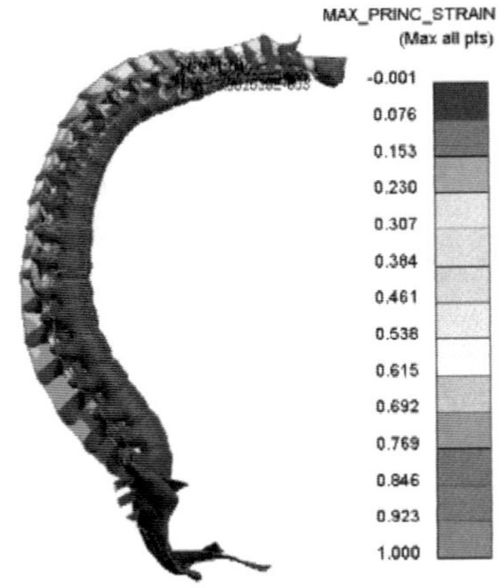

Figure 11. Strain of ligament

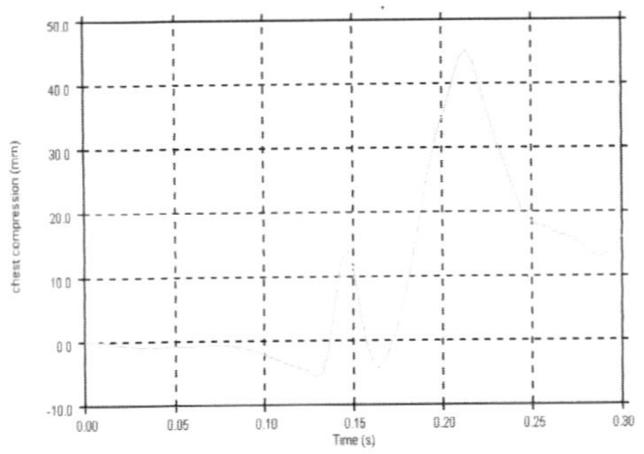

Figure 12. Compression quantity of Chest.

In secondary impact, abdominal diaphragm is squeezed by the console plate. Liver moved up because abdominal muscle tissue squeezed it. In addition, lung, heart and blood vessels are also squeezed in different extent. Nevertheless, the middle and lower part of abdomen is most serious among them.

The stress of lung is shown in Fig. 13. The maximum stress value of lung is 454kPa and the minimum is -178kPa, which is beyond the recommendation value ±10kPa, as shown in Table1. It resulted in pneumothorax and deadly injury possibly.

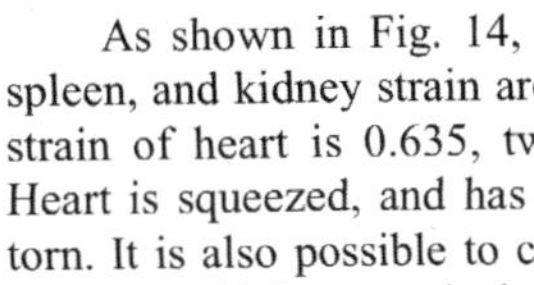
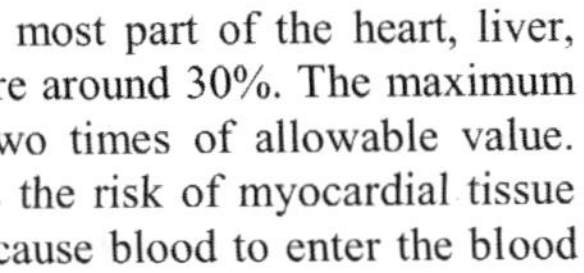

Figure 13. Lung stress **Figure 14.** Abdomen' organs strain

As shown in Fig. 14, most part of the heart, liver, spleen, and kidney strain are around 30%. The maximum strain of heart is 0.635, two times of allowable value. Heart is squeezed, and has the risk of myocardial tissue torn. It is also possible to cause blood to enter the blood vessels, which caused the rupture of the aorta. The maximum strain of liver tissue is 0.98, and the tissue may be torn, which caused the injury of liver. The maximum strain of the spleen is 0.673, which is also beyond the allowable value. The maximum strain of the kidney is 0.836, which is above the limit of 0.3.

The blood vessels of the human body have blood flow, especially the effect of the blood flow of the heart pump. The expansion rate is larger, and the strain limit is 100%. Similarly there is liquid diet in gastrointestinal tract. Especially the small intestine deep-rooted is very soft and flabby, and the damaged strain is 120%. The injury of the blood vessel and the digestive tract is shown in figure 15. The maximum strain in the small intestine is 177%, which is above 120%. So, it may appear small intestinal laceration, intestinal bleeding and other damage. The major blood vessels in the chest and abdomen are basically less than 100%, and the blood vessels are less damaged.

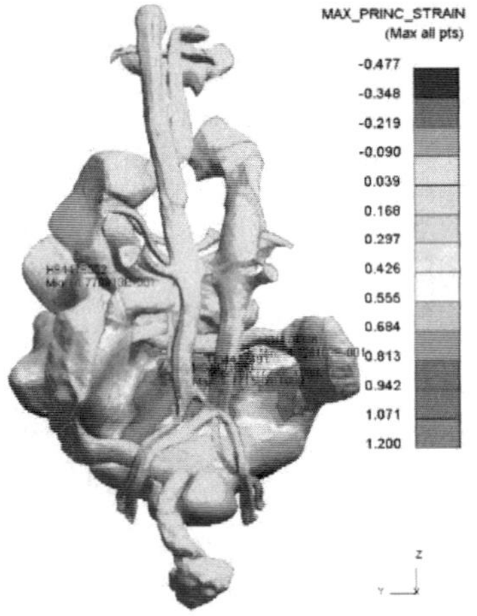

Figure 15. Strain of blood vessels and digestive system.

Simulation results showed that the THUMS dummy has lots of injuries including diffuse contusion of brain, three fractures, hemothorax, pneumothorax , different damage of internal organs and avulsion injury in the intestines. It could be determined that the wounded need emergency treatment.

Comparing with the general rigid dummy, more simulate results, such as the deformation of the skeleton, the strain of the internal organs and the damage of blood vessels can be gotten with THUMS dummy. The simulation results are more realistic than general rigid dummy. It is proposed that in the future study, THUMS dummy should be used widely in train occupant safety simulation.

From the simulation results shown, the injury is serious. The key interior part- cab console should be optimized to reduce injury risk for the driver secondary impact in the future work.

4 Conclusions

This paper describes the application of THUMS dummy in train driver secondary impact. Comparing the simulation result of THUMS dummy and the general rigid dummy, it is proved that the damage of THUMS is more detail and comprehensive in secondary impact simulation. THUMS dummy could be applied to the safety research of crashworthiness of train driver cab in the future work.

Acknowledgements

The authors would like to thank the Ministry of Science and Technology of China, the research work presented has been substantially supported by the project "The key technology research of next generation metro train and the prototype manufacture" (project No.2015BAG12B01).

References

1. Sutton A, "The development of rail vehicle crashworthiness," IMechE, Part F, vol. 216, 97-108 (2002)
2. Wang Wenbin, Kangkang, Zhao Honglun. Joint simulation of crashworthy train set based on non-linear finite element and multi-body dynamic. Journal of Tongji University, vol.39, 1552-1556 (2011)
3. Wang, Wenbin. The Crashworthiness Design of Railway Vehicle and Passive Safety of Occupant. Doctoral Dissertation, Tongji University, Shanghai, (2006)
4. Hechao Zhou, Wenbin Wang, Markus Hecht. Three-dimensional override analysis of crashed railway multiple units .Vehicle System Dynamics, vol.50, 663-674 (2012)
5. Hechao Zhou, Wenbin Wang, Markus Hecht. Three-dimensional derailment analysis of a crashed city tram, Vehicle System Dynamics, vol.51,1200-1215 (2013)
6. Xie SC, Tian HQ. Dynamic simulation of railway vehicle occupants under secondary impact. Vehicle System Dynamics. Vol.51, 1803-1817 (2013)
7. Hyung-Suk Han, Jeong-Seo Koo, "Simulation of Train Crashes in Three Dimensions", Vehicle System Dynamics, vol. 40, 435-450 (2003)
8. Lu G, "Energy absorption requirement for crashworthy vehicles," IMechE ,Part F, vol. 216, 31-39 (2002)
9. Xue X, Schmid F, Smith RA. A study of modelling approaches for rail vehicle collision behavior. International Journal of Crashworthiness, vol.9, 515-525 (2004)
10. ATOC. Vehicle interior crashworthiness. AV/ST9001. London, (2002)
11. European Standard EN 15227:2010. Railway applications-Crashworthiness requirements for railway vehicle bodies, (2010)
12. Tyrell David, "US rail equipment crashworthiness standards," IMechE, Part F, vol. 216, 123-130 (2002)
13. Tyrell David, "Rail passenger equipment accidents and the evaluation of crashworthiness strategies," IMechE, Part F, vol. 216, 131-147 (2002)
14. Kristine,Severson. Research and Development of A Safety Standard for A Safety standard for Workstation Tables in the United States. Proceeding of International Symposium on Passive Safety of Rail Vehicle, Berlin, 107-121 (2013)
15. Tyrell DC. US rail equipment crashworthiness standards. Proc Inst Mech Eng F, J Rail Rapid Transit. vol.216, 123-130 (2002)
16. oão F. Milho , Jorge A.C. Ambrósio & Manuel F.O.S. Pereira .A multibody methodology for the design of anti-climber devices for train crashworthiness simulation, International Journal of Crashworthiness, vol.7, 7-20 (2002)

Association Rule Mining on Five Years of Motor Vehicle Crashes

Jean Raymond Daher[1], Satish Chilkaka[1], Abdunnaser Younes[2], and Khaled Shaban[2]

[1]Department of Computer science and Software Engineering Concordia University, Montreal, Canada.
[2]Qatar University, Doha, Qatar

Abstract. Every year, road accidents kill more than a million people and injure more than 20 million worldwide. This paper aims to offer guidance on road safety and create awareness by pinpointing the major causes of traffic accidents. The study investigates motor vehicle crashes in the Genesee Finger Lakes Region of New York State. Frequency Pattern Growth algorithm is utilized to cultivate knowledge and create association rules to highlight the time and environment settings that cause the most catastrophic crashes. This knowledge can be used to warn drivers about the dangers of accidents, and how the consequences are worse given a specific context. For instance, a discovered rule from the data states that 'most of the crashes occur between 12:00 pm and 6:00pm'; hence, it is suggested to modify existing navigation application to warn drivers about the increase in risk factor.

1 Introduction

Road safety is a major concern all over the world. Approximately 1.3 million people die every year, and 20 to 50 million people are non-fatally injured [1]. In particular, the United States is annually marked with over 37,000 road accident fatalities and 2.5 million injuries

This paper aims to provide suggestions to improve road safety. It outlines the major causes of road accidents by applying data mining algorithms to data collected from past accident records of the Genesee Finger Lakes Region of New York State, where thousands of motor vehicle crashes occur every year. The data was posted in March 2015 by Socrata, a reliable source for analyzing large scales of recorded data [2]. The dataset encompasses 36097 records with 33 attributes. After data preprocessing, the cleaned data is analyzed and mined using the Fp-Growth data mining technique.

The rest of the paper is organized as follows: Section 2 discusses related works. Data pre-processing techniques are discussed in Section 3. Section 4 presents and discusses the results obtained. Finally, Section 5 concludes the paper and outlines potential future work.

2 Literature review

Vehicle accidents have been studied extensively. Xu *et al.* [3] Used Classification and Regression Tree (CART) and found that crash occurrences depend on the roadway characteristics, traffic and environmental factors. However, the dataset is now obsolete and covers three months only.

Fogue *et al.* [4] proposed the Automatic Accident Notification and Assistance System to estimate the severity of traffic accidents and concluded that vehicle speed is the primary factor in head-on collisions. Their findings could be used to improve emergency services; however, the authors did not offer explanations for the causes of crashes.

Krishnaveni *et al.* [5] compared Naïve Bayes, AdaBoostM1, PART, J48 and Random Forest Classifier based on injury severity. They found that weather, vehicle kinematics, vehicle class and driver's age are responsible for crashes.

Nayak *et al.* [6] used text mining to identify the causes of about 20000 motor accidents in Queensland, Austria, collected in 2004-2005. The authors asserted that most crashes are rear-end collisions at intersections and due to speeding. No other important crash factors were considered.

Beshah and Hill [7] studied 18,288 traffic accidents and achieved comparable accuracy using Decision Tree, Naive Bayes and K-Nearest Neighbors.

Zhang and Fan [8] studied data spanning 20 years in Saskatchewan, Canada, using decision trees, and showed that accidents are likely to happen to infirm drivers violating traffic rules, intoxicated drivers, and inexperienced drivers during poor weather conditions. However, vehicle factors were not considered in the study.

Jianfenget *al.* [9] analyzed over twenty thousand accidents in China, seeking a relation between the driver's, vehicle, environment and road conditions and the occurrence of crashes.

Using factor analysis, Haixia and Zhihong [10] identified the most important crash factors in the categories of road and driver conditions. However, their dataset was very small (372 crashes).

Emerson *et al.* [11] used decision trees and regression trees to establish a benchmark for surface skid resistance versus crash probabilities, for all the crashes.

Ramani and Shanthi [12] suggested designing education campaigns for parents and generating awareness to limit the number of fatalities.

Using Classification and Regression Trees (CART), Tree Net and Random Forest Beshah et al. [13] concluded that human behavior is a major factor in road accidents but did not provide any suggestions to improve road safety.

Using multinomial logistic regression, Pakgohar et al. [14] concluded that 98% of crashes happened due to human factor, 70% due to the environmental factor and 32% due to the vehicle factor. However, the environmental factors were not clearly identified.

T. Dipo Akomolafe, A. Olutayo [15] used several classifiers and concluded that tire burst, broken shaft and speeding are the most culpable collision factors. However, other factors that contribute to crashes were not considered.

In retrospect to the findings of the examined literature, we suggest the use of a larger dataset that considers more factors and spans longer periods of time to yield more comprehensive, more reliable and more conclusive collision factors.

3 Data pre processing

The accident data records have two different categories of attributes: qualitative and quantitative [16]. However, most of the attributes are qualitative, which include nominal, binary, and ordinal attributes. The quantitative attributes include interval-scaled attributes, and do not have continuous attributes.

Only 21 attributes of the dataset are considered in this study; the remaining 10 attributes are removed because they have little significance. The dataset contains 450 undefined values, which are filled using random attribute filling [16]. Missing attribute values are chosen randomly between their minimum and maximum limits. Driver age outliers that are less than 15 or more than 95 are ignored. The total number of removed records constitutes only 3.2% of our dataset; hence, the data remains cohesive.

Finally, the five years of accident data records is transformed into numerical records for plotting purposes.

Table 1. Investigated attributes

Attribute	Description	Domain (possible values)
Collision Type	The other party involved in the collision, e.g. another vehicle, a train, a structure.	34
Weather Condition	The weather condition during the collision, e.g., clear weather, cloudy, rainy.	9
Road Surface	The condition of the road on which collision takes place, e.g., dry, muddy, icy.	8
Lights	The road illumination during the collision. e.g., dark, daylight, dawn,	6
Damage	The seriousness of the collision, e.g., fatal, injury, property damage,	6
Traffic Control	Traffic control signs in the accident location, e.g., stop sign, flashing signal, school zone.	19
Collision Location	How the collision occurs, e.g., head on, rear end, sides swipe.	11
Vehicle 1 type	These two attributes describe the vehicles involved in the collision, e.g., a bus, a car, a bicycle.	8
Vehicle 2 type		
Factor Driver 1	The two attributes describe factors or reasons that contributed to the collision, e.g., a defective brakes, alcohol, distraction.	55
Factor Driver 2		
Intersection	Whether one or two roads intersecting.	2
Day of the week	The day of the week on which the collision happens	7
Time	The time rounded to hours at which the collision happens	24
Month	The month in which the collision happens	12
Year	The year in which the crash happens	5

4 Data mining process

The data mining process is a combination of classification of the attributes and their corresponding values and an association mining rule using Fp-growth algorithm [16]. Approximately 2000 rules are generated from the dataset. These rules are then inspected for significance. The first step is eliminating rules that have a support less than 40%. Then, rules that have a confidence less than 70% are eliminated. The confidence is set to 70% to ensure reliability of rules. However, the support is only 40% because some important rules are only present in a small part of the dataset. Rules with a lift measure equal to one are eliminated, as it indicates a lack of correlation

DOI: 10.1051/matecconf/20168102017

between the attributes. From there, almost 700 rules remain and the manual interpretation starts for each rule. The rules contribute to refining the knowledge discovered, thus improving the meaningfulness of those results.

4.1 Preliminary analysis

Rapid Miner is used to evaluate the distribution of the dataset values. Other tools, such as Microsoft Excel are used to count the specific values of interest.

The total number of crashes studied is 34945. The most recorded crashes are with other motor vehicles: 21918 crashes. Most of the crashes are serious and record property damage and injury: 24174 crashes.

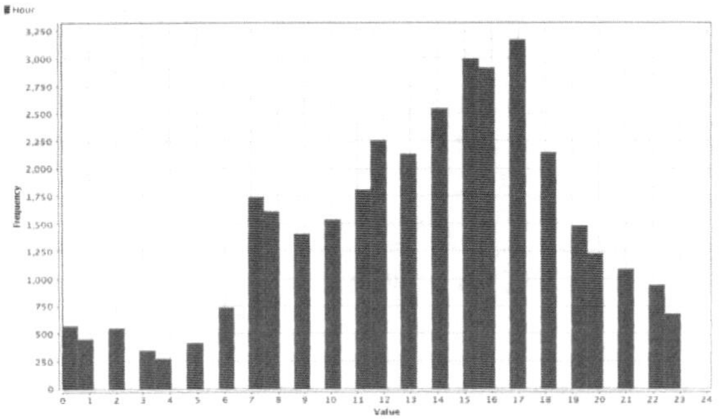

Figure 1. Effect of the hour of the day on crashes distribution

The effect of the hour of the day on the number of crashes is shown in Fig. 1. It can be seen that crashes are considerably more during daytime. More specifically, the number of crashes is the highest in the afternoon between 12:00 and 6:00 pm; in fact, nearly half the accidents (18139) occur during this period, with the peak at 5:00pm (3168 crashes)

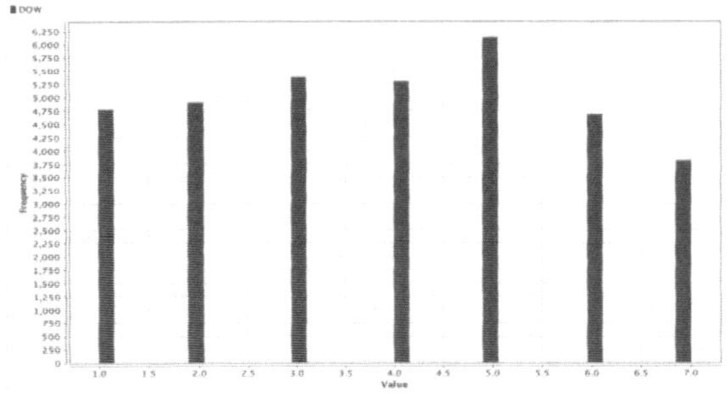

Figure 2. Effect of the day of the week on crashes distribution.

The effect of the day of the week on the number of crashes is shown in Fig. 2. Daily crashes tend to be constant. However, Saturday and Sunday have the lowest number of crashes, which can be explained by the reduction in traffic during these holidays. On the other hand, Fridays have the peak number of crashes, since people tend to stay out late at this day; thus, more traffic volume and the greater possibly of intoxicated drivers result in more accidents during this day.

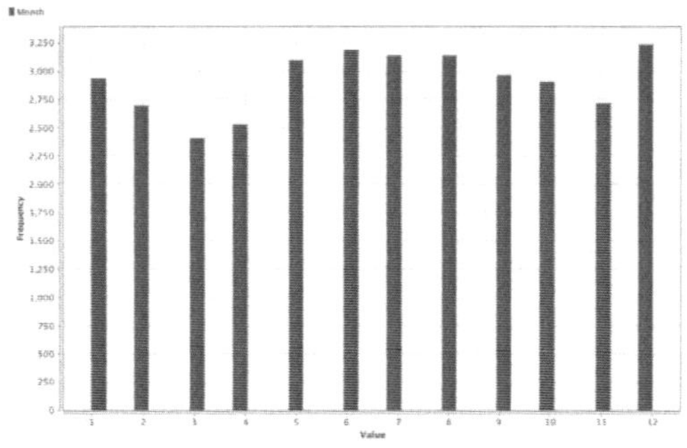

Figure 3. Effect of the month of the year on crashes distribution

In addition, Fig. 3 shows that December records 3337 crashes, the highest number compared to other months. December's records can be attributed to poor weather conditions and also widely celebrated holidays such as Christmas and New Year's Eve during which people consume more alcohol. This factor will be studied later in conjunction with seriousness of crash and weather conditions.

More accidents happen when the road surface is wet (3623, 10%) than when it is snowy (6787, 19.4%); it can be inferred that people are more cautious when it is snowing but ignore the dangers of slippery wet roads. However, most accidents are recorded in dry conditions: 24167 crashes.

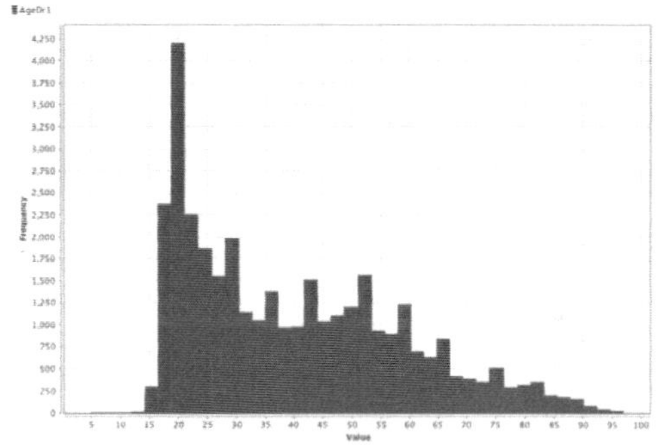

Figure 4. Age of Driver 1 involved in the crash

The effect of the driver's age on the number of crashes is shown in Fig. 4, where it is clear that drivers between the ages of 19 and 23 are the ones accountable for most accidents totaling 6455 accidents, that is, 18% of all crashes.

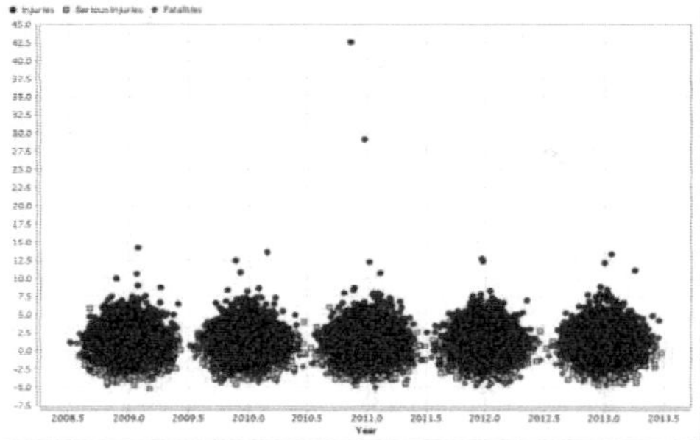

Figure 5. Visualization of occurrence according to severity of crash (injuries/serious injuries/ fatalities)

Fig. 5 shows the crashes over a five-year span along with their severity. Jittering is introduced in the plot, by adding random noise to the data to alleviate the problem of over-plotting similar or neighboring points, and thus better visualize the close values of data. The figure shows that, the crash severity is slightly declining with time. The figure is not conclusive but helps visualize the decreasing trend in the number of crashes and their severity over the years.

Table 2. Discrete count of occurrence according to severity of crash (injuries/serious injuries/ fatalities)

Year	Injuries	Serious Inj	Fatalities	Crash Count
2009	3632	281	33	7202
2010	10012	802	100	7432
2011	9650	811	103	6986
2012	8957	871	115	6691
2013	8855	836	86	6631

Table 2 shows that the number of injuries and fatalities decreased between the years 2010 and 2013. However, the year 2009 has the least severity of crashes.

4.2 Discovered association rules

In this section, more knowledge from the data is extracted using an association data mining technique [16]. The data mining process consists of four main steps illustrated in Fig. 6: Crash data retrieval, nominal to binomial conversion, Fp-Growth Algorithm and association rules creation.

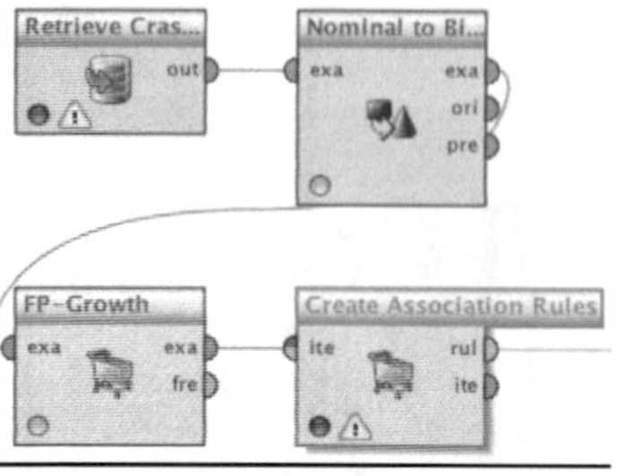

Figure 6. Visualization of occurrence according to severity of crash (injuries/serious injuries/ fatalities)

To generate association rules, the Fp-Growth Algorithm [16] is run on the data fifteen times; every time the properties of the Fp-Growth algorithm and association rule generation are modified to tune the parameters to generate slightly different rules. Each iteration is a search refinement towards more useful data. The challenge appears in interpreting the rules, since many rules are logical but do not contribute towards any knowledge discovery.

The algorithm generates 2247 rules; however, only a few of these rules are useful and non-trivial. Some of the meaningful rules discovered by this study are given below along with their support, confidence, and lift measures:

Rule 1: Crashes in daylight have no fatalities or serious injuries. (Sup:0.6335 Conf:0.8980 Lift: 1.1154)

Rule 2: Crashes in daylight involving two vehicles have no fatalities. (Sup:0.4093 Conf:0.8390 Lift:1.5644)

Rule 3: Crashes in daylight in which the second vehicle involved is a car, van or pickup do not cause serious injuries. (Sup:0.4244 Conf:0.9196 Lift:1.5930)

Rules 1, 2 and 3 indicate that accidents in daylight have little chance of being catastrophic. The lift shows that there is a positive correlation between the lights, and the fatalities and serious injuries. The second and third rules clearly assert that a crash between motor vehicles in daylight will not lead to fatalities or serious injury.

Rule 4: If the road surface is dry then there are no fatalities or serious injuries. (Sup:0.60523 Conf:0.8751 Lift:1.1021)

Rule 5: If the road surface is dry and two motor vehicles are involved in a crash then there are no fatalities. (Sup:0.4020 Conf:0.8389 Lift:1.5641)

Rules 4 and 5 show that the severity of crash decreases when the road surface is dry. The lift measure shows a positive correlation between road surface, fatalities and serious injuries. In Rule 5 it is demonstrated that when a crash occurs between two vehicles and the road surface is dry, the number of deaths is 0. The 1.56 value of lift indicates the strength of this correlation

Rule 6: Crashes in clear weather have no fatalities. (Sup:0.5206 Conf:0.9862 Lift:0.8990)

This rule states that when the weather is clear, no deaths occur with a confidence of 99%. Thus, it is concluded that in daylight, dry road surface and clear weather conditions, crashes are less severe and tend to have no fatalities. A negative correlation between weather and fatalities is present; thus lift <1.

Rule 7: If there are no fatalities and the weather is clear and the second involved vehicle is a car, van or pickup then the road surface is dry. (Sup:0.4102 Conf:0.9362 Lift:1.3538)

This rule shows a positive correlation between the two members. This helps us understand the impact of the conditions on the fatalities. If no one dies in a clear weather crash then the road must be dry.

Rule 8: If no one dies, then no one is seriously injured. (Sup:0.8843 Conf:0.8959 Lift:1.001)

Rule 9: If no one is seriously injured, then no one dies. (Sup:0.8843 Conf:0.9882 Lift:1.001)

These rules tell us that if a crash occurs and no one dies, then the chance of serious injuries is also minimal. Also, if a crash occurs and no one is seriously injured then the chance of fatalities is minimal. These rules help us classify a crash as serious or non-serious, where serious is leading to death. However, the lift measure indicates that fatalities and serious injuries are independent, and there is no correlation between them.

Rule 10: If there are no fatalities and the second vehicle involved is a car, van or pickup then there are no serious injuries. (Supp:0.5411 Conf:0.9150 Lift:1.5850)

Rule 11: If there are no serious injuries and the second vehicle involved is a car, van or pickup then there are no fatalities. (Supp:0.5411 Conf:0.9374 Lift:1.5850)

It seems that Rule 8 and 9 are misleading because the lift was equal to one. After Rule 10 and 11 are analyzed it is understood that when a crash occurs between two motor vehicles, being cars vans or pickups and the fatalities are zero then serious injuries are also equal to zero. The principle also works vice versa. The lift is very elevated indicating the positive correlation between each of these rules.

5 Conclusion

This study uses a new and comprehensive dataset encompassing several years. A methodical examination of all factors went into the study, followed by the omission of all factors found to be outside the scope of the study's objective; factors determined to be consequential to the study include those pertaining to weather, road, and vehicle conditions in precise time frames.

From our established findings on the categorical factors of collisions, we have developed a guideline to encourage drivers to take pre-emptive safety measures on the road:

1. Be more vigilant in the afternoon, between 12 and 6 pm, when most crashes occur.
2. Young drivers should drive responsibly since most people crash their cars in their early twenties.
3. Be extra cautious on Fridays because most accidents occur on Fridays.
4. Do not go over the speed limit.
5. Do not tailgate.
6. When making a right turn, make sure to check your blind spot.
7. Be on the lookout for traffic control signs; when there are none, be extra cautious.
8. When the road conditions are not ideal (clear, dry and daylight), accidents can be very dangerous.
9. If possible, drive during daylight, clear, dry days to reduce possibilities of collisions.

We could embed the results in some navigation software, such as Google maps, so that they would be highlighted in colored indicators reflecting the probability of crash (red for high probability, green for low probability and yellow for intermediate). The probability would be based on the checklist. For example, when the road conditions are dangerous and the timeframe is between 12 and 6 pm, the area would be highlighted in red. This would create awareness for the drivers and possibly reduce the number of crashes and deaths. The speed of the driver could also be tracked and the indicator would be updated accordingly.

Future work would generate new and improved safety practices to be appended to the current checklist based on the data analyzed and possibly finding another, more up to date, dataset to study. Keep in mind that our dataset was last updated in 2013. A new dataset would work on improving the quality of the checklist and possibly adding more points of interest.

References

1. Association for safe international road travel [http://asirt.org/].
2. https://opendata.socrata.com/Education/Five-Years-of-Motor-Vehicle-Crashes/tbyb-ykru
3. X. Xu, Ž. Šarić, A. Kouhpanejade, Traffic&Transportation, **Vol. 26**, (2014), No. 3, 191-199 191.
4. M. Fogue , P.Garrido, F J. Martinez, Juan-Carlos Cano, Carlos T(2012), Advances in Intelligent Systems and Computing, pages 37-46.
5. S.Krishnaveni , Dr.M.Hemalatha(2011), International Journal of Computer Applications, **Volume 23– No.7**
6. R. Nayak, N. iyatrapoomi, J. Weligamage, Proceedings of the 4th World Congress on Engineering Asset Management (WCEAM 2009), (28-30 September 2009).
7. T. Beshah and S. Hill (2010), Proceedings of AAAI Artificial Intelligence for Development (AI-D'10).
8. X F Zhang; L. Fan, Electrical and Computer Engineering (CCECE), 2013 26th Annual IEEE Canadian Conference on, **vol., no., pp.1,4,** 5-8 (May 2013).
9. Xi Jianfeng; C. Xiaodong; W. Shuangwei; Zhurong Tao, Computer Application and System Modeling (ICCASM), 2010 International Conference on , **vol.13, no., pp.V13-230,V13-233,** (22-24 Oct. 2010).
10. Y. Haixia; N. Zhihong, Computer Science and Education (ICCSE), 5th International Conference on, **vol., no., pp.1355, 1358,** (24-27 2010)
11. D. Emerson, R. Nayak, J. Weligamage, N. Piyatrapoomi, Proceedings of the 5th World Congress on Engineering Asset Management (WCEAM 2010).
12. R. G. Ramani,. S. Shanthi, Computational Intelligence & Computing Research (ICCIC), IEEE International Conference on, **vol., no., pp.1, 4,** (18-20 Dec. 2012).
13. T. Beshah, D. Ejigu, A. Abraham, V. Snasel, Pavel Kromer,"Mining Pattern
14. A. Pakgohar, R. Sigari Tabrizi, M. Khalili, Alireza Esmaeili, World Conference on Information Technology, **Vol 3**, (2011), Pages 764–769.
15. T. Akomolafe, Akinbola Olutayo, American Journal of Database Theory and Application (2012), **1**(3), pages 26-38.
16. J. Han, M. Kamber, J. Pei, *Data Mining: Concepts and Techniques*, 3rd Edition, (June 2011).

Comprehensive Evaluation of Entropy-hierarchical Grey Correlation Analysis for Highway Safety Life Protection Engineering

Jin Shuxin, Cao Xudong

School of Highway, Chang'an University, Xi'an, China

Abstract. Different highway safety life protection engineering decision-making have important meaning. The achieving goals and optimal highway safety life protection engineering scheme can not only improve the function of the highway facilities and service level, still can reduce the traffic accident, which caused by the imperfect highway facilities. Different highway safety life protection engineering decision-making is a multiple targets, multi-layers and multi-schemes system evaluation problem. With regard to lack of concrete data on multiple targets, multi-layers and multi-schemes system evaluation problem, make analytical hierarchy process combined with the entropy value analysis into the grey relational comprehensive evaluation method, and then get entropy-hierarchical grey correlation analysis method. This method is a qualitative and quantitative decision method, which combine comparison principle of analytic hierarchy process (AHP) and the entropy principle of entropy value analysis method to determine the relative weight of various indexes between factors layer-by-layer. Then using grey relational analysis by low-layer to high-layer step by step in the possible scheme and referenced scheme. Finally, calculating the comprehensive correlation degree between the possible scheme and referenced scheme, the best plan which has maximum grey correlation degree can be selected.

1 Introduction

With the rapid development of the national economy, china's highway construction has entered a glorious development period; the road traffic safety situation is getting more and more serious. China's road traffic accident death absolute toll has been the highest in the world for many years and with a higher proportion of growth, so the implementation of highway safety life protection project is particularly urgent. The scientific and reasonable decision of this important construction project will indirectly affect the local living standard of happiness and the quality of the national economy, but also has a direct impact on the accident mortality, the incidence of serious accidents and traffic safety level.

The concept of road traffic safety evaluation originated from Britain, which research work started earlier, and then successively developed the relevant content in this field [1]-[4]. Main methods of highway traffic safety evaluation in foreign countries are mostly established on the basis of a large number of traffic accident statistics to analysis for safety influence factors qualitatively [5]. It through the mathematical theory established quantitative relationship between the accident and affecting factors, so as to reflecting the level of road traffic safety. Recent research trends have shifted from absolute indicators to relative indicators as well as the studies of evaluation system and model [6], [7].In addition, some researchers also through analyze and study the changes in amount of traffic accidents before and after the reconstruction of highway structure to put forward a reasonable transformation scheme of highway system and evaluate the effectiveness of transformation, and finally established the relevant evaluation system [8].

Although the research on traffic safety evaluation started relatively lag in our country, but the relevant experts and scholars have also obtained the valuable research results to a certain extent. Wang Yuanyuan [9] structured mountainous highway accident index system based on road alignment index by using negative binomial model and carried out the traffic safety evaluation of the whole highway by evaluating the safety level of road unit. Xu Hongguo [10] discussed fuzzy set-valued statistical method to evaluate the road traffic safety, which can effectively overcome the problems that accuracy is not high caused by non quantitative indicators.

The evaluation methods of highway safety life protection engineering all have certain application scopes and defects [11-16]. Most of the quantitative models only consider the design speed test, and rarely take into account the other factors, especially when they cannot be quantified. The characteristics of various evaluation methods are compared in Table 1.

In view of the above-mentioned evaluation method of highway safety life protection engineering is lack of objectivity, this paper combine subjective weight index determined by AHP and objective weight determined by entropy method into comprehensive weight, which can not only take full consideration of the experience and intuition thinking, but also reduce the human uncertainty factors in the evaluation process, which both has the standard of comprehensive decision making method and reflects the higher problem solving efficiency. Then using grey relational decision method to evaluate the highway safety life protection projects. Finally, the relative optimal engineering scheme can be obtained.

Table 1. Comparison of comprehensive evaluation methods of highway safety life protection engineering

Evaluation method		Characteristics
1	Principal component analysis method	1) To eliminate the correlation effect between evaluation index and reduce the workload of index selection; 2) Computational process can be simplified and accurate evaluation results can be efficient accessed; 3) Can obtain an overall ranking by evaluating multiple comparison objects , cannot evaluate a single object.
2	Fuzzy evaluation method	1) It has certain advantages in the multi factors comprehensive evaluation; 2) Evaluation results be expressed by evaluation object's membership grade to the fuzzy subset, which can be more intuitive and clear to interpret and evaluate; 3) The weight of the index is determined in accordance with the subjective sense of the people, lack of effective objectivity.
3	Grey correlation degree method	1) It has reliable theoretical foundation, clear algorithm meaning and strong practicality; 2) It can clearly and objectively reflect the pros and cons of the safety level of system factors and the integrated security level; 3) The process of evaluating is simple, with good visibility.
4	Artificial neural network evaluation method	1) Can not only realize the multi-factors comprehensive evaluation, but also realize the dynamic evaluation; 2) Algorithm can be realized by computer, with scientific and efficient; 3) When the sample data is more abundant, the accuracy of the model will have much room for improvement; 4) The limited network training samples can lead to the low training accuracy or the unstable network.
5	Analytic hierarchy process method	1) It is a kind of effective decision-making method, which can be simple and easy to grasp and has wide application scope; 2) Combining the quantitative and qualitative analysis, it using numerical to show people's subjective judgment for reduce defects caused by personal factors and make the evaluation results more credible.
6	Matter element analysis method	1) It solves the problem of incompatibility and variability between the evaluation indexes; 2) It helps to judge the things from the angle of change; 3) The index weight is determined by expert opinion, which has certain subjectivity and limitation.

In information theory, entropy is a measure of the uncertainty of things and disorder degree. The correlation degree is a measure of the correlation between the factors, which quantitatively describes the relative change between the factors. The change of the two factors' situation (direction, size, speed, etc.) is basically the same, and then the correlation between them is larger, whereas the correlation degree is smaller [17]. Grey evaluation method is using grey relation analysis theory to judge the relevance level of each evaluation index and evaluation object, so as to achieve the objective and accurate to understand the influence degree of evaluation index affected by the evaluation objects [18].

In general, the optimal solution of a multi objectives decision problem is not existed. By analytic hierarchy process (AHP) method to analyze the weight vector of decision index, and further application of the entropy technology, they can make full use of decision matrix to provide information to correct the priority weights determined by decision maker, thereby helping to form the correct decision [19-22]. On the basis of determining a reasonable weight, the optimal scheme is determined by the method of multi hierarchies grey correlation analysis according to draw up an ideal reference sequence. It can be seen that the entropy-hierarchical grey correlation analysis method is a better method to solve the multi objectives, multi layers and multi schemes evaluation problems.

2 Selection of evaluation indicators of highway safety life protection schemes

According to the characteristics of different highway safety protection projects, the following key characteristics can be chosen:

1) **Safety facilities adaptive indicators**

The more perfect of the safety facilities, the higher value of the index evaluate.

① The protection facilities; ② The sign; ③ The marking; ④ The sight guidance facilities; ⑤ The emergency lane; ⑥ The deceleration facilities; ⑦ The passing track.

2) **Pavement treatment technology indicators**

①The disposal of the pavement diseases; ②Preventive maintenance of pavement; ③Regeneration technology of road surface; ④Pavement reinforcement of reconstruction and extension.

3) **daily safety and defense technical indicators**

① The daily security defense technology for traffic participants; ②Security management of vehicles.

4) **geological hazard prevention and control indicators**

① Prevention and control measures of geological engineering ; ②The optimization of traffic safety facilities under geological hazard.

5) **indicators of safety life protection facilities under the condition of danger**

The more perfect of the safety life protection facilities, the higher value of the index evaluate.

①The safety measures of dangerous section; ②Safety protection measures under accident conditions; ③Special weather safety protection measures; ④Security measures of lane assignments.

3 Entropy-hierarchical grey correlation analysis model

Assuming there are feasible schemes $A_1, A_2, \cdots, A_i \cdots A_m$, and there are two layers of evaluation indicators, the first layer index is recorded as V_p, $p = 1,2,\cdots,r$, the second layer index is recorded as U_j, $j = 1,2,\cdots,s$. x_0 as a reference sequence, x_i is the i program sequence. The problem of system evaluation is transformed into selecting scheme which has the maximum correlation value. Specific steps as follows.

1) **Determination of analysis sequence**

The intermediate state comments set as a (very good, good, general, poor, poor, and very poor), and its membership grade is a collection (0.8, 0.7, 0.6 and 0.5, 0.4, 0.3 and 0.2). According to the evaluation opinions of the expert group, the form of analysis sequence is:

$$(x_0, x_1, \cdots x_m) = \begin{pmatrix} x_0(1) & x_1(1) & \cdots & x_m(1) \\ x_0(2) & x_1(2) & \cdots & x_m(2) \\ \vdots & \vdots & & \vdots \\ x_0(r\times s) & x_1(r\times s) & \cdots & x_m(r\times s) \end{pmatrix}_{r\times s\times m}$$

Assuming all indicators are benefit type, the reference sequence is $x_0 = (0.8, 0.8, \cdots 0.8)$.

2) **The determination of indicator weight in all layers of analytic hierarchy process method**

Using weight determination method of analytic hierarchy process——the root mean square method to get the weight of each indicator. The weight of j evaluation indicator in the second layer is w_j. Similarly, calculate the weight w_p which between the P indicators of the first layer.

3) **The determination of indicator weight in all layers of entropy analysis method**

The entropy of the j evaluation indicator of the second layer is defined as:

$$H_j = -\frac{1}{In\ s}\left(\sum_{i=1}^{m}\frac{x_i(j)}{\sum_{j=1}^{s}x_i(j)}In\frac{x_i(j)}{\sum_{j=1}^{s}x_i(j)}\right) \quad (1)$$

$j = 1,2\cdots s\,; i = 1,2,\cdots m$

Definition of entropy weight v_j as follows:

$$v_j = \frac{1-H_j}{s-\sum_{j=1}^{s}H_j} \quad j = 1,2\cdots s \quad (2)$$

In the same way, entropy and entropy weight between the P indicators of the first layer are:

$$H_p = -\frac{1}{In\ r}\left(\sum_{i=1}^{m}\frac{\sum_{j=1}^{s}x_i(j)}{\sum_{j=1}^{s\times r}x_i(j)}In\frac{\sum_{j=1}^{s}x_i(j)}{\sum_{j=1}^{s\times r}x_i(j)}\right) \quad (3)$$

$j = 1,2\cdots s\,; i = 1,2,\cdots m$

$$v_p = \frac{1-H_p}{p-\sum_{p=1}^{r}H_p} \quad p = 1,2\cdots r \quad (4)$$

4) **The calculation of comprehensive weight**

Calculating the subjective weight w_j and w_p by the AHP and finding the objective weight v_j and v_p under the theory of entropy. Combination of them will get the comprehensive weights λ_j and λ_p. The comprehensive weights λ_j which of the j evaluation indicator of the second layer can be expressed as:

$$\lambda_j = \frac{w_j \times v_j}{\sum_{j=1}^{s} w_j \times v_j} \quad (5)$$

The comprehensive weight λ_p which between the P indicators is:

$$\lambda_p = \frac{w_p \times v_p}{\sum_{p=1}^{r} w_p \times v_p} \quad (6)$$

5) **System correlation analysis of candidate schemes: grey correlation analysis of the second layer**

The equation for calculating the grey correlation degree $\xi_j(i)$ which of second layer is as follows:

$$\xi_j(i) = \frac{\min_j\min_i|x_0(i)-x_j(i)| + \rho\max_j\max_i|x_0(i)-x_j(i)|}{|x_0(i)-x_j(i)| + \rho\max_j\max_i|x_0(i)-x_j(i)|} \quad (7)$$

The ρ represents the discrimination coefficient, general takes 0.5.

6) **System correlation analysis of candidate schemes: grey correlation analysis of the first layer**

The formula for calculating the grey correlation coefficient $\theta_p(i)$ which of first layer is as follows:

$$\theta_p(i) = \sum_j \lambda_j \bullet \xi_j(i) \quad (8)$$

Take the maximum value of a set of rows in $\theta_p(i)$ as reference sequence $\theta_0(i)$. The formula for calculating the grey correlation degree $\varsigma_p(i)$ which of first layer is as follows:

$$\varsigma_p(i) = \frac{\min_i|\theta_0(i)-\theta_j(i)| + \rho\max_i|\theta_0(i)-\theta_j(i)|}{|\theta_0(i)-\theta_j(i)| + \rho\max_i|\theta_0(i)-\theta_j(i)|} \quad (9)$$

7) **The calculation of correlation degree between the i scheme and the reference scheme**

The formula for calculating the grey correlation degree γ_{0i} which between the i scheme and the reference plan is as follows:

$$\gamma_{0i} = \sum_p \lambda_p \bullet \varsigma_p(i) \quad (10)$$

 DOI: 10.1051/matecconf/20168102018

8) **Search for the greatest scheme correlation degree**

$\max(\gamma_{0i})$'s scheme is the best solution.

4 Application examples

It is assumed that there are 3 highway traffic safety and life protection engineering schemes, denoted as scheme A, scheme B and scheme C. According to the expert group's evaluation opinion, get the quantification matrix of each index factor:

$$(x_0,x_1,x_2,x_3)=\begin{pmatrix} & \text{Reference sequence} & \textit{Scheme A} & \textit{Scheme B} & \textit{Scheme C} \\ V_1 & 0.8 & 0.7 & 0.6 & 0.8 \\ & 0.8 & 0.4 & 0.5 & 0.7 \\ & 0.8 & 0.7 & 0.6 & 0.6 \\ & 0.8 & 0.4 & 0.5 & 0.6 \\ & 0.8 & 0.5 & 0.7 & 0.4 \\ & 0.8 & 0.6 & 0.5 & 0.7 \\ & 0.8 & 0.7 & 0.4 & 0.8 \\ V_2 & 0.8 & 0.8 & 0.6 & 0.5 \\ & 0.8 & 0.8 & 0.7 & 0.7 \\ & 0.8 & 0.6 & 0.8 & 0.4 \\ & 0.8 & 0.6 & 0.7 & 0.4 \\ V_3 & 0.8 & 0.7 & 0.6 & 0.7 \\ & 0.8 & 0.6 & 0.5 & 0.8 \\ V_4 & 0.8 & 0.6 & 0.8 & 0.4 \\ & 0.8 & 0.5 & 0.8 & 0.4 \\ V_5 & 0.8 & 0.8 & 0.7 & 0.4 \\ & 0.8 & 0.4 & 0.6 & 0.7 \\ & 0.8 & 0.7 & 0.8 & 0.5 \\ & 0.8 & 0.6 & 0.5 & 0.6 \end{pmatrix}$$

The calculation results of weight of each indicator of each layer are shown in Table 2 and Table 3 through the above steps.

Table 2. The weight of each indicator in second layer

The judgment matrix of indicators in the second layer	AHP weight	Entropy weight	Comprehensive weight
$c_{v1}=\begin{pmatrix} 1 & 5 & 3 & 2 & 4 & 1/2 & 1/2 \\ 1/5 & 1 & 4 & 1/3 & 1/3 & 2 & 1/3 \\ 1/3 & 1/4 & 1 & 1/5 & 1/5 & 1/8 & 4 \\ 1/2 & 3 & 5 & 1 & 3 & 3 & 5 \\ 1/4 & 3 & 5 & 1/3 & 1 & 1/2 & 1/2 \\ 2 & 1/2 & 8 & 1/3 & 2 & 1 & 3 \\ 2 & 3 & 1/4 & 1/5 & 3 & 1/3 & 1 \end{pmatrix}$	0.1998	0.1333	0.185
	0.0821	0.1482	0.084
	0.0514	0.1382	0.049
	0.2823	0.1508	0.297
	0.1041	0.1481	0.107
	0.1826	0.1413	0.179
	0.0977	0.1402	0.095
$c_{v2}=\begin{pmatrix} 1 & 8 & 5 & 2 \\ 1/8 & 1 & 4 & 1/3 \\ 1/5 & 1/4 & 1 & 1/5 \\ 1/2 & 3 & 5 & 1 \end{pmatrix}$	0.5340	0.2459	0.522
	0.1141	0.2198	0.099
	0.0565	0.2632	0.059
	0.2955	0.2711	0.318
$c_{v3}=\begin{pmatrix} 1 & 6 \\ 1/6 & 1 \end{pmatrix}$	0.8572	0.4774	0.845
	0.1429	0.5226	0.154
$c_{v_4}=\begin{pmatrix} 1 & 1/5 \\ 5 & 1 \end{pmatrix}$	0.1667	0.4799	0.155
	0.8333	0.5202	0.844
$c_{v5}=\begin{pmatrix} 1 & 3 & 5 & 1/2 \\ 1/3 & 1 & 1/2 & 3 \\ 1/5 & 2 & 1 & 1/3 \\ 2 & 1/3 & 3 & 1 \end{pmatrix}$	0.3576	0.2469	0.351
	0.2081	0.2689	0.222
	0.1597	0.2277	0.144
	0.2746	0.2565	0.280

Table 3. The weight of each indicator in first layer

The judgment matrix of indicators in the first layer	AHP weight	Entropy weight	Comprehensive weight
$C=\begin{pmatrix} 1 & 1/4 & 1 & 1/2 & 1/2 \\ 4 & 1 & 3 & 2 & 1/2 \\ 1 & 1/3 & 1 & 3 & 3 \\ 2 & 1/2 & 1/3 & 1 & 2 \\ 2 & 2 & 1/3 & 1/2 & 1 \end{pmatrix}$	0.1493	0.1571	0.116
	0.2525	0.1819	0.228
	0.2199	0.2237	0.244
	0.1892	0.2552	0.239
	0.1892	0.1821	0.171

The calculation results of correlation degree of each layer of each scheme are shown in Table 4 and Table 5 following the specific step. Among them, $\rho=0.5$.

In accordance with the correlation degree formula, we can get the calculation results of the scheme's correlation degree as shown in Table 6.

Table 4. The correlation degree of the second layer of each scheme

Indicator in the first layer	Correlation degree		
	Scheme A	Scheme B	Scheme C
Safety facilities adaptive indicators V_1	0.6667	0.5000	1.0000
	0.3333	0.4000	0.6667
	0.6667	0.5000	0.5000
	0.3333	0.4000	0.5000
	0.4000	0.6667	0.3333
	0.5000	0.4000	0.6667
	0.6667	0.3333	1.0000
Pavement treatment technology indicators V_2	1.0000	0.5000	0.4000
	1.0000	0.6667	0.6667
	0.5000	1.0000	0.3333
	0.5000	0.6667	0.3333
Daily safety and defense technical indicators V_3	0.6000	0.4286	0.6000
	0.4286	0.3333	1.0000
Geological hazard prevention and control indicators V_4	0.5000	1.0000	0.3333
	0.4000	1.0000	0.3333
Indicators of safety life protection facilities under the condition of danger V_5	1.0000	0.6667	0.3333
	0.3333	0.5000	0.6667
	0.6667	1.0000	0.4000
	0.5000	0.4000	0.5000

Table 5. The correlation coefficient and correlation degree of the first layer of each scheme

First layer	Correlation coefficients of each candidate scheme			Correlation degree of each candidate scheme		
	A	B	C	A	B	C
V_1	0.4807	0.4458	0.6668	0.3747	0.3353	1.0002
V_2	0.8111	0.5993	0.4014	0.9998	0.4917	0.3334
V_3	0.5735	0.4139	0.6617	0.5843	0.3333	1.0001
V_4	0.4156	1.0000	0.3333	0.3632	0.9999	0.3334
V_5	0.6629	0.6030	0.4641	0.9996	0.6239	0.3333

Table 6. The correlation degree of each scheme compared the idea scheme

Scheme	A	B	C
Correlation degree	0.6727	0.5793	0.5739
Sorting	1	2	3

Therefore, the scheme A is the optimal scheme.

5 Conclusion

In this paper, we put forward the entropy—hierarchical grey correlation analysis method, which is a decision method of multi objectives, multi layers and multi schemes that lack of specific data, to solve the highway safety life protection engineering schemes decision problem. Its core is to use subjective comparison principle of AHP and objective entropy theory of entropy analysis method to calculate the comprehensive weight of each indicator attribute, and calculate the correlation degree of scheme by layer to rank and select the alternatives. The method not only has the characteristics of clear concept and simple calculation, but also has more objectivity than other methods. By combining the analytic hierarchy process with grey relational analysis, the decision-making of different highway safety life protection projects is more reasonable and scientific. But this is only a preliminary study to solve decision-making system problems of highway safety protection projects. Compared with other schemes, the non substitution of this method and quantitative analysis of evaluation results remains to be further in-depth study.

Acknowledgment

This research is supported by National High Technology Research and Development Program of China（Project No. 2007AA11Z248）and the Special Fund for Basic Scientific Research of Central Colleges, Chang'an University (Project No. 310821153402) and Cultivation Project for Excellent Doctorial Dissertations of Central Colleges, Chang'an University(Project No. 310821165021) and Traffic Science and Technology Project of Shaanxi Provincial Transport Department (Project No. 15-26K).

References

1. S. Protor, M. Belcher, "The use of road safety audits in Great Britain," Traffic Engineering and Control, **31**(1993) :460-465.
2. AASHTO, Highway Safety Design and Operations Guide 1997. Washington D.C, AASHTO,(2001).
3. A. Jerry Reagan, "The Interactive Highway Safety Design Model: Designing for safety by analyzing road geometrics," Public Roads, **63**(1994) :215-219.
4. AUSTROADS, *Road safety audit*, Austroads Publication Number AP-30/94, Sydney Australia, Printed by Currency Productions, AUSTROADS, (1994).
5. D. Fambro, K. Fitzpatrick and R. Koppa, "Determination of Stopping Sight Distances," NCHRP Report 400, National A-cademy Press, Washington DC, (1997).
6. K. You, L. Sun and W.Gu, "Reliability-based risk analysis of roadway horizontal curve," Transportation Engineering, **125**(2012) : 137-142.
7. H. C.Chin and S. T.Quek, "Measurement of traffic con-flicts," Saf.Sci, **26**, 3(1997) :169-185.
8. G. Davis, J. Hourdos and H. Xiong, "Outline of a causal theory of traffic conflicts and collisions," Proceedings of 87th Annual TRB Meeting, Washington D.C,(2008).
9. W. Yuanyuan, "Research on the accident causations of mountain road," Beijing: Beijing Univ. Of Technology, (2008).
10. X. Hongguo, L. Zhaohui and W. Chao, "Road safety level evaluation based on grey fixed weight clustering model and factors analysis," Journal of Traffic and Transportation Engineering, **7**, 2(2007) :94-98.
11. L. Huichun, D.Hongxing, "A comparative study of urban expressway traffic safety evaluation method," Journal of Heilongjiang Vocational Institute of Ecological Engineering, **23**, 6(2010) : 3-6.
12. Z. Wenfeng, S. Zhongke, "Application of principal component analysis in evaluation of city road traffic safety," Computer Engineering and Applications, **43**, 36(2007) : 246-248.
13. C. Dongdong, Z. Jinfa, "Fuzzy comprehensive evaluation method of traffic safety system," Hunan Communication Science and Technology, **29**,3(2003) :116-117.
14. H. Wenjie, "The research of urban road traffic safety evaluation based on the BP neural network," Science Technology and Engineering, **9**, 18(2009) :5607-5609.
15. P. Yanrong, Z. Changxu and Z. Shunying, "Road traffic safety evaluation based on the grey clustering theory and artificial neural network technology," Journal of Chongqing Jiaotong University, **24**, 2(2005) :101-105.
16. W. Zijuan, Z. Jianhu and L. Jianfeng, "Study of fuzzy evaluation for urban road traffic safety based on AHP method," Journal of Minjiang University, **30**, 5(2009) :106-109.
17. Y. Jiazu, *The grey system theory and application, Beijing*, Printed by Science Press, (1991) :214-236.
18. T. Yuejin, C.Yingwu and Y.Jinxian, *Systematic engineering theory*, Changsha, Printed by National University of Defense Technology Publishing,(1999) :38-54.
19. L. Enke, M. Yuxiang, "Grey and Analytic Hierarchy Process for Comprehensive Evaluation of Information Systems," Journal of intelligence, **20**, 4(2001) : 416-420.
20. M. Rui, Z. Jiatai, "Research of hierarchy synthetic evaluation based on grey relational analysis," System Engineering Theory and Practice, **10**(2008) :125-130.
21. F. Zhiping, Z.Jin and M. Jian, "An integration method to determine the weight of Multiple attribute decision making," Journal of Management Science, **1**, 3(1998):50-53.
22. L. Jian, "Using basic points to calculate the weighting coefficients in multiple objective decision making," System Engineering Theory and Practice, **4**(2001) :27-30.

A Sensor-based System for Monitoring Hard-shoulder Incursions: Review of Technologies and Selection Criteria

Paraskevi Michalaki[1], Mohammed Quddus[1], David Pitfield[1], Mark Mageean[2] and Andrew Huetson[2]

[1]*School of Civil and Building Engineering, Loughborough University, Loughborough, Leicestershire LE11 3TU, UK*
[2]*Connect Roads, Balfour Beatty, Spen Common Lane, Tadcaster LS24 9NS, UK*

Abstract. According to safety observations from motorway operators in the United Kingdom, the hard-shoulder is occasionally violated by road users travelling in the nearside lane. These unintentional movements (hard-shoulder incursions) can impose risk to operatives performing activities on the network. To further investigate these events, a sensor-based system can be used for monitoring them and collecting related data such as severity of incursion and vehicle classification. A review of vehicle detection technologies that could be applied for this purpose is presented, along with the criteria for selection of the most suitable technology and implementation sites. Two potential non-intrusive systems are also described, a laser- and a radar-based systems, which provide different levels of flexibility and data.

1 Introduction

The hard-shoulder of a motorway is normally used by the public in the event of an emergency or by emergency services as a through route to incidents. The hard-shoulder is also used for_operation and maintenance activities. Thus, vehicles of motorway operators often remain on this lane for short durations whilst personnel perform work under live traffic conditions. Even though interest has been raised on the risk of having an accident on the hard-shoulder [1-7], this has not been investigated through rigorous research.

Safety observations from motorway operators in the UK have shown that vehicles travelling on the nearside lane of the main carriageway on occasion violate the hard-shoulder (Fig. 1). Considering that this could occur at any time an operator or emergency vehicle is on the hard-shoulder, it is important to investigate and further understand these events in order to reduce exposure under high-risk conditions. The aim of this study is to review the existing technologies that could be used to track hard-shoulder incursions and the criteria used for the selection of a system for this purpose.

To collect data of hard-shoulder incursions, a system installed on the network would need to fulfil the following goals: detecting, measuring, identifying and recording/storing incursion-related data.

- ***Detecting***: The system needs to detect the violations of the raised rib edge-line of the main carriageway (commonly referred to as 'rib-line'), which is the line separating the hard-shoulder from adjacent nearside running lane. These violations are caused by vehicles that are not positioned in the middle of the lane and move onto the hard-shoulder while crossing the rib line. These violations are considered to be inadvertent, unless the road user intends to stop on the hard-shoulder in the case of an emergency or vehicle breakdown.
- ***Measuring***: While detecting hard-shoulder incursions, the system should also measure the severity of incursion, which can be defined by the width (D) and the length (L) of the violation (Fig. 2). The width of an incursion represents how far over the rib-line the vehicle has transgressed. Having the trajectory of the vehicle could provide this information, as well as the length which the vehicle travelled over the rib-line.
- ***Identifying***: In order to explore high risk driver groups, these incursions need to be linked to the identity of the vehicle, i.e. the classification of the vehicle and its speed. The classification of the vehicles is in two main categories (e.g. car or truck) or in multiple categories (e.g. car, van, bus, small truck, large truck, transporter or caravan). There are several methods of classifying vehicles, such as number of axles and space between them, total length/ size of the vehicle and vehicle weight.

Figure 1. Hard-shoulder incursion of a truck on the motorway (slight/serious). (Source: P. Michalaki)

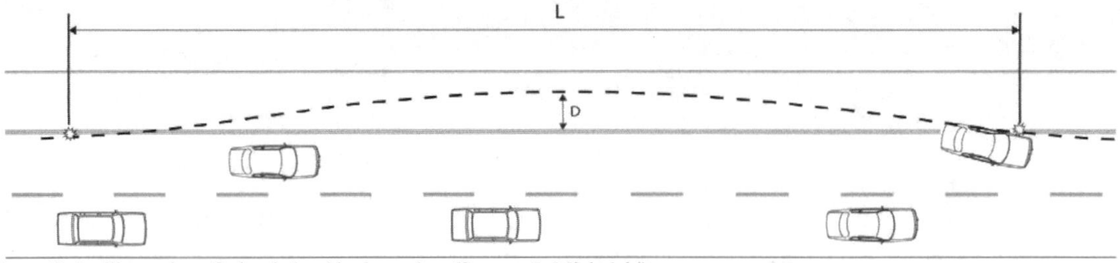

Figure 2. An illustration of a hard-shoulder incursion. (Source: P. Michalaki)

The speed of the vehicle, according to the technology used, might be directly measured or estimated.

- ***Recording and storing***: Finally, all raw data received by the system must be recorded and stored to be readily available for further review and analysis.

2 Review of technologies

In this section, potential sensor-based technological solutions for setting up a hard-shoulder incursions monitoring system are discussed, along with the advantages/disadvantages of each (Table 1).

2.1 Intrusive and non-intrusive technologies: advantages and limitations

The main categorisation of vehicle detection and traffic counting technologies is in two groups: *intrusive* and *non-intrusive* devices. The distinction is based on the level of disruption to normal traffic and the road, as well as the risk to the operator while installing, operating and maintaining the equipment. By definition, non-intrusive devices do not require installation in or on the road. Thus, intrusive devices tend to be permanently installed, while non-intrusive are preferable in temporary, short-term surveys. Both types of technology present several advantages and drawbacks. According to the purpose of the survey, the conditions under which this is performed and the requirements of the motorway operator, it can be decided which device would be most appropriate. Some aspects of these devices are discussed below:

- ***Safety***: Safety of road workers and the public is a key priority. As there is little or no direct interaction with live traffic, non-intrusive devices are generally safer to install. In addition, as the pavement is not disrupted, the carriageway is not adversely affected. However, over-roadway sensors in specific, which are mounted over traffic lanes, can disrupt traffic and pose a safety risk to the installers as the installation may require for vehicles to be parked on the road.
- ***Security***: A concern for non-intrusive devices is security. As they can be installed at the side of the carriageway, they are exposed to risk of accidental damage or vandalism. To secure the equipment against damage or theft, the operator must keep workers informed and updated about the presence and scope of the equipment as well as protect it against theft.
- ***Portability***: Due to the way intrusive devices are fixed on a network (e.g. under the pavement), they do not allow for easy transfer from site to site, which may be a requirement especially due to cost constraints, while non-intrusive can be more flexible.

Table 1. Advantages and disadvantages of sensor devices.

Sensor type		Characteristics	
		+	-
Intrusive	inductive loop	Well established. Insensitive to weather. Flexible design.	Does not provide information on the exact position of the vehicle.
	pneumatic road tube	Low acquisition cost. Easy to install even though intrusive.	May be displaced resulting in loss of data. Snow plows can damage road tubes. If installed in a perpendicular direction to the traffic, the severity of incursion cannot be measured. A parallel installation is examined as an alternative.
	piezo-electric	It measures weight and speed.	Does not provide information on the exact position of the vehicle. Requires working within the traffic lane and is time consuming. If placed on road surface, may be displaced resulting in loss of data.
	magnetic	Insensitive to weather. Ease of deployment.	Difficulty in discriminating longitudinal separation between closely spaced vehicles.
Non-intrusive	infrared	Accurate measurement of vehicle position and speed. Multiple lane operation available.	Operation may be affected by low visibility. Installation and maintenance require lane closure. Requires regular maintenance. High cost.
	radar	Insensitive to weather. Direct measurement of vehicle speed.	High cost. Side-fire installation classifies vehicles based only on length.
	ultrasonic	Low cost.	It can be affected by environmental conditions.

- ***Cost***: Due to traffic management and increased labour, the cost of installing intrusive devices is generally higher. In addition, due to the lack of portability, the amount of locations where the devices can be permanently fixed is reduced.
- ***Environmental conditions***: Both intrusive and non-intrusive devices can be affected by environmental conditions, though not always in the same way. Thus, intrusive devices, such as piezo-electric cable, are more sensitive to wet road conditions, as well as snow and frost. On the contrary, non-intrusive devices, such as cameras, tend to be affected by low visibility (e.g. fog).

2.2 Examples of intrusive technologies

2.2.1 Inductive loops

Inductive loops are one of the most common vehicle detection and traffic counting methods used worldwide [8]. An inductive loop system consists of a cable, which is embedded into or under the roadway in a roughly square configuration, a cable extension and a counting device placed on the side of the road [9, 10]. The loop is based on the principle that a magnetic field introduced near an electrical conductor causes an electrical current to be induced. A metal vehicle passing over the loop acts as the magnetic field and the inductive loop as the electrical conductor [10]. The data supplied by conventional inductive loop detectors are vehicle presence, classification, count and occupancy. Loops cannot directly measure speed; however this can be estimated using a difference loop configuration [11].

Advantages of the inductive loops are its well-known and understood operation as it is mature technology. Also, the equipment cost is relatively low in comparison to other sensor technologies. The same does not apply though to the life-cycle cost of the sensors, as the installation cost is significant. Also, the installation and repair require disruption of traffic and the carriageway pavement. The wire loops are also subject to the stresses of traffic and temperature [8]. In addition, for this study a new configuration would need to be set up, as the existing loops in their normal set-up would not be able to detect incursions.

2.2.2 Pneumatic tubes

Pneumatic road tube technology uses hollow rubber tubes to detect vehicles by the change in air pressure generated when a vehicle tyre (or one axle) passes over a tube [12, 13]. The road tube is normally installed across traffic lanes, perpendicular to the traffic flow direction [11]. A device is placed at the side of the road and while connected to the tubes, it records the vehicle axles by the change in pressure. When activated, the device notes the time of the event [12, 13]. The pneumatic road tube sensor is portable, using rechargeable batteries as a power source [11].

There are several road tube configurations that can be used by a counter depending on the number of lanes and the data needs [12]. Axle counts can be converted to count, speed, and/or classification depending on how the configuration is structured [13]. Looking at the pattern of the compression times (for instance, the time interval between two activations of the counter caused by a vehicle's axles), each compression event is matched to a specific vehicle based on a set vehicle classification scheme. Two tubes attached to the same counter placed a set distance apart on the carriageway can determine the vehicle speed by measuring the time interval between an axle hitting the first and second tubes.

Advantages of pneumatic tube sensors are the low power consumption and the quick installation; this however, requires some intrusion into the flow of traffic [13, 14]. Road tube sensors are usually low cost and simple to maintain. Sensor manufacturers often supply software packages to assist with data analysis. Disadvantages include limited lane coverage and inaccurate axle counting when multiple axle vehicles' volumes are high or when vehicles are travelling at low speed and/or high proximity. Also, accuracy is subject to weather and temperature conditions [10, 11, 12].

For this study, the tubes could either be installed on the hard-shoulder perpendicularly to the traffic flow direction, as normal, or alongside the traffic flow, on the rib-line. In both cases, only one, or the two nearside tyres, would be detected, which could affect the classification and speed estimation. In addition, in the second configuration, the risk of the tubes being detached from the pavement due to traffic (and especially trucks) would be higher. Traffic management would be required for the installation and maintenance of pneumatic tubes (closure of the nearside traffic lane).

2.2.3 Piezoelectic

These sensors convert kinetic energy to electrical, generating a voltage when subjected to mechanical impact or vibration. The voltage is proportional to the vehicle weight [11]. Piezoelectric sensors can also classify vehicles by axle count and axle spacing and measure vehicle speed, when multiple sensors are deployed. Piezo-electric sensors can either be imbedded in the roadway, which requires cutting into the pavement surface, or placed on the road surface [13]. An advantage of this technology is the differentiation of individual vehicles axles with high precision. However, the accuracy of sensors is subject to pavement temperature and vehicle speed. As with pneumatic tubes, they can be displaced if on the road surface, while these can also fail due to poor road surface [11].

2.2.4 Magnetic

Magnetic sensors indicate the presence of a metallic object by the disruption it causes in an induced (active) or natural (passive) magnetic field [15]. The vehicle detection criterion is for the voltage to exceed a predetermined threshold. In order for the sensor to detect the change in the magnetic field, it must be relatively close to the vehicles; thus it is normally installed under or on top of the pavement. Magnetic sensors can be used to collect count, speed, and classification data; however, in operating conditions the sensors have difficulty differentiating between closely spaced vehicles [10, 13] and generally cannot detect the perimeter of a vehicle accurately, which is a requirement for this study. An advantage of the magnetic detectors in comparison to the inductive loops is that they cause less disruption on the road as they cover a smaller area. Also, they are less sensitive to stresses of traffic [11].

2.3 Examples of non-intrusive technologies

2.3.1 Infrared

There are two types of infrared technology: active and passive. The former is a device emitting a laser beam at the road surface, measuring the time for the reflected signal to return. The transmitted beam is interrupted by the vehicles, thus the reduction in time indicates this presence. The mounting position for active infrared sensors can vary from over the lane to roadside. Active infrared sensors can potentially provide information on the presence, the volume, the length, the speed and the number of axles. If the laser wavelength is between 400nm and 700nm, the beam is visible to the human eye and the sensor is called visible red light. The second infrared sensor type (passive) does not emit energy of its own and detects a vehicle by measuring infrared energy radiating from the detection zone. The infrared energy naturally emanating from the road surface is compared to the energy radiating when a vehicle is present. Passive infrared detectors are typically mounted over the traffic lane (e.g. on a gantry or bridge) or alternatively on a pole at the roadside. Both active and passive infrared devices can be used to record count, speed, and classification data and work well under day and night conditions. The main drawbacks are the performance during bad weather, and limited lane coverage [8, 10, 13].

2.3.2 Radar

Radar (RAdio Detection And Ranging) technology detects distant objects and is able to determine their position and speed. With vehicle detection, a device transmits high frequency radio waves at the roadway to determine the time delay of the return signal, thereby calculating the distance to the detected vehicle. In opposition to magnetic sensors, radar can also detect stationary vehicles, while at the same time, it can provide, if preferable, all the vehicle information per section of the detection area. An advantage of the radar is its insensitivity to inclement weather conditions and its capability to operate during day and night. [8, 13]. The radar sensor may be mounted over the carriageway (for single-lane monitoring) or at the side, off the hard-shoulder (for detection across several lanes). Side-mounted radars provide data corresponding to several traffic lanes, though generally not as accurately as in the over-looking position. However, side-mounted radars have also been used for single-lane detection [8].

2.3.3 Ultrasonic

The ultrasonic sensors transmit sound waves at a selected frequency between 20 and 65 kHz, to detect vehicles by measuring the time for the signal to return to the device. The preferred configurations for presence sensors are downward and side viewing, while for speed-measuring sensors is forward-looking. The transducers in both presence and speed-measuring devices convert the sonic energy received back into electrical energy that is then processed. Even though the implementation cost is relatively low, a disadvantage of the ultrasonic sensors is that they can be affected by inclement weather (e.g. low temperatures, snow) [8, 10].

3 Criteria for selection of technology

In order to select the most suitable technology for the hard-shoulder data collection, a series of criteria are applied.

3.1 Ability to collect data required

First and foremost, the equipment selected should be able to collect the data required as accurately as possible. Two potential limitations are the influence of vehicle speed and weather conditions. The technology selected must be effective in high speed conditions (the speed limit for general traffic on UK motorways is 70mph (112km/h) and 56mph for trucks (90km/h) [16]. Other potential issues are vehicles in close proximity, which might not be accurately classified as they are not always recognised as two or more separate vehicles, and stationary or slow-moving vehicles, which cannot be recorded by some technologies.

In terms of the environmental conditions, the performance of some types of technologies is influenced by weather events, such as fog, heavy rain and sub-zero temperatures, all of which are often encountered in the area of study. This also includes the detection of dark objects, which, depending on the abilities of the equipment, may be limited.

3.2 Cost

The overall cost of the use of equipment, which influences which type is used, includes the following:

- ***Procurement***: the cost of purchasing/renting the equipment, as well as all the accompanying elements (e.g. data logger, power supply establishment of internet connection for remote data collection).
- ***Installation***: the cost of the equipment to be deployed on the network. It includes the number of man-hours, the cost for traffic management (e.g. lane closure) and – if required and the possible interventions to the pavement or the existing infrastructure (e.g. gantries, lighting columns) that need to take place, especially if the technology is intrusive.
- ***Operation***: the cost of the energy consumed by the system and the cost for actions required during operation

(e.g. manual data collection, systematic cleaning of the equipment).

- ***Maintenance***: this can be divided into the routine and emergency costs; it is the cost (including man-hours) for inspecting and repairing, as well as the cost of any traffic management requirements.
- ***Replacement/ Complete removal***: this cost depends on the actions required and the possible repairing that the road or the surrounding infrastructure might need, as, for example, in the case of inductive loops.

3.3 Installation, operation and maintenance requirements

It is desirable to have minimum installation, operation and maintenance requirements, as this reduces, apart from the cost, the risk to operatives and disruption for road users. For this, side-fire mounting is preferred to over-head as it is simpler to install (no lane closures required). In this context, it is generally preferable for the equipment to be non-intrusive. This means that it causes minimum disruption to normal traffic and the road. If non-intrusive, the equipment is also portable. This is a major advantage in cases that it is necessary to collect data from several locations and there are financial constraints at the same time (See Table 2).

Table 2. Equipment attributes desirable to the road users and the operator.

	INSTALLATION	OPERATION - MAINTENANCE
ROAD USER	Safe installation. No lane closure or speed restrictions. Minimum disruption.	Safe operation and maintenance. No distraction to drivers.
MOTORWAY OPERATOR	Minimum time on the network for safety and cost. Minimum disruption to the public as a good service should be provided. No lane closure or Traffic Management to be installed. Non-intrusive and/or portable. Use of existing infrastructure as much as possible.	Minimum maintenance required. Reliability (minimum cost for inspection and repairs). Remote data collection. Energy efficiency.

3.4 Health, safety and security

For equipment to be installed and to operate on the network, health and safety requirements must be met. Generally, when a new piece of equipment is installed, a risk assessment and method statement are prepared to recognise and quantify the risks, mitigate or eliminate them and to determine a safe manner to carry out the installation and ongoing maintenance. Competence is key for health and safety for all phases and training is undertaken if required.

3.5 Overall

Overall, it is preferable that the equipment is transferable, i.e. it is possible to move it to another location under different conditions, as it gives the ability to collect data from a greater variety of locations. In addition, if the equipment has been used widely in the past, it means that it should be more reliable and any problems that might appear, as well as the limitations, are generally known. On the contrary, the use of a more contemporary system may offer a better perspective for alternative future uses.

4 Selection of implementation sites

In cases where it is not feasible to continuously monitor an area for incursions, mainly due to cost constraints, the sub-sections monitored should be carefully selected in order to cover an extensive variety of conditions, in terms of road construction characteristics (slope, curvature, hard-shoulder/main carriageway lane width, pavement type, street lighting, etc.) and other conditions, if these are known to change significantly (e.g. traffic volume, vehicle classification proportion). Proximity to a junction may also affect driving behaviour. In addition, the resources available on-site are investigated. These include power supply from existing electric cabinets, existing infrastructure that can be used for mounting the equipment (e.g. bridges, gantries), vehicle restraint system (VRS) which provides safer conditions for installation, maintenance and operation. It is also necessary to comply with health and safety legislation and the specification requirements of the Highways England in terms of installing new equipment on the motorway network, including the following:

- ***Structural Review***: A review of an individual or a group of structure (e.g. a bridge or a gantry) to confirm the load carrying capacity. This is conducted when the need arises, such as the installation on additional equipment on the structure, which increases the operational load [17].
- ***Technical Approval***: The submission of a Proposal, the acceptance of which confirms that the design, assessment, specification or construction works complies with the agreed design and specification certificates [18].
- ***VRS Length of Need***: That is the total minimum length of full height, full containment VRS as required in advance of, alongside, and after a hazard to protect this hazard, plus the additional lengths that are declared by the manufacturer to ensure full containment. Thus, any new equipment installed should not affect the VRS' functionality in case of an impact [19].
- ***Working Width***: This is the maximum lateral distance between any part of the barrier and the maximum dynamic position of it. If the vehicle body deforms around the road VRS so that the latter cannot be used for the purpose of measuring the working width, the maximum lateral position of any part of the vehicle is taken as an alternative. The working width need to remain clear from any installations [19].
- ***Structure Free Zone***: This is a buffer zone adjacent to the paved carriageway and beneath a Structure that

reduces the risk of errant vehicle impacts by providing an appropriate value of headroom [20].

- ***VRS Set-back***: Obstructions immediately adjacent to the edge of the paved carriageway result in drivers reducing speed and positioning their vehicles away from the obstruction. The purpose of the set-back is to provide a lateral distance between the VRS and the carriageway which reduces the effect of the safety barrier on driver behaviour and shyness. Any relaxations must consider the effects on the ability of occupants of stopped vehicles to leave via the nearside doors and the possibility of increased risk due to parking closer to live traffic [20].

5 Potential hard-shoulder incursions monitoring systems

After a range of sensors that can be altered to accommodate the study's needs are taken into consideration and the implementation sites are investigated, two potential solutions are suggested: a laser- and a radar-based system. The former is more flexible and easily transferable, while the latter has the potential to provide more detailed information.

5.1 Laser-based system

This system comprises of the laser sensors, the data logger and power supply. The laser sensor selected is provided by Micro-Epsilon (ILR1030-15), and has the ability to detect dark objects at a sufficient distance and a measuring frequency high enough to capture high-speed objects (Fig. 3) [21]. The sensors, while mounted at the side of the motorway pointing towards it perpendicularly, are measuring the distance to the overpassing vehicles. Due to the small size of the sensors (55x26x102mm – 90grams), they are not considered a risk to road users. As the laser is class 2, it poses no risk in case of eye contact. The sensors are placed inside a standard motorway marker post and are mounted to the rear of the VRS. The marker posts where the laser sensors are installed are additional to the existing ones. The sensors aim at an approximate height of 600mm above the carriageway at the outside edge of the rib-line and are levelled when installed.

Campbell Scientific provides a suitable data logger, also of a small size (241x104x51mm – 0.7kg), to store data from up to three laser sensors (Fig. 4) [22]. The logger has the ability to store data every 10^{-1} seconds, which allows the detection vehicles travelling at a speed of 40m/s or 150kph (90mph). Using the software provided with the data logger, the device is programmed so as whenever a vehicle crosses the rib-line, the measurement is recorded. The system can be powered either by using a battery, or by connection to the mains (where electric cabinets are present). The laser sensors are connected to and powered via the data logger. In order to protect the cables from damage the cables are carried in ducting.

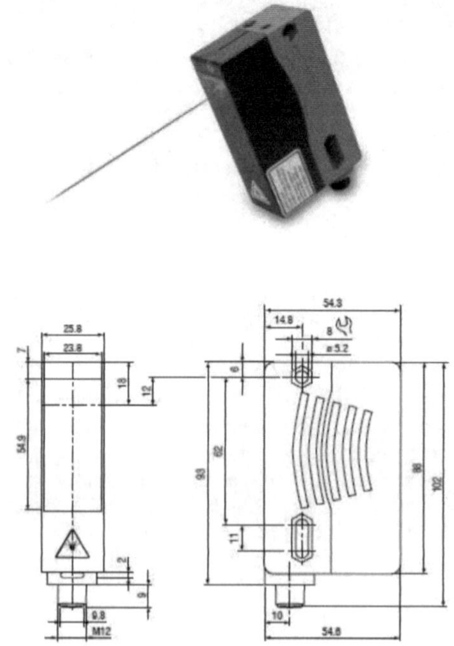

Figure 3. ILR1030-15 Laser sensor by Micro-Epsilon and dimensions (Source : Micro-Epsilon)

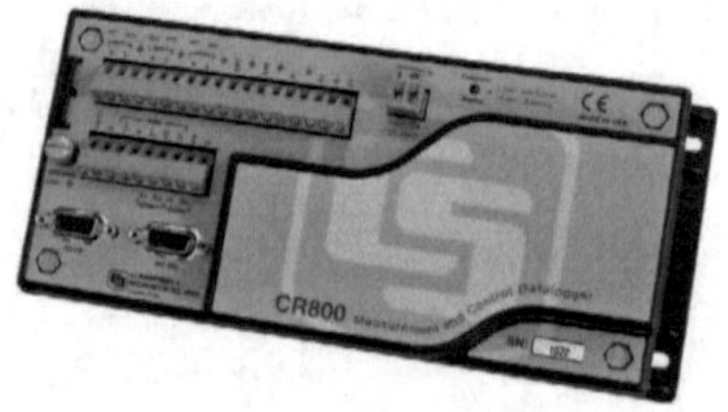

Figure 4. CR800 data logger (Source: Campbell Scientific)

5.2 Radar-based system

The alternative suggested method that can provide more detail is radar, which can monitor a continuous section of the motorway. A potential solution is provided by NavTech Radar [23]. The radar sensor is installed at the side of the motorway, while the mounting height can vary between 2m and 8m. For the hard-shoulder monitoring, a minimum height of 2m could be applied. However, to ensure security of the equipment, a mounting height of at least 3m is preferred. The higher the radar is mounted, the longer distance it can cover, whilst the blind area underneath the radar increases. The radar sensors are positioned to provide the optimum 'line of sight'. In order to check what can be seen by the radar one has to stand at ground level where the radar is mounted, or ideally, at the height of the radar. Whatever is visible to a person is also visible to the radar. Street furniture, such as lamp posts, can create visual obstacles to the radar, especially further away where they appear to be closer together. In a straight line the radar can detect a person located between 20 and 350m away, depending on the road geometry and mounting height.

DOI: 10.1051/matecconf/20168102019

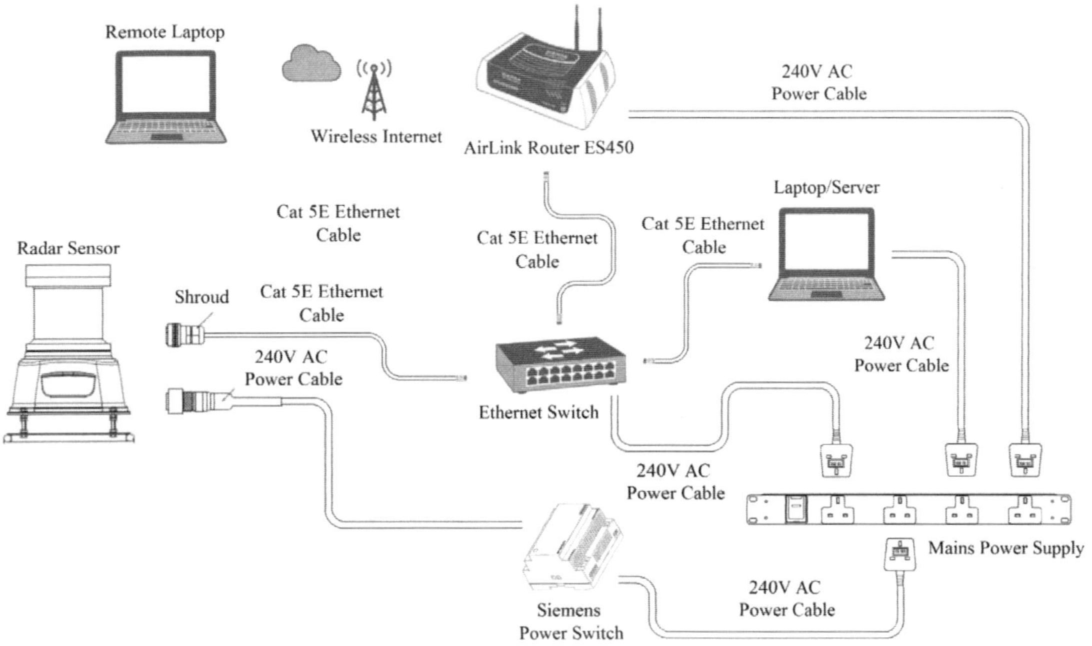

Figure 5. Set up of all the components of the radar system. (Source: P. Michalaki, created using NavTech Radar, Siemens and Sierra Wireless drawings)

The radar sensor may be mounted on dedicated posts, or various other structures – such as walls and gantries – using brackets. It is fitted to a plate on top of the post or the bracket using nuts and bolts which allows the tilt adjustment (levelling). On the M1-A1 Link Road, there are a series of structures that could be used for the radar mounting, such as A-Frame concrete gantries, bridges and variable message signs (VMS). While following the requirements of Highways England in the UK, several mounting solutions are investigated:

- **Custom-made metal bracket**: The bracket can be installed either at the side of an A-Frame gantry support (by drilling and fixing into the concrete) or on its horizontal beam. The main advantages of the metal bracket are the relatively small size, which would not be a distraction to the drivers, and the low manufacturing cost. However, this solution is intrusive to the gantry, would require rectification afterwards, and a full design would be required, which significantly increases the cost. In addition, specialist installation resources and traffic management would be required for the installation of the metal bracket (closure of the hard-shoulder and the nearside traffic lane) resulting in higher disruption for the road user and also higher costs.
- **Pole**: The manufacturing cost for a pole is considerably higher than that for a bracket and its installation would also require disruptive traffic management, further increasing the cost.
- **Trailer Mast System**: A trailer system that can be used for mounting the radar sensor is also considered. The approximate dimensions of the trailer are 5.0x2.0x2.0m and the approximate weight 500kg. The extended height can reach up to 15m. Even though the trailer could operate independently power-wise (solar power facility), which provides more flexibility, its size considerably restricts the locations where it can be used. This solution has higher maintenance requirements as well as certifications for its safety prior to the transfer.
- **Scaffold**: A scaffold is a temporary structure that can provide a stable base for the radar sensor. This is the solution considered the most suitable for this installation. The base plate is fastened to the scaffold and the radar sensor is positioned on top. The advantage of using a scaffold for this purpose is the non-intrusiveness to the existing structures on the network. In addition, the scaffold is set up 'around' a concrete gantry, thus the structure does not carry any additional weight, but does provide added restraint. In order for the scaffold to be installed, a hard-shoulder closure is required. After the scaffold is set up and the desired height is ensured (around 3.50m) using a simple distance measurement laser, the base plate is secured on top of the scaffold and the radar installed and levelled so it is parallel to the motorway surface. The levelling is achieved by fine adjustment of the nuts and bolts of the base plate. The base plate is first levelled to 0° and then adjusted according to the motorway slope. After installation access to the scaffold is prohibited to prevent unauthorised use or tampering with the equipment.

Each radar sensor requires a power and data connection via an Ethernet cable. Remote access minimises the necessity for site visits. In order to setup the radar system, several additional components are required. Fig. 5 shows all the connections of the radar system.

- **Power supply**: The radar uses 24V DC voltage and is powered via a transformer from the 240V AC mains (electric cabinet on-site). The power cable is provided with a military specification connector for the sensor connection and a bare end at the switch connection [24].
- **Ethernet connection**: An Ethernet cable is connected to the sensor for data transfer. A military specification shroud ensures a suitable environmentally protected network connection. At the other end, the Ethernet cable is connected to an Ethernet switch, which is also powered via the mains supply.
- **Laptop/Server connection**: A laptop or server is connected to the radar via the Ethernet switch. All components of the radar's software are installed and running constantly on this device, which is required to have certain specifications and is selected accordingly.
- **Router connection**: In order to have remote access to the laptop/server on-site, a router connection is set up. The IP addresses of all components (router, router's SIM card, on-site laptop/server, remote laptop) are synchronised for the system to function. In order to connect remotely to the laptop/server on-site, remote access software is used [25].

6 Conclusions

In this paper, the safety issue of motorway hard-shoulder incursions is approached. Since little is known about how these accidental movements occur, it is crucial to identify methods to monitor them so they can be further investigated. With no working or similar modifiable system available, alternative vehicle detection methods are considered. A review of intrusive and non-intrusive devices provided the advantages and disadvantages of each technology. Looking at the practical whole life process, from procurement and deployment to operation and removal, the considerations in order to employ the most suitable systems were described including basic criteria for equipment and site selection, such as safety, ease of deployment and cost. Based on these factors, two potential systems, a laser-based and a radar-based are suggested. Further research on field tests is being conducted in order to fully evaluate the suitability of each system and gain more feedback on the system parameters and performance.

References

1. Survive Group, *Surviving the Hard Shoulder.* (2006) See http://www.survivegroup.org/download_files/Leaflet%2022-06-07.pdf (accessed 01/02/2016).
2. Safer Monitoring, *Staying Safe on the Hard Shoulder.* (2012) See http://www.safermotoring.co.uk/SafetyOnHardShoulder.html (accessed 01/02/2016).
3. Safer Monitoring, *What to do if You Breakdown*. (2013) See http://www.safermotoring.co.uk/BreakdownWhatToDo.html (accessed 01/02/2016).
4. Department of Transport, Driver and Vehicle Services, *Drive Safe*. (2014) Government of Western Australia. See http://www.transport.wa.gov.au/mediaFiles/licensing/LBU_DL_B_DriveSafeFull.pdf (accessed 01/02/2016).
5. ACT – Australian Capital Territory, *Road Rules Handbook.* (2015) See http://www.rego.act.gov.au/__data/assets/pdf_file/0009/565029/2015-Road-Rules-Handbook-Accessible-Version.pdf (accessed 05/04/2016).
6. RoSPA – The Royal Society for the Prevention of Accidents, *General Driving Policy Statements*. (2005) See http://www.rospa.com/roadsafety/policy/statements/general-driving.aspx (accessed 01/02/2016).
7. Brake, Driver advice: vehicle maintenance and breakdowns. (2015) See http://brake.org.uk/info-resources/info-research/advice/21-facts-a-resources/resources/291-breakdown (accessed 01/02/2016).
8. L.A. Klein, M. K. Mills, D.R.P. Gibson, *Traffic Detector Handbook. Federal Highway Administration*, **I** (2006)
9. MarshProducts, *The Basics of Loop Vehicle Detection*. (2000).
10. G. Leduc. *Road Traffic Data: Collection Methods and Applications*. Joint Research Centre, European Commission. (2008)
11. L.E.Y. Mimbela, L.A. Klein, *A Summary of Vehicle Detection and Surveillance Technologies used in Intelligent Transportation Systems*. (2007)
12. P. McGowen, M. Sanderson, *Accuracy of Pneumatic Road Tube Counters*. (2011)
13. S.L. Skszek, *" State-of-the-Art " Report on Non- Traditional Traffic Counting Methods*. Prepared for Arizona Department of Transportation, Phoenix, Arizona. (2001).
14. Smith, D., & McIntyre, J. (2002). Traffic Volume Counts. In *Handbook of Simplified Practice for Traffic Studies*. Center for Transportation Research and Education, Iowa State University.
15. L.A. Klein, M.R. Kelley, *Detection Technology for IVHS Volume I: Final Report Detection Technology for IVHS,* **I**. (1996).
16. Department for Transport, 2014. Consultation Response Report Examining the Speed Limit for HGVs over 7.5 tonnes on Dual Carriageway Roads: Summary of Responses and Government Response, London.
17. Highways Agency, Transport Scotland, Welsh Government & Department for Regional Development Northern Ireland, 2011. Design Manual for Roads and Bridges (DMRB) Highway Structures Inspection and Maintenance, **3**, Section 4 Part 22 BD 101/11 Structural Review and Assessment of Highway Structures.
18. Highways Agency, Scottish Executive, Welsh Assembly Government & Department for Regional Development Northern Ireland, 2005. Design Manual for Roads and Bridges (DMRB) Highway Structures: Approval Procedures and General Design, **1**, Section 1 Part 1 BD 2/05 Technical Approval of Highway Structures.
19. Highways Agency, Transport Scotland, Welsh Government & Department for Regional Development Northern Ireland, 2006. Design Manual for Roads and Bridges (DMRB) Highway Structures Design (Substructures & Special Substructures), Materials, **2**, Section 2 Part 22 TD 19/06 Requirement for Road Restraint Systems.
20. Highways Agency, Scottish Executive, Welsh Assembly Government & Department for Regional Development Northern Ireland, 2005. Design Manual for Roads and Bridges (DMRB) Road Geometry, **6**, Section 1 Part 2 TD 27/05 Cross-Sections and Headrooms.
21. Micro-Epsilon, See http://www.micro-epsilon.co.uk/displacement-position-sensors/laser-distance-sensor/optoNCDT_ILR_1030_1031/index.html
22. Campbell Scientific, See https://www.campbellsci.com/cr800
23. NavTech Radar, See http://www.navtechradar.com/automatic-incident-detection
24. Siemens, See http://w3.siemens.com/mcms/power-supply-sitop/en/logo-power/pages/default.aspx
25. Sierra Wireless, See https://www.sierrawireless.com/products-and-solutions/gateway-solutions/airlink/

A Partial Proportional Odds Model for Pedestrian Crashes at Mid-Blocks in Melbourne Metropolitan Area

Alireza Toran pour[1] , Sara Moridpour[1] , Richard Tay[2] and Abbas Rajabifard[3]

[1] *RMIT University, School of Engineering, Civil, Chemical and Environmental Engineering Discipline, Melbourne, Australia*
[2] *RMIT University, School of Business, IT and Logistics, Melbourne, Australia*
[3] *The University of Melbourne, School of Infrastructure Engineering, Melbourne Australia*

Abstract. Pedestrian crashes account for 11% of all reported traffic crashes in Melbourne metropolitan area between 2004 and 2013. There are very limited studies on pedestrian accidents at mid-blocks. Mid-block crashes account for about 46% of the total pedestrian crashes in Melbourne metropolitan area. Meanwhile, about 50% of all pedestrian fatalities occur at mid-blocks. In this research, Partial Proportional Odds (PPO) model is applied to examine vehicle-pedestrian crash severity at mid-blocks in Melbourne metropolitan area. The PPO model is a logistic regression model that allows the covariates that meet the proportional odds assumption to affect different crash severity levels with the same magnitude; whereas the covariates that do not meet the proportional odds assumption can have different effects on different severity levels. In this research vehicle-pedestrian crashes at mid-blocks are analysed for first time. In addition, some factors such as distance of crashes to public transport stops, average road slope and some social characteristics are considered to develop the model in this research for first time. Results of PPO model show that speed limit, light condition, pedestrian age and gender, and vehicle type are the most significant factors that influence vehicle-pedestrian crash severity at mid-blocks.

1 Introduction

1.1 Background

In order to reduce air pollution and obtain better public health outcomes, efforts to encourage non-motorized transport modes have increased in recent years (Wey & Chiu 2013). To increase the number of walking trips, concerns about pedestrian safety has to be addressed. Pedestrians are more likely to be harmed or killed in traffic crashes. They are 23 times more likely to be killed than vehicle occupants [1] and more than 22% of traffic deaths in the world are pedestrians [2]. Every year, 34 pedestrians are killed in traffic crashes in Melbourne metropolitan area, representing 24% of the total traffic fatalities. Mid-block crashes account for 46% of total pedestrian crashes in Melbourne metropolitan area and 49% of pedestrian fatalities occur at mid-blocks.

Many studies have been conducted to examine the factors contributing to frequency and severity of vehicle-pedestrian crashes [3-6] .Whereas many of the studies have chosen to focus on crashes at intersections [7], there are limited studies on vehicle-pedestrian crashes at mid-blocks. Since the factors contributing to vehicle crashes at intersections and mid-blocks are significantly different [8-13], more research needs to be done to develop a model for vehicle-pedestrian crashes at mid-blocks.

Identifying and predicting trends in crash severity is one of the main steps to improve pedestrian safety. In this approach crash severity modelling assist in distinguish influencing variables on pedestrian crashes. Review of literature shows many different statistical models have been applied in crash studies. PPO model that is one of regression models, allows covariates to affect different levels of severity injuries. Therefore, the proportional odds assumption will be relaxed for independents variables that have different effects on different level of injuries.

1.2 Objective and scope of study

The main objective of this research is to identify the factors contributing to the severity of vehicle-pedestrian crashes at mid-blocks. Whereas previous studies have mainly focused on pedestrian crashes at intersections or applied a model for crash risk at mid-blocks for special group of pedestrian (e.g. children) or specific study area (e.g. pedestrian crossing), this research will examine all vehicle-pedestrian crashes in Melbourne metropolitan area. In addition, this research will consider different social-economic variables, such as place of birth, level of pedestrian education and percentage of labour force participates. Also, the distance of crash location from/to public transport stops is another variable used in this research as a novel measure to identify the influence of public transport stops on pedestrian crashes. This research apply PPO model in pedestrian accident at mid-blocks to explore influencing variables in this type of crashes.

This paper begins with a review of previous studies on pedestrian crash modelling and then presents the methodology of this research. Description of the variable data is presented then. It is followed by presenting the model development in this research. Afterward, the results are presented and discussed. Finally, the outcomes are summarized and direction for future research is presented.

2 Literature review

The published literature on pedestrian crash modelling shows that binary models, ordered discrete models and unordered multinomial discrete models are three main statistical techniques which have been widely used to study pedestrian crash severities. In binary models for crash severity, outcomes are injury against non-injury crashes or fatal against non-fatal crashes. These studies used common discrete models such as the binary Logit and binary Probit models.

In recent studies, Sarkar et al. [14] and Ballesteros et al.[15]developed a logistic regression model for pedestrian crashes to predict influencing variables for this type of crashes. However, according to the ordinal nature of crash severity, ordered probability models are popular in traffic crash studies. Lee and Abdel-Aty used this approach to estimate the likelihood of pedestrian injury severity at intersections [7]. In ordered logit and ordered probit models, it is assumed that the parameter estimates are constant across severity level [16].Eluru et al.[6] developed a generalized ordered probability model to examine pedestrian and bicyclist injury severity levels in traffic crashes in the USA.

Limitation of the traditional ordered Logit and Probit models leads to developing unordered models to find important variables in traffic accidents. MultiNomial Logit model (MNL), Mix Logit model and random parameter Logit model are most common unordered models that are used in many pedestrian crash studies [17-20] . However, these models have similar limitations to logit models. For instance, all of explanatory variables must be independent to each other.

The ordered nature of the traffic crash severity cannot be ignored completely. Furthermore, it is not appropriate to assume all of predictors have same effect on different levels of crash severity. PPO model proposed by Peterson and Harrell [21] combines the ordered arrangement in ordinal models and allows certain independent variables to affect different levels of dependent variable differently [22]. Because, PPO models allow a combination of the two ordered and multinomial modelling frameworks, the proportional odds assumption will be relaxed for independent variables that show different effects on different severity level. Sasidharan and Menedez [22] compared PPO models, ordered logit models and multinomial logit models for pedestrian crash injury severities in Switzerland. They found that PPO models have better results than two other models type based on different evaluation criteria.

In summary, there are many studies that analysed pedestrian crash severity. However, there is no study about pedestrian crash at mid-blocks. In addition, independent variables in most of studies were limited to only personal or traffic characteristics. The objective of this paper is to analyse vehicle-pedestrian crash severity at mid-blocks in Melbourne metropolitan area using PPO models. Furthermore, to explore the effects of different variables on this type of crash different socio-economic, traffic and environmental characteristics are analysed in this study.

3 Data description

CrashStats is a database contains road crash statistics of Victoria, Australia from 1987 onwards. The dataset includes crashes where at least 1 person was injured. This dataset includes personal characteristics such as driver and pedestrian age and gender, vehicle characteristics like vehicle type and weight, road and environment condition, including surface, light and pavement condition, and temporal parameters such as date, day and time of the crashes. In addition, this study considers the influence of social and economic characteristics on pedestrian crashes. These social and economic variables include religion, income, occupation type, percent of labour force participation in suburbs, level of education and place of birth that are extracted from Australian Urban Research Infrastructure Network (AURIN). AURIN is the largest single resource for accessing diverse types and sources of data, spanning the physical, social, economic and ecological aspects of Australian cities, towns and communities. Also, AURIN dataset is used to extract geospatial data, such as road network, elevation (which is used to find the road slope), public transport stops, and other data like traffic volume and land-use categories.

To investigate the variables influencing the pedestrian crash severity, data for all pedestrian crashes on public roadways in the Melbourne metropolitan area from 2004 to 2013 are used in this research. A total of 12,279 pedestrian crashes were recorded in Melbourne metropolitan area over the 10-years period where 5,760 of them occurred at mid-blocks. Traffic volume is an influencing variable on accident analysis and many studies emphasized the influence of traffic volume on pedestrian crashes [1, 23, 24]. However, this variable is ignored in many pedestrian crash studies, especially in crash severity models. To consider traffic volume in this research the location of accidents are joined to the Melbourne road network using ArcMap GIS.

In CrashStat, crash severity is divided into three levels (base on Victoria police data source): fatal crashes, serious injury crashes and other injury crashes. A fatal crash refers a crash that at least one person died immediately or death from injuries within 30 days of a collision. In serious injury accidents, at least one person is sent to hospital. In addition, other injury crashes refer to crashes that at least one person injured in accident but

 DOI: 10.1051/mateconf/20168102020

it was not major injury and did not send to hospital. Fatal and serious injuries comprised 51.2% of total pedestrian crashes (4.2% for fatal and 47% for serious injuries). Also, 48.8% of pedestrian crashes were other injury severity. Table 1 to 3 show a summary of the classified data used in this study for different types of pedestrian crash severity between 2004 and 2013. The continuous data such as traffic data and social economic variables which are used in this study are shown in Table 4. In this study, 38 explanatory variables are classified into 6 categories describing the accident time, personal characteristics, traffic and road variables, environment and land use specifications and socio-economic characteristics for each recorded crash. However, lack of access to some pedestrian characteristics parameters data such as pedestrian volume or speed cause to the impact of these characteristics not considered in this paper.

Table 1. Time and personal characteristics variables

Category	Variable		(%)
Time	Season	Springs	23.9%
		Summer	24.0%
		Fall	25.8%
		Winter	26.3%
	Time of day	Morning Peak (7:00 -	13.2%
		Afternoon Peak (16:00 – 18:00)	3.7%
		Off-Peak (10:00 –	35.8%
		Other	47.4%
	Day	Weekdays	74.9%
		Weekend	25.1%
Personal	Pedestrian Gender	Female	42.8%
		Male	55.4%
		Unknown	1.8%
	Driver Gender	Female	29.4%
		Male	59.6%
		Unknown	11.0%
	Pedestrian Age	0-14 (Children)	8.4%
		15-65 (Working age)	72.8%
		> 65 (non-working age)	15.8%
		unknown	3.0%
Personal	Driver Age	< 18	0.4%
		18 - 65	79.9%
		> 65	7.5%
		Unknown	12.1%

In Table 2 occupation types are divided to three categories, White-collar work is performed in an office, cubicle, or other administrative setting.in addition, in this categories Pink-collar worker, whose labour is related to customer interaction, entertainment, sales, or other service-oriented work and Blue-collar refers to manufacturing, construction, mining and agricultural businesses.

Table 2. Traffic, road and environmental and social

Category	Variable		(%)
Traffic and road	Vehicle Type	Light Vehicles	78.4%
		Taxi and Van	9.0%
		Heavy Vehicles	0.7%
		Buses	1.6%
		Motor and Bike	4.4%
		Tram	3.3%
		Train	2.6%
		Other	0.0%
	Traffic Control	No control	78.9%
		Stop go light and	6.1%
		Pedestrian light and cross walk	7.8%
		Unknown	7.1%
	Surface Condition	Dry	82.8%
		Wet, Muddy, snowy	12.9%
		Other	4.3%
	Divided road type	Divided Double line	9.2%
		Divided Single centerline (DS)	26.1%
		Not Divided (ND)	40.0%
		Unknown (U)	24.7%
Environment	Light Condition	Day	62.1%
		Dusk/Dawn	5.6%
		Dark street light on	26.6%
		Dark no street light	4.4%
		Other	1.3%
	Atmosphere Condition	Clear	85.9%
		Raining and snowing	9.0%
		Fog, smoke, dust and strong winds	0.6%
		other	4.6%
	Land Use	Residential	22.8%
		Commercial	29.3%
		Industrial	7.1%
		Community and Educational	2.7%
		Sport, Recreation and parks	4.9%
Social	White Collar workers in Suburb	< 50%	39.4%
		50-80%	59.6%
		> 80%	1.0%
	Pink Collar workers in Suburb	50-80%	46.5%
		> 80%	53.5%
	Blue Collar workers in Suburb	> 50%	8.2%
		50-80%	88.9%
		> 80%	2.9%
	Income (suburb)	Low income (< $600 P/W)	2.3%
		Middle income ($600-$2,499 P/W)	94.2%
		High income (> $2,500 P/W)	3.5%

Table 3. Descriptive statistics for continues variables

Category		Variable	S.D	Mean
Traffic and road		Average daily traffic (ADT)	10151.4	13250.8
		Average road slope	3.0	1.3
		Distance to public transport stops	392.3	134.3
Social[a]	Age and gender	Gender ratio	0.1	1.0
		Median age	4.7	35.4
	Religious affiliation	Christian	6.2	48.9
		Islamic	5.1	3.7
		Other religion	0.0	11.2
		No religion	8.9	26.2
	Place of birth	UK	2.1	3.9
		Southern and eastern of Europe	0.2	0.2
		Middle-East	1.6	0.8
		Other place than Australia	6.9	43.6
		Asia	12.9	15.7
		Indigenous persons	0.4	0.4
	Level of education	With a degree or higher qualification	15.7	45.6
		With a certificate, diploma or advanced diploma	13.1	34.6
	Labor force	Labor force participant rate	8.3	60.0
	population	Suburb population density	2506.7	3276.3

[a] The present of social variable are extracted for each suburb

4 Methodology

In this study vehicle-pedestrian crash severity is categorized into 3 levels; fatal crashes, serious injury crashes and other injury crashes. The probability of crash severity level (j) for a given crash (i) can be specified as [22, 25]

$$P(Y_i > j) = P_{ij} = \frac{EXP(\alpha_j + X_i\beta_j)}{1 + EXP(\alpha_j + X_i\beta_j)} \quad j = 1,2,...,J-1 \tag{1}$$

where X_i is a vector of explanatory variables, β_j is a vector of estimable parameters, and α_j and $\alpha_{(j-1)}$ are the upper and lower thresholds for injury severity j. The difference between this equation and the General ordered logit model (Gen) is that β_j is free to vary across severity levels [16]. In proportional models (PO) parallel- lines assumption may be violated by one or more variables and in PPO models these variables are specified by β_j which differs across equations.

The PPO model for this study is fitted using SAS 9.3. To interpret the results of PPO models it should be cautious that the sign of β does not always determine the direction of the effect of the intermediate outcomes [26].

In these models the marginal effects are useful for interpretation of variables. The marginal effects show how any change in the independent variables can effect on the dependent variable. Figure 1 shows the framework for this study. The final outcomes of this model identify the variables that influence on vehicle-pedestrian crash severity at mid-blocks in Melbourne.

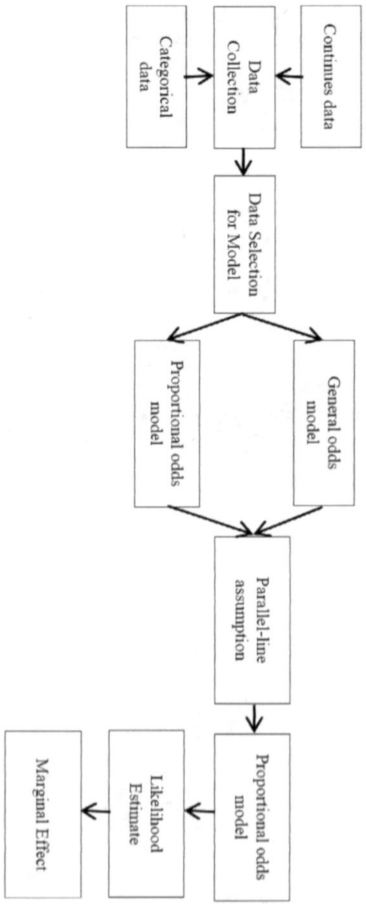

Figure 1. PPO model framework.

5 Results

This section shows the results of general logistic regression and PPO model for pedestrian crash severity at mid-blocks. Moreover, the marginal effect of PPO model illustrates to show influence of variables on crash severity. Table 3 shows the results of likelihood ratio tests for General model (Gen), Proportional Odds model (PO) and PPO. According to this table models reject the proportional odds model in favour of both the non-proportional odds model and the partial proportional odds model. In addition, Table 5 indicates that the partial proportional odds model fits as well as the non-proportional odds model.

Table 6 shows the goodness of fit for tree statistical tests. The goodness of fit of a statistical model describes how well it fits a set of observations. According to Table 4, Null Hypothesis is rejected for all testes, thus the PPO model is fitted to all of observation well (ChiSq < 0.0001). Also, Table 7 shows the hypothesis tests for all variables in the model individually. The chi-square test statistics and associated p-values shown in the table indicate that Speed, Light Condition, Pedestrian Gender, pedestrian age, vehicle type, traffic volume, per-cent of Islamic affiliation in suburbs, distance of public transport stops to location of accidents, and suburb's population density in the model significantly improved the model fit.

Table 4. Likelihood ratio for three Gen, PO and PPO models

Test	ChiSq	DF	Pr>ChiSq
Gen vs PO	113.40	81.00	0.01
PPO vs PO	16.97	3.00	0.00
Gen vs PPO	96.43	78.00	0.08

Table 5. Model fir statistic

Testing Global Null Hypothesis: BETA=0			
Test	Chi-Square	DF	Pr > ChiSq
Likelihood Ratio	648.68	85	<.0001
Score	623.49	85	<.0001
Wald	540.33	85	<.0001

Table 6. Hypothesis test for each variable in the model

Type 3 Analysis of Effects				
	Effect	DF	Wald ChiSq	Pr> ChiSq
1	Season	3	0.57	0.90
2	Time	3	6.8	0.07
3	Day	1	1.68	0.19
4	Speed	3	53.58	<.0001
5	Light Condition	4	47.64	<.0001
6	Pedestrian gender	2	8.33	0.01
7	Driver gender	2	3.63	0.16
8	Pedestrian age	3	93.79	<.0001
9	Driver age	3	5.28	0.15
10	Vehicle type	6	21.29	0.00
11	Traffic Control	5	4.45	0.48
12	Surface Condition	4	4.06	0.39
13	Atmosphere Condition	6	4.84	0.56
14	Traffic volume	1	6.41	0.01
15	Male to Female ration	1	0.01	0.94
16	Median Age in suburb	1	0.00	0.99
17	Christian affiliation	1	1.00	0.32
18	Islamic affiliation	1	5.82	0.02
19	Other religion affiliation	1	0.75	0.39
20	No Religion affiliation	1	0.57	0.45
21	White Collar participants	2	0.88	0.64
22	Pink Collar participants	1	0.60	0.43
23	Blue Collar participants	2	1.64	0.43
24	Born in UK	1	9.43	0.00

Table 7. Hypothesis test for each variable in the model (continue)

Type 3 Analysis of Effects				
	Effect	DF	Wald ChiSq	Pr> ChiSq
25	Born in SE of Europe	1	0.01	0.90
26	Born in Mealiest	2	6.91	0.03
27	Born in overseas	1	0.84	0.35
28	Born in Asia	1	0.29	0.58
29	Indigenous	1	2.01	0.15
30	Higher Qualification	1	1.07	0.30
31	Certificate diploma	1	2.57	0.11
32	Labour force participant	1	0.15	0.69
33	Suburb average Income	2	0.75	0.69
34	Divided road type	3	2.71	0.44
35	Average road slope	1	0.81	0.37
36	Distance to public transport	1	5.61	0.02
37	Land Use	5	28.80	<.0001
38	Population Density	2	18.20	0.0001
39	Existence of Median	1	2.43	0.12
21	White Collar participants	2	0.88	0.64

Table 8 shows results of maximum likelihood estimation and marginal effect analysis for explanatory variables. This table displays that roads with low speed limitation increase the probability of fatal or serious vehicle-pedestrian crashes at mid-blocks more than roads with high speed limits. In addition, roads with speed limit higher than 70 km/h increase the probability for sever pedestrian crashes. Road with speed limits under 60 km/h are local roads and in these roads drivers drive with speed higher than speed limits. Furthermore, in these roads people use more jay walk to cross the roads and these cause to increase the probability of mid-block crashes. Moreover, in roads with high speed limits, vehicle move faster and the braking distance is higher than low speed limits road. In addition, intensity of crash is higher in vehicle with high speed. Thus, the severity of pedestrian-crash in road with high speed limit is usually high.

Table 8 reveals that accident at day condition decrease the severity of pedestrian crashes (marginal effects -0.06). In nights traffic is less and speed is higher than days. In addition, darkness could impacts on vision and reaction of pedestrian and drivers. Therefore, vehicle-pedestrian crashes during the nights are more sever. Furthermore, the results indicate that pedestrian gender and age could be an influencing parameter in pedestrian crashes. Table 8 shows that the probability of sever crash is decreased for female pedestrian (marginal effect -0.02). Usually females drive carefully and with lower speed than males, thus the crash severity of female driver could be lower than male drivers. Against vehicle crash could be another reason for high severity crash of these pedestrian.

Table 8. Maximum likelihood estimation and marginal effect analysis.

Parameter		Severity	DF	Estimate	Standard Error	Wald Chi-Square	Pr > ChiSq	Marginal effect
Intercept		Fatal	1.00	0.14	2.05	0.01	0.94	
Intercept		Serious	1.00	2.93	2.06	2.03	0.15	
Speed	60-70		1.00	-0.19	0.07	8.05	0.00	-0.02
	Other speed		1.00	0.56	0.12	19.96	<.0001	0.07
	Under60		1.00	0.27	0.08	11.42	0.00	0.03
Light Condition	Dark light off		1.00	-0.10	0.17	0.33	0.57	-0.01
	Dark light on		1.00	-0.32	0.11	9.19	0.00	-0.04
	Day		1.00	0.49	0.11	20.74	<.0001	-0.06
	Dusk		1.00	0.06	0.14	0.22	0.64	0.01
Pedestrian gender	Female		1.00	-0.13	0.12	1.34	0.25	-0.02
	Male		1.00	-0.30	0.12	6.23	0.01	-0.04
Pedestrian age	15-65		1.00	0.08	0.08	1.01	0.32	0.01
	Unknown		1.00	0.45	0.20	5.17	0.02	0.06
	Under 15		1.00	0.31	0.12	6.94	0.01	0.04

Moreover, Table 8 reveals that pedestrian crashes with light vehicle (e.g. passenger cars or vans) and heavy vehicle (e.g. trailer or trucks) are more sever and the probability of these crashes to be sever are more than other type of vehicles. Light vehicles usually move in road with speed close to speed limit therefore the pedestrian-vehicle crash in this situation is more likely to be sever. Furthermore, braking distance and intensity of crash could be the reasons of this level of severity for with heavy vehicles-pedestrian crash.

The results of maximum likelihood estimation analysis show that suburbs social characteristics can be an influencing variable in pedestrian crashes. Culture can impact on walking and traffic behaviour. People's traffic behaviour may vary among cultures and this study consistent with the results from previous researches that showed culture and family can be an important factor influencing on traffic crashes [27, 28].

Furthermore, this result shows that land use is a significant variable in pedestrian crashes at mid-blocks. Table 8 indicates that commercial land use increase the likelihood of an accident to be a sever crash. In this land use pedestrian jaywalk and has less attention to traffic condition when want to cross the roads. This lack of attention to traffic condition and vehicles could be one reason that increasing the severity of pedestrian crashes in these areas.

Finally, this table reveals that suburb's population density is a significant explanatory variable in pedestrian crashes at mid-blocks. Since this variable was significant in Linear Hypothesis Results (Pr less than 0.1) in the PPO model is allowed to be vary across the response function. It means that the impacts of this variable on fatal and serious crashes were different. Thus, there are 2 probability results for population density. Table 5 shoes that population density has no significant impact on fatal crashes; however increasing population density in suburbs can increase the probability of serious injury crashes for pedestrian at mid-blocks. It is clear that increasing the population in suburbs increase the traffic and the probability of pedestrian crashes, however this density can decrease the vehicle speed and probability of fatal crashes.

6 Results

PPO model is a type of logistic regression that allows certain individual predictor variables that meet the proportional odds assumption to affect different level of response variable with the same magnitude, while other predictor variables that do not meet the proportional odds assumption can have different effects on different level of response variable. To analysis the likelihood estimation of explanatory variables in pedestrian crash severity at mid-blocks pedestrian crashes on public roadways in the Melbourne metropolitan area from 2004 to 2014 used in this research. This research applied crash severity (fatal, serious injury and other injury crashes) as target variables and 38 different socio-economic variables (such as population, income, religion, occupation, and etc.),environment variables (such as light condition, land use, surface condition and etc.), traffic and road characteristics (including road slope, vehicle type, traffic volume, distance from public transport stops and etc.), personal characteristics (like age and gender) and temporal variables (such as time, day and date of the crashes) were used to develop the model. Results of proportional odds assumption analysis indicated that some culture variable and population density in suburbs reject the proportional odds assumption. Thus, in final model these two variables allowed to be vary across the model response. This study showed that Speed, Light Condition, Pedestrian Gender, pedestrian age, vehicle type, traffic volume, culture and distance of public transport stops to location of accidents, and suburb's population density were the variables that significantly improved the model fit. However, in maximum likelihood estimation analysis the Speed, Light condition, Pedestrian characteristics, culture, distance of accidents to public transport, land use and population density were influencing variables. This research indicated that risk of severity crashes at roads with high speed limit, and day condition are more than other roads. Furthermore, personal characteristics of pedestrian had significant impacts on severity of crashes. This study showed that pedestrian under 15 years old is more likely to be killed or harmed in crashes. In addition, this study revealed that commercial area increase the probability of sever crash for pedestrian at mid-blocks. Also, population density in suburbs had two different impacts in crash severity model. Increasing the population density increased the probability of serious crashes, however this variable decrease the likelihood of fatal pedestrian crashes.

This study showed the ability of PPO model in severity crash analysis for pedestrian crashes. However, it can be useful to compare the results of this model and other models and techniques.

References

1 L.F. Miranda-Moreno, P. Morency, and A.M. El-Geneidy, Accident Analysis & Prevention, (2011). **43**(5): p. 1624-1634.
2 WHO, *Pedestrian Safety, A road safety manual for decision-makers and practitioners*. (2013), World Health Organization.
3 T.K. Anderson, Accident Analysis & Prevention, (2009). **41**(3): p. 359-364.
4 J.-K. Kim, et al., Accident Analysis & Prevention, (2008). **40**(5): p. 1695-1702.
5 D. Dai, Journal of Transport Geography, (2012). **24**(0): p. 206-214.
6 N. Eluru, C.R. Bhat, and D.A. Hensher, Accident Analysis & Prevention, (2008). **40**(3): p. 1033-1054.
7 C. Lee, and M. Abdel-Aty, Accident Analysis & Prevention, (2005). **37**(4): p. 775-786.
8 B. Roudsari, R. Kaufman, and T. Koepsell, Traffic Injury Prevention, (2006). **7**(3): p. 283-289.
9 A.S. Al-Ghamdi, Accident Analysis & Prevention, (2003). **35**(5): p. 717-724.
10. S.S. Pulugurtha, and V.R. Sambhara, Accident Analysis & Prevention, (2011). **43**(1): p. 439-446.
11 A. Lightstone, et al., Injury Prevention, (2001). **7**(2): p. 155-160.
12 D.N. Moore, et al., Accident Analysis & Prevention, (2011). **43**(3): p. 621-630.
13 S.A. Bennet, and N. Yiannakoulias, Accident Analysis & Prevention, (2015). **78**: p. 94-103.
14 S. Sarkar, R. Tay, and J. Hunt, Transportation Research Record: Journal of the Transportation Research Board, 2011(2264): p. 128-137.
15 M.F. Ballesteros, P.C. Dischinger, and P. Langenberg, Accident Analysis & Prevention, (2004). **36**(1): p. 73-81.
16 P.T. Savolainen, et al., Accident Analysis & Prevention, (2011). **43**(5): p. 1666-1676.
17 D. McFadden, and K. Train, Journal of applied Econometrics, (2000). **15**(5): p. 447-470.
18 K. Kim, and E.Y. Yamashita, Journal of Advanced Transportation, (2007). **41**(1): p. 69-89.
19 Y.-I. Kim, S.H. Park, and S.-Y. Kho. *A Case Study of 'Continuous Risk Profile' Approach for Hotspots Identification on Korean Expressways*. in *17th ITS World Congress*. (2010).
20 H.A. Aziz, S.V. Ukkusuri, and S. Hasan, Accident Analysis & Prevention, (2013). **50**: p. 1298-1309.
21 W.A. Harrell, The Journal of Social Psychology, (1991). **131**(3): p. 367-372.
22 L. Sasidharan, and M. Menéndez, Accident Analysis & Prevention, (2014). **72**: p. 330-340.
23 M. Wier, et al., Accident Analysis & Prevention, (2009). **41**(1): p. 137-145.
24 C.D. Cottrill, and P. Thakuriah, Accident Analysis & Prevention, (2010). **42**(6): p. 1718-1728.
25 R. Williams, Stata Journal, (2006). **6**(1): p. 58-82.
26 S.P. Washington, M.G. Karlaftis, and F.L. Mannering, *Statistical and econometric methods for transportation data analysis*. (2010): CRC press.
27 R. Factor, D. Mahalel, and G. Yair, Accident Analysis & Prevention, (2008). **40**(6): p. 2000-2007.
28 P.F. Agran, et al., Inj. Prev., (1998). **4**(3): p. 188-93.

MATEC Web of Conferences **81**, 02021 (2016) DOI: 10.1051/matecconf/20168102021
ICTTE 2016

Type Classification of Motorized Vehicles Red Light Running and Their Characteristics Analysis

Ya Ping Zhang[1], Yan Liu[1], Lan Liu[2] and Chuan Yun Fu[3]

[1]*School of Transportation Science and Engineering, Harbin Institute of Technology, Harbin, China*
[2]*School of Safety and Environment Engineering, Hunan Institute of Technology, Hengyang, China*
[3]*College of Transportation and Logistics, Southwest Jiaotong University, China*

Abstract. In order to study types of motorized vehicles red light running and their characteristics, the concept of critical distance was introduced, combining with the start and end of red light, to classify motorized vehicles red light running into three categories: deliberate and unintentional red light violation at red onset and red light violation at red end. Based on video observation data, the number of motorized vehicles that commit any of the afore-mentioned three types of red light violation was statistically analyzed, and then their characteristics under various traffic control and management conditions were explored using z-test method. The results show that the majority of red light runners at red start are deliberate, while the unintentional red light violators also share some proportion. At the end of red light, there are a few vehicles violating red light as well. Additionally, the rates of red light running are obviously varying between those three classifications of red light running under the conditions of countdown timer, automatic red light camera, green flashing device, different speed limits and peak period.

1 Introduction

Red light running is a common traffic violation, but also one main reason to lead to the intersection traffic accident. According to statistics, the traffic accident caused by running a red light accounts for 2.5% of the total number of intersection accidents[1]and the frequency of this behavior is about one time per every 3.5 minutes[2-3]. An American survey shows that 1.3 times red light running will occur by every 1000 cars going through intersections with the frequency of 3 times every hour[4].Similar research found that the number ranges from 0.45 to 38.5[5-6]. Running a red light often causes very serious consequences, especially loss of life and property. Therefore, it is of great significance to classify the type of red light running analyzing their characteristics, and put forward the corresponding intervention measures to reduce the occurrence of this behavior.

Predictors of red light running include being under the age of 30, having poorer driving records, driving smaller and older cars, not wearing a safety belt, getting closer to the vehicle in front, angrily gesturing to other drivers and drinking alcohol prior to the violation[7]. By establishing the matched group, a study of 2015 showed that local, male drivers and passenger vehicles are more likely to run red light and the probability of red light running increases when traffic volume is higher[8]. Lum et al [9]compared the number of red light running vehicle before and after the installation of countdown timer. They found that it reduced by 65% within the first two months and later gradually restored to the level before the installation. Qian et al [10-11] found that the countdown will give the drivers a signal to accelerate through the intersection with an increase of red light behavior.

Kejun Long et al [12] argued that the countdown could effectively prevent radical acceleration and sudden stop, but could not eliminate the phenomenon of red light running. Helai Huang et al [13] noted that the dilemma zones would cause red light running, which the green signal countdown timer would lead to. Some scholars have studied the influences of red light camera, and different conclusions were made. Retting et al. [14-16] studied the impact of the red light camera, the results showed that red light camera could significantly reduce the rate of red light running. Another study [17] showed that the rate of running a red light had reduced by 69% in three months of the installation of red light camera and the number was 38% at the 6th month. While Sze et al [18] think there is no obvious connection between red light running rate and red light camera.

In addition, Yang[19] and other research indicates that, the behaviors of red light running have different characteristics under different traffic flow. Elimitiny et al [20] pointed out that with the increasing of traffic volume, red light running rate would be higher. Morning and evening peak will also increase the red light behavior, working day, the speed limits, vehicle speed on the yellow onset and distance to the stop line affected red light running behavior significantly.

Various studies have focused on the behavior of motorized vehicles red light running and their influence factors Little literature has classify the type of red light running at the beginning of red light, and the red light running at red end has not been explored. In view of this, this paper derives and calculates the critical distance based on the observation data, classify the type of red light running and analyzes the characteristics of every type.

2 Data collection

2.1 Study sites

Seven signal intersections in Harbin city are selected as the video observation sites. The main selection principle of observation sites are excellent sight on the target lanes, no shelter, pedestrian overpass. According to the research purpose of this study, the selection of the intersection should have different signal control scheme, management facilities and speed limit. Table 1 shows the detailed information of the observation sites. For the convenience of records, the following sections of this study use the abbreviation names of each road.

2.2 Video observation

Data collection has been carried out during the period between 1st July to 31st July, 2015. All traffic surveys were carried out during working days of the week (Saturday to Wednesday), during afternoon period (2.00 pm to 6.00 pm).We defined that the first two hours non-peak hour and the latter two hours peak hour. Every observation site were recorded three days totally. Illustration of video recording is showed as figure 1.

Video observation including the following steps:

(1)Set up the camera on the pedestrian overpass, close to import road in the middle lane, adjust the camera lens and focal length and fixed camera.

(2)Before the shooting, mark 12 pairs of reference points every 10 meters at the entrance lane on both sides from the stop line

(3)After marking the end of the reference point, establish a benchmark at each of the reference points. Record the whole process until the last pair of points.

(4)Turn off the camera after the completion of the steps 3.Start the camera to shoot the traffic flow and traffic signal control on the import road at 2 pm.

(5)In the process of shooting, the staff measured and recorded the geometric parameters of the intersection and traffic management facilities, etc.

Through the video recording of seven signal intersections around Harbin urban area for, a total of 84 hours video data were obtained.

Table 1. Detailed information of observation sites

intersection	Cycle/s	Yellow interval/s	Countdown timer	Red light camera	Green flash device	Speed limit（km·h^{-1}）
ZhengYi Road-XueFu 3rd Street	111	3	yes	yes	no	50
ZhengYi Road-BaoJian Road	183	4	yes	no	no	70
HongQi Street-HuangHe Road	167	4	yes	yes	no	60
XianFeng Road-HongQi Street	161	4	yes	no	no	60
ZhongShan Road-GeXin Street	142	3	yes	yes	no	60
DaXin Street-GongYu Street	119	3	no	yes	no	60
HuangHe Road-HuaShan Road	177	3	no	yes	yes	50

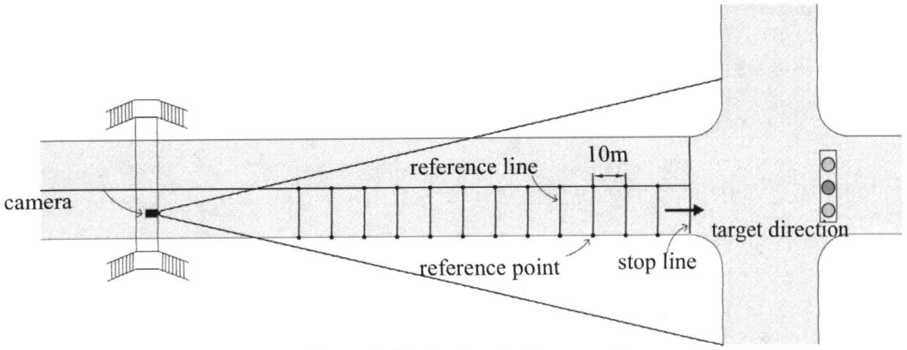

Figure 1. Illustration of video recording

2.3 Data extraction

Use Potplayer to play the video data, draw the reference lines in the video on the screen with the drawing software. Then play the video of the yellow light duration and red start and red end period frame by frame. Data extracted include: moment of turning to yellow light, distance between target car to stop line on the yellow onset, distance between rear wheel of target car to upstream recent reference line on yellow onset, vehicle type, time headway, moment of braking lights turning on and the number of red light running at red end, etc.

In this article, target vehicle means the first stop vehicle during the yellow light, vehicles running yellow light running red light at both red onset and offset. After data extraction, exclude some useless data including the incomplete data caused by the bad observation sight. Data that driver's perception reaction time is negative should be eliminated as well. Table 2 shows the sample size at each observation site.

Figure 2. Scenes for data collection

Table 2. Sample size at each observation site

intersection	First-stop-car at yellow light	Running yellow light	Running red light
ZYXF	90	351	37
ZYBJ	86	159	40
HQHH	216	378	68
XFHQ	168	371	36
ZSGX	265	324	21
DXGY	133	275	46
HHHS	278	474	77
Number of observation	1236	2332	325

Usually case, during the red light, the red light behavior occurs very rarely, in the process of video observations also found no the behavior. Therefore, the statistics of vehicle running the red light include the red onset and red offset. This study only focuses on this two conditions.

3 Type classification of motorized vehicles red light running

3.1 Critical distance

In this paper, the critical distance is used as the criterion for the classification of the behavior type of the vehicle running red light. We assume that there exists a critical distance (d) perceived by a normal driver at each intersection when noticing the beginning of a yellow phase.[21]. Drawing on the ideas of literature [21], the derivation of critical distance is as follows:

Logit model is applied to analyze the decision of driver's driving behavior, that is, choose to continue driving or stopping at the yellow onset. Driving behavior decision is a binary dependent variable y, $y=1$ means the choice of continue to drive, $y=0$ means the choice of stopping. Independent variables include two continuous variables and four discrete variables.

Driving behavior decision is a binary dependent variable y, $y=1$ means the choice of continue to drive, $y=0$ means the choice of stopping. Independent variables include two continuous variables and four discrete variables. Continuous variables refers to the distance of the vehicle to stop line on the yellow onset x_1 and the corresponding speed x_2, the discrete variables refers to the countdown timer x_3, the camera x_4, green flash device x_5, and peak time x_6. Among them, countdown positive $x_3=1$, countdown negative $x_3=0$, similar to x_4 and x_5, peak period $x_6=1$, non- peak period $x_6=0$

Then, the binary dependent Logit model is as follows:

$$Logit(P_{y=1}) = \ln(\frac{P_{y=1}}{1-P_{y=1}}) = \alpha + \beta_1x_1 + \beta_2x_2 + \beta_3x_3 + \beta_4x_4 + \beta_5x_5 + \beta_6x_6 \quad (1)$$

where

$P_{y=1}$	probability of continue to drive
$Logit(P_{y=1})$	logarithm of probability of continue to drive and stop
α	constant
$\beta_1, \beta_2, ..., \beta_6$	coefficient

Match the observation data using R and calibrate model parameters. The results show that only the two continuous variables significantly affect red light violation statistically significant at the 5% level. Furthermore, the speed at the yellow onset is a function of distance, it can be implied that the distance from the target car to the stop line at the yellow onset is major factor that affect diver's driving decision to pass or stop. Therefore, Eq.(1) becomes:

$$Logit(P_{y=1}) = \ln(\frac{P_{y=1}}{1-P_{y=1}}) = \alpha + \beta x \quad (2)$$

where

x distance from the target car to the stop line at the yellow onset

Change the form to the probability of driving choice:

$$P_{y=1} = \frac{1}{1+e^{\alpha+\beta x}} \quad (3)$$

Now, to derive the critical distance to the stop line, d, for a driver either to clear the intersection or to make a stop, we can consider $P_{y=1}$ as a statistical density function for a random variable, D. Then, the critical distance to the stop line is defined as the median of this random variable, D, that is, d is the value of D, for which

$$\int_0^d P_{y=1} dx = 0.5 \quad (4)$$

Finally, the parameters are estimated by a maximum likelihood technique with the likelihood function:

$$L(\alpha,\beta) = \sum_{d=1}^{n}\left[\ln(\frac{1}{1+e^{\alpha+\beta x_d}}) + \alpha + \alpha y_d + \beta x_d - \beta x_d y_d\right] \quad (5)$$

where $y_d = 1$ if a driver accepted a distance to pass and 0 if a driver rejected a distance to make a stop, n=number of observed decisions (pass or stop), and x_d =a vehicle's distance to the stop line when the yellow phase starts.

The maximization of $L(\alpha,\beta)$ reveals values for α and β in Eq.(3). Since this is the distribution function of a logistic distribution, Eq.(4) can be solved for d as the mean of this distribution, which is: $d = \frac{\alpha}{\beta}$

3.2 Type of red light running

Calculate each critical distance of intersection with the method of 2.1 based on the observation data. We can classify the red light running at red onset into two types:

① Deliberate red light running: passing the intersection if the current location to the stop line at the start of yellow phase is longer than the critical distance

② Unintentional red light running: passing the intersection if the current location to the stop line at the start of yellow phase is less than or equal to the critical distance

In addition, red light running at the red end has been researched in this study.

(1)Red light onset

According to the critical distance, the statistical results of red light running vehicle are as table 3. As shown in table 3, the critical distance ranges from 30m to 60m and the majority are at the range of 40m to 50m.As far as the number is concerned, deliberate red light running is higher than unintentional red light running. It is worth noting that the total violation at the ZSGX intersection were deliberate red light running, however, the deliberate red light violation is less than the unintentional violation at the ZYXF intersection.

Thus it can be seen that the majority of red light runners at red start are deliberate, while the unintentional red light violators also share some proportion. Short yellow duration, the dilemma zone and drivers' distraction may cause some unintentional red light running.

(2)Red light end

Table 4 shows the statistical results of red light running vehicles at red end. It indicates that this type of red light running occurred a little during the observation period. Among all of the observations, it scarcely happened at ZYXF, DXGY and HHHS intersection. It can be seen that red light running at the red end is less than the red light running at the red onset comparing the results in table3 and table 4. Some time is needed to start a car at the red light end from steady to activation. On the contrary, it is easier for a car at the red light onset to run a red light.

Table 3. Statistics of red light running vehicles at red onset

intersection	Critical distance/m	Red light running number	Deliberate red light running	Deliberate red light running percentage/%	Unintentional red light running	Unintentional red light running percentage/%
ZYXF	59.2	36	15	41.7	21	58.3
ZYBJ	48.6	22	12	54.5	10	45.5
HQHH	36.4	57	35	61.4	22	38.6
XFHQ	41.5	17	12	70.6	5	29.4
ZSGX	41.0	11	11	100	0	0
DXGY	50.5	45	26	57.8	19	42.2
HHHS	45.5	77	44	57.1	33	42.9
Mean	—	37.8	22.1	63.3	15.7	36.7

Table 4. Statistics of red light running vehicles at red end

intersection	ZYXF	ZYBJ	HQHH	XFHQ	ZSGX	DXGY	HHHS	Mean
Red light running number	1	18	11	19	10	1	0	8.6

4 Red light running characteristics analysis

4.1 Method

It is difficult to compare the characteristics of every type of red light running under different conditions using the red light running rate only. Hence, Z test is employed for the significance analysis of red light running rate. The method of Z test is as follows：

$$Z_{stat} = \frac{p_1 - p_2}{\sqrt{p(1-p)}\sqrt{\frac{1}{Q_1} + \frac{1}{Q_2}}} \quad (6)$$

Where, $p_1 = \frac{N_1}{Q_1}$, $p_2 = \frac{N_2}{Q_2}$, $p = \frac{N_1 + N_2}{Q_1 + Q_2}$.

Z_{stat} Z value

p_1, p_2 Red light running rate

N_1, N_2 Number of red light running

Q_1, Q_2 Traffic volume

The critical value of Z test at 95% confidence level is ± 1.96, which means that it is significant statistically when $Z_{stat} > 1.96$ or $Z_{stat} < -1.96$.

4.2 Countdown timer

Divide the intersections based on having countdown timer or not, calculate the red light running rate of three classifications, respectively. Take the deliberate red light running as example, calculate Z value under with and without countdown timer condition.

$$p_1 = \frac{69}{43858} = 1.573 \times 10^{-3} \quad p_2 = \frac{85}{132394} = 6.420 \times 10^{-4}$$

$$p = \frac{69 + 85}{43858 + 132394} = 8.737 \times 10^{-4} \qquad Z_{stat} = 5.72$$

Similarly, calculate the Z value of unintentional behavior and the one at the red end. Table 5 shows the results under different conditions.

Z test results indicate that countdown timer is significant variable for three red light running, which means that the countdown timer affect red light running behavior significantly. In terms of the red light running rate, the 2 types of red light running at red light onset are significantly less than the no countdown condition, while the red light end is opposite. This shows that the countdown timer can effectively reduce the red light running at the red onset, but to a certain extent, will increase the rate at the end of the red light.

Table 5. Results of z-test for red light running rate under with and without countdown timer conditions

Red light running type		Red onset deliberate rate/‰	Red onset unintentional rate/‰	Red end running rate/‰
condition	No countdown timer	1.57	1.19	0.02
	Countdown timer	0.64	0.44	0.45
Z value		5.72	5.43	-4.16
significance		√	√	√

√: significant, ×: not significant.

4.3 Red light camera

Divide the intersections based on having red light camera or not, calculate the red light running rate of three classifications, respectively. Z-test results are as showed in table 6.

Table 6 shows that under red light camera, the red light running rates at red light onset are higher than the ones without red light camera. But deliberate running a red light rate does not have statistical significance. It means that red light camera has no significant effect on deliberate red light running on the red onset and the unintentional red light running will be affected by the device. In addition, there is significant difference on the red light running rate under with and without red light camera. And the rate under no camera condition is higher than the one when the camera exists. The results show that red light camera can reduce the red light violation on the red end and increase the unintentional red light running. But it has no significant effect on the deliberate red light running.

Table 6. Results of z-test for red light running rate under with and without automatic red light camera conditions

Red light running type		Red onset deliberate rate/‰	Red onset unintentional rate/‰	Red end running rate/‰
condition	No red light camera	0.62	0.39	0.96
	red light camera	0.95	0.69	0.17
Z value		-1.91	-2.11	7.43
significance		×	√	√

4.4 Green flash device

Divide the intersections based on having green flash device or not, calculate the red light running rate of three classifications, respectively. Z-test results are as showed in Table 7.

Table 7 shows that there is a significant difference under with and without green flash device conditions, which indicates that green flash device affect the behavior of the red light significantly. Under green flash condition, red light running rate at red onset is higher than the ones without the device, and the influence at the end of the red light is contrary. The results indicate that green flash easily lead to the occurrence of running a red light when turning to red light, but can effectively reduce violation at the end of red period.

Table 7. Results of z-test for red light running rate under with and without green flash device conditions

Red light running type		Red onset deliberate rate/‰	Red onset unintentional rate/‰	Red end running rate/‰
condition	No green flash device	0.72	0.50	0.39
	green flash device	1.96	1.51	0
Z value		-5.83	-5.59	2.92
significance		√	√	√

DOI: 10.1051/matecconf/20168102021

4.5 Speed limit

There were three different limit speed at the observation sites, 50km/h, 60km/h and 70km/h. Table 8 shows the red light running rate under different speed limit. Z-test results are as showed in Table 9.

Table 8. Red light running rate under different speed limits

Red light running type		Red onset deliberate rate /‰	Red onset unintentional rate /‰	Red end running rate /‰
Limit speed/ (km·h⁻¹)	50	0.96	0.87	0.02
	60	0.90	0.50	0.44
	70	0.56	0.46	0.83

Table 9. Results of z-test for red light running rate under different speed limits

Compared condition	Z value(significance) deliberate red light running	unintentional red light running	red light running at red end
50 km/h-60 km/h	0.33 (×)	2.88 (√)	-4.97 (√)
60 km/h-70 km/h	1.60 (×)	0.19 (×)	-2.28 (√)
50 km/h-70 km/h	1.74 (×)	1.88 (×)	-6.84 (√)

From table 8 we can see that red light running rate at the red onset decreases with the increase of the speed limit. and it has an opposite effect at the red end. However, z test results shows that there are no significant difference of deliberately running red light under different limit speed. The unintentional red light running rate under 60km/h is less than the rate under 50km/h, and there are significant difference of red light running rate at red end under different limit speed. The above results indicate that the speed limit value has no effect on the behavior of running a red light at the red light onset, and the increase of the speed limit will significantly increase the occurrence of red light running at the end of the red signal.

4.6 Peak period

Calculate the red light running rate of three classifications under peak and non-peak period, respectively. Z-test results are as showed in table 10.

Table 10. Results of z-test for red light running rate under different periods

Red light running type		Red onset deliberate rate/‰	Red onset unintentional rate	Red end running rate/‰
condition	Non-peak period	0.94	0.65	0.38
	Peak period	0.82	0.60	0.30
Z value		0.86	0.36	1.02
significance		×	×	×

From table 10, we can see that the red light running rates under non-peak period are greater than the peak period But from the Z test results, this difference is not significant statistically. Therefore, peak hour can not affect the red light running behavior significantly.

5 Conclusions

Motorized vehicles red light running is one of the important causes for frequent traffic accidents at signalized intersection. This study classified the type of red light running according to red start and red end based on the video. Based on the critical distance, red light running was divided into three types combined with the red start and red end: deliberate and unintentional red light violation at red onset and red light violation at red end.

Video observation and statistics found that, majority of red light runners at red start are deliberate(63.3%), while the unintentional red light violators also share some proportion(36.7%). The video observation also showed that drivers would run a red light at red end but significantly less than the number of running a red light at red onset.

Through the analysis of the characteristics of the red light running under different conditions, it can be seen that the influence of different traffic control and management measures on the behavior of red light is different. Countdown timer reduces red light violation at red onset, but the green flash device produces the opposite effect. Red light camera increases unintentional red light violation at red onset but has no significant effect on deliberate behavior. Raising speed limit has no significant effect both on deliberate and unintentional red light violation at red onset. Running a red light at the red end will be decreased significantly under red light camera and green flash device conditions. But the countdown timer and higher speed limit will produce the opposite effect. The three types of red light running will not be influenced significantly by the peak hour.

It is necessary to explain that the results of this study are based on a small sample size because of the experimental conditions, time and other restrictions. Further research will expand the scope of investigation and extend the survey time at same signal intersection.

Acknowledgments

This paper is supported by Open Project of Key Laboratory of Ministry of Public Security for Road Traffic Safety, China (2015ZDSYSKFKT06-1).

References

1. QIU Chen-lu, XU Hui-ying, SHAO Zhi-hua, TANG Li-hu. China Pub Sec **34**,84-87(2014)
2. Mohamedshah Y M, Chen L W, Council F M. Report No. FHWA RD-00-112, FHWA, DOT, U.S., (2000).
3. Milazzo J S, Hummer J E, Prothe L M. Ins Transp Res Edu, NCSU, U.S., (2001).

4. Retting R A, Williams A, Greene M. Transportation Res Rec, **1640**,23-26 (1998).
5. Kamyab A, McDonald T, Stribiak J. Center Transp Res Edu, ISU, U.S., (2000).
6. Retting R A, Greene M., Transportation Res Rec 1-7 (1997).
7. Retting R A, Williams A F. J Safety Res, **27**,9-15 (1996)
8. WANG Xue-song, LI Jia, ZHONG Chu-jun. Urb Trans of China, **13**, 73-79 (2015).
9. Lum K M, Harun H. Journal of Transportation Res F, **9** 29-41 (2006)
10. QIAN Hong-bo, HAN Hao. China Safety Sci J, **20** 9-13 (2010)
11. CAO Yi, YANG Zhong-zhen, ZUO Zhong-yi. China Safety Sci J, **25** 77-82 (2015)
12. Long K, Liu Y, Han L D. Safety Sci, **54** 8-16 (2013)
13. Huang H L, Wang D, Zheng L, Li XiaoQi. Acc Ana Prev, **71** 248-260 (2014)
14. Retting R A, Ferguson S A, Hakkert A S. Traffic Inj Pre, **4** 17-23 (2003)
15. Porter B E, Johnson K L, Bland J F. Acc Ana Prev, **50** 1104-1111 (2013)
16. McCartt A T, Hu W. J Safety Res, **48** 57-62 (2014)
17. Chen P L, Pai C W, Jou R C, Wafaa Saleh, Ming-Shin Kuo, Acc Ana Prev, **75** 128-136 (2015)
18. Sze N N, Wong S C, Pei X, Choi, Y.K.Lo. Acc Ana Prev, **43** 265-271 (2011)
19. Yang C Y D, Najm W G. J Safety Res, **38** 311-321 (2007)
20. Elmitiny N, Yan X D, Radwan E, Acc Ana Prev, **42** 101-111 (2010)
21. Liu Y, Chang G L, Yu J. J Transp Eng, **138** 31-42 (2011)

Evaluating the Road Safety Design through High Resolution Satellite Image: A Case Study of Karachi Metropolitan

Salman Zubair and Lubna Ghazal

Department of Geography, University of Karachi, Karachi 75270, Pakistan

Abstract. Humanity is suffering from numerous natural, technological and health related hazards. Urban Road crash is one of the growing health issues these days in both developed and developing countries. Pakistan stands 1st in Asia and 48th in the world in this regard. Similarly, the metropolitan city of Pakistan, Karachi; ranks fourth in the list. Various reasons are responsible for these crashes in Karachi. Around 34% of crashes in the city were accounted due to errors in road geometry. In this study use of high resolution satellite imagery made it possible for identifying geometrical errors at the U-turns on major arteries of the city. It was also recognized that most of the U-turns were built on the fastest lane of the roads with average distance of 1.1 Km apart, are marked as vulnerable for considerable number of severe injury and fatal crashes. Moreover, inlet wall of all median U-turns were found broken, suggested that the car crash had occurred at least once. To cross check this observation, nearly 120 U-turns were surveyed and marked on the satellite imagery based on convenience. Trained professionals interviewed the people working/living nearby the U-turns. Out of 120 U-turns studied, 72.5% were without wall/median and 27.5% were with wall/median. Average number of people got injured or died due to crashes were statistically significant ($p<0.05$) between the above mentioned types of U-turns. In order to reduce geometrical errors use of RS (Remote Sensing) and GIS (Geographical Information System) techniques are strongly suggested to be incorporated while planning road design in the city. This would certainly save the resources particularly the lives of the people.

1 Introduction

The Global outbreak of road crashes mortalities and morbidities have emerged as a serious health issue. It costs more than 1.3 million lives and 20-50 million people are either being disabled or injured every year across the globe [1]. Multiple factors are responsible to vehicle collision but inappropriate road designs coupled with speedy vehicle are proven to be the key factors for serious injuries or deaths caused by road crashes [2]. Urban road environment is subjected to severe injury/fatal road crashes. This might be due to the complexity in road design and higher number of road users. Similarly, fast driving/riding in urban metropolitan could be more risky because it reduces the response time to any hazard on road, making it difficult for rider/driver for subsequent control over the vehicle.

However, controlled speed could be beneficial to road users in many ways. It does not only reduce the number as well as the severity of injury caused by the crashes but also increases the response time to road users [3]. Similarly, reduced vehicle speed could also check the fuel consumption, vehicle emission, noise and vehicle functioning cost. It is proved in various studies that controlled speed could reduce the number as well as severity of road crashes irrespective of road design in some instances [4]-[6]. Conversely, increased in travel speed is responsible to increase in the chances of crashes as well as deaths [4], [7]. Some of the studies revealed that speed limit of 14-15km/h is responsible to increase the chances of fatal road crashes to 15% and 26% respectively [8]. However, chances of fatality are 20% less for an adult if collided by a car moving with utmost speed of 50km/h. Therefore, the percentage of death may increase to 60% if hit at the speed of 80 km/h [9]. In UK, decrease in 1.6km/h average speed could reduce the chances of collision by 6% [10], [11]. Considering Urban environment, vehicle speed is subjected to road condition as well [12]. Smoother and wider roads compel drivers to move faster and vice versa [13], thus responsible to higher number of road crashes [14]. In Urban Environment, roads are embellished with multiple accessories. U –turn is an important feature that maneuvers the vehicle to 180 degree turn to direct the travel towards the opposite direction [15]. U-turns are considered as possibly highly dangerous if not situated correctly and could be responsible to greater number of vehicle crashes [16]. Globally, U-turns are considered as most unsafe road feature that is susceptible to severe injury road crashes [17]. Similarly, in Thailand U-truns are considered as one of the key contributors in road crashes and are responsible to increases the crash risk specifically on major highways of the country. In Karachi more than 30% of car crashes are accounted for error in road design and presence of U- turns are an important reason in this regard [18]. High speed of vehicles could also be responsible for higher chances of fatality to the pedestrian. If a pedestrian is being hit by a vehicle with an average speed of 60km/h, 50km/h or 30km/h than he/she would be susceptible to die with the proportion of 90, 50 and 10 percent respectively. Likewise, chances of severe injury also yield greater if a pedestrian is hit with the following proportion of increasing speed, 65km/h, 70km/h, 75km/h and 80km/h would double, quadruple, ten times and thirty two times more the probability of severe crash [1]. Vehicular speed law is an important parameter to evaluate the occurrence of road crashes. Unfortunately, Pakistan lies among those countries where there is neither any speed limit followed nor any high speed law exists in the urban areas 'Figure 1', making difficult for the road users who face higher probability of road crashes.

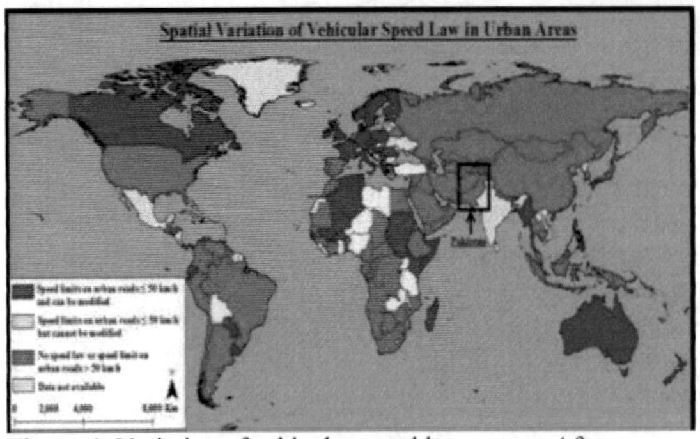

Figure 1. Variation of vehicular speed law source: After Toroyan, 2015.

2 Methodology

Following work flow "Figure 2" was adopted to carry up this research.

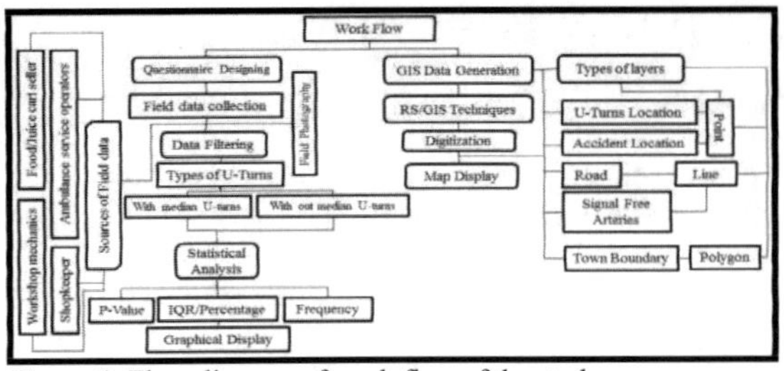

Figure 2. Flow diagram of work flow of the study.

2.1 Field data collection

Questionnaires were designed that hold information regarding various aspects of road crashes that include number of people injured or killed due to crash, time of crash, types of collision, days of accidents with high or low frequency. A cross-sectional interviewer-based survey was conducted to various parts of Karachi by trained investigators in year 2015. Nearly one-twenty U-turns were surveyed and minimum three people working in the close vicinity of each U-turn were interviewed. These include shop keepers, fruit cart seller, workshop mechanics, ambulance service providers who were more exposed to U-turns nearly 10 hours daily.

2.2 Use of RS/GIS techniques

Initially layers of road, towns and accident locations were digitized and Geo-coded data of car crashes of different years were overlaid on high resolution satellite imagery. Later on density mapping was performed to figure out the hotspots of crash areas. It was revealed that most of the U-turns proximity were showing high cluster of crashes. Similarly, with median U-turns had broken wall which is also visible on high resolution imagery, suggesting that the car crash is very common on such U-turns. That inflicted the idea of inquiring about the impact of U-turns on collision cases in Karachi. Figure 4 shows different segments of road in the city. Movement of vehicles is slowest on left lane whereas, it is fastest at the right lane. Presence of median wall reduces the space for vehicles that not only cause consistent traffic jam in rush hours but also result to serious crash cases. Insufficient road accessories made visibility of median wall extremely poor for the vehicle runner thus, ended up with collision.

3 Results and analysis

Data was analyzed to multiple statistical tests using *Statistica Six Sigma*. Firstly, data was put on two-sample Wilcoxon rank-sum test to assess the association between the average numbers of road accidents per week, average number of people got injured or died per week and the two different types of U-turns. Statistical significance was tested by two tailed test with P values of less than 0.05.

Secondly, Pearson's rank correlation was applied to check the ranking order of two different types of samples. It was revealed in all the surveyed U-turns that only one correlation was found positive i.e average number of crashes per week while, rest of the two conditions were found negatively correlated. Similarly, those U-turns situated on Signal Free Arteries positive correlation were found on average number of fatalities per week. Rest of the other two conditions was negatively correlated of the given condition.

Thirdly, Chi-Square test was applied on two types of sample data sets. In this regard it was found out that all the three conditions were found statistically insignificant. Same test was also applied to the dataset extracted for Signal Free Arteries. The test showed significant relationship between the numbers of crashes took place on SFA whereas, rest of the other two conditions exhibited non-significant relationship.

Finally, Geoffrey H. Havers (GEH) test was applied to check the association of crash data. Except one variable of Morning/Evening & Night crashes, all the other variables showed the resultant value of less than 5 that suggest the data collected for this research had a good match. However, the said variable showed the value of 6 may not be a good association of the data.

Out of those 120 U-turns survey locations, seventy-three percent were without wall/median and twenty-seven percent were with wall/median. Similarly, Thirty percent of all the surveyed U-turns lied on SFA and rest of the seventy percent were situated other than SFA of Karachi. Mean values computed for the responses based on numerical dataset were average while frequencies were chalked out of categorical dataset that included type of severity, the time and days and types of vehicle involved in those accidents. P-values were calculated for mean-weekly number of different types of road crashes i.e. only injury reported accidents and fatal accidents. The mean-weekly number of accidents found statistically insignificant with p-value is greater than 0.05. However, mean-weekly number of injury based crashes and fatal crashes were statistically significant having P-value less than 0.05. With-wall U-turns caused more injury based and fatal crashes than without-wall U-turns. Night time and evening hours were found out as most common times for road crashes of both types of U-turns. On without wall U-turns, occurrence of road crashes were highest on week-days i.e. (41.4%) from Monday to Friday. However, occurrences of road crashes were observed high at all the seven days of week i.e. (48.6%). Percentage involvement of loading pick-up, rickshaw and truck on other than SFA showed the value of 2% each. Rest of the vehicles involvement in road crashes was very meager in terms of percentage. Different types of vehicles were involved in road crashes near U-turns of both types. Involvement of cars/bike depicted the highest percentage i.e. (27% and 37%) on with wall and without wall U-turns. Similarly, involvements of bike on two different types of U-turns were 21% and 25% while, involvements of Bike/Car/rickshaw were 24% and 8%. Whereas, cars contribution in road crashes on both types of U-turns were computed 9% and 1.1% respectively. Percentage of bike/rickshaw on without wall U-turn were 12.6 and involvement of rest of the other types of vehicles in road crashes were less than 5% near both the types of U-turns.

Table 1. Profile of road safety status.

Profile of road safety status by type of U-turns.							**Profile of road safety status by type of U-turns on SFA.**					
Questions	U-turn with Wall (N=33) Median Frequency (IQR/Percent)	U-turn without Wall (N=87) Median Frequency (Percent)	*P-Value	Pearson Correlation (r)	Chi-Square Test	GEH Value	U-turn With Wall (N=25) Median Frequency (IQR/Percent)	U-turn Without Wall (N=11) Median Frequency (IQR/Percent)	*P-Value	Pearson Correlation (r)	Chi-Square Test	GEH Value
Avg. No. of crash/week	4.2 (3.6)	3.5 (4.2)	0.2702	0.117	0.109	2.529	6.3 (4.4)	6.3 (9.4)	0.008	-0.286	0.016	2.612
Avg. No. people injured/week	4.0 (3.5)	2.3 (3.5)	0.0043	-0.040	0.664	2.348	5.3 (4.5)	5(4.3)	0.01	-0.043	0.191	2.039
Avg. No. people dying/week	0.0 (1.0) @	0.0 (0.0) @	0.0047	-0.188	0.994	1.814	12 (87)	1.75 (24)	0.03	0.336	0.146	1.215
Record of All U-turns.							**Record of U-turns on SFA.**					
Most frequent time of crashes	U-turn with wall (Percent)	U-turn without wall (Percent)	GEH Value				U-turn with wall (Percent)	U-turn without wall (Percent)	GEH Value			
Morning	4 (12 %)	7 (8.1 %)	1.27				4(28.5%)	1(7.1%)	1.89			
Morning/Afternoon	1 (3 %)	0 (0 %)	1.41				1(7.1%)	1(7.1%)	0.00			
Morning/Evening & Night	3 (9.1 %)	27 (31 %)	6.19				1(7.1%)	5(35.7%)	2.30			
Afternoon	3 (9.1 %)	6 (6.9 %)	1.41				2(14.2%)	1(7.1%)	0.81			
Evening & Night	12 (36.4 %)	30 (34.5 %)	3.92				1(7.1%)	1(7.1%)	0.00			
Afternoon/Evening & Night	5 (15.2 %)	5 (5.7 %)	0.00				2(14.2%)	2(14.2%)	0.00			
Anytime (Morning/Afternoon/Evening & Night)	5 (15.2 %)	12 (13.8 %)	2.40				3(21.4%)	3(21.4%)	0.00			
Most frequent day of roadside crashes												
Working days of week	11 (33.3 %)	36 (41.4 %)	5.01				10(71%)	3(42.8%)	2.74			
Weekends	4 (12.1 %)	10 (11.5 %)	2.26				2(14%)	1(14.2%)	0.81			
Monday	1 (3 %)	1 (1.1 %)	0.00				0(0.0%)	0(0.0%)	-			
Wednesday	0 (0 %)	1 (1.1 %)	1.41				0(0.0%)	0(0.0%)	-			
Thursdays	1 (3 %)	0 (0 %)	1.41				0(0.0%)	0(0.0%)	-			
Friday	0 (0 %)	2 (2.3 %)	2.00				1(7.1%)	1(14.2%)	0.00			
Every day of week	16 (48.6 %)	35 (40.3 %)	3.76				1(7.1%)	2(28.5%)	0.81			
No answer	0 (0 %)	2 (2.3 %)	2.00				0 (0.0%)	0 (0.0%)	0.85			
Record of All U-turns.							**Record of U-turns on SFA.**					
Most Frequent vehicle involved in road crash	U-turn with wall (Percent)	U-turn without wall (Percent)					U-turn with wall (Percent)	U-turn without wall (Percent)				
Bike	7(21.2%)	22(25.2%)					5(20.8%)	1(11.1)				
Bike/Car	9(27.2%)	33(37.9%)					8(33.3%)	5(55.5%)				
Bike/Car/Rickshaw	8(24.2%)	7(8.0%)					6(25%)	1(11.1%)				
Car	3(9.0%)	1(1.1%)					3(12.5%)	0(0.0%)				
Rickshaw	1(3.0%)	2(2.2)					1(4.1%)	0(0.0%)				
Loading pickup/bus	2(6.0%)	2(2.2)					0(0.0%)	1(11.1%)				
Taxi	0(0.0%)	1(1.1%)					0(0.0%)	1(11.1%)				
Any	1(3.0%)	1(1.1%)					1(4.1%)	0(0.0%)				
Bike/Rickshaw	0(0.0%)	11(12.6%)					0(0.0%)	0(0.0%)				
Bike/others	0(0.0%)	5(5.7%)					0(0.0%)	0(0.0%)				
Truck	2(6.0%)	2(2.2%)					0(0.0%)	0(0.0%)				

* P-values are based on two-sample Wilcoxon rank-sum (Mann-Whitney) test, Pearson Correlation and Chi-Square Test

@ Wide variation was observed with mean and standard deviation of 0.47 (SD: 0.78) and 0.12 (SD: 0.36) for U-turns with and without median, respectively.

3.1 Analysis of U-turns on signal free arteries (SFA)

From the same data set multiple queries were applied to filter out the crash cases incurred on SFA only. On U-turns situated on SFA, number of Road crashes at with-wall U-turns and without-wall U-turns were 25 and 11 respectively proved that with-wall U-turns are more susceptible to road crashes than without wall U-turns. Averaged numbers of accidents on both types of U-turns were same i.e 6.3. Similarly, average numbers of injury cases were nearly same i.e. 5.3 and 5.0. However, number of fatal crash showed substantial difference i.e 12 reported by the respondent near with wall U-turns, suggested that faster road are more vulnerable to road users. While, number of fatal accidents were only reported 1.75 on without wall U-turns on same arteries, proved that this type of U-turn has lesser susceptibility of collision occurrence because of their median free design. One possibility of higher number of road crashes on with wall U-turns is that median wall is situated on fastest traffic lane that reduces the reflex time of the drivers, thus increases the probability of road crashes. Responses of above three questions specifically on Signal Free Arteries were statistically significant with P-value less than 0.05.

At the U-turns (with wall) morning hours were considered the most susceptible to road crashes (i.e. 28%). However, on without median U-turns morning/evening and night time cases showed the highest proportion of 35%. Afternoon and evening hours depicted 14% of crashes on both the types of U-turns on SFA. Unidentified times of the day were reported as 21%. While, rest of the times of the day reported 7% crashes on both the types of U-turns. Different types of vehicle were involved in both types of U-turns (i.e with wall and without wall U-turns). Involvement of bike alone cause 21% and 11% crashes along with cars and rickshaw is 58% and 66% crashes on two different types of U-turns respectively. Cars alone constituent 12.5% of road crashes on with wall U-turns whereas, no case of car crash was reported on without wall U-tunrs. Loading pickup and taxi combine to cause 11% road crashes on without wall U-turns. Figure 4 shows some of the with-median U-turns situated on different arteries of Karachi. Different colored arrows show different segments of road and arrow direction show the direction of traffic flow. Yellow arrows depict that the left lane which is the slowest traffic lane in the city. Green arrow show right most lanes in the city. It is noteworthy that most of the with-median U-turns are situated on right lane which is also the fastest traffic lane in the city that reduced the reflex time of the driver and increases the certainty of crashes. In all three figures red arrows highlight the broken walls of the with-median U-turns. It is clear from the field pictures that considerable part of the median wall has been broken which proves that the car crashes is frequent on such spots. Furthermore, consistent traffic jam is also observed on such location because median wall reduce the space of forward moving vehicles. This result to higher number of Property Damage Only (PDO) crashes.

3.2 GIS work

Geo-coding of 120 surveyed U-turns were done by taking their coordinates while doing the field work. Town map of Karachi was made with the help of scanned maps through different government sources. However, road network and Signal Free Arteries (SFA) were also made by digitizing on high resolution satellite imagery. It is noteworthy that SFAs initiated from peripheral parts of the city, passing through medium to high population density towns (Figure 3-b) making higher chances of road accidents within the city. Inverse Distance Weighted (IDW) technique was applied on U-turns layer in Arc GIS 10.1 environment to figure out the concentration of U-turns in Karachi (Figure 3-c). Results imply that most of the U-turns are situated on roads where fast flow of traffic is observed; therefore, increasing the risk of road crashes coupled with the injury and fatality cases. On contrary central parts of the city have fewer U-turns that reduce the chances of severe injury and fatal road crashes within the city.

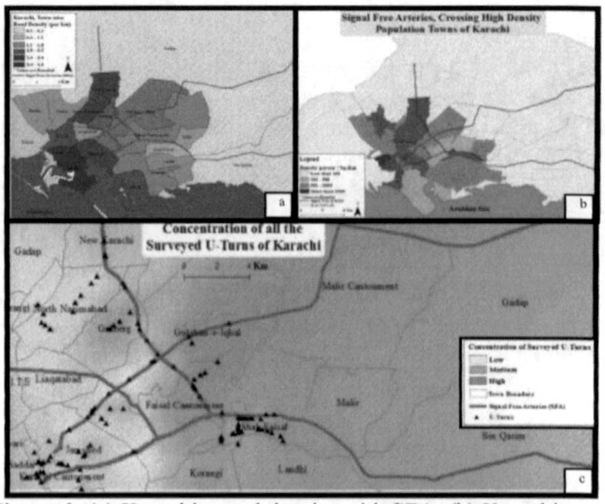

Figure 3. (a) Karachi - road density with SFA, (b) Karachi-population density, (c) concentration of U-turns in Karachi

4 Discussion

Concentrating on driver skill alone for the complex issue of acquiring road safety would smog the distinction between infrastructure design and its influence on drivers. Globally, more than 1.2 million people are killed every year. This problem is the ninth leading cause of death for all-aged group people and is the leading cause of death among the group of 15-29 years age people, spoiling up to 5% GDP share in low and middle income group countries of the world. Pakistan stands first in Asia [24] in terms of leading RTA death conceiving countries of the world with more than 8,000 casualties reported in 2013 by Police. Likewise, Karachi ranks fourth [25] in this regard with more than 1,000 road accident causalities in the year 2012. It is estimated that the actual cases are much higher than the reported figures. It is because, every road accident does not get registered by the traffic police, as parties involved in such accidents resolve the conflicts on their own [26]. Results from the study suggest that with wall U-turns are more likely to encounter higher number and severe intensity of road crashes thus, causing to higher number of fatalities. Further, same U-turns are more suitable for 180 degree maneuvering of vehicle on other side of the road as compared to without wall U-turns. In Karachi most of the with wall U-turns are located on Signal Free Arteries aimed for faster traffic movement and are responsible to higher number of mortality cases. However, without wall U-turns experience lesser frequency of fatal crashes.

DOI: 10.1051/matecconf/20168102022

Source: Authors

Figure 4. High resolution satellite images showing placement of U-turns.

It was postulated that faster vehicular traffic are more susceptible to severe injury and fatal road crashes due to the presence of two different types of U-turns on selected arteries of the city. This shows a significant positive relationship with road crashes hence this particular feature (i.e. U-turn) on road compromise the road safety within the city.

5 Conclusion

It was concluded that this is the first study conducted in this region that computed the impact of U-turn on road crashes. Various statistical tests were executed then it is finally suggested that presence of U-turns especially (with-wall) is a pitfall on major arteries of the city. However, safer road design should be the top most priority for the road designers. Planners can incorporate the use of latest technologies and recent research work before taking final decision regarding changes in the infrastructure. Strong and timely coherence among different departments such as health, planning and development, traffic and communication department and academician is needed to promote the culture of productive research and concrete planning. Mass transit systems should be launched as at the earliest due to growing demand of population in Karachi. This would reduce the consumption of valuable resources as well as the adverse impact on the-environment and will certainly help to get better solutions of the real world problems like Road crashes.

References

1. T. Toroyan, "Global status report on road safety 2015," *Supporting a decade of action. Geneva: World Health Organization, Department of Violence and Injury Prevention and Disability,* (2015)
2. P. Thomas, A. Morris, R. Talbor, H. Fargerlind, "Identifying the causes of road crashes in Europe," *Annals of advances in automotive medicine,* vol. **57**, p. 13, (2013)
3. A. Habibovic, J. Davidsson, "Causation mechanisms in car-to-vulnerable road user crashes: Implications for active safety systems," Accid Anal Prev, vol. **49**, pp. 493-500, (2012)
4. R. Elvik, T. Vaa, A. Erke, M. Sorensen, *The handbook of road safety measures*: Emerald Group Publishing, (2009)
5. A.K. Lund, J.K. Wells, H.M. Baum, "Motor Vehicle Crash Fatalities and 65 Mph Speed Limits on Rural Interstates in 40 States," (1989)
6. L. Mourad, L.-b. Patrick, V. Francoise, "Impact of Speed on Road Safety: A Times Series Analysis Approach," in *Proceedings of the Road Safety on Four Continents Conference*, (2005)
7. P.P. Maglio, S. Srinivasan, J.T. Kreulen, J. Spohrer, "Service systems, service scientists, SSME, and innovation," *Communications of the ACM,* vol. **49**, pp. 81-85, (2006)
8. R. Elvik, T. Vaa, A. Hoye, M. Sorensen, "Factors contributing to road accidents," *The Handbook of Road Safety Measures. Oxford,* pp. 29-79, (2004)
9. E. Rosen, H. Stigson, U. Sander, "Literature review of pedestrian fatality risk as a function of car impact speed," Accid Anal Prev, vol. **43**, pp. 25-33, (2011)
10. M. C. Taylor, D. Lynam, A. Baruya, *The effects of drivers' speed on the frequency of road accidents*: Transport Research Laboratory Crowthorne, (2000)
11. Brake, "Go 20 towards changing the default speed limit to 20mph," (2015)
12. J. Edquist, C. Rudin-Brown, M.G. Lenne, "Road design factors and their interactions with speed and speed limits," *Monash University Accident Research Centre,* p. 30, (2009)
13. M. Martens, S. Compte, N. A. Kaptein, "The effects of road design on speed behaviour: a literature review," (1997)
14. B. Fildes, S. Lee, "The Speed Review: Road Environment, Behaviour, Speed Limits, Enforcement and Crashes. Monash University Accident Research Centre, Prepared for Road Safety Bureau," *New South Wales, and Federal Office of Road Safety, Canberra,* (1993)
15. BMV. (2016), *Chapter Five Safe Vehicle Operation*. Available: http://www.in.gov/bmv/ files/Drivers_Manual_Chapter_5.pdf
16. I. Meel, U. Brannolte, A. Vesper, "Road Traffic Safety Evaluation At Thai U-turns Using Severity Conflict Indexes."
17. Unnamed. (2016, 05/01/2016). *The U-turn as a Threat to Road Safety*. Available: https://arrivealive.co.za/The-U-turn-as-a-Threat-to-Road-Safety
18. S. Zubair, S. J. H. Kazmi, "Spatial Framework for the Assessment of Road Traffic Accidents in Karachi," *JBAS,* vol. **9**, p. 525, (2013)
19. A. Hasan, M. Mohib, "The case of Karachi, Pakistan," (2003)
20. S. Zubair, M. A. Shaikh, "U-turns and road safety-perspective from Karachi," *JPMA,* vol. **65**, pp. 84-86, (2015)
21. R. Jooma, "Using data to save lives: Road Injury Surveillance in Karachi," RTI&PC, Road Injury Research & Prevention Centre, Karachi, Karachi (2013)
22. T. Mahmood, A. Wasim, A. Amir, "Title," unpublished
23. W. A. Khan. *Urban Resource Centre*. (2008)
24. S. Reporter. (2012, 23/09/2013). *jang.com*. Available: http://e.jang.com.pk/11-18-2012/karachi /page42.asp (2012)
25. Waheeduddin. (2010, 2/6/2014). *The Express Tribune*. Available: http:// tribune.com. pk/ story/ 26389/karachi-is-number-4-in-world-stats-of-highest-road-fatalities/ (2010)
26. M. Shaikh, A. Shaikh, Z. Siddiqui, "Road rage and road traffic accidents among commercial vehicle drivers in Lahore, Pakistan," (2012)

Application of IC Card License for Road Transportation in Commercial Vehicles Supervision and Service

Weiwei Li and Jingyan Gu

Research centre of Transport and Logistics, Research Institute of Highway Ministry of Transport, No.8 Xitucheng Road, Haidian District, Beijing 100088, People's Republic of China

Abstract. IC card electronic license for road transport includes the IC card commercial vehicle's certificate and IC card practitioner's qualification certificate. In China, the IC card electronic license for road transport is the electronic ID card which must be carried by each commercial vehicles and practitioners. This paper briefly introduces the basic situation, data format and security keys architecture of IC card electronic license for road transportation of China. In order to strengthen the supervision and service of commercial vehicles, this paper puts forward the overall application framework of IC card electronic license for road transport. The application examples of IC card license in the supervision of passenger station, dangerous goods transport management, governance overload and logistics park and port area management are discussed. The practical application results show that the application of IC card electronic license for road transport is an important technical means to improve the supervision ability and service quality of the road transportation industry.

1 Introduction

Commercial vehicles are vehicles engaged in road transport business operations. Road passenger transport management of taxi, tour bus, class line of passenger cars, and road freight transport business trucks, etc. all belong to the commercial vehicles. In road traffic, due to the wide operating range of commercial vehicles, day long operation time, a large number of passenger traffic and a huge amount of logistics operations, commercial vehicles are prone to traffic accidents. Especially the operation of passenger cars, dangerous chemicals transports vehicles and overloaded trucks often cause serious consequences once the accident happened.

In China, the existing commercial vehicles are more than 14 million vehicles, and practitioners are more than 20 million people. With the rapid growth of commercial vehicles and practitioners, the lack of timely and accurate grasp of the operation status of the vehicle, it has been difficult to adapt to the needs of modern road transport management. In addition, the development of China's road transport is still relatively extensive, large-scale, low degree of organization, market concentration is not high. In all the more than seven million and five hundred thousand business households for road freight of the whole country, with five cars below accounted for about 98%, with more than ten cars accounted for less than 1%. Passenger car has five households accounted for about 85% of the following, with more than ten cars accounted for about 13.5% [1]. The characteristics of the large scale of China's road transport industry and the dispersion of business operations resulted in a high degree of market operating order specifications, but also which is for the higher management and services requirements to the commercial vehicles.

The information and intellectualization has become the main developing direction of modern traffic safety supervision system and smart card system, global position system (GPS), geographic information system (GIS), computer, wireless communication and Internet technology have been widely integrated application, which is to achieve integrated management of vehicle operation and quality service provides new techniques. Among them smart card system is utilized in many ways in many countries [2]. At present, the Ministry of transport of China is to gradually implement the national IC card electronic license for road transport. IC card electronic license for road transport includes the IC card commercial vehicle's certificate and IC card practitioner's qualification certificate. IC card as the operator carry electronic archives, by virtue of its security to read and write, easy sharing, facilitate the automatic data entry and other advantages, to enhance the level of industry supervision and service provides powerful technical means.

To ensure for realizing IC card for road transport of electronic documents in the provinces and cities a common card in the whole country, the Ministry of transport of China issued relevant standard [3]. The IC card electronic license for road transport must be the key security system, unified appearance style and anti-counterfeiting scheme and unified data format unification, unified key equipment and technology. At the same time, the Ministry of transport of China unified leadership and co-ordination arrangements for the construction and application of IC card electronic license for road transport.

This organization of this paper is as follows: In section 2, we briefly introduce the technical characteristics of IC card electronic license for road transport in China. After that, in section 3 the general application framework of IC card electronic license for road transport is proposed. In section 4 we discussed the typical application scenario and the application effect is analysed and summarized. Finally, section 5 gives the conclusion with the prospect of further work.

2 Technical characteristics of IC card electronic license for road transport of China

2.1 General technical introduction of IC card electronic license for road transport

IC card electronic license for road transport in China select non-contact type CPU card. The non-contact CPU card features include: First, the chip and the COS security technology for the CPU card provides the dual security guarantee, with its own operating system CPU card on the computer network system requirements are low, can be achieved offline. IC card electronic license for road transport can be read and write under the network environment, which can be guaranteed under the offline state of a variety of applications. The second is to achieve a real sense of one card and more applications. Each application is independent of each other, and is controlled by the respective key management system. In the IC card electronic license for road transportation set aside the area for other management departments need to read and write. Other management departments can use the original hand-held reading and writing system for IC card electronic license for road transport to read and write, so as to achieve the IC card electronic license for road transport cross sectorial applications. The third is the card storage capacity is large, can provide 1K~64K bytes of data storage. The fourth is the card is with an independent security module. Encryption, decryption, and transaction processing are implemented by using the corresponding entity SAM card key, thereby completing the security authentication between the user and the user card.

2.2 Data format of IC card electronic license for road transport

The data format of the IC card electronic license for road transport is in accordance with the relevant requirements of the transportation industry standards in China the IC card for road transportation certificate applied unified national style and the unified data format. The information contents of the IC card electronic license for road transport mainly include the basic information file, the basic directory file, the national main application area and the local main application area [3]. Data storage space is 8954 bytes. The basic information file mainly includes the provincial administrative division code identification, vehicle territorial administrative division code, card version number, card initialization date, the card initialization authority and card number and other information. The basic directory file includes the contents of the various types of applications included in the card. The national main application areas include the province of annual audit cases accumulated data file, outside the province inspection application information file information such as the information. The local main application areas include two types of data: one is static data, the other is scalable data. Static data is with the vehicle and personnel basic information about the data, such as vehicle license, road transportation certificate number, license plate number, business qualification, scope, secondary maintenance, technical grade status and owner's name, identity card number and other basic situation and vehicles and the personnel of the picture file. The scalable data is generated in the offline management information, such as fees to pay, annual audit records, records of violation, administrative penalties and other information data.

2.3 Security keys architecture of IC card electronic license for road transport

In order to strengthen the security and confidentiality of IC card electronic license for road transportation and key management, IC card electronic license for road transportation key system adopt ministerial and provincial two level management models. Ministerial key includes the ministerial road transport of electronic license business root key and the ministerial road transport of electronic license business master key. Ministerial road transportation electronic certificate business root key can be controlled to produce the ministerial road transport electronic license business master key, card control key, etc. The master key of the ministerial road transport of electronic business license includes road transport of electronic license the read and writes the key, such as IC card for road transportation electronic certificate authentication key, IC card for road transportation electronic license outside the province inspection business application key. Provincial root key is the key of provincial local application. Provincial business key mainly includes local application related key [3]. At present, China's Ministry of transport has established the IC card electronic certificate for road transportation of ministerial key system, to ensure that the IC card electronic license for road transport can achieve to be read and written across the whole country.

3 General application framework of IC card electronic license for road transport

In this paper, we offer the overall framework of IC card electronic license for road transport management and application according to the above IC card technology characteristics. IC card electronic license for road transport management and service system is mainly composed of ministerial level and provincial level two system. The framework is illuminated in Figure 1.

The ministerial level is responsible for the unified provincial level management of the whole country. The ministerial level key management system is mainly to achieve the management of road transport industry the root key, the generation and management of the national unified reading and writing key, and the secure distribution of provincial key. Relying on the ministerial and provincial road transportation information system network and through the ministerial level road transportation information service system of IC card management function expansion, the Ministry of transport can achieve relevant information about the IC card electronic certificate for road transportation issuing, tracking and summary.

The provincial level is responsible for the issuance and application of IC card electronic license for road transport. Relying on available provincial road transport management system expansion development of IC card issuing and management subsystem, the provincial level key management system can realize the establishment of the IC card system initialization, key management system and IC card application system [3]. IC card initialization system mainly uses the equipment to realize the IC card initialization and initialization key filling, IC card quality inspection and sorting, file structure and the basic information data write functions.

In addition, the ministerial level can rely on the ministry-province networking data exchange platform to build the national IC card electronic license for road transportation black list data exchange and query functions, which can achieve cross regional and trans-departmental linkage inspection, collaborative supervision to provide technical supporting platform.

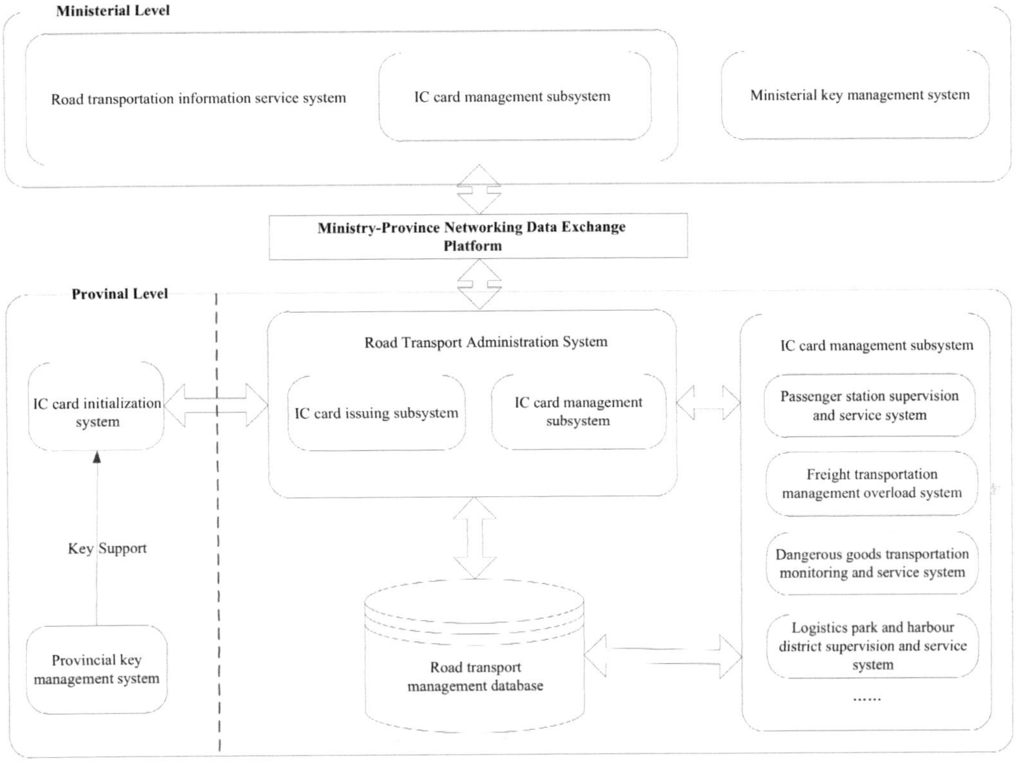

Figure 1. Application framework of IC card electronic license for road transportation.

4 Typical application scenario analysis of IC card electronic license for road transport

IC card electronic license for road transportation has been in the passenger vehicle station management, dangerous goods transport management, governance overload, Logistics Park and port area management and other aspects of a wide range of applications. The typical application scenario analysis is as follows.

4.1 Passenger station supervision and service system

The IC card electronic license for road transportation as a carrier, by recording bus stop check, security check, inspection certificates, sign up for a class, check the business aspects of management information station, the realization of the entrance and safety supervision station and sign up for a class of whole business process for passenger vehicles. At the same time, IC card as an authoritative carrier of security cases and the entrance and exit the station, but also provides strong evidence for road traffic inspection and safety inspection [4].

4.1.1 Passenger vehicles enter the station verification

When passenger vehicles enter the antenna reading area, the IC card can be identified by the antenna, and through the background of the vehicle library data to determine the conditions of the vehicle. According to the electronic data inspection results are divided into two categories, in line with the conditions of vehicles to enter the station without stopping. The vehicles which do not meet the conditions will be turned to manual control.

4.1.2 Car wash, security and class report management

After allowing the vehicle to enter the station, in the car wash disinfection, routine security checks, classes and other aspects of the use of IC cards to record business information in real time. In addition, the system provides the third party application access interface such as the fingerprint, face recognition, alcohol detection system to effectively monitor the vehicles travel safety. When vehicle class the scheduling system and documents will verify the system data to determine the execution system automatically to sign up for a class of results and push the result to the station LED display. The driving personnel can pass through the LED display clear the contents of the subsequent execution of the task. The document verification system is the automatic information collection and verification of the five kinds of documents, which includes the driver's license, qualification certificate, driving license, and vehicle certificate and line card.

4.1.3 Passenger vehicle departure verification

When the passenger vehicle is ready to leave the station, once the vehicle enter the antenna area and the antenna read the IC card, all the information includes the car flight number and number of passengers and other information through the antenna will be written into the IC card. All the implementation results will be displayed in the LED display of the station. At the same time, the system allows the vehicle to leave the station without a stop. Exit of station will write the vehicle trips, and number of passengers and other information into the IC card through the antenna, which will facilitate management by mobile terminals.

4.2 Freight transportation overload management system

The focus of freight transportation overload management is to focus on the source and road control. The IC card electronic license for road transport can be used as the sole identity card of the freight transportation vehicle. The road transportation certificate, qualification certificate, driving all the cards, card of drive a vehicle, weighing information such as information entry in the IC card, can realize multi card into single unity as the basis for supervision [5].

4.2.1 Supervision of the source of cargo transport

IC card electronic license for road transportation is the permit document for source of freight transport. Cargo transport vehicles must hold IC cards to enter the source enterprise loading. In the source of freight source goods enterprise will be responsible for writing the loading of information in road transportation IC card. All freight source units on government publicity will be all included in the road transportation IC card source overload management system. In the scattered goods transport is relatively concentrated area the goods concentrated weigh point will be set up to implement IC card management to achieve full coverage of the source.

4.2.2 Roadway checks at toll station or overload inspection station

The control system of overload and highway toll system for docking can be in the toll station to achieve the regulation of vehicles. When the overload vehicles passed the weighing system, the software can display the information overload and overrun. At the same time, the display LED set on the highway toll plaza will show the ban line information to the driver. When the overloaded vehicles entered into the toll lanes, the toll collection system will be on the toll again prompted to limit information. If the toll collectors insist on forcing the pass card, then the system will automatically upload the vehicle information to the monitoring system.

4.3 Dangerous goods transportation monitoring and service system

Combined IC card electronic license for road transportation with the record risk goods transportation electronic waybills, transport of dangerous goods vehicles running state, loading dangerous goods of the type and number, driving route, and the driver, supercargo and information management for process monitoring, the transport of dangerous goods from the vehicle scheduling, loading, transportation and unloading will realize the closed-loop management of whole process, which will reduce the safety hidden danger to the greatest extent [6].

4.3.1 Controllable source and traceable responsibility

Before the transportation, the carrier online fill in the waybill information and automatically download to the road transportation IC card through reading and writing device or vehicle terminal. Prior to the shipment of the goods the shipper will verify the transport vehicles and personnel qualification of the transport of dangerous goods qualification enterprises, vehicles and personnel engaged in the transport of dangerous goods by reading the IC card electronic license, which will prevent the illegal operation behaviour on the bills to fill in the content confirm. In addition, through the system you can check whether the source of the company commissioned a carrier of the carrier's transport of goods to provide strong evidence of responsibility after the accident occurred.

4.3.2 Real time monitoring and reliable emergency response

Combined with the electronic waybill information and the real-time vehicle route and track, management personnel can know the real-time dangerous goods of each type, weight and objective, which will provides effective means in real time for industry statistics analysis and emergency disposal. When unexpected events occur, industry management department query electronic waybills, obtain goods category, the number of information, and combining with GIS intuitive understanding of dangerous goods transport vehicles location of the surrounding topography, lakes, villages and other information, for emergency and accident rescue plan is formulated to provide effective technical support.

4.4 Logistics Park and harbour district supervision and service system

4.4.1 Freight market credit information service

IC card electronic license for road transportation is an important support in the transport market for the commercial vehicles and practitioners of the identity of the qualification certification. In logistics parks and port freight yards, by authorization authentication terminal, national vehicle and personnel database and IC card electronic license for road transportation by CA authentication technology, provide transport practitioners member, vehicle certification service for the public through IC card authentication and network authentication mode. The system will be convenient for users to check the authenticity and legitimacy of the relevant documents, and select the normal operation of the vehicle and professional personnel commissioned by the transport business. This service will effectively regulate the transport market, maintaining the normal order of the transport market.

4.4.2 Park and station management and service

On national highway transport freight hub station field, logistics park and key port deployed the card reading device, and in the IC card electronic license for road transportation to increase field management and service function, can realize field station one card multi-service system, so as to facilitate the vehicle certification and in and out of the station management. Logistics Park and port will be as a national freight service system of merchants, for the cardholder to provide parking, accommodation, catering, repair, and picking and preferential, consumption points and other value-added services, cardholders with in each station field of the harbour and cumulative frequency of consumption can enjoy more preferential incentives.

5 Conclusion and prospect

Overall the application of China's IC card electronic certificate for road transportation is still in its early stage, which is mostly concentrated in the industry management. This study summarizes innovative thinking and means in IC card electronic certificate for road transportation in the supervision of passenger station, dangerous goods transport management, governance overload and logistics park and port area management. IC card electronic certificate for road transportation is an important carrier of vehicle and personnel credit information. The credit information management is the key for vehicle and personnel management and service. The future research is more concentrated in the field of industry service is actively exploring using by IC card. For example, by IC card electronic license for road transportation as the carrier, the establishment of national freight service alliance, and oil retail companies, vehicle maintenance and rescue companies, telecommunications companies, insurance companies and other value-added service providers identified for the cardholder to provide the relevant preferential policies and enjoy the group purchase benefit. In addition, it can also be linked to the transport business households and practitioners credit record and enjoy the highways, transport station, value-added service providers to provide preferential service standards and credit records of good cardholders will enjoy greater discount preferential service. Otherwise it has no right to enjoy, and even be included in the blacklist to accept severe punishment. This will effectively promote the transport industry practitioners to regulate business, honest and trustworthy, truly scientific management and services to achieve the perfect unity.

References

1. X. Liu. Transport Business China **5**, *Adhere to Market Leading to Promote the Intensive Development of Road Transport Industry*. 18-20 (2014)
2. W. Kou, S. Poon, E. M. Knorr. Payment Technologies for E-Commerce, Smart Cards and Applications. 95-126, (2003)
3. Ministry of Transport of the People' s Republic of China. China Communications Press, *JT/T 825-2012 IC card license for road transportation*. (2012)
4. X. Zhang, L. Liu, L. Zhang, China ITS Journal **9**, *Research and application of IC card technology in transportation*. 131-134, (2012)
5. L. Wang, X. Yang, N. Shen, L. Li. Highway & Automotive Applications **2**, *Research on the application of IC card in Shanxi Province*. 79-81, (2013)
6. X. Wei. Modern Industrial Economy and Informationization **46**, *The Role of IC Road Transportation Certificate in Management of Highway Transportation Industry*. 90-100 (2013)

Analysis of Urban Rail Transit Seamless Transfer Standard

Liu Li Fen[1,2] and Wang Wen[1]

[1]*School of Highway, Chang'an University, Xi'an, China*
[2]*Xi'an municipal design and Research Institute Co., Ltd, Xi'an, China*

Abstract. With the improvement of urban rail transit network and traveler requirement of travel , the concept of rail seamless integration has been proposed. This paper analyzes the meaning of integration and seamless integration and the connotation and characteristics of urban rail transport were seamlessly transfer analysis. On this basis, the organization of urban rail transit traffic in different ways are discussed.According to Brighton G (Bou ladon G) hypothesis, tranfer distance is devided into ideal, desirable, acceptable, tolerable, intolerable five levels and their corresponding standard values are quantized, getting the standard of transfer time and distance to evaluate the quality of rail integration and Seamless Transfer of proposeing the standards of integration time and distance.

1 Introduction

With the development of the city, rail transportation network development is an inevitable direction to play the maximum benefit rail transportation. Transfer trip promotes integrated development of urban public transportation and plays an important role in the rail transport and urban public transport operations. Transfer the merits of transfer system directly affects the rail can play a role, but also directly reflects the rail transportation efficiency and service levels.In the urban rail transit development and travelers for travelers quality improvement background,,urban rail transit seamless transfer concept has been applied in the design of rail transportation, scholars have studied urban rail hubs, city passenger hubs and cohesion layout theoretical model and evaluation of transfer efficiency [1]-[7]. Now the citys of construction rail transit are considering combining with neighboring transit system to facilitate the transfer, to achieve the integration of public transport and seamless transfee, saving transfer time.

2 The concept, connotation and characteristic of Rail seamless transfer

2.1 The concept of Rail seamless transfer

Urban rail transit transfer system is large and complex, but also provides great convenience to the traveling public. With urban development and travel demand increasing, travel transfers is inevitable. From the perspective of travelers from the starting point to the end to complete a travel, the transfer is the same mode of transport between different vehicles or different modes of transport conversion process in the process of travel.

The transfer of Rail transit network between different line train is the process from one train to another train.Under normal circumstances, for the train and change into the train runs along a fixed line and stop at different stations (except platform transfer), so you need to transfer the vast majority of passengers on foot to achieve. Swapped out trains and swapped trains runs along a fixed line and stop at different stations (except platform transfer), The vast majority of transfers to passengers are achieved by foot.

The transfer between rail transit and external traffic mainly refers to the convergence transfer of rail transit with airports, the city's railway station. The transfer between rail transit and other modes of transportation Mainly involves between conventional bus, taxi, private car, bicycle, walking and other modes of transport interchange.

From the perspective of travelers from the starting point to the end to complete a travel, the transfer is the same mode of transport between different vehicles or different modes of transport conversion process in the process of travel.

So "seamless transfer" refers to the use of the facilities of convergence integration, Minimizing the consumption of transfers between different modes of transportation, including time, money, etc. However, the "seamless" is just an idealized state, even rail traffic in developed countries and regions, can not be fully realized "zero distance" between the modes of transportation. Therefore, from the view of space and time analysis, seamless transfers is to minimize walking and waiting time of passengers and achieve vehicle stable, punctuality. Seamless Transfers is to achieve seamless space, but also time to do seamless.Since walking distance can be measured by time, so seamless transfer, the most important is the time to do a seamless, namely to minimize passenger walking time and passenger transfer waiting time.

The most important indicator of evaluating the effectiveness of rail transfer system is to save time, reduce fatigue residents ride line. Transfers efficiency of urban rail transportation hub directly affect the traffic shortcut, and the convenience of travel to hundreds of thousands or even millions of passengers. Therefore, reasonable planning and construction of transfer hub then improving its function, is of great significance.

2.2 The connotation of urban mass transit seamless

Urban rail transit seamless transfer refers to the seamless operation phase (such as convergence channel, line, transfer and transfer hall building and the like), but to achieve this convergence must be from the planning, design, construction, operation and manage the entire process to reflect, involving effective coordination of multiple modes of transportation in the city between, seamless transfers hub connotation analysis from the following six aspects.

2.2.1 Rational layout

Hubs layout is more to consider when planning the hub from the start thinking, rational layout is more in terms of social, environmental and economic considerations, the layout is more macro level to consider the hub its configuration and coordination with neighboring facilities.

2.2.2 The capacity matching

Hubs for point, line, surface configured to consider the integrated transport hub transport capacity through the other modes of transportation unified arrangted and operated on the both in time and space efficient configuration, while providing passengers with a full range of real-time, accurate "seamless" of information services to achieve efficient transport can match.

2.2.3 Ride facilities are available

Through scientific and comprehensive system planning and design, and to make the track as the backbone of other modes of transport interchanges pick layout optimization, functional, structured, joint transport network can match the transfer, as a starting point to consider changing take facilities hub configuration, the specific configuration of the facility should be modern information technology and modern management tools based on intelligent infrastructure, automated to meet the travel needs of the urban social and economic development and residents.

2.2.4 Conducive to the development of surrounding land

Hub from the planning, it should be with the overall urban planning, network planning, landscape layout to match that transfer hub from the site will begin to take into account the level of land use, which also reflects a seamless transfer is a kind of convergence continued hubs. Energy saving and environmental protection.

First, the entire process from design to operation, it should be considered hubs throughout the layout, the size of the size of the transfer station, transfer facilities and coordinate the relationship between the environment; secondly, from a technical perspective, the need to consider sustainable, low-carbon technologies in the convergence transfer application; and finally, through the development of relevant local laws, regulations and standards to control hubs carbon-related indicators.

2.2.5 Human services

Hubs from the entire stages of planning, design, construction, operation and management have to take into account the service of humanity, that is, to consider the convergence of facilities, such as humanized landscape layout, during the operational phase to be considered intelligent infrastructure and staff quality of service, vehicle facility itself user-friendly configuration.

Rational layout, capacity matching, cohesion interchange facilities, conducive to the surrounding land development, environmental protection and human services seamlessly connotation six aspects, including the transfer of transport capacity and efficiency analysis, at the same time reflects the "low-carbon economy, smooth and safe" and other targets.

2.3 The characteristic of urban rail transit seamless

2.3.1 systemic

Urban rail transit Transfer System is a comprehensive and coordinated system consisting of urban rail transport and other transport modes in two ways by the constitution, but also deals with other outside urban transportation modes of coordination, in addition, other external factors , and these factors mutual restraint, mutual influence, must use viewpoint systematic approach to resolve.

2.3.2 Integration and hierarchy

Urban rail transit and other transportation mutual coordination and unified and effective convergence constitutes a seamless transfer, which reflects the overall characteristics; they are both point (hub coordination), line (line network coordination), face (macro-coordination) layer progressive presence in urban public transport system, with a clear hierarchy.

2.3.2 Dynamic and comprehensive

Metro convergence is not a static state, is a comprehensive, real-time changes in the state, according to changes in demand requirements, regulatory control parameters, function, structure, layout, economic, policy, and so the whole internal and external the system is in a virtuous cycle. So that all aspects of the organic unity of coordination.

3 Analysis of rail transportation organization

3.1 Inter rail transportation organization

3.1.1 Non-paid area transfer

Passengers need to transfer the ticket again, increasing time into the outbound zone transfers formalities unpaid no dedicated transfer facilities, walking long distance passenger and transfer passenger flow lines and other intertwined. Non-paid area transfer, often caused by poor planning, reflecting the mass transit system itself defects, should be avoided.

3.1.2 Paid district transfers

Transfer passengers without tickets again, including the platform transfer, the transfer node, the transfer station hall and transfer channels in several ways.

Platform transfer: directly transfer the passengers get off to walk to the other side of the platform, waiting for the train.

Node Transfer: Transfer passengers get off the escalator to reach the floor, use another platform waiting.

Transfer station hall: the transfer of passengers to get off after the first arrival transfer hall, waiting to go to another site.

Channel Transfer: Transfer passengers get off the transfer channel to the other using the platform waiting.

Platform transfer can greatly simplify the transfer flow line is the ideal way to transfer, but the requirements for network planning, circuit design, train operation organization of high and expensive. Node Transfer、channel transfer and transfer station halla currently common rail transfer organization, convenience descending order as follows: Node Transfer> channel transfer> transfer station hall. A variety of transfer within the organization can be used in Metro Station.

3.2 Rail transit and ground conventional public transfer organization

Conventional Public carrying capacity is relatively small, but the line flexible, broad coverage, applicable to passenger rail hub. Rail and bus transfer organization focuses on: Continuous transfer space and information.
Making mass transit hub design, should be considered a valid ground feeder bus and rail transit, rail transit transfer station near the bus station or the bus station to introduce rail station building, in order to form a comprehensive Hubs hall.

3.3 Rail Transit and private transport interchange organization

Private transport including bicycles, private car traffic in several ways. Transfer traffic organization within the rail transportation hub focused on bicycle, private car parking reasonable, including determining the size of parking, planning and service facilities.

High bicycle usage in our country, many cities of the reference residents travel survey, rail transportation hub in densely populated areas, non-motorized transfer is second only to walk the second largest rail transit transfer travel. Therefore, in the planning and design of rail transportation hub, it should consider certain non-motor vehicle parking space, if floor space is limited, you can set up an underground car park. It should also create lanes and facilities equipped with the appropriate model to ensure travelers' safety[8].

For private cars, the parking requirements for higher mass transit hub in urban areas often do not set conditions, can be built in the car parking area is located at the junction of the suburban rail transportation hub, encourage car users to transfer rail into the city, that park and Ride mode (P + R).

3.4 Rail and train transfers organizations

Ideally, if the inner rail transport hub comprising train transportation, urban rail transit should be shared with the train station hall, channels, squares and other facilities, in order to simplify the transfers streamline, improve facility utilization [9]. However, due to differences in vote system, security and other aspects, we can not be fully integration of two modes of transportation. The current urban rail transit train interchange organization there are three main forms:

3.4.1 Under the square in front of the train station separate building urban rail transit station.

Rail alignments may be parallel to the railway line, orthogonal, oblique and other two modes of transportation use entrance mutual convergence. The advantage is available Railway Station Square, and cost savings. The disadvantage is that long distance transfers, Station Square Traffic organizational complexity.

3.4.2 in building urban rail transit station and the railway station, the use of integrated hub transfer hall organization transfer.

This method will use a variety of line traffic hub complex in construction, shorten the distance passengers to transfer to simplify plane traffic organization, widely used in the world's major cities.

3.4.3 transit selected ground-mounted or high shelf, or placed on the side of the railway station square in front of the road and the railway station are arranged in parallel.

Commonly used method is to organization transfers rail station and the train station connected by channel or overpass. The program covers a smaller rail transit station, but traffic interference of different nature, affecting the landscape, transfers mode is not ideal.

4 Rail seamless transfer standard quantitative research

Rail transit transfer system to transfer as the main service purpose, namely to ensure that in the case of various modes of transport cross and mixed to achieve passenger flow in the most convenient transfer in the transfer system. As travelers between the same or different mode of transport interchanges generally realized by walking, although sometimes in the civil aviation and subway ride through the track, elevators and other mechanical joint transfer on foot, but by human foot to achieve interchange objectives is essential.

Therefore, walking distance and walk-time transfer function is theoretical basis to quantitative research hubs transfer system.

For this reason, Brighton G (Bouladon G) assumptions is introduced[4]. The function of Walking distance and feel comfortable walking time can be represented as:

$$T = Kd^{r} \tag{1}$$

T—transfer walking time

d—Transfer walking distance

k,r—coefficient(according to the mode of transportation survey data analysis calculated)

Since travelers inherent individual differences, to transfer distance, transfer time requirements have different standards, so at present, there is no standardized results to comprehensive transportation hub transfer distance and transfer time. In General, taking into account the variety of modes of transportation converge more experience-oriented, but travelers will, cost control, ease of travel into account, always want less time to transfer, transfer distance as short as possible. According to a function within walking distance and walking time, the transfer distance control is divided into five grades,as shown in Table I.

Table 1. Transfer distance classification

grade	A	B	C	D	E
description	ideal	desirable	acceptable	tolerable	Intolerable

Table 2. k,r,s values of different modes of transport

Mode of transportation	aviation	railway	coach	Rail	taxi	bus	Private car	bicycle
K	25	28.48	23.82	23.46	20.77	20.18	25.26	19.16
r	1.32	1.25	1.3	1.55	1.22	1.33	1.46	1.51

Normally, traveler requirement for transfer distance and the hub transfer service level evaluation is ambiguous, after the grade division, the transfer distance will be relatively objective and reasonable results. Considering the passenger flow characteristics of different modes of transport, including traffic distribution, passenger walking speed and walking space, whether passengers carrying luggage and other factors, combined with the comprehensive passenger transport hub status quo at home and abroad research data [10], k,r,s values of different modes of transport is initialized, as shown in Table 2.

Some well-known foreign transportation hub in dealing with transfer time, more control in 1min completed, or will be controlled within transfer distance 60 m. Learn from the successful experience of foreign countries, combined with the special characteristics of pedestrian traffic that, according to the curve and function of pedestrian flow and density in a horizontal channel of the transport hub to transfer time for standards[6], corresponding to A, B, C, D, E, different levels of traffic transfer time is respectively 90, 90-150, 90-300, 300-480, >480, as shown in Table 3.

Table 3. Transfer time classification

transfer Level	A	B	C	D	E
description	idael	desirable	acceptable	tolerable	Intolerable
Transfer time(s)	90	90-150	150-300	300-800	>480

According to the characteristics of our country's special pedestrian traffic and the relational model of pedestrian flow parameters in transport hub within different facilities, the transfer system is Classified by transfer time. In the function of walking distance and feel comfortable walking time as the theoretical basis, transfer distance can be obtained under different levels,as shown in Table 4.

Table 4. The rail transit transfer distance under different level

	Transfer mode					
level	Coach (m)	Rail (m)	Taxi (m)	Bus (m)	Private car(m)	Bicycle (m)
A	100	100	80	90	110	50
B	100-500	100-150	80-150	90-150	110-180	50-100
C	500-700	150-220	150-220	180-240	180-230	100-150
D	700-900	270-370	200-300	240-350	230-300	150-220
E	>900	>370	>300	>350	>300	>220

Rail seamless transfer request is a short transfer time and transfer passengers short walking distance. Through the above research, passengers seamless transfer the requirement for seamless transfer is at least an acceptable state and transfer graded should be in c and c and above. Therefore, the interchange time standard of rail seamless transfer should be controlled in T≤300s, transfer distance criteria in Table 5.

Table 5. Seamless transfer distance standard

Mode of transportation	coach	Rail	taxi	bus	Private car	bicycle
Walking distance(m)	<700	<220	<220	<240	<230	<100

It must be noted that, due to the different ways of transfer walking distance standard is not the same, and transfer distance can probably calculate the walking time, transfer time can not be drawn. Also within walking distance of the transfer will be affected by other factors lead to an increase walking time,such as traffic, channel congestion, wicket mechanical failure and so on. Transfer time includes the walking time and reflects the size of transfer distance, but also can directly describe the ease of transfer.So seamless transfer standard generally use transfer time to reflect.

5 Conclusion

With the development of rail transportation and the improvement of people's quality requirements for travel, seamless rail concept has been applied to transportation planning and design. Accroding to application of Brighton hypothesis, transfer distance standards can be divided into five grades, according to the related research, it is concluded that the transfer time and distance under different level standards and Seamless transfer concept. In the final analysis the deficiency of the transfer distance as standard,transfer time should be presented as the standard of seamless transfer.

Acknowledgment

This work is supported by National High Technology Research and Development Program of China（Project No. 2007AA11Z248）and Chang'an University(Project No. 310821165021) and Traffic Science and Technology Project of Shaanxi Provincial Transport Department (Project No. 15-26K).

References

1. J. Wang, "Key Issues on Transfer Organization of City Passenger Traffic Hubs," Beijing, Beijing Jiaotong University, 10-32, (2006).
2. D. Li, "Study on Planning and Design Theory for External Passenge Transport Hub of Metropolis," Nanjing, Southeast China University, (2006)
3. J. Su, S. He, "Evaluation of Transfer Efficiency for Urban Rail Transit," Agricultural Equipment & Vehicle Engineering **12**, 22-24, (2009).
4. Y. Wang, "Design Seamless Transfer between Metro Station and Bus:Case Study of on Financial Hi — Tech Zone Station on Gu hou — Foshan Line," Tunnel Construction **35**, 354-358, (2015)
5. D. Yang, L. Gao, "Exploration of the Three-dimensional Traffic Zero Transfer Design of Xi,an Textile CityPassenger Stations," Urbanism and Architecture **33**, 5-6, (2014)
6. X. Pan, Z. Ke, "A Study on Transfer D istance and Construction Indexes of Passenger Transit Hubs," Communications Science and Technology Heilongjiang **1**, 103-104, (2008).
7. H. Jia, L. Yang, M. Tang, "Pedestrian Flow Characteristics Analysis and Model Parameter Calibration in Canprehensive Transport Teaminal," Journal of Transportation Systems Engineering and Information Technology **5**, 117-123, (2009).
8. J. Xu, "Study on Seamless Connection Evaluation of Urban Rail Transit and Bus," Chongqing Jiaotong University, (2014)
9. X. Lu, W. Jiang, "Seamless Connection and Functionalities of Passenger Transport Terminals," Urban Transport of China **12**, 1-4, (2014)
10. D. Wu, "Research on Construction of Comprehensive Transportation Hub in the Economic Zone on the West Side of the Straits," Fujian, Fujian Provincial Department of transportation, (2008)

MATEC Web of Conferences **81**, 03003 (2016) DOI: 10.1051/matecconf/20168103003
ICTTE 2016

Traffic Conflict Technique-Based Setting Method of Freeway Climbing Lane

Xin Gu, QiaoJun Xiang and Shen Li

Southeast University, School of Transportation, 210096 Si Pai Lou #2, Nanjing, China

Abstract. Speed of heavy vehicles is greatly reduced on upgrade sections due to their insufficient climbing capability, thus having great impact on traffic safety. Setting climbing lanes can solve this problem effectively. Considering China's traffic composition and operational status, this paper proposed the traffic conflict technique-based setting method of freeway climbing lanes in order to set foundations for climbing lanes with the goal of promoting traffic safety. Based on large freeway upgrade section data and actual traffic characteristics, the paper built simulation models with VISSIM. 220 schemes were simulated by setting different road and traffic conditions of models. The paper then used Surrogate Safety Assessment Model (SSAM) to calculate traffic conflict, analyzed the influence of slope, slope length and mixed rate of heavy vehicles on traffic safety on upgrade sections. Finally, taking the reduction extent of the collision rate after setting climbing lanes as an indicator, this paper used K-means to classify the effectiveness level of the climbing lanes. The paper found the setting position and basis of the climbing lanes. The findings of this paper can provide support to effective decision making with regards to constructing future climbing lanes.

1 Introduction

Heavy vehicles usually have a great speed reduction on upgrade sections, due to their insufficient climbing capability, which has effect on the traffic safety. Setting climbing lanes on the upgrade sections would be an effective strategy to lessen this negative impact. However, since the construction of climbing lanes entails large difficulties and high construction cost, it is necessary to demonstrate the rationality of setting climbing lanes on upgrade sections of freeways.

Currently much research on setting methods and effects of setting climbing lane has been studied. Hesham Rakha and Bin Yu utilized a linearly increasing variable power vehicle dynamics model to develop the TRUCKSIM software for the modeling of truck acceleration behavior along grade sections. Using the software, the sensitivity of truck performance curves to roadway and truck characteristics are quantified [1]. Bin Yu's research also demonstrates that the vehicle weight-to-power ratio, vehicle power, and pavement condition are critical variables in designing truck climbing lanes [2]. Other Researchers also propose climbing lane setting methods based on traffic capacity [3]-[5]. Venkataraman Shankar evaluated the effect of climbing lane from the speed difference [6]. Abishai Pomst's results show that setting climbing lane can improve traffic flow, safety and capacity [7]. LIANG Guohua analyzes the influence of number of lanes, slope, slope length, speed and mixed rate of heavy vehicles on traffic capacity in climbing section to test the effectiveness of building climbing lane [8].

Other studies show the different climbing lane setting standards. American studies show the deviation between running speed of heavy vehicle and the average running speed on a given longitudinal gradient section is bigger, the chance of accidents occurring is higher. Therefore, AASHO requires that climbing lanes should be set on sections where the speed of large vehicles is reduced by 16km/h than its initial speed from the start of the gradient [9], [10]. In the Investigation and Research Report on Climbing Lanes, Japan adopts a simulation method to determine the appropriate sections and conditions of setting climbing lanes based on the aspects of traffic service and traffic safety [11]. China Academy of Transportation Science proposes the maximum limits on the slope and length of longitudinal gradient of all levels of highways, based on the vehicular speed differences and the requirements of satisfying the highway service level demands [12]-[14].

From the literature review, there seems to be three main issues in China: 1) most calculation methods used in setting conditions and sections of climbing lanes in China are based on the research results from other countries. The fact, traffic composition and the dynamic performance of the large vehicles in China are greatly different from those of other countries, is not taken into account. This may cause improper climbing lanes setting; 2) most studies only consider capacity rather than traffic safety; 3) the current foundations of setting climbing lanes are not qualified in detail. So in practice, decisions on constructing climbing lanes are based on the subjective judgments of the designers.

In order to address these issues, this paper proposes the traffic conflict technique-based setting method. This method perfects the foundations of climbing lanes setting with regard to traffic safety. Traffic conflict technique is an approach that overcomes the lack of reliable accident records, relying instead on observations of conflicts [15]. Microscopic simulation models are also seen as promising tools to evaluate road safety levels of existing and future infrastructures [16]. The VISSIM simulation model and Surrogate Safety Assessment Model (SSAM) have been proven to provide reasonable estimates for field measured traffic conflicts [17].The main object of this study is to find suitable climbing lane setting standard for China in terms of traffic safety. Actual typical upgrade section data has been surveyed in order to obtain actual traffic characteristics. The research can provide effective decision support for managers to set up climbing lanes.

2 Method

2.1 Data Collection

The traffic survey was conducted at an upgrade section of Nanjing-Hangzhou Highway (Section No. K2092+000～K2093+800), which has slope length of 1800m, gradient of 2.9%, vertical curve radius of 20000m, one-way for two lanes and one lane width 3.75 m. We divided the upgrade section into ten point to collect traffic data, such as the volume of different vehicles, the running speeds on different sections etc. (Fig. 1).

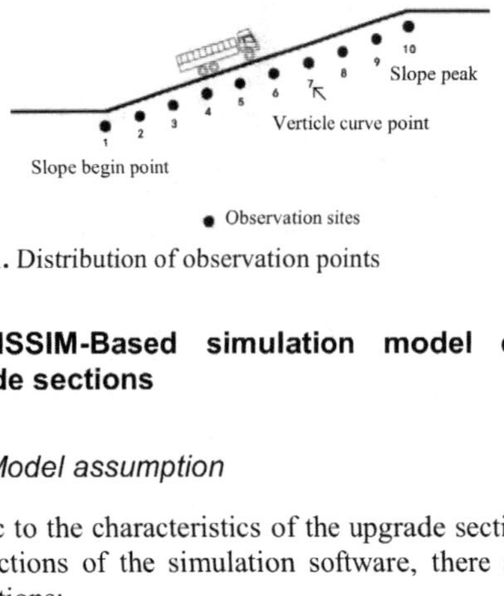

Figure 1. Distribution of observation points

2.2 VISSIM-Based simulation model of the upgrade sections

2.2.1 Model assumption

Specific to the characteristics of the upgrade sections and the functions of the simulation software, there are four assumptions:

- Neglect the influence of horizontal curve and simplify the horizontal alignment of upgrade sections into straight lines. Just consider the geometrical parameters such as slope, length and width of the lane in the design of the simulation experiment.
- Take the vertical curve on the top of slope as straight line of the upgrade sections. Neglect the running speed difference caused by the difference of the lengths of vertical curve.
- Neglect the influence of the lateral clear width on the operation characteristics of the traffic flow.
- The design speed is 100km/h.

2.2.2 Parameter calibration

Traffic Parameters Calibration of Representative Vehicles :In the simulation, we categorize vehicles into two categories, small cars (Car) and heavy vehicles (HGV), one have large flow and another have poor climbing performance. Meanwhile, combining the analysis of vehicle performance and driving behavior, we choose VW PASSAT as the representative for Cars and DFM EQ5208XXY2 as the representative of HGV.

Calibration of the Drivers' Behavior: We calibrate the drivers' behavior parameters from the following three models: the car-following model, the lane-changing model and the transverse behavior model [18]. In this paper, we adopt the Wiedemann99 Car-following model which can simulate the traffic flow characteristics of highway sections.

Simulation Time Period and Number of Times: Traffic collision is randomly time-varying. To eliminate the influence of time, this paper sets the simulation period as 43,200s, i.e. 12 hours. Meanwhile, to eliminate the influence of random factors in single simulation, we simulate each assumption 10 times and adopt the average value of the total 10 simulations.

2.2.3 Model verification

Table 1. Speeds and traffic flow test.

	Vehicle Types and Sections		Actual Value(km/h)	Simulation Value(km/h)	E (%)
Speed Test	Cars	Section 1	93.43	92.2	1.32
		Section 4	90.51	90.7	0.21
		Section 7	88.87	86.4	2.78
	HGVs	Section 1	70.1	72.5	3.42
		Section 4	60.39	49.7	1.14
		Section 7	42.43	44.5	4.88
	Time Interval (min)		**Actual Observed No. of Vehicles**	**No. of Vehicles in simulation experiments**	**E (%)**
Traffic Flow Test	0～5		55	53	3.64
	5～10		47	46	2.13
	10～15		50	48	4.00
	15～20		55	57	3.64
	20～25		42	44	4.76
	25～30		44	42	4.76

We carry out the simulation by setting up a simulation network of upgrade sections in the VISSIM, and taking the values of traffic flow, ratio of heavy vehicles etc. as input parameters, which are obtained from the practical survey and statistical results. Test simulation model by two indicators, average speed of each section and the traffic flow. The verification result is shown in Table 1. In addition, we also use root-mean-square percent error (RMSP), correlation coefficient R and Hill inequality coefficient U as the testing indexes to verify the effect of the model. Error E<5%; it is within the normal range of error and the other three indexes also satisfies the requirements RMSP<15% , r>0.8 , U<0.3 . Therefore the traffic simulation model is valid.

2.2.4 Simulation scheme

On upgrade sections, factors which influence the traffic flow mainly include road conditions (slope of gradient, length, radius of curve, number of lanes, width of lanes etc.), traffic conditions (volume of traffic, traffic composition etc.), environmental conditions (weather conditions, environments along the roads etc.).

This paper establishes kinds of simulations by inputting different traffic conditions and road conditions. We can get different simulation scenarios to analyze traffic operation under different conditions. The specific conditions of simulation schemes are shown in Table 2. In analyzing the combination of different road traffic conditions, a total of 220 schemes have been simulated.

Table 2. Specific values of road conditions and traffic conditions.

Road Conditions	**Slope (%)**	3	4	5
	Length(m)	800 1000 1200 1400 1600 1800	600 800 1000	600 800
	No. of lanes	Bi-directional four lanes		
	Width of Lanes(m)	3.75		
Traffic Conditions	**Volume of Traffic(veh/h)**	600、800、1000、1200		
	Ratio of large vehicles (%)	10、20、30、40、50		

2.3 SSAM-based conflict analysis

SSAM operates by processing data describing the trajectories of vehicles driving through a traffic facility, which provided by VISSIM in this paper, and identifying conflicts. For each vehicle-to-vehicle interaction SSAM calculates surrogate measures of safety and determines whether or not that interaction satisfies the criteria to be deemed as conflicts [19]. TTC is a critical parameter, which is defined as the minimum time to collision value observed during the interaction of two vehicles on collision course. If at any time step the TTC drops below a given threshold (2s in this paper), the interaction is tagged as a conflict. In addition, we don't take PET into account in this paper, so we use its default value. We also take the default value in the setting of the angular range of collisions; it means that when the angle of the collision vehicles satisfies this requirement, the incident will be defined as side collision.

2.4 K-means-Based safety assessment

In this paper, we take the reduction of traffic conflict rate as the index to assess the effect of the climbing lanes. And we adopt the K-means algorithm to process the data, since the data are in large amount and discrete. The K-means algorithm is a typical clustering algorithm based on distances and easy to use. Firstly we analyze the data with SPSS to determine the center of the primary categories , then use the K-means algorithm to categorize the data into four categories, i.e. Excellent, Good, Middle and Poor to express the effectiveness of setting climbing lane.

3 Results

3.1 Setting position of the climbing lane

The scenario whose road with the gradient of 3%, slope length of 1600m, traffic volume of 1000veh/h and heavy vehicle ratio of 30% is taken as an example. The optimal setting position of climbing lane is determined through analyzing the traffic collisions under different start-end points.

When the climbing lane has not been set, the speed variation tendency of heavy vehicles after entering the upgrade section is obtained through simulation (Fig. 2). As illustrated in Fig. 2, the heavy vehicles have different speed reductions at different sections after entering the upgrade section. Specifically, the sections respectively corresponding to the speed reductions 15km/h, 20km/h, 25km/h and 30km/h are set as the start points of the climbing lane, and the speed recovery positions are respectively set as the end points of the climbing lane. The simulation models of the climbing lane are thus established under the above four conditions. Therein, the transition lengths of the branching section and the confluence section are respectively set as 100m and 200m. The section adjacent to the longitudinal slope is a straight slope, so the length of the additional section is set as 150m. The traffic conflict distribution under the above four conditions before and after setting climbing lane is obtained through simulation (Table 3).

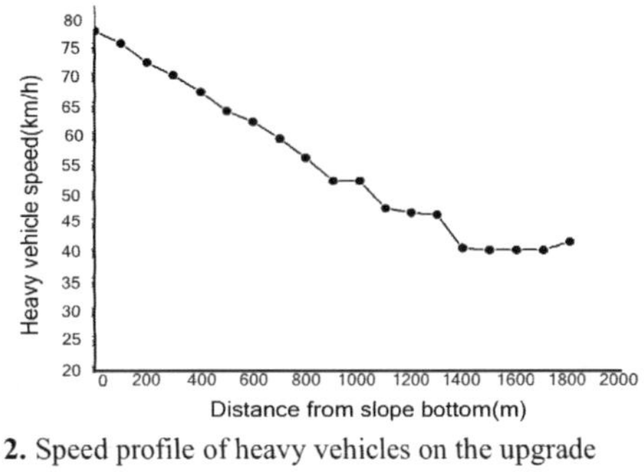

Figure 2. Speed profile of heavy vehicles on the upgrade section.

According to the simulation result, the number of rear-end collisions on the upgrade section is obviously reduced after setting the climbing lane. While the number of side impact collisions is increased, the total number of collisions is reduced. Moreover, when the start point of the climbing lane is set on the section corresponding to the speed reduction 20km/h (namely 700m), the total number of the collisions and the number of rear-end collisions are minimized. The cause analysis is as follows:

Table 3. Traffic collision distribution under the conditions of different start points of the climbing lane.

Setting Condition \ Collision Type		Number of Rear-end Collisions	Number of Side Impact Collisions	Total Number of Collisions
No Climbing Lane		300	108	408
With Climbing Lane	15km/h Speed reduction	80	219	299
	20km/h Speed reduction	24	228	252
	25km/h Speed reduction	58	253	311
	30km/h Speed reduction	126	224	350

When the start point of the climbing lane is set on the section corresponding to the speed reduction 15km/h (namely 400m), the heavy vehicle needs to have additional speed reduction to travel to the climbing lane, thus causing more rear-end collisions with the vehicles on the main lane. When the start point is set at the section corresponding to the speed reduction 25km/h (namely 900m), at the middle position of the upgrade section, such climbing lane setting can effectively reduce the number of the collisions between the middle and top upgrade section.

When the start point of the climbing lane is set at the section corresponding to the speed reduction 30km/h (namely 1000m), the speed of the large vehicle has been reduced to the balanced speed. The speed difference between small vehicles and large vehicles is then gradually stabilized, such that the climbing lane can no longer effectively reduce the number of the traffic collisions.

According to above analysis, when the section corresponding to the speed reduction 20km/h of the heavy vehicles is set as the start point of the climbing lane, the number of the traffic collisions can be effectively reduced, thus achieving the optimal effect. Therefore, the speed reduction 20km/h is set in the article as the basis for setting the start point of the climbing lane.

3.2 Effect analysis for the setting of the climbing lane

The reduction extent of traffic conflict rate after setting climbing lane on the freeway upgrade section is related to the slope length, the gradient, the traffic volume and the large vehicle rate, as shown in Fig. 3 (a~ d).

According to Fig. 3 (a), under the same ratio of large vehicle, the reduction extent of collision rate is overall slightly increased along with the increment of the gradient; according to Fig. 3 (b), under the same traffic volume, the reduction extent of collision rate is increased along with the increment of the slope length.

According to Fig. 3 (c), under the same ratio of large vehicle, the collision rate reduction extent of the upgrade section is 10%~20%, and significantly influenced by the traffic volume. Specifically, when the ratio of large vehicle is relatively small (less than 30%), the reduction extent of collision rate is firstly increased and then reduced along with the increment of the traffic volume. When the traffic volume is 1200veh/h, the reduction extent of collision rate reaches to the maximum value. When the ratio of large vehicle is relatively large (more than 30%), the reduction extent of collision rate reaches to the maximum value at the traffic volume of 1000veh/h.

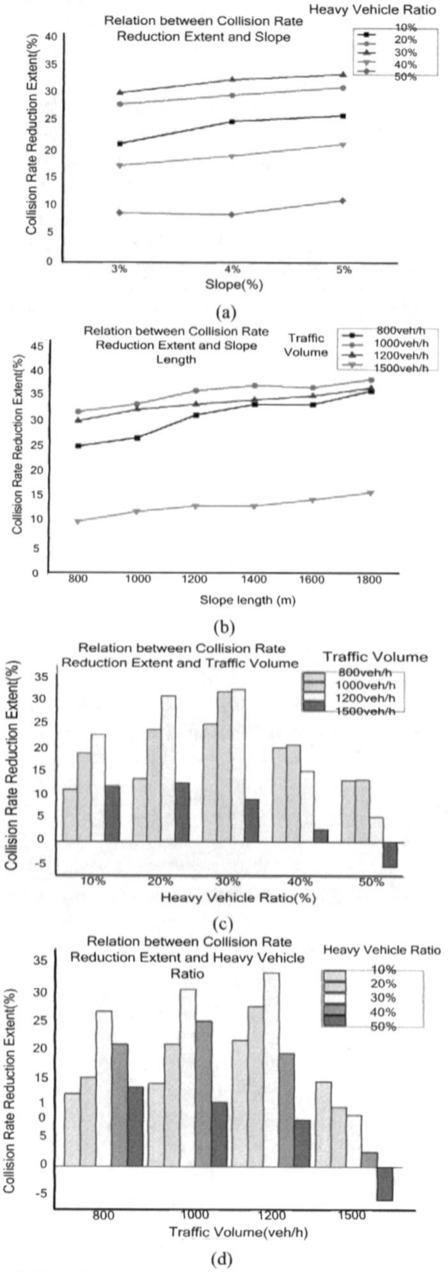

Figure 3. Results.

The analysis is as follows: when the traffic volume is relatively small, vehicles have high running freedom and small mutual influence. So the setting of the climbing lane cannot obviously improve traffic safety. Along with the increment of the traffic volume, large vehicles gradually significantly influence small vehicles, and the small vehicles need to find suitable opportunities to change lane for overtaking. At this moment, if the climbing lane is set, most of the large vehicles will be separated to the climbing lane, thus effectively reducing the influence on small vehicles, reducing the speed change requirements of small vehicles for lane change. Finally this can reduce vehicle collisions and improve traffic safety.

When the traffic volume reaches to a certain value, the traffic flow on the upgrade section is under following driving state. Small vehicles rarely have space to change lane for overtaking, so the vehicle collision is small. Large vehicles are also not the main factor influencing the traffic safety of the upgrade section. So setting climbing lane in this occasion only has small effect for traffic safety improvement, and may increases the lane change requirements of some large vehicles, thus increasing rather than reducing the traffic collision rate.

According to Fig.3 (d), when other conditions are not changed, the reduction extent of collision rate presents a parabola form along with the change of the large vehicle ratio. Under the condition that the traffic volume is not more than 1200veh/h, when the large vehicle rate is 30%, the reduction extent of collision rate reaches to the maximum value. When the large vehicle ratio is less than 30%, the reduction extent of collision rate is increased along with the increment of the large vehicle ratio. When the large vehicle ratio is more than 30%, the reduction extent of collision rate is reduced along with the increment of the large vehicle ratio.

Under the condition that the traffic volume is more than 1200veh/h, when the large vehicle ratio is 10% and 20%, the reduction extent of collision rate is relatively large. When the large vehicle ratio is more than 20%, the reduction extent of collision rate is reduced along with the increment of the large vehicle ratio.

3.3 Effectiveness evaluation

K-mean is adopted for the effectiveness evaluation of the data obtained thereby. The final clustering result obtained from ten iterations is shown in Fig. 4.

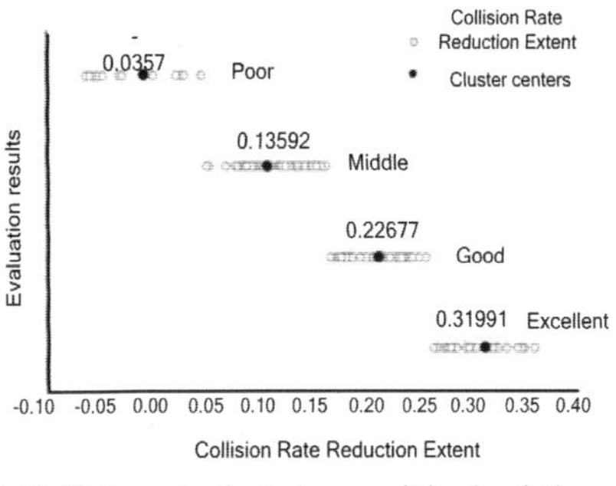

Figure 4. Collision rate clustering result in simulation experiment.

According to the cluster results, we classify the effectiveness of the climbing lanes into four levels. If the reduction extents of traffic collision rate (CR) > 0.31991, it's excellent; if 0.22677<CR<0.31991, it's good; if 0.13592<CR<0.22677, it's Middle; if CR<0.13592, it's poor.

The upgrade section with slope length of 1200m, gradient of 3%, traffic volume of 800veh/h ~ 1500veh/h, and large vehicle ratio of 10%~50% is taken as an example to compute the collision rate. And we use reduction extents of traffic collision rate to analyze the effect of the setting of the climbing lane, as shown in TABLE4.

Table 4. Effect Analysis for the Setting of the Climbing Lane with Slope Length of 1200m and Gradient of 3%

Traffic Volume (veh/h)	Large Vehicle Ratio	Collision Rate Reduction Extent	Effectiveness Evaluation Result
800	10%	11.11%	Middle
	20%	21.43%	Good
	30%	31.25%	Excellent
	40%	23.81%	Good
	50%	16.67%	Middle
1000	10%	23.53%	Good
	20%	26.32%	Good
	30%	36.00%	Excellent
	40%	26.47%	Good
	50%	13.51%	Middle

According to TABLE4, when the large vehicle ratio is 10% and the traffic volume is 1000veh/h~1200veh/h, setting climbing lane has good effect. When the traffic volume is relatively small or large, setting climbing lane has middle effect. The reasons are following: when the traffic volume is relatively small, the traffic flow is under free flow state and large vehicles slightly influence small vehicles. However, when the traffic volume is relatively large, the traffic flow is under saturated state, and setting climbing lane cannot effectively separate the large vehicles to reduce traffic collisions.

When the large vehicle ratio is 20%~40% and the traffic volume is not more than 1200veh/h, setting climbing lane has excellent or good effect. This indicates that setting climbing lane has relatively good effect for improving traffic safety.

When the traffic volume is more than 1200veh/h, setting climbing lane has poorer effect along with the increment of the large vehicle ratio. This indicates that setting climbing lane cannot significantly improve the traffic safety. When the large vehicle ratio is increased as 50%, setting climbing lane cannot improve the traffic safety of the upgrade section, but oppositely increase the traffic collision rate and reduce the traffic safety.

4 Conclusion

The effect of setting freeway climbing lane and the basis for the setting's specifics were systematically researched in this paper with the goal of traffic safety. The authors aimed to provide suggestions regarding the setting position and the setting method of the climbing lane. However, this study was limited to analyzing data from the angle of road and traffic condition. In reality, there are other influencing factors of the upgrade section such as traffic control condition, environment condition, and more. Future studies may look into including traffic control and environment conditions as variables to be considered for modeling. The main suggestions that can be drawn from this study are summarized by the following points.

Setting Position of the Climbing Lane: The start point of the climbing lane should be set at the position where the speed reduction of large vehicles is at 20km/h. The end point needs to be set at the speed recovery position. It is also necessary to comprehensively consider the relation between the start-end points to ensure that these points are set at the positions with the features of good inter visibility, easy identification, and smooth transition. Moreover, the setting effect and the investment level must also be considered. A minimum length of 500m for the climbing lane is suggested. When the calculated length for the climbing lane is less than 500m (using the methods aforementioned), the setting of a climbing lane is not advised.

Setting Basis of the Climbing Lane: When the traffic volume is less than 1200veh/h, when the large vehicle ratio is less than 20%, or when the large vehicle ratio is more than 40%, setting a climbing lane on the upgrade section cannot improve traffic safety. So it is not advisable to set climbing lane although it is necessary to adopt such measures as lane limitation to improve the traffic safety of upgrade section;When the traffic volume exceeds 1200veh/h, the traffic flow on the upgrade section becomes saturated. In this case, the setting of the climbing lane can improve the traffic capacity but cannot effectively improve the traffic safety of the upgrade section. Furthermore, the setting of the climbing lane may increase the traffic collisions among vehicles, so it is not advisable to set the climbing lane under such conditions.

When the traffic volume does not exceed 1200veh/h and the large vehicle ratio is 20%~40%, the setting of the climbing lane on the upgrade section can effectively improve the traffic safety; especially, when the large vehicle ratio is about 30%, the positive effect is most obvious.

References

1. H. Rakha, B. Yu, Journal of Transportation Engineering **130**, 6, 753-767, (2004)
2. B. Yu, Faculty of Virginia Polytechnic Institute and State University, (2005)
3. Q. Yang, Y. Zhang, Highway **9**, 98-101, (2005)
4. G.O.N.G. Ni-na, Communications Standardization **12**, 193-194, (2008)
5. L. Wang. Research on Setting Climbing Lane on Steep Sections of Mountain Road. Shan Dong University, (2010)
6. V. Shankar, F. Mannering, W. Barfield, Accident Analysis and Prevention **27**, 3, 371-389, (1995)
7. A. Pomst, I. Reshetnik, Transport Research Part A **21**, 6, 401-410, (1987)
8. G. Liang, R. Ma, X. Shen, Y. Liang, Journal of Chang'an University (Natural Science Edition) **34**, 1, 23-30, (2014)
9. American Association of State Highway and Transportation Official. A Policy on Geometric Design of Highways and Streets. Washington DC: AASHTO, (2011)
10. R. Pal, K. C. Sinha, Transportation Research Record: Journal of the Transportation Research Board, No. 1640, Transportation Research Board of the National Academies, Washington, D. C, 1-9, (2007)
11. Japan Highway Public Corporation. Japan Highway Design Essentials. Shaanxi Travel & Tourism Press, China, (1991)
12. Guidelines for Safety Audit of Highway. Publication JTG/T B05-2004.Ministry of Transport of the People's Republic of China, (2004)
13. Technical Standard of Highway Engineering. Publication JTG B01-2003. Ministry of Transport of the People's Republic of China, (2004)
14. Design Specification for Highway Alignment. Publication JTG D20-2006. Ministry of Transport of the People's Republic of China, (2006)
15. M. Parker Jr, and C. V. Zegeer, Observers Manual. Publication FHWA-IP-88-027. FHWA, U.S. Department of Transportation, (1989)
16. A. L. Vasconcelos, A. Seco, A. B. Silva, et al., Transportation Research Record: Journal of the Transportation Research Board, No. 2432, Transportation Research Board of the National Academies, Washington, D. C., 1-9, (2014)
17. J. So, I.K. Lim,Y.J. Kweon, Transportation Research Board of the National Academies, Washington, D. C., (2015)
18. PTV VISSIM User's Manual 5. 0. Planung Transport Verkehr AG, D-76131 Karlsruhe, Germany, (2007)
19. D. Gettman, L. Pu, T. Sayed, and S. G. Shelby. Surrogate Safety Assessment Model and Validation: Final Report. Publication FHWA-HRT-08-051. FHWA, U.S.Department of Transportation, (2008)

MATEC Web of Conferences **81**, 03004 (2016) DOI: 10.1051/matecconf/20168103004
ICTTE 2016

Design of Intelligent Transportation Inquiry System Based on MapX in the Environment of VC++

Juan Cheng[1] and Lu Sun[2]

[1]*School of Transportation, Southeast University, No.2 Sipailou, Nanjing, China*
[2]*Department of Civil Engineering, Catholic University, Washington, DC, US*

Abstract. This paper applied MapInfo, the professional soft ware tool of GIS, integrated secondary exploiture combining with electronic maps, and made use of the exploiture flat roof Visual C++ as the tool of visualize development, transferred MapX, a control of MapInfo, integrated them. The paper designed the Inquiry System in Intelligent Transportation, which including query system of road information, query system of bus information, query system of district information. It can be carried out space analysis and query function based on GIS. Adopted SQL Server manage attribute data, by data binding, attribute data in SQL Server and victor picture data were combined.

1 Introduction

Transportation is the link and lifeline of urban economic activity, which plays an important role on the sustainable development of city and the improvement of people's living standards. With the rapid development of the national economy, in many cities of the world the phenomenon of traffic congestion appeared which caused the traffic accident, traffic noise, traffic pollution and so on. In order to improve the traffic condition in a region, the intelligent transportation system (ITS) has emerged. ITS is a comprehensive, real-time, accurate, safe and efficient transportation system with advanced electronic technology, information technology, sensor technology and system engineering technology [1]. Geographical Information System (GIS) is an important part of the intelligent transportation system, which is based on its outstanding spatial information management capability.

2 Overview of related technologies

2.1 Geographical Information System for Transportation (GIS-T)

GIS-T is a computer software and hardware system for collecting, storing, managing, analyzing, processing spatial information and traffic information, which is the extension of GIS in the traffic field. In a word, GIS-T is an integrated technology of GIS for transportation [2].

2.2 Database

SQL Server is a relational database management system for C/S model [3]. SQL Server is data storage and retrieval model in the middle of 1980s and has continued to develop to nowadays, which tightly integrate with Windows NT operating system [4].

3 Design of intelligent transportation inquiry system

3.1 System requirement analysis

Intelligent transportation information inquiry system using graphics display function of GIS is a graphical operating platform, which provides users with interface who can easily query and management [5], [6].

This paper uses Visual C++ and MapX to develop, using MS SQL Server as background database and using SQL language through the ADO database access interface to control the database.

3.2 Overall design of system function

3.2.1 Overall system

The system which makes full use of the map display function of GIS displays the different geographical elements and supports the management and function of data.

3.2.2 System requirements

(1) The system operation must be simple and practical.

(2) Security of information system.
(3) Stability and reliability of the system.

4 Realization of the system function

4.1 Realization of the system database

The composition of the database of intelligent transportation information inquiry system is shown in Fig. 1.

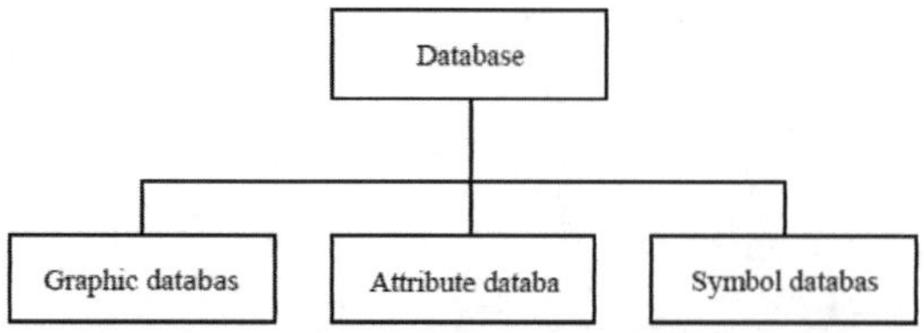

Figure 1. the Construction of the System Database Diagram

Database table is an important part of the system. During the formation of each map layer, MapInfo create a database table for each layer automatically. Processing of map layer by MapInfo is actually a map layer that corresponds to a database table processing [7].

4.1.1 Graphic database

Graphics database of intelligent transportation information inquiry system is scanned according to the map obtained, which can only be used as a base map. Other layers are obtained by two methods. One is the use of AutoCAD, which is shown in Fig. 2. Another, it should draw a map layer with MapInfo when the system requires a new layer. MapInfo will generate database table and the corresponding file automatically.

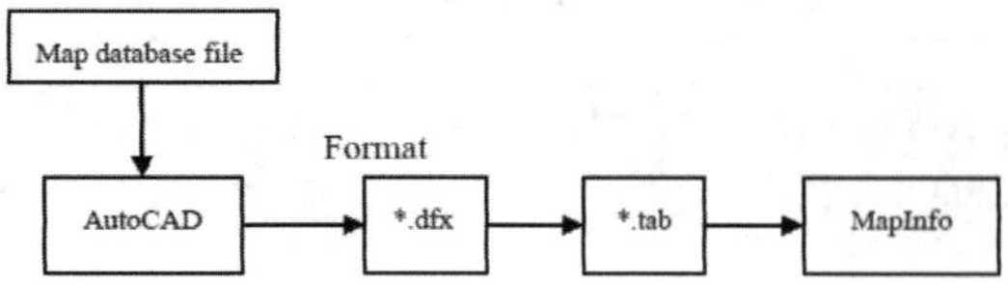

Figure 2. Format conversions of MapInfo and AutoCAD diagram

4.1.2 Attribute database

The relational data structure is used in the MapInfo attribute database. The attribute database of main layers of the system is shown in Table 1.
where:

RoadId is the number of the road.
RoadName is the name of the road.
RoadLength is the length of the road.
RoadWidth is the width of the road.
RoadRank is the level of the road.

Int is integer. The storage size of this type is 4 bytes and the range of the value is -2^{31}~2^{31}-1.

Varchar is character and the format is varchar[(n)]. The storage size is the actual length of the input character.

Float is floating-point and the format is float[(n)].N is used to store the mantissa of scientific notation where the range of the value is 1~53.

Table 1. Structure of road information table

name of the row	date type	length	allow null
RoadId	int	4	
RoadName	varchar	10	✓
RoadLength	float	8	✓
RoadWidth	float	8	✓
RoadRank	varchar	10	✓

4.1.3 Symbol database

MapInfo system has its own symbol database shown in Fig. 3.

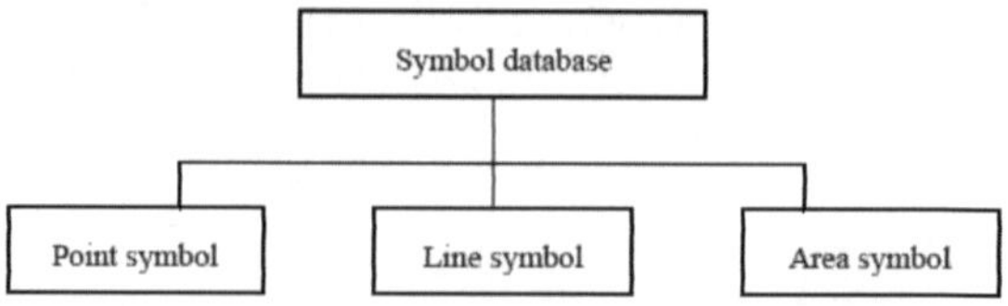

Figure 3. The symbol database of system diagram.

4.2 Realization of the system function

4.2.1 Security module

The system is set up a table to store user names and passwords in the database. The system compares the user name and password entered by the user with the information existed in the database, if the user exists and the password is correct, then enter the main interface. Landing fenster display error messages when user name or password is incorrect.

User types are divided into system administrators and ordinary users. The system administrator can change the administrator's password and modify information of general user, while ordinary users can modify their own user name and password.

User table structure is shown in Table 2.

Table 2. Structure of User Table

USERID	USERNAME	USERPWD	USERTYPE
1	admin	*******	system administrator
2	april	*******	ordinary users
3	fly	*******	ordinary users
....	....	....	....

In order to protect the security of the database, in addition to setting password, we take a database backup function. When the destruction in the database, we can restore the database.

4.2.2 Operation module

(1) Basic Operations Modulz

Basic operations module includes arrows, zoom in, and zoom out, pan, full map display, central display. These functions can be found on the system function navigation bar

(2) Advanced Operations Module

Advanced operations module includes selection, rectangular selection, round selection, polygon selection and ranging

(3) Inquiry Module

• Road information inquiry

We can inquiry the length, width and grade of the road. We can modify information of each road and save the changes to the database at the same time.

• Traffic districts information inquiry

We can inquiry the range, area and population of the traffic zone. We can modify information of each traffic zone and save the changes to the database at the same time.

• Bus information inquiry

We can inquiry peak headway, flat peak headway, early morning departure time, evening departure time, fares, starting station and terminus of each line. We can modify information of each line and save the changes to the database at the same time.

• Optimal path algorithm of least transfer

According to travel habits of people, we choose the bus immediately if there is a direct line from station A to station B. If there is more than one direct line, then considering the route of the distance, we choose the nearest bus scheme. If there is no direct bus, we will consider changing a bus. If there is a station C which is the public station of station A and station B, we will change the bus at station C. If there is no station C, we will consider changing a bus in station D which is the public station of station C and station B. If there is no station D, then we will change bus more than three times to the destination. If there is more than one choice in the above case, we will consider the distance and will choose the shortest travel plan.

Transfer plans of bus lines are shown in Figure 4.

• Shortest Path Analysis

Determine bus changing plan of the shortest path between the two places of city.

(4) Analysis Module

• Road information inquiry

The system shows the number of all main road and inquiries the length of the each main road which can navigate to the corresponding location in the map.

• Positioning traffic districts

Select a traffic zone, the system can be positioned to the map and highlight it.

• Ratio of main road, secondary road and branch road in urban road network

$$\gamma=\sum_{i=1}^{m} a_i : \sum_{j=1}^{n} b_j : \sum_{k=1}^{p} c_k \quad (1)$$

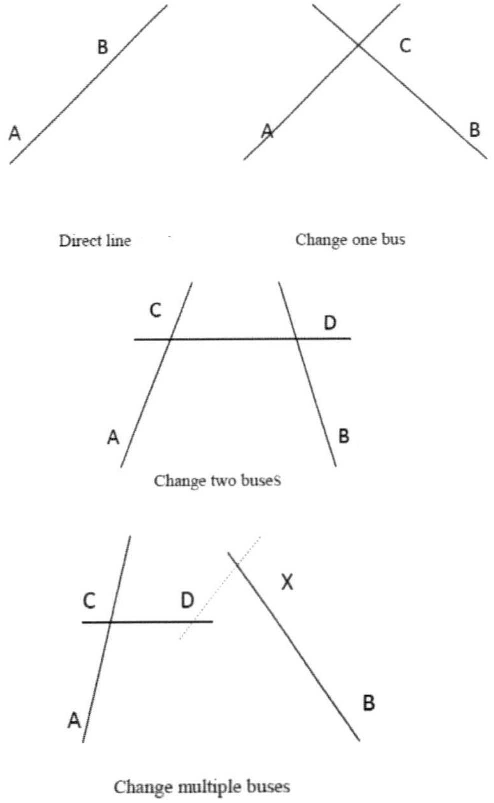

Figure 4. Transfer plans of bus lines.

In equation (1), γ is the ratio of main road, secondary road and branch road in urban road network. a_i is the mileage of main road in road network where i=0,1,2,…,m. b_j is the mileage of secondary road in road network where j=0,1,2,…,n. c_k is the mileage of branch road in road network where k=0,1,2,…,p.

• Thematic Map

For a large number of digital data, the user may not value the specific figures, but want to get the overall distribution of the situation, For example, population density thematic map. If users need to get the distribution of the total number of male and female population in each area, they can choose the number of male or female population as a special variable and create a pie chart, etc.

5 Example

According to the method above, Intelligent Transportation Inquiry System of X city is constructed in this paper. Intelligent Transportation Inquiry System of Nanchang city uses Visual C++ and MapX to develop, using MS SQL Server as background database and using SQL language through the ADO database access interface to

DOI: 10.1051/matecconf/20168103004

control the database. The main functions of the system are shown as follows.

• Login System Diagram is shown in Fig. 5.

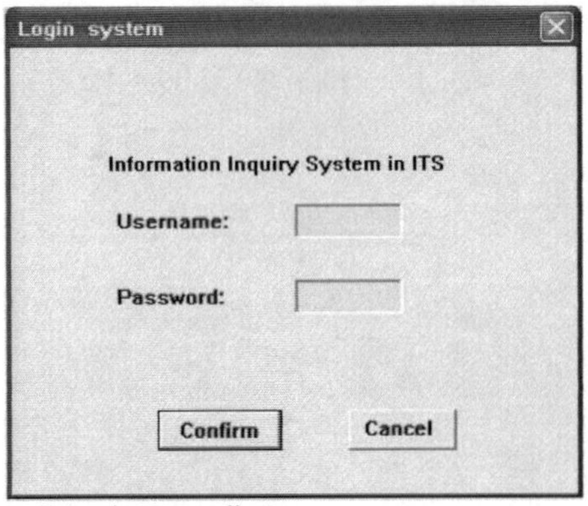

Figure 5. Login system diagram.

•Main Operating Diagram is shown in Fig. 6.

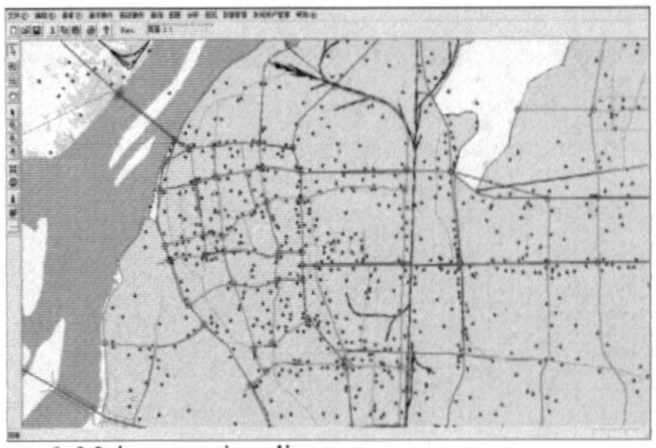

Figure 6. Main operating diagram

6 Conclusion

This paper describes the traffic geographic information system and database technology, and also introduces the function design of the intelligent transportation inquiry system in detail. Using Visual C++ programming language combined with GIS component MapX and Server SQL database, this article developed intelligent transportation inquiry system and completed each module of the system. Finally, the paper users Intelligent Transportation Inquiry System of Nanchang city to verify.

Acknowledgment

This study is sponsored in part by the National Science Foundation under Grants CMMI-0408390 and CAREER award CMMI-0644552, by the National Natural Science Foundation of China under grants 51250110075, 513111130and U1134206, by Ministry of Communication of China under grant 0901005C, and by Jiangsu Natural Science Foundation under Grant SBK200910046, to which the authors are very grateful. The authors would also like to thank anonymous reviewers for their very helpful comments and constructive suggestions, with help us to improve the content and presentation of the manuscript.

References

1. National ITS Architecture for America (Version 4.0), CD-ROM, (2002).
2. S. B. Miles, Application and Issues of GIS as Tool for Civil Engineering Modeling, Journal of Computing in Civil Engineering, **13**(3): 144-152, (1999).
3. H. J. Miller, Yi-Hwa Wu, GIS Software for Measuring Space-Time Accessibility in Transportation Planning and Analysis, GeoInformatica, **4**(2), 141-159, (2000).
4. R. Beheshti, R. Miehels., The global GIS: a ease study, Civil Engineering Informatics, Faculty of Civil Engineering and Geosciences, Delft University of Technology, Netherlands, (2001).
5. Jianjun Xiao, Research and application of Transportation Geographic Information System, Central South University, (2005).
6. F. Kurt, W. Lotharand , A Hybrid Expert System, GIS and Simulation Modeling for Environmental and Technological Risk Management, Computer-aided Civil and Infrastructure Engineering, (1999).
7. D Xiong, H Lin, Spatial Data Handling for ITS: Perspective, Issues and Approaches, GeoInformatica, **4**(2), 215-230, (2002).

MATEC Web of Conferences **81**, 03005 (2016) DOI: 10.1051/matecconf/20168103005
ICTTE 2016

Managing Transportation Megaproject Schedule Risks Using Structural Equation Modelling: A Case Study of Shanghai Hongqiao Integrated Transport Hub in China

Wei Peng [1,2] and Liangfa Shen [1,3]

[1] *School of Economics & Management, Tongji University, Shanghai, China*
[2] *Bank of China Investment Management Co. Ltd, Shanghai, China*
[3] *Construction Science, Texas A&M University, College Station, Texas, USA*

Abstract. Tight schedules, multifunctional scopes, and colossal sizes usually characterize transportation megaprojects as challenging tasks for completion. In order to address these situations, a schedule risk management method was developed in this paper based on the structural equation model. In the proposed method, risk identification, evaluation and response were arranged as a sequence, and the expert elicitation technique was adopted in order to quantify the schedule risk status. To demonstrate the applicability of the proposed model, a megaproject case in China, the Shanghai Hongqiao Integrated Transport Hub (SHITH), was chosen. Information within the expanded risk register was collected including the probability and consequence of risk events, the complexity of risk responsible owners, the reaction time, and the time lasting for risk countermeasures. Final risk control results showed that the method could not only address the schedule risks correlations effectively, but also maintained the simplicity for construction management practices.

1 Introduction

Transportation megaprojects have currently addressed more and more public attentions due to large investment, advanced technologies and long-time construction. Megaprojects could be defined as projects costing more than $1 billion and lasting more than five years in construction duration [1]. The typical transportation mega projects can be found all around the world such as in Europe, the Øresund Bridge between Denmark and Sweden, the Vasco da Gama Bridge in Portugal, and the German Maglev Train between Berlin and Hamburg; in Asia, Hong Kong's Chek Lap Kok Airport, China's Quinling Tunnel, Japan's Akashi Kaikyo Bridge, and Malaysia's North–South Expressway etc. The important role of transportation megaprojects has been revealed in the change of globalization and civilization [2]. Compared to ordinary projects, transportation megaprojects bear more attributes in social contexts, such as the inadequate previous experience, the economic and environmental effects, the technology innovations, the political and cultural effects, and the safety issues.

Due to the substantial impacts on communities and environment, it is important to identify and assess the schedule risks for building megaprojects [3]. For megaprojects, aligning with the original timeline and scopes could not only guarantee the missions that megaprojects hold to complete on time, but also save the public resources and protect civilian benefits [4]. As a result, transportation megaprojects are usually characterized by tight schedules, multifunctional scopes, and colossal sizes. However, the large-scale engineering, construction and infrastructure projects are usually complex and notoriously difficult to manage [5]. Schedule overrunning frequently occurs during the planning and construction phases, named the "megaproject paradox" [1]. For instance, in the United States, the Boston Central Artery/Tunnel Project, commonly known as the Big Dig, projected the official planning phase to start in 1982 and was originally scheduled for completion in 1998. Finally, however, the project was concluded on December 31, 2007, with nine-year delay.

Therefore, improving the schedule techniques of megaprojects with risk analysis has become a prospective field for transportation construction management. The remainder of the paper is structured as follows: Section 2 reviewed the existing schedule risk assessment literature in transportation project, which is set up as a steady basis for current research. Section 3 introduced the modified risk register and structural equation modeling in detail, as the two basic methods adopted in the schedule risk management model. In order to demonstrate the applicability of the proposed method, an in-depth analysis of a real-life megaproject, the Shanghai Hongqiao Integrated Transport Hub (SHITH), is presented in Section 4. The project schedule risk identification and ranking approach is outlined prior to the presentation of the schedule risk measurement model. The case study concludes with a discussion of SEM outputs interpretation and application in decision-making. Section 5 summarized the paper.

2 Related studies

Project risk is defined as an uncertain event or condition that, if occurs, has a positive or negative effect on one or more project objectives such as scope, schedule, cost and quality [6]. Risk management can thus be critical, since risks are underlying throughout the whole project life cycle. Generally, risk management is the life cycle process that includes planning, identification, analysis, response planning, and controlling risk on a project [6]-[8]. The objectives are to increase the likelihood and impact of positive events, and decrease the likelihood and impact of negative events in the project [6]. From a risk management perspective, construction risks such as adverse weather, poor subcontractor performance, safety and environmental risks, resource availability, and coordination among various programs need to be carefully monitored [8].

As a foundation of the risk management process, risk registers are usually used to house a complete list of risks that may be faced in one place. With dynamic documentation, the risk register is updated throughout the project as old risks are closed out and new risks are identified [6]. Traditionally, two variables of the risks, the likelihood and the consequences, were adopted in order to track the changes quantitatively [9], [10]. However, limitations exist in real application. For instance, the inner correlation or clustering between risk events generally got lost in a risk register. A warning was set against the dangers of one-sidedly using risk registers and related practices in containing management risk [11]. As a result, varied risk register improvements occurred in the risk management. Based on the standard risk register, a more stable risk prioritization measure was proposed by fusing elements of different strategies from several matrix-based schemes [12]. A combination of these schemes led to a cubic-structure, rather than a matrix, as risks are evaluated along three dimensions: impact, probability, and discrimination. The ranking model was similar to the generalized business impact analyses proposed by [13]. The first dimension, impact, was taken directly from the research of [14] and it used a five-point scale for evaluating risk impact. The second dimension, probability, was based on [15]. Also, a PIS (probability, impact and significance) method was proposed by [16], which encompassed the significance criterion into risk quantification.

Multiple risk sources might result in transportation project delay such as organization and management, economic, technique, and environment [17]. In order to address the characteristics of these underlying sources, various schedule risk management methods were adopted in the previous research. Probability theory was used to get a best-fit distribution of schedule overruns occurring in Australian constructions and engineering projects [18]. The identified probability distribution function enabled a realistic estimate of schedule overrun in schedule risk determination. A schedule risk analysis methodology was also proposed with activity-risk factor matrix and risk correlation coefficient matrix, in order to find out the pessimistic project completion time, critical activities and risk elements [19]. When coming to more complicated correlations, networks could also be adopted for analyzing the schedule risks [20]. Other hybrid risk analysis models involve the integrated risk management and proactive scheduling [21], project tracking [22], and schedule assessment and evaluation [23].

Transportation megaprojects hold close relationships with the level of political centralization and economic development, which functioned as a safety valve in society [24]. Distinguished from the previous schedule risk research, this paper presents a statistical schedule risk management method adopted by the SHITH project. In order to enhance the transportation megaproject performance, two main virtues of the method are demonstrated thusly: 1) schedule risk characteristics are addressed such as the correlations between risk events, and 2) the simplicity of methodology is maintained, which is important for real construction application. Thus, transportation megaproject owners or managers from different sectors could effectively use and implement the method in real life practice.

3 Methodology

3.1 Modified risk register

Risk Breakdown Structure (RBS) is an approach that could be considered as a hierarchically organized depiction arranged by risk categories. Following the concept of Work Breakdown Structure (WBS), RBS provides a means for the risk manager to structure the risks being addressed or tracked. Based on RBS, risk could be built as a central repository for all risks identified by the project or organization and, for each risk, includes information such as risk probability, impact, counter-measures, and so on. Concerning the schedule risks for transportation megaprojects, three major factors are included in the proposed risk register that evaluate the risk score for transportation megaprojects: 1) probability of risk events P, 2) consequence of risk events C, and 3) urgency of risk events U. For the urgency of risk events, three sub-factors are used for evaluation including: 1) complexity of risk responsible owners X; 2) reaction time of each risk responsible party; and 3) time lasting for risk countermeasures. The modified risk register framework is demonstrated in Table 1 below for megaproject schedule risks.

Table 1. Modified risk register framework.

Risk item	Probability	Consequence	Urgency				Risk score
			Complexity	Reaction time	Countermeasure		
K_1	P_1	C_1	X_1	R_1	L_1	U_1	S_1
K_2	P_2	C_2	X_2	R_2	L_2	U_2	S_2
...	...	...	...	...	...	...	...
K_j	P_j	C_j	X_j	R_j	L_j	U_j	S_j
...	...	...	...	...	...	...	...
K_n	P_n	C_n	X_n	R_n	L_n	U_n	S_n

Note: Stands for the risk index and n stands for the total number of risks identified in RBS, where. $j \in \{1,...,\text{n}\}$ For each of the above variables, survey data is collected from project managers and engineers from respective points of view. Here, m stands for the number of surveyed experts. Datasets are represented as .

$$P_j = [p_j^1, p_j^2, ..., p_j^m], C_j = [c_j^1, c_j^2, ..., c_j^m], X_j = [x_j^1, x_j^2, ..., x_j^m], R_j = [r_j^1, r_j^2, ..., r_j^m], \ L_j = [l_j^1, l_j^2, ..., l_j^m] \cdot$$

3.2 Structural equation modeling

Structural equation modeling (SEM) technique is an extension of the general linear model that was originally developed by sociologists and psychologists. Attractive virtues of SEM include: 1) assumptions underlying the statistical analyses are testable giving the investigator full control and enabling furthering understanding of the analyses; 2) regression coefficients, means, and variances can be compared simultaneously; and 3) ability to fit non-standard models, including longitudinal data, databases with time series analysis, and databases with non-normally distributed variables [25]. In the construction management, SEM was used for describing and quantifying the influence of situational factors in project environments [26-27]. The SEM approach provides theoretical insight into how individual factors work together to enable the modeling of multiple latent variables that cannot be measured directly. Four geometric symbols are schematically used in SEM–a circle or ellipse, a square or rectangle, a single-headed arrow, and a double-headed arrow, each of which represents an important component in the analytic process. For instance, the circle means a latent variable, while the square symbol represents an observed variable. The single-headed arrow implies the impact of one variable on another. The doubled-headed arrow means a covariance or correlations between pairs of variables. Each variable is connected with another by single or double headed arrow.

Combining the SEM method with the modified risk register, the hypothesized second-order model sample of factorial structure for the SHITH schedule risk is represented in Fig. 1. The SHITH schedule risk model presented hypothesizes a priori that: 1) the sample schedule risks in SHITH can be explained by two factors S1 and S2; 2) each item-pair measure has a nonzero loading on the schedule risk factors that it was designed to measure (termed a target loading), and a zero loading on all other factors (termed non-target loadings); 3) S1 and S2 are correlated; 4) errors associated with each measure are uncorrelated. In more details, K1 and K2 are chosen to include the observed variables of (P1, C1, X1, R1, L1) and (P2, C2, X2, R2, L2). Based on the observed variables, structures of the latent variables (U1, S1) and (U2, S2) are framed. Other latent variables e1~e12 represent the measurement and residual errors. Causal effects are represented by the single-headed arrows in the path diagram. For instance, X1, R1 and L1 lead to U1, while U1, P1 and C1 lead to S1. By contrast the bidirectional arrow between S1 and S2 represents the correlations and covariance between the two risks without an explicitly defined causal direction.

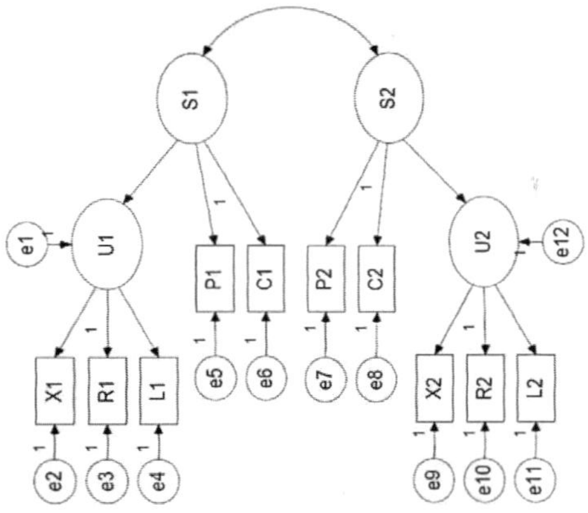

Figure 1. Example of confirmatory factor analysis measurement model for the SHITH schedule risks.

4 Case study

4.1 Modified risk register

The Shanghai Hongqiao Integrated Transport Hub (SHITH) is a super-large type of world-class transport hub that is located in west Shanghai with a total area of 26.2 square kilometers and an investment of more than $12 billion. As a major urban infrastructure project started during the eleventh "Five-Year Plan" of China (2006-2010), the SHITH serves as the transit center to

DOI: 10.1051/matecconf/20168103005

effectively connect Shanghai with other cities in the Yangtze delta area, which also supports the city's future development as an international trade, financial, shipping center. Due to the Shanghai World Expo 2010 from May to August, the SHITH project is also an important setup for facilitating the transportation service during that time. According to Fig. 2, the Shanghai Hongqiao Airport Terminal 2 SHAT2) and the Traffic Center (TC) are the two major civil projects of SHITH. For TC, there are four main subprojects: west traffic square, metro station, maglev station, and east traffic square. In general, a variety of modern transportation forms are involved, including aviation, high-speed rail lines, intercity rail lines, maglev, bus rapid transit, metro lines, etc. The railway station takes a total area of 1.3 million square meters, which is the largest railway station in Asia. Also, the Hongqiao Maglev Station is built with a scale of 10 sets and 10 lines, which is the first large maglev passenger station in China. All the facilities among the SHITH constitute as a cohesive whole. If any part was defected, the SHITH system cannot operate properly.

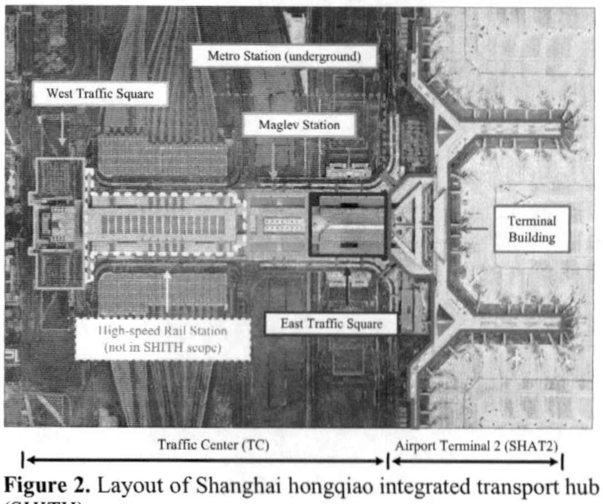

Figure 2. Layout of Shanghai hongqiao integrated transport hub (SHITH).

The overall construction work of SHITH needed to be completed within 38 months and put into public use right before the expo. In general, the SHITH project followed the original schedule plan that the SHAT2 project started in February 2007 and completed in March 2010, while the TC project started in July 2007 and completed in April 2010. Aligning with the schedule plan was thus one of the highest priorities in the SHITH project objectives. However, multiple project participants are involved making the working interfaces complex. The whole SHITH program was delivered in the form of traditional design-bid-build with 18 general contractors and 20 operating companies in total. Also, 12 different state-owned companies were involved in the co-finance activities. As a result, overlaps frequently occurred on both the construction and investment interfaces. Taking the traffic center as an example, $2.5 billion have been spent by Shanghai Rainbow Investment Company who works as the main responsible party on the civil work, while the Shanghai Shentong Metro Group was also responsible for the construction of two metro stations. Furthermore, the construction and operation work were mostly carried out by subcontracting. It is likely that the project investors had more interfaces with the subcontractors or sub-operators, and more integration problems within a project to deal with. As a result, difficulties were added in to maintain the project schedule.

4.2 Schedule risk management process

In order to address the dynamic construction progress, effective risk management is a multiple-stage process through the whole construction phase. In the proposed method, three major steps are represented in Fig. 3: 1) risk identification, 2) risk evaluation, and 3) risk response, which build up updating loops for every evaluation period. In general, both Step I Risk Identification and Step II Risk Evaluation compose the risk assessment process. Risk treatment is composed by the Step III Risk Response and information updating. Two research methods are adopted so that the expert elicitation technique is used for data collecting and the SEM method is used for data analyzing.

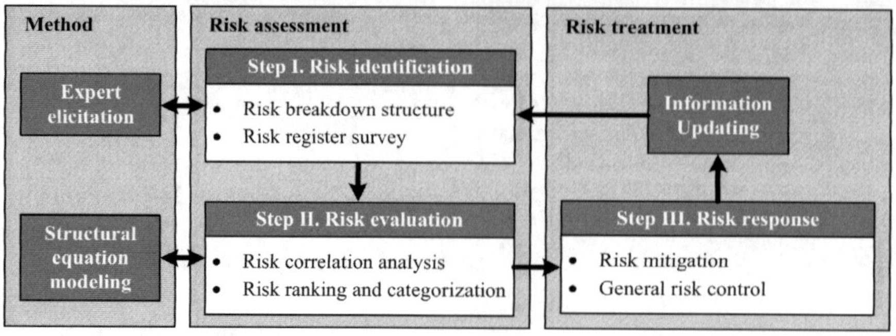

Figure 3. Conceptual framework for the SHITH schedule risk management.

4.3 Risk identification process

The schedule risk management model for transportation megaprojects starts with risk identification. RBS and risk register are built by categorizations in order to identify the potential schedule risks and quantify the corresponding characteristics for a certain evaluation period. The expert elicitation technique was adopted for building the RBS and risk register in accordance with the SHITH construction execution plans. First of all, experts involved were widely spread in order to cover the whole schedule risk opinions from different project perspectives, including officials from the engineering management and financing department of Shanghai Rainbow Investment Corp (the investor and owner of TC), department leaders from the financing, engineering, planning, IT service, and aviation sections of airport construction headquarter (airport project management department), senior project managers from construction management firm (Shanghai construction group), as well as the general contractors, designers, and supervisor relative to the SHITH construction. Other interested parties were also invited to participate in revising the drafted RBS and risk register in the last round of schedule risks investigation. Finally, the modified risk register was reported in detail to the top managerial committee of project stakeholders, such as the SHITH construction authority. Feedback was confirmed and final agreement was then reached.

The SEM program standard errors are computed under the assumption of large sample sizes. Considering the real situation of SHITH, 12 experienced experts were invited for workshop discussions and questionnaire surveys. Based on their monthly risk evaluations with repetitive but updated risk items, 40 results were collected for each item with the truncations of extra data, where m=40. All the invited experts have more than 15-year construction experience and 11 of them have participated similar large-scale projects such as Shanghai Pudong International Airport project, Guangzhou Baiyun International Airport, thus experienced with the construction, commissioning and testing work in SHITH. The estimates on the schedule risks could be deemed as relatively highly precise though subjective. Aligning with the time table of SHITH, modified risk registers were developed respectively for SHAT2 and TC according to the primary risk sources. Taking SHAT2 for example, risks were divided into five categorizations: 1) management and organization risks; 2) technique risks; 3) economic risks; 4) law risks; and 5) safety and environmental risks, similar to TC. As shown in Table 2, 30 risk items are identified based on the expert elicitation, where n=30.

For the modified risk register, each expert is required to provide the numeric values concerning the probability, consequence, and urgency of risks events in the questionnaire. In order to be consistent in the evaluation scale for all experts, a Likert Scale is used and tailored to the SHITH project situation. The definition of each level in the scales is presented in Tables 3 shown below.

The first two dimensions in risk registers are regular risk probability and consequence. In order to fit the schedule risk characteristics of transportation megaprojects, a third dimension, urgency, is added in for evaluation. Three different aspects of urgency, as the complexity of risk responsible owners, the reaction time of each responsible party, and the time lasting for risk countermeasures are classified in the register. Since the numbers of responsible parties regard risk management, they normally range from 1 to 5 in the SHITH project, which is the same number adopted as in the evaluation scale. For example, stands for only 1 party involved in construction.

Based on the proposed evaluation criteria, experts gave the opinions on five aspects of the risk events: the probability, the consequence, the complexity, the reaction time, and the time lasting for risk countermeasures. From the result, unique characteristics are presented for different risk items in the risk register.

Table 2. Rbs of shat2 with five categorizations.

Code	Risk events
1 Management and organization risks	
K_1	Temporary headquarter staffs are not dedicated to current management work for the possible personel shake-up
K_2	General project management company's insufficient supervision towards design changes due to lack of qualified reviewer
K_3	The operation company is not well adapted for the new technical management mode at the beginning of operation
K_4	General construction management contactor's poor management and coodination
K_5	Conflict in project construction and operation interface
2 Technique risks	
K_6	Issues in outer ring land buying
K_7	Lack of meteorological observatory demolition plan
K_8	North slide road acceptance work is disturbed by surrounding road pavement
K_9	Delay in internal area decoration
K_{10}	Main building is not synchronized with ancillary works in progress
K_{11}	Limited working spacingfor terminal equipment installation and decoration
K_{12}	Slow execution of construction site turnover
K_{13}	Operation Management Center (OMC) project started ate
K_{14}	Design institute site service failed to meet the project schedule requirement due to the designer's lack of experience and authority
K_{15}	Design company and industry supervision office being unfamiliar with new aviation design standard
K_{16}	Many new design changes after planed deadline
K_{17}	Lack of confirmation on terminal building decoration design scheme
K_{18}	Delay in foreign designer's decoration delivery
K_{19}	Airport terminal equipment and information system installation and debugging works are heavy and complex
K_{20}	Lack of coordination in display device installation work
K_{21}	Delay in the airport operation manager raining
K_{22}	Uncertain period for energy center and water supply system combined testing work
K_{23}	Operation company is not well adapted for the new automatic information system
3 Economic risks	
K_{24}	Contractor lengthening work period when receiving building material cost pressure
K_{25}	Project acceptance is not synchronized with other civil works in progress
4 Law risks	
K_{26}	Project design review and document approval lays behind
K_{27}	Hidden fire risk resulting from the flammable decoration materials
K_{28}	Theft construction sites
5 Safety and environment risks	
K_{29}	Hidden safety problems resulting from the above and in-ground cross construction
K_{30}	Disturbance to runway construction due to flight operations

4.4 Risk evaluation process

Based on the survey results, SEM is framed in order to get the risk score and the corresponding correlations for risk categorization and ranking. Causations and correlations among risk events are analyzed that attempt to link the certain risks to others.

Table 3. Likert scales for measurement of the risk events.

	Level	Score	Definition
Probability of risk events(Pj)	Impossible	1	Very exceptional circumstances with 20% or less chance
	Hardly	2	Unlikely to occur with 20% to 40% chance
	Possibly	3	Sometimes with 40%~60% chance
	Very possible	4	In most circumstances with 60% to 80% chance
	Almost definite	5	Almost certain to and expected to occur with 80% or greater chance
Consegience of risk events(Cj)	Very low	1	No delays or within 5days
	Low	2	Slight delay within 6-15days
	Medium	3	Moderate delay within 16-30 days
	High	4	Considerable delay within 31-60 days
	Very high	5	Serious delay with more than 61 days
Complexity of risk responsible owners(Xj)	Very simple	1	Single party involved
	Simple	2	Single parties involved
	Medium	3	Single parties involved
	Complex	4	Single parties involved
	Very Complex	5	Single parties involved
Reaction time of each responsible party(Rj)	Very short	1	No waiting time or within 2 days
	Short	2	Waiting time within 3-7days
	Medium	3	Waiting time within 8-15days
	Long	4	Waiting time within 15-29days
	Very long	5	Waiting time with more than 31 days
Time lasting for risk countermeasures(Lj)	Very short	1	No waiting time or within 2 days
	Short	2	Waiting time within 3-7days
	Medium	3	Waiting time within 8-15days
	Long	4	Waiting time within 15-29days
	Very long	5	Waiting time with more than 31 days

We state that the management and organization schedule risks model presented in Fig. 4 as an example, which hypothesizes a priori that: 1) management and organization schedule risks in SHITH can be explained by five factors: S1, S2, S3, S4 and S5; 2) each item-pair measure has a nonzero loading on the schedule risk factors that it was designed to measure (termed a target loading), and a zero loading on all other factors (termed nontarget loadings); 3) the five schedule risk factors are correlated; 4) errors associated with each measure are uncorrelated. Numerous iterations are conducted to achieve the inner correlations for schedule risk analysis with the support of expert investigation data as input. In order to illustrate how confirmatory factor analysis assists risk evaluation in detail, risk items are listed in the first section of RBS in Table 2, the category of management and organization risks, is chosen as an example. AMOS 17.0 software is used to quantify the relationship levels. In the simple case, 65 variables are arranged through the confirmatory factor analysis that including 25 observed variables and 40 latent variables, as Figure 4 shown below. Within the 40 latent variables, 35 of them are exogenous latent variables, and , and 5 of them, , are endogenous latent variables. Note that, the standardized estimated parameters are shown along with the arrow line. The final model concretizes the conceptual model that revealing the logical impacts that all observed and latent variables listed have on the risk scores.

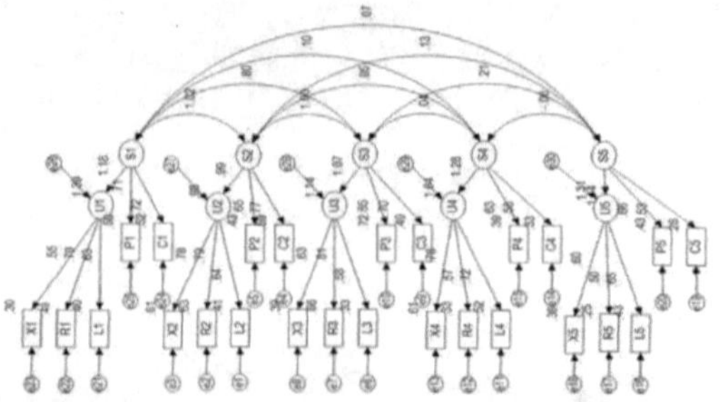

Figure 4. Output of confirmatory factor analysis measurement model for the management and organization schedule risks in SHITH.

Taking the risk item as an example, among the three observed variables having influence (coefficient = 0.70) set the most significant effect, then followed by (coefficient = 0.63) and (coefficient = 0.55). Similar to , (coefficient = 1.18) is the most significant factor setting influence, then followed by (coefficient = 0.72), and (coefficient = 0.71). The same analysis could also be applied to the risk items , and . Detailed correlation analysis is shown in Table 4 and the estimations are demonstrated to be statistically significant. Estimation of the Regression Weights/ Result of Factor Analysis.

Correlations among different risk items are also shown in Table 5. Taking the covariances of S1 and S2 as examples, the estimate is 1.065 and the standard error is 0.322. Since C.R. = 3.313 that is greater than 2, and $p<0.001$, the correlation between S1 and S2 is very significant at the 0.0001 level. Thus, risk items could be taken care at the same time. Based on the analysis, and are closely correlated, while and are relatively independent.

Table 4. Correlations among different risk items.

Latent variable		Observed	Estimate	S.E.	C.R.	P
	U1->	X1	0.994	0.260	3.822	***
	U1->	R1	0.935	0.206	4.541	***
	U1->	L1	1.000			
S1->	U1->		0.859	0.184	4.657	***
S1->		P1	0.724	0.164	4.419	***
S1->		C1	1.000			
	U2->	X2	1.297	0.281	4.622	***

	U2->	R2	1.375	0.293	4.690	***
	U2->	L2	1.000			
S2->	U2->		0.798	0.176	4.543	***
S2->		P2	0.930	0.197	4.727	***
S2->		C2	1.000			
	U3->	X3	1.006	0.296	3.402	***
	U3->	R3	1.304	0.323	4.032	***
	U3->	L3	1.000			
S3->	U3->		0.955	0.261	3.667	***
S3->		P3	1.159	0.229	5.065	***
S3->		C3	1.000			
	U4->	X4	1.166	0.248	4.712	***
	U4->	R4	0.712	0.211	3.423	***
	U4->	L4	1.000			
S4->	U4->		1.449	0.518	2.797	.005**
S4->		P4	0.890	0.241	3.688	***
S4->		C4	1.000			
	U5->	X5	0.883	0.291	3.038	.002**
	U5->	R5	0.794	0.306	2.593	.100**
	U5->	L5	1.000			
S5->	U5->		1.416	0.596	2.376	.015**
S5->		P5	0.965	0.338	2.854	.004**
S5->		C5	1.000			

Note: *p<0.1; **p<0.05; ***p<0.001. S.E. stands for the approximate standard error; C.R. stands for the critical ratio, which is the parameter estimate divided by an estimate of its standard error.

Table 5. Estimation of the Risk Score Covariance.

Latent variables			Estimate	S.E.	C.R.	P
S1	<--->	S2	1.065	0.322	3.313	***
S1	<--->	S3	0.765	0.267	2.862	0.004**
S1	<--->	S4	0.083	0.126	0.655	0.513
S1	<--->	S5	0.045	0.125	0.359	0.720
S2	<--->	S3	0.912	0.283	3.226	0.001***
S2	<--->	S4	0.037	0.121	0.306	0.760
S2	<--->	S5	0.083	0.127	0.653	0.514
S3	<--->	S4	0.026	0.108	0.243	0.808
S3	<--->	S5	0.127	0.122	1.040	0.298
S4	<--->	S5	-0.030	0.081	-0.365	0.715

Note: *p<0.1; **p<0.05; ***p<0.001. S.E. stands for the approximate standard error; C.R. stands for the critical ratio, which is the parameter estimate divided by an estimate of its standard error.

In SEM applications, the overall fitness of the structural model is also needed to assess the goodness-of-fit tests in order to demonstrate the statistical reliability. The results of the schedule risks management model are discussed below. First, a recommended level of /df, where df is the degree of freedom, is usually greater than 5. The value /df = 1469.9/260 = 5.65 in this case demonstrates a good match. P-value = 0.52 > 0.05 suggests the statistical result is no significant and the proposed model is consistent with the observed data. Chi-square test is found to be extremely sensitive to the sample size. As a result, a variety of alternative goodness-of-fit indices have been developed to supplement the chi-square statistic. The Goodness-of-Fit Index (GFI) = 0.953, the Comparative Index (CFI) = 0.976, and the Root Mean Square Error of Approximation (RMSEA) = 0.014 which are greater or less than suggested acceptable levels of >0.95, >0.9, <0.05, respectively, implying that the significance of variables to the model is augmented. Then, each cause of delay has a strong connection with its grouped class and significant influences on project schedule delay and other causes of delay.

In this case, the confirmatory factor analysis demonstrates a satisfied hypothesis testing result. However, possibilities still exist that the fitness tests might reject the original hypotheses in similar practices. In that case, the model fitting adjustment needs to be conducted. If still not satisfied, exploratory factor analysis could be conducted in order to identify potential underlying risk factors.

4.5 Risk response process

Understanding the direct or indirect relationships between risk factors would help practitioners take effective risk response actions. Based on the results of confirmatory factor analysis, several risk countermeasures have been adopted in order to address the schedule risks. From the project point of view, the corresponding risk countermeasures are conducted respectively for each risk item, and the general risk control is intended for arranging all the risk mitigation measures in an effective way. Updating for the risk management loop is conducted periodically, usually by month, and the final risk management result is thus well-controlled.

By taking the management and organization risks section from the SHAT2 risk register for an example,K1, K2 and K3 are addressed simultaneously due to their strong inner correlations. Corresponding risk countermeasures focus on doing good personnel arrangements. The airport headquarters need to pay more attention on the working staffs' personal goal achievements to make sure understanding their personal development plan and aligning to be consistent with the direction of the company and personal development arrangements. For K4 and K5 which have to be treated respectively, communications with the external interface and the general contracting office need to be maintained smoothly during the period of construction. Coordination between the main contractors keeps information flow spreading among the construction sites and, as a result, the human resources department could consider assign certain number of engineers working on site in order to identify and solve problems in time.

In general, schedule risks management results for both SHAT2 and TC are shown in Figure 5 that demonstrates the proposed risk management method could identify, evaluate and response the schedule risks effectively toward megaprojects.

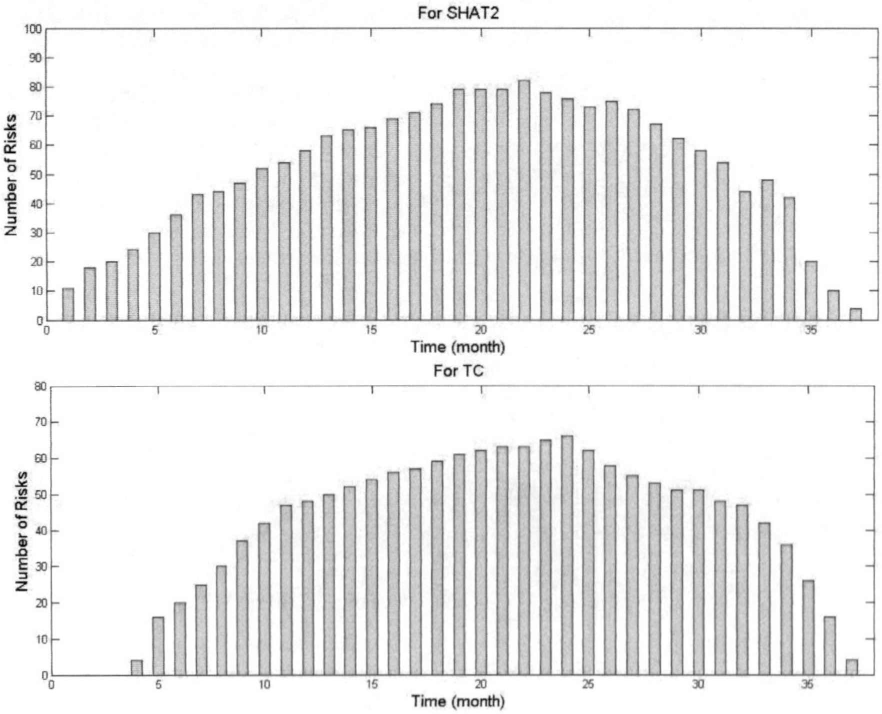

Figure 5. OutRisk control results for the Shanghai hongqiao airport terminal 2 and Traffic Center.

5 Conclusion

This study mainly focused on developing an efficient risk management method to address the schedule delays problems for transportation megaprojects. Taking the inner correlation and dependencies of megaproject schedule risks into consideration, this paper develops a statistical schedule risk management model for transportation megaprojects. Within the process, two techniques, the risk register and the structural equation modeling (SEM), are integrated for describing and quantifying the scores for each risk item. Survey results are collected from five aspects such as probability of risk events, consequence of risk events, and complexity of risk responsible owner's structure, reaction time, and time lasting for risk countermeasures. Then, two subprojects of the Shanghai Hongqiao Integrated Transport Hub (SHITH) are selected to demonstrate the applicability of the proposed model. Lessons learned from the proposed megaprojects demonstrate that the proposed risk management method could effectively identify, evaluate and response the schedule risks for megaprojects. Correlations among schedule risk characteristics are addressed, and the application simplicity of methodology is maintained.

Based on the SEM results, this study described and quantified the different aspects of schedule delay risks for transportation megaprojects. However, one limitation of this study is the relatively small sample size, as the structural equation modeling is a large sample technique. Nevertheless, this is the first study to our knowledge that examines the interrelationships existing within observable variables affecting project planning in order to understand how the individual variables work together in influencing planning effectiveness. Furthermore, despite the relatively small size of the sample, the hypothesized model provided a good explanation of how the project planning process interacts with its environment, and how these interactions influence project-planning effectiveness. Further work will focus on cross-validating the developed covariance structure model using data drawn from a larger sample, and data collected from other geographical locations.

References

1. B. Flyvbjerg, N. Bruzelius, & W. Rothengatter, *Megaprojects and risk: An anatomy of ambition*, Cambridge University Press, Cambridge, (2003)
2. M. Pacione, Dubai, Cities, **22**, no. 3, pp. 255-265, (2005)
3. K. Beck & C. Andres, *Extreme programming explained: Embrace change*, Addison-Wesley Professional, Boston, (2004)
4. M. Giezen, Keeping it simple? A case study into the advantages and disadvantages of reducing complexity in mega project planning, International Journal of Project Management, **30**, no. 7, pp. 781-790, (2012)
5. W. R. Scott, R. E. Levitt, & R. J. Orr, *Global projects: Institutional and political challenges*, Cambridge University Press, Cambridge, (2011)
6. PMI, A guide to the project management body of knowledge (5th ed.): Project Management Institute, Inc., Newtown Square, (2013)

7. H. P. Berg, Risk management: Procedures, methods and experiences, Rsik Management, **1**, no. 6, pp. 79-95, (2010)
8. KPMG, How to successfully manage your mega-project, KPMG International Cooperative, http://www.kpmg.com
9. /us/en/issuesandinsights/articlespublications/pages/how-to-manage-mega-project.aspx. Accessed March 5, (2013)
10. D. Cooper, S. Grey, G. Raymond, & P. Walker, *Project risk management guidelines: Managing risk in large projects and complex procurements*, John Wiley & Sons, Hoboken, (2005)
11. O. Renn, The role of risk perception for risk management, Reliability Engineering & System Safety, **59**, no. 1, pp. 49-62, (1998)
12. H. Drummond, MIS and illusions of control: an analysis of the risks of risk management, Journal of Information Technology, **26**, no. 4, pp. 259-267, (2011)
13. H. F. Cervone, Project risk management, OCLC Systems & Services, **22**, no. 4, pp. 256-262, (2006)
14. C. Traeger, Conducting a business impact analysis, Faulkner Information Services, Docid: 0018494, (2005)
15. Z. F. Lansdowne, Risk matrix: an approach for prioritizing risks and tracking risk mitigation progress, Proceedings of the 30th Annual Project Management Institute, Philadelphia, PA, October, pp. 10-16, (1999)
16. T. Kendrick, *Identifying and managing project risk: essential tools for failure-proofing your project*, AMACOM Div American Mgmt Assn., New York, (2003)
17. S. H. Han, D. Y. Kim, H. Kim, & W. S. Jang, A web-based integrated system for international project risk management, Automation in Construction, **17**, no. 3, pp. 342-356, (2008)
18. V. T. Luu, S. Y. Kim, N. V. Tuan, & S. O. Ogunlana, Quantifying schedule risk in construction projects using Bayesian belief networks, International Journal of Project Management, **27**, no. 1, pp. 39-50, 2009
19. P. Love, C. Sing, X. Wang, D. Edwards, & H. Odeyinka, Probability distribution fitting of schedule overruns in construction projects, Journal of the Operational Research Society, **64**, no. 8, pp. 1231-1247, (2013)
20. A. Öztaş, & Ö. Ökmen, Judgmental risk analysis process development in construction projects, Building and Environment, **40**, no. 9, pp. 1244-1254, (2005)
21. Ö. Ökmen & A. Öztaş, Construction project network evaluation with correlated schedule risk analysis model, Journal of Construction Engineering and Management, **134**, no. 1, pp. 49-63, (2008)
22. D. Schatteman, W. Herroelen, S. Van de Vonder, & A. Boone, Methodology for integrated risk management and proactive scheduling of construction projects, Journal of Construction Engineering and Management, **134**, no. 11, pp. 885-893, (2008)
23. M. Vanhoucke, On the dynamic use of project performance and schedule risk information during projecttracking, Omega, **39**, no. 4, pp. 416-426, (2011)
24. S. F. Moosavi & O. Moselhi, Schedule assessment and evaluation, Construction Research Congress 2012: Construction Challenges in a Flat World, West Lafayette, pp. 535-544, (2012)
25. G. Jia, F. Yang, G. Wang, B. Hong, & R. You, A study of mega project from a perspective of social conflict theory, International Journal of Project Management, **29**, no. 7, pp. 817-827, (2011)
26. B. M. Byrne, *Structural equation modeling with AMOS: Basic concepts, applications, and programming*, Routledge, (2013)
27. M. Mainul Islam & O. O. Faniran, Structural equation model of project planning effectiveness, Construction Management and Economics, **23**, no. 2, pp. 215-223, (2005)
28. K. Molenaar, S. Washington, & J. Diekmann, Structural equation model of construction contract dispute potential, Journal of Construction Engineering and Management, **126**, no. 4, pp. 268-277, (2000)

MATEC Web of Conferences **81**, 03006 (2016)
ICTTE 2016
DOI: 10.1051/matecconf/20168103006

Anticipatory Adaptation of Signalisation Based on Traffic Flow Forecasts within a Self-organised Traffic Control System

Matthias Sommer and Jörg Hähner

Organic Computing Group, University of Augsburg, 86150 Augsburg, Germany

Abstract. Autonomously adapting signalling strategies to changing traffic demand in urban areas have been frequently used as application scenario for self-adapting systems. Striving for the ability to cope with the dynamic behaviour of traffic and to react appropriately to unforeseen conditions, such solutions dynamically adapt the signalisation to the monitored traffic demands. The Organic Traffic Control (OTC) system is one of the most prominent representatives in this domain. OTC implements a multi-layered observer-/controller architecture. In this paper, we extend OTC's observer with a time series forecast component to create forecasts of future traffic developments for turning movements. These forecasts are then used to proactively adapt signalisation parameters. We demonstrate the benefit of the developed approach in terms of reduced travel times, and vehicle emissions within near-to-reality simulations of realistic traffic conditions from Hamburg, Germany.

1 Introduction

The vehicular traffic domain is a vivid research field, both for industrial and academic research institutions. Novel trends, such as self-driving cars [1], car-to-car communication [2], and traffic-adaptive control systems [3, 4], have the overall goal to optimise the existing traffic infrastructure towards a more efficient utilisation of road networks. However, we have to face certain negative impacts on the environment due to the increase of mobility, especially in urban areas. Consequently, this leads to an increase in pollution, a rising number of incidents, and an inefficient use of the transportation system.

Urban road networks are characterised by a large amount of signalised intersections in vicinity that need to be monitored and optimised. Due to the mathematical complexity, the online optimisation of a single signalised intersection is not feasible with analytical methods. Above all, the dynamic characteristics of traffic, the unpredictable behaviour of humans, and the highly complex dependencies between several streams throughout the road network make it an interesting and challenging field for self-adapting and self-organising solutions.

Traffic management systems can be divided into two categories: fixed-loop systems and adaptive systems. The first category relies on static time-based signalisation with a limited set of predefined signal plans for each signalised intersection. These signal plans are developed by traffic experts based on historic data. In consequence, fixed-time signal plans are not responsive to dynamic and changing traffic demands, and become outdated over time. In contrast, adaptive systems alter the signal timing based on the observed state of traffic. The traffic demands are monitored in real-time by sensors along the road, and serve as input for the signal time optimisation at each intersection. Some traffic management systems react to the traffic measured in the previous interval [4] while others adjust the signalisation proactively to the estimated traffic demand in the near future [5]. This continuous adaptation leads to a better system performance, even in case of changing traffic demands.

We propose and evaluate a highly self-adaptive technique to relieve an engineer from complex, design-time-situated tasks related to signal plan design, forecast selection, and configuration. In this work, we use time series forecasts of traffic flows to proactively adapt the signal plans at signalised intersections to current and future demands. The forecasting process follows the idea of multi-model ensemble forecasting, combining individual forecasts of several independent forecast methods.

The remainder of this paper is structured as follows. First, we provide a brief overview of the related work in the fields of traffic control and short-term traffic forecasting. We move on, presenting OTC; a fully decentralised, self-organised traffic control system. Furthermore, we introduce our forecast module for time series ensemble forecasting. We then elaborate on our approach using short-term forecasts of traffic flow to adapt the green times of traffic lights. We present the evaluation of our approach that is based on near-to-reality

DOI: 10.1051/matecconf/20168103006

simulations of an intersection in Hamburg, Germany. We conclude with a summary of our findings.

2 Related work

Our approach to predictive traffic control introduced in this paper touches a number of topics of different research disciplines.

2.1 Traffic-adaptive systems

In the last decades, several traffic-adaptive systems have emerged. Traffic-adaptive control systems such as OPAC [6], SCATS [7], and SCOOT [8] adapt green times, cycle times, and/or phase sequences according to the recently monitored traffic demands. Often, the adjustment is executed every five minutes, respectively after one to three full cycles [5]. This description of the current traffic demands can be represented by the number of vehicles passing a sensor at the respective signalised intersection, the occupancy level of a detector, or the current queue length in front of a traffic light. These responsive adaptive systems [3] adjust signal timings based on recent data collected over a certain time span (several minutes or cycles) before implementing a timing chance. This responsive process has the disadvantage that the adaptations lag behind current traffic demands [3]. In contrast, we try to omit this drawback by making forecasts of the short-term developments of traffic flow, and by adapting signal timings to future demands.

2.2 Short-term traffic forecasting

Making forecasts of future traffic developments is an important aspect of Intelligent Traffic Management Systems. Typically, forecast techniques take a time series of the recent traffic flows to forecast the estimated traffic conditions of the near future. A time series can be formalized as $ts = X_1, \dots, X_n$, where n determines the length of ts. Forecasting strives to look ahead the current value of ts and to predict the value m steps in the future X_{t+m}. There are also some techniques that rely on historical information (e.g. Daily Load Curves [9]) or a combination of both. Furthermore, there are advanced forecast techniques that allow to add more input variables such as traffic density, weather conditions, or data from adjacent roads (e.g. Kalman filters [10] or Artificial Neural Networks [11]). A collective consideration of several forecast models has proven to be a more powerful approach than just relying on one individual technique [12]. Consequently, we follow the idea of ensemble forecasting, combining the forecast of several independent techniques.

3 Architecture

This section presents the Organic Traffic Control (OTC) system for the self-optimising and self-adaptive control of signalised intersections [4]. Furthermore, we briefly introduce our forecast module for time series forecasting [13] which we later use to make short-term forecasts of traffic flows of turning movements.

3.1 Organic traffic control

Most urban road traffic networks work with fixed-time signal plans defining the green times for each turning movement. A traffic engineer has to define a signal plan for each signalised intersection. This time-consuming task needs a fair amount of expert knowledge. Additionally, these static signal plans are prone to obsolescence due to changing traffic demands [14], and cannot offer ideal signal timing for every situation and traffic demand. Learning the best situation-to-signalisation mapping replaces the problem of which technique to prefer by the question of how to describe the particular traffic condition, and how to automatically learn this mapping.

The self-organised traffic control system OTC follows this idea. OTC creates and selects signal plans matching demands without human interaction, following a reinforcement learning approach. Fig. 1 illustrates the architecture of the OTC system.

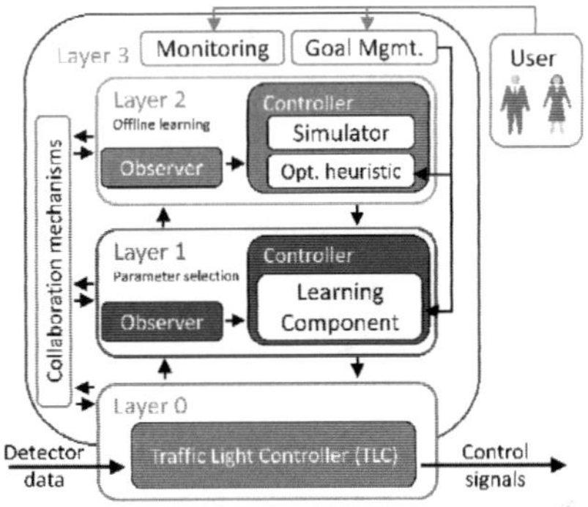

Figure 1. Multi-layered architecture of the OTC system.

An industry-standard, parametrisable traffic light controller (TLC) is extended by the OTC logic. The TLC is responsible for adapting phase durations (i.e. green times of traffic lights) according to the currently observed traffic conditions at the particular intersection. OTC works in fully decentralised mode – meaning that one individual instance of OTC deployed on a TLC controls one intersection only. Thereby, a safety-oriented self-learning concept is applied to allow for a continuous self-optimisation process, while simultaneously staying within controllable boundaries of system behaviour.

This optimisation is done in a two-step fashion. Layer 1 is responsible for the online learning process. The Layer 1 observer processes the input received from sensors at the local intersection. The observer passes a situation description of the current traffic conditions on to the Layer 1 controller which is represented by a learning component. Here, this component is realised as a learning classifier system. In case that no matching rule was found (rules resemble mappings of situation descriptions to signal plans), the offline learning process on Layer 2 is activated. Here, new rules (signal plans) are generated, simulated, and evaluated. The plan with the lowest estimated average delay is returned to the controller on Layer 1. The simulation is executed by the traffic simulator Aimsun [15], combined with an evolutionary algorithm, based on the intersection's actual topology, and the current traffic demands.

3.2 Time series forecast module

The observer's task on Layer 1 (Fig. 2) is to create a situation description of the current traffic demands of the underlying intersection. First, raw data recorded by sensors is received by the *Monitor*. Raw data is often noisy and might be subject to disturbances (e.g. because of malfunctioning sensors). Therefore, the received raw data is pre-processed, e.g. it can be filtered or smoothed. Second, the processed information is passed on to the *Situation Analyser* which builds a situation description representing the current system conditions. Therefore, the measurements from the sensors are converted into a rolling detection interval. Third, the *Performance Analyser* retrieves statistical information about the system's performance, derived from the current monitored data and the history of previous measurements. Fourth, the *Forecast Module* generates forecasts based on historical and recent data for future points in time. Several forecasting techniques each make a forecast, which are combined by a combination strategy into one comprehensive forecast. Fifth, the *Situation Descriptor* receives the current situation description, and the forecasts for future time steps, and creates one overall situation description. Finally, this description is passed on to the controller, where it serves as basis for the signal plan selection process.

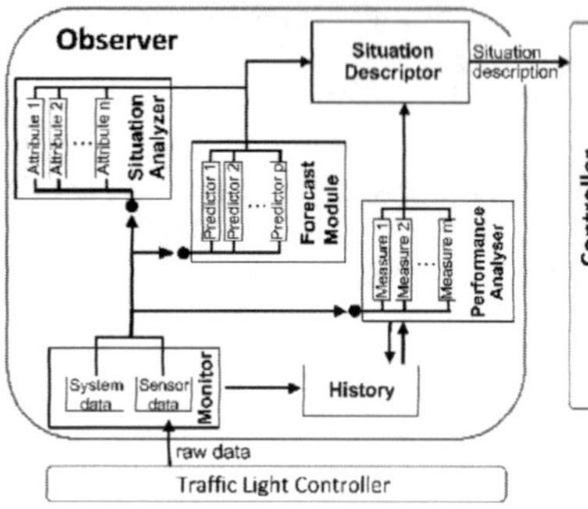

Figure 2. Detailed view of the observer on Layer 1.

4 Proactive adaptation of green times

OTC runs in cycles – meaning that a review of the current control strategy takes place at the beginning of each second or third cycle of the currently active phase plan (due to safety reasons and to allow the current strategy to show its performance). Taking into account a standard duration of such a phase plan of about 90 seconds, OTC decides about an adaptation in intervals of three to five minutes. Hence, an appropriate forecast would alter the adaptation strategy of OTC: Instead of selecting the most promising signal plan for the current traffic conditions, selecting the one for the estimated traffic demands in the near future can lead to a better anticipatory adaptation strategy. Historical and current traffic flows of turning movements are used to estimate the most probable traffic state at a certain future point in time. This process involves the following four steps:

1) Create a vector describing the current traffic demands.
2) Create short-term forecasts for future traffic demands.
3) Combine these vectors into one situation description.
4) Use the new vector to select the next signal plan.

4.1 Representation of the current traffic demands

We estimate the intersection's traffic flows based on a real-valued vector $\vec{M} = (M_1, \dots, M_n)$ containing the estimated hourly flows for each of the intersection's signal groups. We make no limitations concerning the type of detection hardware to measure these traffic flows. Fig. 3 depicts a class diagram with the entities involved in the creation of $\vec{M}$ and their relations among each other.

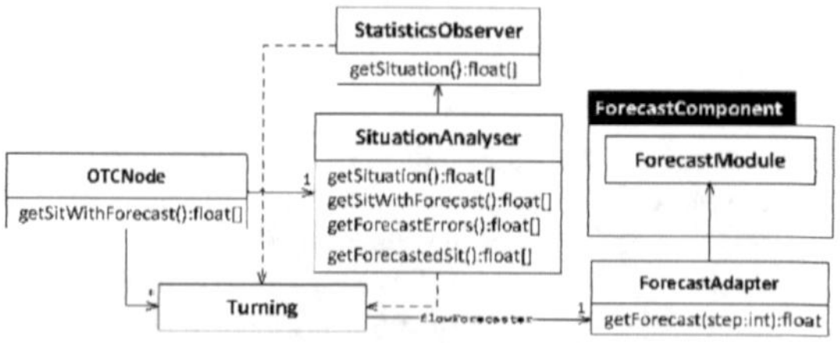

Figure 3. Class diagram showing a partial view of OTC.

An *OTCNode* resembles OTC's abstracted internal model of the physical intersection. Therefore, it knows all its related turnings and detectors, its approaching and outgoing sections, and the current signal parameters. A description of the currently monitored traffic flows for each turning of the respective intersection can be derived via the associated *SituationAnalyser*. Each turning's traffic demands are available as time series via a *StatisticsObserver* which retrieves its data from physical sensors, such as inductive loop detectors.

4.2 Forecasting the traffic flows of turning movements

Analogously to vector $\vec{M}$ describing the current traffic demands, we have to derive forecasts for each of the intersection's turning movements. Therefore, we extend each *Turning* by a *ForecastAdapter* which triggers the *ForecastModule* to generate forecasts of the future traffic flow based on time series of the recently monitored traffic flows (Fig. 3).

Our *ForecastModule* follows the idea of ensemble forecasting, resembling a weighted combination of forecasts calculated by several independent forecast methods. First, several forecasting methods P_1 to P_n independently generate forecasts based on their individual model, and historic and recently monitored data. The forecasts are generated for the point in time t, when the

DOI: 10.1051/matecconf/20168103006

next signal plan change can be executed. These forecasts F_1 to F_n are then combined based on the selected combination strategy by the *Forecast Combiner*. This module calculates the accumulated forecast as $\hat{F} = f(\vec{F},\vec{w}) = \sum_{i=1}^{n} F_i * w_i$. The according weights $w_1, \dots, w_n$ can be calculated by a basic approach, such as the simple average of all forecast values, or more complex ones, such as machine learning techniques that learn the optimal weights over time. The associated weights are managed by the *Weight Learner*. They are iteratively updated based on the forecast errors from the previous time step $t - 1$ derived from the actual value v_{t-1} and the respective forecasts. Finally, the combined forecast $\hat{F}$ is calculated based on $\vec{w}$ and the respective forecasts $\vec{F}$.

4.3 Weighted Combination of Traffic Flows and Forecasts

Having generated a vector $\vec{F} = (F_1, \dots, F_n)$ consisting of the forecast for each of the intersection's turning movements, we combine $\vec{F}$ with the current traffic flow $\vec{M}$ as

$$\vec{\mathbf{C}} = \alpha * \overrightarrow{\boldsymbol{M}_t} + (1-\alpha) * \vec{\mathbf{F}}_{\mathbf{t+m}} \quad (1)$$

One crucial point is the selection of the optimal weight factor α. This can be a static weight or it can be dynamically calculated based on the accuracy of the latest forecasts. Consequently, a performance measure is needed that defines the forecast accuracy of each individual turning's forecast module. The higher the accuracy, the more likely it is that future forecasts will be equally accurate. Therefore, higher accuracy leads to more influence of the forecasts on the combined situation vector.

Algorithm 1 depicts the combination process. The vector *sit* describes the current traffic conditions at the local intersection (one value for each turning) and *F* contains a forecast value for each turning. The forecasts are created with the previously introduced forecast framework. These two vectors are combined based on the vector of forecast errors *e*. Each forecast error is calculated independently for each turning's forecast component in a sliding window fashion with the Mean Absolute Scaled Error (MASE). The resulting MASE depicts a reliable forecast accuracy, in case its value is lower than one. The lower the MASE value, the higher the influence of the forecasts on the final combination (see lines 5 to 7). For example, if the MASE is zero for a certain turning, the current value is neglected and only the respective forecast value is used at the according position in sit_{adj}. In case the MASE is higher than a previously defined maximal error, the currently monitored sensor values are taken without adaptation (line 9). Otherwise, the forecast error accounts proportionally to the final result. Finally, the adjusted situation vector sit_{adj} is returned. The result is an adjusted traffic flow vector combining the forecast and the current traffic flow of each turning according to the estimated forecast accuracy of the respective forecast module.

Algorithm 1. Build a situation description combining traffic flow forecasts and the current situation.

Data: situation sit, forecast vector F, forecast error vector e
Result: Vector with adjusted situation sit_{adj}

```
1   sit_adj ← Ø;
2   error_max = 1.2;
3   for i = 0;i < sit.length; i++ do
4     error = |e[i]|
5     if error ≤ error_max then
6         α = error/error_max
7         sit_adj[i] = α * sit[i] + (1 − α) * F[i]
8     else
9         sit_adj[i]=sit[i]
10    end
11  end
12  return sit_adj
```

Figure 4. Algorithm 1

4.4 Representation of the current traffic demands

Based on this weighted combination of the current and the forecasts of the turning's future traffic flows, the new situation is then passed-on to the controller on Layer 1. Finally, the controller selects a new signal plan matching the input vector provided by the observer. The new signal plan is activated once the currently active signal plan's cycle is finished.

5 Evaluation

In order to demonstrate the benefit of our approach, we conduct a simulation study to evaluate the impact of traffic flow forecasts within the learning-based traffic control strategy of OTC. The evaluation is done with the traffic simulation toolkit Aimsun 8.0 [15] which is a widely used tool in the field of professional traffic engineering. We compare three approaches: a) the reference solution running the fixed-time control strategy installed in reality, b) the standard OTC system as introduced in Section 3.A, and c) the new proactive adaptation of the signal timings as introduced in Section 4. We perform three independent runs for each approach and simulation model.

5.1 Experimental setup

1) Traffic Simulation: We developed near-to-reality simulation models of urban areas in Hamburg, Germany that reflect the actual topology, traffic conditions as derived by a census, and the actual control strategies running in reality. Due to space restrictions, we only present the results of one intersection. K3 (Fig. 5) is a four-armed intersection with eleven turnings and part of the federal highway 433 connecting Alsterkrugchaussee, Deelboge, and Borsteler Chaussee. K3 has 25 signal groups, ten of which serve motorised traffic movements. Six inductive loop detectors serve as input to the traffic-actuated control.

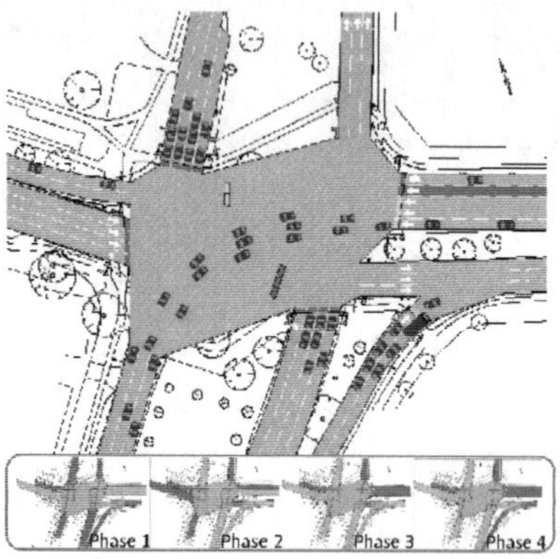

Figure 5. Aimsun's simulation model of intersection K3 at Hamburg, Germany.

The traffic data was collected at Tuesday, May 4 2004. In this census, cars and trucks passing the intersection were counted with a time resolution of 15 minutes. The simulation duration ranges from 5.30 a.m. to 12.30 p.m. including a typical morning rush hour scenario (6 a.m. to 8 a.m.). OTC starts without further knowledge, only the standard fixed-time signalisation is known in advance.

2) Traffic Demand: Fig. 6 illustrates K3's traffic profiles for three different replications with different random simulation seeds. The horizontal axis shows the time, the vertical axis depicts the traffic flow in vehicles per hour. The simulated period starts with low traffic density before it drastically raises up to 8000 vehicles per hour between 6 a.m. and 8 a.m. After 9 a.m., the traffic density slowly decreases to a medium level. The ratio of passing trucks to cars lies between 2% and 3%.

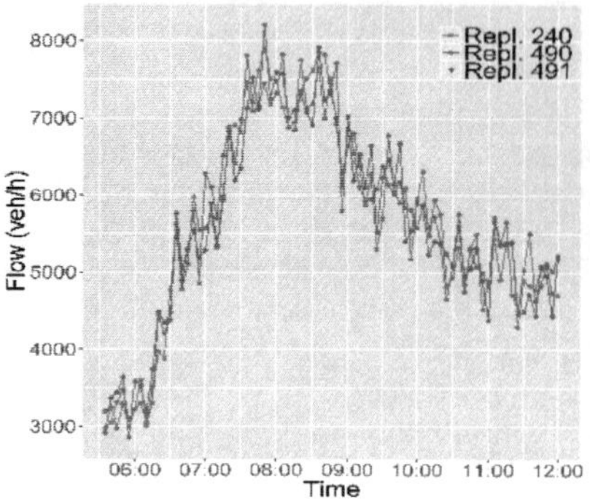

Figure 6. Vehicular traffic flow profile for K3.

3) Signal Plan: K3's simulation model (Fig. 5) depicts the real topology including the actual fixed-time signal programs that have been developed by traffic engineers. Table I shows the static fixed-time signal timings for the deployed traffic light controller which is later-on used as a reference controller in the simulation study. The signal plan has four phases and defines the green times and the interphase durations in between green phases. The duration of yellow light is three seconds, and the signal plan's cycle time is 90 seconds.

Table 1. Fixed-time signal timings in seconds (IP=Interphase, GT= Green time)

Phase 1		Phase 2		Phase 3		Phase 4			Cycle time
IP	GT	IP	GT	IP	GT	IP	GT	IP	
10	33	13	9	4	3	7	9	2	90

4) Forecast Module: Parameters for the forecast module are set as follows. The forecast methods used are Cubic Spline, Croston, Double Exponential Smoothing (DES), and Exponential Smoothing (ES). They are provided by the forecast package implemented in R [16]. They calculate their forecasts based on the last 10, 15, 15, and 10 observations. DES and ES automatically estimate their smoothing value α. Croston's α parameter is set to the default value 0.1. The simple average is used as combination strategy.

5.2 Experimental results

We evaluated our approach by means of several metrics:

- Mean queue of vehicles in front of red traffic lights.
- Average stop time in seconds per kilometre.
- Vehicle's emissions: mono-nitrogen oxide (NOx), carbon dioxide (CO_2), and particulate matter (PM).

The emission of these three pollutants has been estimated with the help of Aimsun's internal pollution emission model which is based on [17]. Each vehicle's emission is estimated based on its current speed and acceleration using non-linear multiple regression techniques. To remove random factors, the results are averaged over three independent replication runs with different seeds.

1) Forecast Accuracy: Fig. 7 presents the time plot of the actual and the forecasted traffic flow for a random turn of K3. The forecast horizon equals twice the signal plan's cycle time, here 2*90s=180s. The horizontal axis displays the simulated time step, the vertical axis the vehicular flow in vehicles per hour. Although the actual traffic flow (dotted line) shows heavy fluctuations, the forecasts (solid line) are rather precise. The time plot shows that the forecasts tend to be too low in certain situations. Usually, this happens when traffic flow decreases.

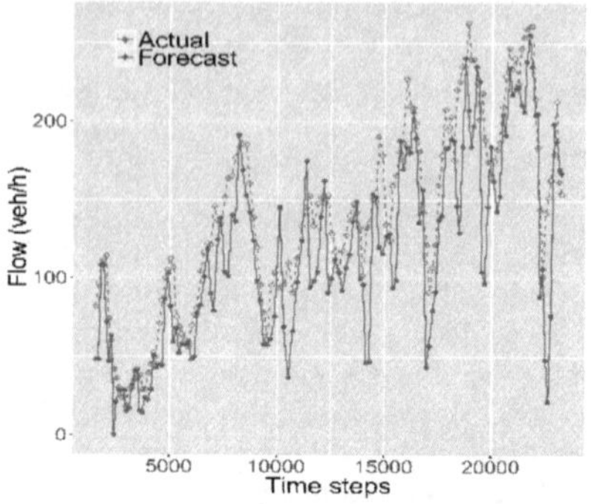

Figure 7. Actual vs. forecasted traffic flow in vehicles per hour.

The histogram of the forecast errors in Fig. 8 supports this observation. Still, this plot shows that the distribution of forecast errors is roughly centred around zero, more or

DOI: 10.1051/matecconf/20168103006

less normally distributed, and slightly skewed to the left compared to a normal curve (solid line).

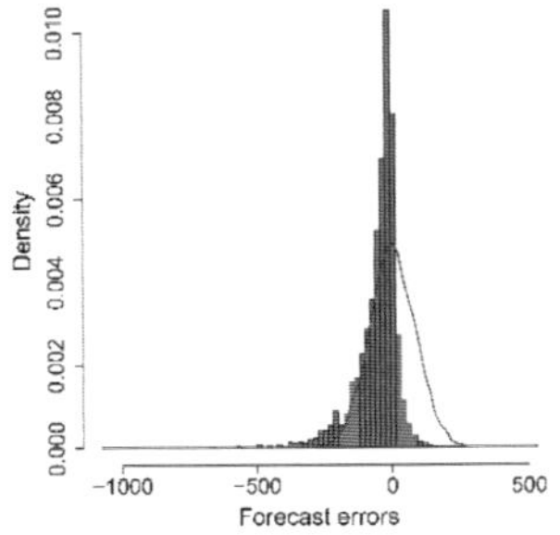

Figure 8. Forecast error histogram for 3 replications.

2) Simulation Results: The results of the simulation runs are given in Table 2. Compared to the standard OTC (fixed-time signal plan), the forecast-augmented OTC-control significantly reduces the mean queue by 5.3% (16.3%), and the average stop time by 3.4% (12.0%). Consequently, OTC with forecasts reduces vehicle emissions, such as CO_2 (FTC: 1.4%, OTC: 0.5%), and NOx (2.5%, 0.6%), having slightly higher values for PM (-2.7%,-0.6).

Table 2. Average simulation results (3 replications)

	FTC	OTC	OTC+Forec.
Mean queue [veh]	52.4	46.4	43.9
Stop time [s/km]	152.4	137.0	131.6
CO_2 [g/veh]	565.3	560.2	557.0
PM [g/veh]	0.329	0.336	0.338
NOx [g/veh]	1.60	1.57	1.56

Fig. 9 shows the time series for the total average delay in seconds per kilometre. Using the static fixed-time signal plans, vehicles are faced with an average delay of 171.7 seconds per kilometre. The use of OTC-adapted signal plans reduces this value by 8.9% to 156.3 seconds. Compared to the standard OTC (fixed-time signal plan), the additional consideration of forecasts results in further improvement by 3.4% (12.0%), reducing the average delay to 151.0 seconds. It flattens the steeply rising peak in the morning that occurs from 6 a.m. to 8 a.m. before developing to a normal level until 10 a.m. The time, until a normal level of delay is reached, can be reduced.

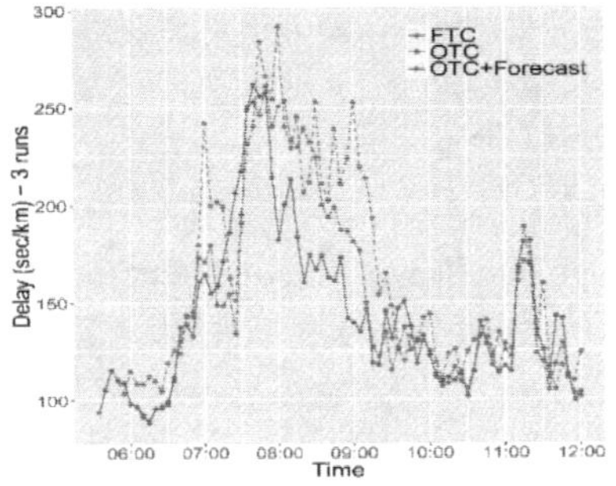

Figure 9. Average network-wide delay (3 replications).

6 Conclusions

The Organic Traffic Control (OTC) system is a self-adaptive feedback controller for adapting the green durations of traffic lights at intersections to the changing demands of traffic behaviour. We compared this approach to our forecast-augmented solution and to a real-world controller that is actually implemented in Hamburg, Germany. Using this setup, we demonstrated that our developed technique outperforms the standard OTC and the reference solution, especially in highly demanding conditions such as the morning rush-hour. This is especially promising since peak load poses the most severe challenge to traffic control.

References

1. M. Birdsall, *Google and ITE: The road ahead for self-driving cars*, ITE Journal **84**, 5, pp. 36–39 (2014)
2. N. Navet, Y. Song, F. Simonot-Lion, C. Wilwert, *Trends in automotive communication systems*, Proc. of the IEEE **93**, no. 6, pp. 1204–1223 (2005)
3. M. Selinger, L. Schmidt, *Adaptive traffic control systems in the U.S.: Updated summary and comparison*, HDR Engineering, Tech. Rep. (2010)
4. H. Prothmann, *Organic Traffic Control*. KIT Scientific Publishing (2011)
5. A. Stevanovic, National Research Council (U.S.), *Adaptive traffic control systems: domestic and foreign state of practice, Synthesis of highway practice* (2010)
6. N. H. Gartner, *OPAC Strategy for demand-responsive decentralized traffic signal control*, in Control, Comp., Communic. in Transp. (1989)
7. A. G. Sims, K. W. Dobinson, *The Sydney coordinated adaptive traffic (SCAT) system – Philosophy and benefits*, IEEE Trans. Veh. Techn. **29**, 130-137 (1980)
8. D. I. Robertson, R. D. Bretherton, *Optimizing networks of traffic signals in real time – the SCOOT method*, IEEE Trans. Veh. Technol. **40**, 11–15 (1991)
9. R. Chrobok, O. Kaumann, J. Wahle, M. Schreckenberg, *Different methods of traffic forecast based on real data*, European Journal of Operational Research **155**, 3, pp. 558–568 (2004)
10. S. Garside, K. Lindveld, and J. Whittaker, *Tracking and predicting a network traffic process*. Intern. Journal of Forecasting **13**, pp. 51–61 (1997)
11. M. S. Dougherty, M. R. Cobbett, *Short-term inter-urban traffic forecasts using neural networks*, Intern. Journal of Forecasting **13**, 1, pp. 21–31 (1997)
12. R. Adhikari, R. K. Agrawal, *Performance evaluation of weights selection schemes for linear combination of multiple forecasts*, Artif. Intell. Rev., 529-548 (2014)
13. M. Sommer, S. Tomforde, J. Hähner, D. Auer, *Learning a Dynamic Re-combination Strategy of Forecast Techniques at Runtime*, Proc. of IEEE Int. Conf. Autonomic Computing, pp. 261–266 (2015)

14. M. C. Bell, R. D. Bretherton, *Ageing of fixed-time traffic signal plans*, 2nd Int. Conf. On Road Traffic Control, pp. 77–80 (1986)
15. J. Barcelo, J. Casas, *Dynamic network simulation with AIMSUN*, Int. Symp. on Transp. Simulation. Kluwer, pp. 1–25 (2002)
16. R. J. Hyndman, Y. Khandakar, *Automatic time series forecasting: the forecast package for R*, J. Statistical Software **26**, 3, pp. 1–22 (2008)
17. L. I. Panis, S. Broekx, R. Liu, *Modelling instantaneous traffic emission and the influence of traffic speed limits*, Science of The Total Environment **371**, pp. 270–285 (2006)

Management of Truck Loading Weight: A Critical Review of the Literature and Recommended Remedies

Mozhgan Aliakbari, Sara Moridpoure

RMIT University, Melbourne, Australia

Abstract. Traffic accidents involving heavy trucks have social and economic effects on society. However, little research has focused on the influence of heavy truck specifications such as weight. Apportioning the maximum permissible gross weight of trucks allows trucking companies/owners to consolidate loads, and therefore reduce the vehicle-kilometres required to collect and distribute a given amount of goods/material. While drivers/managers are responsible for ensuring that trucks are loaded appropriately and in compliance with regulations, some may take chances and overload vehicles. This increases the need for formal and documented inspections, in order to reduce traffic hazards on public roads due to overweight loading. According to a New South Wales Centre for Road Safety report in 2014, crashes involving heavy trucks often result in serious road trauma outcomes. When a heavy truck is involved in a crash, the vehicle mass raises the crash forces involved and hence increases the severity of the crash. Therefore, interventions should be established to mitigate or prevent these crashes from occurring. Currently, weight checks are required for trucks and truck drivers must drive to a weighbridge for a weight check. Since this is a random process, truck drivers may take the risk of driving an over-loaded truck on some occasions. This paper reviews existing studies concerning safe system interventions in relation to truck gross weight management and a framework is presented to effectively manage truck loading weight. The result may be a reduction of injuries and fatalities involving heavy trucks.

1 Introduction

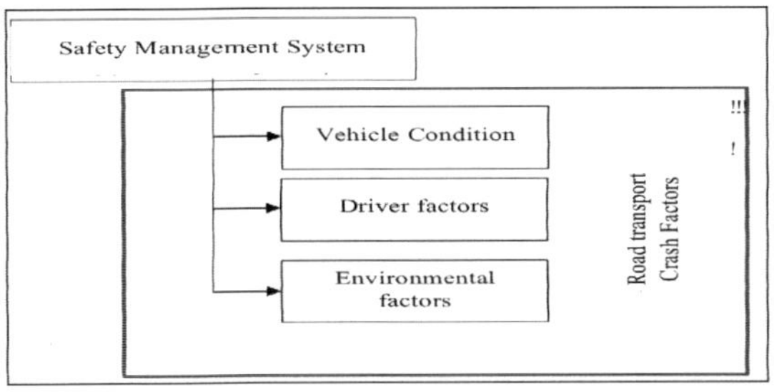

Figure 1. Inter relationship of truck crash factors

Traffic accidents are common on urban highways and rural roads in Australia, resulting in economic and social effects. The main factors responsible for traffic accidents fall into into three classes: driver performance, vehicle and equipment factors, and road and climatic conditions [1]. Fig. 1 illustrates the complex interrelationship between these factors. Vehicle and equipment factors refer to truck specifications such as weight, size and mechanical and operational failure. This is important due to the impact of truck crashes on accident costs. For instance, according to an analysis by the Bureau of Infrastructure, Transport and Regional Economic (BITRE) in 2006, human expenses (production losses) accounted for 32% of the standard cost of highway collisions [2]. Vehicle repair followed this cost at 24%, human costs (non-pecuniary) 10%, human costs (medical and others) 5% and insurance administration at 8% of the total average cost of A$ 3,180,598, including A$326,869 for truck fatalities and serious injuries [3].

2 Influence of truck gross weight loading on the severity of road crashes

The number of trucks on Australian roads has increased greatly to 3.3% of total vehicles. This increase poses safety challenges for road users in terms of possible accidents, and trucks are involved in almost 15% of all fatal accidents in Australia [1]. However, the implementation of the Safety Management System reduced the number of fatal crashes involving trucks from 276 in 2005 to 220 in 2014.

According to the aggregate fatal crash data in Australia compiled by BITRE (2014) [4], trucks in Australia are classified as either rigid or articulated (see Fig. 2).

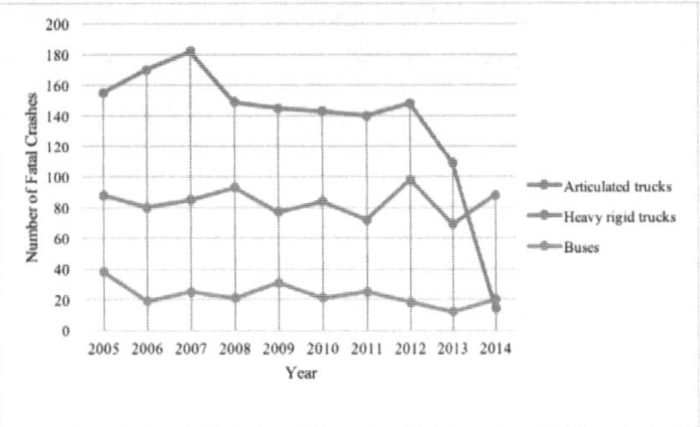

Figure 2. Fatal crashes by type of truck.

The BITRE report (2014) [4] further shows that two thirds of the articulated trucks and one third of the rigid trucks were involved in fatal accidents, as presented in Table 1.

TABLE 1. Number of Fatal Crashes by Type of Truck Involved, Australia 2005-20104

Year	Type of trucks		
	Articulated trucks	Heavy rigide trucks	Buses
2005	155	88	38
2006	170	80	19
2007	182	85	25
2008	149	93	21
2009	145	77	31
2010	143	84	21
2011	140	72	25
2012	148	98	18
2013	109	69	12
2014	14	88	20

Several factors have been cited for this phenomenon, with truck gross weight being the most significant [5]. The loading weight plays a significant role in determining the severity of traffic accidents involving trucks. Collisions involving heavy vehicles frequently lead to serious disturbances to roads, because when a heavy articulated vehicle is in an accident, the mass of the vehicle increases the accidental forces and consequently amplifies the seriousness of the collision [5]. The present crash rate in Australia is 2.5 victims per 100 million truck kilometers traveled [6]. The severity of accidents involving heavy trucks is high because of the high impact magnitude as a result of trucks' gross weight (refer to Fig. 3).

The effect of the gross load weight of trucks involved in accidents is more pronounced for long trucks with several combinations, such as tractor semi-trailers plus trailers or tractor semi-trailers alone. These truck combinations have double the crash rate of rigid trucks. For instance, there is a high fatal crash rate of 8.6% per hundred million vehicle miles for truck combinations compared to rigid trucks [7]. This shows that truck gross loading weight affects the severity of accidents.

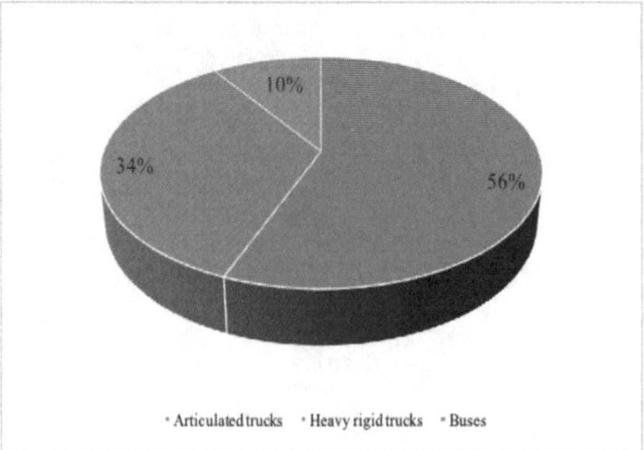

Figure 3. Percentage of fatal crashes by type of truck involved, Australia 2005-20104 [4].

Studies of the influence of truck loading weight on the severity of accidents have reported similar trends. For instance, truck load combinations involving unfilled and near-empty trucks registered relatively elevated collision participation rates with a factor of two compared to loaded trucks [7]. It is important to note that the impact of truck gross load on accidents is dependent on the exposure of unfilled trucks and their participation in these accidents [8].

Other studies elsewhere have reported low accident involvement rates for heavily-loaded trucks, with empty or partially-empty trucks reporting high accident involvement rates [7]. These findings are supported by engineering assumptions that empty trucks are dynamically less stable compared to loaded trucks and have an inherent inability to generate braking force compared to high weight trucks [6]. In addition, unloaded trucks are characterized as having longer braking distances in comparison with loaded trucks. High crash rates have been reported for empty combinations of single-unit trucks due to premature wheel lock-up and associated stability issues, because truck brakes characteristically work more efficiently under loaded conditions [8]). These findings show the relationship of increased gross truck weight to braking performance. It has been shown that truck gross weight and braking performance are inversely proportional, implying that for these trucks to attain acceptable levels of brake performance, they should have certain weight loadings, thus minimizing the severity of truck accidents [9].

In addition, the number of truck victims in Australia showed a wide variation from 2005-2010. In 2005, fatalities were 235, escalating to 272 in 2008 and reducing to 216 for the duration of the global financial crisis of 2009. The number increased further to 237 in 2010. Based on the truck type, a similar trend for crashes was observed with two-thirds articulated trucks and one-third rigid trucks accounting for these fatalities. However, fatal deaths and crashes involving trucks have remained relatively constant since 1991, with the lowest number of cases reported in 2003 [10]. Over the same period, the number of trucks on Australian roads increased by 18%

with a 37% increase in the kilometres traveled and a 70% increase in tonne-kilometres [4], as summarized in Figure 2. This phenomenon can be attributed to improvements in truck safety since 2002 (see Fig. 4)

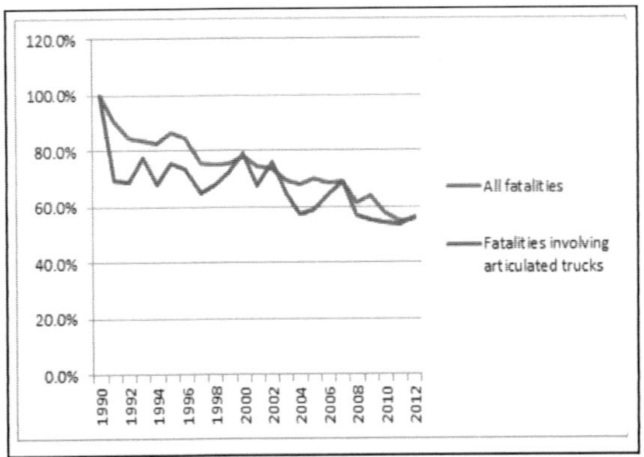

Figure 4. Total road deaths compared to those involving articulated trucks, 1986-2012 [4].

3 Existing safety system interventions with regard to truck gross weight management

3.1. The use of truck tracking technology

Technology has been adopted since the 1980s to manage the risk of overloaded trucks and to collect heavy vehicle data on highways 11]. Weigh-In Motion (WIM) sensors on highways (such as piezoelectric, bending plate, load cell, and quartz sensors) are employed to measure the weight of axle(s), and communication systems are used for data transmission to a database. The latest wave of electronic technology is telematics systems to bring more quality, effectiveness and improved ability to detect overloaded vehicles [12].

The existence of new technologies reduces the gap between the available safe system measures and measures taken to detect overloaded trucks and reduce the risk to traffic safety. However, the use of WIM is sophisticated and expensive because of Australia's vast distances, dispersed roads and highways. Furthermore, the use of telematics is disputed within the trucking industry and between the industry and the government in relation to whether it should have a regulatory function in order to make the data collected available to regulators to check industry compliance with the rules [11] [12] [13].

3.2 Government safety system interventions

To ensure road safety, the Australian Government has regulated the gross weight loading of trucks plying urban highways and rural roads. The maximum allowable truck sizes and weight limits for these trucks are specified as follows: 80,000 pounds gross weight and axle weight of up to 20,000 and 34,000 pounds for single and double axles and size limits of 102 inches (width) by 48 feet (length) [14]. To enforce the law on maximum permissible truck loads, the government through the Australian Transport Council (comprising of ministers in charge of transport from the Australian federal government, states and territories and observers from local governments) developed *the National Heavy Vehicle Safety Action Plan 2003-2005* [14] and the *National Heavy Vehicles Safety Strategy 2003-2010* [9]. The strategy complements *the National Road Safety Strategy 2001-2010* by focusing on critical impacts in an attempt to lessen deaths and injuries relating to heavy trucks. The objective of the heavy vehicle stratagy is to facilitate the implementation of a 40% mortality reduction in the national policy target, thereby reducing the Australian heavy vehicle crash rate to low global rates. To achieve this, the strategy highlights eight objectives: extra effective rate management, reduced driver injury, safer heavy vehicles, augmented seat belt employment by drivers, safe roads, effective enforcement, enhanced driver and business administration and improved investigation and education. Other safety system interventions geared towards minimizing the severity of heavy truck crashes include the adoption of modern state-of-the-art technologies such as adaptive cruise control, installation of collision warning and avoidance systems and speed limiters to help truck drivers avoid risky situations that could result in truck crashes and deaths. These complement the government regulation of the maximum and minimum gross weight of these trucks. It has been observed that regulations on the highest and lowest gross weight alone without supportive measures will not assist in reducing the number and severity of accidents arising from heavy trucks on Australian roads [15]. This combined effort will help in attaining internationally recommended and acceptable crash rates.

To effectively address the increasing number of fatal accidents in Australia, other factors affecting the safety of heavy trucks should be addressed. These include, but are not limited to, lane-departure and curve speed warnings, electronic stability control and braking, and trucks' mechanical condition in terms of brake efficiency. The government road safety strategy has addressed these factors adequately through improved maintenance of road conditions, improved compliance with existing rules and extending and refining enforcement programs. These provide additional safety for trucks [16]. To ensure the effectiveness of the road safety strategy, educational training and awareness programs play an important role, by training truck drivers on the adverse effects of not following traffic rules strictly. Drivers are also reminded periodically of the need to load their trucks within the recommended gross weight range to reduce fatal truck crashes in Australia [17].

The government of Australia through the relevant bodies has instituted rigorous roadside inspection programs with the objective of ensuring that the trucks plying both urban and rural roads adhere to various requirements such as the minimum and maximum gross weights. Secondly, these inspection programs serve to ensure that heavy trucks carry the necessary safety devices such as speed limiters and adaptive cruise control [17].

Truck design is an important safety system intervention in addressing truck crashes and deaths in

Australia due to its impact on the severity of truck collisions with other road users such as light vehicles on urban highways and rural roads [16]. Truck design affects truck stability in terms of weight distribution. In the case of an error in truck design, there will be increased instances of truck accidents [18] [27].

4 Possible remedies and framework

The Australian government through its agency the Australian Transport Council should increase its efforts in dealing with crashes and deaths involving trucks with excess loads. The current accident prevention measures in Australia focus on three issues: drivers, roads, and vehicles. For instance, in one study of major fatal truck deaths and crashes, driver fatigue accounted for 10% of crashes and speeding was responsible for 32% [18] [19]. In relation to vehicle factors, current control measures emphasize truck design, management of speed of trucks and the introduction of engineering measures such as black spot programs in addition to large-scale implementation of cost- effective measures in the heavy truck sector with the objective of improving road safety. These measures aim for the improvement of the road network all over Australia.

However, these measures have proven inadequate, since they have failed to address accidents involving heavy trucks. The main reason for the inadequacies in the current system is overlooking the impact of gross weight on crashes and deaths involving trucks [20]. This has led to the establishment of weighbridges on major highways to check whether trucks adhere to the permissible gross weights advised by the law. However, some drivers take the risk of either driving overloaded vehicles or avoiding these weight checks since in the current system, weight checking is a random exercise, allowing some overloaded trucks to go unchecked [20].

In the United States, there have been proposals on truck size and weight liberalization to eliminate the heavy five-axle tractor semi-trailers to reduce the weight of trucks in an attempt to minimize fatalities and accidents involving these trucks. This culminated in the removal of the Federal bridge formula's 80,000 pound restriction on truck loads. The removal of this cap enabled weight reduction of trucks, thus reducing truck gross loading weight and the associated accident severity [21]. In Brazil, the authorities have established minimum truck size-and-weight restrictions through compulsory weight checks with the objective of minimizing truck accidents. However, some truck drivers and owners/companies still overload their trucks, as reports show that 20% of the trucks passing through weigh stations are overloaded [20]. The existence of gaps in current laws have made them ineffective, as truck accidents have been on the rise with 60% of the trucks involved in accidents being overloaded [23]. There is a need for the authorities to formulate complementary regulations to complement these laws, thus managing truck gross weight loading and the severity of truck accidents [24].

The study of existing control measures as explained above suggests that the risk of catching an overloaded vehicle is relatively low, at least when the route is outside the areas where weigh-in-motion systems are employed or where the presence of enforcement personnel is expected [24] [25]. Therefore, for planning purposes, comprehensive traffic and weight data are essential to advance the system and collect reliable forecasts of road users both on highways and in rural areas [27]. A combined traffic data collection and enforcement regime is possible using the framework shown in Fig. 5.

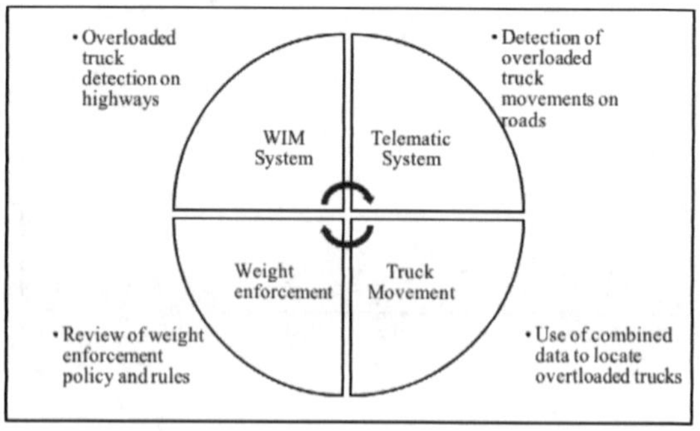

Figure 5. Overload truck management framework enforcement effectiveness.

The combination of the information collected by the traffic data collection system will lead to continuous reviewing of road traffic policies and rules on weight enforcement and make a significant contribution to road transport safety. However, the weight enforcement and traffic data collection activities may not be reduced. In fact, the overall coverage and effectiveness may significantly increase [28].

5 Conclusion

To improve the effectiveness of weight checks, the Australian government should adopt a compulsory as opposed to random formal and documented inspection process for all trucks. This will involve the certification and stamping of official documents as proof that the trucks have undergone weight checks, with the particulars of the inspection stations and authority personnel who checked these trucks, and endorsements as to whether or not the truck gross weight is within the recommended values. This should be followed by confirmation of whether or not the trucks went through weight checks on different locations along the road by confirming and checking details and particulars obtained from weigh stations as contained in relevant official documents. The Road Safety Authority should also enact laws to deal with truck drivers and owners/companies found driving overloaded trucks. This will help in managing truck gross weight loading by eliminating overloading of these trucks, hence minimizing the severity of accidents involving trucks.

References

1. Trucking Association, 2011-2013 Strategic Plan. Forrest Canberra, Australia 2603. (2010)

2. Deighton-Smith, R., Review of the Road Safety Remuneration System Final Report , Australia , pp 10-101, (2016)
3. BITRE (Bureau of Infrastructure, Transport and Regional Economics), Road crash costs in Australia 2006, Report 118. Canberra, Australia, (2009)
4. BITRE (Bureau of Infrastructure, Transport and Regional Economics), Road Deaths Australia 2011 statistical summary, Canberra , Australia, (2014)
5. New South Wales Centre for Road Safety, Road traffic crashes in New South Wales: Statistical Statement for the year ended 31 December 2013, Government of New South Wealth, Australia, (2014)
6. Driscoll, O. P. , Major Accident Investigation Report 2013.National Truck Accident Research Centre. Brisbane, Australia, (2013)
7. af Wåhlberg, Anders E. "Meta-analysis of the difference in accident risk between long and short truck configurations." Journal of Risk Research **11.3**, pp 315-333, (2008)
8. National Heavy Vehicles Safety Strategy. (2001),Retrieved January 11, 2013, from National Heavy Vehicle Safety Strategy 2003–2010, Australian Transport Council, (2001) Available www.ntc.gov.au/DocView.aspx?page=A02208316301560020
9. National Heavy Vehicle Safety Action Plan. (2001, September 3). Retrieved May 5, (2010) from Australian Transport Council, Available www.ntc.gov.au/DocView.aspx?page=A02208507300550020, 2010.
10. Wilson, L.J. and T.W. Horner, Data Analysis of Tractor- Trailer Drivers to Assess Driver's Perception of Heavy Duty Truck Ride Quality, Final Report, No.: DOT-HS•805-139~ PB-8•0-**118144**, (2009)
11. O'Brien, Eugene J., et al. "Direct measurement of dynamics in road bridges using a bridge weigh-in-motion system" pp 263-270, (2013)
12. Mikulski J. Introduction of telematics for transport. InELEKTRO, 2012, pp. 336-340. IEEE, (2012)
13. Jarossi, Linda, Anne Matteson, and J. Woodrooffe. "TRucks involved in Fatal Accident Fact Book 2006, University of Michigan Transportation Research Institute." Retrieved September 8 (2008)
14. National Heavy Vehicle Safety Action Plan. (2001, September 3). Retrieved May 5, 2010, from Australian Transport Council,(2010) Available www.ntc.gov.au/DocView.aspx?page=A02208507300550020
15. Raftery, SJ; Grigo, JAL; Woolley, JE. , "Heavy vehicle road safety: A scan of recent literature". Journal of the Australasian College of Road Safety, **22**(3), pp 18-24, (2011)
16. Scott-Parker, B., Goode, N., & Salmon, P., The driver, the road, the rules… and the rest? A systems-based approach to young driver road safety. Accident Analysis & Prevention, **74**, pp 297-305, (2015)
17. Newton, J., Howard, E., & Wishart, D. , A Guide to Applying Road Safety within a Workplace: A Bilateral Approach to Organisational Road Safety in Australia and New Zealand, (2013)
18. de Pont, J., 2014, October. Prescriptive limits or performance based standards?. In HVTT13: International Symposium on Heavy Vehicle Transport Technology, 13th, San Luis, Argentina, (2014)
19. Edgar, J., Regulating heavy vehicle safety in New Zealand using performance standards. In Fourth International Symposium on Heavy Vehicle Size and Weight. Ann Arbor, MI. International Symposium on Heavy Vehicle Weights and Dimensions. Michigan, pp. 115-119, (1995)
20. Tziotis, M., Recent Australian Road Research Board research on heavy vehicle safety, Journal of the Australasian College of Road Safety, **22,**(3), pp 22(3), (2011)
21. Woodrooffe, J., & Blower, D. , Survey of the status of truck safety: Brazil, China, Australia, and the United States, (2012)
22. Ribeiro, Sandra Furlan, and R. R. José. "Road accidents in Brazil." IATSS research 29.2 , pp 68-70, (2005)
23. Blower, D., & Woodrooffe, J. , Survey of the status of truck safety: Brazil, China, Australia, and The United States, University of Michigan Transportation Research Institute, (2012)
24. Smith, R.D, Review of the road Safety Remuneration System, Jaguar Consulting Pty Ltd ,Australia, pp10-65, (2014)
25. Sharwood, L.N.; Elkington, "Large Truck Accident Causation, "National Highway Traffic Safety Administration Technical Report No. DOT HS-806-300, (2012)
26. Franke, R., & Shinstine, D. S. , Effects of Truck Traffic on Crash Injury Severity on Rural Highways in Wyoming using Bayesian Binary Logit Models 2. Transportation Research, pp 42-43, (2014)
27. BITRE(Bureau of Infrastructure, Transport and Regional Economics), Truck productivity: sources, trends and future prospects, Report 123, Canberra, Australia Canberra, (2011)
28. Castillo-Manzano, J. I., Castro-Nuño, M., & Fageda, X., Exploring the relationship between truck load capacity and traffic accidents in the European Union. Transportation research part E: logistics and transportation review, **88**, pp 94-109, (2016)
29. Steadman, L. A., & Bryan, R. J., Cost of Road Accidents in Australia. Canberra, Australia, (2013)

Travel Mode Detection Exploiting Cellular Network Data

Arash Kalatian and Yousef Shafahi

Sharif University of Technology, Department of Civil Engineering, Tehran, Iran

Abstract. There has been growing interest in exploiting cellular network data for transportation planning purposes in recent years. In this paper, we utilize these data for determining mode of travel in the city of Shiraz, Iran. Cellular data records -including location updates in 5minute time intervals- of 300,000 users from the city of Shiraz has been collected for 40 hours in three consecutive days in a cooperation with the major telecommunications service provider of the country. Depending on the density of mobile BTS's in different zones of the city, the user location can be located within an average of 200 meters. Considering data filtering and smoothing, data preparation and converting them to comprehensible traces is a large portion of the work. A novel approach to identify stay locations is proposed and implemented in this paper. Origin-Destination matrices are then created based on trips detected, which shows acceptable consistency with current O-D matrices. Finally, Travel times for all trips of a user is estimated as the main attribute for clustering. Trips between same origin and destination zones are combined together in a group. Using K-means algorithm, records within each group are the portioned in two or three clusters, based on their travel speeds. Each cluster represents a certain mode of travel; walking, public transportation or driving a private car.

1 Introduction

Studying users' travel behaviors have always been of paramount importance for transportation planners and policy makers. Transportation modal share, which is defined as the percentage of users using a given mode of transportation between a given origin and destination, is of interest as it gives insight into utilities of different transportation facilities and their changes and trends over time. For this purpose, Surveys (including questionnaires and interviews) have conventionally been conducted to gather sample data required. Being time consuming and expensive are concomitant problems of such methods [1].

Development of location-aware technologies such as body-worn sensors, Bluetooth transceivers, satellite positioning systems and mobile phones in recent decades has led to the introduction of novel alternative approaches for gathering users' data and extracting survey parameters. Determination of Origin-destination matrices, mode of travel, travel time, route choice and activity patterns are some of the main issues that have been discussed in recent years.

Among these new methods, body worn sensors and Bluetooth transceivers need additional facilities and therefore have the lowest penetration rate. GPS-based methods, although precise, requires GPS modules to be turned on, which leads to the reduction in devices' battery usage. Hence, collecting large-scale data for transportation planning purposes passively is not possible [2].

Mobile phone based methods employ GSM signal strength, fluctuation or connections to extract users' positions and movements. No infrastructure is required to obtain records as mobile phone operators already acquire such information, often as Call Detailed Records, for their own purposes. Because of the inherent inexpensiveness and high penetration rate of these methods, this paper aims to exploit cellular network data to estimate modal split. In order to achieve this goal, origin-destination matrices are also developed and travel time between different zones is estimated using the same data.

This paper is organized as follows. Section 2 is a review of previous works on the subject. The data set and the preparation process which involves handling ping-pong handovers and smoothing data are described in Section 3 and 4 respectively, while the process to distinguish stay locations from moving locations and development of origin-destination matrices is discussed in Section 5. In Section 6, travel times and speeds between different origins and destinations are estimated and by clustering, modes of travel for number of trips is identified. Conclusions and future research plans are outlined in Section 7.

2 Related works

Most of the existing works have used GPS tracking data or GSM data along with external data obtained from additional modalities such as body sensors or accelerometers to acquire users' travel information, which can only be applied to a limited number of users [3]-[5].

Among those who have utilized cellular data information, the main attitude toward solving the problem consists of clustering/classification of users' trips based on features like travel time and speed along with additional infrastructural information to infer transportation mode of travel. In [6] for instance, GSM data were merged with real-time location of buses and taxis in the city of Rome and developed a software to monitor real-time urban mobility. The addition of location of buses and taxis helps to understand their trajectory which can be used as an additional feature to detect the mode of transportation.

Another relevant example is [1], in which authors coupled anonymized call detailed records with spatial coverage of mobile network cells and route map information of main transportation modes and inferred users' trajectory and mode of transport. For validation purposes, Kernel density path generation was used.

K-means clustering algorithm and hidden Markov model were implemented to infer transportation mode in [7]. Users' activities were smoothed out by knowledge of "normal" behavior. By using five prior states and implementing Hidden Markov Model, the most likely state of the users is predicted.

While fine-grained sampling may not always be available, in [8], the authors used coarse-grained mobile phone call detailed records to infer transportation mode. Travel times between pairs of defined origins and destinations were estimated -for weekdays and weekends separately- and travelers were clustered into three subgroups using k-means algorithm: walking, public transit and driving cars. However, the variation of travel times in different times of the day has not been considered by the authors. In addition, since clustering requires sufficient data to be meaningful, trips within origins and destinations with a small number of users traveling in between, have been neglected.

3 Dataset description

Anonymized spatiotemporal data of 300,000 mobile phone users in the city of Shiraz, Iran have been collected for approximately 40 hours in three consecutive weekdays, in a cooperation with Hamrah-Aval, Iran's leading telecommunication service provider. The data set covers about 1/3 of Shiraz's mobile phone users.

Unlike Call Detailed Records Data, in which location of each user is updated when the user makes a call, sends an SMS, connects to internet or changes cell area, records for each individual user is updated in 5-minute time intervals. Thus, the data set consists of about 144 million records, each containing anonymous user ID, time of the day and coordination (Latitude and Longitude) of the transceiver the user is connected to.

For this research, 24-hour records of the second day were extracted as the main data. Other records can be used for validation purposes. In order to have a strong focus on trips within the urban area zones, users whose cell phones were never turned off or dropped signal from 6 a.m. to 9 p.m., and spent the whole period in urban areas of Shiraz were extracted. About 50,000 users fulfilled the criteria.

4 Data preparation process

After the extraction of users meeting the criteria defined previously, anomalies were detected and removed or replaced. In the second step in smoothing out data, bouncing of cells back and force between two or three base stations while the user is stationary, was handled. This phenomenon, known as ping-pong handover effect, occurs because of fluctuations in the received signal strength and causes changing in cells IDs while the user is not actually moving [9]. An algorithm introduced in [10] was modified slightly and implemented to distinguish movements due to ping-pong H/O effect from actual mobilities. The slightly modified algorithm is as follows:

Step 1- A cell graph is defined for each user with vertices of all observed GSM cells. There is an edge between two vertices if and only if a transition occurs between them.

Step 2- Any two or three vertices set a cell cluster if and only if:

A) They form a connected subgraph. This makes sure all cells in a cluster are adjacent to each other.

B) The average time spent at the cluster is larger than the sum of the individual average times spent at each cell within the cluster. This condition is met only if there is a cell fluctuation in the cluster.

Step 3- The clusters with cell(s) in common are merged together.

Step 4- Location of each cluster is represented by the weighted average of its consisting cells coordinations. The weight of each cell coordination is set according to the time spent by the user at that cell, divided by total cluster time.

The above algorithm is at first implemented for two-way ping-pong handovers, meaning that only clusters containing two cells are detected. The algorithm is repeated for the output data of the previous implementation until no significant change occurs in the cell graphs. The next stage would be the detection of three-way ping-pong handovers, which takes into account clusters including three cells. About 65% of movements (not trips) were detected as a result of ping-pong H/O effect. Which shows the necessity of replacing them with reasonable records, weighted average of corresponding cells here. Fig. 1 shows a three-way ping pong H/O effect detected by the algorithm for sample user: the user is bouncing back and force between points A, B, and C. these points are clustered together and replaced with point S, as indicated by a star sign.

DOI: 10.1051/matecconf/20168103008

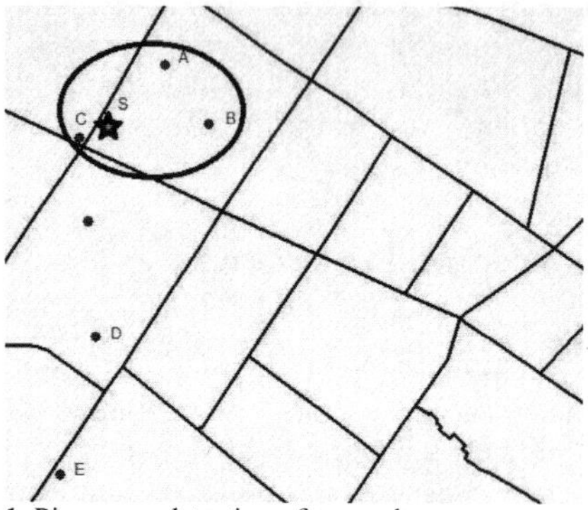

Figure 1. Ping-pong detection of a sample user.

5 Origin-destination matrices

5.1 Stay location detection

In order to recognize trips between different origins and destinations for a user, it is required to know whether a record is an activity point (or stay location, including home, work, school, shopping, etc.) or an on-the-path point (a point user passed by during her trip). For this purpose, a method is proposed as follows:

Step 1- The geographic coverage area of all Shiraz urban area cells is estimated by Voronoi tessellation method. The locations of transceivers are considered as the generating points for the tessellation. There are 544 transceiver antennas in Shiraz urban area, partitioning the urban area into 544 distinct zones. Fig. 2 depicts part of Shiraz traffic zones (thick lines) along with BTS [a] antenna's cover area (thin lines), created with ArcMap Software.

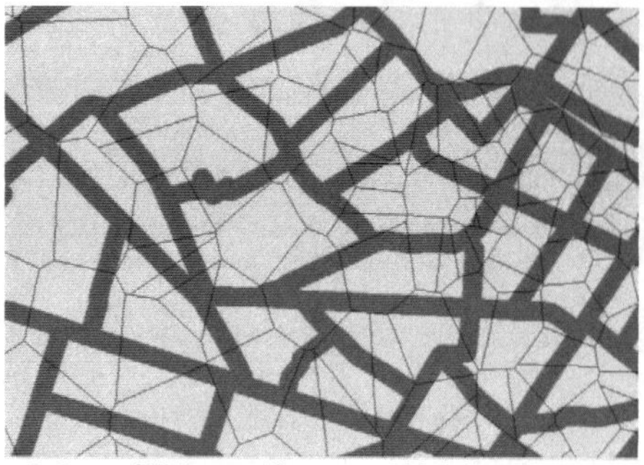

Figure 2. Part of Shiraz traffic zones (thick lines) and transceivers cover area (thin lines).

Step 2- If the largest dimension of the corresponding cell's cover area is smaller than the length the user can cover when she walks during time t, it can be inferred that the user has been stationary during that time. In other words, for a given user u that spends t minutes consecutively at location l in cover area v, if the following condition is true, l is marked as a stay location:

$$t * S_{walk} > D_v \quad (1)$$

In which S_{walk} is the walking speed, approximately 80 meters/minute according to HCM[b] and D_v is the largest diameter of cover area v. When a user meets this condition, even if she walks, the cover area would change. Not changing means that the user has remained still.

It must be noted that (1) is a sufficient but not necessary condition for a record to be marked as stay location. Other methods must be taken into account to identify other records moving/stationary status

Step 3- Trips are investigated with origins and destinations extracted from Step 2, trips with similar origin and destination locations are spotted and converted to two trips, with the furthest point in trajectory marked as destination and origin for each trip respectively. This will make sure that no user makes a trip without a destination, and separate origin and destination pairs are assigned to all trips.

Implementing the methods mentioned above, users' statuses can be identified. The average number of trips per day for all users is computed to be 1.81 trips/day/person (neglecting intrazonal trips), which is acceptable compared to Shiraz City Transportation Studies statistics [11].

5.2 Studying trips

Trips between different traffic analysis zones of Shiraz were aggregated to build the origin-destination matrices for different hours of the day. For each user, a stay location that the user spent at least 3 hours during night hours is marked as home. Each stay location acts as the origin for one trip and the destination for another one, except for those recognized as home.

Total number of inter-zonal trips for each hour during the day is depicted in Fig. 3. As shown in the figure, the A.M. peak hour in Shiraz occurs between 11 A.M. and 12 P.M. and the P.M peak hour occurs between 5 P.M. and 6 P.M., both consistent with previous transportation studies of the city. A relative peak can be observed between 7 A.M and 8 A.M., which is also similar to the one in studies.

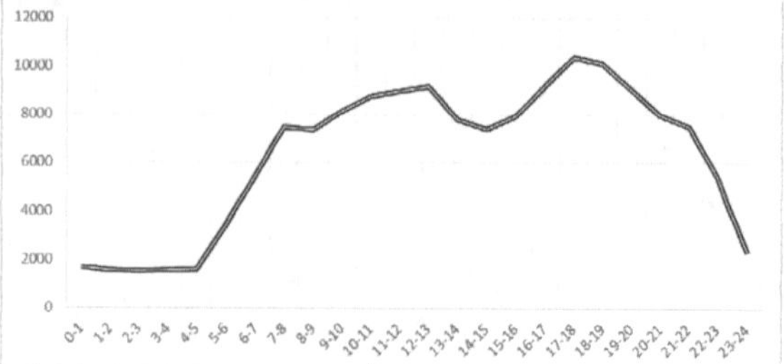

Figure 3. Total number of inter-zonal trips per hour.

6 Determining mode of travel

Our goal is to infer transportation mode of each trip by estimating speed. A first approach is to combine trips between origin O and destination D in hour h in one

[a] Base Transceiver Station

[b] Highway Capacity Manual, 2010

group and cluster this group into three subgroups: walking, driving a private car and public transportation (mainly buses in Shiraz). There will be a group for every origin-destination pairs in every hour. All conditions of trips within a group are similar together that travel speed can be used to determine the mode of the travel. The drawback of this approach is that not enough trips occur between any origin and destination in every hour to partition the records into clusters. The maximum number of trips between an origin and destination in the peak hour appears to be less than 100. More than 95% of origin-destination-hours do not include sufficient records to be clustered.

A second approach to solving the problem is to neglect the effect of time of the day on traffic speed and combine all the records of trips between origin O and destination D for all 24 hours together. Records from different weekdays can also be added as done in [8]. The problem with this approach is the fact that travel speed during different hours of the day varies significantly, especially in central congested zones. This approach may be applicable to trips on not congested roads.

To overcome these issues, we combined trips occurring at similar hours of the day together. Travel speeds for different modes of transport in these similar hours are assumed to be comparable. Take time period of 4 P.M to 9 P.M for instance; all trips from TAZ[c] 88 to TAZ 64 of the city of Shiraz occurring in this time period are merged together in a group. We then partition this group into clusters each representing a certain mode of travel. There is a total of 128 trips in this group.

Using K-means unsupervised learning algorithm, this group is divided based on travel speed into two clusters. The distance between these two zones is less than 2 km at the furthest points, which makes it possible for users to walk between these zones. The centers of clusters are 5km/h and 17 km/h and 29 km/h, representing walking and public transport and private car respectively. The size for each mode of transport is estimated to be 6, 29 and 93, which means the modal share of 4.5% for walking, 22.4% for public transit and 72.1% for driving. The method can be applied to several origin-destination pairs. Although there still remains a number of trips undecided, due to the insufficient number of similar trips.

7 Conclusions and future work

In this paper, we tried to convert big cellular network data to comprehensible transportation trajectories of users. After handling the inherent anomalies of real GSM data, stay locations were detected and origin-destination matrices for different hours of the day were developed. We also proposed a method to estimate modal share. Although we reached sufficient records for clustering in some origin-destination pairs, the majority of trips does not belong to these pairs and therefore their mode of travel cannot be estimated. As this paper presents some parts of an ongoing research, mode detection methods will be extended in several aspects in future works.

A proposed method being conducted for future works is to group trips similar to each other in some defined features together. These defined features are adding a feature to distinguish trips within CBD from other trips and trips between zones that public transit covers from other trips, adding economic level of origin zone, traffic congestion among available routes, etc. these features are assumed to be factors affecting mode choice and travel speed, thus, travel times and speeds of trips with similar features would be more comparable. By taking these features into consideration, a major number of trips would be covered. In the final stage, results would be compared to available transportation statistics.

Another problem that may lead to a decrease in accuracy is assigning each stay location to one TAZ. Some zones, especially in central areas of the city, are fairly small that assigning a BTS Antenna to them may be a reasonable choice. For other zones, this assumption would cause errors due to the large area. To solve this issue, some novel approaches are proposed and being worked on, defining membership functions for TAZs.

As travel speeds for public transit and private cars do not differ significantly especially in congested zones, to better distinguish between driving by vehicle and by bus, an android application is being developed for bus drivers of different bus lines of the city of Shiraz to capture the traces to identify their trajectory. The proximity of the user to these trajectories in different hours can be used as an attribute for detecting trips through public transit. Transportation mode share between different origins and destinations can be inferred thereafter using a large amount of data that can be acquired from cellular network data. Validation of the method is done by comparing the results with surveyed data collected from the city.

Acknowledgement

The authors would like to gratefully acknowledge the support of Hamrah-Avval for providing the dataset used in this research. Colleagues from Shiraz Municipality are also gratefully acknowledged for providing geographical region polygons.

References

1. J. Doyle, P. Hung, D. Kelly, S. McLoone, R. Farrell, *Irish Signals and Systems Conference*, (2011)
2. S. Timothy, A. Varshavsky, A. LaMarca, M.Y. Chen, T. Choudhury, I. Smith, S. Consolvo, J. Hightower, W. G. Griswold, E. De Lara. UbiComp 2006: Ubiquitous Computing, pp. 212-224, (2006)
3. S. Reddy, M. Mun, J. Burke, D. Estrin, M. Hansen, M. Srivastava. ACM TOSN, **6**, no. 2 (2010)
4. L. Stenneth, O. Wolfson, P. S. Yu, B. Xu. *ACM SIGSPATIAL International Conference on Advances in Geographic Information Systems*, pp. 54-63 (2011)
5. H. Xia, Y. Qiao, J. Jian, Y. Chang. Sensors. **14**, no. 11 (2014)

[c]Traffic Analysis Zone

6. F. Calabrese, M. Colonna, P. Lovisolo, D. Parata, C. Ratti. IEEE ITS Transactions, **12**, no. 1, pp. 141-151 (2011)
7. I. Anderson, H. Muller, *ISWC*, pp. 127-128 (2006)
8. H. Wang, F. Calabrese, G. Di Lorenzo, C. Ratti. *IEEE ITSC,* **13**, pp. 318-323 (2010)
9. T. Tettamanti, Z.A. Milacski, A. Lorincz, I. Varga, Periodica Polytechnica. Transportation Engineering, **43**, no. 2 (2015)
10. K. Laasonen, M. Raento, H. Toivonen. Pervasive Computing, pp. 287-304 (2004)

DOI: 10.1051/matecconf/20168103009

Route Choice Estimation Using Cell Phone Data

Homa Taghipour and Yousef Shafahi

Sharif University of Technology, Civil Engineering Department, Tehran, Iran

Abstract. Nowadays development of cell phone network provides huge and ubiquitous data, with wide application in transportation science. One of the most important advantages of these kinds of data is enabling the process of collecting information without any active users' interference. A big data set consisting of 300,000 cell phone users' information in Shiraz are studied. This data set includes spatiotemporal information of travelers for every 5 minutes in a time span of 40 hours in two consecutive days. The spatial part of each user's information contains the position of the BTS (Base Transceiver Station) to which his cell phone is currently connected. Due to the existence of outliers, it is necessary to smooth the data initially. One of the main reasons of errors in the data set is ping pong handover, which leads to false transitions and must be eliminated. After the data preparation, stay locations are determined for each user and a trajectory for each pair of origin and destination is estimated. At this step based on network information of the city, a method to match trajectories with the network map is applied. Finally the obtained results indicate whether travelers choose the shortest path or other possible alternatives.

1 Introduction

Nowadays the advancement of technologies and cell phone network expansion can provide a massive and ubiquitous database. This comprehensive data set that belongs to the majority of people in society, yields a wide application in transportation science. Knowing the trajectories that are chosen by travelers, is really important and useful for practice in certain transportation planning steps, such as traffic assignment. According to the Wardrop's first principle of equilibrium, the travel times in all routes, which are used, are equal and also less than those routes that are not used. Utilizing this principal is completely common in traffic assignment step. However, based on observations and recent researches, this equilibrium does not necessarily occur and people sometimes decide to choose longer trajectories because of a couple of reasons such as the number of stops in a route, the quality of route roads and even the route aesthetic.

There are lots of recent researches conducted in this field, but the major deficiency existed in most previous researches is the lack of sufficient data. The main goal of this research is human trajectory estimation between a pair of origin and destination by using numerous cell phone data information and the result shows that how travelers choose their routes.

There are some kinds of methods to collect data in order to estimate travelers' trajectories. Although one of the most popular methods is using GPS database, collecting these kinds of data encounters some problems. The need for direct cooperation of travelers is one of the main problems, furthermore using of cell phones equipped with GPS is not popular among people and also GPS consumes battery charge of cell phones severely. Therefore, GPS data utilization cannot provide a data set that covers the majority of people travel information. The other common method to collect data is gathering questionnaire which has some imperfections. The high cost of collecting data, the possibility of information inaccuracies and the extremely time-consuming process of preparing data are some of the imperfections in questionnaire gathering method.

In this research frequent spatiotemporal cell phone data set is employed. One of the most important advantages of these kinds of data is enabling the process of collecting information to be accomplished passively and without any active interference of the users. Moreover, a large amount of data, which are attainable, lead to high accuracy. It is also easily possible to collect these appropriate data in a short duration. On the other side, cell phone data include some inaccuracies and errors that must be corrected.

After the data preparation, stay locations are determined for each user and a trajectory for each pair of origin and destination is estimated. At this step based on network information of the city, a method to match trajectories with the network map is applied. Finally the obtained results indicate whether travelers choose the shortest path or other possible alternatives.

2 Literature review

Researchers believe that it is possible to extract travel information from cell phone data, but accuracy and quality of the data depends on the network conditions such as density, situation and power of BTSs (Base Transceiver Station) in the network. There are several researches which study on estimating travel information using cell phone data. In this section some of these related researches are explained briefly.

The appropriate data dramatically affect the outcome of these kinds of researches. Most researchers had access to the event-based cellular data. These kinds of data contain the special information of mobile users at the moments that they are connected to the cellular network, by making a call, sending a SMS or using internet and sometimes when their BTS zone is changed. Authors in [1] used the event-based cellular data collected by Airsage [2]. In this research the spatial information was calculated by triangulation method by means of the Airsage's Wireless Signal Extraction technology to reach more accurate and precise data. The researchers selected the information of those users who had more than 1000 connections during a given day. This selection is biased because of getting limited. In this fashion, only information of special group of society who are more active on using cell phone is acquired. The event-based data are also used in [3]-[7], etc.

Removing errors from data plays a significant role in gaining better results. Three methods to detect outliers from a set of cellular data in order to improve traffic modeling, are discussed by authors in [5]. In this research, the transition between two points is defined as an outlier, if the transition speed is more than 250 Km/h. Using the recursive naive filter, the look-ahead filter and the Kalman filter are the mentioned three methods, each of which is advisable for a special purpose.

The most challenging part of cellular data preparation is to eliminate the false result caused by ping pong handover, which is discussed in many researches. Authors in [8] introduced an algorithm to detect and remove ping pong handover by defining cell graph and cell cluster. Researchers in [9] used a moving average method to smooth out high frequency ping pong handover. Authors in [10] defined the repetitive transitions between two or three antennas as ping pong handover, put those antennas in a cluster and did not count the internal transitions. Researchers in [11] introduced a method by defining a set of recent antennas and a ping pong counter, to filter false transitions generated by ping pong phenomena. Authors in [12] applied an almost similar method to remove ping pong phenomena.

In order to determine the stay locations, each research use its own method. As an instance, authors in [6] determined the stay locations by defining the spatiotemporal threshold. In other words, in a sequence of data records, a point which is closer than one kilometer to its last point and its time stamp is more than 1 hour after its last point, is a stay location.

Author in [13] at map matching step used Hidden Markov Model to match a sequence of points to map, instead of matching them one by one, in order to increase the accuracy. Authors in [1] divided users into sedentary and commuter categories and calculated the error between real trajectory and mobile data trajectory using an interpolation method. Authors in [6] after an initial data smoothing, at map matching step used a method to detect some probable trajectories between each pair of origin and destination based on local transportation network. Afterward, authors selected the closest one to the cellular data trajectory as the correct trajectory. Authors in [4] proposed a method to estimate trajectories, using the information of high probability intermediate areas and weighting streets based on their coverage. They showed that travelers do not choose the shortest path consistently. In order to match trajectory to transportation map, authors in [14] applied a multi-path route generation algorithm to find the set of plausible trajectories and then they chose the best one by using Needleman-Wunsch algorithm. The goal of researchers in [7] was not gaining the maximum likelihood by finding the most likely trajectory. They generated set of possible trajectories based on the right probability distribution. Authors in [3] to map match the trajectories, made a set of probable transportation network trajectories by incremental assignment between each pair of origin and destination, then chose the best trajectory which had the shortest squared deviations between each point of mobile data trajectory to that network trajectory.

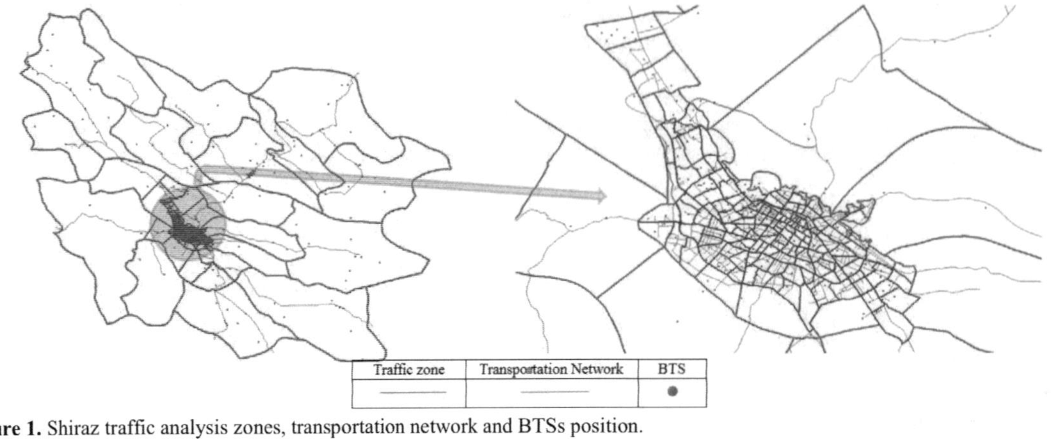

Figure 1. Shiraz traffic analysis zones, transportation network and BTSs position.

3 Data

A big data set consisting of 300,000 cell phone users' information in Shiraz city of Iran are studied to achieve travelers' trajectories between each pair of origin and destination. Shiraz is the fifth most populated city of Iran, with the population of about 1.5 million. The city divided into 183 traffic zones and there are 484 BTSs throughout the city, which are denser in the central zones. The figure 1 depict the study zone. In the most related researches usually cell phone data are recorded at the moment that a person uses his cell phone, in the other word cell phone data consist of phone events, such as phone calls, short messages and connecting to the internet. However, the data set that have been used in this research include frequent periodic spatiotemporal information of travelers for every five minutes in a time span of 40 hours in two consecutive days. The spatial part of each user's information contains the position of the BTS to which his cell phone is currently connected. Due to the existence of outliers, it is necessary to smooth the data initially. One of the main reasons of errors in the data set is caused by ping pong handover, which leads to false transitions and must be eliminated.

4 Methodology

In this section, the successive steps of estimating travelers' trajectories are discussed. The first step of this procedure is to prepare the data. After eliminating all the errors existed in the data, the stay locations are determined. Eventually, by using the BTSs trajectories, it is possible to achieve the main goal of this research. Final step is to map match the BTSs trajectories and reach the correct transportation trajectories. In the following steps, three fundamental assumptions have been applied:

1. The signaling power of all BTSs are the same as each other.

2. Each cell phone connect to the nearest BTS around it.

3. The disruptive effect of spatial objects like tall buildings on power of signaling is neglected.

4.1 Data preparation

These kinds of data are usually error prone and need to get prepared before utilization. Initially, some records which have favorable characteristics are selected. The selected data contain records of the travelers which were entirely in Shiraz city zones in the studied day and their spatiotemporal information in every five minutes of the day are available. Afterward, the raw primary data are converted to the spatial information of the selected travelers in addition to the stay duration of each location.

Commonly, cell phones are consecutively connected to some different BTSs, even if the user of that cell phone is immobile, it is called ping pong handover and this phenomena leads to false movements. At this step it seems necessary to look for a solution in order to modify the ping pong handover. It is possible to detect ping pong transitions by using suggested algorithm of Hong and Kim [11]. Based on mentioned algorithm, a transition to j is a ping pong handover if conditions (1) and (2) meet:

$$j \in Bn_u \tag{1}$$

$$pn_u = pn_{max} \tag{2}$$

When Bn_u is the set of three recent BTSs to which user "u" is connected at its n_uth transition and pn_u is ping pong counter which takes measure in what follows:

0	if	$1B_n(j) = 0$	and	$pn_{u-1} = 0$
$pn_{u-1} - 1$	if	$1B_n(j) = 0$	and	$0 < pn_{u-1}$
$pn_{u-1} + 1$	if	$1B_n(j) = 1$	and	$pn_{u-1} < pn_{max}$
pn_{max}	if	$1B_n(j) = 1$	and	$pn_{u-1} = pn_{max}$

The pn_{max} is the maximum value of pn_u which is equal to 2 based on this research conditions, and the indicator function of $1A_x$ is 0 if $x \notin A$ and 1 if $x \in A$.

After ping pong handover detection, those locations which the ping pong transition occurs between them, are replaced with an expected new location. The expected new location is the weighted average of initial locations, and can be calculated by weighting the locations by the stay duration of each one. Finally, at the end of this step, the BTSs trajectories of the travelers are extracted. A BTSs trajectory is the total of straight lines between each two BTSs that a traveler is connected to them.

4.2 Stay location extraction

Stay locations are the places which travelers have a stop there for a special purpose. These purposes might be going to home, work, shopping, entertainment, etc. In this section the process of determining the stay locations is discussed. One location is defined as a stay location if a traveler's cell phone remain connected to a BTS more than 20 minutes and this duration has been chosen due to the Shiraz local network conditions. Sometimes a cell phone is connected to a special BTS in two consecutive records and the distance between that BTS to its adjacent BTSs is short enough that if the traveler is moving even in walking speed, his cell phone can be connected to the other BTS easily. In these situations the stays are defined as stay locations too.

4.3 Estimationing the trajectory between a pair of origin and destination

After determining the stay locations, for each traveler, there are some stay locations as origins and destinations of his travels and some intermediate points between them. Now it is the time to estimate the network trajectory that each traveler choose to transfer from an origin to a destination.

4.4 Map matching

Conspicuously, cellular data are inaccurate and one of the most important steps to make them qualified enough to use is map matching. Some kinds of data like GPS data approximately give the information of cell phone exact location on the network, but cellular data can give just the position of the BTS to which the cell phone is connected.

So converting BTSs locations to actual locations is really necessary. There are lots of approaches in order to meet this goal. Some of them are categorized into simple search techniques such as point-to-point matching and point-to-curve matching and some of them are the complex techniques and use the theories like probability theory and fuzzy logic theory. These methods are discussed in [15]. In this paper the result of map matching process is to extract network trajectory from BTSs trajectory. The approach is to find some trajectory options initially and then choose the best one as the final transportation trajectory.

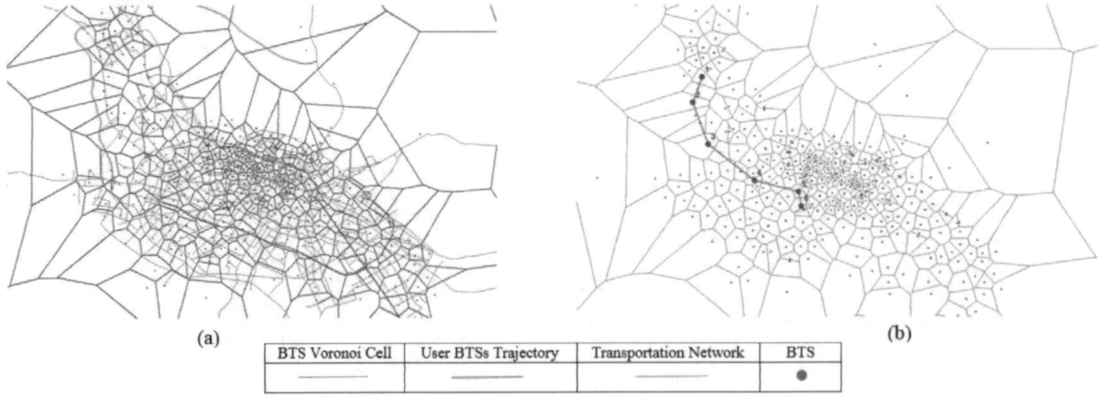

Figure 2. (a) BTSs Voronoi cells, (b) a sample user BTSs trajectory.

In order to determine some trajectory options for each pair of origin and destination, a method almost like the one discussing in [6] is applied. Firstly, it is essential to divide the whole study zone into voronoi cells using BTSs as seeds and keep these cell's characteristics in a set named "V". Voronoi cell is the region consisting of all points which are closer to its seed than the other seeds. At the next step all transportation network links of the study zone are kept in another set named "t". After preparation of these sets, it is possible to make a new set for each transportation network link, named "Vt" and put the information of all cells intersected by that link in the set. Then perform the same process for each BTSs trajectory, in other words, for each BTSs trajectory create a set named "Vb" and store the information of all cells intersected by the BTSs trajectory in that set. Hence, for each BTSs trajectory with the determined Vb set, the selected links to estimate the correct transportation trajectory, are all the t links which there is at least one common element in their Vt set and the Vb set, it means $Vt \cap Vb \neq \emptyset$.

At last, the correct transportation trajectory must be chosen from the set of candidate trajectories for each pair of origin and destination. To reach this goal, total distance between the BTSs trajectory and each trajectory options is calculated and the trajectory with the shortest distance is the correct final transportation trajectory. The total distance is the sum of distances between the intermediate points of BTSs trajectory and each transportation trajectory.

5 Result

This section illustrates the above mentioned methodology that is used to determine the transportation trajectory for one trip of a random user. After preparing data and removing all the aforesaid errors, the methodology is performed by utilizing GIS software. The divided study zone into voronoi cells is stored in the V set. The Shiraz transportation network is also kept in the t set. At this step it is possible to estimate the Vt sets for each transportation network link, using the characteristics of V and t sets, as illustrated in Figure 2.a. For a random user, a trip between two stay locations is considered, this trip comprises six points: one origin, four intermediate points and a destination. The points are obtained after the process of error eliminating, such as ping pong handover. The BTSs trajectory is indicated in Figure 2.b. The Vb set is created by the BTSs trajectory and V set.

Therefore, those transportation links which are candidate to make the set of probable trajectories, based on the methodology, are selected as shown in Figure 3.a. Finally the most acceptable transportation trajectory that is the closest one to the BTSs trajectory is determined. Figure 3.b. indicates the ultimate trajectory for the mentioned trip.

6 Conclusion

In this paper some algorithms have exploited to estimate the transportation trajectory of cellphone users. At first process of data preparation which is really essential to get a better result is discussed. After that, stay locations are determined. Then, map matching is applied to reach the correct transportation trajectory. This paper explicates the fundamental part of research and result can be used in other related researches. Finding trajectories for a group of users between a pair of origin and destination using periodic cellular data might be the subject of next researches.

An appropriate data plays a significant role to reach the goal. A repetitive, periodic and large dataset can easily increase the accuracy and preciseness of the result. Some practical applications of this research is in traffic controlling, urban planning, etc.

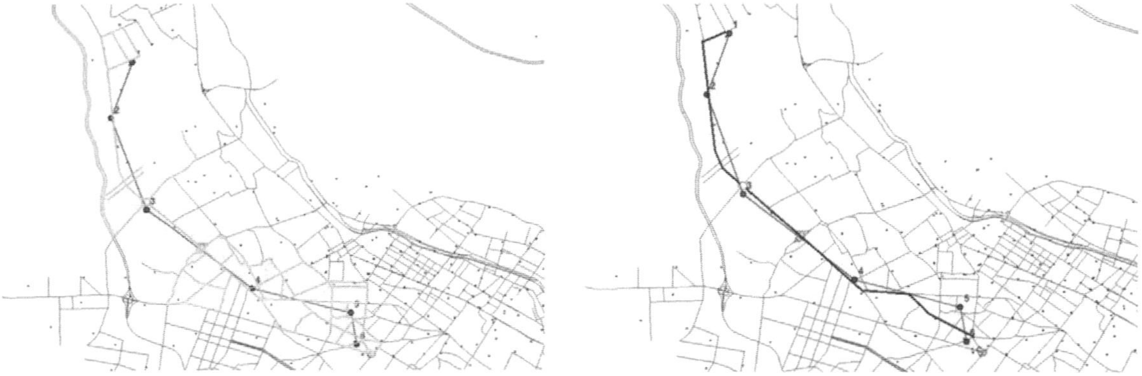

(a) (b)

Candidate Transportation Link	Final Transportation Trajectory	User BTSs Trajectory	Transportation Network	BTS
				●

Figure 3. (a) Candidate transportation links and (b) final transportation trajectory (for sample BTSs trajectory).

References

1. S. Hoteit, S. Secci, S. Sobolevsky, C. Ratti, G. Pujolle, Comput. Netw. **64,** 296-307 (2014)
2. http://www.airsage.com
3. T. Tettamanti, H. Demeter, I. Varga, Acta polytech. Hung. **9**, 207-220 (2012)
4. I. Leontiadis, A. Lima, H. Kwak, R. Stanojevic, D. Wetherall, K. Papagiannaki, *Emerging Networking Experiments and Technologies*, 121-132 (2014)
5. C. Horn, S. Klampfl, M. Cik, T. Reiter, TRR. **2405**, 49-56 (2014)
6. M. Dash, K. Kiat Koo, T. Holleczek, G.Yap, Sh.P. Krishnaswamy, A. Shi-Nash, *Mobile Data Management*, **1**, 243-250 (2015)
7. T. Vajakas, J. Vajakas, R. Lillemets, Int. J. Geogr. Inf. Sci. **29**, 1941-1954 (2015)
8. K. Laasonen, M. Raento, H. Toivonen, Pervasive. Comput., 287-304 (2004)
9. J. Yoon, B.D. Noble, M. Liu, M. Kim, *Mobile systems, applications and services*, 177-190 (2006)
10. J. Lee, J.C. Hou. *Mobile ad hoc networking and computing*, 85-96 (2006)
11. J. Hong, H. Kim, *Computer Communications and Networks,* 1-6 (2009)
12. B. Nunes, K. Obraczka, *World of Wireless, Mobile and Multimedia Networks*, 1-6 (2011)
13. A. Thiagarajan, *PhD diss., Massachusetts Institute of Technology,* (2011)
14. J. Schlaich, T. Otterstätter, M. Friedrich, *transportation research board of the national academies*, (2010)
15. M.A. Quddus, W.Y. Ochieng, R.B. Noland, Transp. Res. Part C. Emerg. Technol. **15**, 312-328 (2007)

MATEC Web of Conferences **81**, 04001 (2016) DOI: 10.1051/matecconf/20168104001
ICTTE 2016

Research on Aerospace Cooperative Continuation Observation Strategy for Maritime Moving Target

Liang Xingxing, Chen Chao, Li Jin, Zhang Mingxing, Cheng Gaungquan and Xiu Baoxin

Science and Technology on Information Systems Engineering Laboratory, National University of Defense Technology, Changsha 410073, P. R. China.

Abstract. Maritime moving targets continuation observation has important implications to maintain the safety of the navigation barrier. Existing methods either rely on satellite observations, or only rely on UAV observations. Even though the two cooperative observations, it confined to a simple task allocation. They did not solve the problem of maritime moving targets continuation observation. On the basis of difference between satellites observation and UAV observation, we constructed the aerospace collaborative continuation observation model for maritime moving target, and put forward aerospace collaborative continuation observation strategy (ACCOS) to reduce the complexity of solving the model. According to the spatial and temporal characteristics of the task, ACCOS extracted five sub problems to solve the model: Satellite planning, UAV flight plan, UAV observation sequence, target potential area and distribution probability density prediction, UAV path planning. In order to realize the optimization goal of the model, the five sub problems are modelled and solved in turn. Finally, the simulation results show that the proposed method can greatly reduce the average observation period of the target, and effectively solve the problem of the continuous observation of maritime moving target.

1 Introduction

Due to the limitation of the horizon, it is difficult to continuously detect the long-range maritime moving target. Only using the upper unmanned aerial vehicle (UAV) observation, it will lead to blindly trap into the search for the lack of a priori information; only rely on satellites, it will get a cover on the target about less 20% observation time. Also, optical satellite usually is been affected by meteorological conditions, light illumination condition; and electronic satellite affected by the outside world is weak, but the positioning accuracy is poor, need to cooperate with other observation equipment.

There are many researches about the satellite observations of the moving target, for example, to study the various algorithms for moving target search by satellite [1], satellite observation area construction [2], the key technology of the multi satellite cooperation of observation planning [3] and so on. Relevant research to solve the problem of more fully, but the study itself is only limited to similar satellite search or to a stationary target observation, to only complete the target observation. They could not complete the continuous near real time observation of maritime moving targets.

Compared with fixed orbit satellite observation, UAV's observation effect is more flexible and better. It can be regarded as an important supplementary resources of satellite. At present, many research focus on the UAV real-time online planning. UAV research focuses on multi UAV cooperative search method description [4], multi UAV search path optimization [5-6], the exchange of information between planning and search of multi UAV [7-9] and so on. These research made the assumption for the distribution of the targets that designated search area at first, and then find regional goals. This model is relatively simple, and could not effectively reflect and solve the real problem. For the sea area which is far from the mainland, existing detection method mostly use other means to obtain the target location, and then sent UAV to closely observe, the specific process is to estimate the potential area after finding the target, then to closely detect the targets. The existing methods are only qualitative analysis, and there is no appropriate model to quantify the analysis.

In conclusion, UAV and satellite observations have advantages and disadvantages, but the two can complement each other. Scholars put forward the multi-platform cooperative evolutionary algorithm, which is used to divide the observation mission [10]; to set up a space cooperation service planning model, the analysis of the aerospace collaborative work principle [11-12]; the key technologies of collaborative of the space for the detailed description [13-14]. But such research also only stay in the assignment of scheduling, they complete the regional division of the platform through the corresponding means, and on this basis, finish task planning of every platform. But their research target is static, and has not formed the true meaning of the joint.

For continuous observation of maritime moving target, this paper constructs the collaborative space continuation observation model, puts forward the aerospace collaborative continuation observation strategy (ACCOS) to solve the model. On the basis of temporal characteristics of tasks, there is five sub problem solving models from ACCOS: Satellite planning, UAV flight plan, UAV sequence, target potential area and distribution probability density prediction, UAV path planning. After solving this five problems, getting the observation window of the satellite observation sequence set, UAV observation sequence, path planning results and trigger event time series. From the result, it reduced the average observation period of observation platform, solving the problem of continuous observation of maritime moving target.

The structure of the paper is as follows. In section II of the paper, maritime moving targets continuation observation model is provided, along with input, output, constraints and optimization objective. Section III outlines the aerospace collaborative continuation observation strategy. Several case studies are illustrated in Section IV, along with their results. Conclusions and future work are then provided in section V.

2 Problem description and model

2.1 Problem description

The problem is described as continuous observation of maritime moving target. The target moves from single platform observation area to multi-platform observation area. The satellite observation information couldn't meet the needs of the target observations, need to send UAV to the target area, together with the satellite to complete continuous observation of moving targets. Therefore, the problem can be described for the moving target set Tar and satellites, according to the actual intention, send UAV to the target's potential regions to closely observe. In the process of observation, satellite sent observed target information to guide UAV, UAV found the optimal observation path according to the optimization objective, to complete the task of continuous observation for the targets.

2.2 Model modelling

Maritime moving targets continuation observation is a complex system engineering. The process involves multiple entity unit, output and constraints. So in this paper, an aerospace collaborative continuation observation model is made to describe the complex problem. Some definition is shown as follow.

2.2.1 Input

$Tar=\{tar_{1},\cdots,tar_{N}\}$: Observation target sets;

(Lon_{Base},Lat_{Base}) : Coordinate of UAV base;

(P_i,V_i) : Position and speed vector of target tar_i;

$Plat(Sat,UAV)$: Observation platform, and $Sat=\{sat_1,\cdots sat_{N_{sat}}\}$ is satellite resource set, $UAV=\{uav_1,\cdots uav_{N_{uav}}\}$ is UAV resource set;

$Vis_{sat_a-i}=\bigcup win_{sat_{\alpha k}-i}$: The observation windows set of satellite sat_α to target tar_i, ordered by time, $win_{sat_{\alpha k}-i}=(ws,we)$ is the start and end time of once observation.

2.2.2 Output

The target's observation windows and location precision set from all satellites:

$$SatVis_i=\{(win_{ik},pre_{ik})\mid k=1,2,\cdots\} \tag{1}$$

The observation sequence of uav_i:

$$Seq_i=\{tar_{i1},\cdots tar_{ik},\cdots tar_{i1}\} \tag{2}$$

The observation path of target tar_i:

$$R_i=(P_1,P_2,\cdots,P_n) \tag{3}$$

The trigger event time series:

$$EVENT=\{(time_i,event_i)\mid i\in 1,2,\cdots\} \tag{4}$$

$time_i$ is occurrence time of event $event_i$.

2.2.3 Constraints and optimization objective

Optimization objective for the continuous observation of maritime moving targets:

$$\min\overline{T}=\frac{\sum_{i=1}^{N}T_i}{N} \tag{5}$$

Constraints is:

$$P_i\notin Area \tag{6}$$

$$\alpha_{sat_i}<\max\alpha_{sat_i} \tag{7}$$

$$\sum_{k=1}^{N_{UAV}}X_i^k=1,x_i^\alpha=\{0,1\} \tag{8}$$

$$\sum_{k\in Seq_i}L(P_k,P_{k+1})<flyD \tag{9}$$

$$\frac{\sum_{k\in Seq_i}L(P_k,P_{k+1})}{flyV}<flytime \tag{10}$$

α_{sat_i} is the sensor's angle of optical satellite; $\max\alpha_{sat_i}$ is maximum sensor's angle of optical satellite; $L(P_k,P_{k+1})$ is the distance between position P_k and

P_{k+1} ; $flyD$ is maximum cruising distance of UAV; $flyV$ is maximum cruising speed of UAV; $flytime$ is maximum cruising time of UAV.

Optimization objectives ensure target the average observation period of target is the shortest; Constraints (6) expressed in the UAV into the search, the target cannot enter one's own space; (7) said the satellite's swing angle can not exceed the maximum swing angle; and (8) said each target needs to be observed and belong to a UAV to observe; (9) said that the UAV cannot exceed the farthest cruising range; (10) said UAV cannot exceed the maximum flight time.

3 Operational research

Aerospace collaborative continuation observation model regards maritime moving target continuation observation problem as mathematical model, but the model itself is complex to solve. In order to overcome this problem, this paper proposes aerospace collaborative continuation observation strategy (ACCOS) to reduce the complexity of the model's solution. This section divided into two parts to study ACCOS: first is the overall aerospace collaborative design framework; the other is the specific description of ACCOS.

3.1 Overall design

Under aerospace collaborative maritime moving targets continuation observation involves multiple element, including the satellite, UAV, control center and so on. A complete task decision process usually is made by observe, orient, decide and act, namely the OODA. As time passing by, it constitutes a dynamic cycle. On a maritime moving target continuation observation also follow the OODA process. The design of the overall framework as shown in Fig. 1.

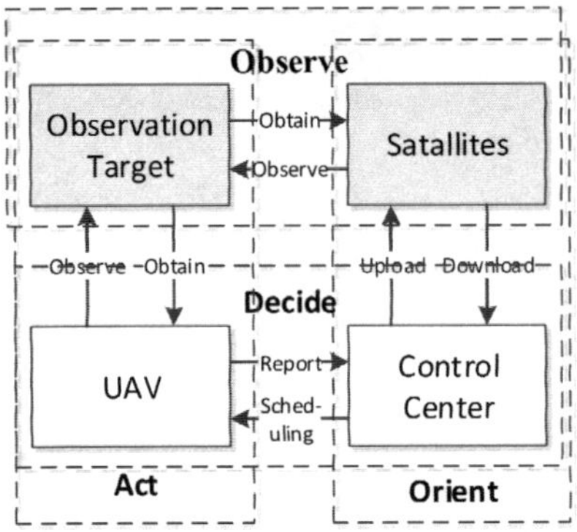

Figure 1. Overall design framework.

Fig. 1 describes the overall framework of maritime moving target continuation observation tasks by aerospace collaborative: 1. electronic observation satellite and optical imaging satellite observe the observation area, transferring the regional information timely to the control center; 2. After satellite's observation information is sent to control center, it identifies suspicious target within the region, analyzes the target intention, judges whether adjust the satellite attitude. According to the relevant constraints, it is necessary to decide whether the UAV should fly closely observation area; 3. Control center send UAV to observation area, setting the UAV track, and planning online; 4. According to the instruction information, UAV fly to the designated location. The cycle is a dynamic closed-loop process, effectively solving maritime moving targets continuation observation problem.

3.2 Detailed design

The aerospace collaborative overall design framework is given for maritime moving target continuation observation in last section, describing the information interaction relation between the various units, but have not yet specifically described strategies for each unit, and couldn't give the clear method to solve the model. Therefore, there is detailed analysis of the design for the ACCOS in this section.

3.2.1 Strategy description

ACCOS describes the strategy of various elements on the task timeline, that is, what the problem is, what strategy is been used.

The specific description of ACCOS is as follows.

(1) First, electronic satellite is used to detect the target. If the target is in the observation window, it detects the target and returns the result; otherwise, it continues to wait, and re-start the task.

(2) Secondly, satellite plans is made, according to the type of the next satellite. If it is optical satellite, we will analyse the obtained information, getting next arrival time interval and target potential area, calculating the swing angle of the optical imaging satellite attitude; otherwise, it continues to wait, and re-start the task.

(3) Next, analyzing the situation is essential while making satellite plans. If the analysis result shows that it is necessary to send UAV, the UAV flight plan should be made, including the number of sending UAV and UAV base; otherwise, it continues to wait, and re-start the task.

(4) When we decided to send the UAV, the UAV flight path should be modified according to satellite observations result. If UAV will arrive the target area, it is necessary to analyze the UAV observation sequence analysis; on the contrary, it is well to keep the cruise state and modify the UAV flight path.

(5) After finishing analyzing the UAV observation sequence, we need to determine whether the UAV is near the target. If it is, it is time to predict the distribution probability density of the target area; if not, it is ok to maintain the cruise state, and revise the route.

(6) Finally, when the forecast is over, UAV turns to the search for the target. When there is target which is still not observed in the observation sequence, UAV will turn to the next target observation; when all targets have been observed, UAV turns into the tracking observation, and repeats prediction - search mode.

3.2.2 Strategy implementation process description

ACCOS describes a complete mission design: from the beginning of the satellite observation, to UAV turning into the target tracking observation.

ACCOS contains three planning: satellite-control centre planning (SCP), control centre -UAV planning (CUP), UAV autonomous planning (UAP). SCP contains the interaction between satellites and control centre; CUP contains the interaction between control centre and UAV; UAP shows UAV autonomous planning.

Along with the time axis, ACCOS decomposes problems and reduces the complexity of the model solution. According to the design of the ACCOS, there is five sub problems in the three kinds of planning: satellite planning, UAV flight plan, UAV observation sequence, potential target regions and the distribution probability density , UAV path planning.

(1) Satellite planning. Satellite planning is the beginning of the task, containing many kinds of satellites and control centre, and each satellite observation window is the starting point of the decision. The process specifically described as follows: electronic satellite finds the target in a large range observation area at first; when finding the target, control centre value the target's observation value, to determine whether optical imaging satellite plan to observe this goal, getting satellite observation result. The planning result of this stage is to get the observation window of the satellite to the target.

(2) UAV flight plan. UAV flight plan is the beginning of the UAV planning, containing control centre and UAV. The distance between the target and the base is far away. When UAV flew near the target, the target distribution has been changed. So the control centre needs to decide how much UAV should been send and gives the UAV cruise path. The results of this stage is the number of UAV and the cruise routes planning.

(3) UAV observation sequence. UAV observation sequence is the beginning of UAV searching. In UAV cruise, control centre constantly revises UAV's flight path according to the received information. When the UAV begin to carry out their tasks, control centre will start planning UAV observation sequence containing target assignment and observation sequence planning: the former divides the target into group according to the number of UAV, and the latter plans to get the observation sequence of each UAV for the assigned target. The result of this stage is the result of the target assignment and observation sequence of UAV.

(4) Target potential area and distribution probability density prediction. Getting potential target regions and the distribution probability density is the beginning of UAV autonomous planning. With the cooperation of the satellite, UVA make a reasonable prediction of the target distribution using the received information. The result of this stage is to get potential target regions and the distribution probability density.

(5) UAV path planning. According to the target distribution information predicted by the previous stage, UAV obtains the observation path of the target combining with the optimization goal. The result of this stage is to get UAV path planning for each UAV.

4 Simulation example and result analysis

At present, less research is in aerospace collaborative continuation observation, while more research is in the collaborative task scheduling. So there is no recognized benchmark test problem set to compare the algorithm running results. In order to verify the applicable and feasible of the method, we uses the STK to set a scene, four sets same orbit altitude electronic satellite and four groups same orbital altitude optical imaging satellite, while the amount of UAV is sufficient.

Algorithm used Visual Studio C + + to achieve and run in microcomputer which is Intel Core 3.40GHz, 4G of memory; Using DBSCAN clustering algorithm with radius parameter $\varepsilon = 3km$ and neighborhood density threshold $MinPts = 1$; target assignment is based on UAV number in the task by K-means clustering; single UAV observation sequence and actual flight path calculation used genetic algorithm with the crossover probability that is 0.8 and the mutation probability that is 0.2 calculated by several times; when the maximum evolution algebra or converge to the global optimum is reached, the algorithm stops.

In simulation process of this experiment, control center decided to send three UAVs based on the observation information, cooperating with the satellite to finish target continuation observation. When the satellite observes the target, it update the information to the UAV. Specific simulation response event schedule is shown in Table 1:

Table 1. The simulation events.

Event	Time(s)	Description
Target run into area	2200	Based on Constraints
Send UAV	2200	Based on DBSCAN
Satellite's Observation	6254	Optical satellite
Target assignability	6637	Based on K-means
Satellite's Observation	8928	Electronic satellite
UAV1 turns into tracing	9462	None
UAV2 turns into tracing	9500	None
UAV0 turns into tracing	9700	None

Table 2. The satellite observation ability.

Number	Observation Window	Duration Time	Positioning Accuracy
1	2181.9-3142.6	960.753	0
3	3363.3-3764.3	400.996	0
6	6254.2-6637.9	383.707	1
1,8	8928.3-9624.0	695.679	0.07
5,3	9632.6-10051.2	418.575	0.41
7	12252.9-12634.7	381.886	1
3	16511.9-17188.4	676.512	0
2	19822.0-20615.2	793.277	0
2	26412.2-27359.6	947.407	0
Average Duration Time		639.57s	
Average Interval Time		2425s	
Average Positioning Accuracy		0.33	

In this paper, we use multiple types of satellite observations for the target, and the duration is about 26400s (7h), from the beginning of satellite observation to track the target by UAV. The observation results are shown in Table 2. Number 1-4 represents satellite electronic observation satellite, and 5-8 means optical imaging satellite; Observation Window is said that target could be observed by satellites in this time; Duration Time represents the size of the observation window; Positioning Accuracy means satellite's capacity where optical is 1 and electronic is 0.

Through the analysis of the results in Table 2, we could see the lack prediction of satellite observations. By only relying on satellites, coverage of satellite observation is only 20.8%. In this experiment, Target's speed is 50km / h. In the blank space of observation, target's movement range is 1122.2km2, and moving target range is very wide. Therefore, it is essential to use UAV cooperating with satellite for achieving target continuation observation.

After UAV taking-off, according to the target distribution and real distance, control center determine whether to start planning for UAV observation sequence. The simulation of the observation sequence for UAV is shown in Table 3: Number means the UAV number; Observation Sequence represents UAV observations sequence, and the serial number of the first goal means the UAV search first goal; Flying Time represents the time interval for the target observation.

Table 3. UAV observation sequence.

Number	Observation Sequence	Optimization Objective	Flying Time
0	9-10-9	240.13km	86.4s
1	1-5-6-2-0-1	248.43km	172.8s
2	8-3-4-7-8	245.37km	130.46s

As we can see from the above analysis, due to the different target distribution and UAV observation sequence, observation gaps isn't same and maximum flying time is 130.46s. Compared to satellite observations, UAV observation's accuracy and level of detail has also been improved.

Based on the above analysis, in space cooperation, the ability of two types of resources has been fully applied, improving the average observation period from 2425s to 141.7s, solving the problem that target run into area by only UAV observation.

5 Conclusion

Because of the different focus of observation type, continuous observation of maritime moving target hasn't been efficiently solved. This paper constructs the maritime moving target continuation observation model, puts forward the aerospace collaborative continuation observation strategy (ACCOS) to solve the model. On the basis of temporal characteristics of tasks, getting sub problem solving models from ACCOS: Satellite planning, UAV flight plan, UAV sequence, target potential area and distribution probability density prediction, UAV path planning. After solving this five problems, we get the observation window of the satellite observation sequence set, UAV observation sequence, path planning results and trigger event time series. From the result, we found that it improved the average observation period.

Acknowledgment

This project was supported partially by National Natural Science Foundation of China (71471174).

References

1. Y. Ci, Y. Xu, Y. Tan. Comparison of Several Algorithms for Maritime Moving Target Search by Satellite[J]. Acta Armamentarii, **1**: 029, (2009).
2. C. Ran, H. Wang, G. Xiong, et al. Research on Mission-Planning of Ocean Moving Targets Imaging Reconnaissance Based on Improved Genetic Algorithm [J]. Journal of Astronautics, **2**: 025, (2010).
3. Y. Guo. Research on key techonology research for mulit-type satellitestask planning [D]. National University of Defense Technology, (2009).
4. Y. Altshuler, V. Yanovsky, I. A. Wagner, et al. Efficient cooperative search of smart targets using uav swarms[J]. Robotica, **26**(04): 551-557, (2008).
5. P. B. Sujit, D. Ghose. Search using multiple UAVs with flight time constraints[J]. Aerospace and Electronic Systems, IEEE Transactions on, **40**(2): 491-509, (2004).
6. Y. Zhang. The UAV formations Research based on probability calculations and state predicted [D]. Dalian University of Technology, (2013).
7. Y. Xuan, C. Huang, W. Wu, et al. Coverage search strategies for moving targets using multiple unmanned aerial vehicle teams[J]. Systems Engineering and Electronics, **35**(3): 539-544, (2013).
8. J. Tisdale, Z. Kim, J. K. Hedrick. An autonomous system for cooperative search and localization using unmanned vehicles[C]//AIAA Guidance, Navigation and Control Conference and Exhibit, Honolulu, HI: 18-21, Aug. (2008).
9. J. George, P. B. Sujit, J. B. Sousa. Search strategies for multiple UAV search and destroy missions[J]. Journal of Intelligent & Robotic Systems, **61**(1-4): 355-367, (2011).
10. G. Bai, L. Xing, R. He, st al. Scheduling multi-platforms collaborative observation based on cooperative evolution[J]. Journal of National University of Defense Technology, **35**(4): 182-188, (2013).
11. J. Li, Z. Zhong, W. Hu, et al. A service model based on SWE for space-aeronautics cooperation earth observing operations[J]. Journal of National University of Defense Technology, **35**(3): 108-113, (2013)
12. J. Li, Z. Zhong, N. Jing, et al. Space-air Resources Multi-phase Cooperation Task Planning Approach Based on Heterogeneous MAS Model[J]. Acta

Aeronautica ET Astronautica Sinica, **34**(7): 1682-1697, (2013).
13. J. Yu. Research on Key Technologies of Cooperative Task Scheduling for Airborne and Spaceborne Earth Observing Assets[D]. National University of Defense Technology, (2011).
14. J. Bai, Y. Wen. Multi-sensor Cooperative Detection Technology for Sea Surface Targets[J]. Radio Engineering, **45**(3): 10-13, (2015).

MATEC Web of Conferences **81**, 04002 (2016)
DOI: 10.1051/matecconf/20168104002
ICTTE 2016

Assessing the Possibility of Presenting a Semi-Stochastic Speed-Density Function

Mohammad Maghrour Zefreh, David Baranyai and Adam Torok

Budapest University of Technology and Economics, Department of Transport Technology and Economics, Muegyetem rkp 3, Budapest, Hungary

Abstract. Underlying a fundamental diagram is a relation between traffic speed and density which roughly corresponds to drivers' speed choices under varying car-following distances. Stochastic and deterministic models are mainly two different categories of speed-density models. The advantages of deterministic models are their mathematical simplicity and analytical tractability though their results will show just the average parameters. Although the stochastic models may represent more accurate results taking uncertainty into account, they are often hard to use and analytically not tractable. The aim of this paper is to investigate the possibility of presenting a model which is neither completely deterministic nor completely stochastic but easy to use and understand which incorporates different traffic conditions and speed distributions. Monte Carlo Method has been used to generate different speed distributions based on different traffic conditions and consequently generating their relevant densities. Surveying the relation between the mentioned speed distributions and the obtained densities kept the chance of presenting a model which is neither completely deterministic nor completely stochastic but easy to use and understand which incorporates different traffic conditions and speed distributions.

1 Introduction

Flying birds, running water, electric current, internet packets, and moving vehicles can all be considered as flows, yet each exhibits distinct characteristics. One of the attributes that distinguishes vehicular traffic flow from other kinds of flow is the so-called fundamental diagram – the relationships among traffic flow characteristics (e.g. flow q, speed v, and density k) which are typically represented graphically [1].

Fundamental relations of traffic flow have historically been established either empirically or derived from car-following models. Generally, they come in two flavors: speed-density relations and flow-density relations, which, respectively, express the mean speed and the mean flow rate as functions of traffic density. The speed-density relationship serves as the basis to understand system dynamics in various disciplines [2]. It can be used as a tool to study on moving objects (or particles) in many scientific areas: pedestrians [3, 4], conveyors, network information packages [5], crowd dynamics [6], molecular motors, and biological systems [7]. It has been almost 75 years since Greenshields' seminal paper Study of Traffic Capacity in 1935 [8]. Attaching empirically derived curves to a fitted linear model of the speed-density relationship started a new era of transportation science and engineering. Due to its strong empirical nature, the efforts to find a perfect theory to explain these particular shapes mathematically never cease, but they always achieve limited success. There is a fairly large amount of effort devoted to revising or improving such an over-simplified relationship. These efforts include single-regime models: Greenberg's Model [9], the Underwood Model [10], Northwestern [11], Drew [12], and the Pipes-Munjal Generalized Model [13].There are also multi-regime models which include: two-regime models such as Edie Model [14], multi-regime model by cluster analysis [15], two-regime model [16], modified Greenberg, and three-regime models [16, 11].

A generalized deterministic form of speed-density relationship (so-called regression models or fitting models) is written in Eq. 1.

$$v_k = v(k) + \varepsilon_k \qquad (1)$$

where, v(k) is the deterministic speed-density relationship in which given density k there is one fixed corresponding speed value and ε is the measurement error [2]. Deterministic models are justified on the basis that they describe the average value of the dependent variable given the independent variable inputs [17].

The advantages and drawbacks of deterministic single- and multi- regime speed-density models have been well-documented in a multitude of publications [16, 18, 19]. The advantages of deterministic models are their mathematical simplicity and analytical tractability. The main drawback of deterministic models is that they describe average system behaviors. However, the mean alone may not be capable of describing the entire traffic conditions. It should be underlined that traditional deterministic families of models which essentially describe a 'pair-wise' relation between speed and density are not sufficient to include the traffic speed variance.

In the stochastic modeling framework of speed-density relationship, traffic speed v is a random function of density k and random variable ω. In this type of modeling, the random process will be discretized by an appropriate expansion technique which characterizes second-order statistics of the random process using uncorrelated random variables and the deterministic orthogonal set of functions. A simple illustration of the deterministic and stochastic model is shown in Fig. 1.

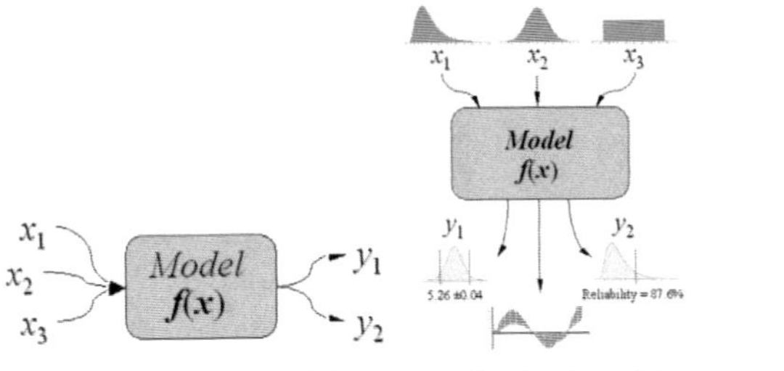

Figure 1. A simple illustration of deterministic and stochastic model [20]

To sum up, stochastic models are hard to use and deterministic models are not enough accurate. The aim of this paper is to investigate the possibility of presenting a model which is neither completely deterministic nor completely stochastic but easy to use and understand which incorporates different traffic conditions and speed distributions. Practically this paper describes the possibility of presenting a model that has deterministic input parameters (desired deterministic density) and stochastic output parameters (related speed distribution) considering different traffic conditions (Referring to Fig. 1.). The main contribution of this paper is to survey on the relation of speed-density parameters to see whether it is possible to derive speed parameter capable of presenting behaviour of traffic dynamics rather than presenting just the average speed of the entire traffic flow.

2 Methodology

Since there is a distribution of traffic speed at a certain density level due to the stochastic nature of traffic flow, automatically 'pairwise' pattern from deterministic models will be in contrast with the nature of traffic flow. What if in deterministic speed-density models a certain level of density will be incorporated with one or more part of different speed distributions related to that specific traffic condition?!

The development of mathematical tools focused on the modelling of the speed distribution in a traffic flow is widely reported in the scientific literature [21-24]. Many papers concern this problem since vehicles speed distribution is an important input parameter in lots of issues, such as kinematical traffic simulation model, road design, speed limit evaluation, road traffic noise prediction, traffic safety evaluation, bicycle performance evaluation, analysis of pedestrian walking, road transport emission estimation [25-33], [46] etc. Since 1940 majority of papers considered speed distribution as Gaussian or Normal distribution (Fig. 2) [34] and only extreme situations are considered as the traffic volume exceeds the practical road capacity, the speed distribution may become so heavily skewed toward the higher speeds that all semblance of normality is lost.

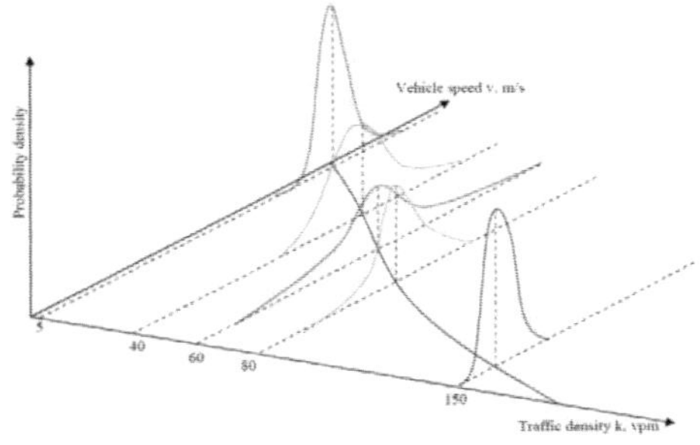

Figure 2. A simple illustration of stochastic speed-density model [2]

If the traffic conditions are not homogeneous, the normal distribution is not well suitable anymore. There is a lot of research that has examined distribution models for motorized vehicle speed data, such as normal distribution [35-37], lognormal distribution [38, 39], skew-t distribution [40, 41], gamma distribution [42], Beta distribution and Chi-Square distribution [25], as well as composite distribution [43]. Overall, roadway sections may have more than two modes, e.g., uncongested speed range, interim speed range which lies between uncongested and congested conditions, and congested speed range. Jun [43] mentioned if the characteristics of speed data are homogeneous (or unimodal distribution), the speed data of interstate freeways can be generally described by conventional probability density function (pdf) such as normal distribution, gamma distribution, lognormal distribution etc. However, if speed data follow a bimodal distribution (or a mixed distribution), there is no specific distribution function available [22].

The conditions that turn to a different speed distribution are quite often realized in non-highway or urban roads, where, in general, the traffic stream is much more complicated. This assumption is supported by Harmonoise [44], where different speed distributions are related to different traffic situations. This large spread of typologies, together with other parameters such as the presence of traffic lights and the road surface conditions, leads to a significant deviation of the speed distribution curve from the generally accepted unimodal normal distribution [25]. In the present research, the traffic condition has been assumed mixed containing the all possible traffic conditions and consequently the all possible speed distributions in order to cover all the possible scenarios.

In the current paper, Monte Carlo Method has been used to build different speed distributions and generate a link between them and their related density. Monte-Carlo methods (or Monte-Carlo simulations) are computational algorithms that rely on repeated random sampling [45].

 DOI: 10.1051/matecconf/20168104002

This method generates suitable random numbers of parameters or inputs to explore the behaviour of a complex system or process. The random numbers generated follow a certain probability distribution function (PDF) that describes the occurrence probability of an event. Typically simulations run many times in order to obtain the distribution of an unknown probabilistic parameter. It should be underlined that the number of vehicles, their average speed, and travel distance are remain constant in all cases as similar inputs in Monte Carlo Simulation, but the outputs (speed and density) were totally different based on the nature of traffic flow. Generated speed distributions by Monte Carlo Method have been shown in Fig. 3.

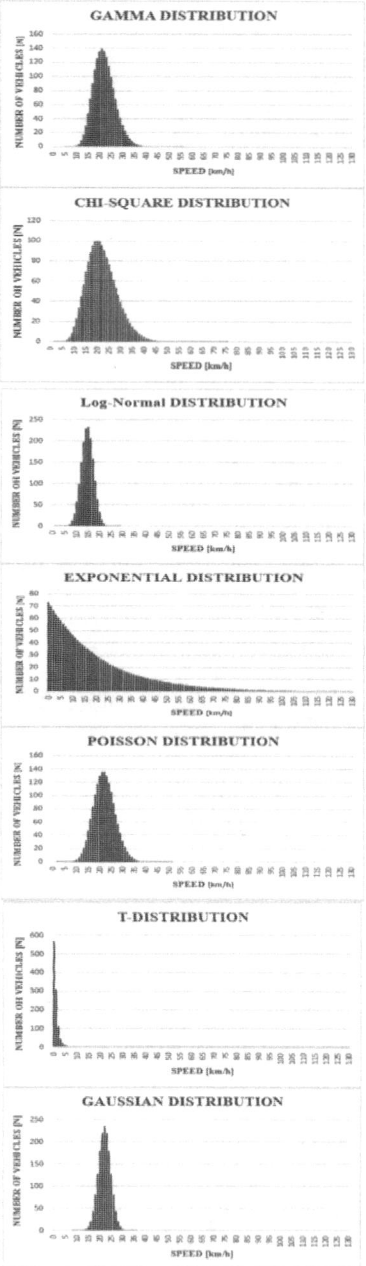

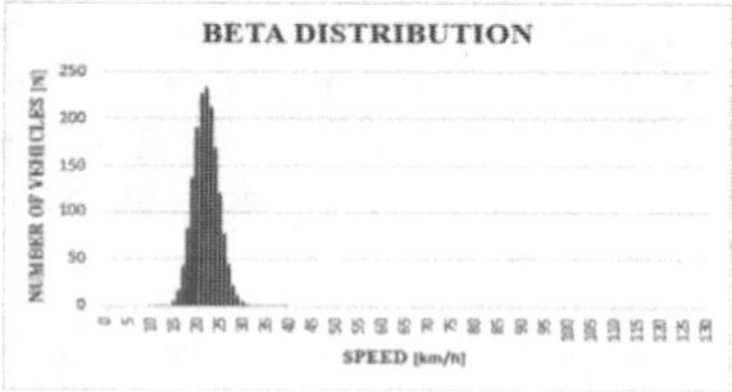

Figure 3. Different speed distributions generated by Monte Carlo Method

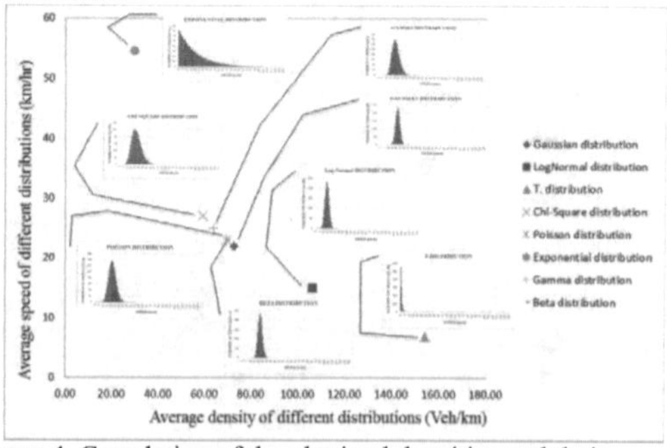

Figure 4. Correlation of the obtained densities and their relevant speed distribution

After running the Monte Carlo Method, generated speed distributions were converted to their relevant densities and the Average and Variance of the obtained density related to each and every speed distribution were computed. These different densities related to different speed distributions (different traffic condition) of the sample highway were shown in Fig. 4.

3 Results

By surveying different densities, it was found that in some cases different speed distributions will describe better the traffic conditions and some speed distributions were roughly covered each other (e.g. Poisson and beta distribution). In this paper based on the international literature overview, the different traffic scenarios were identified and described. Schematic fitting of the mentioned correlation in Fig. 4. by generally accepted deterministic speed-density models may imply the possibility of capturing the chance of presenting the semi-stochastic speed density model which deals with different traffic conditions rather than the average speed (see Fig. 5.).

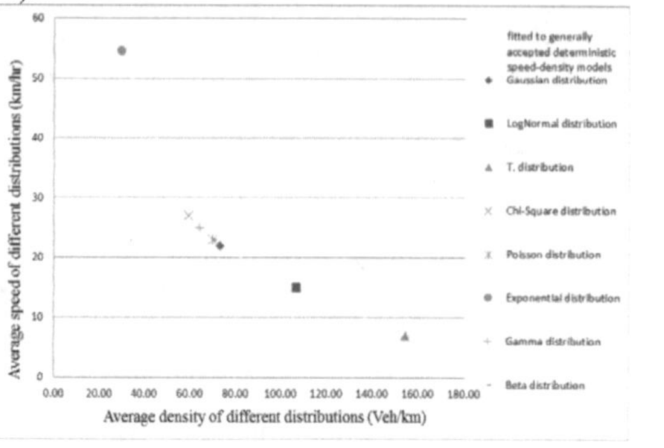

Figure 5. Schematic fitting of the Average speed-density diagram

It should be highlighted that these schematic fitting (Fig.5.) is based on the relevant average speed and density of different speed distributions. It should be noted that these different speed distributions may have the intersection in some certain densities with each other with different weights. For instance, although the average of the exponential speed distribution is around 55 km/hr, its domain is between zero km/hr and around 110 km/hr with more weight toward zero km/hr (see Fig.4.). Since these different speed distributions are based on different traffic conditions, this fact implies that the output of the deterministic speed-density function cannot be a single average speed but should be the function of speed composed of different relevant speed distributions.

4 Conclusion

In this paper, the authors tried to provide a method in which traffic speed variance will be incorporated by traffic density. The aim of this paper was to investigate the possibility of presenting a model which is neither completely deterministic nor completely stochastic but easy to use and understand and incorporates different traffic conditions by determining different speed distributions. The main contribution of this paper is to survey on the relation of speed-density parameters to see whether it is possible to derive speed parameter capable of presenting behaviour of traffic dynamics rather than presenting just the average speed of the entire traffic flow. The results showed that different speed distribution based on different traffic conditions may have the intersection with each other- however their averages were fitted to the generally accepted deterministic speed-density models- and this fact implies that discrete modelling is not viable in this case. The scope for future research of the authors will be applying the fuzzy theory to help the composition of the complex speed-density probability function taking intersections into account. The final complex semi-stochastic model will be validated by the traffic data of Budapest, Hungary compared to other density-velocity models.

References

1. Wang, Haizhong, Jia Li, Qian-Yong Chen, and Daiheng Ni. 2011. Transportation Research Part A: Policy and Practice 45 (6): 554–66. doi:10.1016/j.tra.2011.03.010
2. Wang, Haizhong, Jia Li, Qian-Yong Chen, and Daiheng Ni. 2009. The 88th Transportation Research Board (TRB) Annual Meeting. Washington, DC. http://people.umass.edu/ndh/Publications/C21.pdf
3. Jr., S. J. Y. Smith, J. M. Modeling Circulation Systems in Buildings Using State Dependent Queueing Models,Queueing Syst., 1989, 4, pages: 319-338.
4. Rajat, J. Smith, J. Modeling Vehicular Traffic Flow using M/G/C/C State Dependent Queueing Models. Transportation Science, **31**, pages: 324-335, (1997).
5. Gabor S, C. I. The analogies of highway and computer network traffic. Physica A, 2002, 307, pages: 516-526.
6. Helbing, D. Traffic and Related Self-Driven Many-Particle Systems. Rev. Mod. Phys., 2001,73,1067-1141.
7. Chowdhury, D. Schadschneider, A., Nishinari, K. Physics of Transport and Traffic Phenomena in Biology: from molecular motors and cells to organisms. Physics of Life Reviews, (2005), 2, 318.
8. Greenshields, B. D. A Study in Highway Capacity. Highway Research Board Proceedings, 1935, 14, pages: 448-477.
9. Greenberg, H. An analysis of traffic flow. Operation Research, (1959), **7**, pages: 79-85.
10. Underwood, R. T. Speed, Volume, and Density Relationship. Quality and Theory of Traffic Flow, Yale Bur. Highway Traffic, New Haven, Connecticut, (1961), pages: 141-188.
11. J. S. Drake, J. L. Schofer, A. D. May. A Statistical Analysis of Speed Density Hypotheses. Third International Symposium on the Theory of Traffic Flow Proceedings, Elsevier North Holland, Inc. New York, (1967).
12. Drew, D. R. Traffic Flow Theory and Control, McGraw-Hill Book Company, 1968, Chapter 12.
13. Pipes, L. A. Car-Following Models and the Fundamental Diagram of Road Traffic. Transportation Research, 1967, 1, pages: 21-29.
14. Edie, L. C. Car-Following and Steady-State Theory for Noncongested Traffic. Operation Research, (1961), **9**, pages: 66-76.
15. Lu Sun, J. Z. Development of Multiregime Speed-Density Relationship by Cluster Analysis. Transportation Research Record, (2005), Volume **1934**, pages: 64-71.
16. May, A. D. Traffic Flow Fundamentals, Prentice Hall, Englewood Cliffs, (1990).
17. Douglas I. Rouse. Stochastic Modeling of Plant Disease Epidemic Processes, Handbook of Applied Mycology Sail and Plants, Vol. 1 edited by Dilip K. Arora etc, (1991).
18. Hall, F.L., Hurdle, V.F., Banks, J.H. A synthesis on recent work on the nature of speed flow and flow-occupancy (or density) relationships on freeways. Transportation Res. Rec. 1365, pp. 12-18, (1992).
19. James H. Banks. Review of Empirical Research on Congested Freeway Flow. Transportation Res. Rec. 1802, pp. 225-232, (2002).
20. J.W. Wittwer, 'Monte Carlo Simulation Basics ', 2004. [Online]. Available: http://www.vertex42.com/ExcelArticles/mc/MonteCarloSimulation.html. [Accessed: 03- March- 2016].
21. Castro M., Sanchez J.A., Vaquero C.M., Iglesias L., Rodriguez-Solano R., (2008), Automated GIS-based system for speed estimation and highway safety evaluation, Journal of Computing in Civil Engineering, **22**, 325-331.
22. Dey P.P., Chandra S., Gangopadhaya S., (2006), Speed distribution curves under mixed traffic conditions, Journal of transportation engineering, 132, 475-481.

23. Fitzpatrick K., Carlson P.L., Wooldridge M.D., Brewer M.A., (2000), Design factors that affect driver speed on suburban arterials, Technical Report Documentation, Texas Transportation Institute, Report n. 1769-3.
24. Trozzi C., Vaccaro R., Crocetti S., (1996), Speed frequency distribution in air pollutants emissions estimate from road traffic, Science of the Total Environment, 189/190, 181 – 185.
25. Iannone, G., Guarnaccia, C., & Quartieri, J. (2013). Speed distribution influence in road traffic noise prediction. Environmental Engineering And Management Journal, **12**(3), 493-501.
26. Berry, Donald S., Daniel M. Belmont, and others. 1951. "Distribution of Vehicle Speeds and Travel Times." In Proceedings of the Second Berkeley Symposium on Mathematical Statistics and Probability, 589–602. The Regents of the University of California. http://digitalassets.lib.berkeley.edu/math/ucb/text/math_s2_article-43.pdf.
27. Lin S, He M, Tan Y, et al. Comparison study on operating speeds of electric bicycles and bicycles: experience from field investigation in Kunming, China. Transp. Res. Record (2008); 2048: 52–59.
28. Wang D, Zhou D, Jin S, et al. Characteristics of mixed bicycle traffic flow on conventional bicycle path. Presented at 94th annual meeting of the transportation research board, Washington, DC, 11–15 January (2015).
29. Vadeby, Anna, and \AAsa Forsman. 2014. Transport Research Arena (TRA) 5th Conference: Transport Solutions from Research to Deployment. http://tra2014.traconference.eu/papers/pdfs/TRA2014_Fpaper_18042.pdf.
30. Maurya, A.K., Das, S., Dey, S. (2015), 11th International Conference of Eastern Asia Society for Transportation Studies (EASTS 2015) in Cebu City, Philippines during September 11-14, 2015.
31. Hustim, M. Fujimoto, K. (2012) Road Traffic Noise under Heterogeneous Traffic Condition in Makassar City, Indonesia. Journal of Habitat Engineering and Design, 4(1), 109-118.
32. B., Aly, S.H., Ramli, M.I, Sumi, T. (2012) Driving Cycle of Passenger Cars on Heterogeneous Traffic Situations: Case Study on an Urban Road in Makassar, Indonesia. Proceeding of the 8th International Symposium on Lowland Technology.
33. Chandra, Satish, and Anish Kumar Bharti. 2013. "Speed Distribution Curves for Pedestrians During Walking and Crossing." Procedia - Social and Behavioral Sciences 104 (December): 660–67. doi:10.1016/j.sbspro.2013.11.160.
34. Loutzenheiser, D. W., & Greenshields, B. D. (1940). Percentile Speeds on Existing Highway Tangents. In Highway Research Board Proceedings (Vol. 20).
35. Leong HJW. The distribution and trend of free speeds on two-lane two-way rural highways in New South Wales. In: Proceedings of the 4th Australian road research board conference, part 1, Melbourne, VIC, Australia, 14–16 October 1968, pp.791–808. Vermont South, VIC, Australia: Australian Road Research Board.
36. McLean J. R. (1978): Observed Speed Distributions and Rural Road Traffic Operations. 9th Australian Road Research Board Conference Proceedings., Part 5, Australian Road Research Board, Vermont South, Victoria, Australia, 235-244.
37. Minh, C. C., Sano, K., and Matsumoto, S. (2005) The speed, Flow and Headway Analyses of Motorcucle Traffic. Journal of the Eastern Asia Society for Transportation Studies, 6, 1496 – 1508.
38. Wang, Y., Dong, W., Zhang, L., Chin, D., Papageorgiou, M., Rose, G., Young, W. (2012) Speed modeling and travel time estimation based on truncated normal and lognormal distributions. Journal of the Transportation Research Board, No. 2315, 66-72.
39. Gerlough DL and Huber MJ. Traffic flow theory: a monograph. Special report 165. Washington, DC: Transportation Research Board, National Research Council, June, 1975.
40. Zou, Y. (2013) A Multivariate Analysis of Freeway Speed and Headway Data. Texas A&M Transportation Institute.
41. Zou, Y., and Zhang, Y. (2011) Use of Skew-Normal and Skew-t Distributions for Mixture Modeling of Freeway Speed Data. Journal of the Transportation Research Board, No. 2260, 67-75.
42. Haight F.A., Mosher W.W., (1962), A practical method of improving the accuracy of vehicular speed distribution measurements, HRR 341, Highway Research Board, Washington, D.C., 92116.
43. JUN, J.. Understanding the variability of speed distributions under mixed traffic conditions caused by holiday traffic. Transportation Research Part C-Emerging Technologies, **18**, 599-610, (2010).
44. Harmonoise. (2004), Imagine project funded by EC under the sixth framework program, On line at: http://www.imagine-project.org.
45. Torok, Adam. 2015. International Journal for Traffic and Transport Engineering **5** (3): 278–85. doi:10.7708/ijtte.2015.5(3).05.
46. A. Török and M. M. Zefreh, Acta Tech. Jaurinensis, **9**, no. 2, p. 118, Apr. (2016)

Intelligent Traffic Information System a Real-Time Traffic Information System on the Shiraz Bypass

Amir Sodagaran[1], Narjes Zarei[1] and Zohreh Azimifar[2]

[1]*Shiraz Traffic Control Centre, Shiraz Municipality, Shiraz, Iran*
[2]*Electircal and Computer Engineering, Shiraz University, Shiraz, Iran*

Abstract. Real-time traffic information system is an Intelligent Transportation System (ITS) that allows commuters to make their traveling plan better. In this regard, an intelligent and real-time traffic information system was developed based on the video detection and an image processing algorithm was applied to measure traffic-flow according to the average speed of vehicles. Then, traffic status of each pass way is broadcasted to the electronic boards installed on all decision making entrance / exit. Different levels of congestion related to the routes ahead are shown on the boards with different colors in order to assist commuters. This system was implemented on the Shiraz Dry River's bypasses which account as vital routes to moderate traffic of city center. Experimental results are promising due to the proximity of determined traffic status by the system compared to the detection done by traffic experts. Average speed improvement is another result of using this system. This intelligent system developed and implemented in Shiraz city for the first time in Iran.s.

1 Introduction

Nowadays, travelers desire to be aware of real-time traffic flow which allows them to optimize the routes according to their destination. This issue becomes more significant in the roads with limited entrances and exits without secondary paths such as tunnels and bridges. Indeed, in such roads, if the drivers enter the passage without knowledge of traffic condition, they should tolerate the condition until next exit which result in a long delay and travelers inconvenience. Shiraz Dry River's bypasses are example of such special roads. More precisely, Shiraz bypasses pass diagonally from the middle of the city and connect northwest to southeast and vice versa. They accounts as a minor but vital pathway that passes traffic from the city center especially in peak hours. According to nature of Shiraz bypass with high distance between most entrances and exits, real-time traffic information is greatly valuable for bypasses' traveler decision making.

In this regard, Intelligent Transportation systems (ITS) suggest real-time traffic status systems as an effective solution to apply traffic control strategies which provides drivers with traffic information. Obviously, a reliable real-time traffic status information system is obtained through three steps: data collection, processing and dissemination in which a great variety of technologies has been developed for each step [1].

There is a variety of sensing technologies for detecting vehicles in the roadway such as inductive loops, vehicles probes, radar sensors, aerial surveillance and cameras. Although collected data using embedded detectors is accurate, maintenance cost of employing these technologies such as wiring and digging is the main challenges. In addition, regard to their limited detection area, gathering the data need employing many detectors. Recently, video detection is the most commonly used method regard to its wide-area detection. At the following, collected data such as occupancy, volume and speed of passed cars received from videos and other detectors should be carefully e integrated and translated to provide useful traffic information which is primary task of traffic control centers (TCCs).

In the current work, optical flow is used to detect vehicle motion in the received images which is defined as pattern of apparent motion of objects, surfaces, and edges in a visual scene caused by the relative motion between an observer and the scene [2], [3]. Recently, the term optical flow is also used by roboticists, encompassing related techniques from image processing and control of navigation including motion detection, object segmentation, time-to-contact information and motion compensated encoding [4], [5]. Various methods are proposed to compute optical flow and motion detection [6] - [9].

Finally, after intelligent detection, real-time traffic information should be disseminated to the road users through various technologies such as television, telephone, radio, boards, internet, cell phones and navigation devices [1]. Recently, tendency of methods is toward using variable message signs (VMS) which are the most easily accessible one for the most travelers.

Indeed, VMSs are easily used by all road user types with different facilities and educational level.

As an example of such real-time traffic condition information, we can refer to real-time of traffic information in Shanghai. In 2003, the Shanghai Urban Construction & Communications Center, Shanghai Electrical Apparatus Research Institute and Tongji University developed 'urban intelligent traffic navigation system' to serve the drivers with real-time traffic information of shanghai's highways. In this regard, 196 electronic traffic information boards were installed to show different levels of congestion with three colors; red, yellow and green." Red means blocked (speed less than 20km/h), yellow indicates jammed (speed between 20km/h and 45km/h), while green shows smooth traffic (speed of more than 45km/h). Some boards can also provide estimated arrival times to the next exit." Data collection in this system is done through 3,278 underground sensors installed at 300-400m intervals on the main roads. As mentioned above, although collected data by these methods is accurate, digging, wiring and maintenance cost are the main challenges of the embedded detectors. In addition, regard to limited detection area, gathering the data need employing many detectors. At the other hand, received data from many detectors should integrated and translated to provide traffic information [10].

Another popular ITS method to extract and process the current state of the roads is Automatic Number Plate Recognition (ANPR). For example, Traffic Scotland has implemented ANPR cameras to record the Vehicle Registration Numbers (VRN) of passing traffic. These are then encrypted at a computer outstation and stored alongside the related time stamp and camera location. By matching the encrypted VRN between camera locations, a journey time can be derived between those two ANPR locations. This information is then collated by the Traffic Scotland Control Centre, and will be used in the future to inform drivers using variable message signs (VMS). Although, by this method an accurate journey time and average speed is estimated for long distance journeys, there are some delays in real-time dissemination of traffic information and the price of ANPR cameras makes this solution as an expensive one [11].

The reminder of this paper is organized as follows: section II describes the specification of Shiraz Dry River's bypasses while a short description of ITS technology is presented in section III. Proposed method is then explained in section IV. Finally, experimental results and conclusion are presented in section V and VI, respectively.

2 Shiraz dry river's bypasses

Shiraz is sixth most populous city of Iran and accounts as a east-west city. Shiraz Dry River's bypasses are the roads which connect northwest to southeast and vice versa with length of 15 kilometers, width of 8 meters and 21 entryways and exits.

The most important challenge related to these bypasses is that as the cars entered the track, there is no alternate route until the next exit and since there is no prior information about the traffic status of bypasses, if the drivers faced with congestion, they should tolerate the condition until the next exit which results in a long delay and inconvenience. In addition, entry of other vehicles in heavy traffic status, double the problem and result in traffic anomalies and dangerous violation such as circumvent the ban and reverse gear.

At the other hand, given the vital role of the bypasses as a route which helps to evacuate traffic of city center especially in the morning and evening peak hours, awareness of traffic flows ahead in the decision making entrance/exit can be so useful. Therefore, providing drivers with real-time traffic information is purpose of this article..

3 Intelligent transportation system

Nowadays, it is believed that solving the problem of traffic congestion, rely not only on improving transportation facilities but also on employing different technologies for more efficient use of existing infrastructure. In this regard, IT technology has been increasingly applied in the transportation industry and allows the elements in this domain to use microchips and sensors in order to communicate with each other. The efforts in this regard led to formation of Intelligent Transportation Systems (ITS) which are commonly grouped within different categories, such as; Advanced Traveler Information Systems (ATIS) includes the applications such as Real-time Traffic Information Provision, Route Guidance/Navigation Systems, Parking and Weather Information Systems. Some other systems which try to apply traffic control strategies are placed in the Advanced Transportation Management Systems (ATMS) category such as Ramp Metering, Dynamic Traffic Signal Control that are mostly managed in the Traffic Control Centers [1].

To provide real-time traffic information system, different technologies are applied such as Global Positioning System (GPS), Dedicated-Short Range Communications (DSRC), wireless networks, mobile telephony, radio wave or infrared beacons, roadside camera recognition and probe vehicles or devices [1]. Indeed, the underlying technologies are used in collection, management and broadcasting traffic information data on online maps, cell phones or variable message signs (VMS) in the roads to help drivers choosing an optimal route with less congestion.

4 Proposed method

To come up with a precise explanation of developed intelligent traffic status information system, proposed method is described below through three phases: (1) Data collection, (2) Data aggregation and translation and then (3) information dissemination.

4.1. Data collection

Among different sensing technologies for detecting vehicles, tendency of recently methods is toward employing cameras which provide efficient wide-area detection besides giving a full monitoring facility that result in extracting more information about traffic events, accident and also weather conditions. Data collection is done through 21 pan/tilt/zoom cameras installed at 500 meter intervals on the roads and transmitted to the traffic control center (TCC) via fiber and wireless networks. Due to the capability of cameras for defining multi-presets, more than 40 critical segments (points) are covered by these cameras. Indeed, for each camera, two or more presets are defined and every two minutes, the cameras rotate once upon these points. Thus, one segment may be covered by more than one view delivered by different presets of different cameras. In this case, the traffic status determined for that segment is combination of the status detected from different views. More precisely, the worst case is reported as the traffic status of that segment.

4.2 Image processing algorithm

At the following step, received real-time videos from cameras should be processed and translated to deliver traffic status information. To come up with this, an image processing algorithm is applied through 4 steps: (1) filter out background, (2) calculate optical flow and then (3) measure the displacement after noise removing and then finally (4) estimate the velocity of vehicles to determine volume of traffic which is determinative of traffic status.

4.2.1 Filter out the background

As it is obvious, the images taken from the cameras included a lot of unrelated objects such as trees and landscapes with plenty of edges. Since the applied image processing algorithm is based on the detected edges, omitting unrelated ones result in increase of system accuracy besides frame rate increment. To achieve this, a filtering step is done at the beginning of system configuration to determine region of interest (ROI). It is worth to mention that, since the presets of cameras are pre-defined, there is no need for automatic recognition of ROIs and the traffic expert determines the region of interest for all presets of cameras.

4.2.2 Calculate optical flow

After filtering the background, remaining regions are considered in the calculation. In this regard, vehicles motion in the videos is calculated using optical flow algorithm. More precisely, after converting videos to continuous frames, a corner detection algorithm is applied on each frame and flow of detected corners is recognized through ongoing frames. Then, the distance of selected points is calculated in two continuous frames. In this order, the displacement traveled by the vehicles is measured.

4.2.3 Remove noise and background

Due to the noise-sensitivity of optical flow algorithm, the computed displacement may be calculated with error. To decrease the error, we model all calculated displacement using a Gaussian function.

Since there is no sudden jump in the vehicles motion, the fourth quartile which relate to the high displacement is omitted. On the other hand, motionless points which belong to the background should also be ignored. First quartile indicates displacement of these points and should be omitted. Average speed is then calculated based on the displacement of remaining points. After omitting displacement of error points, vehicles speed should be calculated based on the remaining displacement which is computed based on dividing distance with time as in "Eq. (1)"

$$v = d/t \tag{1}$$

4.2.4 Velocity estimation

At the following, average speed of displacement which is actually average vehicles speed is calculated for each segment and real-time traffic status determined based on it. In the proposed method, traffic level was measured by transportation engineers according to level of service (LOS) scale defined in the Highway Capacity Manual publication. In our qualitative classification, green color represents A (free flow), B (reasonably free flow) and C (stable flow, at or near free flow) levels, yellow color represents D (approaching unstable flow) and E (unstable flow, operating at capacity) levels and red color represents F (forced or breakdown flow) level [12].

(a) (b) (c)

Figure 1. Examples of intelligent traffic status detection of developed system in different status: (a) free flowing (b) slow (c) congested status indicated by green, yellow and red respectively.

Indeed, regard to the specification of passages and also knowledge of traffic experts, different average speed intervals are assigned to one level of congestion. More specifically, in the developed system, average speed of more than 40 km/h is assigned to the free flowing traffic status (A). If the calculated speed is less than 20 km/h, the related traffic status is considered as congestion or breakdown flow (F). Unstable flow as D is assigned to the average speed between 20 km/h and 40 km/h. Different real-time traffic status detected intelligently by the developed system is presented in "Fig. 1".

4.3 Information Dissemination

Finally, the real-time traffic status determined for each segment should be disseminated for public use. Among different types of information dissemination, dynamic message sign(VMS) is the most easily accessible one for the dominant travelers. Therefore, the real-time traffic information is passed digitally to 21 boards installed before all entrance and exit of bypasses. For more convenient usage, the digital boards show real-time traffic status by three colors; green, yellow and red. Green shows free flowing traffic status while yellow indicates slow traffic status and red means heavy traffic or congestion. "Fig. 2" represents an example of digital board installed before an entrance of Shiraz bypasses which show real-time traffic status related to four segments ahead. In addition, real-time traffic status information of bypasses is also shown on an online map as in "Fig. 3" and can be used via cell-phones and internet and etc.

Figure 2. An example of digital board installed before an entrance of Shiraz bypasses which show real-time traffic status related to four segments ahead.

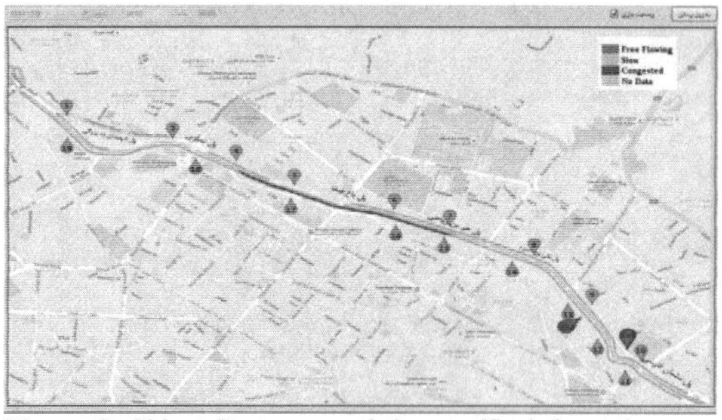

Figure 3. Shiraz Bypasses real-time traffic information map. Red, yellow and green colors show traffic status of flow flowing, slow and Congestion, respectively.

5 Experimental results

To evaluate the accuracy of implemented system, a comparison of manual state and auto is done for period of nearly two months. More precisely, traffic status determined intelligently by the system was inserted in the database and at the same time, traffic experts were employed to consider a full observation of traffic conditions and record their traffic status detection. . Accuracy of the system was computed by considering the differences between auto and manual detections. "Fig. 4" is related to a segment of bypasses with the worst accuracy. This specific segment experiences highly variable traffic status during a day and is one of the most challenging bypasses.

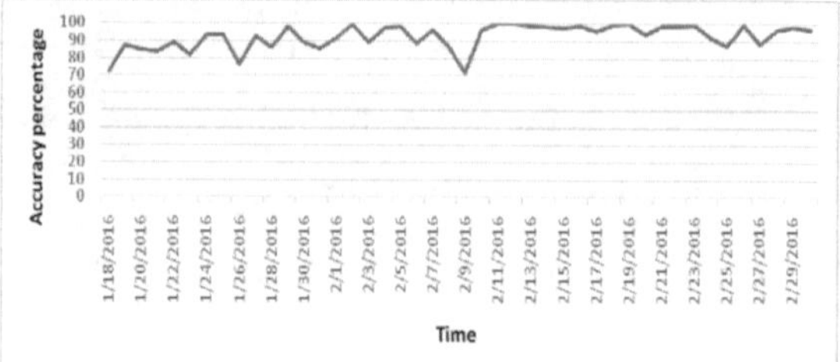

Figure 4. The worst accuracy of intelligent traffic information system in one of the most challenging segment of the Shiraz bypasses.

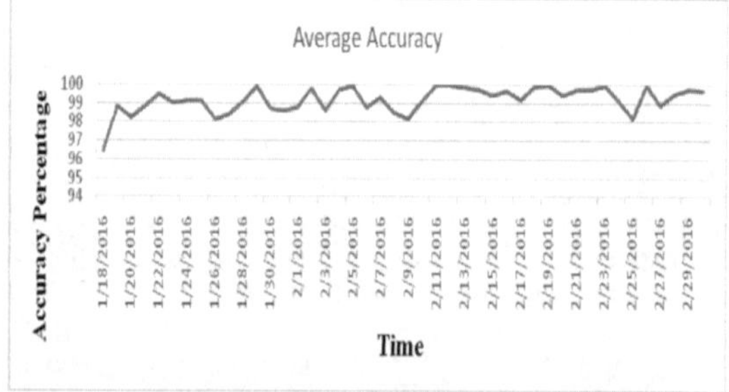

Figure 5. Average accuracy of intelligent traffic information system for all the segments of Shiraz bypasses.

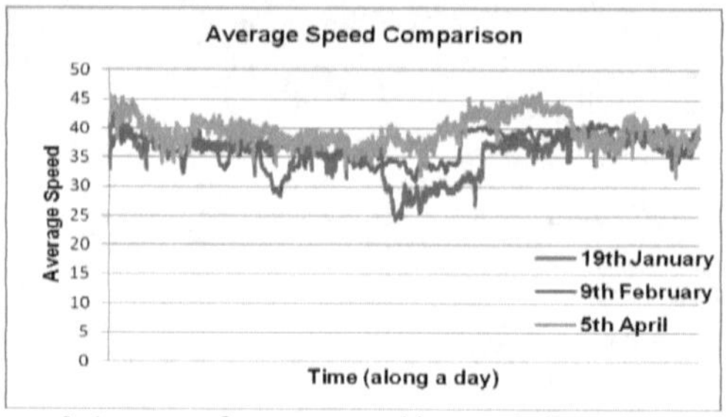

Figure 6. Average of average speed between all segments on 3 Tuesdays in January, February and April.

In "Fig. 5", average accuracy of the system for all the segments of bypasses is indicated. As it is shown, in both figures, the accuracy of intelligent real-time traffic status is improved through period of two months.

As another effect of employing this intelligent real-time traffic information, average speed of vehicles in the bypasses is reported in "Fig. 6". Given that a greater number of citizens become familiar with the system over the time and rely on it to plan their trip based on the real-time traffic information, we are witnessing an improvement in average speed as in "Fig. 6". In this figure, average speed of all segments is averaged on three Tuesday selected from three months: January, February and April.

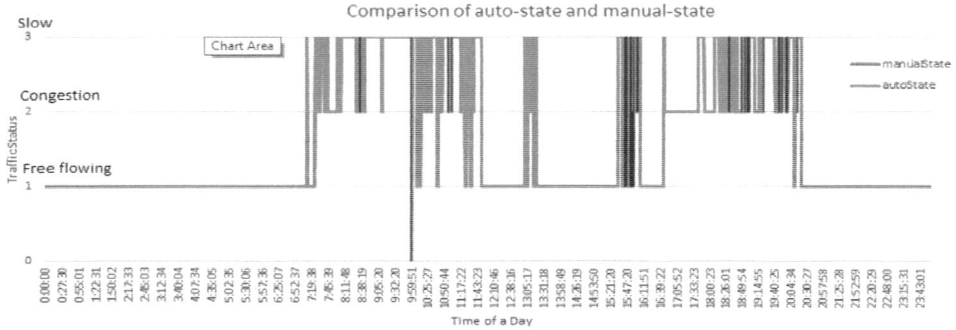

Figure 7. Comparison of intelligently determined traffic status with the detection done by traffic experts. Number 0 in the figure indicated off-camera.

For further investigation on the miss detection of system, a sample day in a week, specifically, Tuesday is considered. As we know, Tuesday is commonly selected in traffic domain, because it is a day with normal traffic state without special traffic behavior. In "Fig. 7", a clear comparison is considered between traffic status determined intelligently and by traffic experts. The points with blue color are the miss detection of system. Obviously, the beginning and end of chart indicate free flowing traffic status which is related daybreak hours.

6 Conclusions and future work

It is widely accepted that, awareness of traffic flows ahead result in better route choosing for travelers with less congestion. However, providing real-time traffic information and the way of sharing the information is so important. The implemented system employs cameras and image processing algorithm to provide accurate traffic status information on digital boards on the Shiraz Dry River bypasses.

Reducing staffing operations which result in reduction of human errors and providing 24-hours continuous functionality are the main obtained advantages.

Extension of similar intelligent traffic information systems to other passages of Shiraz city in order to detect real-time traffic status is considered on the agenda. In addition, accident detection and also congestion detection based on occupancy are the two options considered to be added the in future work.

On the other hand, due to nature of Shiraz bypasses and proximity of them to the river, one of the important usages of the developed system is the time of heavy rain. Currently, in such situations, the system is manually set to blue color for flooded roads to avoid entrance of travelers. Intelligently detection of flooded roads is another option which is planned to be added to the system.

References

1. Stephen Ezell. "*Intelligent Transportation Systems*". The Formation Technology & Innovation Foundation (ITIF) (January 2010).
2. Burton, Andrew, and John Radford, eds. *Thinking in perspective: critical essays in the study of thought processes*. Methuen, (1978).
3. Warren, David H., and Edward R. Strelow, eds. *Electronic Spatial Sensing for the Blind: Contributions from Perception, Rehabilitation, and Computer Vision*. Springer Science & Business Media, **99**, (2013).
4. Aires, Kelson RT, Andre M. Santana, and Adelardo AD Medeiros. "Optical flow using color information: preliminary results." *Proceedings of the 2008 ACM symposium on Applied computing*. ACM, (2008).
5. Beauchemin, Steven S., and John L. Barron. "The computation of optical flow." *ACM computing surveys (CSUR)* 27.3 (1995): 433-466.
6. Tao, Michael, et al. "SimpleFlow: A Non-iterative, Sublinear Optical Flow Algorithm." *Computer Graphics Forum*. Vol. 31. No. 2pt1. Blackwell Publishing Ltd, (2012).
7. Zach, Christopher, Thomas Pock, and Horst Bischof. "A duality based approach for realtime TV-L 1 optical flow." *Pattern Recognition*. Springer Berlin Heidelberg, (2007).
8. Wei, Shui-gen, et al. "Motion detection based on optical flow and self-adaptive threshold segmentation." *Procedia Engineering* 15 (2011): 3471-3476.
9. Klappstein, Jens, et al. *Moving object segmentation using optical flow and depth information*. Springer Berlin Heidelberg, (2009).
10. http://www.traffictechnologytoday.com/news.php?NewsID=3624
11. http://www.transport.gov.scot/road/technology-and-traffic-scotland.
12. Manual, Highway Capacity. "Transportation research board." *National Research Council, Washington, DC* 113 (2000).

MATEC Web of Conferences **81**, 04004 (2016)
ICTTE 2016
DOI: 10.1051/matecconf/20168104004

Research on Prediction Method of Freeway Operation Situation Based on Short-Term Traffic Flow Multi-Parameters Regression

Xiaodan Zhang, Liande Zhong and Yongsheng Chen

Research Institute of Highway, Ministry of Transport, 8 Xitucheng Road, Haidian District, Beijing 100088, China

Abstract. In order to realize the short-term prediction of traffic flow operation security situation, so as to enhance the operation safety of freeway, based on the traffic flow detection data and traffic accident data of Beijing section in Jing-ha freeway, the security situation prediction model based on short-term multi-parameters was established in the paper. Firstly, we extracted all the traffic flow data of microwave coils and videos, as well as the traffic accidents after updating the electromechanical system of Beijing section in Jing-ha freeway, so as to establish the risk prediction database, and developed the pre-analysis software of basic data; secondly, the traffic flow data of 30 minutes prior to the time of the accident were divided into six slices at 5-minute level, and the volume, speed, occupancy as well as their statistical parameters were selected as the chosen parameters of the model; next, single-parameter Logistic regression analysis was carried out in each time slice respectively; finally, based on the results of the correlation analysis of parameters and the single-parameter modeling significance, the multi-parameters Logistic regression model was established in each time slice respectively, thus obtaining the short-term prediction model of the traffic flow operation security situation. The results indicate that, the modeling fitting effect of slice 1 is the best, that is the change of the traffic flow parameters and their statistics in 5 minutes prior to the time of the accident can effectively predict the possibility of the accident, in which the average speed has a significant impact on the risk of the accident.

1 Introduction

Intelligent transportation system (ITS) is widely used in the global scope, which makes the traffic managers have large real-time traffic situation data. Many researchers and practitioners have been fully aware that all the advantages of ITS will not be recognized without realizing the ability of the traffic flow short-term prediction [1] (Brian L. Smith, 2002). The traffic flow prediction model can provide such a kind of ability, and can provide forward-looking traffic management as well as comprehensive travel information service.

At present, the object of traffic forecast is mainly urban road, and it focuses on the unblocked reliability and efficiency [2]-[7], including the assessment of traffic operation situation, short-term traffic forecast, road traffic condition judgment and so on, but the researches on the traffic flow operation security situation prediction are rare.

In China, due to the lack of detailed accident data and microscopic traffic flow data as support, resulting in a serious shortage in the real-time traffic flow security analysis theory, leading to the current freeway security management lag behind the real-time traffic situation prediction in our country. The United States and Canada began to carry out researches on the traffic accident detection algorithms and the traffic flow harbinger characteristics before the traffic accident from 1990s [8-15]. Among them, Chris adopted the speed differences between upstream and downstream, and the variances of the cross-section speed as the characterization factors of the traffic flow real-time risk discrimination [9], the results of which was referenced by Kansas state highway agency of USA in 2006. However, the main shortcomings were unknown of the risk reason and subjectivity of the risk rank assessment, and its scientificity needs to be further examined.

So, the prediction method of freeway operation situation based on short-term traffic flow multi-parameters regression is researched in this paper, thus achieving the short-term prediction of the traffic flow operation situation, and the results are helpful to reduce the risk of accident, decrease traffic accidents, and improve the operation security of freeway.

2 Data processing

Jing-ha freeway (Beijing section) is about 39.7 kilometers, two-direction of which has 20 microwave detectors (they can detect the traffic flow data of two directions), 16 video vehicle detectors (they can detect data of one direction), the average spacing of detectors toward Beijing direction is

 DOI: 10.1051/matecconf/20168104004

1.32 kilometers, and the average spacing of detectors far away from Beijing direction is 1.53 kilometers. All the traffic flow data of microwave coils and videos, as well as the traffic accidents after updating the electromechanical system of Beijing section in Jing-ha freeway are extracted in this paper, in which, the traffic accidents include time, location, type, reason, and et. al., and the traffic flow data are speed, volume and occupancy of divided-lane at 1-minute level.

First, we select the nearest detector data within 2 kilometers prior to or back to the location of accident, in order to screen data for first round, in which, the choice of accidents should avoid these caused by external factors such as weather, linear conditions, drivers and vehicles as much as possible, only in this way we can accurately excavate the rule of the traffic flow fluctuation affecting on the accidents, and in consideration that the relationship between the causes and effects of single-vehicle accidents and the traffic flow may not be strong under high service level, so the select of accident samples is more emphasis on the multi-vehicle accidents in large traffic volume (under C-classical service level); and then, we screen data for second round: check the quality of the traffic flow data, including deleting and processing the singular values such as speed with 0 km/h, diagnose the outliers by the data spatio-temporal graph method and the statistical method, and correct the abnormal data values by the simple difference method and the filtering method, in order to improve the accuracy and reliability of the analytical results, thus selecting the speed, volume and occupancy of detectors with better data quality to pair with accidents, and extract the traffic flow data of the control groups by 1:4 ratio.

The control group data meet the following requirements: the date is different with the corresponding accident; the time, week, and location are the same as the corresponding accident; the control group has no accidents at this location at the same day. Then, we select the control groups with better data quality according to the same method mentioned above, so as to establish the database required in this paper.

Because the location of traffic accident recorded by the police department is a cross-sectional stake number, therefore, it is necessary to aggregate the divided-lane traffic flow data into a cross-sectional traffic flow data, thus taking the cross-sectional data as the foundation of the research, so, the divided-lane data are transformed into the cross-sectional data by weighting method: $q_s=\sum_{i=1}^{n}q_i$, $k_s=\frac{\sum_{i=1}^{n}q_i\cdot k_i}{\sum_{i=1}^{n}q_i}$, $v_s=\frac{\sum_{i=1}^{n}q_i\cdot v_i}{\sum_{i=1}^{n}q_i}$, in which, q_i, k_i, v_i are respectively volume, occupancy, speed of divided-lane, n is the number of lanes.

We chose a tall building on the southwest corner of the intersection and used a video-based system to record traffic conditions. Then, we selected 150 typical crossing cases, including two straight-moving vehicles from the orthogonal direction, as shown in Fig. 1(b). To avoid interference from other vehicles to drivers' crossing behavior, we selected only the simplest crossing cases. In each case, one straight-moving vehicle encountered another straight-moving vehicle and no other object.

3 Mathematical model

3.1 Logistic regression model

Binary Logistic regression model is commonly used to quantitatively analyze the impact of explanatory variables on binary dependent variables, also can be used to estimate the occurrence probability of a category of the dependent variables, and from the traffic flow operation results, the dependent variables just can be divided into two categories: accidents and non-accidents. The probability of the accident corresponding to one sample data is:

$$P(x_i)=\frac{1}{1+e^{-x_i'\beta}},\quad i=1,2,\cdots,n \tag{1}$$

The linear expression after logit transform is:

$$\ln\frac{P(x_i)}{1-P(x_i)}=x_i'\beta,\quad i=1,2,\cdots,n \tag{2}$$

where, $P(x_i)$ represents the probability of traffic accident; $x_i'\beta$ represents the linear combination of explanatory variables: $x_i'\beta=\beta_0+\beta_1x_{1i}+\cdots+\beta_kx_{ki}$, x_{ki} is the value of variable k in i-th sample; β_0 is the regression intercept; $\beta_1,\beta_2,\cdots,\beta_k$ are the regression coefficients of explanatory variable x_{ki}; $\beta_0,\beta_1,\beta_2,\cdots,\beta_k$ can be calculated by the maximum likelihood estimation method: $\ln L(\beta,x_i)=\sum_{i=1}^{n}[\beta_0+\beta_1x_{1i}+\cdots+\beta_kx_{ki}-\ln(1+e^{\beta_0+\beta_1x_{1i}+\cdots+\beta_kx_{ki}})]$.

3.2 Logistic model testing

In Logistic regression, likelihood ratio test, Akaike information criterion (Akaike Information Criterion, AIC) and Schwarz criterion can be used to reflect the goodness of the model fitting.

We adopts AIC to reflect the fitting effect of the final model in this paper, that is $AIC=-2LL+2(K+S)$, in which, K is the number of independent variables of the model, S is the total number of response variable categories minus 1, the range of $-2LL$ is 0 to infinity, which is the smaller the better. Under the same conditions, the smaller value of AIC indicates the better model fitting.

The prediction accuracy for classification is usually used to reflect the prediction accuracy of the model. Using Logistic model to predict the classifications needs specifying a threshold of probability, that is, when the probability calculated by Logistic model is greater than a specified threshold, it is discriminated as the traffic accident, and when the probability is less than a specified threshold, it is discriminated as the security state with non-accident. The threshold value decides the forecast accuracy of each category and the total samples, the current researches commonly use the proportion of a category in the whole samples as the threshold value of this category prediction. Because of researching the prediction method of freeway traffic operation situation short-term in this paper, the proportion of accidents in the whole samples is adopted as the threshold value.

4 Research results and analysis

4.1 Data preparation

In order to predict the traffic accident in advance, the calibrated traffic flow data within 30 minutes prior to the accident is extracted in this paper, meanwhile, the traffic flow data of the control group in the corresponding period is extracted for each accident. Through screening of data quality, we ultimately retain 112 groups of accident samples and 448 groups of non-accident samples as the control groups for modeling research, and divide them into 2 categories: accident and non-accident, that is the value of dependent variable is 1 indicating accident and 0 meaning non-accident.

Firstly, the original traffic flow data of divided-lane are aggregated into a cross-sectional traffic flow data, secondly, in order to avoid the data noise due to the short acquisition interval, the data are converged with 5-minute level in order to get averages and standard deviations, thus for 5-minute aggregation half an hour period is divided into 6 time slices: the interval between time of accident and 5 minutes prior to the accident is named as slice 1, interval between 5 to 10 minutes prior to the accident as slice 2, interval between 10 to 15 minutes prior to the accident as slice 3 and so on, interval between 25 to 30 minutes prior to the accident as slice 6.

4.2 Model parameters selection

We extract 9 statistical parameters as the chosen model parameters: the average values of speed *as*, the average values of volume *av*, the average values of occupancy *ao*, the standard deviations of speed *ss*, the standard deviations of volume *sv*, the standard deviations of occupancy *so*, the coefficients of variation of speed *cvs* ($cvs = \frac{ss}{as}$), the coefficients of variation of volume *cvv* ($cvv = \frac{sv}{av}$), the coefficients of variation of occupancy *cvo* ($cvo = \frac{so}{ao}$) at 5-minute level.

4.3 Modeling steps

The binary Logistic model of the freeway traffic security operation situation deduction based on short-term traffic flow multi-parameters is established by using statistical analysis software with R programming language in this paper, the methods are as follows:

(1) We use the correlation analysis method to select parameters in each time slice, which makes the highly correlated variables be not into the Logistic model;

(2) For each parameter, we carry out a binary Logistic regression model in each time slice (single-variable Logistic regression), and select parameters which have significant impact on the traffic accident risk of freeway as the candidate variables of the subsequent steps;

(3) Combine with the analysis results of step (1) and (2) to select parameters for entering the final model, and then select the reasonable explanatory variables for modeling according to the backward selection method of the Logistic regression, thus obtaining the final models of six slices, the basic steps are as follows:

a) First, all the variables are contained in the model;

b) Then, calculate the z test values of all variables, and get the corresponding P values;

c) Last, find the largest P value, if the P value is greater than the significance level α_{out} , this variable is eliminated;

d) Go back to step b) for the next round of elimination.

Among them, the significance level of the reserved values is set as: including variables is $P \le 0.05$, excluding variables is $P > 0.1$.

4.4 Logistic model regression and analysis

Firstly, we use the correlation analysis method to select parameters in each time slice, the results of slice 1 are shown in Table 1.

Generally, if the correlation coefficient between two parameters is > 0.6 or < -0.6, they are strongly related to each other, and they can not enter the model at the same time. Through the experimental results, it can be seen that, for slice 1:

Cvv and *cvo*, *sv* and *av*, *ao* and *as*, *cvs* and *ss* are strongly related to each other, so they can not enter the model at the same time.

The results of the other slices are obtained by the same method.

Slice 2: *Cvv* and *cvo*, *sv* and *av*, *ao* and *as*, *ao* and *so*, *cvs* and *ss* are strongly related to each other, so they can not enter the model at the same time.

Slice 3: *Cvv* and *cvo*, *sv* and *av*, *ao* and *as*, *ao* and *so*, *cvs* and *ss* are strongly related to each other, so they can not enter the model at the same time.

Slice 4: *Cvv* and *cvo*, *ao* and *as*, *ao* and *so*, *cvs* and *ss* are strongly related to each other, so they can not enter the model at the same time.

Slice 5: *Cvv* and *cvo*, *sv* and *av*, *sv* and *so*, *ao* and *as*, *ao* and *so*, *cvs* and *ss* are strongly related to each other, so they can not enter the model at the same time.

Slice 6: *Cvv* and *cvo*, *sv* and *av*, *ao* and *as*, *ao* and *so*, *cvs* and *ss* are strongly related to each other, so they can not enter the model at the same time.

Secondly, for accurately analyzing the impact of a single traffic flow parameter on the probability of accident (single-factor analysis), for each parameter, we carry out a binary Logistic regression model in each time slice, for 5-minute aggregation there has 6*9=54 single-variable model, due to the same modeling method, the Logistic regression results of *as* as the independent variable are only shown here in Table 2:

It can be seen, for parameter *as*, the significance levels of parameters in six slices are all $P<0.001$, indicating that the average value of speed at 5-minute level has significant effect on the traffic accident risk of freeway, which is selected as the candidate variable of the subsequent modeling process.

Table 1. Correlation coefficients of variables

	cvv	*sv*	*av*	*cvo*	*so*	*ao*	*cvs*	*ss*	*as*
cvv	1	0.123703	-0.5397	0.74723	0.115362	-0.26757	0.385528	0.406205	-0.0469
sv	0.123703	1	0.634693	-0.06899	0.394076	0.278207	-0.07135	-0.06005	0.081788
av	-0.5397	0.634693	1	-0.54284	0.249592	0.507524	-0.26532	-0.31465	-0.0451
cvo	0.74723	-0.06899	-0.54284	1	0.236815	-0.34532	0.364011	0.499118	0.032488
so	0.115362	0.394076	0.249592	0.236815	1	0.589883	0.405378	0.058607	-0.50241
ao	-0.26757	0.278207	0.507524	-0.34532	0.589883	1	0.191952	-0.27208	-0.69086
cvs	0.385528	-0.07135	-0.26532	0.364011	0.405378	0.191952	1	0.731012	-0.45599
ss	0.406205	-0.06005	-0.31465	0.499118	0.058607	-0.27208	0.731012	1	0.074069
as	-0.0469	0.081788	-0.0451	0.032488	-0.50241	-0.69086	-0.45599	0.074069	1

Table 2. Regression analysis results of *as*

Slice	Coefficient (β_i)	Standard error	z value	Significance level ($\Pr(>\lvert z\rvert)$)	$Exp(\beta_i)$	$Exp(\beta_i)-1$
1	-0.05770	0.00926	-6.231	4.62e-10	0.943933	-5.6%
2	-0.058400	0.009493	-6.152	7.66e-10	0.943272	-5.7%
3	-0.05357	0.00891	-6.012	1.83e-09	0.94784	-5.2%
4	-0.046271	0.008381	-5.521	3.37e-08	0.954783	-4.5%
5	-0.043120	0.008094	-5.327	9.97e-08	0.957796	-4.2%
6	-0.046069	0.008469	-5.439	5.34e-08	0.954976	-4.5%

Odds ratio of the traffic flow parameters can be used to quantify the impacts of different traffic flow parameters on the risk of accident. The odds of an event is defined as the ratio of the probability with occurrence and the probability without occurrence, therefore, for the Logistic regression model:

$$\log[\frac{\Pr ob(event)}{\Pr ob(nonevent)}]=\beta_0+\beta_1X_1+\cdots+\beta_kX_k \quad (3)$$

$$Odds=\frac{\Pr ob(event)}{\Pr ob(nonevent)}=e^{\beta_0+\beta_1X_1+\cdots+\beta_kX_k} \quad (4)$$

It can be seen that when the *i*-th independent variable changes a unit, the change of odds is $Exp(\beta_i)$, that is:

$$Odds\ ratio=\frac{Odds(X_i+1)}{Odds(X_i)}=Exp(\beta_i) \quad (5)$$

If the coefficient of an independent variable is positive, it means that odds will increase, and this value will be greater than 1; if the coefficient of an independent variable is negative, it means that odds will decrease, and this value will be less than 1; when the coefficient of an independent variable is 0, this value is equal to 1. The change percentage of odds ratio is $Exp(\beta_i)-1$.

From Table 2, the odds ratios of *as* in six slices are all less than 1, and with the farther away from the time of accident, the value is increasing, which shows that the probability of accident will decrease with the increase of *as*.

The Logistic regression analysis results of the other independent variables are obtained by the same modeling method: for parameter *av*, the independent variables are not significant in slice 1~6, which shows that the correlation between the change of volume and the accident is relatively low; for parameter *ao*, the significance levels of parameters in six slices are all $P<0.001$, indicating that the average value of occupancy at 5-minute level has significant effect on the traffic accident risk of freeway, which is selected as the candidate variable of the subsequent modeling process, and the odds ratios of *ao* in six slices are all greater than 1, and with farther away from the time of accident, the value is decreasing, which shows that the probability of accident will increase with the increase of *ao*; for parameter *ss*, the independent variables are not significant in slice 1~6; for parameter *sv*, the independent variables are significant in slice 1 and slice 4, the significance levels of them are $P<0.05$ and $P<0.1$ respectively, the other slices are not significant, and the odds ratios in slice 1~6 are all less than 1, which shows that the probability of accident will decrease with the increase of *sv*; for parameter *so*, the significance levels of parameters in slice 1~6 are all $P<0.01$, indicating that the standard deviation of occupancy at 5-minute level has significant effect on the traffic accident risk of freeway, which is selected as the candidate variable of the subsequent modeling process, and the odds ratios of *so* in slice 1~6 are obviously greater than 1, showing that the probability of accident will increase with the increase of *ao*; for parameter *cvs*, the significance levels of parameters are more higher in slice 1, 2, 3 and 6, the significance levels are $P<0.01$, the significance levels are both $P<0.1$ in slice 4 and 5, and the odds ratios are far greater than 1, thus the probability of accident will increase significantly with the increase of the coefficient of variation of speed; for parameter *cvv*, the independent variables are significant in slice 3 and slice 4, the significance levels of them are

$P<0.1$ and $P<0.05$ respectively, the other slices are not significant, and some odds ratios are far less than 1 and some odds ratios are slightly greater than 1, which shows the impacts on the possibility of accident may be not consistent; for parameter *cvo*, the independent variables are all not significant in slice 1~6.

In summary, in slice 1, the modeling effects of parameters *sv*, *so*, *ao*, *cvs*, and *as* are the most significant; in slice 2, the modeling effects of parameters *so*, *ao*, *cvs*, and *as* are the most significant; in slice 3, the modeling effects of parameters *cvv*, *so*, *ao*, *cvs*, and *as* are the most significant; in slice 4, the modeling effects of parameters *cvv*, *sv*, *so*, *ao*, *cvs*, and *as* are the most significant; in slice 5, the modeling effects of parameters *so*, *ao*, *cvs*, and *as* are the most significant; in slice 6, the modeling effects of parameters *so*, *ao*, *cvs*, and *as* are the most significant.

Combining with the correlation analysis results of parameters and the single-parameter modeling significance results, the variables which have significant affect on the accident risk of freeway and are uncorrelated with each other are selected for modeling, so the final variables entering the Logisitic regression model of muti-parameters in each slice are respectively: for slice 1, *cvv*, *sv*, *so*, *cvs*, *as*; for slice 2, *sv*, *cvo*, *so*, *cvs*, *as*; for slice 3, *cvv*, *sv*, *so*, *cvs*, *as*; for slice 4, *cvv*, *sv*, *so*, *cvs*, *as*; for slice 5, *av*, *cvo*, *so*, *cvs*, *as*; for slice 6, *sv*, *cvo*, *so*, *cvs*, *as*.

The final regression results of model are shown in Table 3.

Table 3. Comparison of regression analysis results of slice 1~6

Slice	Variable	Coefficient (β_i)	Standard error	z value	Significance level (Pr(>|z|))	$Exp(\beta_i)$
1	Constant	3.21024	0.74016	4.337	1.44e-05	24.78503
	as	-0.05770	0.00926	-6.231	4.62e-10	0.943933
	The fitting parameters of the model: sample size=560, $-2LL=504.15$, AIC=508.15					
2	Constant	3.285394	0.761098	4.317	1.58e-05	26.71951
	as	-0.058400	0.009493	-6.152	7.66e-10	0.943273
	The fitting parameters of the model: sample size =560, $-2LL=506.65$, AIC=510.56					
3	Constant	2.89824	0.71468	4.055	5.01e-05	18.14219
	as	-0.05357	0.00891	-6.012	1.83e-09	0.94784
	The fitting parameters of the model: sample size =560, $-2LL=512.45$, AIC=516.45					
4	Constant	2.836702	0.735927	3.855	0.000116	17.05941
	cvv	-1.804372	0.893609	-2.019	0.043467	0.164578
	as	-0.046800	0.008598	-5.443	5.23e-08	0.954278
	The fitting parameters of the model: sample size =560, $-2LL=518.77$, AIC=524.77					
5	Constant	3.033395	0.923545	3.285	0.00102	20.76762
	cvo	-1.535999	0.932708	-1.647	0.09960	0.215241
	av	-0.030455	0.014048	-2.168	0.03016	0.970004
	so	0.346117	0.183208	1.889	0.05886	1.413568
	cvs	-6.078032	3.376873	-1.800	0.07188	0.002293
	as	-0.041821	0.009812	-4.262	2.02e-05	0.959041
	The fitting parameters of the model: sample size =560, $-2LL=520.45$, AIC=532.45					
6	Constant	2.319736	0.683262	3.395	0.000686	10.17299
	as	-0.046069	0.008469	-5.439	5.34e-08	0.954976
	The fitting parameters of the model: sample size =560, $-2LL=524.32$, AIC=528.32					

Table 4. Comparison of regression model prediction accuracy of slice 1~6

Model	Actual Classification	Forecast Classification		Total (rate)
		Accident (rate)	Non-accident (rate)	
1	Non-accident	135 (30.13%)	313 (69.87%)	448 (100%)
	Accident	68 (60.71%)	44 (39.29%)	112 (100%)
2	Non-accident	141 (31.47%)	307 (68.53%)	448 (100%)
	Accident	65 (58.04%)	47 (41.96%)	112 (100%)
3	Non-accident	144 (32.14%)	304 (67.86%)	448 (100%)
	Accident	64 (57.14%)	48 (42.86%)	112 (100%)
4	Non-accident	149 (33.26%)	299 (66.74%)	448 (100%)
	Accident	59 (52.68%)	53 (47.32%)	112 (100%)
5	Non-accident	150 (33.48%)	298 (66.52%)	448 (100%)
	Accident	59 (52.68%)	53 (47.32%)	112 (100%)
6	Non-accident	159 (35.49%)	289 (64.51%)	448 (100%)
	Accident	57 (50.89%)	55 (49.11%)	112 (100%)

From the results, the regression variable of the optimal final model in slice 1, 2, 3 and 4 is only *as*; the optimal model of slice 4 retains two variables: *cvv* and *as*; the final model of slice 5 retains five variables: *cvo*, *av*, *so*, *cvs* and *as*. In that case, *as* (the average value of speed at 5-minute level) is the most closely related to the probability of accident, and the significance level of each slice achieves to $P<0.0001$, indicating that *as* has the remarkable influence on the traffic accidents risk of freeway. Odds ratios of the traffic flow parameters are also given in this table to quantify the impact of different traffic flow parameters on the risk of accident, it can be seen that, the odds ratios of *as* in six slices are all less than 1, and with farther away from the time of accident, the value is increasing, which shows that the probability of accident will decrease with the increase of *as*. By comparing the values of AIC in final model of each slice, we get that, the farther away from the time of accident, the value of AIC is increasing, illustrating that the fitting effect of the final model is the better the closer to the time of accident. So, the change of the average speed in 5 minutes prior to the accident is the most effective to predict the probability of accident, which is used as the risk characterization factor of real-time safety assessment for the traffic flow operation.

The final model of the security situation deduction is as follows:

$$P(x_i) = \frac{1}{1+e^{-x_i\beta}} = \frac{1}{1+e^{-(3.21024-0.0577as)}} \quad (6)$$

where, as is the average speed of 5 minutes prior to the accident.

After specifying a reasonable threshold value, the calibrated model can predict the risk of freeway traffic accidents in real-time. Because in the total samples, the proportion of accidents is 20%, so the threshold is set to 0.2 in this paper, namely when the probability of the model output is greater than 0.2, it is discriminated as the traffic accident; and when the probability of the model output is less than 0.2, it is discriminated as the safety state with non-accident. The comparison of the model prediction accuracy of slice 1~6 is shown in Table 4.

In slice 1, it can predict 60.71% of accidents and 69.87% of non-accidents, the total prediction accuracy is 68.04%; in slice 2, it can predict 58.04% of accidents and 68.53% of non-accidents, the total prediction accuracy is 66.43%; in slice 3, it can predict 57.14% of accidents and 67.86% of non-accidents, the total prediction accuracy is 65.71%; in slice 4, it can predict 52.68% of accidents and 66.74% of non-accidents, the total prediction accuracy is 63.93%; in slice 5, it can predict 52.68% of accidents and 66.52% of non-accidents, the total prediction accuracy is 63.75%; in slice 6, it can predict 50.89% of accidents and 64.51% of non-accidents, the total prediction accuracy is 61.79%. It can be seen that, the farther away from the time of accident, the total prediction accuracy is decreasing, and for the accident prediction accuracy, the prediction accuracy of slice 1 is significantly higher than that of the other slices, so the accident risk prediction model established in slice 1 is capable of using the real-time traffic flow data to predict the freeway traffic accident risk in real-time well.

5 Conclusions

The traffic flow data of the nearest detectors prior to or back to the location of the typical accidents of Beijing section in Jing-ha freeway are extracted in this paper, on the basis, we select the binary Logistic regression method to establish the security risk prediction model and realize the real-time prediction of the probability of accident by using the average speed at 5-minute level, which providing the more scientific support for the traffic control and the traffic emergency management decisions of freeway.

Acknowledgment

Fund: Applied Basic Research Project of Ministry of Transport (2013319223170).

References

1. Beijing Traffic Research Center. *TMC Beijing Live-presentation Meeting Information* **12**, 15-18 Beijing, (2005).
2. G. Wang, X. Guo, Y. Jiang, J. Hou. Journal of Highway and Transportation Research and Development **27**(4), 155-158, (2010).
3. Y. Liang, Y. Chen, F. Ren. Journal of Highway and Transportation Research and Development **22**(12), 105-108, (2005).
4. K. Chen, W. Zou, X. Li. Highways & Automotive Applications **3**, 59-61, (2012).
5. J. Chen, W. Zhou. Journal of Chang'an University (Natural Science Edition) **22**(4), 52-54, (2002).
6. H. Guo, H. Shi. Journal of Highway and Transportation Research and Development **22**(8), 102-105, (2005).
7. Q. Xu, R. Yang. Journal of Highway and Transportation Research and Development **22**(12), 131-134, (2005).
8. Weil R, Wootton J, Garcia-ortiz A. Mathematics Computer Modeling **27**(9), 257, (1998).
9. Lee Chris Choongho. *School of Civil Engineering, The University of Waterloo*, Ontario, (2004).
10. Abdelaty M, Uddin N, Abdalla F, et al. Transportation Research Record **1897**, 88-95, (2004).
11. Abdelaty M, Pande A. Journal of Safety Research **36(1)**, 97-108 (2005).
12. Abdelaty M, Uddin N, Pande A. Transportation Research Record, **1908**, 51-58, (2005).
13. Lee C, Hellinga B, Saccomanno F. Transportation Research Record, **2749**, 67-77, (2003).
14. Lee C, Saccomanno F, Hellinga B. Transportation Research Record ,**1784**, 1-8, (2002).
15. Hossain M, Muromachi Y. *The 89th Annual Meeting of the Transportation Research Board*, 1-15 Washington, D. C., (2010).

MATEC Web of Conferences **81**, 04005 (2016) DOI: 10.1051/matecconf/20168104005
ICTTE 2016

Construction and Application of Passenger Flow Simulation Evaluation Index System in Urban Rail Transit Transfer Stations

Xiaohong Fang[1], Aijun Feng[2]

[1] *Beijing Rail and Transit Design & Research Institute CO., LTD, Beijing, China*
[2] *Beijing Urban Construction Design & Development Group CO., LTD, Beijing, China*

Abstract. To popularize the efficient application of the passenger flow simulation technology in urban rail transit transfer stations, and make the simulation evaluation systematic, comprehensive and reasonable, a passenger flow simulation evaluation index system of transfer stations was constructed, based on the characteristics of transfer stations. Then the calculation method, evaluation scope and evaluation standard of each evaluation index were proposed. Finally, through Anhuaqiao station of Beijing urban rail transit, the evaluation index system and evaluation standards were verified feasible, comprehensive and effective.

1 Introduction

Transfer station of urban rail transit is the important place, whose efficiency affects the ability of the whole network and overall service level. At present, passenger flow simulation technology has gradually been used in the design and operation optimization of many urban rail transit stations, as a new technology to guarantee the construction and traffic organization scheme meet the demand of passenger flow distribution ability, and make sure the operation of the station safe and efficient. But it has not formed a set of complete evaluation index system, which causes differences in the selection of indicators of most applications due to the difference of cities, research units and the objects of study, makes it impossible to form example of the outcome, and also set up obstacles to pedestrian simulation technology application in a wider range. Therefore, the evaluation index system and evaluation standards of passenger traffic simulation in urban rail transit transfer stations, which could provide criterion for reasonable and effective evaluation of the transfer stations, is badly needed.

2 Evaluation index system design principles

Construction of a complete evaluation index system is complex system engineering, some principles must be considered:

• Systemic principle [1]: to reflect the status, characteristics and the dynamic change of evaluation objects from all aspects, to summarize each link of the whole object systematically and comprehensively, avoid one-sidedness.

• Objectivity principle: to ensure objectivity and impartiality of the evaluation index system, to ensure the comprehensiveness, reliability and accuracy of the data source.

• Scientific principle: the selection, definition, calculation method and measure of indicators must base on scientific theories, and could widely use in the service field of rail transit stations and the evaluation of pedestrian traffic facilities.

• Operability principle [1]: the indicators selected can be obtained by direct or indirect method, which have strong operability.

• Coordinative principle: the indicators in the meaning, calculation method and other various aspects should be coordinated with other related factors index system, which can avoid conceptual confusion.

3 Hierarchical division of evaluation index system

Traditional station hub designs mainly refer to code for metro design and other architectural design, and usually use static and average calculation method, which consider microcosmic behavior characteristics of pedestrian and emergency in the station inadequate, and unable to accurately judge the movement efficiency of passengers in stations [2]. Therefore, dynamic simulation evaluation must pay attention on "average" and "transient" at the same time, and also give consideration both to service level and emergency evacuation. Passenger flow simulation evaluation system in urban rail transit transfer stations was established by four types of indicators as capacity adaptation, instantaneous impact, comfort and emergency evacuation capacity, and then the types were subdivided into specific evaluation indicators.

Hierarchical division of evaluation index system was shown in Fig. 1.

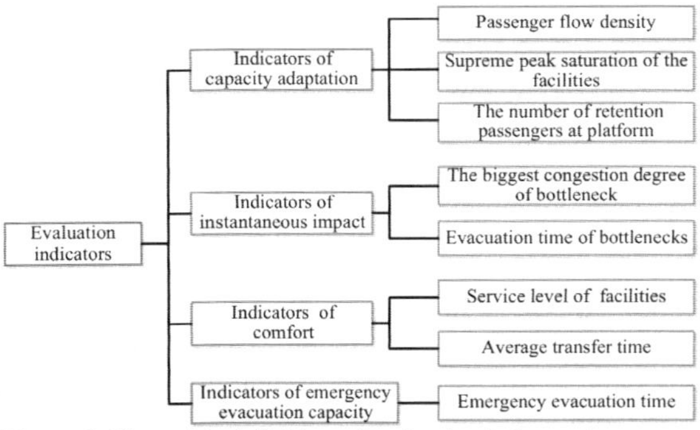

Figure 1. Hierarchical division of evaluation index system.

4 Analysis of the evaluation indicators

4.1 Indicators of capacity adaptation

4.1.1 Passenger flow density

The passenger flow density of urban rail transit refers to the number of people in each unit area. This indicator directly reflects the safety and comfort of the rail transit station area [3]. In the dynamic analysis, the traffic density is changing along with the passenger movements, which makes the value more in line with the actual operation state, including spatial and temporal distribution of the passenger flow density. Spatial distribution of the passenger flow density includes average passenger flow density distribution and instantaneous maximum passenger flow density distribution; the temporal distribution of the passenger flow density refers to the duration that various time points reach or exceed a certain density value (could be seen as the critical density) in station during the peak hours.

4.1.2 Supreme peak saturation of the facilities

The utilization efficiency ratio of the facilities reflects the level of coordination by the volume of passenger flows and the maximum capacity of traffic facilities. The ratio, denoted by θ , can be expressed as a quotient of the throughput during supreme peak hours and the capacity of the facilities.

$$\theta = \frac{Q_h * \rho}{C} \quad (1)$$

In this equation, Q_h represents throughput during peak hour; C represents maximum capacity of traffic facilities; ρ represent coefficient of supreme peak.

4.1.3 The number of retention passengers at platform

To insure the normal operations in transfer stations, all transfer demands should be satisfied, otherwise, passengers will be delayed and wait on the platforms. When the number of retention passengers at platform exceeds the capacity, the platform will get crowded and potential risks will appear. The accordance of the line capacity and the transfer passenger flow is represented by the number of the retention passengers on the platform. The retention passengers are rail transit riders who are waiting for the approaching train but fail to board the train because of crowd.

$$M_Z = Q_h - C_f * \beta \quad (2)$$

In the equation, M_Z denotes number of retention passengers at the platform; Q_h denotes number of people arriving at the platform; C_f denotes available space of the arriving train compared to the capacity. β denotes crowding coefficient (The value of crowding coefficient is related with the degree of the patience for passengers in subway cars).

4.2 Indicators of instantaneous impact

4.2.1 The biggest congestion degree of bottleneck

Passengers of rail transit arrive at stations with no continuity and equilibrium, but the traffic capacity of facilities is limited and fixed. Therefore, when large numbers of passengers reach a rail transit station within short a time, they may make a significant impact on facilities. This impact effect causes traffic congestion at different areas in the station. When too many passengers delay at one particular location, a bottleneck of traffic flow appears in the station. The bottleneck will affect other functions of the station and bring safety problems as well.

The bottlenecks often appear at stairs, escalators, ticket brakes, security checks and transfer channels. Sometimes, bottlenecks occur at waiting areas on the platform, passageways between stairs and the platform edges and queuing-up-for-checking areas outside of the station. The maximum congestion level of the bottleneck is reflected by the maximum number of assembling passengers or the maximum density of the bottleneck area in every passenger flow aggregation or evacuation process.

4.2.2 Evacuation time at bottlenecks

Because of the conflict between the dynamic passenger arrival rate and fixed transit capacity, when lots of passengers reach a rail transit station in a short time, the station becomes crowded. Moreover, the crowds cause bottlenecks of passengers' evacuation.

The evacuation time of a bottleneck is the time duration from when a bottleneck occurs to when the bottleneck disappears.

$$T_b = T_{start} - T_{stop} \quad (3)$$

In the equation, T_b represents the evacuation time at a bottleneck; T_{start} represents the time when a bottleneck appears and T_{stop} represents the time when the bottleneck disappears.

4.3 Indicators of comfort

4.3.1 Service level of facilities

Service level is a comprehensive evaluation of passenger evacuation efficiency and walking comfort of different facilities. Indicators [2] of evaluating level of passenger service usually includes occupied area per passenger, maximum walking speed, freedom degree of pedestrians, the possibility of outpacing and crossing other passengers as well as the level of security and comfort. The service level of various facilities can be analysed and concluded automatically from the simulation software and the results are denoted by different colours. The performance of services is usually represented by cumulative time percentage within the simulating time. The output can be data table and diagrams.

4.3.2 Average transfer time

Transfer times refers to the time in a transfer station between a passenger getting off a train of a certain line and the passenger getting to another line he/she is about to get on. The average transfer time is the most intuitive indicator to reflect connection convenience. It typically includes walking time during transfer and waiting time during transfer. The transfer time a passenger spends to transfer in station n can be expressed as follows.

$$t_n^h = t_n^b + t_n^w \tag{4}$$

In the equation, t_n^b represents walking time during transfer and t_n^w represents waiting time during transfer.

4.4 Indicators of emergency evacuation

Indicators of emergency evaluation are measured by the minimum evacuation time needed for all the people in the station to evacuate to safety regions in emergency. According to the current specifications, such as Code for Metro Design and Code for Design of Urban Rail Transit, emergency evacuation time usually refers to the minimum required time [4,5] to evacuate all the persons (passengers on the incoming train, passengers and staff on the platform) from the train/platform to the station hall in peak hours.

The time to evacuate passengers in emergency circumstances can be calculated using the following equation.

$$T_e = t_a + t_r + t_m \tag{5}$$

In the equation, T_e represents the total time needed for evacuation; t_a represents the time needed for detecting emergencies; t_r represents reaction time of passengers and t_m represents the time of evacuating passengers.

5 Evaluation scope and evaluation standard

5.1 Basic principles of evaluation standard

1) If there are specific provisions in the relevant laws and regulations, standard, government documents of urban rail transit, the evaluation standard executes unified standard if the documents in accordance with the same standard, else executes the strictest standards if the file standard have conflicts.

2) Although there are no clear stipulations, but the evaluation criteria has be approved scientific and reasonable in the industry or through any form of expert review.

3) The evaluation criteria could reach the domestic and even international advanced level.

4) If there have not been any related standards, the evaluation criteria should be proved reasonable and reliable by qualitative or quantitative methods.

5. 2 Summary evaluation scope and evaluation standard

The summary of simulation evaluation indicators and evaluation standards are shown in Table 1.

6 Instance application

6.1 Station introduction

Anhuaqiao station is the transfer station of Line8 with Line12 in Beijing metro. The station of Line 12 is lay out east-west trending along the north third ring road, and the station of Line8 north-south trending along drum tower street [6].

The station of Line 12 is a three-floor underground island station, and the station of Line8 is a two-floor underground island station. The two lines interchange passengers through transfer channels in Anhuaqiao station. Design scheme and Passenger flow organization scheme are shown in Figure 2 and Figure 3.

6.2 Simulation test and scheme optimization

According to the station planar construction scheme and passenger flow data prediction, the station passenger flow simulation model at morning peak hour is set up. Through simulation of the original station design model, it occurs serious congestion running to 7:25 am (morning peak hour 7:00-8:00), and the simulation is forced to stop.

In the original station design scheme, the facilities and transfer channel of Line12 can meet the demand of passenger flow in forward peak hour, but the number of transfer passengers between Line8 and Line 12 in forward peak hour is large, and streamline crossover before stars and escalators in south station hall of Line8 is serious. That cause serious congestion happened before two groups of stairs (escalators) of Line8, and make its capacity can't meet the traffic demand in forward peak hour.

Table 1. The summary table of simulation evaluation indicators and evaluation standard

Types of evaluation	Evaluation Indicator	Implication	Evaluation scope	Evaluation standard	Evaluation level
Capacity adaptation	Passenger flow density (person/ m^2)	Reflecting the traffic congestion degree of effective area(safety and comfort of the area)	Side of the platform	1.33～3 person/m^2	-
			Middle walk area of the platform	≤1.2 person/m^2	-
			Station hall	0.8 person/ m^2-1.0 person/ m^2	-
			Effective area of the platform	<0.6	Good
				0.6～1.0	General
				>1.0	Bad
	Supreme peak saturation of the facilities	Reflecting the utilization of facilities under the condition of unbalanced time distribution of passenger flow	Security checks, ticket brakes, stair, escalator, channels, etc.	<0.8	Good
				0.8～1.0	General
				>1.0	Bad
	The number of retention passengers at platform (person)	Reflecting the retention on platform when a batch of passenger flow come in super hour	Upside and downside platform	<0	Good
				0～292	General
				292～400	Not good
				>400	Bad
Instantaneous impact	The biggest congestion degree of bottleneck	Instantaneous passenger flow impact causes station distribution bottleneck, reflecting the congestion degree of the bottleneck(show as congestion density)	Stair and escalator	<2.0	Acceptable
			Walk channel facilities, such as transfer channel, stair and escalator	<2.17	Acceptable
			Queue facilities, such as the area before stairs, tickets brakes, security checks, waiting zones	<3.33	Acceptable
	Evacuate time of bottleneck(min)	The duration from the bottleneck appear to disappear	Bottleneck facilities	Evacuated before the arrival of next train	-
Comfort	Service level of facilities[7]	Comprehensive evaluation of the pedestrian evacuation efficiency and comfort level on facilities(shown as cumulative time proportion)	Platform, security check, tickets brakes, stair, escalator, channels, etc.	Cumulative time proportion achieves grade C and above in queuing space and grade D and above in walking space: ≥70%	Good
				50%～70%	General
				<50%	Not good
	Average transfer time(min)	Reflecting the impact of transfer mode and facilities perfect degree to transfer time	All transfer directions	<3	Good
				3～5	General
				>5	Bad
Emergency evacuation	Emergency evacuation time(min)	Reflecting the least time for passengers and staff to evacuate to safe areas	From platform to station hall(or ground)	6	-

Note:
1) Main standards come from "*Code for Design of Metro*", "*Code for Design of Urban Rail Transit*", "*Construction Standards for Urban Rail Transit Project*", and "Fruin service level evaluation standard"; 2)The other standards were got from calculation and deduction based on the above reference.

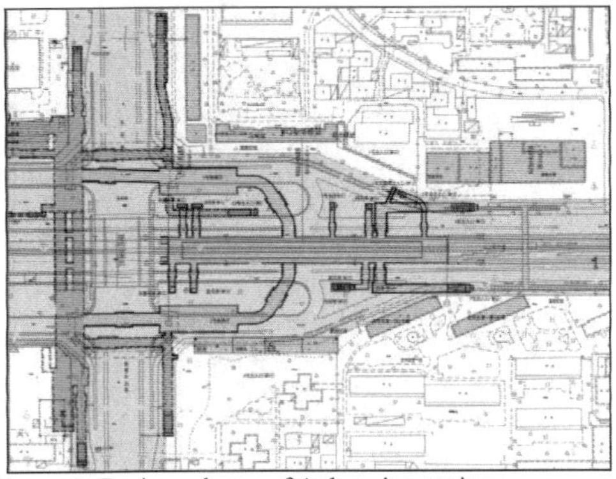

Figure 2. Design scheme of Anhuaqiao station.

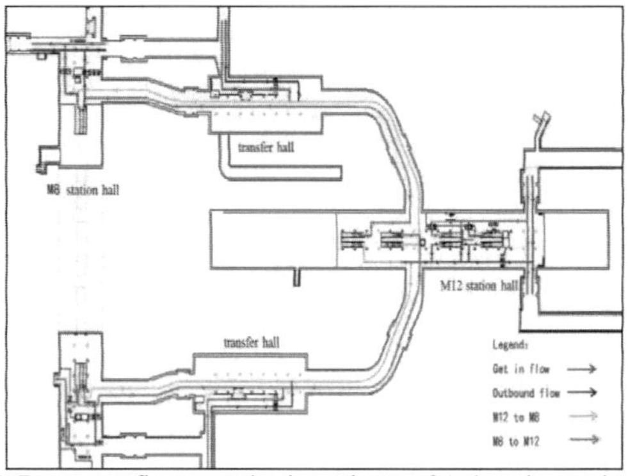

Figure 3. Passenger flow organization scheme of Anhuaqiao station.

Therefore, the optimized station design increases two groups of ascending stairs and one-way transfer channel from Line8 to Line12 in the middle of the Line 8 station, change the original two-way transfer channel to one-way transfer channel from Line12 to Line8, and change the running direction of two escalators in the south of Line8, which can reduce streamline crossover, and alleviate the passenger flow pressure of the stairs (escalators) on both north and south sides of Line8. Through simulation of the optimized station design model, it can run smoothly from 7:00 am until 8:00 am. The congestion bottlenecks relieve obviously or disappear, and the station simulation operates in good condition.

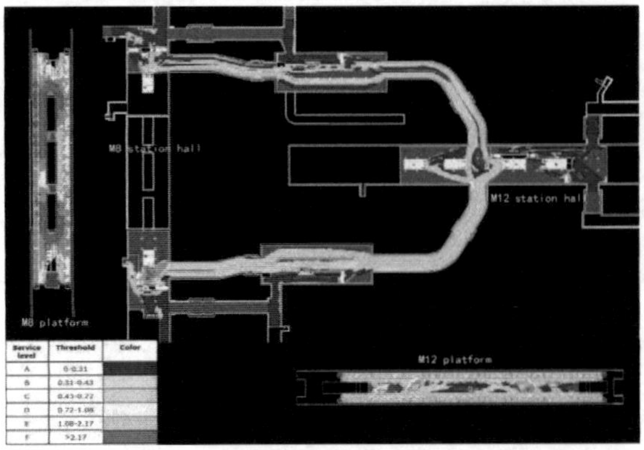

Figure 4. The average density figure of Anhuaqiao station original design scheme.

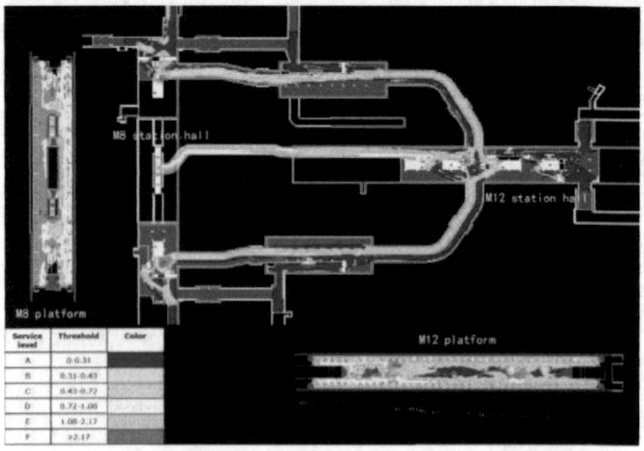

Figure 5. The average density figure of Anhuaqiao station optimal design scheme.

6.3 Contrast of evaluation indicators

According to the passenger flow simulation evaluation index system and evaluation standard proposed before, the instance chose some typical evaluation indicators, and make an effective assessment of the optimization effects through comparing the indicators before and after optimization.

6.3.1 Passenger flow density of the side platform

Table 2. Indicator contrast figure of passenger flow density of the side platform

Test position	Applicable standard	Test result of original scheme	Test result of optimization scheme	Indicator contrast
The downside platform of Line8	1.33～3 person/ m2	Maximum 3.10 person/ m2	Maximum 2.22 person/ m2	Passenger flow density of the downside platform of original scheme is bigger than standard, but the optimization scheme is in the scope of the standard value.

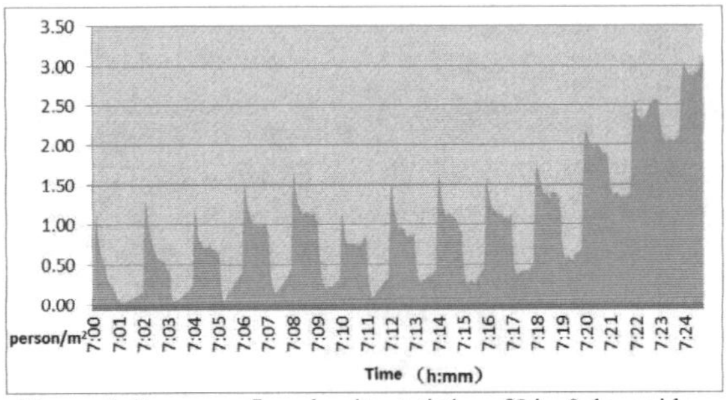

Figure 6. Passenger flow density statistics of Line8 downside platform of original scheme.

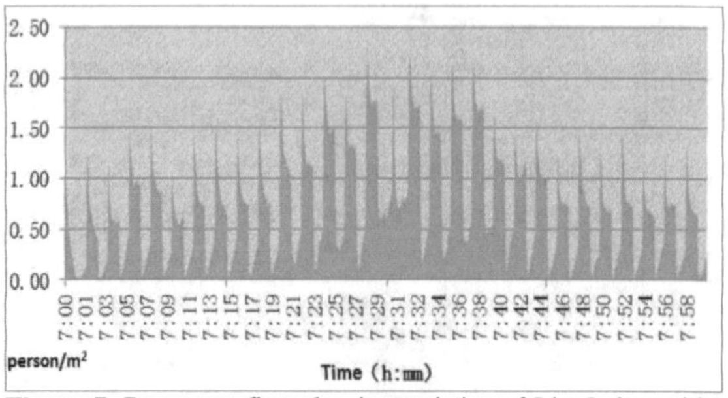

Figure 7. Passenger flow density statistics of Line8 downside platform of optimization scheme.

6.3.2 Supreme peak saturation of the facilities

Table 3. Indicator contrast figure of supreme peak saturation of the facilities

Test position	Applicable standard		Test result of original scheme	Test result of optimization scheme	Indicator contrast
The south upward and downward escalators of Line8	<0.8	Good	Upward escalator: saturation1.13; Downward escalator: saturation0.99.	Upward escalator: saturation0.78; Downward escalator: saturation0.75.	The capacity of upward and downward escalators of original scheme reaches saturation, so adaptability is worse; but the capacity of upward and downward escalators of optimization scheme is enough, so adaptability is good.
	0.8～1.0	General			
	>1.0	Bad			

6.3.3 Facilities service level

Table 4. Indicator contrast figure of facilities service level

Test position	Applicable standard			Test result of original scheme	Test result of optimization scheme	Indicator contrast
Queen area before the south stairs and escalators of Line8 platform	Cumulative time percentage achieves grade C and above in queuing space and grade D and above in walking space	≥70%	Good	Cumulative time percentage achieved grade C and above is 6%	Cumulative time percentage achieved grade C and above is 71%.	Queen area service level of optimization scheme improves obviously compared with the original scheme.
		50%～70%	General			
		<50%	Not good			

6.3.4 The biggest congestion degree of bottleneck

Table 5. Indicator contrast figure of the biggest congestion degree of bottleneck

Test position	Applicable standard	Test result of original scheme	Test result of optimization scheme	Indicator contrast
Queen area before the south stairs and escalators of Line8 platform	<3.33 person/m2	Maximum 4.16 person/ m2	Maximum 2.41 person/ m2	Queen area traffic density is greater than the standard in original scheme; but within the scope of the standard value in optimization scheme.

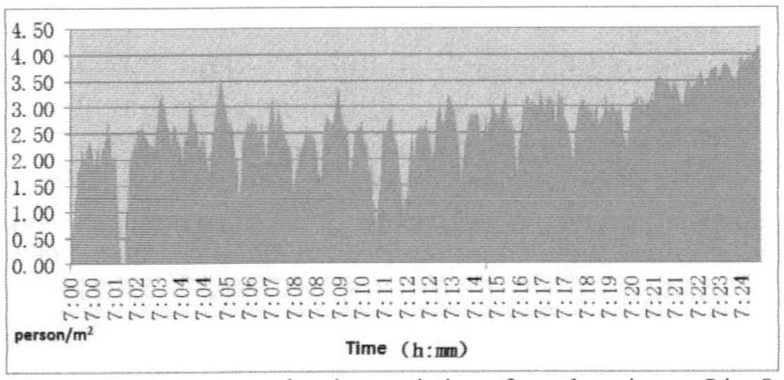

Figure 8. Queen area density statistics of south stairs on Line8 platform of original scheme

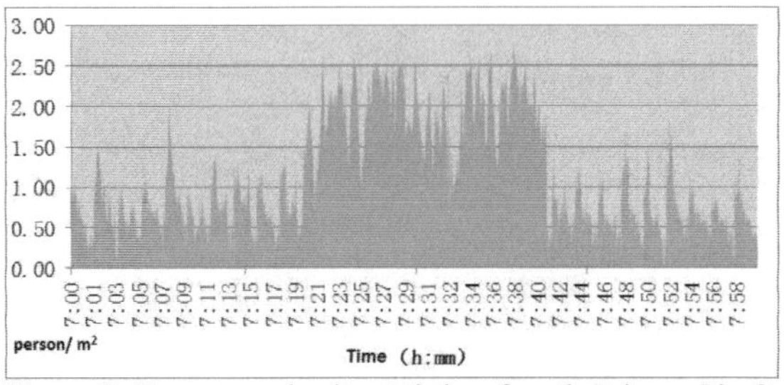

Figure 9. Queen area density statistics of south stairs on Line8 platform of optimization scheme

6.3.5 Evacuate time of bottleneck

Table 6. Indicator contrast figure of evacuate time of bottleneck

Test position	Applicable standard	Test result of original scheme	Test result of optimization scheme	Indicator contrast
The platform of Line8	Evacuate completely before the arrival of next train	There are a large number of passengers stranded on the platform	There are a little passengers stranded on the platform	Platform evacuate time is much greater than departure interval in original scheme, but less than departure interval in optimization scheme.

DOI: 10.1051/matecconf/20168104005

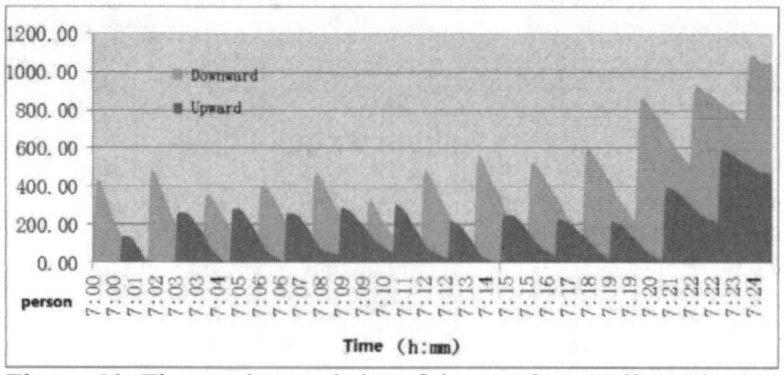

Figure 10. The number statistics of the people get off on Line8 platform of original scheme.

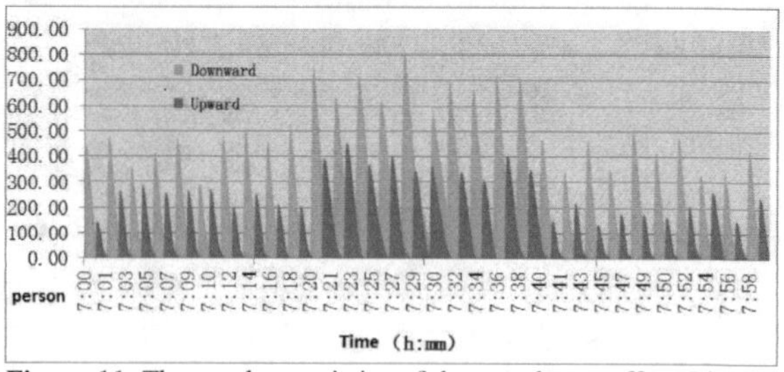

Figure 11. The number statistics of the people get off on Line8 platform of optimization scheme.

From the contrast of evaluation indicators, the optimized scheme is improved significant at passenger flow density, the degree of congestion, service level, and relief time of bottleneck compared with the original plan, and convenience and conform at transfer live up to the corresponding requirements. The instance verified the high feasibility, accuracy, efficiency and completeness of the passenger flow simulation evaluation index system and evaluation standards in urban rail transit stations proposed in this paper.

7 Conclusion

The evaluation index system covered four types of indicators of capacity adaptation, instantaneous impact, comfort and emergency evacuation capacity was proposed, based on multiple scientific applications of passenger traffic simulation in urban rail transit transfer stations. Then the evaluation index system was subdivided into eight specific indicators, such as passenger flow density, supreme peak saturation of the facilities, and service level of facilities and so on. These indicators reflected the possible problems of the transfer stations from different aspects. And the evaluation standard of each indicator was set. We hope that through more station simulation analysis, the evaluation index system and evaluation standard could be constantly revised perfect, and they can provide scientific decision support for the design and operation of transfer stations in urban rail transit.

Acknowledgment

This work is supported by Beijing Nova Program: Z141106001814027. And we thank the simulation project team of Beijing Rail and Transit Design & Research Institute CO., LTD. for sharing the instance date of Anhuaqiao station.

References

1. Chu Dongzhu, Lin Yanyu, Wei Shuxiang. Application in Influenced Urban Realm Around Rail Transit Station Microscopic Pedestrian Simulation Method and Urban Design. Chongqing, Chongqing University Press, pp.58-59 (2015).
2. Li Dewei. Modeling and Simulation of Microscopic Pedestrian Flow in MTR Hubs. Beijing, Beijing Jiaotong University (2007).
3. Wu Yongyan, Liu Xiaoming, Wei Zhonghua, Rong Jian, Sun Lishan. "Simulation Evaluation on Pedestrian Facility Layouts of Transfer Station". Journal of Transport Information and Safety, vol. **30**, pp. 74-77 (2012).
4. GB 50157-2013 Code for Design of Metro. Beijing, China Planning Press (2013).
5. DB 11/995-2013 Code for Design of Urban Rail Transit. Beijing, Beijing Municipal Commission of Urban Planning and Beijing Municipal administration of Quality and Technology Supervision (2013).
6. Beijing Rail and Transit Design & Research Institute CO., LTD. Research on Passenger Flow Simulation Technology in Urban Rail Transit and Construction of Evaluation Index System. Beijing (2014).
7. J. J. Fruin, "Pedestrian Planning and Design", Metropolitan Association of Urban Designers and Environmental Planners, New York, NY, USA (1971).

An Analysis to Validate the Travel Time in Order to Develop an Animated Simulation to Reproduce the Congestion Situation during a Massive Downpour

Mohammad Hannan Mahmud Khan, Motohiro Fujita and Wisinee Wisetjindawat

Scientific and Engineering Simulation Department, Nagoya Institute of Technology, Nagoya, Japan

Abstract. It is undesirable to face traffic congestion during an emergency. Congestion does not only cause delay to returning home commuters but also impede emergency vehicles. It is necessary to identify the congestion location as well as congestion severity in order to find the causes and countermeasures, furthermore setting priorities among remedial steps to alleviate the severity. In this study, we prepare an animated simulation of returning home behavior of commuters during a downpour. A taxi probe data was used to prepare the simulation. However, a direct visualization of the raw data does not produce a comprehensive representation of the real traffic situation. This study aims at reproducing efficiently the real traffic situation under such condition that can be easily recognized by a simple and quick glance. This paper discusses the method to reproduce the data as well. The proposed method provides a better realization of the traffic condition. The animated simulation that would be built from a preferable pattern can be used to explore the level of congestion in numerous locations and to evaluate the performance of countermeasures to overcome the congestion.

1 Introduction

In recent years, coastal cities including Nagoya, Toyota, and Ise have been facing a series of downpour disaster quite frequently. For example, on 20th September 2011 Nagoya city experienced a devastating downpour caused by Typhoon Roke.

Strong winds and driving rain led to the cancellation of train services in Nagoya. Moreover, several companies were forced to close the facilities as a precaution. This situation increased the number of commuters bounded for home. Consequently, they chose to travel by cars as the only available option for returning home. This produced an additional traffic load. Flash floods and extreme river waves covered roads and bridges. It reduced the road capacity and generated a further dimension of congestion severity.

During an emergency, severe congestion is not welcomed since it related to life and death of victims. Therefore, alleviating congestion is one of the greatest challenges in traffic operations. As a result, this study tries to establish a traffic simulation to inspect the performance of countermeasures to alleviate traffic congestion.

Most of the recent studies focus on pre-disaster evacuation and post-disaster network reconstruction [1]. However, quite a few studies worked on the situation during a disaster, which is further subtle on typhoon as a catastrophe. Y. Hiroi, N. Sekiya, R. Nakajima, S. Waragai, and H. Hanahara [2] conducted a questionnaire survey to study and discuss the usage of smart technology while returning home during an earthquake. T. Kazuyuki, F. Makoto, S. Shigeki, and O. Yusuke [3] developed models to describe the preference for returning home during an earthquake. Though these works investigate on the returning home passengers, yet lack considering downpour situation.

Study on downpour situations, such as M. Fujita, M. Jun, and K. Suzuki [4] focused on individual attitude and awareness of typhoon evacuation among public transport users for any future downpour using a questionnaire survey. However, an analytic approach is absent there even though it is among quite a few studies focusing typhoon.

In addition, M. H. M. Khan, W. Wisetjindawat, M. Fujita, and K. Suzuki [5] investigated the GIS approach using taxi probe data and examined the travel time accuracy. Even it discussed the potential for using probe data over traditional survey data, the development of a validation system was lacking there. Moreover, M. H. M. Khan, W. Wisetjindawat, and M. Fujita [6] analyzed the congestion situation through GIS approach utilizing taxi probe data, yet that was lack of a proper justification.

The previous literatures reveal that no existing analysis focused on the commuters returning home through roadway while no other possible travel mode during a heavy torrential rain event. They also lacked the presence of an animated real-time simulation in this context. Our simulation will be helpful in overcoming this lack of a well-developed simulation. It could find applications that include exploration of level of congestion intensities in several locations and also can be used to evaluate the performance of countermeasures.

2 Outline of taxi probe data

We used taxi car as probe vehicle in which the data was received from a Taxi Probe Research Center. The vehicle is capable of sending the data of Positioning with time, Speed, Angle and Vehicle status. A good number of GPS device installed cars (taxis) was left moving in a traffic stream.

We have received the recorded data for three consecutive days, 19th to 21st of September 2011 for 72 hours long. The recording interval of recording the data was an average of 200m traveled path and 20s of travel duration. The location for recording is Aichi prefecture with most of the vehicles moving in and around Nagoya city. As our probe vehicle is a taxi car therefore, vehicle status data such as with/without passenger were important concern in the data preparation and analysis as well.

3 Creating the basic animated simulation during downpour disaster

In this section, we construct the basic animated simulation using taxi probe data to represent the congestion severity around Nagoya city during downpour disaster.

The purpose of this simulation is to communicate with drivers for a better recognition of the congestion severity during disaster. In general, the number of probe data during disaster is inadequate to represent to real traffic situation. Especially, when the congestion level is worsening, the probe cars cannot enter to the traffic; hence the number of probe data becomes much smaller. During severe traffic congestion, the probe vehicles are also strictly running along with other vehicles in the traffic stream. It is possible to consider that other vehicles exist in between the probe vehicles. This study prepares a visualization of the probe vehicles with a proper size and color for a better representation of the congestion level. In addition, using the probe vehicles cannot represent other vehicles in between the probe cars. Thus, the probe data are expanded to represent the other following vehicles in the traffic stream.

GIS is used to prepare the simulation. In Figure 1, a lot of circles with difference in size and colors are shown. These circles denote probe data for that time window. Time window is the particular period of time, for which length the data are plotted on a single frame. As in the legend, the size and color vary according to the speed classes. The slower vehicles are with larger sizes and the most red in color.

GPS data was recorded as the vehicle changed its position for a particular distance. It resulted in fewer data for the lower speed vehicles. Therefore, representation of unmodified data would obviously underestimate the real situation. For this reason, we require augmenting the lower speed values to the extent of real life data volume.

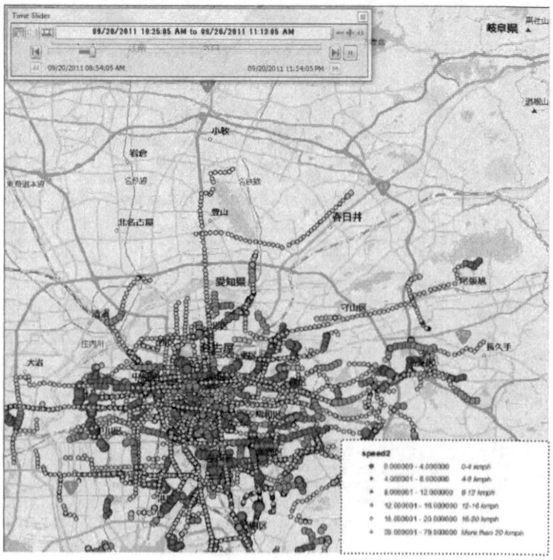

Figure 1. Simulation output window from basic animated simulation.

Table 1. Factors for different speed classes.

Speed Class	Multiplication Factor
0 – 4 km/h	4
4 – 8 km/h	4
8 – 12 km/h	4
12 – 16 km/h	1

We proposed the multiplication factors as in Table 1, for different speed classes. We achieved this factor after experimenting with a number of patterns. We could not validate the output at that time.

Therefore, in this study we would do the evaluation to propose a well suited pattern of multiplication factor for the better portrayal of the actual situation.

4 Methodology

4.1 Data transformation

The data recorded are simply comprised different location with time-tagged information along with point speed recording. In general, GPS contains huge dataset to track the vehicle down along its route that makes the dataset large in volume. To process the data, we transformed the raw data into desired indicators. We used Java platform to facilitate data transformation.

4.1.1 Preparation of probe data

Data reduction is one of the essential parts of the research. As GPS data is the continuous trajectory for a vehicle for the specific duration of recording, therefore, data reduction process is introduced to identify OD trips and desired travel indicators. It needs to cut the continuous trajectory into trip ends primarily by vehicle status. The origin of any trip is defined by the change of another vehicle status to occupied status contrary to determine the destination of that trip. The trips without passengers were not included in the data, as there are chances for any unusual driving behavior. In this research, we proposed vehicle status as an indicator to define origin and destination of a trip.

It seems difficult to differentiate between a genuine trip end and just stopping there while a taxi waits at a signalized intersection for a longer duration. As congestion situation is observed in this research, therefore appropriate care is required to handle these data. As long as there is a passenger in the car has an indication of not waiting for any unusual reasons but to follow the travel path. So, long stops for any journey, carrying passengers played a substantial role to determine travel time.

This study concentrated on the return home behavior. For this reason, the trips outbound from the center of Nagoya city is only filtered out from the whole dataset based on their positioning information and approaching direction. We used the probe data as a benchmark dataset to perform the experiment. Probe data consisted of travel information for a number of taxis' wherein every Taxi runs multiple trips. We stored travel time and distance traveled for every change in position for each trip by the difference in the time stamp and by determining linear distance respectively.

4.1.2 Preparation of simulation data

In order to mimic the actual situation, we proposed several patterns of data expansion. We factored the number data regarding their speed classes to prepare the simulation environment, including the pattern we chose so far, as in Table 2.

In the earlier study, we proposed to set the time window for 48 minutes therein. We further augment the dataset to make the dataset similar to the simulation presented in the study.

4.2 Data analysis

We have analyzed the data for the day of 20th September 2011 as the downpour was the most severe on the same day in Nagoya city. Simulation output showed a continuous trajectory of multiple frames for a particular time window. Therefore, firstly, we need to begin the analysis by filtering out the data of a given time window.

We considered each test probe data as a square object with a particular area. The initial area of that object was set as the size of the largest circle in the simulation, as the largest one would draw the attention while displaying the simulation output. In this study, we examined three different sizes to determine the most suitable one for the development of the new animated simulation. In Table 3, we can observe the pattern development. Firstly, the variation in multiplication factor, from Table 2. Secondly, the variations in setting the initial area of the test probe object.

The object traveled through its path and explore for any simulation speed data inside the boundary based on positioning data. The observer would store the top most data in case of multiple data. Concurrently, the observer varies its area while the speed data is not from the lowest speed class. It modified its area for three instances up until to achieve the lowest speed data. On the contrary, it stored a value of 20 km/h as a speed data. This approach is unique, as the probe observer itself was picking up the speed data that is supposed to be displayed.

The concept of changing area of the square observer was very essential. It helped to facilitate lower speed classes by varying the area.

So, the observer recorded the speed value which would be displayed in the simulation. Then, we used the distance data of the test probe object to determine the travel time by the speed distance relationship. Next, the data were grouped by vehicle and particular trips to achieve the overall travel time for any given trip. Finally, both of the travel time, traveled in actual by probe data and data recorded by the Observer probe were compared.

The whole process was repeated several times for patterns from 1A through 5C (Table 3). Then, the performance indices are determined for all of the patterns to propose the most effective pattern for the development of the animated simulation.

Table 2. Pattern distribution for simulation data

Speed Class	Pattern 1	Pattern 2	Pattern 3	Pattern 4	Pattern 5
0 – 4 km/h	2	3	4	6	12
4 – 8 km/h	2	3	4	5	12
8 – 12 km/h	2	2	4	4	8
12 – 16 km/h	1	1	1	1	1

Table 3. Detail distribution of various patterns

Speed Class	Size of the Observer (m)	Patterns														
		1A	*1B*	*1C*	*2A*	*2B*	*2C*	*3A*	*3B*	*3C*	*4A*	*4B*	*4C*	*5A*	*5B*	*5C*
		52	*78*	*104*	*52*	*78*	*104*	*52*	*78*	*104*	*52*	*78*	*104*	*52*	*78*	*104*
0 – 4 km/h		2			3			4			6			12		
4 – 8 km/h		2			3			4			5			12		
8 – 12 km/h		2			2			4			4			8		
12 – 16 km/h		1			1			1			1			1		

5 Results and discussions

We will briefly discuss the statistical output from the analysis for the time window of 1400 hours to 1448 hours. We chose the following graphs to report the outcome of the experiment. Fig. 2 is the output of pattern 1A as described in Table 3.

We observe in Figure 2 that most of the trip time data are close to each other. Trip time data is the travel time required for each trip for the probe and simulation data as well. Figure 3 displays the output distribution of pattern 1C. The trip time data are quite confined, which provided a good fit line. It also seems that the simulation trip time is quite similar to the probe output.

The root mean square error (RMSE) defined by equations (1) is used to judge the proficiency:

$$RMSE = \sqrt{\frac{1}{N}\sum_{i=1}^{N}(x_i - \hat{x}_i)^2} \quad (1)$$

where, N is the total number of testing cases, $\hat{x}_i$ the predicted simulation trip time and x_i the probe trip time.

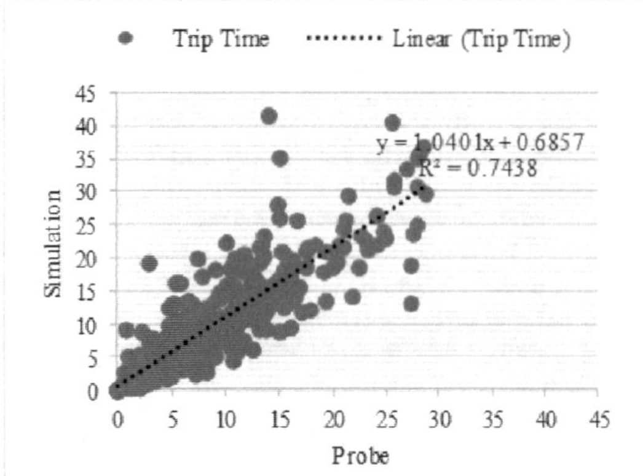

Figure 2. Travel time for different trips (Minutes)

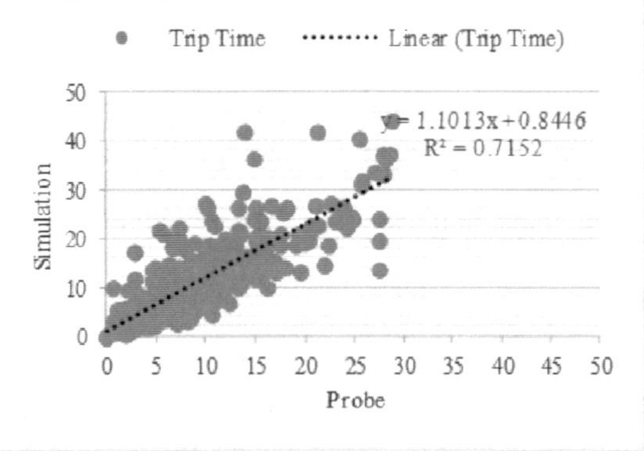

Figure 3. Travel time for different trips (Minutes)

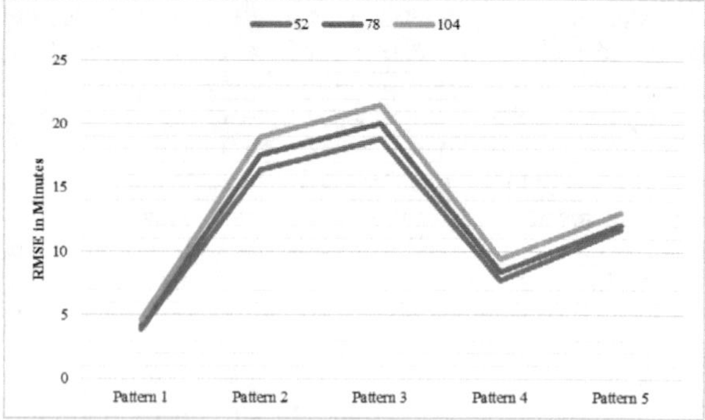

Figure 4. Variations of RMSE (Minutes)

In Fig. 4, the RMSE values are drawn for three different sizes of the probe observer. According to the RMSE value, we find the prediction for Pattern 1 and Pattern 4 is better than any other patterns. We can conclude that, pattern 1A as in Table 3 can be the most preferable pattern to develop the simulation.

6 Conclusion

A well representative animated simulation is essential to communicate with drivers for a comprehensive perception of the congestion severity during a disaster. With this animated simulation, the congested locations can be efficiently recognized in order to anticipate proper countermeasures to mitigate the congestion severity. However, the available probe data was insufficient to represent the real situation. Therefore, the probe data should be expanded to the extent of the real traffic volume for a better portrayal of actual situation. This study has simulated the returning home commuters in different patterns and compared the travel time with the actual movement of the probe vehicle during the downpour.

From the results, pattern 1A is the most preferable pattern to prepare the simulation. From the simulation we can perceive that, the central Nagoya was heavily congested and the roads with bridges and tunnels were congested as well. This simulation will work as a tool to communicate with the drivers for a comprehensive understanding of the traffic situation for any particular time. However, this paper could not include the consequences on the driver's decision to depart for the journey. This study still require a proper justification for model calibration and validation, which would be considered in the future work. As a transferability of this work, the prepared simulation can be demonstrated to the drivers in order to collect their opinions and these opinions can be used for a further analysis on departure choice. Our future work will also focus on how to use this simulation for forecasting the travel time for commuters simply by adjusting the severity of downpour.

References

1. H. Yusuke, K. Masao, *Traffic Monitoring immediately after a major natural disaster as*

revealed by probe data – A case in Ishinomaki after the Great East Japan Earthquake, Tr. Res. Part A: Pol. And Prac., **75**, 1-15 (2015)

2. Y. Hiroi, N. Sekiya, R. Nakajima, S. Waragai, H. Hanahara, *Questionnaire survey concerning stranded commuters in metropolitan area in the East Japan Great Earthquake,* J. Ins. Soc. Saf. Sc., **15**, 343-353 (2011)
3. T. Kazuyuki, F. Makoto, S. Shigeki, O. Yusuke, *Behavior analysis of the people having difficulty in going home in Tokyo metropolitan area-the 2011 off the pacific coast of Tohoku Earthquake,* Proc. of 15th World Conference on Earthquake Engg. (2012)
4. M. Fujita, M. Jun, K. Suzuki, *Behavioral analysis of homebound public transport users during downpour conditions,* J. of the As. Tr. St., **1(3)**, 318-333 (2011)
5. M. H. M. Khan, W. Wisetjindawat, M. Fujita, K. Suzuki, *An analysis of Probe data on traffic congestion during the Typhoon using GIS application,* Proc. of the Est. As. Soc. for Tr. St. **9** (2013)
6. M. H. M. Khan, W. Wisetjindawat, M. Fujita, *Analysis of traffic congestion due to Typhoon using taxi probe data,* Proceedings of JSCE, **47** (2013)

Prediction of Ship Traffic Flow Based on BP Neural Network and Markov Model

Pengfei Lv, Yuan Zhuang and Kun Yang

College of navigation, Wuhan University of Technology, 1178 YouYi Road, Wuchang District, Wuhan, China

Abstract. This paper discusses the distribution regularity of ship arrival and departure and the method of prediction of ship traffic flow. Depict the frequency histograms of ships arriving to port every day and fit the curve of the frequency histograms with a variety of distribution density function by using the mathematical statistic methods based on the samples of ship-to-port statistics of Fangcheng port nearly a year. By the chi-square testing：the fitting with Negative Binomial distribution and t-Location Scale distribution are superior to normal distribution and Logistic distribution in the branch channel；the fitting with Logistic distribution is superior to normal distribution，Negative Binomial distribution and t-Location Scale distribution in main channel. Build the BP neural network and Markov model based on BP neural network model to forecast ship traffic flow of Fangcheng port. The new prediction model is superior to BP neural network model by comparing the relative residuals of predictive value, which means the new model can improve the prediction accuracy.

1 Introduction

Ship traffic flow is composed of ships and other water transports. In order to establish the actual vessel traffic flow characteristics model to reveal the effects of changes of regularity and other factors, it's important to research the basic characteristics with time and space. The distribution regularity of ship arrival and departure is the basis research of vessel traffic flow, channel capacity, vessels accident analysis and whether the channel dimension is reasonable. The distribution regularity research of Tianjin port main channel and Xijiang River waterway obey normal distribution [1-2]. Studies on container ship to the port and Jinzhou Yangtze River Bridge waterway fit Poisson distribution [3-4]. ZHANG Huai Hui demonstrates that the distribution law of ships to port obeys the binomial distribution by mathematical statistics, the mathematical and physical meanings are clear and afford to the objective law, the binomial distribution matches the truth, the Poisson distribution is a similar method of the binomial distribution [5]. The distribution model varied because of characteristics of port and waterways .Currently there is no a uniform distribution model to describe the ship regularity. In this paper, we fit the data using different statistical distribution models and finally get the optimal distribution model.

Accurate forecast of ship traffic flow is the key to maritime traffic control and induced. How to accurately forecast traffic flow has been a hot research scholars. For years, experts and scholars established a variety of forecasting models and methods, such as the historical average, time series, regression analysis, Karman filtering, etc. [6].These traditional prediction methods are basically carried out by ways of mathematical statistics and the prediction accuracy is low and does not have the self-learning ability. BP artificial neural network is a network in a classical algorithm and it itself has a strong nonlinear mapping and self-learning ability to improve prediction accuracy. However, this model has two drawbacks. First, it's easy to fall into local optimal solution. The second is the convergence speed is slow, which will lead to decline the prediction accuracy. Scholar modelled models to improve Prediction accuracy including the methods of BP (Back Propagation) neural network， optimized BP neural network based on modified particle swarm optimization algorithm and nonlinear time series based on BP neural network [7-9].

Markov chain is a stochastic discrete process in mathematics which has the Markov properties. With a given knowledge or information, the past (previous status) is irrelevant to predict the future (future status).Scholars established gray Markov chain model to predict traffic volume, which could get a better prediction accuracy to meet the requirement of short-term forecast [10-11]. LI Huai Jun combined exponential-smoothing-Markov short-term traffic flow forecast method, which could solve the problems existing in exponential smoothing and improve forecast accuracy [12]. Yan Qi proposed a stochastic approach, Hidden Markov Model (HMM), for short-term freeway traffic prediction [13]. Some researcher combined BP neural network theory and Grey neural network with Markov theory to improve the prediction of annual runoff variation [14-15]. WANG Yi Fan incorporated the Markov chain concept into fuzzy stochastic prediction [16].

However, the gray model is only available for a strong exponential sequence, which can only describe monotonous and mutative process, it does not apply to non-monotonic sequence or swing development saturated S-shaped sequence. The size of the ship traffic is closely related to the weather, freight volume, sea conditions and so on, so the trend of ship traffic flow will not necessarily exhibit exponentially. In this paper, BP neural network model and Markov chain model will be combined to improve the short-term ship traffic flow.

2 Distribution model and hypothetical test and forecasting model

2.1 Distribution model and hypothetical test

The regularity distribution model divided into normal, Poisson, binomial, negative binomial, logistic, gamma (Γ distribution), t-Location Scale, Waybill and so on.

2.2 Forecasting model

2.2.1 BP neural network

BP network is based on Back Propagation algorithm for Multi-layer feed forward neural network（FNN）, which was first proposed by Rumelhart and McClelland in 1985. It is currently the most widely used and the most powerful of an artificial network. The algorithm learning process is consists of the forward propagation and reverse propagation. In the forward propagation process, beginning from the input layer input information pass to the layers, after treatment then produces an output, and get an error of the output to the desired output. Another again to reverse the spread of computing, from the output layer to the input layer, using forward error propagation obtained connection weights adjusted layer by layer, making the desired network output close to sequential output.

2.2.2 BP neural network modelling steps

1) Determine the network structure: Determine the network layer, the number of neurons within each layer and excitation function.

$$f(x) = 1/(1+e^{x}) \tag{1}$$

Input layer is i, Hidden layer is j and output layer is k.

2) Initial values to the connection weights w_{ij}, w_{jk} and neuron threshold θ_j of Neural Networks. At the same initialization sample data, set the minimum error E, learning rate η, the maximum number of training n.

3) Calculate the hidden layer network output and actual output according to the input samples.

Hidden node output:

$$O_j = f\left(\sum_i w_{ij}x_i - \theta_j\right) \tag{2}$$

Output node output:

$$y_k = f\left(\sum_i w_{jk}x_k - \theta_k\right) \tag{3}$$

4) Calculate the total error of actual output and the desired output:

$$E = \sqrt{\frac{1}{N}\sum_{u=1}^{N} e_u^2} \tag{4}$$

where, N is the number of samples, e_u is the error of actual output and the desired output, u is the sample number. If the error meet the requirements, then end the training, or continue.

5) Repeatedly adjust the weighting value of each neuron, making E to achieve the error range requirements by repeat steps 3 to 5.

2.3 Markov prediction model

Markov chain predictions is a method which based on the initial status probability vector and the status transition probability matrix to speculate about the future status of a certain period of a variable. This theory is based on Markov process, which describes the dynamic process of a random time series.

2.3.1 Markov prediction model status division

According to Markov chain sequence data into a plurality of different statuses, which are represented with E_1, $E_2 \cdots E_m$. *The* status transition only occurs in t_1, $t_2 \cdots t_m$.

2.3.2 Status transition probability matrix

The transition probability of Markov chain with the status of E_i through k step transition to status E_j is represent by $p_{ij}^{(k)}$.

$$p_{ij}^{(k)} = \frac{m_{ij}^{(k)}}{M_i} \tag{5}$$

where: M_i is the total number of status E_i appears, $m_{ij}^{(k)}$ is the times of status E_i through k step transition to status; m is the number of status division. Then the first step in the status transition probability matrix as follows:

$$\boldsymbol{P}^{(1)} = \begin{bmatrix} p_{11}^{(1)} & p_{12}^{(1)} & \cdots & p_{1m}^{(1)} \\ p_{21}^{(1)} & \cdots & \cdots & p_{2m}^{(1)} \\ \vdots & \cdots & \vdots & \vdots \\ p_{m1}^{(1)} & p_{m2}^{(1)} & \cdots & p_{mm}^{(1)} \end{bmatrix} \tag{6}$$

DOI: 10.1051/matecconf/20168104007

Repeated use of C-K equation (Chapman Kolmogorov equation), then the k-step transfer probability as follows.

$$P^{(k)}=(P^{(k)})^k \quad (7)$$

Set the initial vector of a variable in the initial status E_i as $P^{(0)}$, the status vector after k steps transfer as follows.

$$P_k= P^{(0)}\times P^{(k)}= P^{(0)}\times(P^{(1)})^k \quad (8)$$

2.4 BP neural network and markov forecasting model

Firstly, instruct the BP neural network model, reach conclusion of prediction sequence derived from the actual sequence, and get the Relative residuals. After the residuals is classified to [0,1], the classified relative residuals is placed under the status division, and then calculating the status transition probability matrix, therefore predicting the future trends.

Table 1. The statistical characteristics of each vessel traffic flow of Fangcheng port

	Into port				Out port			
	Data 1	Data 2	Data 3	Data 4	Data 5	Data 6	Data 7	Data 8
Min	0	0	2	3	0	0	1	2
Max	19	70	51	102	22	66	52	104
Means	2.77	21.39	24.17	48.33	2.78	21.05	24.31	48.14
S.d	3.20	11.14	7.34	16.00	3.23	11.19	7.22	16.00

2.4.1 Modelling steps

1) Instruct the BP neural network model.

2) Calculate the classified relative residuals.

$$x(t)=\frac{\hat{X}(t)-X(t)}{X(t)} \quad (9)$$

$$\hat{x}=\frac{x-x_{min}}{x_{max}-x_{min}} \quad (10)$$

where, $X(t)$is the actual sequence of values, $\hat{X}(t)$ is the predicted value sequence, $X(t)$ is the relative residuals. X is an element of the $x(t)$, x_{min} is the minimum element of the $x(t)$, x_{max} is the maximum element of the $x(t)$, $\hat{x}$ is an element of the $x(t)$ which is classified.

3) Calculate the status transition probability matrix.

4) Prediction. The status vector after k steps can be obtained by Equation (8), in order to determine the status interval in which the k steps is, thereby obtaining a final prediction.

3 Experiment

3.1 Analysis of distribution regularity of ship arrival and departure

3.1.1 Data sources

Gather the number of ships entering and leaving the Fangcheng port channel from October 2014 to September 2015, and use the samples to analyse the regular pattern of vessel entering the port of Fangcheng port.

3.1.2 Statistical analysis

The statistical characteristics of each vessel traffic flow of Fangcheng port are shown in Table 2.

Table 2. The parameter of each distribution model

	Distribution Model	Into port	Out port
Channel 1	Negative Binomial	r = 0.97, p = 0.26	r = 1.11, p = 0.29
Channel 2	Normal	μ=21.39, σ=11.14	μ=21.05, σ=11.19
	Negative Binomial	r = 3.77, p = 0.15	r = 3.67, p = 0.15
	t-Location Scale	μ=20.64, σ=9.28, v=6.38	μ=20.05, σ=8.70, v=4.83
	Logistic	μ=20.71, s=6.09	μ=20.26, s=6.00
Channel 3	Normal	μ=24.17, σ=7.34	μ=24.31, σ=7.22
	Negative Binomial	r = 18.55, p = 0.43	r = 20.63, p = 0.46
	t-Location Scale	μ=23.88, σ=5.39, v=3.94	μ=23.86, σ=4.50, v=2.73
	Logistic	μ=23.96, s=3.90	μ=24.03, s=3.70
Main Channel	Normal	μ=48.33, σ=16.00	μ=48.14, σ=16.00
	Negative Binomial	r = 8.89, p = 0.16	r = 8.72, p = 0.15
	t-Location Scale	μ=48.45, σ=13.73, v=7.31	μ=48.21, σ=13.60, v=6.89
	Logistic	μ=48.46, s=8.81	μ=48.20, s=8.79

3.1.3 Hypothetical test

Less than three vessels berthing in Channel 1 everyday, the data of into port obey negative binomial distribution(r=0.97,p=0.26). The data of out port obey negative binomial distribution(r=1.11,p=0.29).

There are about 21 vessels inbound and outbound in channel 2 everyday. When inbound , its distribution obey t-Location Scale Distribution(u=20.64,σ=9.28,v=6.38) and Logistic Distribution(u=23.96,s=3.90),but its divergence 14.741<16.938,So t-Location Scale Distribution is superior to Logistic Distribution, when outbound, its distribution obey t-Location Scale distribution (u= 20.05,σ=8.70, v=4.83).

There are about 24 vessels inbound and outbound in channel 3 everyday. When entering the port, its distribution obey t-Location Scale distribution (u= 23.88, σ= 5.39,v= 3.94) and Logistic Distribution (u = 20.71,s = 6.09). But its divergence 13.171< 22.108 , So t-Location Scale Distribution is better than Logistic Distribution, when outbound, its distribution obey t-Location Scale distribution(μ=23.86,σ=4.50,v=2.73).

About 48 vessels inbound and outbound in main channel every day. When entering the port, the distribution obey the normal distribution (μ=48.33,σ=16.00), t-Location Scale distribution (μ= 48.45, σ=13.73,v=7.31) and Logistic Distribution (μ= 48.46,s=8.81), but its divergence 7.366<10.685<16.269, So Logistic distribution is the best. When departing, the distribution obey the normal distribution (μ= 48.14, σ= 16.00), t-Location Scale distribution (μ=48.21,σ=13.60, v=6.89) and Logistic Distribution (μ=48.20, s=8.79), but its divergence 9.520<9.947<20.623,so Logistic distribution is the best.

3.2 Prediction of ship traffic flow

For convenience , the data of the first 12 days of 12 months are used to evaluate the performance of the proposed vessel traffic flow forecasting scheme. The first 11 data points are used as the training sample, while the remaining 1 data points are employed as the testing sample for measuring forecasting performance of the proposed model. Data are shown in Table 3.

3.2.1 BP neural network forecast

Set up BP neural network forecast model, input layer i=24, hidden layer j=10 and output layer k=12, Training times is 5000 and the error is 5%.The result of prediction see Table 4.

3.2.2 Markov forecast model

(1) Status Division. With the method of Golden Ratio [9], the relative residuals classified can be divided into three statuses in Table 4, which are [0, 0.286], (0.286, 0.748], (0.748, 1]. The relative residuals can be divided into three statuses according to Formula (9), which are E_1[-29.32%,-2.56%], E_2(-2.56%, 40.66%], E_3(40.66%, 64.23%].

From Table 4, the initial status of No.10 is E_1,which means initial vector $\boldsymbol{P}^{(0)}$=[1 0 0].Revert the value of X(t) according to Equation (11). $\hat{X}$ (11)=77.16, the probability of next status is E2, the value interval of X(11) is [75.23,130.03] and the average value is 103. Similarly, after two-step transfer, $\hat{X}$ (12)=44.98, the most possible status of two-step transfer is E_2 or E_3, the value interval of X(12) is [43.86, 75.80] or [75.80,113.1]and the average of the two- interval is 77. The comparison of each model's prediction value see Table 5.

Table 3. The data of the first 12 days of 12 months

	Days											
	1	2	3	4	5	6	7	8	9	10	11	12
10	52	47	42	31	33	40	53	70	73	73	77	64
11	58	55	39	58	60	51	54	46	51	55	59	68
12	37	30	30	27	46	61	34	27	46	71	62	38
1	62	46	49	55	42	34	49	54	49	37	27	37
2	39	59	73	73	60	54	53	48	59	70	55	45
3	26	29	37	49	49	47	47	45	47	70	72	58
4	56	63	59	45	32	26	27	56	82	91	69	54
5	60	40	16	46	52	52	56	56	50	46	51	45
6	40	47	49	51	33	49	37	31	41	46	59	59
7	45	71	71	46	36	33	58	56	57	56	56	65
8	37	54	58	55	52	48	45	52	42	52	80	61
9	38	36	33	39	45	46	42	40	42	71	96	73

Table 4. Result of BP neural network prediction model

No.	M	A.v	P.v	R	R.r	R.r.c	S
1	1	38	31.12	-6.88	-18.11%	0.120	E_1
2	2	36	47.27	11.27	31.31%	0.648	E_2
3	3	33	47.19	14.19	43.00%	0.773	E_3
4	4	39	64.05	25.05	64.23%	1.000	E_3
5	5	45	57.28	12.28	27.29%	0.605	E_2
6	6	46	54.99	8.99	19.54%	0.522	E_2
7	7	42	39.26	-2.74	-6.52%	0.244	E_3
8	8	40	41.78	1.78	4.45%	0.361	E_2
9	9	42	43.29	1.29	3.07%	0.346	E_2
10	10	71	50.18	-20.82	-29.32%	0.000	E_1
11	11	96	77.16	-13.84	-15.21%		
12	12	73	44.98	-23.02	-38.38%		

where :M means month,P.v means Predictive value,R means Residual,R.r.c means Relative residuals,S means Statues

4. Discussion

4.1 Analysis of distribution regularity of ship arrival and departure

By chi-square test, the data of Channel 1 obey negative binomial distribution, the data of Channel 2 and 3 obey t-Location Scale distribution, the data of main channel obey Logistic distribution.

Table 5. The comparison of each model's forecast value

No.	A.v	BP P.v	BP R.r	BP & M P. v	BP & M R. r
11	96	77.16	-15.21%	103	7.29%
12	73	44.98	-38.38%	77	5.48%

where : A.v means Actual value, BP P.v means BP Predictive value, BP R.r means BP Relative residuals, BP & M P. v means BP and Markov chain Prediction values, BP & M R. r means BP and Markov chain Relative residuals.

4.2 Prediction of ship traffic flow

Table 5 shows that the use of the prediction model of BP neural network and Markov chain to forecast vessel short-term traffic flow, which can reduce the relative residuals from -15.21% and -33.38% to 7.29% and 5.48%. The absolute value of the prediction error is less than 10%.The model can improve prediction accuracy.

5. Conclusion

The results of Fangcheng port shows that the distribution regularity of ship arrival and departure may obey few of distribution models. In this paper, we use a variety of distribution models to fit the data, avoiding the error caused by one fitting.

The prediction accuracy of short-term ship traffic flow is great significance to ensure the safety of navigation, efficient use of resources and reduce maritime accidents. This paper setup a prediction model based on BP neural network and Markov Chain to improve the prediction accuracy of short-term ship traffic flow. The empirical study indicates that using this method can achieve forecasting accuracy improvement feasible. This paper only forecast the ship traffic flow from the macro and did not consider the impact of other factors, such as weather, sea conditions, and policies and so on. In the future studies, we can put these factors into consideration to design a more efficient traffic flow forecasting model.

References

1. J.X. Liu, Y.B. Li, J. WHUT(T.S.&E), *Distribution Regularity Analysis of Ship Arrival and Departure at the Tianjin Port Main Channel*, **32**(2), 351-353(2008)
2. J. Yu, W. Zhang, J.H. Jiang, P. Liao, J.T &T.E, *Probability distribution of vessel traffic flow in Xi Jiang waterway*, **6**(2), 88-93(2006)
3. X.L. Han, Z.Q. Lu, L.F. Xi, Eur.J.O.R, *A proactive approach for simultaneous berth and quay crane scheduling problem with stochastic arrival and handling time*, **207**, 1327-1340(2010)
4. Z.R. Tan, X.P. Yan, L. Liu, J.T.I&S, *Examination of Vessel Traffic Flow Distribution Model in Yangtze River Bridge Area*, **28**(2), 70-73(2010)
5. H.B. Zhang, D.H. Jiang, J.P&W.E, *Distribution law of ships to port and calculation of reasonable number of berths*, **9**, 51-56(2014)
6. Wang,L.J.Xu,N.Q.Song,T.Xu,J.DMU, *Distribution function and empirical study of container liner's arrival discipline*, **39**(4),107-110(2013)
7. Y.H. Tian, J.B. Chen, J.S&O.E,*Vessel Traffic Flow Prediction Based on BP Neural Network*, **39**(1), 122-125(2010)
8. S. Li, L.J. Liu, M. Zhai, J. Systm, E.T&P,*Prediction for short-term traffic flow based on modified PSO optimized BP neural network*, **32**(9), 2045-2049(2012)
9. J.G.Zhai, T.F.Yan, X.P.Yan, J.SHMU, *Prediction of vessel traffic flow based on BP neural network and residual analysis*, **34**(1), 19-22(2013)
10. Z.M.Liu, Z.X.Jia, X.L.Li, J.ECJTU, *Traffic Volume Forecast Based on Gray Markov Chain Model*, **29**(1), 30-34(2012)
11. C.M.Kong, Y.Han, J.F&E, *Prediction of Volatile Traffic Volume Based on the Grey-Markov Model*, **31**(1), 92-96(2015)
12. H.J.Liu, Z.Gao, Z.X.Wang, J.Zhang, J.S&T(Algorithm), *Short-term Traffic Flow Forecasting Based on Exponential Smoothing and Markov Chains*, **22**(12), 132-135(2013)
13. Q.Yan, I. Sherif, J.T&R, Part C, *A Hidden Markov Model for short term prediction of traffic conditions on freeways*, **43,** 95-111(2014)
14. Y.M.Wang, X.J.Yu, J.X.Chang, Q.Huang, J.EofWH, *Pr-ediction of runoff based on BP neural network and* Markov model 41(5), 14-17(2008)
15. Y.P.Jing, X.Zhang, Y.Luo, J.N.A&F.U(Nat. Sci. Ed.), *Forecasting of urban water demand based on combining Grey and BP neural network with Markov chain model*, **39**(7), 229-234(2011)
16. Y.F.Wang, S.M.Cheng, M.H.Su, J.A.soft.C, *Incorporating the Markov chain concept into fuzzy stochastic prediction of stock indexes*, **10**, 613-617(2010)
17. G.G.He, Y.Li, S.F.Ma, J.Systm, E.T&P, Discussion on short-term traffic flow forecasting methods based on mathematical models, **20**(12), 51-56(2000)

MATEC Web of Conferences **81**, 04008 (2016) DOI: 10.1051/matecconf/20168104008
ICTTE 2016

Evaluation and Modelling of Traffic Noise on the Asian Highway in Golestan National Park, Iran

Shiva Gharibi[1], Mozhgan Aliakbari[2], Abdolrassoul Salmanmahiny[1] and Hossein Varastehe[1]

[1]*Gorgan University, Gorgan, Iran,*
[2]*RMIT University,Melbourne ,Australia*

Abstract. The increasing number of vehicles on Iran's highways and major roads has led to an increase in noise levels. As a result, traffic is now considered a main source of noise pollution. This paper reports on the modelling of traffic noise levels in Golestan National Park, Golestan using vehicle data and other environmental features. For the evaluation of noise and the recording of independent environmental variables, Sampling stations were selected using a systematic-random method at 76 points at various distances and between 0-250 meters from the road. At each sampling point, traffic flow (number and speed of vehicles, number of horn beeps) was measured for 15 minutes from 8 am to 8 pm. Simultaneously other environmental variables were assessed, including the geometry of the road surface and location conditions .The best multivariable regression based on the correlation coefficient (R) and the coefficient of determination (R2) was achieved. The R-square (73%) and the adjusted R-square (68%) of the regression equation were 73% and 68% respectively. The results of modelling show that the most important variables affecting noise pollution are distance from the road, roughness coefficient, speed of medium-weight vehicles, relative humidity, and height and number of light vehicles. There is a negative correlation with distance from the road and noise pollution.The accuracy of the model was found to be about ±5 dB. Therefore, the model is suggested for the prediction of traffic noise on the Asian Highway in Golestan National Park.

1 Introduction

Environmental noise is a worldwide problem, and noise is becoming an increasingly omnipresent, yet unnoticed, form of pollution, even in developed countries [1]. The sources of noise are broadly divided into two parts: industrial and non-industrial. Noise originates from human activities, especially the development of transport and industry. Road traffic noise is a non-industrial noise that is the most widespread source of noise in all countries and the most prevalent cause of annoyance and interference. It is directly proportional to the volume of vehicles. According to Fig. 1, noise generated by vehicular traffic is the most important source of noise [1].

However, noise from road traffic can be measured or predicted by using a common model and prediction model [2] .Both models are helpful tools to evaluate traffic noise conditions and traffic noise analysis while prediction models are important tools for planning and design of new roads/highways. Traffic engineers often compare traffic to a fluid, assuming that a certain volume must flow through the road system. Traffic noise prediction models in the 1950s and 1960s were designed to predict a single-vehicle sound pressure level (Lp) at the roadside. Later models were not intended to predict single vehicle levels [3]. Traffic noise can be either measured or predicted in the case of noise pollution management. Ref [2] Since noise measurement is complicated and time-consuming, and moreover impossible at the design stage , traffic noise prediction models are required for the design of new highways and other roads or the redesign of traffic flow on existing roads in order to achieve comfortable traffic noise conditions. In the last few years, a number of prediction tools have been created to estimate noise levels in several countries [4].

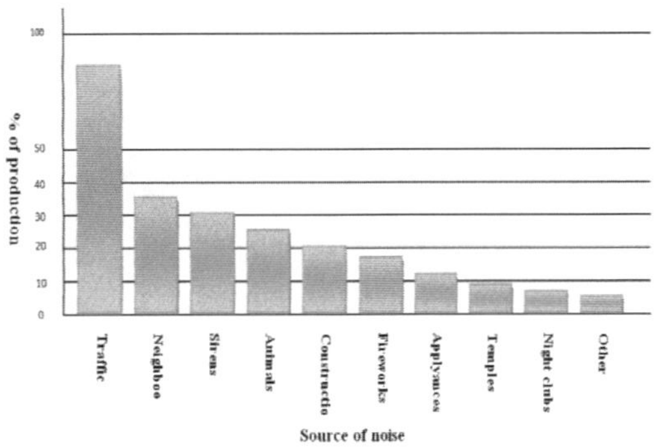

Figure 1. The most important sources of noise [1].

2 Different prediction models

Traffic noise is a component of environmental noise and is the result of the sum of various types of noise from traffic sources (e.g., cars, buses, trains, etc.). There are many parameters directly or inversely correlated with traffic noise levels [5]. For example, increased traffic volume, in terms of vehicles per hour, causes an increase in the noise level [6].The models based on the correlation between traffic and geometrical parameters, with the *Leq (equivalent continuous sound pressure level in dB)*measured in experimental studies, generally have as input the traffic flow, the composition and the average speed of the vehicular current, slope, the ratio between the height and the distance of the barriers and sometimes the road bed conditions [7].

The most important variables considered in the development of mathematical models are traffic flow rate and composition [1].The earliest road traffic noise model was that given in the 1952 Handbook of Acoustic Noise Control. It was offered for speeds of 35±45 mph and distances greater than 20 feet [3]. Lam and Tam (1998) in Hong Kong developed a noise prediction tool using the Monte-Carlo technique Ref [4] taking into account traffic flow, traffic speed and traffic composition in terms of percentage of heavy vehicles.

Reed [8] used a SPreAD-GIS tool for modelling spatial patterns of anthropogenic noise propagation in natural ecosystems. The SPreAD-GIS incorporates commonly available datasets on land cover, topography, and weather conditions to calculate noise propagation patterns. Ref [5] describes the development of a model for assessing TRAffic Noise EXposure (TRANEX) in an open-source geographical information system and high correlation was seen between modelled and measured L_{eq}.

Therefore, since each of the current noise prediction models only a few variables and there is no model for the natural environment in Iran, a model is required to predict traffic noise on the Golestan Road taking into account all variables affecting the noise of traffic.

3 Methodology

Figures and tables, as originals of good quality and well contrasted, are to be in their final form, ready for an operational process to develop a prediction model based on characteristics influencing the equivalent noise level has been defined and is shown in Figure 2.

3.1. Study area

Golestan National Park is the first national park in Iran and in the classification of protected areas in the World Conservation Union for the conservation of nature and natural resources it is in second place [9]. This park was determined as a National Park in 1967 and was selected as a biosphere reserve in 1976 [10]. It is located in the northeast of Iran and between the 37° 16' 43″ - 37° 31' 35″ of north and the 55°43'25″- 56°17'48″ of east longitude, and covers 91.895 hectares. The park is located between three provinces: Golestan, Khorasan and Semnan [11]. Fig. 3 shows a view of the study area.

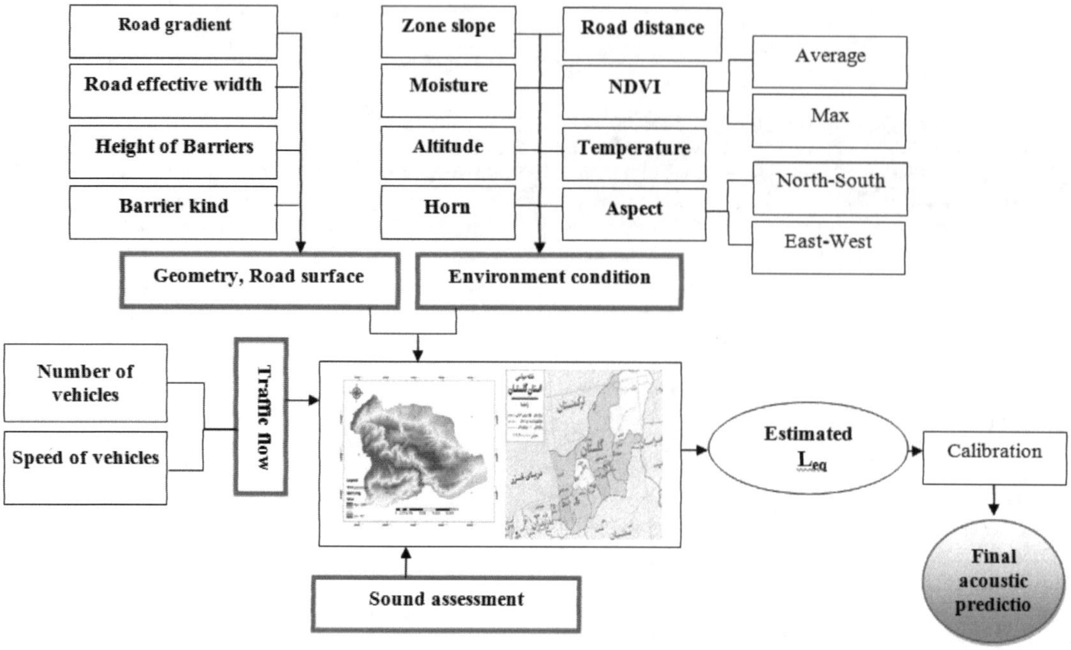

Figure 2. Flowchart of the operative process.

Figure 3. A view of the study area (Shiva Gharibi, 2011).

3.2 Technique

This study was conducted in September 2012, along the Asian Highway in Golestan National Park. The study was conducted during a week of the year with the heaviest traffic flows at various times from 8 am to 8 pm. At first the GIS software used to determine the unique area based on Digital elevation model (Dem), slope, aspect and Normalized Difference Vegetation Index (NDVI). Then 250 meter buffers were applied on two sides of the road and this zone contained 11 homogeneous regions. Using IDRISI software, sampling stations were selected in this area at 76 points using a systematic-random method at various distances between 0-250 metres from the main road (Fig. 4) and covered the entire Golestan road (26.51 km).

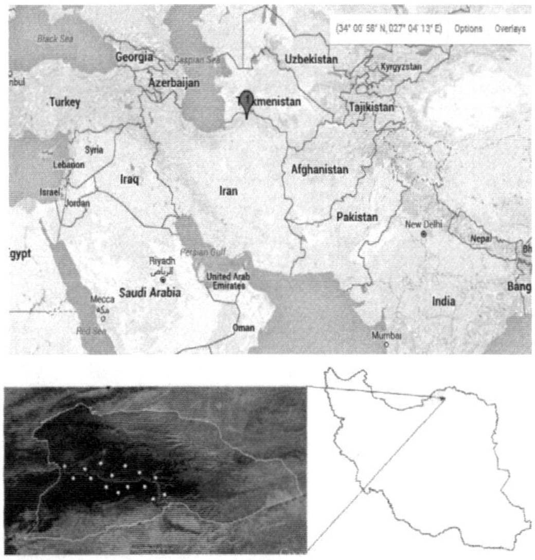

Figure 4. Location of Golestan national park and sampling stations.

Ref [12] at each sampling point, measurements of sound and other variables were taken at 15 minute intervals from 8 am to 8 pm. Sound level meter devices model CEL-450 were used for evaluating the dependent variable or Leq. Simultaneously 19 independent variables affecting the sound were also measured. The developed model has as many Independent variables as possible for estimation traffic noise (Leq). Three groups of variables were considered in the model, as shown in Table 1.

The number of vehicles was recorded using a video camera, and the speed of vehicles was measured using EDIUS 6 software and then physical formulations. By multiplying this numbers at 4, number of vehicles and horns was obtained in 1hour. Measurement and data collection was performed in lack of rainfall and wind environmental conditions and each variable was measured within a circular plot and the centrality of the sampling point and others were measured at roadside. Fig. 5 shows some of the variables measured. Each of these variables makes a specific contribution to sound emission, but some may have a much larger role in the in spread of sound and some may have little role. Fig. 5.

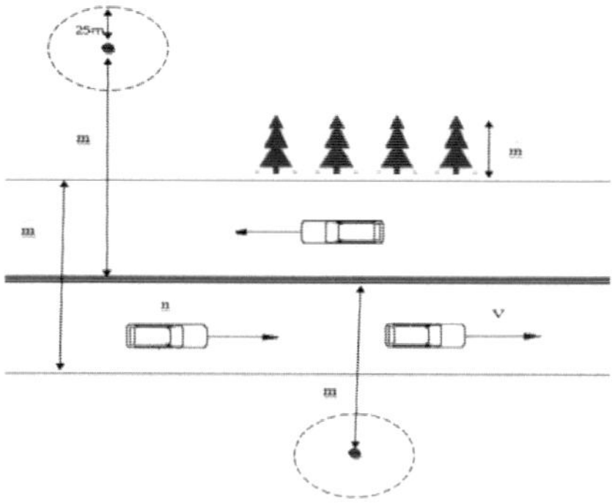

Figure 5. Graphic form of sampling plot.

3.3 Model development

Modelling based on L_{eq} (1h) as the dependent variable and 19 other variables as independent variables was performed using SPSS software. To provide the best multivariate regression, the stepwise regression method was used with a correlation coefficient (R) and the coefficient of determination (R^2). In the modelling, due to the high correlation coefficient L_{eq} (15) instead of L_{eq} (60) can be used [15]. Therefore, a model was developed for equivalent sound level (1h) and to evaluate the final model, data from 26 samples out of 76 samples were used for model validation.

Table I. Model inputs (Independent variables affecting the sound)

A. Traffic flow	B. Geometry of road surface	C. Location conditions
1. Number of light vehicles	8. Road slope (per cent)	11. Distance from road (m)
2. Number of medium-weight vehicles	9. Road effective width (m)	12. Slope of sampling plot
3. Number of heavy-weight vehicles	10. Noise barrier height	13. Altitude
4. Speed of light vehicles		14. Average of NDVI *
5. Speed of medium-weight vehicles		15. Temperature (°C)
6. Speed of heavy - weight vehicles		16. Moisture (per cent)
7. Number of horn beeps		17. Roughness coefficient (per cent)
		18. North-South aspect
		19. East-West aspect

*Normalized Difference Vegetation Index (NDVI)

DOI: 10.1051/matecconf/20168104008

4 Results

The results showed that the noise level at all stations based on the equivalent sound level was 51-79 dB. The values of equivalent sound levels at all stations are shown in Fig. 6.

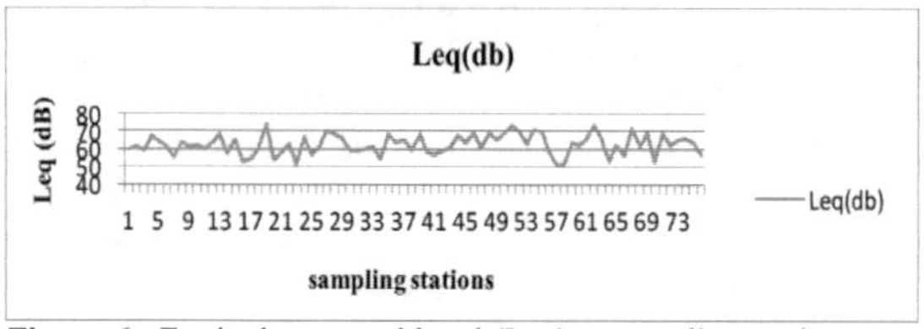

Figure 6. Equivalent sound level (Leq) at sampling stations

For the modelling, according to the goodness-of-fit test results, the best linear relationship between variables was determined. The result of multivariate linear regression analysis of the Leq (1h) model is obtained with the correlation coefficient (R) and the coefficient of determination (R2) that is given as follows (Equation 1).

$$L_{eq}(1h)= 75.61- 14.2Log(Road_{dis})+ 0.654Log(rugged)+ 0.018\ (V_m)+ 0.106\ (Mois)+ 0.006\ (z)+ 0.005\ (Nl) \quad (1)$$

where,

$Road_{dis}$: distance from a traffic line to receiving point (m)
Rugged: Ground roughness coefficient (per cent)
V_m: Speed of medium-weight vehicles
Mois: Moisture (percent)
Z: Altitude (m)
Nl: Number of light vehicles

Table I shows that the R-square and the adjusted R-square of the regression equation were 73% and 68% respectively and Fig. 7 shows a comparison between measured Leq and predicted values based on the proposed model.

Table 2. Details of predicted model

Model	R	R Square	Adjusted R Square	Std. Error of Estimate
1	0.825	0.726	0.683	3.082

Details of the statistical analysis as descriptive statistics (Table 2) and distribution of errors (Figure 7) are shown in Table 3.

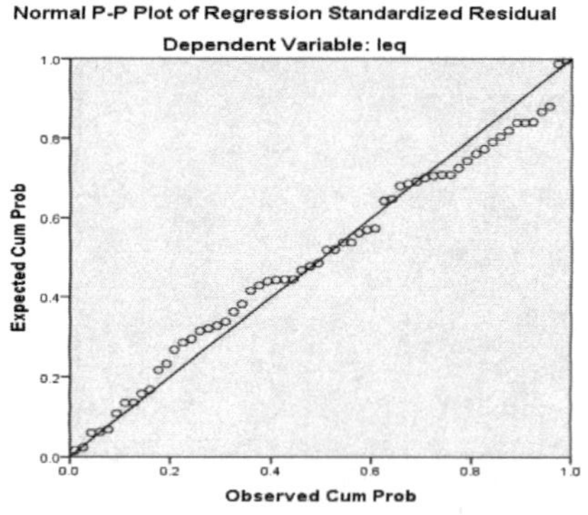

Figure 7. Scatter plot of predicted and measured values of Leq

Table 3. The details of predicted model

Variables	Mean	Standard deviation	Number of sampling points
Equivalent sound level	62.27	5.472	60
Temperature	32.3512	4.599	60
Relative humidity	40.2072	15.1526	60
Road effective width	10.39	3.624	60
Altitude	749.78	166.364	60
Aspect	126.02	121.472	60
Average of NDVI	147.7856	49.69	60
Road slope	27.57	13.308	60
Noise barrier height	7.17	4.555	60
Number of light vehicles	1035.50	196.086	60
Number of medium-weight vehicles	99.61	28.045	60
Number of heavy-weight vehicles	67.97	22.065	60
Speed of light vehicles	165.53	94.211	60
Speed of medium-weight vehicles	138.23	83.841	60
Distance from road (m)	1.7412	0.29943	60
Slope of sampling plot	1.1401	0.46416	60
Speed of heavy - weight vehicles	1.7641	0.13691	60
Number of horn beeping	1.4122	0.36726	60
Roughness coefficient	0.1605	1.98613	60

Based on the results, an acoustic zoning map was obtained by using the interpolation method for the Asian Highway in Golestan National, Park Fig. 8.

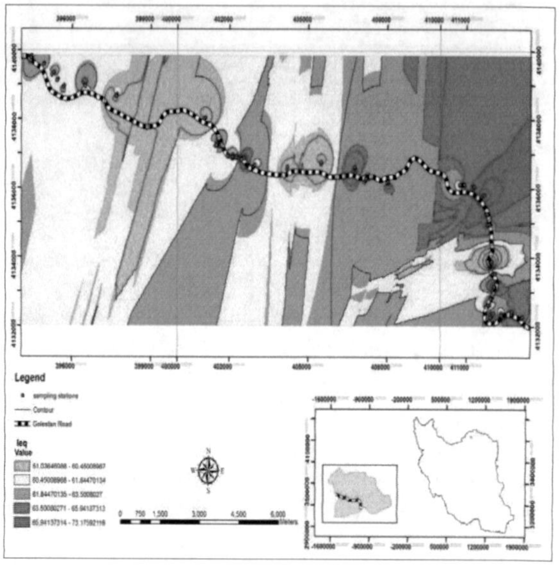

Figure 8. Zoning map of equivalent sound levels (Leq) in Golestan National Park

5 Conclusion

The present study has introduced a model for the prediction of road traffic noise in Golestan National Park. This paper is the result of a study that was performed during 2012 in Iran. The data set consisted of 76 samples, utilized to develop a new model for national park conditions using regression analysis. The measurements show that there is serious noise pollution due to traffic in the park which is not in the acceptable range, according to policies of the Environmental Protection Agency of Iran. The results of modelling showed that the most important variables affecting noise pollution are distance from the road, roughness coefficient, speed of medium-weight vehicles, relative humidity, and height and number of light vehicles. There is a negative correlation with distance from the road and noise pollution. Environmental noise levels can vary over a wide range as a result of the diversity of site conditions and activities occurring during field measurements. However, this prediction model has been developed to predict the noise in a standard situation where the road surface is dry and there is no snow or ice on the terrain.

6 Recommendation

Highways and roads are one of the most important anthropogenic impacts on the environment and the noise generated by the speed of medium-weight vehicles and the number of light vehicles is the most annoying noise. It is an increasing environmental problem in terms of nuisance and annoyance to species in the communities around the highways in Golestan National Park and the increase in number and speed of vehicles on the roads has caused increasingly greater highway noise problems to the surrounding areas. Some bird species such as woodpeckers, nuthatches, robins, chaffinches and blackbirds are most sensitive to sound levels, but since the noise pollution is a slow killer for the community of birds, very little effort has been made to ameliorate the problem. Therefore, the developed regression model can be applied for all protected areas in the country, because the suggested model has a high coefficient of determination (73%), which indicates the adequacy of the model.

References

1. Zannin, P. H. T.; Calixto, A. D. and Ferreira, J. A. , A survey of urban noise annoyance in a large Brazilian city. The importance of a subjective analysis in conjunction with an objective analysis, Journal of Environmental Impact Assessment Review, **Vol 23**, no 2, Pages 245–255, (2003).
2. Bendtsen, H. 1999. The Nordic prediction method for road traffic noise. Journal of Science of The Total Environment, **Vol 235**, no 1–3, 1, pp 331–338, September (1999)
3. Campbell, S.A. 2001. Critical Review of Some Traffic Noise Perdiction Models, Journal Applied Acoustics. **Vol 62**, no 3 , pp 271–287, (2001)
4. Gundogdu, O., Gokdag, M., Yuksel, F. 2005. A traffic noise prediction method based on vehicle composition using genetic algorithms. Journal of Applied Acoustics, **Vol 66,** no **7,** pp 799–809 , (2005).
5. Gulliver, J., Morley, D., Vienneau, D., Fabbri, F., Bell, M., Goodman, P., Beevers, S., Dajnak, D., Kelly, F. J. and Fecht, D. ,Development of an open-source road traffic noise model for exposure assessment, Journal of Environmental Modelling & Software, **Vol 74**, pp 183–193, (2015)
6. Paz, E. C., Ferreira, A. M. C., and Zannin, P. H. T. ,Comparative study of the perception of urban noise. *Journal of Public Health,* **Vol 39**, no 3, pp 467–472 , (2005)
7. Vincenzo, B., Crocco, F. and Mongelli, D.W.E., A *mathematical model for traffic noise prediction in an urban area.* 4th WSEAS International Conferense on Urban Planning and Transportation UPT2011." Corfu Island, Greece , pp 405-410, (2011)
8. Reed, S., Boggs, J. L. and Mann, j. p., A GIS tool for modeling anthropogenic noise propagation in natural ecosystems. Journal of Environmental Modelling & Software, **Vol 37**, pp 1–5, (2012)
9. Majnonian, H.; Dehzad, B.; Farhang Dareshori, B and Maigoni, H. , *Certificate of Golestan National Park*. (in Persian.), (2000).
10. Darvish Sefat, A., *Atlas of Protected Areas of Iran.* University of Tehran Publications(Persian.), pp 5-157, (2007)
11. Sohrabi, M. Lichens from Golestan National Park (Iran). Journal of Folia Crypotg, Fasc, **Vol 41**, pp 105-108, (2005).
12. Peris, S. J. and Pescador, M. 2004. Effects of traffic noise on passerine populations in Mediterranean wooded pastures. Applied Acoustics, **Vol 65**, pp 57-366, (2004).
13. Samadyar, H, Effective Elemnts of Traffic Noise Control in the Urban Environment , II National Conference of Air plloution and Environmental Health (2007), Iran
14. Singh, N. and Davar, S. C. , Noise Pollution- Sources, Effects and Control. Journal of Hum. Ecol., **Vol 16,** no 3, pp 181–187, 2004
15. Zaree Chahoki, M. A., 2010, *Data analysis in natural resources research using SPSS software.* University of Tehran Press, Iran,1st ed, (2010).

 DOI: 10.1051/matecconf/20168104009

Analysis of Traffic Capacity on Mountainous Two Lane Highway Adding Climbing Lane

Yunwei Meng[1,2], Hailing Li[2], Hejun Chai[2] and Zongling Yan[2]

[1]*The Key Laboratory of Road and Traffic Engineering, Ministry of Education, Tongji University, 201804,Shanghai, China*
[2]*China Merchants Chongqing Communications Technology Research & Design Institute Co., LTD, 400067,Chongqing, China*

Abstract. During the construction of mountainous double lane highway, climbing lane plays a certain role for enhancing the traffic capacity. In order to explore traffic capacity improvement effect for different conditions of climbing lane, 20 representative models are chosen, which contain various combinations of alignment and traffic parameters. The changes of traffic capacity, average speed, delays, and saturation in models are obtained before and after the use of a climbing lane by means of numerical simulation. The results show that the use of a climbing lane could improve traffic capacity, average speed, while reducing delays and saturation. The improvement effect is different according to different combinations of alignment and traffic parameters. The research could provide a reference for mountainous climbing lane construction intend.

1 Introduction

Because of the limitation of geology, topography and engineering investment, two-way two lane highway is common form in mountainous area of China. Two lane highways in mountain area is the main artery, assuming the dual role of the internal and external communication. Its adequate traffic capacity, high safety performance, is particularly important to regional social and economic development. In addition to low alignment parameters, mountainous two lanes are also traffic accident-prone sections, especially in uphill ones. Due to lower travelling speed of trucks, creating a large speed difference between cars, it will significantly reduce the traffic capacity of the road, thereby inducing safety accidents. Adding climbing lane is an effective, economic, environmental and feasible solution to eliminate the phenomenon.

2 Literature review

It will form a larger travel impact on the entire road when the length of the continuous uphill road section and the proportion of truck in traffic volume reach peak value. More serious congestion will occur in the impact area, while car adventure temporarily occupy the opposite lane, it will be more likely to cause traffic accidents [1]. Climbing lane is set on a section with severely reduced traffic capacity, which is a special lane for truck travelling. By means of physical separation [2], a lane is added to the right of the main road for continuous uphill section. After trucks traveling some distance, let it back into the main road, which could improve the traffic capacity of the road and traffic safety [3]. Some research found that it is possible to generate traffic bottlenecks in the lane separation and convergence position. It can effectively ease traffic congestion by extending the length of the transition section [4]. Setting up a reasonable climbing lane, it can slow down the occurrence of traffic congestion [5]. The weight power ratio of the truck is the factor that should be considered when setting up the climbing lane, and it is directly related to the climbing performance of the vehicle. Therefore, the weight power ratio of the truck is an important basis for the calculation of longitudinal slope and slope length of the climbing lane [6].

All of the above studies show that setting up the climbing lane is beneficial to improve the traffic capacity of the section. This paper intends to make a comparative analysis of traffic capacity, average speed, delay and saturation before and after setting up the climbing lane. This paper is intended to evaluate the effect of setting up the climbing lane in mountainous highway.

3 Numerical simulation process

According to the theory of dynamics and kinematics, the running characteristic of each kind of vehicle can be obtained by calculating. When the vehicle is running on longitudinal section, the vehicle is subjected to the combined action of traction and resistance, and the resistance includes air resistance, rolling resistance, slope resistance and inertia resistance. Before reaching the equilibrium speed, the running process of the vehicle can be simplified as a uniformly deceleration motion.

VISSUM is used for simulation. It is a kind of traffic simulation modeling tool, which is microscopic and based on time interval and vehicle driving behavior. It can be used to simulate and analyze the traffic operation condition of various road traffic conditions, and it is an effective tool to evaluate road traffic engineering design and road traffic planning. VISSIM simulation system is composed of many modules, and these modules have different functions.

3.1. Determination of representative vehicle

According to the survey data combined with the development truck industry, the representative model of the two-lane highway truck in the mountainous area of China is chosen as EQ1228v19D2 of Dongfeng truck. Representative vehicle is used in the process of simulation.

3.2 Simulation scheme

The research object of this simulation is the climbing lane of the two lane highway in mountainous area. In general, when the slope of the road is more than or equal to 3%, climbing lane could be proposed.

According to the "Technical Standard of Highway Engineering" (JTG B01-2014), the design speed of the secondary highway is generally 80km/h, 60km/h, and the longitudinal gradient is generally not more than 6%. Taking into account the characteristics of the mountain road, the simulation of the longitudinal slope gradient could be chosen as 3%, 4%, 5%, 6%, and 7% respectively.

At the same time, the design speed of the secondary highway is 60km/h, the maximum slope length of the longitudinal slope is 1200m, and combined with the time complexity of computer simulation, the slope length of the simulation are chosen as 200m, 300m, 600m, 900m, 1200m respectively.

In this simulation experiment, the mixing ratio of the truck in traffic volume can be chosen as 10%, 20%, 30%, 40%, 50%, and 60% respectively.

Combined with the actual traffic of the secondary highway in mountain area and the time complexity of computer simulation, the simulated traffic can be used 1000veh/h, 1500 veh/h, 2000 veh/h, 2500 veh/h respectively.

According to the possible value of above parameters, there are 600 kinds of simulation combination schemes in theory. By consulting the design engineers, and according to the principle of the priority of traffic capacity, 20 groups are selected as the combination of the simulation experiments, as shown in Table 2. The 20 combinations are simulated in the condition of a climbing lane or no climbing lane. The test of effectiveness of setting up the climbing lane is conducted through the comparative analysis.

Table 1. Design parameters combination of simulation

Group number	Longitudinal slope (%)	Slope length (m)	Length of climbing lane (m)	Proportion of trucks (%)	Traffic (veh/h)
1	3	900	300	10	1000
2	4	600	300	10	1000
3	5	300	200	10	1000
4	6	600	300	10	1000
5	7	300	200	10	1000
6	3	1200	300	10	1500
7	4	600	300	10	1500
8	5	600	300	10	1500
9	6	300	200	10	1500
10	7	300	200	10	1500
11	3	1200	300	10	2000
12	4	600	300	10	2000
13	5	600	300	10	2000
14	6	600	300	10	2000
15	7	300	200	10	2000
16	3	300	200	10	2500
17	4	600	300	10	2500
18	5	300	200	10	2500
19	6	300	200	10	2500
20	7	300	200	10	2500

Adding a lane in the climbing section of the secondary two-lane highway, it is used for large vehicles climbing, that is the climbing lane. According to the speed prediction model of truck in steep slope, the change of the running speed of the truck is determined. VISSIM traffic simulation software is used to simulate and analyze the traffic state. The changes of traffic capacity, average speed, and travel delay and service level are studied, which provide the basis for the optimization of the length,

DOI: 10.1051/matecconf/20168104009

slope and other parameters of the climbing lane. Among them, the traffic capacity is classified implementation according to the "Technical Standard of Highway Engineering" (JTG B01-2014).

VISSIM software is applied to simulate the micro simulation of the uphill road section of two lane highway. The specific steps are as follows: (1) Building the simulation road network of the uphill section. The simulation model of road network is established in VISSIM, and the numbers of road lanes, lane width, slope, length and so on are set up. (2) Setting traffic flow parameters. According to the experimental scheme, the road traffic volume, the proportion of vehicles, the expected vehicle speed are input to the road network model. (3) Setting traffic control parameters. Priority of traffic conflict point, give way rules, the deceleration area and speed decision-making point are carried out. Setting the climbing lane is to make fast vehicles and slow vehicles moving apart. Therefore, the traffic management and control module of the separated and convergence from main road are needed. (4) Setting evaluation module, and simulation runs and output evaluation index. (5) Analyzing the output of the evaluation parameters.

4 Result analysis

According to the 20 groups of combination design in Table 1, simulation analysis is carried out. The traffic capacity, driving speed, delay, saturation values of uphill section are obtained before and after the setting climbing lane. Only figures of group 1-5 are shown, while figures of other groups are similar.

4.1 Traffic volume=1000veh/h, group1-5

In the condition of road traffic volume 1000veh/h, the truck proportion 10% and the slope gradient 3%, 4%, 5%, 6% , 7%, the parameters of the slope length and the length of climbing lane are obtained by the five groups, namely group 1-5. The traffic capacity of section, the average speed and the delay of the car and the saturation are obtained through the numerical simulation, as shown in Fig. 1.

The simulation result analysis shows that, after setting the climbing lane, traffic capacity has increased significantly. Climbing lane enable truck and car lane driving respectively, improving the degree of freedom of the car, so the vehicle's average speed increases and the vehicle delay is low. Moreover, after setting climbing lane in the given traffic flow conditions, the degree of dispersion has a larger decline. It shows that the setting the climbing lane can increase the speed of the car, and the speed is more stable, which can effectively improve the level of climbing. After addition a climbing lane, the saturation of the main road sections are reduced to some extent, enhancing the road service level, improving the service quality of the road. It is improved from the original fourth service levels to three levels of service.

4.2 Traffic volume=1500veh/h, group 6-10

In the condition of road traffic volume 1500veh/h, the truck proportion 10% and the slope gradient 3%, 4%, 5%, 6%, 7%, the parameters of the slope length and the length of the climbing lane are obtained by the five groups, namely group 6-10.

The simulation result analysis shows that, after setting the climbing lane, traffic capacity has increased significantly. It is improved from the original fourth, fifth service level to fourth service level.

4.3 Traffic volume=2000veh/h, group11-15

In the condition of road traffic volume 2000veh/h, the truck proportion 10% and the slope gradient 3%, 4%, 5%, 6%, 7%, the parameters of the slope length and the length of the climbing lane are obtained by the five groups, namely group 11-15.

The simulation result analysis shows that, after setting the climbing lane, traffic capacity has increased significantly. It is improved from the original fifth service level upgrade for the fourth service level.

4.4 Traffic volume=2500veh/h, group 16-20

In the condition of road traffic volume 2500veh/h, the truck proportion 10% and the slope gradient 3%, 4%, 5%, 6%, 7%, the parameters of the slope length and the length of the climbing lane are obtained by the five groups, namely group 16-20.

The simulation result analysis shows that, after setting the climbing lane, traffic capacity has increased significantly. It is improved from the original fourth, fifth service level upgrade to fourth service level.

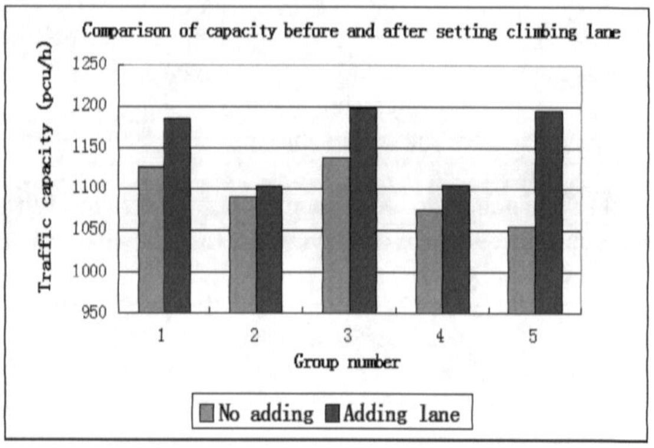

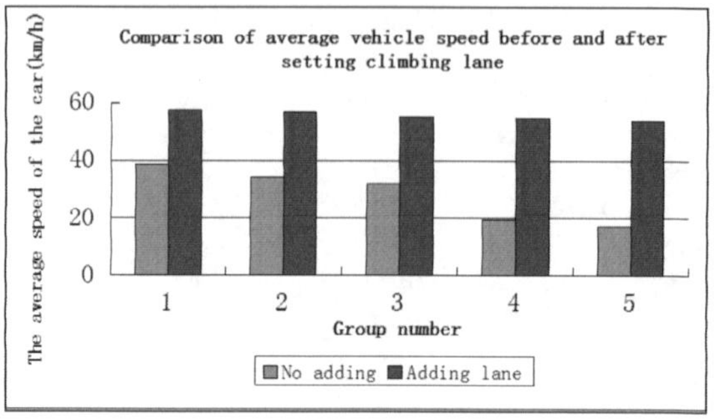

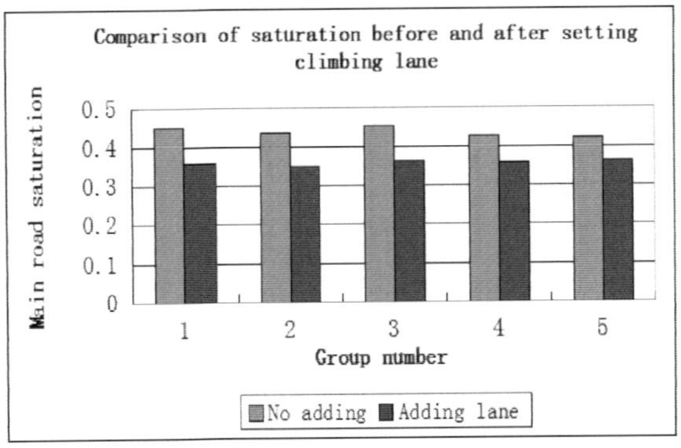

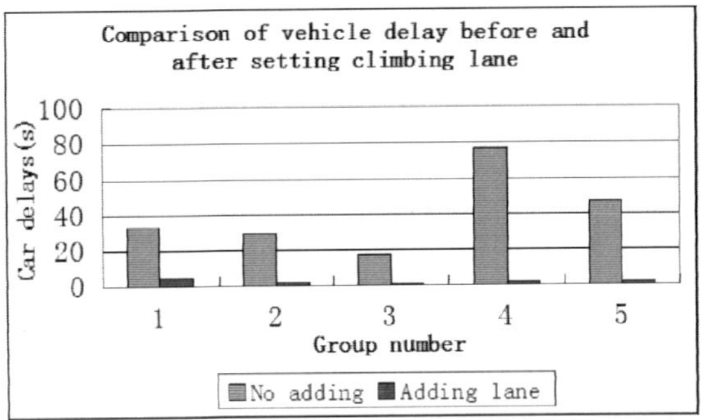

Figure 1. The comparison parameters between before and after setting climbing lane (group 1-5)

5 Conclusion

From numerical simulation, after setting climbing lane to mountainous double lane highway, the traffic capacity of the section, average speed increase while delays and saturation are decreased, especially in the car delays, and the level of service has increased. The results show that adding climbing lane could improve the traffic capacity, while it is conducive to road safety. It should also be noted that the effect of adding a climbing lane is not the same for different alignment and traffic combinations. As a combination of 1-5, the traffic capacity increases by 1% -12% while it reaches maximum when longitudinal slope is 7%, average speed increases by 49% -218% while it reaches maximum when longitudinal slope is 7%, delays decreases by 85% -97% while it reaches maximum when longitudinal slope is 6%, saturation decreases by 14% - 21% while it reaches maximum when longitudinal slope is 7%.

6 Discussion

Factors that affect the traffic capacity on climbing section can be considered from two aspects of road and traffic. Road factors include lane number, slope gradient and slope length, while traffic factors include traffic volume, vehicle speed and proportion of trucks in whole traffic volume. Number of lanes is two in this study.

Slope gradient: The slope gradient of the road has a great influence on the traffic capacity of climbing sections. In the fixed slope length, the larger climbing gradient, the lower speed of the vehicle and traffic capacity is.

Slope length: The slope length has less influence on traffic capacity of the climbing section relative to the slope gradient. Under fixed slope gradient, the longer climbing length, the lower speed of the vehicle and traffic capacity is.

Traffic volume: In the climbing section, whole traffic volume has a certain impact on its capacity. When traffic volume is small, traffic density is low as a result of fewer vehicles in the driveway, the number of vehicles through the climbing section is less; if traffic volume is larger, there are more vehicles in the driveway in the certain lane numbers, which result in high traffic density, even reaching the saturation flow, while it could induce traffic jams seriously.

Speed: The influence of vehicle speed on traffic capacity of climbing section is mainly expressed in the dispersion of speed. The greater the dispersion of the speed, the more unstable the traffic flow is, which could result in disorder of vehicles and less traffic capacity.

Proportion of trucks: The influence of the proportion of trucks on the traffic capacity of climbing section is great. When there are more cars and fewer trucks, it will result in less lane changing behavior and stable vehicle speed, so traffic capacity of climbing road will not reduce obviously. If proportion of trucks is bigger, it will result in lower speed of climbing section and less traffic capacity. When proportion of trucks lying between of the above two cases, on the one hand, because of truck speed is low in the climbing section, it will reduce the traffic capacity, on the other hand, speed dispersion will increase due to the lane changing behavior occurs frequently, which will also cause reduced traffic capacity.

20 simulated combination models containing alignment and traffic parameters are chosen to explore the effect of climbing lane. Adding climbing lane could effectively increase traffic capacity and average speed while decrease delays and saturation. Different combination has different effect. The length of climbing lane is contained in the slope length, so various proportions of climbing lane result in much simulated results. While the length of the climbing lane is related to that how much trucks could travel on the climbing lane at the same time. The position of climbing lane in the uphill highway, separated and convergence from main road, which are also factors influencing traffic capacity, requires further study.

Acknowledgment

This work was funded by the National Science and Technology Support Program of China (No. 2015BAK09B00, 2015BAK09B01, 2015BAK09B02), Chongqing Science & Technology Commission (No. cstc2014yykfb30001, cstc2014jcsf30001, cstc2015shmszx30007). It was also funded by Guizhou Transportation Science and Technology Project of Guizhou Provincial Transportation Department (No. 2015-121-035).

References

1. R.R. Tian, Q.J. Xiang, S.T. Hu. Research on Traffic Safety of Freeway Upgrade Section. Procedia - Social and Behavioral Sciences, **96**, 548–556, 2013.
2. M.J. Cassidy, K.Kim, W.Ni, W.H. Gu. A problem of limited-access special lanes. Part I: Spatiotemporal studies of real freeway traffic. Transportation Research Part A, **80**, 307–319, 2015.
3. S. Labi. Efficacies of roadway safety improvements across functional subclasses of rural two-lane highways. Journal of Safety Research, **42**, 231–239, 2011.
4. M.J. Cassidy, K.Kim, W. Ni, W.H Gu. A problem of limited-access special lanes. Part I: Spatiotemporal studies of real freeway traffic. Transportation Research Part A, **80**, 320–329, 2015.
5. Y.R. Kim, S.K. Lee, J.P. Moon, J.H. Jeong, J.S. Choi. A Study on Appropriate length of Passing Lane in Consideration of Traffic Characteristics of Two-Lane Roads in South Korea. Procedia - Social and Behavioral Sciences, **43**, 671– 680, 2012.
6. C.Y. Zhuang. *Design of climbing lane in mountainous freeway based on vehicl dynamic performance.* Xi'an, Chang'an University, 2006.

Tour Planning and Ride Matching for an Urban Social Carpooling Service

J. Jamal, A.E. Rizzoli, R. Montemanni and D. Huber

IDSIA USI/SUPSI, Manno, Switzerland

Abstract. SocialCar is a research project aiming at integrating carpooling with existing transportation systems in urban and peri-urban environments. The system, that also takes benefit from social network interactions, is based on tour planning and ride matching algorithms to suggest the users the available alternatives for their trips. In this paper we overview some approaches to model general public transportation networks and carpooling services. A route planning algorithm, able to integrate different types of public transportation, bike rentals and carpooling services is finally presented.

1 Introduction

Commuting for short or long distances has become an everyday activity for most people worldwide. For some people the preferred mean of transportation is a car, in such case one can easily find the shortest path from source to destination using a navigation system. On the other hand, if one prefers to use public transportation, there are many websites and mobile applications that are used to find available connections. In some cases it is advantageous to use multiple means of transportation for a single trip, such as driving to a railway station and using public transportation to reach the desired destination. Still, most car transports are performed by a single driver, which is rather inefficient in terms of resource usage. Carpooling is another effective mean of transportation, which is usually defined as the cooperation of two or more persons regarding the use of a single vehicle to meet their mutual commuting needs.

Carpooling has many benefits, such as reducing the number of vehicles participating in the transportation system, and it also benefits the individuals participating in the carpool. These benefits include reduced fuel costs, reduced toll costs, and potentially reduced driving stress for the passengers in the vehicle, and possibly some benefits in the social aspect.

In some countries such as the United States and parts of Europe, carpooling is encouraged by the governments. However, research has shown that the attractiveness of carpooling has been decreasing [1] because of low fuel prices, and strong incomes [2]. The psychological barrier of riding with strangers also plays a role [3].

SocialCar is a research project that seeks to assimilate carpooling into existing mobility systems by means of powerful planning algorithms and integration of big data from public transport, carpooling, and crowd sourcing. The project's mission is to design, develop, test, and roll out a service that simplifies the travel experience of citizens in urban and peri-urban areas.

The SocialCar service differs from existing carpooling services because it does not simply offer a peer to peer taxi and it is not a commercial service. However, SocialCar users - that is both passenger and driver, can also benefit from existing carpooling services (where available). These commercial services are interfaced with SocialCar so that the data on carpool offers and requests becomes available to the wider SocialCar user community. It is expected to be able to gather reliable real-time information on the traffic by the SocialCar community, and therefore being able to monitor and update the initially planned trip in real-time, and planning the trip based on the current state of traffic.

2 SocialCar network models

As transportation networks are becoming more complex and means of transportation are increasing, route planning is gaining more and more importance. The problem of route planning involving different means of transportation is called *multi-modal route planning*. The goal of the problem is very simple. Given a source and target in the road network, and a departure time, with the desired means of transportation, the route planner should return an optimal route with respect to travel time that shows us what roads and which forms of transportation to use.

SocialCar aims to solve a multi-modal route planning problem. In order to solve such a problem, different transport options need to be structured as a multi-layer temporal network, as described in [4]. The multi-modal network model is strictly related to the algorithms that can be used to solve the planning problem, so its design and implementation depends on the specific algorithms to be applied to it.

2.1 The multi layer temporal networks

The algorithm needs an underlying network in order to create alternative route solutions for users willing to move from origin to destination selecting alternative transport means.

The network must be made available as a *graph*, composed by *nodes*, which are connected by *edges*. The nodes represent the junctions in the network, from where links depart. A link connects two nodes.

The network is composed of *layers*. Each layer represents a different transport mode. Nodes that are present in multiple layers simultaneously represent intermodal connection points. There are also interlayer links which represents the travelling time (by foot) required to transit from a transport mode to another. The nodes where a modal change can take place have been defined as *switch points* in the work discussed in [5].

The layers associated with scheduled transport are defined as *temporal networks* [6]. Each edge in the graph modeling the network represents a segment in a route, while the nodes are the stop/stations.

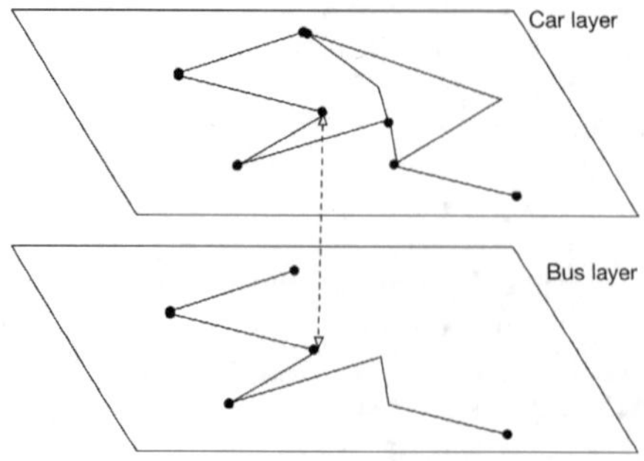

Figure 1. Two different layers contain the graphs for car and bus routes. The dashed line represents an interlayer link, which allows a modal change across a selected route.

2.2 Modelling public transport as temporal networks

Modeling public transport network requires an additional level of complexity with respect to the road networks, as routes are possible only when a service is scheduled.

In order to represent the transport services offered on top of the transfer network, each node in the public transportation network is annotated with a list of departure times containing all the routes departing from that node.

The timetable information (bus number, arrival/departure time, and destination) is used to generate the temporal network representation. Each ride from a stop/station to the next one is represented with a directed edge with duration equal to the difference between the arrival time of the destination node and the departure time from the origin node. The resulting network is called a *Time-Expanded Graph*.

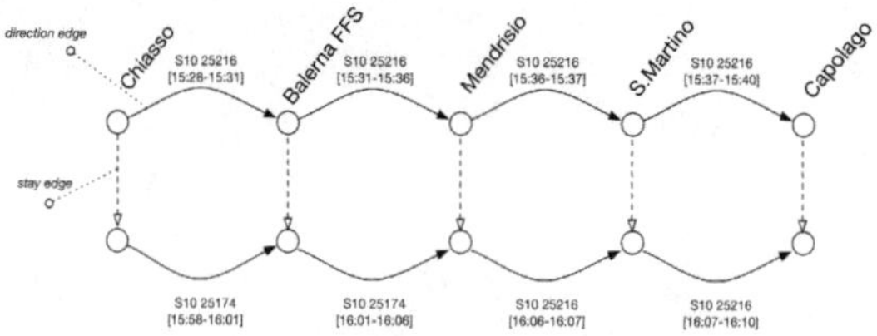

Figure 2. A timetable converted into a time-expanded directed graph. Each edge has the departure and arrival time between the consecutive nodes.

The main advantage of the above representation is that standard algorithms for shortest path work with little to no modifications, as the resulting network is static. A drawback is the large size of the resulting network (even if sparse), which can lead to longer computation times with respect to the *time-dependent* case.

Time-dependant graphs are another approach to modeling public transit networks. This representation does not require proliferation of graph nodes in order to represent different departure times from a station. On the other hand, it uses information contained in the timetables to generate the arc weights "on the fly" [7].

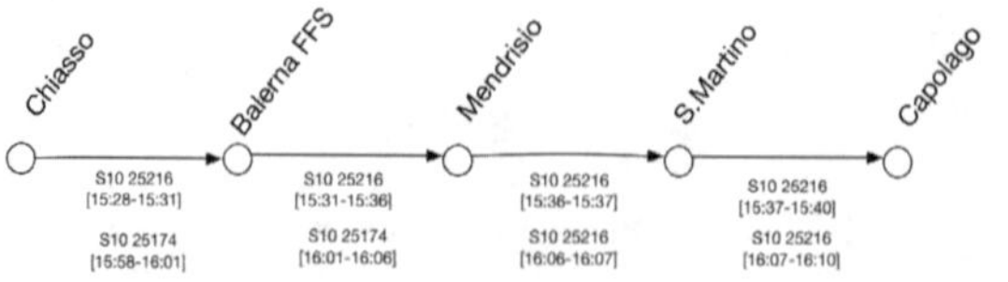

Figure 3. A timetable is used to dynamically calculate weights (travel times) on the graph.

In a multilayer network, each transport mode is represented by a different layer, and each transport line can be also represented as a different layer. Solutions where the user can switch modes and lines have been defined as *switch points* in [5]. Such a solution is described in Fig. 4.

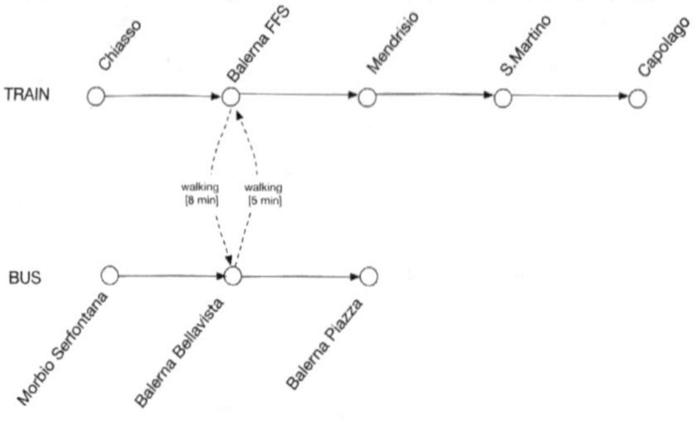

Figure 4. In Balerna the passenger can switch to the bus line. Balerna is therefore a switch node

The road network is therefore considered as another layer on top of the previous ones, but no time constraints are fixed on the graph edges. In SocialCar the public transportation network is modeled as a time-dependant graph.

2.3 Modelling carpooling services

SocialCar also represents car pooling services, which are a hybrid transport mode halfway between public transport lines (car poolers driving to work tend to stick to fixed schedules) and private transport (car poolers are not bound to a specific route, they can change the route dynamically depending on traffic conditions).

For the above reason, car pooling services can be represented as public transport services, while deviation will be represented using real-time feed. The real-time feed contains the essential information that is needed to describe alterations to planned routes, such as cancellations, delays, and updated routes. Notifying users about pick up points (stops) that have moved, as well as deviation in time from the expected schedule, is essential and very important for SocialCar. Another essential element is the availability of places on the car (residual capacity). Unlike public transportation, cars have very limited capacity, and as soon as all places are taken, the car is no longer available as a service; therefore the real-time feed must also be enriched with such information.

In order to represent this information, data of the car pooling service must be organized to the segments which connect pick-ups and drop-offs.

In Fig. 5 a typical situation is represented. The car starts the trip at home with the driver and one passenger. At the next stop, two passengers are picked up and then the car is fully occupied, meaning that there is no residual capacity to serve other passengers. This information is used by the route planning algorithm so that this particular car pooling service is not chosen in a potential solution by the algorithm.

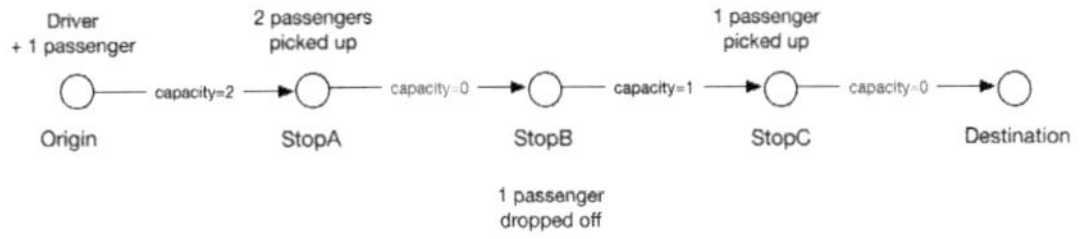

Figure 5. A carpooling service with pick-ups and drop-offs.

3 Route planning and carpool matching

One of the main challenges of the project is to develop a fast and efficient method able to integrate route planning and carpool matching into a single algorithm. This can be achieved by specializing algorithms of the Dijkstra family [8]. Dijkstra's algorithm computes a shortest path in graph from a given source to all other nodes, in route planning the graph is a road network. For road networks, there exist many algorithms to reduce the time of calculating a shortest path from one location to another. The work of Hart [9] allows the usage of optimality preserving heuristics to decrease the number of iterations necessary to compute the distance to a given selected target node. One of the heuristics used in such a context are discussed in [10], and [11]. A substantial improvement in computing shortest paths on road networks has been provided in [12]. In our implementation we used such techniques. A further improvement, not implemented yet, will be to deal with uncertainty in travel times, that can potentially have a strong impact on a multimodal journey. In order to do so, we will use a strategy similar to that described in [13] and [13].

A *Global Carpooling Matching Service* was designed by [15], where the sevice advises registered candidates how to combine their trips by carpooling. In their work they model the periodic trips as nodes in the graph, and the edges are labeld with the probability of scucess calculated by a learning mechanism.

In general, we can consider two alternative approaches to matching passengers to cars during the ride matching process: in the first approach, the car does not detour from its original path, and passengers need to travel to the closest possible stop. As described in Fig. 6, the car travels from origin to destination along a pre-determined optimal route. The passengers can be picked up at intermediate stops which coincide with public transport stops in order to favor modal changes.

The second approach, shown in Fig. 7, changes the car's route in order to pick up passengers at their preferred location. This approach involves solving a routing problem, and it might be sub-optimal for the driver, as the expected travel duration can be considerably extended.

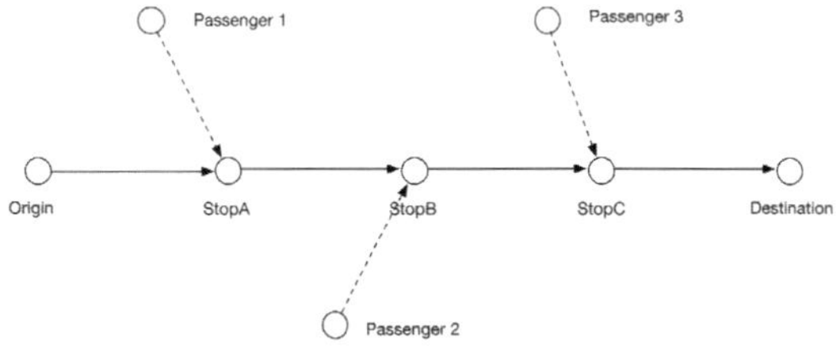

Figure 6. The car poolers are picked up at pre-determined stop points, along the optimal path from origin to destination.

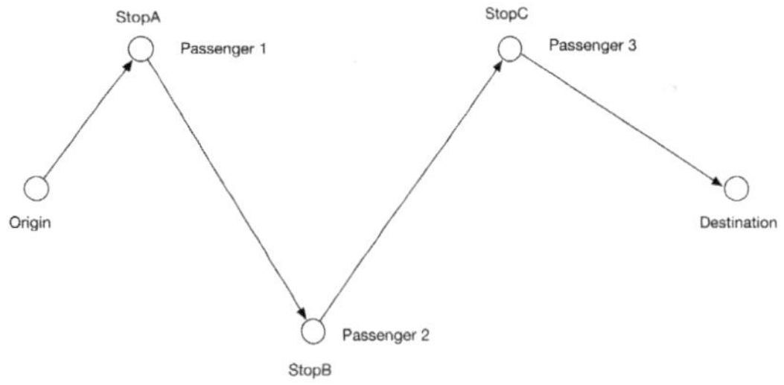

Figure 7. The car modifies the original route in order to pick up the passengers.

In SocialCar we tend to privilege the first approach, as it strengthens the similarities between car pooling services and public transport and is more realistic for a driver, but at the same time small detours will be allowed.

In particular, the main work will be in the preparation of the graph representing the network on which the algorithm itself will be run: with appropriate combinations of the layers described in Section II, it is possible to use the same approach both for route planning and carpool matching,

The final network of the algorithm is shown in Fig. 8 and consists of three layers. The first layer contains the road network, the second layer contains the public transit network, and the third layer contains the carpools. The second and third layers are connected through the fixed stops (public transit stations), and every carpooler with no capacity left at some specific stop, is not considered by the algorithm.

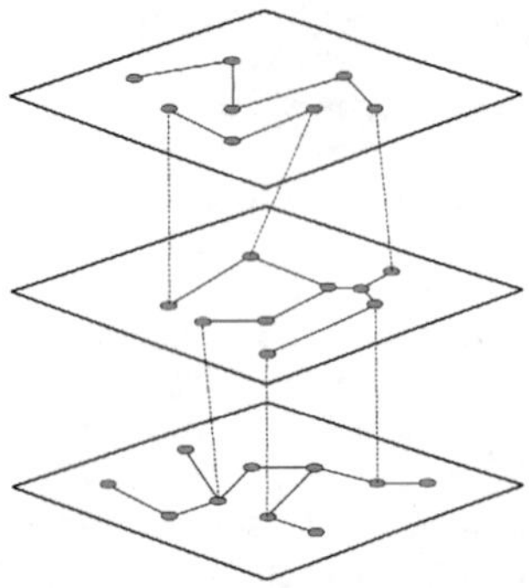

Figure 8. A network with three layers for different transport modes. The bottom layer is the road network, the middle layer is the public transit network, and the top layer is the carpool network. Additional layers could be introduced to model time-dependency.

Initially the starting point of the passenger is mapped to the closest vertex residing in the road network, the mapped vertex is the source vertex used by the shortest path algorithm. The algorithm is executed on the the three layer network to find a trip combining both public transit and carpooling.

Notice that the links between the middle and top layers are undirected connections, meaning that the passenger can travel both ways between the two modes of transportation. Therefore, it would be a possibility to use public transportation to reach a pick up point on the route of carpool A and use public transportation to reach another pick up point on the route of carpool B to reach the final destination of the passenger.

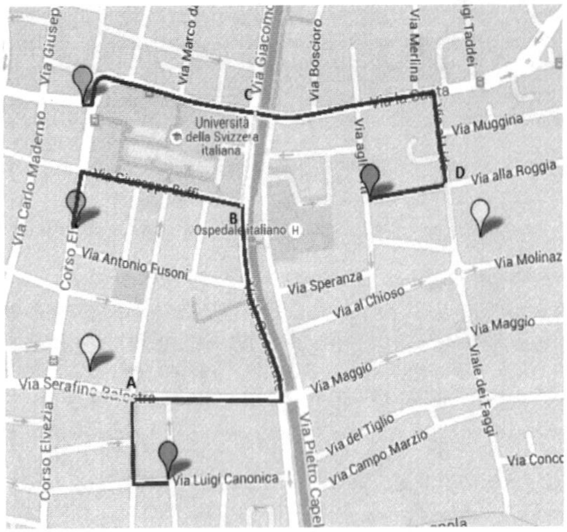

Figure 9. A screenshot of a possible tour planning and ride matching solution using SocialCar.

Fig. 9 shows a screenshot of a possible solution of a trip plan using carpooling. The green and red markers show the start and finish point of two journeys of drivers registered to the car pooling service, willing to carry passengers and with a time schedule feasible with the needs of the passenger interested in using the services provided by the system. The bottom left yellow marker shows the starting point of the passenger, and the top right yellow marker shows the final destination of the passenger. The passenger starts his trip by travelling from the starting point to the nearest pickup point. The passenger is picked up at point A to reach point B and leaves the carpool. The passenger then proceeds by walking to point C and is picked up there by another car pool to be dropped off at point D. The trip of the passenger ends by walking from point D to the desired final destination. Notice that time and the related synchronization elements are omitted in the figure, together with capacity constraints.

4 Conclusion

In this paper, after having introduced the project SocialCar, we have shown that using a time-dependent representation of the road and transit network is suitable for the needs of SocialCar, being it designed for (peri-) urban areas.

For carpooling matching, two approaches have been considered, and the approach where passengers are picked up from a pre-determined route is currently adopted, as it strengthens the used of both public transportation and carpooling services, and also represents more realistically the behavior of a driver in a carpooling system.

Future work will mainly concern the integration of uncertainty about travel times and public transportation timetables into our model. A further step will be about the use of tailored social network information to further improve the reliability of the planner.

Acknowledgment

The work was co-funded by the European Union's Horizon 2020 Research and Innovation Program under grant agreement no 636427: "SocialCar: open social transport network for urban approach to carpooling".

References

1. ND. Chan, SA. Shaheen. "Ride Sharing in North America: Past, Present, and Future", Tran. Rev. **32**, 93 (2012).
2. H. Park, R. Gebeloff, "Carpooling Declines as Driving Becomes Cheaper", NY Times (2011).
3. Gonçalo Correia, Jose Manuel Viegas, "Carpooling and carpool clubs: Clarifying concepts and assessing value enhancement possibilities through a Stated Preference web survey in Lisbon, Portugal", Transportation Research Part A, 45, (2011), 81-90.
4. R. Gallotti, and M. Barthelemy, "The Multilayer Temporal Network of Public Transport in Great Britian", Sci. Data, **2** (2015).

5. L. Liu, "Data models and algorithms for multimodal route planning with transportation network", PhD thesis, Technische Universitaet Muenchen, (2010).
6. P. Holme and J. Saramäki, "Temporal networks", Phys. Rep., **519**, 97 (2012).
7. E. Pyrga, F. Schulz, D. Wagner, C.D. Zaroliagis, "Efficient models for timetable information in public transportation systems", ACM J. Exp. Alg. **12** (2007).
8. E.W. Dijkstra, Mathematik, "A note on two problems in connexion with graphs", **1,** 269 (1959).
9. P.E. Hart, N.J. Nilsson, B. Raphael, "A formal basis for the heuristic determination of minimum cost paths", IEEE Trans. Sys. Sci. CybT. **4**, 100 (1968).
10. RH. Möhring, H. Schilling, B. Schütz, D. Wagner, T. Willhalm, "Partitioning graphs to speed up dijkstra's algorithm", J. Exp. Algo. **11**, 189 (2006).
H. Bast, S. Funke, D. Matijevic, P. Sanders P, and D. Schultes, "In transit to constant time shortest-path queries in road networks", *Proceedings of the Meeting on Algorithm Engineering & Expermiments*, 46 (2007).
11. R. Geisberger, P. Sanders, D. Schultes, D. Delling, "Contraction hierarchies: Faster and simpler hierarchical routing in road networks", In WEA, 319 (2008).
12. R. Montemanni, L.M. Gambardella, "An exact algorithm for the robust shortest path problem with interval data", Comp. and OR **31,** 1667 (2004).
13. R. Montemanni, L.M. Gambardella, "The robust shortest path problem with interval data via Benders decomposition", 4OR **3**, 315 (2005).
14. L. Knapen, A. Yaser, S. Cho, D. Karen, A. Abu Dbai, T. Bellemans, D. Janssens, G. Wets, A. Schuster, I. Sharfman, K. Bhaduri. J. Amb. "Exploiting graph-theoretic tools for matching in carpooling applications", Inteli. Huma. Comp. **5**, 393 (2013).

 DOI: 10.1051/matecconf/20168104011

Estimating Activity Patterns Using Spatio-temporal Data of Cellphone Networks

Seyedmostafa Zahedi and Yousef Shafahi

Sharif University of technology, Civil and environmental engineering department, Azadi Ave., Tehran, Iran

Abstract. The tendency towards using activity-based models to predict trip demand has increased dramatically over recent years, but these models have suffered insufficient data for calibration. This paper discusses ways to process the cellphone spatio-temporal data in a manner that makes it comprehensible for traffic interpretations and proposes methods on how to infer urban mobility and activity patterns from the aforementioned data. Movements of each subscriber is described by a sequence of stays and trips and each stay is labeled by an activity. The type of activities are estimated using features such as land use, duration of stay, frequency of visit, arrival time to that activity and its distance from home. Finally, the chains of trips are identified and different patterns that citizens follow to participate in activities are determined. The data comprises 144 million records of the location of 300,000 citizens of Shiraz at five-minute intervals.

1 Introduction

Transportation planners need to understand human movements to better design networks that suit citizens' needs and their demand for travel. There are two main ways to model trips in an urban area; trip-based models and activity-based models. The required data to calibrate such models are normally gathered by interviews or surveys. These traditional methods of getting data have several inconveniences. The process of sending out questionnaires and recollecting them is quite costly. Entering or importing the data into an appropriate database is very time-consuming. The infeasibility of surveying large percentages of the society has introduced errors into engineer's analyses. Moreover, not all the citizens are comfortable with such detailed questions on their daily trips and they might answer the questionnaires incorrectly. Some people do not even take the time to answer the interviews or questionnaires. Usually, the educated portion of the society understands the importance of these surveys and cares to participate in the procedure, which leave us with biased data. All of these inconveniences and problems, hinder the frequent use of such traditional methods and have prompted scholars to come up with better ways to acquire sufficient data to calibrate the models. Between the two prevalent models, the trip-based ones have long been used and calibrated by the survey data. The activity-based models however, face problems when used with survey data. Their multi-level nested nature along with their numerous alternatives demand for huge data if they are to correctly model and predict trips. As the location-aware technologies grew and developed, scientists paid more and more attention to the feasibility of using them to observe urban mobility. Among these technologies, cellphone networks stand out as a more promising way of collecting data. Cellphone networks have a built-in capability of recording the location of their subscriber's cellphone without the need of any additional infrastructure. They try to always be aware of the whereabouts of their subscribers to be able to calculate the cost of making a call and maintain the readiness for a fast connection. There are four events that result into registering a cellphone's location. When a cellphone leaves a location area, during a call, when a cellphone is turned on and a periodic location update. Cellphone networks are capable of tracking a large portion of citizens for as long a period as required and for as many times as needed. The location updates are passive and the subscribers do not take any actions to make it happen. When we use the data obtained by cellphone networks to represent citizens' movements the participant are unaware of being in the sample which increases the reliability of the data. On the other hand, some issues arise when we try to elicit traffic-related information from cellphones spatio-temporal data. Location accuracy might be insufficient. The accuracy by which networks estimate a subscriber may vary between 50 meters to several kilometers. In addition, it is assumed that the cellphone network records the location of the nearest BTS (Base Transceiver Station) to the subscriber's cellphone which is not necessarily always true. A cellphone which is holding still may connect to several nearby BTSs and have the network record different locations for it while it has been stationary the whole time. A phenomenon known as the Ping-Pong handover. Scholars have tried to overcome these issues and provide frame works for practical use of these spatio-temporal data. Next section presents a review of the literature on this topic.

2 Related work

Scholars have made efforts to develop methods to use cellphone data to understand and estimate urban mobility and trip patterns. Studies were initially simple and were mainly addressing the feasibility of using mobile phones to measure traffic variables. Ygnace et al. (2000) tried to find out if it is possible to use cellphones as a probe to estimate travel times [1]. Cayford and Johnson (2003) studied the effective parameters on practical use of cellphone traces for generating traffic information [2]. One of the obvious outcomes was spatial accuracy. Afterwards Asukara and Hato undertook a more complicated research to elicit travel behavior from cellphone data [3]. They tried to actually use this data for traffic interpretations. They enumerated some advantages for using cellphone data; advantages such as higher accuracy in comparison to traditional methods, all-day-long coverage of cellphone networks, reachability of data in all sorts of weather conditions and etc. However their main contribution was developing a label setting algorithm that distinguishes stay locations from on-move locations. Recent studies are more precise in their objectives. Yang Xu et al. (2015) tried to understand aggregate human mobility patters using passive mobile phone location dataset from Shenzhen, China [4]. Their study presented a home-based approach to find human movements patterns that considered homes of individuals as anchor points and references to analyze those individuals' activities. Then, they categorized people based on their approximate home locations to obtain aggregate mobility patterns for each BTS. Finally they used a multilevel hierarchical clustering algorithm to classify regions that show similar mobility patterns. Mahdieh Allahviranloo et al. (2015) conducted a research to mine activity pattern trajectories and allocate activities in different parts of the network [5]. Although their data was obtained from GPS, the method they proposed can be used for mobile phone data as well. They tried to infer the types of activities in which each individual has engaged at different locations by features such as duration of stop and distance from home. Later they used a Markov chains with conditional random fields to find the relations between individual's socio-economic attributes and activity sequencing and spatio-temporal trajectory of activities. Again in the same year Peter Widhalm et al. made an effort to discover urban activity patterns in cell phone data [6]. In order to do so they developed a two-staged method. In the first stage, they detected stays and extracted geocoded time stamps that form trip chains. In the second stage they combine stays with land use data to cluster activities. They model the dependencies, activity type, trip scheduling, and land use type via a relational Markov network. They tested their method by a CDR (see section 4) dataset of the Boston city and claimed that the results agreed with the city surveys. Shan Jiang et al. used data of Singapore to infer activity-based human mobility pattern from mobile phone data [7]. The patterns are supposed to be suitable for the activity-based approach. By parsing trajectories to extract stays and monitoring the most frequently communicated tower during night they were able to detect homes of the individuals. They used the concept of "motifs" as a representative of the complex activity pattern and chains of trips which can go into numerous combinations and provided an algorithm to identify the daily motifs for each individual. Motifs are very simple sketches of human movements. Several different movement patterns can be shown by one motif. Later they interpreted the results and demonstrated the spatial patterns of human mobility for that data.

3 Data

Previous scholarly works have been more or less with Call Detailed Records. CDR data are location plus time records that happen during a call. When a subscriber receives or makes a call his or her location is recorded in the networks database along with the time of the event. The problem with CDR data is that they are very sparse and sporadic. Not everyone makes or receives enough phone calls during a day that their movements can be observed and even if they do there is no certainty that the data catches all the participated activities. There might be stays and activity locations that remain hidden because there has been no record for them. Our data however, are records of periodic update. Three hundred thousands of citizens of Shiraz have been tracked every five minutes and their location have been recorded for 40 hours which is nearly one fifth of the city's population for almost two days and nights and comprises 144 million records. These records are anonymous so that the privacy of the citizens are not violated. The precision by which the location is recorded in the network differs regarding the density of BTSs in various parts of the city, however, in crowded zones with dense presence of BTSs the network can locate its subscribers with an error of less than 50 meters.. The 300,000 participants' cellphones were scanned the night before to make sure that they reside within the urban areas of the city, however, some of them made trips outside the city in the 40 hour period. We used ArcGIS to filter those who had spent a fair share of their time outside the city. Other than cellphone data, 4 percent of the citizens were surveyed in a day during the 40 hour period. Twenty thousand questionnaires were handed to random household and their daily chains of trips were gathered by these questionnaires and can be used to verify the methods presented here in this paper.

4 Methodology

The data is intractable and quite difficult to handle. Different phenomena such as Rayleigh fading effect and Ping-Pong hand over distort the image seen by the network of the subscriber's movements. Therefore, some preprocessing needs to be done in order to convert the data into meaningful trips and stay locations that show movements and activity participation. Peter Widhelm et al [6] used a low-pass filter to eliminate the outliers and smooth the movements with a velocity higher than an acceptable range. Then they use an incremental clustering algorithm to detect stays and to convert raw cellphone track into a sequence of visited places. Jiang et al [7] applied an outlier detection algorithm based on time intervals and distances between consecutive points to eliminate the two types of noises that exist in the raw data. Other than eliminating signal jumps and outliers they agglomerated points that are spatially close but not necessarily adjacent in temporal sequence to obtain one unique location that represents an activity location. These methods may work well for CDR, but are not practical for PLU (periodic location update) data. In CDR data there might be several records for a single user in a matter of minutes but on the other hand the network may be unaware of its subscriber for hours. That is when a clustering algorithm comes in handy. But in PLU data, where there are records in predefined intervals (in our case five minutes) Ping-Pong hand over and frequent signal jumps between towers dominate the data and insert fake movements in the trajectory of subscribers. Since the distance between the towers that Ping-Pong occurs varies significantly, a clustering algorithm that uses distance as a similarity feature is not effective. In the following sections we describe how to distinguish Ping-Pong handover from real displacements and then provide stepwise instructions to discern stays that were motivated by an activity from passing-by points.

4.1 Distinguishing Ping-Pong handover from real movements

Cellphones connect to the BTS from which they receive the strongest signal. Since signal strength decays exponentially as the distance between cellphone and tower increases, it seems a logical assumption that cellphones connect to the nearest tower which will later inform us of the whereabouts of the subscriber. However, this is not always the case. In telecommunication science there is a phenomenon known as the Rayleigh fading effect. When a signal is generated and propagates through an environment, due to the reflections and refractions that happen in its way, the signal power strengthens and weakens randomly. These random fluctuations in strength change the BTS that the cellphone connects to and therefore, changes the location that represents the cellphone in the network database. As a result, within consecutive records a cellphone may be represented by different locations while its actual location has not changed. Sometimes presence of tall buildings block the way and the cellphone connects to a more accessible tower. And after a slight displacement the obstacle is removed and cellphone changes the connecting BTS to a nearer one. Sometimes a tower is fully loaded and a nearby cellphone is not welcome and has to connect to another BTS While shortly afterwards load decreases and the former tower is suddenly able to accommodate the cellphone. Sometimes the topology of the region is in a way that cellphones tend to connect to farther towers that are in similar altitude as the cellphone device. All of these examples cause frequent back and forth movements of cellphones between towers that resembles a game of Ping-Pong and therefore, is called the Ping-Pong handover. The Ping-Pong handover exists all over the data and mistakenly increases displacements of each and every individual. It assigns invalid movements to cellphone users, even those that are not making any moves. There is no discernable pattern for Ping-Pong handover. It can happen between any numbers of towers for a single cellphone. It can happen every now and then at unknown intervals. Sometimes towers hand over a cellphone every minutes or so and sometime a cellphone lingers on a tower for hours.

Despite the fact that Ping-Pong handover does not follow any regular pattern it defies logical human movements. It is very unusual for a subscriber to oscillate between two or more points for a notable period. Humans follow a purpose when they make a trip and try to minimize the trips necessary to achieve these purposes. Lasonen et al. [8] used these oscillations and came up with three conditions that detects if Ping-Pong handover is occurring among a set of towers. The first condition requires that all towers in a set are close to each other. The second condition checks if oscillation is happening and that is if the average time spent visiting a set is larger than the sum of the individual times on each cell. The third condition makes sure that no subset of the towers satisfy the second condition and minimizes the number of members in each set. The problem with this algorithm is that if a cellphone is oscillating among three towers say A, B and C and then moves to another point and starts Ping-Ponging among different cells that has an intersection with the previous ones, for example C, D and E this algorithm will combine all of the towers and outputs a single location for the cellphone with the meaning that no actual movement has occurred, which is not true. Hong and Kim notice that the same Ping-Pong handover happens in WLAN traces and that 90% of transitions are irrelevant to actual user movements [9]. They try to filter false transitions. They consider a transition to be Ping-Pong if it satisfies two conditions. A) The transition should be among l recently associated BTSs. B) There should be at least p_{nu} current transition among these l BTSs. Readers can refer to [9] for a more complete explanation. We have employed the method provided by Hong and Kim to distinguish ping pong hand over from real movements. We chose l to be three and p_{nu} to one. However, their method had some deficiencies. First, their conditions detect Ping-Pong hand over with a delay. It takes at least three hand overs for their method to detect Ping-Pong. Second, there were finite number of situations where their method couldn't detect a Ping-Pong hand over. We completed their method, added those undetected conditions, removed the delay and inserted some modification to better distinguish Ping-Pong hand over from real movements.

4.2 Distinguishing stays and activity locations from on-move points

After eliminating false transitions, it is now time to determine at which points a user has stopped to make an

DOI: 10.1051/matecconf/20168104011

action and at which points was moving to get to the places he or she wanted to visit. Correctly detecting stays is of paramount importance when it comes to extracting activity patterns; it is crucial to other transportation uses of cellphone data such as mode choice or route assignment. Failing to detect a stay will change the patterns and trip chains dramatically. For example it can drop a three-node activity pattern into a two-node one and mislead us to erroneous results. Besides, by missing a stay point the path that the traveler moves along is unusually long. Hardly exists a logical explanation as to why the traveler uses such path. As discussed earlier Widhelm et al. used an incremental clustering algorithm to detect such stays. In the time-stamped location sequence, Shang et al. clustered the points which are spatially close (within the threshold of Δd) and temporally adjacent. But, as mentioned earlier these methods are not efficacious for PLU data. Dash et al. [10] used two thresholds to detect stays. They postulated that a person is staying in a location and performing an activity if its cellphone remains within a radius of Rd during a time limit of Td. They ran some sensitivity analyses and determined these thresholds in such a way that trips rates and number of stays matched the data from surveys. However this does not sound like an appropriate procedure because the objective of using cellphone data is to derive traffic-related without the need of surveys. Their method makes cellphone data dependent to survey data for calibration and practical use.

The method proposed in this paper considers all the records of a user after applying the Ping-Pong preprocessing and decides if each record is a stay or not. Since our data is in five-minute-intervals a subscriber may spend 1 second to 10 minutes in a given locations so it is not easy to tell if a location was recorded during a move or during a stay. By considering factors other than duration of stop it is possible to robustly detect a stay. Here are three conditions that if any of them is met, the point under consideration is highly likely to be a stay.

1- If a cellphone connects to a tower (or remains within a cluster of Ping-Ponging towers) for more than 20 minutes. It can surely be stated that the user of that cellphone is purposefully staying near that tower. Because the configuration of the city and the BTSs inside it are in a way that a moving cellphone is not likely to connect to one tower (or remains within the cluster of the towers with Ping-Pong handover) for more than 20 minutes. In other words four consecutive repetition of the same location in the time-stamped sequence indicates a stay

2- For the points with two or three consecutive repetition of the same location. If the cover range of the tower or towers representing that location is fairly small it can be assumed that the user has stopped in that location to perform some activity.

Fig.1 illustrates how the second condition works. If the area that a BTS tower covers is small or if there are several towers in close vicinity of it; a moving cellphone can easily leave the tower's territory and connect to other BTSs. Hence, if the cellphone remains connected to one tower or in other words its representative location repeats two or three times consecutively in the sequentially time-stamped records, it means that the subscriber holding this cellphone has intentionally stopped near that BTS and is not moving. By small cover area, we mean an area that a walking passenger can cross it within the expected amount of time spent on that tower. For example if the data of the i_{th} user has recorded tower j for three consecutive times we expect that the user's cellphone has been connected to that tower for 15 minutes (which can be between 10 to 20 minutes). Now, if we assume that the speed of an average passenger is 2 meters per second, a cellphone that is moving at least as fast as a walking passenger can move 2*15*60= 1800 meters. Which is the diameter of a circle with an area of 2.54 km squared. If the area belonging to a certain tower is notably smaller than 2.54 km^2 and a cellphone is connected to this tower for 10 or 15 minutes, it indicates that the subscriber was not trying to move away from that tower. The cover range of each BTS in the city is obtained from a Thiessen polygon.

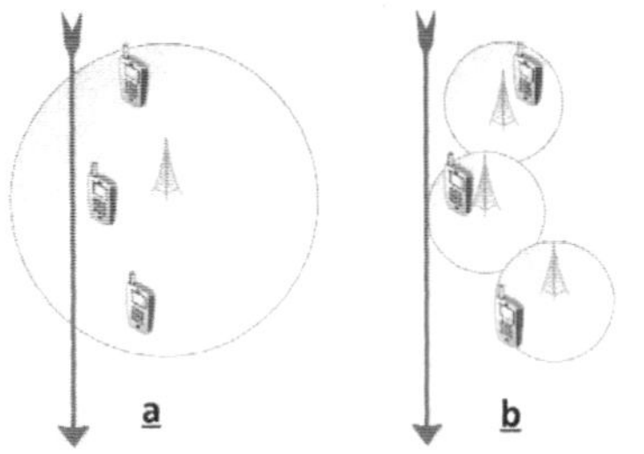

Figure 1. - a) if a cellphone is passing by a tower with a large coverage area, only one location may be recorded for the cellphone in consecutive time stamps b) if the cellphone is moving in a zone that BTSs are placed closely to each other, different locations will be recorded in consecutive time stamps for a moving cellphone as it connects to different BTSs.

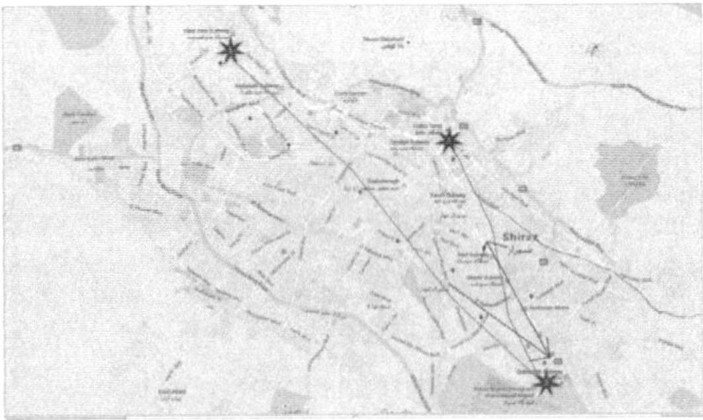

Figure 2. Significant change in the moving direction of a passenger and how our method detects this change and finds the stay location

3- Major direction change

Significant changes in overall trend and direction of a person is motivated by activities. People tend to use the routes that can more or less directly take them from an origin to a destination. The paths on which subscribers move are normally smooth, especially when seen by a network that catches their location only every five minutes and the sharp U turns and left turns on intersections are neglected. Hence, if a person deviates considerably from the overall trend of his or her movements; there is a reason behind it and the reason is

 DOI: 10.1051/matecconf/20168104011

to make a stop to participate in an activity. Consider Fig.2. A tour guide leaves his home at Gasr dast (marked by 1), goes to pick up a tourist at the airport (marked by 2) and finally takes him to the Hafez Tomb (marked by 3). Since the pick-up may take less than ten minutes, our first two conditions are highly likely to fail to detect a stop at the airport. The third condition however, notices that the traveler was initially moving south and at the airport he changes his direction and goes north which indicates that the traveler had something to do at the airport and he was not merely passing by it. In order to find out whether or not a point meets the third condition, the overall direction and moving trend before that point is compared to the trend after it and if these trends have notable differences in direction the point understudy is considered to be a stay location. The trend before a point is defined as the unit vector that represents the direction of the line that best fits the two anterior points (recorded BTS locations) and the current one. Like-wise the trend after the point is the vector that represents the direction of the line that best fits the two posterior points and the current one. The angle between these two vectors are computed and if it exceeds 130 degrees the point under study is probably a stay location.

4.3 Feature extraction to detect activity types

When a stay is detected, it means an activity is being performed during that stay. The next step would be to identify in which sort of activity is the traveler participating. Each type of activity has several characteristics that help to its recognition. For example the activity type of "work" usually happens in working hours. A pick-up/drop off activity usually takes no longer than several minutes. In order to determine the type of activities a person performs in a day, each stay should be enriched with features that contribute to identifying the type of activities during that stay. Our method proposes 8 features for a given stay. The arrival time to the stay, the departure time from the stay, duration of the stop, sequence of stay in the chain of trips, frequency of visits to that stay in the trip chain, distance of the stay to home, distance from previous stay and the dominant land use type of the location of the stay.

4.4 Assigning activity types to stays

When a stay is detected it means that the traveler has made a stop to participate in some activity. The problem here is to identify what sort of activity was the traveler doing during that stop. [5] Used adaptive boosting algorithm to infer activity types. The adaptive boosting algorithm is a set of classifiers that needs train data which we do not. We used K-mean clustering algorithm to classify activities to six groups and then manually label them to "Work", "School", "Shop", "Personal business", "Recreation" and "drop off or pick up". All different stays of all the people in the dataset has been aggregated and each stay was assigned features that were previously extracted except the sequence of stay in the chain of trips which will later be used to find the patterns of activity participation. Features have been normalized in a way so that they can be compared to each other. After running extensive sensitivity analyses 11 seeds were inserted in the model that resulted in to 11 clusters. One for work, one for school, two for personal business, two for shopping, two for recreation, and three for pick up/ drop off.

5 Concluding remarks

The methodology presented in this paper gets the raw PLU data from cellular phone network as input and gives a series of activities that the subscriber has performed during a day as output. After finding the types and timings of activities that each citizen has participated in, the next step would be to extract patterns that are similar in the citizens' chains of activities. This step is still understudy and will be presented in the future publication of the authors.

References

1. J.-L. Ygnace, C. Drane, Y. Yim, and R. De Lacvivier, PATH, (2000)
2. R. Cayford and T. Johnson, *Geography compass,* **8**, 49-62, (2003)
3. Y. Asakura and E. Hato,*Transportation Research Part C: Emerging Technologies,* **12**, 273-291, (2004).
4. Xu, Yang, Shih-Lung Shaw, Ziliang Zhao, Ling Yin, Zhixiang Fang, and Qingquan Li. *Transportation* ,**42**, 625-646, (2015)
5. Allahviranloo, Mahdieh, and Will Recker. *Transportation*, **42**, 561-579 ,(2015)
6. Widhalm, Peter, Yingxiang Yang, Michael Ulm, Shounak Athavale, and Marta C. González. ,*Transportation*, **42**, 597-623, (2015)
7. J. Shan, J. Ferreira Jr, and M.C. González. In *Int. Workshop on Urban Computing,* (2015)
8. Laasonen, Kari, Mika Raento, and Hannu Toivonen. In Pervasive Computing. Springer Berlin Heidelberg, 287-304., (2004)
9. J. Hong, H. Kim, Computer Communications and Networks, *Proceedings of 18th International Conference on. IEEE*, (2009)
10. M. Dash, K.K. Koo, T. Holleczec, G. Yap, et.al.*(MDM), 2015 16th IEEE International Conference* ,**1**, 243-250, (2015)

Neuro-Fuzzy Prediction of Cooperation Interaction Profile of Flexible Road Train Based on Hybrid Automaton Modeling

Lejla Banjanovic-Mehmedovic[1], Nermin Delic[2], Ivana Butigan[3], Suad Kasapovic[1] and Ivan Bosankic[4]

[1]*Faculty of Electrical Engineering, University of Tuzla, 75000 Tuzla, Bosnia and Herzegovina*
[2]*Magicsoft, 75000 Tuzla, Bosnia and Herzegovina*
[3]*Leftor, 75000 Tuzla, Bosnia and Herzegovina*
[4]*IVL, 75000 Tuzla, Bosnia and Herzegovina*

Abstract. Accurate prediction of traffic information is important in many applications in relation to Intelligent Transport systems (ITS), since it reduces the uncertainty of future traffic states and improves traffic mobility. There is a lot of research done in the field of traffic information predictions such as speed, flow and travel time. The most important research was done in the domain of cooperative intelligent transport system (C-ITS). The goal of this paper is to introduce the novel cooperation behaviour profile prediction through the example of flexible Road Trains useful road cooperation parameter, which contributes to the improvement of traffic mobility in Intelligent Transportation Systems. This paper presents an approach towards the control and cooperation behaviour modelling of vehicles in the flexible Road Train based on hybrid automaton and neuro-fuzzy (ANFIS) prediction of cooperation profile of the flexible Road Train. Hybrid automaton takes into account complex dynamics of each vehicle as well as discrete cooperation approach. The ANFIS is a particular class of the ANN family with attractive estimation and learning potentials. In order to provide statistical analysis, RMSE (root mean square error), coefficient of determination (R2) and Pearson coefficient (r), were utilized. The study results suggest that ANFIS would be an efficient soft computing methodology, which could offer precise predictions of cooperative interactions between vehicles in Road Train, which is useful for prediction mobility in Intelligent Transport systems.

1 Introduction

Cyber-physical systems (CPSs) feature a tight combination and coordination between the system's computational and physical elements and integration of computer and information-centric physical and engineered systems [1]. Intelligent Transport systems (ITS) fall in the framework of cyber-physical systems due to the interaction between physical systems (vehicles) and a distributed information gathering and dissemination infrastructure. The transportation sector is responsible for a large part of the world's energy consumption and greenhouse gas emissions [2]. This is especially important if it is expected that the freight transport will increase considerably in the coming years. This provides a strong motivation for developing efficient road transportation sector, especially a cooperative approach, since road transportation is currently mainly done by individual vehicles.

The ITS rely on understanding, predicting and affecting the interactions between vehicles. The cooperative intelligent transport system (C-ITS) includes two or more actors working towards a common or mutually beneficial goal, purpose, or benefit; enabled by interaction and information exchange between the actors. Within C-ITS, information is shared between many actors such as vehicles, infrastructures, cloud services, etc. However, only sharing information is not enough to be considered a C-ITS, cooperation, and interaction between the actors in the system is also required [3].

The vehicle platoon is an example of cooperative intelligent transport system, where the desired behaviour of a vehicle is generally defined by a desired distance to the previous vehicle in the platoon [4]. Platoon is a control system, in which one vehicle - a leader, gives instructions to other vehicles - followers in form of an acceleration, deceleration, or joining/leaving a Platoon. A vehicle platoon can be defined as a set of vehicles that move together while keeping a predefined geometrical configuration without any material coupling. The most widely studied platoon configuration is the column, also known as train configuration, Road Train (RT) [5]. Road Train has many advantages, from increasing traffic density and safety, to simultaneously decreasing fuel consumption and driver tiredness [6].

A very basic Road Train is equipped with an Advanced Cruise Control (ACC) system, and the Co-operative Adaptive Cruise Control (CACC) systems are developed as an extended cruise control. These systems are able to react to vehicles which are located in front of them, and are capable to communicate with the vehicles located in front and in the back [7]. However, both ACC and CACC systems are never designed to function as a part of a Road Train or to increase road capacity. The longitudinal algorithms used in CACC systems are not well suited for Road Train: shock wave effects and string instability are likely to occur. In order to tackle this problem, Road Train systems were designed so that they would be capable to communicate not only with the vehicle located in front of them, but so that they could also gather information about the behaviour and aspects of the platoon as a whole. The final *flexible* Road Train can overcome these problems, whereas smart vehicles with artificial intelligence could merge and leave or enter the platoon whenever the driver wants.

Road Train models can be found in many systems [8], ranging from systems which do not include communication between the vehicles to systems which use full communication between the vehicles. There exists physics-inspired model of the Road Train: [8] it represents the interactions as virtual spring-damper systems, [9] it considers the Road Train as a multi-agent system, which mimics animal interaction behaviour, while [10] modelling the forces between the vehicles as Newton forces.

There is a lot of research in the field of traffic information predictions such as speed, flow and travel time [11]-[13]. Vehicular velocity data have proven to be very useful in many contexts, including the understanding of road usage patterns in urban areas [14].

Various techniques, e.g. neural networks, wavelet, neuro-fuzzy could be adopted to develop a parameter estimation model. The artificial neural network (ANN) can be used as an analytical method for various prediction purposes because it provides the benefit of independency on the knowledge of internal system parameters, compressed compact solution in terms of multi-variable problems and rapid computation [15]. The neural network prediction of the speed profile is presented in [11], [12]. The speed profile neural network based prediction for selected agents using vehicle-to-vehicle (V2V) communication is presented in [13].

Amongst the many neural network systems, one of the most commonly used and powerful ones is the adaptive neuro-fuzzy system (ANFIS) [16]. ANFIS is a hybrid intelligent system that increases the capability of automatically learning and adapting.

The goal of this paper is to introduce the novel cooperation behaviour profile prediction through the example of the flexible Road Train, as useful road cooperation parameter which contributes to the improvement of traffic mobility in Intelligent Transportation Systems. Our methodology consists of flexible Road Train cooperation modelling based on hybrid automaton and ANFIS predictive model which has proven to have the quality for prediction of cooperative behaviour profile of the flexible Road Train.

This paper is organized as follows: Section 2 presents proposed hybrid automaton modelling framework of the flexible Road Train; Section 3 presents the cooperation interactions profile, using proposed behaviour pattern coding scheme. Section 4, presents the neuro-fuzzy prediction of the cooperation interactions profile. Finally, we give conclusions and directions for future work.

2 Flexible road train modeling

The Road Train is often modelled as a platoon of identical vehicles. We assume that each vehicle has the ability to communicate with other vehicles and is able to send and retrieve data content. The Road Train is formed so that each agent can be in one of the next modes: leader or follower in Road Train. The autonomous vehicles can use three manoeuvres: join, merge and leave the Road Train.

We model the Road Train as a hybrid system which exhibits a combination of discrete and continuous behaviour [17]. Hybrid automaton takes into account complex nonlinear dynamics of each vehicle and discrete cooperation approach. Model of a flexible Platoon consists of the following modules, presented on Fig. 1: Leader Vehicle Module, Cooperative Module, Control Module of Followers and Behaviour Scenario Module.

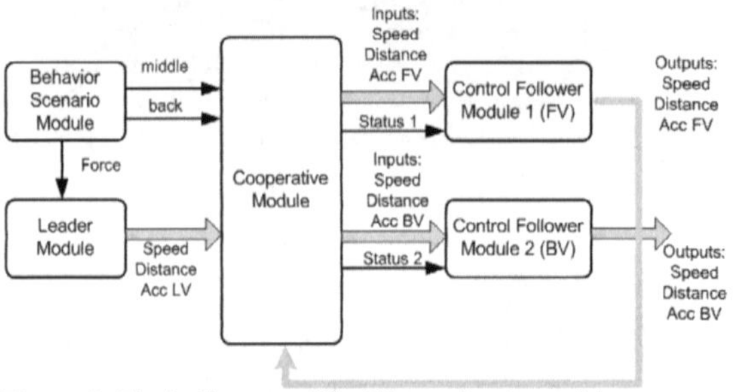

Figure 1. Block-diagram of flexible road train model.

Leader Module consists of model leader's vehicle dynamic. *Cooperative Module* is realized through the finite state machine (FSM) and together with *Control Modules of Followers* defines a hybrid approach in vehicle dynamics system. *The Behaviour Scenario Module* is realized using finite state machine, and it selects the parameters for different behaviours of the vehicles in the Road Train, as well as the speed changing. This module gives different case scenarios for the possibilities for joining, merging and leaving the follower vehicles from the Road Train in time.

2.1 Vehicle modeling

When it comes to vehicle modeling, we assume mathematical model of longitudinal motion of the vehicle [6], with two coordinated systems (vehicle-fixed coordinate system, $B(C; x, z)$, and Earth-fixed coordinate system, $E(O; x_0, z_0)$.

Application of Newton's second law on vehicle moving in the x and z directions, effects of aerodynamic force D_A and the rolling resistance force R_x, give the next equation:

$$m\dot{u} = F_x - mg\sin\theta - f_r mg\cos\theta - \frac{1}{2}C_{air}(u + u_w)^2 \quad (1)$$

where m is the vehicle mass, g is the gravitational constant; $G=mg$ is the weight of the vehicle; F_x is the tractive force, u is the vehicle-forward velocity, u_w is the wind velocity, θ is inclined upon angle with respect to horizontal plane (slope of the road) and constant C_{air}

(depends of the air density and the drag coefficient). The subscripts r refer to the rear tire-reaction forces. Module leader is the block that represents the implementation of leader vehicle dynamics. The input variable is a force and the outputs are the speed, the acceleration, and the velocity.

2.2 Control module of follower vehicles

Under assumption that the dynamic behaviour of the object with respect to the nominal trajectory is linear, we can apply methods for synthesis developed for linear systems: proportional-integral-derivative (PID) controller design, Linear Quadratic Regulator (LQR), adaptive optimal control etc. In this paper, a (PID) control design approach is used in scope of Control Modules of Followers: first (FV) and second vehicle (BV). We generate reference acceleration a_0 , velocity v_0 and position x_0 of the Road Train leader. These values go through Cooperation Module and then to the PID controller where they are processed according to [6]:

$$u = \Delta F_x = K_P(x_0 - x) + \frac{K_I}{s}(x_0 - x) + K_D(v_0 - v) \quad (2)$$

where x and v are position and speed of each follower vehicles and K_P, K_I, and K_D are proportional, integral and derivative gains of the PID controllers for follower vehicles.

The movement of vehicles is in the absolute coordinate system G (O; x_o, y_o), which is fixed to the road with origin in the starting point, O. Positions x_i, velocities v_i, and accelerations a_i are measured with respect to G $(O;\ x_o, y_o)$. Coordinate system L $(L;\ x_L, y_L)$ is fixed to the vehicle-leader with origin in the centre of its mass. Using vehicle model (1), $\theta = 0$ and $u_w = 0$, we can find acceleration of the vehicle in this form:

$$\dot{u} = a = \frac{1}{m}\left(F_x - f_r mg - \frac{1}{2}C_{air}u^2\right) \quad (3)$$

$$F_x = \Delta F_x + F_{x0} \quad (4)$$

Control force F_x is determined by a PID controller, i.e. with equation (2). Substituting (2) in (3), it is possible generate the acceleration for the i-th vehicle. The following equations represent linear state space model of the i-th vehicle in the platoon:

$$\dot{x}_i = v_i \quad (5)$$

$$\dot{v}_i = a_i \quad (6)$$

$$\dot{a}_i = \frac{1}{m}\begin{bmatrix} K_{Ii}(x_{i-1} - x_i - hd_i) + K_{Pi}(v_{i-1} - v_i) + \\ K_{Di}(a_{i-1} - a_i) - C_{air}u^0 a_i \end{bmatrix} \quad (7)$$

where hd_i is constant distance between i-1-th and i-th vehicle, variables $x_{i-1}, v_{i-1}, a_{i-1}$ of the previous (i-1-th vehicle) are input variables for the i-th vehicle.

2.3 Cooperative module

Cooperative Module gets the following as inputs: parameters of the joining, merging and/or leaving of individual follower vehicles from Behaviour Scenario Module, and speed and distance of the leader and first follower. Based on this data, this module makes the settings of status flag and relative distance, speed and acceleration of individual followers by specific joining, merging and/or leaving manoeuvres.

The finite state machine (FSM) of cooperation between the leader and two followers in the Road Train is presented in Fig. 2. It consists of 5 states: *Only LV on*, *FV on*, *FV on BV on*, *FV off BV on* and *FV joining BV on*.

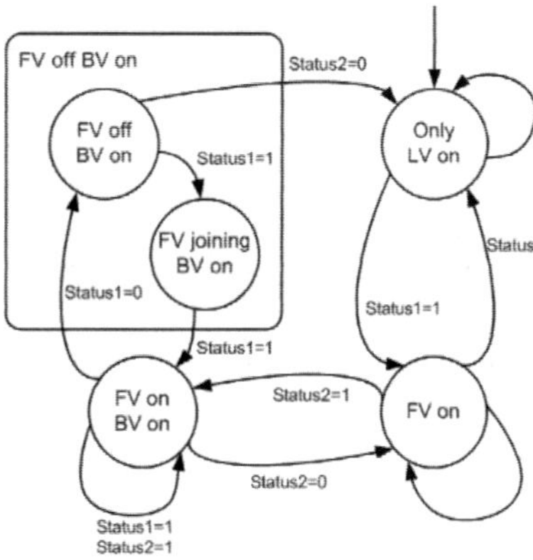

Figure 2. FSM of flexible road train.

The transitions between states are defined with value flags Status 1 and Status 2. The first state is the default state (*Only leader on*), where the only leader vehicle is in the Road Train. From this state, it is possible to make transition with new event to the state, where the first follower joins the Road Train (*FV on*). The Road Train (RT) remains in this state until it gets a signal for the inclusion of another follower or the exclusion of the first follower vehicle. In the first case, the transition leads to the state (*FV on BV on*) in which both follower vehicles are in the Road Train. In the second case, the transition leads to the state (*Only leader on*), where both follower vehicles are excluded. From the state (*FV on BV on*), the two transitions with new events are possible toward other states: toward the state (*FV off BV on*) which excludes the first follower, and toward the state (*FV on*), where the second follower is excluded. From the state (*FV off BV on*), there are two possible transitions: excluding of another vehicle, transition to the state (*Only LV on*), or merging the FV vehicle to the Road Train, transition to the state (*FV joining BV on*) and (*FV on BV on*).

3 Cooperation interaction profile

We propose the coding of cooperation interactions between the vehicles in the Road Train toward cooperation profile. The coding scheme transforms more behaviour patterns in time into specific sum of activities in Road Train, presented in Table 1. In this way, we describe the complex system interactions only with one variable, which we use for tracking and predicting mobility in ITS. In case of more vehicle agents in the Road Train, we can use another coding scheme, for example binary coding.

Table 1. Coding of cooperation interactions in road train.

Notation of behaviour	Sum of behaviour activities in time	Explanation
1	1	Only LV included
3	1+2	LV and FV included
4	1+3	LV and BV included
5	1+4	LV deceleration
6	1+2+3	All vehicles included
7	1+2+4	LV and FV deceleration
8	1+3+4	LV and BV deceleration
11	1+2+8	LV and FV acceleration
12	1+3+8	LV and BV acceleration
10	1+2+3+4	All vehicles; deceleration
14	1+2+3+8	All vehicles; acceleration

The Road Train (RT) control and cooperation model presented in this paper are demonstrated by means of simulation scenarios, which consider several situations that the vehicles in the RT may commonly encounter. All vehicles are the same with numerical values: $u^0 = 20\frac{m}{s}$, $\theta^0 = 0,\ m = 1000kg, \rho = 1{,}2\frac{kg}{m^3}\ ,\ A_f = 1{,}2m^2,\ C_d = 0{,}5,\ f_r = 0{,}01, g = 9{,}81\frac{m}{s}, u_w = 0, C_{air} = 1{,}2$.
The desired distances among vehicles are $d_{xi0} = 100m$. Parameters of PID controllers are: $K_{Pi} = 100, K_{Ii} = 400, K_{Di} = 200$.

The vehicles speed profiles and generated cooperation profile of joining/merging/leaving of followers vehicles in the RT are highlighted in Fig. 3 respectively. On the basis of the road cooperation behaviour profile, we can track the activities of all the vehicles in the RT in time.

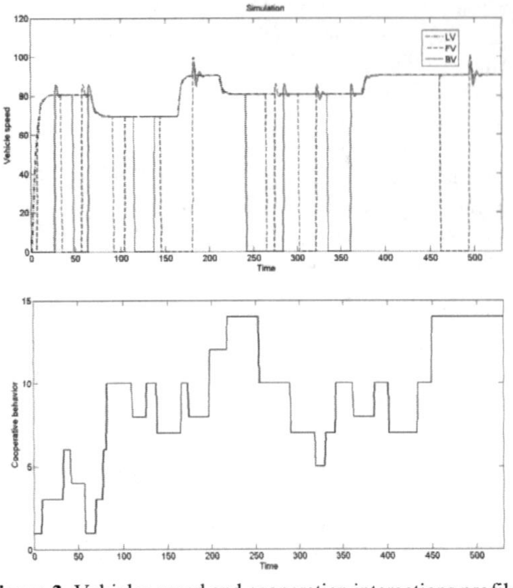

Figure 3. Vehicles speed and cooperation interactions profiles.

4 Adaptive neuro-fuzzy prediction

In this study, the ANFIS model was used to estimate the cooperative interactions profile in relation to the speeds of the leader and first and second follower vehicles. The main core of the ANFIS is the fuzzy inference system (FIS). The FIS is funded on the basis of expertise described in the 'IF–THEN' rules. The fundamental organization of FIS includes: a rule base of a fuzzy rules selection, a database that establishes the membership functions (MFs) employed in the fuzzy rules and the reasoning mechanism to extract a rational output product. The method can be utilized to estimate the characteristics of various systems incorporating the uncertainty. The ANFIS combines the power of the FIS with a neural network back-propagation learning algorithm. The ANFIS structure consists of 5 layers. Fig. 4 shows the ANFIS structure with three inputs.

The first-order Sugeno model, including the fuzzy IF–THEN rules of the Takagi and Sugeno's class, and three inputs were all employed in our study:

$$\text{if } x \text{ is } A \text{ and } y \text{ is } C \text{ and } z \text{ is } E \text{ then } f_1 = p_1x + q_1y + r_1z + t$$

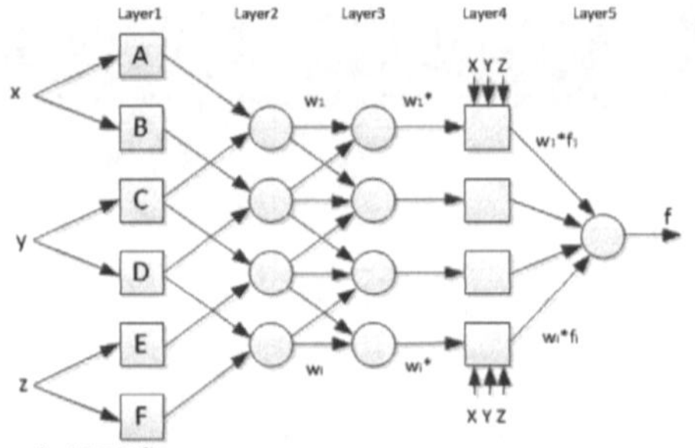

Figure 4. ANFIS structure.

The 1st layer is comprised of input parameters membership functions (MFs). This layer provides the input quantities to the proceeding layer. Within this layer, each node is considered as adaptive with node function, $O = \mu(i)$, where $\mu(i)$ is the membership function. The parameters of this layer are designated as *premise parameters*. The 2nd layer is also termed as the membership layer, and it monitors the MFs' weight. Each node in this layer is considered as non-adaptive. This layer acts as a multiplier for the receiving signals and it produces outcomes $w_i = \mu(i)_i\mu(i)_{i+1}$. Each node output represents the firing strength of a weight or rule.

The 3rd layer is known as the rule layer. Here, all neurons act as the pre-condition matching the fuzzy rules. The nodes in the 3rd layer are considered non-adaptive. Each of the nodes computes value of the rule's firing strength over the sum of all rules' firing strengths in the form of $w_i^* = \frac{w_i}{\sum_i w_i}$. The outcomes are refered to as the normalized firing strengths. The 4th layer (known as the defuzzification layer) is responsible for providing the output values as a result of the inference of rules. Every 4th layer node is an adaptive node having the node function: $O_i^4 = w_i^* \cdot f = w_i^*(p_ix + q_iy + r_iz + t)$. In this layer, the $\{p_i, q_i, r_i, t\}$ is the variable set. The variable set is designated as *the consequent parameters*. The 5th and final layer is known as the output layer. It adds up all the receiving inputs from the preceding layer. Thereafter, it converts the fuzzy classification outcomes into a binary (crisp). The single node of the 5th layer is considered non-adaptive. This node calculates the total output as the whole sum of all receiving signals

$$O_i^5 = \sum_i w_i^* \cdot f = \frac{\sum_i w_i f}{\sum_i w_i} \quad (8)$$

In this prediction application, hybrid learning algorithms were employed to identify the variables within the ANFIS model. In the forward pass of the hybrid learning algorithm, the functional signals are passed forward until Layer 4. Moreover, consequent parameters are calculated using the least squares estimate. In the backward pass, the error rates propagate backwards and the premise parameters are updated by the gradient descent.

The input and output values are gathered according to the experimental scenarios in Road Train model. In order to evaluate the performance of ANFIS as a prediction model, it is necessary to have a rich dataset to experiment with. We used 3201 samples from behaviour scenario of the Road Train model. The leader, first and second follower's speeds from the Road Train model are used as input parameters and the generated cooperation behaviour profile is used as output parameter of ANFIS. In this study, 50% of the data was used to test the samples, while the other 50% was used for samples training.

As a data-driven model, the ANFIS ability for rational estimations is typically reliant on the choice of the ANFIS parameters. Therefore, the factors controlling the system (training epoch number, training error goal, step size rate, number and shapes of MFs) should be considered in the developing of a reliable network.

There are various ways to assess the performance of a prediction model. For the evaluation of the quality of predictors, we used several measures: RMSE (root mean square error), coefficient of determination (R^2) and Pearson coefficient (r). The best result of s prediction is using R^2=0.99, RMSE=0.155 and MSE=0.02 for 7 Gauss membership functions for each input variables in 10 epochs. The comparison of predicted and cooperative behaviour from Road Train model is presented on Fig. 5.

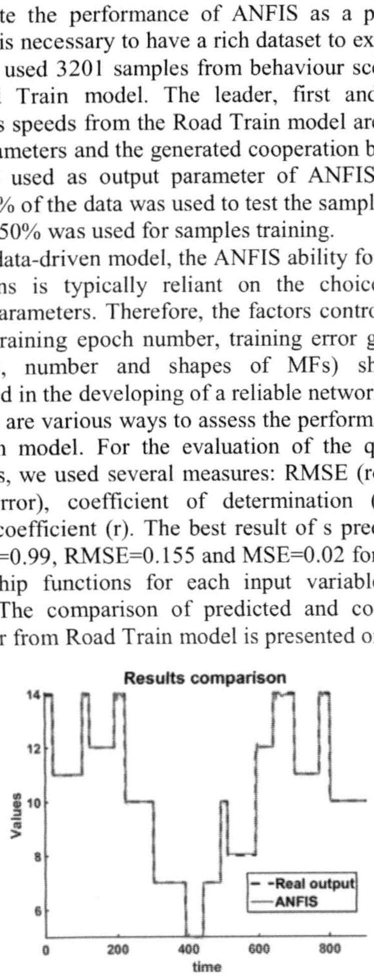

Figure 5. Prediction results of cooperation interactions profile.

5 Conclusion and future work

In this study, the Road Train hybrid automaton model was developed to simulate control and cooperation interactions between the vehicles. We introduced the coding scheme of behaviour patterns in the Road Train, for a specific cooperation interaction profile, in order to describe the complex system interactions only with one variable, which are useful for the prediction of traffic mobility in ITS. The results from ANFIS prediction model of cooperative interactions in Road Train showed the feasibility of this approach in terms of performance evaluation in the prediction method.

Future work will be devoted to exploring different prediction models, in order to test their performance in terms of influence on the vehicle speed profiles in the cooperation profile of the Road Train.

Acknowledgment

The authors express their sincere thanks for the funding support they received from Ministry of Civil Affairs of Bosnia and Herzegovina.

References

1. I. Stojmenovic, "Machine-to-Machine Communications With In-Network Data Aggregation, Processing, and Actuation for Large-Scale Cyber-Physical Systems", IEEE Internet of Things Journal, **1**(2), (2014)
2. B. Besselink, V. Turri, S.H. van de Hoef, K.-Y. Liang, A. Alam, J. Martensson, K.H. Johansson, "Cyber-physical Control of Road Freight Transport", (2015)
3. M. Aramrattana, T. Larsson, J. Jansson, C. Englund, "Dimensions of Cooperative Driving, ITS and Automation", *IEEE Intelligent Vehicles Symposium (IV), Korea,* (2015)
4. A. Ali, G. Garcia, P. Martinet, "The Flatbed Platoon Towing Model for Safe and Dense Platooning on Highways", Intelligent Transportation Systems Magazine, IEEE, **7**(1) (2015)
5. M. El-Zaher, B. Dafflon, F. Gechter, J.-M. Contet, "Vehicle platoon control with multi-configuration ability", Procedia Computer Science, **9**, 1503 (2012)
6. Z. Gacovski, S. Deskovski, "Different Control Algorithms for a Platoon of Autonomous Vehicles", International Journal of Robotics and Automation (IJRA), **3**(3), 151 (2014)
7. G. Tillema: "Merging into a Demo 2000 Platoon", University of Twente, The Netherlands (2006)
8. D. Yanakiev, I. Kanellakopoulos, "A simplified framework for string stability analysis in AHS,"*Proc. 13th IFAC World Congr.*, **177** (1996)
9. G. Franck, C. Vincent, C. Francois, "A reactive multi-agent system for localisation and tracking in mobile robotics,"*Proc. 16th IEEE Int. Conf. Tools Artificial Intelligence*, **431** (2004)
10. F. Contet, P. Gruer, A. Koukam, "Application of reactive multiagent system to linear vehicle platoon,"*Proc. IEEE Conf. Tools Artificial Intelligence, Patras, Greece*, **67** (2007)
11. E. Lee, J. Kim, W. Yoon, "Traffic speed prediction under weekday, time, and neighboring links' speed: Back propagation neural network approach". *In-Lecture Notes in Computer Science. Berlin, Germany: Springer-Verlag*, **626** (2007)
12. J. Park, Y.L. Murphey, R. McGee, J.G. Kristinsson, M.L. Kuang, A.M. Phillips, "Intelligent Trip Modeling for the Prediction of an Origin–Destination

Traveling Speed Profile", IEEE Transactions on Intelligent Transportation Systems, **15** (2014)
13. I. Bosankic, L. Banjanovic-Mehmedovic, F. Mehmedovic: "Speed Profile Prediction in Intelligent Transport Systems Exemplified by Vehicle to Vehicle Interactions", Cybernetics and Information Technologies, Special Issue on Control in Transportation Systems, **15**(5), (2015)
14. P. Wang, T. Hunter, A.M. Bayen, K. Schechtner, M.C. Gonzalez, "Understanding road usage patterns in urban areas", Nature Scientific Reports, **2** (2012)
15. X. Yang, M. Behroozi, O. A. Olatunbosun, "A Neural Network Approach to Predicting Car Tyre Micro-Scale and Macro-Scale Behaviour". Journal of Intelligent Learning Systems and Applications, **6**, 11 (2014)
16. V. Nikolic, S. Shamshirband, D. Petkovic, K. Mohammadi, Z. Cojbasic, T. A. Altameem, A. Gani, "Wind wake influence estimation on energy production of wind farm by adaptive neuro-fuzzy methodology", Elsevier Energy, 1 (2014)
17. N. Lynch, R. Segala, F. Vaandrager, "Hybrid I/O automata, Information and Computation" **185**, 105 (2003).

DOI: 10.1051/matecconf/20168104013

Mobile Robot Path Tracking Using Slip Estimation for Kinematics Information

Pornporm Boonporm [1] and Natdanai Tantawiroon[2]

[1]*School of Mechanical Engineering, Suranaree University of Technology Nakhonratchasima, 30000 Thailand*
[2]*Department of Mechanical Engineering, Kasetsart University Kampangsaen campus, Nakhon Phatom, Thailand*

Abstract. This paper presents the advantages of kinematic model over the dynamics one to control the path of autonomous mobile robot in outdoor working environment since the kinematics model is easier to be sensed and calculated the inertial information. Certainly, global navigation satellite system receiver accompanied with conventional inertial measurement unit are used together, so-called INS, in order to compensate the limitation of each other. The slip angle is one of parameters that cannot be direct measured but with the measurement techniques using two antennas receiver, it can be estimated. Fuzzy logic control rule of determination is used for simply heuristic one in order to track the robot path. The results show the satisfactory of tracking path.

1 Introduction

The research in autonomous vehicle has been developed nowadays for the large amount of application especially road safety. The robot systems require a number of capabilities such as environment sensing, localization, offline and online planning, trajectory generation and robust control of path tracking. Trajectory tracking is mentioned about the vehicle is require to track a time parameterized reference path following the vehicle require to converge to and follow a path without any temporal specifications. The major challenge of the tracking process is the modeling as well as the measurement method selection. Some of them tried to model exactly as possible as dynamics aspects in force and torque calculation. There are some disadvantages concerning the parameters estimation and the system nonlinear behavior. From this reason, some researcher use kinematics model to approximate and estimate the physical of the vehicle ride and handling. There are some examples using kinematics. Wang et al [1] demonstrated the trajectory planning algorithm for four wheel steering vehicle based on kinematics Ungoren et al [2] presented the various approach of controlling the vehicle stability and four wheel steering using kinematics relationship of measured signal applied to bicycle model.

The number of studies on vehicle dynamics handling aspects has been increasing. Sideslip is one of the interesting criteria because it concerns the longitudinal and lateral movement and related to the cornering performance but it is difficult to determine directly since it related forces in any directions and cornering stiffness. So that there are many researches involving the slip estimation based on kinematics method like Chaichaowarat et al [3] proposed the algorithm to estimate the slip from the rear wheel drive.

Next requirement of such problems concerns sensor. Inertial Measurement Unit is commonly used to measure acceleration and rate of rotation as well as velocity but there is some limitation about position of mounting, noise, and etc. Thus, GNSS positioning information can also be used for other purposes such as navigation and monitoring of location based on time. In order to reduce the disadvantages of the two types of sensor, the sensor fusion is used and in this case it is called inertial navigation system (INS). Some works are in literature as Preda, et al. [4], introduced the methodologies to use common GPS receiver, it had potential and convenience to install in test vehicle with limitation only the line of slight from navigation satellites. The vehicle kinematics was estimated from GPS trajectory It is proved to be sufficient for vehicle kinematic measurements especially real-road environments. And our previous work, [5] R. Thitipathanapong et al. purposes the aggressive behaviour detection using GNSS. We also evaluated in this paper, the novel consumer grade multi-GNSS navigation receivers were investigated with the driving behaviour detection algorithm with multiple satellite navigation system were applied in this study. The consumer grade satellite navigation system could apply as input probe for vehicle monitoring system. In conclusion inertial sensor and satellite positioning systems have emerged as accurate device with high data rate providing position sensing data in outdoors and now widely integrated for automotive system. These make autonomous vehicles to obtain rapid position feedback in world coordinate during outdoor operation.

To enhance performance of the GNSS based estimation, the cost effective methods for predicting slip has been investigated. An Amount of them tried to apply the low cost GNSS receiver with the implement techniques and algorithm e.g. Bevly [6] presented the achievable performance of the combine GNSS/INS system using covariance analysis from Kalman filter. Further research from Bevly et al [7] introduced two antenna GNSS receiver in conjunction with INS to improve the estimate value of slip angle. Yoon et al [8] have also demonstrated the two antenna receiver for compensating low update rate and tried to overcome the unsynchronized updated between the two signals.

In our study, we are interested in developed the GNSS signal form the consumer grade receiver fused to the inertial measurement unit, then the fuzzy logic control shame is used in consideration with lateral and longitudinal acceleration and no side slip is assumed. It has already published in Boonporm et al., [9]. We found that the more speed does the vehicle have, the more tracking error has occurred so that sideslip angle has been reconsideration. In this paper we'd like to propose the tracking method in case of sideslip. The difficulty arises because we cannot directly measure this value but estimate from the sensor measurement information instead.

2 Kinematics slip model

The bicycle model is considered in order to identify the relationship among the various parameters like lateral, longitudinal as well as yaw dynamics. In Figure 1, X-Y earth fixed coordinate frame is defined in direction like east and north and x-y-z rotating frame at the vehicle's center of gravity. In plane kinematics, one of the rotation angle about z-axis, called yaw angle (ψ) is taken into account.

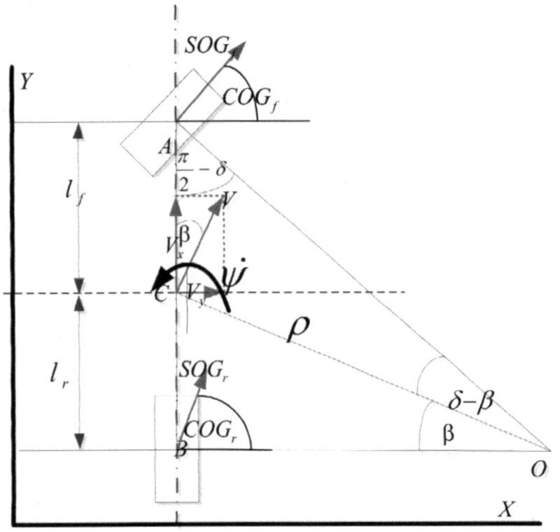

Figure 1. Bicycle model of lateral dynamics.

To define the relationship between the fix frame X-Y and the rotating frame x-y, the acceleration analysis is perform. Let $r = -R\hat{j}$, the acceleration of rigid body is defined as

$$\vec{a} = \dot{\psi}\hat{k}x(\dot{\psi}\hat{k}x(-R\hat{j})) + \ddot{\psi}\hat{k}x(-R\hat{j}) + 2\dot{\psi}\hat{k}x(-\dot{R}\hat{j}) + \ddot{x}\hat{i} + \ddot{y}\hat{j} \quad (1)$$

So that the lateral acceleration is

$$\vec{a}_y = \dot{\psi}^2 R + \ddot{y} \quad (2)$$

The term $\ddot{y}$ is interesting because it shows the vehicle slip so that side slip behavior cannot be negligible.

2.1 GNSS receiver model

In our study, we monitor the robot kinematic behavior in horizontal plane by means of acceleration determination and side slip. The Inertial Measurement Unit (IMU) should be applied to take direct data of acceleration and rotation speed. But from some disadvantages mentioned above, we are also applied the GNSS receiver with speed over ground (SOG) which is the velocity vector and course over ground (COG) at 10 Hz time step. COG denotes the direction of travel (or velocity vector) of the vehicle. Vehicle course is different from vehicle heading or yaw due to vehicle sideslip or lateral velocity.

$$COG = \psi + \beta \quad (3)$$

$$\beta = \tan^{-1}\frac{V_y}{V_x} = \sin^{-1}\frac{V_y}{\sqrt{V_x^2 + V_y^2}} = \sin^{-1}\frac{V_y}{V} \approx \frac{V_y}{V} \quad (4a)$$

or

$$\beta \approx \frac{\dot{y}}{SOG} \quad (4b)$$

There are amount of study concerning the vehicle body slip estimation [6]-[8] and their results proved that the two antenna GNSS receiver is recommended.

The disadvantages when using two-antenna receiver is the time synchronization. The problem of unsynchronized update of two GNSS receivers at front end and rear end when updating the measurements is addressed and solve following Yoon. Equations (5)-(6) are used with the front one as well as the rear end with equations (7)-(8) The idea of asynchronous update is proposed and applied by Yoon stated that the Kalman filter measurement update is executed only for the updated sensor and we also use the recommended measurement matrix with proving in the mentioned paper which are different between front end and rear one.

$$SOG_f \cos COG_f = V_x \cos\psi - (V_y + l_f\dot{\psi})\sin\psi \quad (5)$$

$$SOG_f \sin COG_f = V_x \sin\psi + (V_y + l_f\dot{\psi})\cos\psi \quad (6)$$

$$SOG_r \cos COG_r = V_x \cos\psi - (V_y - l_r\dot{\psi})\sin\psi \quad (7)$$

$$SOG_r s\text{in} COG_r = V_x \sin\psi + (V_y - l_f\dot{\psi})\cos\psi \quad (8)$$

Consequently, V_x and V_y can be expressed as functions of $(SOG_f, COG_f, SOG_r, COG_r)$ under assumption of the same standard deviation of GNSS velocity errors.

2.2 Inertial measurement unit

Governing equations for IMU measurements are

$$a_{x\_imu} = \dot{V}_x - V_y\dot{\psi} + w_x \quad (9)$$

$$a_{y\_imu} = \dot{V}_y + V_x\dot{\psi} + w_y \quad (10)$$

$$\psi_{imu} = \dot{\psi} + w_\psi \quad (11)$$

where w_x, w_y, w_ψ are Gaussian white noise in x,y and yaw rate respectively.

2.3 Data fusion

The GNSS will lost the signal in some case such as under the tree, or driving in urban area with high building, so that the fused data between both signal are helpful for correcting the data. Figure 2 show the algorithm of data fusion. It consist of two sources of them, the former is GNSS receiver that data is collected via serial port and the process the signal and at the same time the acceleration from accelerometer, the rotation rate from gyro and calculated heading from magnetometer are collected and processed. Both of them can fuse together using kalman filter technique.

3 Path tracking control

Path tracking in horizontal plane can be implemented as shown in Figure 2, starting with path planning depended on algorithm and constraint. We have got the desired trajectories in X-Y global coordinate direction and yaw rate with time parameterized ($X_{des}, Y_{des}, \psi_{des}$). After that the profile is calculated and compare with the information from sensor especially longitudinal, lateral acceleration and yaw rate as well as side slip. Then the control algorithm is implemented for the error between them. The control signal is generated for two motor, driving and steering.

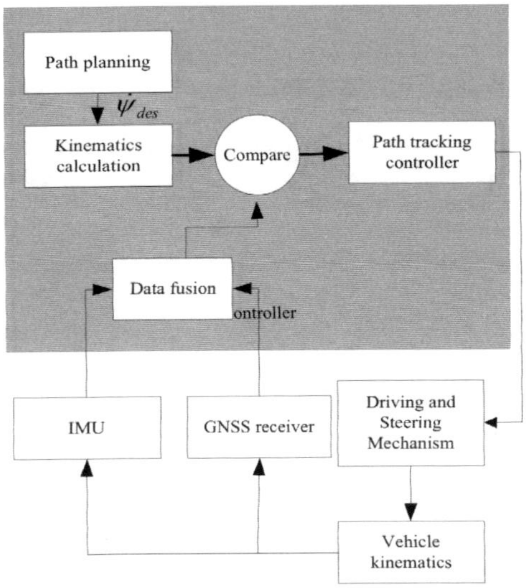

Figure 2. Tracking control algorithm.

3.1 Error model

Aim of control is tracking error that is defined as the shortest distance between the robot and the desired path for e_x, e_y and the yaw error with the difference between yaw rate and the desire one from path planning algorithm calculation.

$$\begin{bmatrix} e_x \\ e_y \\ e_\psi \end{bmatrix} = \begin{bmatrix} \cos\psi & \sin\psi & 0 \\ -\sin\psi & \cos\psi & 0 \\ 0 & 0 & 1 \end{bmatrix} \begin{bmatrix} X_{des} - X \\ Y_{des} - Y \\ \psi_{des} - \psi \end{bmatrix} \quad (12)$$

Refer to bicycle model in figure 1, the error dynamics respect to vehicle frame can be derived and simplified to

$$\begin{bmatrix} e_x \\ e_y \\ e_\psi \end{bmatrix} = \begin{bmatrix} 0 & \dot{\psi}_{des} & 0 \\ -\dot{\psi}_{des} & 0 & 0 \\ 0 & 0 & 0 \end{bmatrix} \begin{bmatrix} e_x \\ e_y \\ e_\psi \end{bmatrix} + \begin{bmatrix} \cos e_\psi \\ \sin e_\psi \\ 0 \end{bmatrix} \dot{x}_{des} - \begin{bmatrix} \dot{x} \\ \dot{y} \\ \dot{\psi} - \dot{\psi}_{des} \end{bmatrix} \quad (13)$$

3.2 Tracking control

In tracking environment, the issue of uncertainty in both model and surrounding are considered and make the model more complex and consequently control problem. One of the effective and ease to implement is the fuzzy logic control since it does not require an accurate mathematical model of the process. We have chosen the simply one, center of gravity method. When used in a control system, FLC is robust since it provides a fast rise time and a small amount of overshoot. However, some difficulties can occur when designing the FLC as stated in [9].

The input variables of acceleration norm as $a = \sqrt{a_{long}^2 + a_{lat}^2}$ which should be not exceed than the threshold in g-g diagram and yaw rate are considered as input. Speed and the command steering angle are classified as output. The results is acceptable but there is error in lateral direction when the vehicle drive faster.

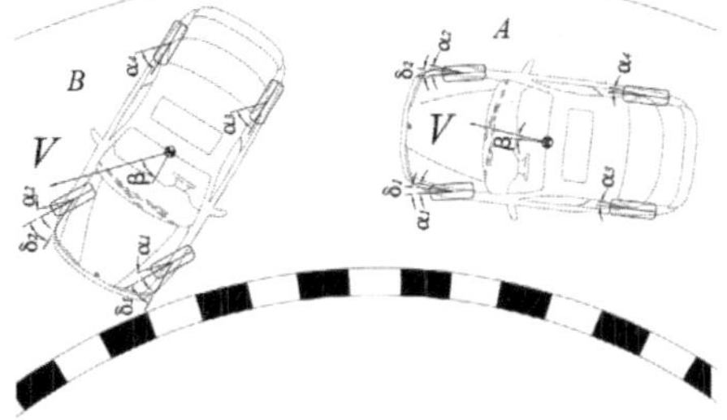

Figure 3. Vehicle driving as low-sideslilp (A) and large sideslip (B) during steady state cornering following Abdulrahim et al [10].

Consequently, the additional control with slip is implemented since we determine its behavior during cornering; the vehicles are subject to centripetal acceleration directed toward the center of rotation. As shown in Figure 3, vehicle A drives at a small sideslip

DOI: 10.1051/matecconf/20168104013

and incurs acceleration that is mostly in the negative lateral direction. Vehicle B is oriented at a large angle relative to the velocity, which is here perpendicular to the turn radius. Functions will be presented to describe slip in term of path curvature and velocity.

The additional control rule is applied as mention in Figure 4. We consider the yaw rate error since the measured yaw rate is not the same as desire one because of side force leading to the lateral movement $\ddot{y}$ and $\dot{y}$ where slip is introduced in equation (4b).

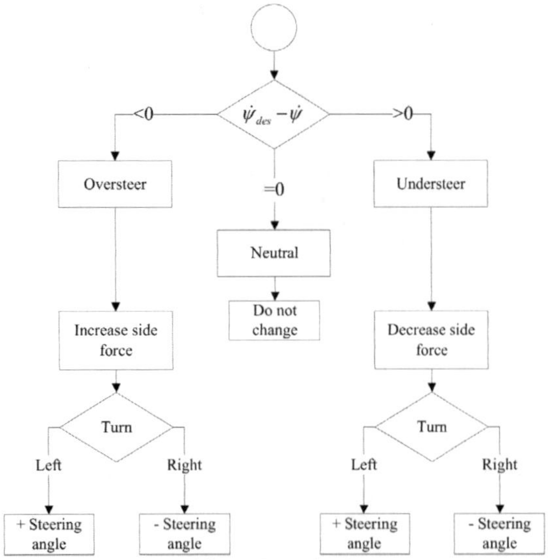

Figure 4. Flowchart of fuzzy determination.

4 Experimental results

The algorithm mentioned above is applied to the middle size vehicle like autonomous robot equipped with two main sensor as U-blox multi GNSS receiver with two antennas and inertial measurement unit both have 10 Hz sampling rate

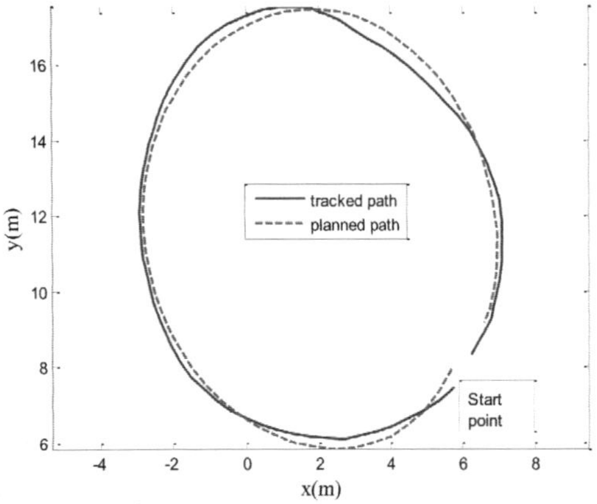

Figure 5. Compare the tracked path to the planed path.

One example of the results during cornering is driving with keeping the constant speed in the trajectory planning, in our sample experiment in Figure 5. we control its speed approximately 5 m/s The yaw rate occur during cornering in left or right turn In the same manner but only plus and minus sign depend on coordinate system. The result of the tracking is demonstrated in Figure 5. The desire path is shown in dash line and the sensor data readings are shown in solid line. The error of path tracking still occurs due to unbalanced of the steering link and some unbalance of the structure, as well as the sensor noise from GNSS receiver and IMU. There still have been error in lateral direction more than longitudinal direction even though the slip is compensated as demonstrated in Figure 6.

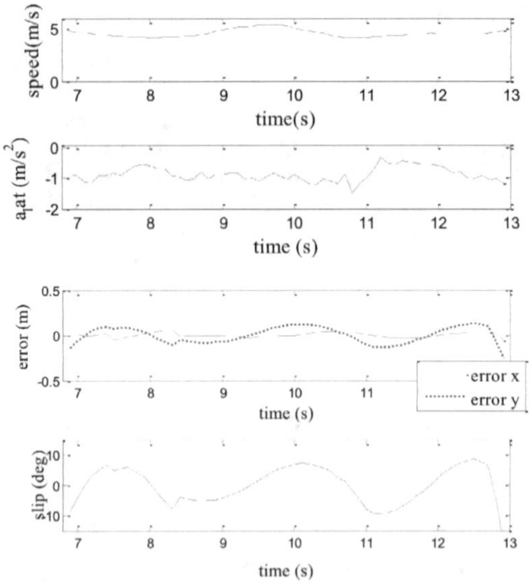

Figure 6. Speed, lateral acceleration, translational error and slip angle versus time from the tracked path's sensor reading.

5 Conclusion

The kinematic model with slip behavior consideration is useful for path tracking problem without knowing its dynamics model and parameters which are difficult to evaluate. In order to measure the kinematics information, we use the information from GNSS receiver with two antenna fused with IMU to overcome some disadvantages. The results are satisfied but there are some position error still occurs but it is smaller. The extend measurement technique and hybrid control are in future consideration.

References

1. D. Wang, F. Qi, *IEEE ICRA* (2001)
2. A. Y. Ungoeren, H. Peng, H. E. Tseng Int. J Vehicle Autonomous Systems, **2** (2004)
3. R. Chaichaowarat, W. Wannasuphoprasit, Int. J. of Automotive Technology, **17** (2016)
4. D. Preda, G. Covaciu, G. Ciolan, D. Dima, *SMAT, Craiova, Romania* (2008)
5. R. Thitipatanapong, P. Boonporm, N. Noomwongs, P. Wuthimanop et al, *SAE Technical Paper* (2015)
6. D. M. Bevly, Int. J. of Dyanmic System Meausement and Control., ASME, (2004)

7. D. M. Bevly, J. Ryu, J. C. Gerdes, *IEEE ITS.*, 7, 4 (2006)
8. J-H. Yoon, H. Peng, *IEEE Vehicular Technology* **63,** 6(2013)
9. P. Boonporm, N. Tantawiroon, Romanian Review, Precision Mechanics, Optics and Mechatronics, INCDTM (2015)
10. M. Abdulrahim, *SAE World Congress* (2006)

The Influence of Weight Distribution on the Handling Characteristics of Intercity Bus under Steady State Vehicle Cornering Condition

Ekalak Prompakdee[1], Pornporm Boonporm[2] and Supakit Rooppakhun[2]

[1]*School of Manufacturing Engineering, Institute of Engineering, Suranaree University of Technology, Nakhonratchasima, 30000 Thailand.*
[2]*School of Mechanical Engineering, Institute of Engineering, Suranaree University of Technology, Nakhonratchasima, 30000 Thailand.*

Abstract. The vehicle cornering behavior is an important performance in handling stability especially for the intercity bus. Accordingly, one of the significant parameters concerns the weight distribution which is affected by the center of gravity (CG). In this paper, the effect of weight distribution while varying the turning radius is compared and it should be expressed, and interpreted by understeer gradient which is influenced by the location of CG. The characteristic of intercity bus was modeled and evaluated using the multi-body dynamic analysis software. The analysis has been conducted under steady state cornering based on a total of three configurations, with front/rear axle weight in percent, as 40/60, 45/55, and 50/50.The results stated that the magnitude of weight distribution on front axle of bus in a range of 40% to 50% caused the incremental value of understeer gradient and it was also increased as three times in each turning radius since the difference lateral slip angles between front and rear are expanded.

1 Introduction

Nowadays, travelling by means of intercity bus in Thailand plays an extremely important role. From the department of land transport statistics, there are more than 35 million passengers for the intercity bus passenger [1]. Consequently, the number of intercity bus accidents should be taken into consideration and the root causes came from its yaw stability during cornering. It required both static and dynamic tests for this safety issue [2]. The static stability test related the vehicle's rollover resistance is a calculation of the center of gravity that affects from the significant parameters of vehicle weight distribution such as the position of engine, transmission, axle and other component [3, 4]. In case of dynamics, the subject deals with the study of vehicle response related to driver input, ability of the vehicle to stabilize its motion against external disturbances and ride characteristics of vehicle [5-7]. The vehicle handling performance including the stability control were the crucial factors involved to the standard tests consisted of lateral acceleration, roll moment, braking performance, and CG position [8, 9].

Generally, the most autonomies safety systems such as antilock braking, traction control and stability control have been introduced in passenger vehicles but in such systems like heavy vehicle (e.g., truck and bus) requires a sound knowledge of the steering response characteristics under both steady-state and transient conditions as well as handling performance [10]. However, the research related to the study of dynamics response during cornering of the heavy vehicle.

Therefore, the study aimed to evaluate the handling characteristics of a typical intercity bus under steady-state cornering condition. Using the multi-body dynamic model and simulation, the effect of weight distribution including various turning radius of intercity bus to handling characteristics (e.g. lateral acceleration, slip angle, and understeer gradient) were conducted. The understeer gradient can be evaluated by operating the vehicle around a constant radius turn and observing steering angle versus lateral acceleration.

2 Steady state vehicle cornering

The well-known bicycle model is considered in order to identify the relationship among the various parameters like lateral, longitudinal as well as yaw dynamics. In figure 1, X-Y earth fixed coordinate frame is defined in direction like east and north and x-y-z rotating frame at the vehicle's center of gravity. Assuming plane kinematics, the rotation angle about vertical axis, called yaw angle (ψ), is taken into account.

To define the relationship between the fix frame X-Y and the rotating frame x-y, the acceleration analysis is perform. Let $r = -R\hat{j}$, the acceleration of rigid body is defined as

$$\vec{a} = \dot{\psi}\hat{k}x(\dot{\psi}\hat{k}x(-R\hat{j})) + \ddot{\psi}\hat{k}x(-R\hat{j}) + 2\dot{\psi}\hat{k}x(-\dot{R}\hat{j}) + \ddot{x}\hat{i} + \ddot{y}\hat{j} \quad (1)$$

So that the lateral acceleration is

$$\vec{a}_y = \dot{\psi}^2 R + \ddot{y} \quad (2)$$

The term $\ddot{y}$ is interesting because it shows the vehicle slip so that side slip behavior cannot be negligible

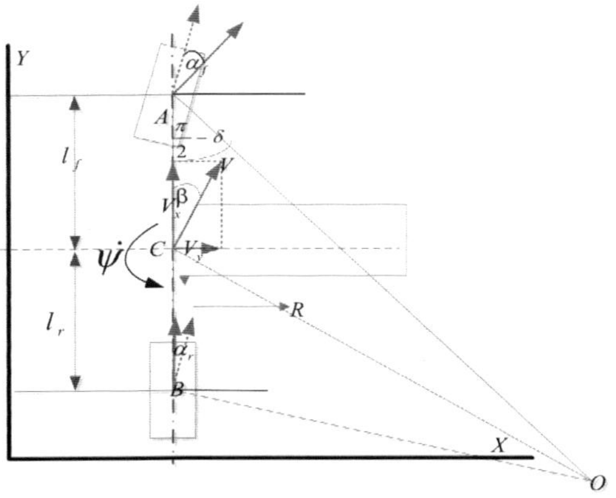

Figure 1. Steady state vehicle cornering condition.

Steady state cornering is the practical test in order to verify the handling characteristics. The aim of this test is to determine the understeer gradient of difference configuration of weigh transfer. The front and rear wheel slip angle denote by α_f and α_r, respectively as shown in figure 1. The instantaneous turn center O of the vehicle is the point at which the two lines perpendicular to the velocities of the two wheels meet. Under the assumption that the road radius is much larger than the wheel base of the vehicle. From the geometry, we have the relationship of the steering angle δ that can be approximated as

$$\delta-\alpha_f-\alpha_r \approx \frac{L}{R} \tag{3}$$

Consider the dynamics equations, α_f and α_r are related to the forces acted on tired and the small slip is assumed so that the lateral tired force is proportional to its slip angle with the constant called cornering stiffness for front C_f and rear C_r

The steady state steering angle in equation 3 can be rewritten to

$$\delta=\frac{L}{R}+\alpha_f-\alpha_r=\frac{L}{R}+\left(\frac{m_f}{2C_f}-\frac{m_r}{2C_r}\right)\frac{V_x^2}{R} \tag{4}$$

or

$$\delta=\frac{L}{R}+K_v a_y \tag{5}$$

In this case of understeer, the understeer gradient K_v should greater than zero due to a larger slip angle at the front tires compared to the rear tires. The relationship between weight transfer and wheel slip angle is interpreted in following equation 6.

$$\frac{m_f}{C_f}>\frac{m_r}{C_r} \Rightarrow \alpha_f>\alpha_r \tag{6}$$

3 Multi-body dynamic simulation

The basic information required to understand multi-body dynamic simulation is its mathematical formulation. It also explains how lateral forces are generated while a vehicle turning, and their relation with the slip angle and lateral acceleration as well as understeer gradient. The intercity bus model is used to conduct for estimating vehicle handling characteristic as shown in Figure 2.

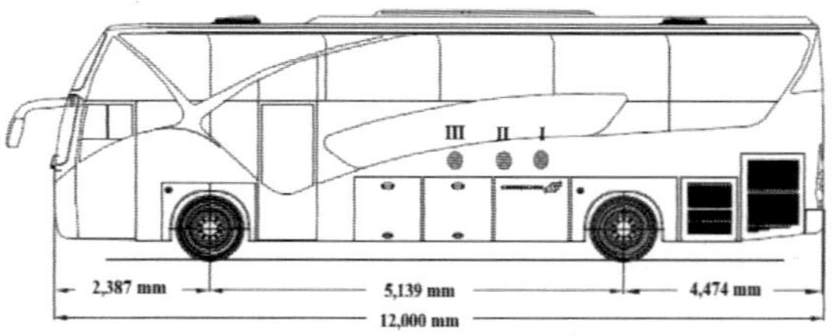

Figure 2. Location of Center of Gravity (CG.)

Table 1. Specification of intercity bus

Parameters		Bus model
Gross Vehicle Weight		1,8000 kg
Overall Length		12,000 mm
Wheelbase		5,139 mm
Track Width	*Front*	2,120 mm
	Rear	1,780 mm
Suspension	*Front*	Coil Spring
	Rear	Coil Spring
Front Axle		Double wishbone
Rear Axle		Rigid axle
Tires		318/80 R22.5

3.1 The virtual intercity bus model

According to the virtual model of intercity bus model, the multi-body dynamic model has been carried in MSC ADAMS/Truck tool (MSC software). Table 1 showed the specificaiton of intercity bus model. The virtual model consised of sub-system as following: chassis body, powertrain, steering system, anti-roll bar, wheel model, front and rear suspension system. Table II display the weight distribution of intercity bus model and the distance of CG position from front axle in this study.

Table 2. The weight distribution of an intercity bus model

Layout	Distance from front axle	% WD (F/R)
I	3,083 mm	40/60
II	2,826 mm	45/55
III	2,570 mm	50/50

3.2 Steady state cornering condition

The vehicle dynamic test under steady state condition was virtually simulated on a total of nine conditions. A three constant cornering radius consisted of 50-m, 75-m, and 100-m was performed in each difference weight distribution. For the test, the vehicle models were maneuvered on the test track at initial longitudinal velocity 12 km/h for left turn. The data were carried out by 0.1 interval time up to 30 km/h under 50 seconds on the flat track.

4 Results and discussion

According to the result, the firstly is investigated the lateral translation or lateral slip $\ddot{y}$ explained in equation 2. Normally, the lateral acceleration rises related the curvilinear motion, because of the change in direction in vehicle rotating frame which should have the value as $\dot{\psi}x(\dot{\psi}xr)$. However, the side slip phenomena occur and lead to instability in the real environment. Figure 3 show the value of lateral slip acceleration $\ddot{y}$ form each layout model, and the result also show that the increased vehicle speed of the bus have the more lateral slip acceleration is arise. Consequently, the bus speed in range of 12 - 30 km/h is taken into consideration.

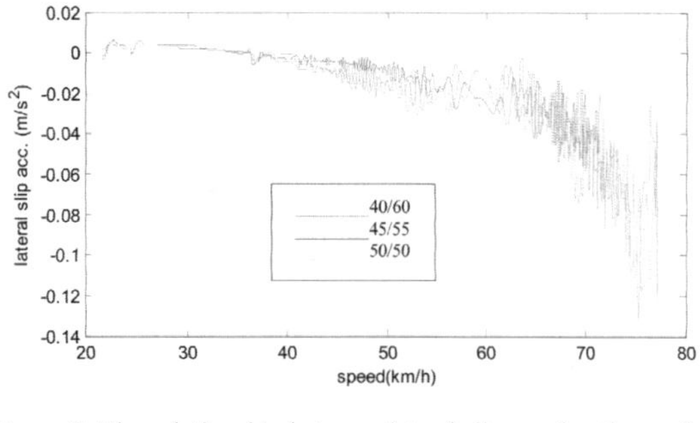

Figure 3. The relationship between lateral slip acceleration and the vehicle speed.

Lateral acceleration versus front wheel side slip is depicted in figure 4. Following equation 4 and 5 the relationship between the steering angle and lateral acceleration is stated and it is shown that the slope of the graph the understeer gradient. But in our cases, constant steering angle input is deployed for three layouts with 75 meter constant turning radius. The steering angle and wheel slip angle are associated and can be approximate via the bicycle model from figure 1 by assuming small rear wheel slip evidenced in figure 5. The slopes of these are summarized in table 3.

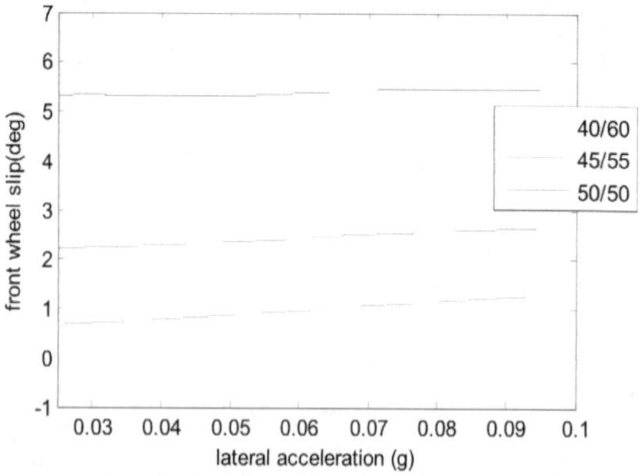

Figure 4. Front wheel slip angle versus lateral acceleration.

Table 3. Understeer gradient of each layout in various turning radius condition

Layout	% WD (F/R)	Understeer Gradient (deg/g)		
		50 m	*75 m*	*100 m*
I	40/60	2.73	2.72	2.83
II	45/55	6.65	6.73	6.93
III	50/50	9.2	9.04	8.98

Table 3 displayed the comparison of weight distribution effective between the various turn radiuses. Clearly, the increase in weight distribution on front axle disclosed the increasing of understeer characteristic. Regarding the magnitude of understeer gradient, the result is closely agree with the N. Yu., et al., [10] which reported the steady state handling behavior of transit bus experimentally. The study revealed that the understeer gradient for the bus left turn was experimentally determined to be 5.1 deg/g. Generally, the understeer gradient of a specific vehicle depend on the several of parameters such as suspension spring stiffness, steering ratio, and wheel alignment conditions, and weight distribution, etc.

In addition, the theoretically understeer gradient is directly proportion to the weight on front axle. That is, the understeer gradient increases with the increase in weight on the front axle. Therefore, with the increase in weight on rear axle, the result exhibit tendency to oversteer behavior [11, 12]. The trends of study results hold well when compared with the theoretical predictions. Thus, the conclusion can be draw that obtained result are reliable.

Figure 5 showed the comparison of the front slip angle and rear slip angle to confirm the understeer characteristics in each configuration referred to equation

 DOI: 10.1051/matecconf/20168104014

6. The variation of slip angle of front and rear wheel seems to be the same over the speed range. The slip angle of front wheel increases when the weight distribution is increased toward the front axle

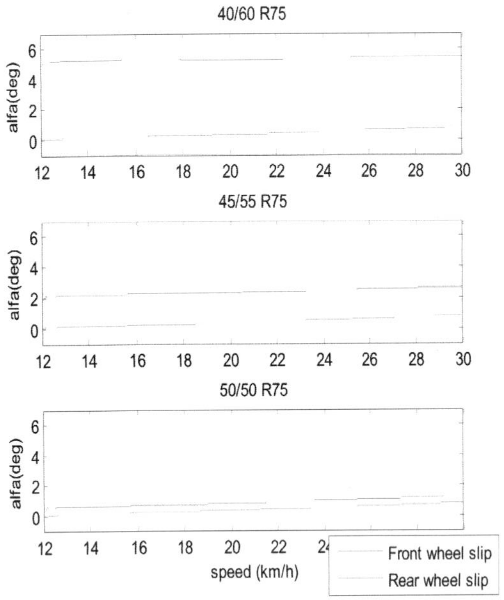

Figure 5. The comparison between front and rear slip in some cases as mentioned.

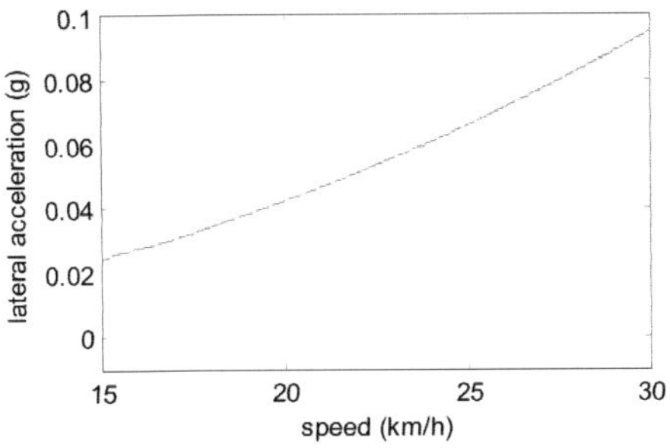

Figure 6. Lateral acceleration determination

One another important parameter ensuring safety during cornering is lateral acceleration as shown in Figure 6 for the worst case. This value is not exceed the threshold value of the heavy duty truck and bus which is approximately less than 0.2g for the heavy duty truck mentioned as our experience. Although this threshold seems to be subjectivity, we did perform the experiment in the passenger car and light pick up as reported in [13]. In our simulation range with maximum speed 30 km/h during cornering, these values are far from the threshold.

Yaw rate evaluation should be mentioned although it does not show directly the handling stability; it played the important in stability control issue. We tried to compare the rate for the layout I by varying the turning radius. The yaw rate signal just slightly exceeds the desire yaw rate which can be derived from lateral acceleration and the velocity as

$$\psi_{des} = \frac{a_y}{\sqrt{v_x^2 + v_y^2}} \quad (7)$$

We found that yaw rate increases when the radius is smaller and it will increase when it drives faster. The error between simulated and desired one is raised concerning the faster speed. The results from other two cases are in the same trend.

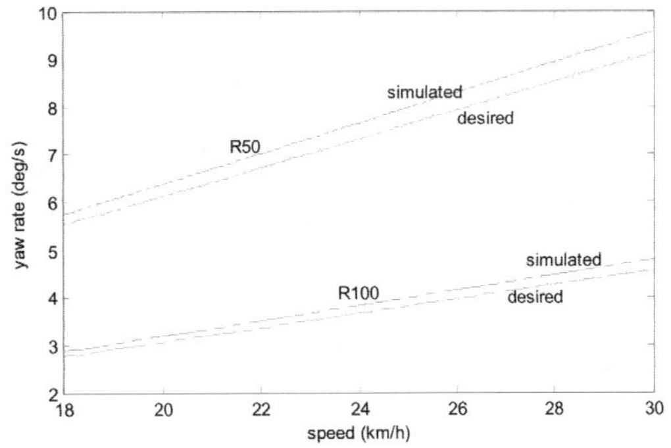

Figure 7. Desired yaw rate and simulated yaw rate in layout 40/60 by varying the turning radius

Theoretically, in curvilinear translation motion the vehicle speed should be in the tangent direction of the travel path or vehicle longitudinal direction, but actually lateral velocity occurs because of side slip. It is obviously that the vehicle moves in an undesired mode. The side slip angle (β) is defined as equation 8.

$$\beta = \tan^{-1}\frac{V_y}{V_x} = \sin^{-1}\frac{V_y}{\sqrt{V_x^2 + V_y^2}} = \sin^{-1}\frac{V_y}{V} \approx \frac{V_y}{V} \quad (8)$$

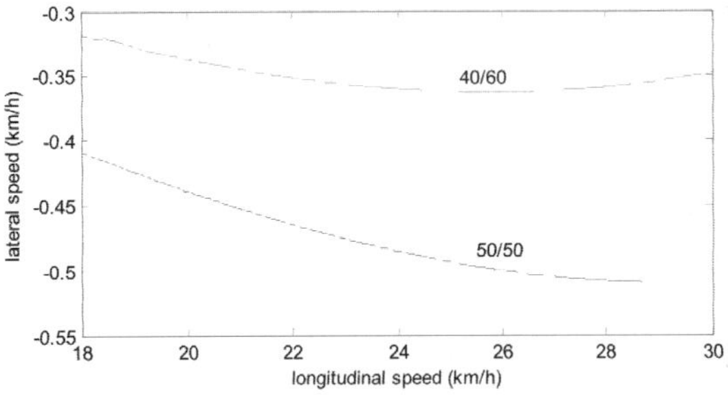

Figure 8. Bus speed in longitudinal and lateral direction using R100.

In our cases of low velocity, mentioned in figure 8 in layout I and III with turning radius 100 m. The lateral speed is near to zero compare to longitudinal one, then the vehicle speed is estimated as the longitudinal speed. In the graph, we can point to side slip angle that layout I has this value less than layout III, consequently, it can be interpreted that model III tends to have more understeer gradient than model I. So that, it is the same way as reported in table 3.

5 Conclusion

Applying the multi-body dynamic simulation, the effect of weight distribution in each layout with three different turning radiuses to evaluate the vehicle handling characteristic of intercity bus was performed. As the results, the magnitude of understeer gradient increased approximately three times when the weight distribution is increased towards front axle. The variation of slip angle of front and rear tires is responsible for the variation of understeer gradient.

Acknowledgment

The authors would like to acknowledge Suranaree University of Technology, and also thank National Metal and Materials Center for the sincere support among the research activities of this article.

References

1. Declare of Department of Land Transport of Thailand, *The Government Gazette*, **129**, 27 (2012)
2. L. Mai, P. Sweatman, *Australian Road Research Board*, (Rept. No. AIR323-2, 1984)
3. A. Reński, Proc. 24th *International Technical Conference on the Enhanced Safety of Vehicles* (Gothenburg, Sweden, 2015)
4. L. Prochowski, K. Zielonka, A. Muszyński, J. K Power and Trans, **19**, 3 (2012)
5. S. Murthy, M. Gowda, H. Venna, *SAE Technical Paper* (2016)
6. B. Ayalew, N. Yu, S. Muthiah, B. T. Kulakowski, ASME Int. Mech. Eng Con and Exp, (2004)
7. J. S. Jo, S. H. You, J. Y. Joeng, K. I. Lee, K. Yi, Int. J. of Auto. Tech **9**, 5 (2008)
8. H. H. Kim, J. Ryu, Int. J. of Auto. Tech **12**, 4 (2011)
9. M. Sherman, G. Myers, *SAE Technical Paper* (2000)
10. N. Yu, S. Muthiah, B. Kulakowski. *9th International Symposium on Heavy Vehicle Weights and Dimensions* (2006)
11. D. Vilela, R. S. Barbosa, J. of the Braz. Soc. of Mech. Sci. and Eng., **33** 4, (2011)
12. T. D. Gillespie, *Fundamentals of vehicle dynamics, Society of Automotive Engineers* (1992)
13. P. Boonporm, *Int. C on Science Ecology and Technology I*, (Vienna, Austria, 2015)

Artificial Neural Network for Transportation Infrastructure Systems

Koorosh Gharehbaghi

RMIT University, Melbourne, Australia

Abstract. Artificial Neural Networks (ANNs) represents the overall interconnection of the systems together with numeric weighting that can be tuned based on experience, system Inputs, Processing and Outputs. Moreover, the real advantage of ANNs is the ability to solve complex system problems such as one which are found within the Transportation Infrastructure Systems. Artificial Neural Networks (ANNs) for Transportation Infrastructure System must incorporate system engineering techniques that will be sustainable for future years and maintained at acceptable levels. Accordingly, this paper will introduce the concept of Artificial Neural Networks (ANNs) and its core functions for the optimization of Transportation Infrastructure Systems in particular the maintenance processes.

1 Introduction

Generally, Artificial Neural Networks (ANNs) are nonlinear models that are non-parametric and are utilized to determine approximate functioning of a System for a real-life application. Commonly, the core action of Artificial Neural Networks (ANNs) tends to fall within the process and control categories [1].
Furthermore, the Artificial Neural Networks (ANNs) also represents the total interconnection of the systems together with numeric weights that can be tuned based on experience, making neural nets adaptive to inputs, and capable of learning [8], [9].

Although, ANN with Back Propagation (BP) learning algorithm is widely used in solving various classification and forecasting problems; streamlined ANNs are employed to treat super complicated problems in which too many variables need to be simplified in a model. ANN are also powerful data-driven, and flexible computational tool having the capability of capturing nonlinear and complex underlying characteristics of any physical process with a high degree of accuracy [8].

Furthermore, ANN are suitable for inverse modeling when the numerical relations between input and output variables are unknown, and cannot be established.

The main advantages of using Artificial Neural Networks (ANN) include ability to [9]:

- Operate large amount of data sets.
- Implicitly detect complex nonlinear relationships between dependent and independent variables.
- Detect all possible interactions between predictor variables.

Moreover, the real advantage of ANNs is the ability to solve complex system problems such as one which are found within the Transportation Infrastructure Systems (TIS). Transportation Infrastructure is seen as the interconnectivity between the physical and tangible assets that is required to support and develop a nation. It is therefore essential to administer the Transportation Infrastructure efficiently in order to provide continuous, sustainable and economic services to the population. A fundamental aspect of this effective administration is the design of Transportation Infrastructure Systems (TIS).

There are many competitive factors currently impacting the Transportation sector, which requires careful system considerations such as; the high capital cost of design and construction, changes to the legislations, and the impact of the increase in demand and usage of Transport Infrastructure [5]. Other areas which require careful system consideration include [4], [5]:

- Transport Infrastructure Planning and Development. Transport Infrastructure Planning and Development involves the various aspects of land use planning, building construction and land subdivision which are administered by governments at all levels. The aim is to achieve high quality and sustainable development outcomes in the urban and rural areas, and taking into account important issues such as preservation of the physical environment as a part of broader Transport Infrastructure Planning and Development assessment.
- Transport Infrastructure Economic Development. The main aim of major cities role as an increasingly dominant regional service centers are underpinned by a strong regional economy based in manufacturing, logistics, wholesale trades etc. Thus indicating the importance of regional areas, since their economy is usually dynamic and vibrant, making it well placed to cope with future economical structural changes resulting from the needs of an ever-changing demographic profile. As a part of a general economic development, regional economic development program for Transport Infrastructure needs

to be created and encompass into any Transport Infrastructure Systems (TIS). The proposed TIS, in an economic sense (as well as overall methodology) is robust, sequential and more adaptive, and considers more parameters (such as Engineering requirements) than the other existing generic models.

• Transport Infrastructure Engineering Requirements [11]. These requirements provide fundamental tools of evaluation of Transport Infrastructure and their performances. These are four fundamental aspects (The Four Core Elements) which need to be built-in into the Transport Infrastructure Systems (TIS). The four Core Elements are:-

1) Infrastructure Asset Rehabilitation and Renewal. This element includes the assessment and renewal of all assets such as bridges, roads, and buildings. As a part of Infrastructure Asset rehabilitation and renewal this first core element initially includes *"Evaluating asset condition and performance"*. Once an extensive and detail asset condition and performance evaluation has been applied, the second phase *"developing an asset renewal strategy"* commences. This involves the identification of future asset renewal costs and addressing the prioritization and funding of the typical liabilities.
2) Structural Performance and Operation. This element includes, damage threshold for structures, performance of structures to extreme events such as act of GOD including earthquakes, and/or strategies for the development of the next generation of standards. This area includes the management and supervision of the Transportation Infrastructure Assets via theories and techniques such as structure performance analysis. The structure performance analysis methods are alternatives to avoid the computational complexity problem associated with other techniques such as discrete event simulation. In this process even if a finished conceptual and technical framework is not yet obtainable, significant advantages have been obtained not only from performance but also from correctness analysis stages.
3) Sustainable and Tangible Materials. This element includes the thorough investigation of utilized materials such as, Concrete, Timber, Iron, Steel and Asphalt. In addition, Sustainable and Tangible Materials approach could include developing new materials processes to enhance sustainability by decreasing pollution, emissions, energy consumption and improving eco-efficiency of materials processing, and modular construction and advanced materials.
4) Emerging Robust Technologies and Innovative Tools. This element includes the use of software such as RIVA (Real-time Asset Valuation Analysis) or Sensor applications in Infrastructure monitoring, and energy efficient structures. Moreover, this element includes the research, recommendation and implementation of the newly created and derived technologies (expertise, equipment and machinery) within the Transportation Infrastructure Assets. The Emerging Robust Technologies and Innovative Tools will be needed for Transportation Infrastructure Optimization in this new high technological era to meet the public's demand for high-quality and convenient transportation. The successful implementation of these practices will encourage and enable responsible authorities to utilize the new technologies appropriately to complete the works quickly with the minimum disruption to traffic, while incorporating quality that will ensure long-lasting, low-maintenance facilities.

All of these three areas need to be part of any system analysis and design process [11]. The Transportation Infrastructure Systems (TIS) provides this integration and a comprehensive system development strategy.

2 Transportation Infrastructure Systems (TIS)

System Engineering not only provides a structure break down but also stage by stage evaluation and modification [2] [10]. This is the most important stage of any system development, and therefore requires utilizing an appropriate Engineering methodology to further enhance the performance of TIS.

Fig. 1 represents the Engineering Mechanism for Transportation Infrastructure Systems (TIS), and its parameters and their relationship.

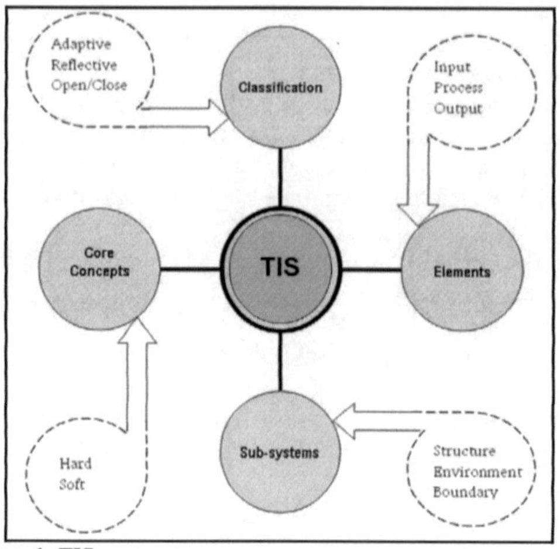

Figure 1. TIS parameters.

A primary objective, when designing the appropriate System parameters is not only to create a System and Sub-system mechanism but also (as demonstrated in figure 4), to establish classification and elements of the System. While the sub-systems typically deal with system environment and structure, system classification and elements provide the additional sub-categorization [6].

On the other hand, while the core concepts of System could be straight forward since it deals with Soft and Hard methodology, the elements is where the essential design questions will be considered. The process of System

Engineering Mechanism for TIS is unique and innovative and thus is the central methodology of this paper. To better design an operative TIS a comprehensive Transportation Infrastructure Management System (TIMS) need to be established.

3 Transportation Infrastructure Management System (TIMS)

As it can notice in Fig. 2, TIMS is a very dedicated process which deals with Varity of Transportation parameters. As a result, TIMS usually covers many subsystems, among which Pavement Management System (PMS) and Bridge Management System (BMS) are the most important.

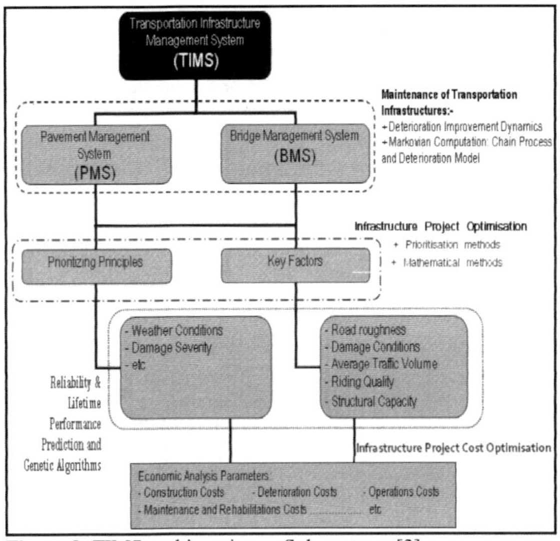

Figure 3. TIMS and its primary Sub-systems [3].

Achieving the ultimate goal of having an effective and successfully operative Transportation Infrastructure Management System (TIMS), dictates the collection of sizable data which relate to the condition of the various transportation infrastructures. This is imperative and necessary to be able to effectively carry out the TIS optimization process.

Furthermore, TIS performance models need to be constantly developed and maintained to assist with the development of priorities. In addition, in determining TIS performance enhancement, all of the TIMS and its subsystems need to be carefully analyzed and integrated. Also, and as part of a systematic approach to optimizing TIS, this paper promotes using a condition enhancement approach, which in turn augments the overall performance of the transportation infrastructure.

Characteristics of the different prioritization methods can be found in Gharehbaghi (2014). The mathematical optimization for TIMS, on the other hand, falls into two categories: static and dynamic. The latter is made possible via Dynamic Programming [5] and [6]. The static and dynamic categories usually use the maintenance strategy together with budgetary constraints as the decision variables. While the *static* category incorporates certain constant variables, the *dynamic* category is the most important and includes maintenance activities.

4 Damage detection and monitoring techniques

As Gharehbaghi (2015) correctly argued, the traditional damage detection strategies consist of visual inspection and localized non-destructive evaluation such as radio X-ray, Radiographic, Eddy current and Ultrasonic techniques. Moreover, the damages are normally modeled by structural parameter reduction (ie Young's modulus) other than establishing the damage (crack) scheme. On the other hand the monitoring techniques aim is to provide a proper structural diagnosis of the 'state' of the constituent materials, of the different parts, and of the full assembled and functioning structures. These two steps are perhaps the most appropriate of ANN application.

5 The use of ANN in damage detection and monitoring techniques

Generally, several damage detection schemes utilize neural networks to detect, localize, and quantify damage in structures. Zapico et al. (2001) utilized number of neural networks for damage assessment, namely the multilayer perceptron (MLP) network with back propagation and the radial basis function (RBF) network. They concluded that the MLP network might be used in connection with vibration-based inspection whereas the RBF network completely failed. Although there are other authors who have instigated various ANN models, none were closely aligned with TIS parameters, which are proposed in this paper.

6 Artificial Neural Network (ANN) for transportation infrastructure systems

The ANNs have significant advantages in particular for data mining prediction and Classification [7]. In addition, ANNs provides specific System tools such as Control devices which once utilized effectively, would be a significant benefits for specific domains such as Transportation Infrastructure. The proposed model (Fig. 3) is a robust prototype and it has additional benefits especially for the Process Control and System Performance predictions. These two (process control and system performance) are fundamental part of Transportation Infrastructure systems, thus any artificial Neural Network collaborations need to incorporates both TIS and TIMS.

ANNs for Transportation Infrastructure System must incorporate engineering techniques that will be sustainable for future years and be maintained at acceptable levels. As it can be noticed, the ANN for TIS is a *multi-layer* and involves traditional Inputs, Processing and Outputs stages. This procedure involves a general system engineering approach:-

$$\therefore\ IX_n \rightarrow \Sigma wq_n \rightarrow Ox_{nq}$$

where:

- IX_1 ...... IX_n = input: Transportation System Parameters = TIS
- W = Weights (TIMS) } • ***Process Control***
- q = Function
- OX_{1q}...... OX_{nq} = output: solution based on each of TIMS functions. ***System Performance***

Of these three system engineering stages, the most involving is the actual process control stage where it incorporates both weights (TIMS) together with the function segmentation. This process stage is a multi-layer segmentation where the system core exists.

The overall performance of this model needs to be constantly developed and maintained to assist with the total development of priorities. In addition, in determining system performance enhancement, all the three main system stages need to be carefully analyzed and integrated holistically.

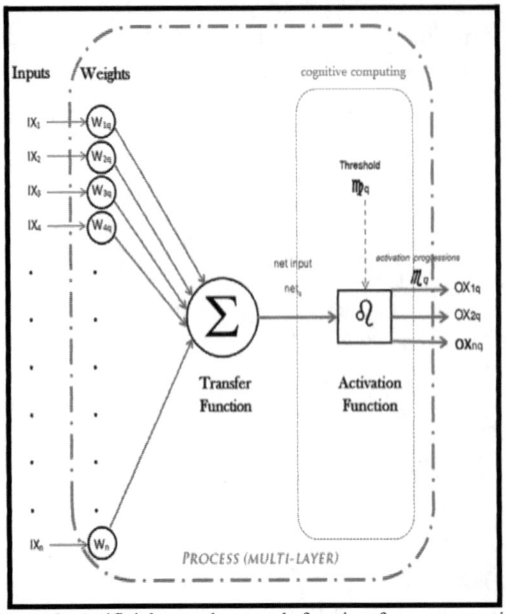

Figure 3. Artificial neural network function for transportation infrastructure systems.

7 Application of ANN for improved maintenance of transportation infrastructure

The inclusive application of ANN and improved maintenance of Transportation Infrastructures can be maximized in particular for the condition assessment of Transportation Infrastructures via the process control and system performance. This condition assessment should primary focus on the evaluation of the state of health, and then on concentrate on the regional or local damage assessment.

The utilization of ANN (multi-layer), involves: setting up the system parameters limits which categorizes the condition status; diagnosing the existing damage including the severity level; and finally measuring the consequence of damage on structural reliability scale. Conclusively, these steps consist with the mentioned damage detection and monitoring techniques.

Presently this model is in its development stage and hence need to combine working examples to further improve the ANN. These working case studies could consist of three main classifications of Transportation Infrastructures: Sea, Road, and Air Transportations Infrastructures. Furthermore, these case studies will be further refined into additional sub-classifications depending on their size, complexity and uniqueness. With the above processes, the proposed model could be refined specifically for an explicit group of Transportation Infrastructure assets.

8 Example of ANN for transportation infrastructure systems

ANN is ideal as a part of the Transportation Maintenance regime. Over time, traffic and environmental effects will damage traditional road surfaces, which require rehabilitation.

Although life expectation of Transportation maintenance projects varies from country to country, a typical expectation of several decades of service can be expected with major rehabilitation efforts performed depending on the planned schedules. Table 1, provides examples of ANN.

Table 1. ANN examples for periodic maintenance

Input (IX_n)	Process (Σwq_n)	Output (Ox_{nq})
Grading (IX_1)	Site preparation (Σwq_1)	Light and Heavy Grading (Ox_{1q})
Gravel Resurfacing (IX_2)	Site set up (Σwq_2)	Re-gravelling (Ox_{2q})
Bituminous Pavement (IX_3)	Repair of the pavement edges (Σwq_3)	Fog Seal (Ox_{3q})
Surface Treatment Resurfacing (IX_4)	Single surface dressing Resurfacing with 30mm asphalt concrete overlay (Σwq_4)	Slurry or Cape seal (Ox_{4q})

As it can be noticed, the examples of ANN are robust and specific especially for the Process Control and System Performance predictions. Accordingly, this table can be utilized (via various ANN software) to further develop and improve specific maintenance regimes, as shown in Fig. 4.

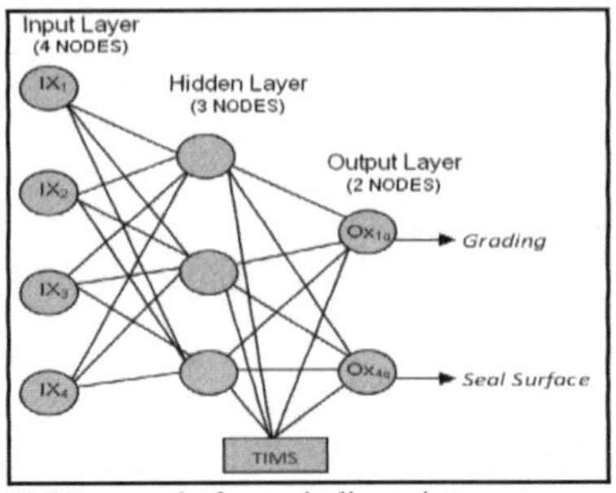

Figure 4. ANN example for periodic maintenance.

As it can be noticed the Fig. 4 consists of 9 interlinks. Although one issue with ANN is its ability to only converge to the local optimization, with the use of TIMS as a subsequent association this can be surpassed. In addition, the output layers can also be altered to represent the actual maintenance procedure more intimately.

9 Conclusion

Artificial Neural Networks (ANNs) represents the overall interconnection of the systems together with numeric weights that can be tuned based on experience, system Inputs, Processing and Outputs. This paper discussed the concept of a robust and sequential Artificial Neural Network (ANN) specifically for the Transportation Infrastructure Systems (TIS). In doing so this paper also argued the importance of ANN specifically as a part of damage detection and monitoring techniques régime.

As this paper demonstrated that ANN has many advantages if the problems cannot be solved by clear algorithm. In addition, ANN has the ability to be instructed to handle large dataset. There are various intelligent algorithms available and accordingly ANN is not a new concept. However the ANN overall ability to solve complex and interchangeable system problems (such as one which are found within the Transportation Infrastructure Systems) is its core advantage. Accordingly, this paper investigated and discussed the ANNs and its core functions for optimization of Transportation Infrastructure Systems.

Finally this paper presented an example of ANN for the Periodic Maintenance of Transportation Infrastructure. This example not only demonstrates the overall benefit of ANN but furthermore validates its process as the basis of the optimized Transportation Infrastructure Systems. Upon successful amalgamation, the proposed model could be specifically refined for definite group of Transportation Infrastructure assets. Lastly, the Fuzzy theories could also be utilized to further enhance the proposed model performance, thus further optimizing the results.

References

1. Anastassiou, George, "Intelligent Systems II: Complete Approximation by Neural Network Operators", Springer, (2016).
2. Biju, Issac, and Nauman, Israr, "Case Studies in Intelligent Computing", CRC, (2014).
3. Gharehbaghi, K and Georgy, M, "Utilization of Infrastructure Gateway System (IGS) as a Transportation Infrastructure Optimization Tool", International Journal of Traffic and Transportation Engineering, SAP, USA, **4**, No.1, (2015).
4. Gharehbaghi, K, "Infrastructure Asset Optimisation in Local Governments: Australia Study", International Journal of Civil, Structural, Environmental and Infrastructure Engineering Research Development, **4**, (6), 33 – 42, Dec. (2014).
5. Gharehbaghi, K, "Transport Infrastructure Development Modelling for Dispersed cities", proceedings of 5th International conference on Civil, Environmental and Medical Engineering, Melbourne, 8 - 10 April, (2015).
6. Gharehbaghi, K. and Raso, V., "Optimisation of Infrastructure Systems for Melbourne", Proceedings of the 2nd Annual International Construction Conference, Athens, Greece, June 18 – 21, (2012).
7. Longbing, Cao, "Metasynthetic Computing and engineering of Complex Systems", Springer, (2015).
8. Mitsuo. Gen; Kuinam J Kim; Xiaoxia Huang; Yabe Hiroshi, "Industrial engineering, Management Science and Applications", Springer, (2015).
9. Powel, Bruce, "Agile systems engineering", Morgan Kaufmann, pp. 23 - 57, (2016).
10. Rainey, Larry, and Tolk, Andreas, "Modeling and simulation support for system of systems engineering applications", John Wiley and Sons, (2015).
11. Sladkowski, Aleksander, and Pamula, Wieslaw, "Intelligent Transportation Systems - Problems and Perspectives", Springer (2015).
12. Zapico, L; Worden, K; and Molina, F, "Vibration-based damage assessment in steel frames using neural networks", smart materials and structures, **10**, Number 3, (2001).

A Comparative Study of Urban Road Traffic Simulators

Mustapha Saidallah, Abdeslam El Fergougui and Abdelbaki Elbelrhiti Elalaoui

Computer Science Department, ReSI team, Moulay Ismail University, Faculty of Science, 50000 Meknes, Morocco

Abstract. Road traffic management has become a worldwide concern. Several traffic simulators have been developed in order to contribute to solving traffic congestion problems. Comparative studies of simulators in this field of activity are concerned with the comparison of the simulation results with the results of the real situation; others are interested in the ability of certain platforms to simulate public transport systems. Our study aims purports to cover carryout of existing simulators in the sense that, on the one hand, poised in eleven major reviews of simulation platforms (commercial and open-source) the most used, given that the existing comparative studies do not cover all simulators we compared. On the other hand, our comparative study takes into consideration new criteria such as use wireless sensors and the ability of simulators to support GIS (Geographic Information System).

1 Introduction

Road traffic becomes increasingly a priority concern, especially in urban areas. We need a simulation of this traffic to better understand the problems of congestion in urban areas.

In general, the simulation is defined as a dynamic representation of some part of the real world through time. It is a tool that is widely used to test or evaluate a plan of action before its implementation.

Several studies have brought about the development of traffic simulation software, and also the comparison between these simulators is the subject of several articles.

The eleven selected simulators in our study are the most used and most mature, and we have compared them according to new criteria such as the use of wireless sensors which provides new opportunities for ITS (Intelligent Transport Systems), as well as the integration of Geographic Information Systems (GIS) which are effective tools for decision support in the field of road traffic, providing accurate data on the real world, and facilitating the establishment of the road network infrastructure.

The paper is organized as follows: Section 2 presents the different simulation software. Section 3 presents the related work on the comparison of urban road traffic simulators. The proposed criteria for the comparative study, a comparative table and a discussion of our comparison are listed in Section 4. In Section 5, a conclusion of our work will be provided.

2 Simulation software

Currently, there are several traffic simulation software, such as SUMO, MATSim, MITSIMlab, AIMSUN, CORSIM, Paramics, SimTraffic, VISSIM, TRANSIMS, etc. Some programs focus on the behavior of the vehicle in detail, the others are not interested in because they insist much on the simulation of a wide area. In simulation area, several tools can simulate a large city like MATSim simulated traffic in Berlin and Zurich.

2.1 AIMSUN (Advanced Interactive Microscopic Simulator for Urban and Non-Urban Networks)

AIMSUN [1], available from TSS-Transport Simulation Systems (Spain), is able to reproduce the real conditions of traffic of any network transport. It is, among others, used to develop and test the traffic control systems, traffic management rules, access controls, the location of tolls, public transport networks, lanes, and can work together with the vehicle guidance systems, and other real-time applications. It has many advantages over other software on the market, it allows notably to model the different network models in the same simulation. AIMSUN is embedded in GETRAM, a simulation environment composed of a traffic network editor (TEDI), a network database, a module for performing simulation and an Application Programming Interface which aims at interfacing to other simulation or assignment models. Through an additional library of DLL functions, the GETRAM extensions, the model is capable of communicating with external user-defined applications such as real-time control logic.

2.2 ARCHISIM

ARCHISIM [2], developed by the team Modeling and Simulation of the National Institute for Research on Transport and Safety (INRETS), is a behavioral simulation model, its implementation follows the multi-agent concepts. Drivers of vehicles are simulated agents. Their functions are based on three main processes: perception, decision and action. In ARCHISIM, there are two types of simulated components: agents and objects. The agents will gather information, strategize and then send the information. The main simulation agents are the drivers and pedestrians. The objects are road signals for example (horizontal or vertical). Each simulated driver is an autonomous software agent which evolves in a virtual environment and interacts with the other agents of the simulation performing its goals according to its skills and the current situation.

2.3 CORSIM (CORridor SIMulation)

CORSIM [3], sponsored and developed by The Federal Highway Administration (FHWA), is a traffic simulation software for signal systems, road networks, and highway systems. It consists of an integrated set of two models that represent the environment of the entire traffic. NETSIM represents traffic on city roads. FRESIM represents traffic on roads and highways. CORSIM provides its own interface and driver software, separate from TShell. In addition to the user interface, the CORSIM driver provides access to a new output data processor. The output processor enables the user to accumulate user-selected statistics and summary data during multiple runs of CORSIM. It writes the collected data to an Excel workbook, a comma-separated file, and/or a tab separated file.

2.4 MATSim (Multi-Agent Transport Simulation)

MATSim [4] developed by the Polytechnic of Zurich, provides a set of tools to implement a very large simulation based agents. It is used to simulate traffic in Zurich (Switzerland), Berlin (Germany), Padang (Indonesia) and Toronto (Canada). It can simulate the traffic of a vast region throughout the day. But it is not interested in retail vehicle behavior. MATSim pursues an activity-based approach to demand generation. Unlike other transportation simulation packages MATSim is throughout agent-based and generates individual activity plans as input to the network loading rather than (time-dependent) origin-destination matrices as typically used in dynamic traffic assignment.

2.5 MITSIMLab (MIcroscopic Traffic SIMulation Laboratory)

MITSIMLab [5], developed at MIT (Massachusetts Institute of Technology), is a traffic simulator that assesses the impacts of alternative designs traffic management systems, information systems for travelers, public transport operations, and various ITS (Intelligent Transport Systems) strategies at the operational level and contributes to their further refinement. It can evaluate systems such as advanced systems for traffic management and road guidance systems. The role of MITSIMLab is to represent "the world". Traffic and network elements are represented in detail in order to capture the sensitivity of traffic flows to the control and routing strategies. MITSIMLab is an open-source application where its core models have been written in C++ and are fully available. It has been successfully applied in several traffic and research studies in the USA, the UK, Sweden, Italy, Switzerland, Japan, Korea, Malaysia and Portugal.

2.6 Paramics (Parallel Microscopic Simulation)

Paramics [6], marketed by Quadstone Paramics (UK), is fully scalable and designed to handle such diverse scenarios such one intersection with a highway congested or modeling of an entire city traffic system. It helps give a realistic representation of the landscape, with 2D and 3D visualization. It takes into account the buses, trams and also simulates pedestrians. It is used in different traffic patterns (urban, inter-urban, tunnel, port, and parking) and in many countries. A Paramics model is represented by a combination of "nodes, links and other associated objects" to replicate real life geometry constraints. Upon release from an "origin zone", each vehicle attempts to complete its journey towards a "destination zone" whilst being bounded by physical and dynamic vehicle parameters. Through the use of micro simulation, Paramics allows users to simulate individual vehicle movements to predict future travel pattern behavior as a result from a change in traffic volume or geometric road layout.

2.7 SimTraffic

SimTraffic [7], marketed by Trafficware (United States), is easy to use and its graphic interface provides the user with an encoding road network time significantly shorter than other models. Even inexperienced users can get a single operational network installation in a very short time. SimTraffic is a link-node model that uses driver behavior and vehicle performance algorithms to simulate individual vehicle movements through a network. The capabilities of SimTraffic were expanded in subsequent versions to model additional features such as freeways, ramps, and roundabouts. A significant disadvantage of SimTraffic is the lack of API functions or supporting detailed output of vehicle-state variable information and automated statistical analysis capabilities of other codes.

2.8 SUMO (Simulation of Urban Mobility)

SUMO [8] was developed at the German Aerospace Centre. In this simulator vehicles can move freely, collision between vehicles and accidents are simulated. Each vehicle has its own road and vehicle routing is dynamic. The vehicle behavior is taken into consideration such as changing lanes. Roads in SUMO are shown as a plurality of lanes. And the width of each lane is fixed.

Also, the vehicle width is fixed. And it does not take into account the different types of vehicles. SUMO allows modelling of intermodal traffic systems including road vehicles, public transport and pedestrians. SUMO can be enhanced with custom models and provides various APIs to remotely control the simulation.

2.9 TRANSIMS (TRansportation ANalysis and SIMulation System)

TRANSIMS [9], developed at Los Alamos National Laboratory (USA), is an integrated set of tools to conduct analysis of a regional transportation system based on a cellular automaton. It uses a new modeling paradigm of individual travelers and their multi-modal transport (This is the carriage of goods by at least two successive modes of transport. Part of the journey can thus be carried by road, the other by sea, river or rail). TRANSIMS differs from previous travel demand forecasting methods in its underlying concepts and structure. These differences include a consistent and continuous representation of time; a detailed representation of persons and households; time-dependent routing; and a person-based micro simulator. These advances are producing significant changes in the travel forecasting process. To date the TRANSIMS models have been tested with data from Dallas, Texas, and Portland (Oregon).

2.10 TransModeler

TransModeler [10], marketed by Caliper Corporation (USA). It can simulate all types of road networks, highways and areas of the city center, and can analyze multimodal networks. We can model and visualize the behavior of complex systems of circulation in an environment of two-dimensional or three-dimensional to illustrate and assess the dynamics of traffic flows, traffic lights, and overall network performance. TransModeler model freeway and urban networks in the same network with driver behavior models that are sensitive to complex inter-vehicle interactions in merging areas and intersections, model high occupancy vehicle (HOV) lanes, bus lanes and toll facilities to better understand their effects on traffic system dynamics and model evacuation plans and scenarios for response to natural disasters, hazardous spills, and other emergencies.

2.11 VISSIM

PTV VISSIM [11] developed by PTV (Planung Transport Verkehr AG) in Karlsruhe, Germany. It is part of the Vision Traffic Suite which also includes PTV Visum (analysis and forecast traffic) and PTV Vistro (optimization of the signal and the impact on traffic). VISSIM is one of the most used simulation software to simulate, evaluate and validate new transport policies and control systems. VISSIM is a multimodal simulator that allows users to define a full range of vehicle types including passenger cars, buses, trucks, and heavy and light rail vehicles as well as pedestrians and cyclists. One very important feature of VISSIM that sets it apart from others simulation models is its component object model (COM) programming interface. The COM interface allows users to develop and implement their own applications on the VISSIM network using computer programming language such as C++, Visual Basic, or Python. The COM interface provides user-developed applications with an access to the network topology, signal control, path flows, and vehicle behavior enabling VISSIM to model complex control logic and sophistical transportation systems and components [12].

3 Related work

Several articles have focused on the comparison of urban road traffic simulators.

In [13] the authors have made a comparison of seven simulators according to their ability to simulate public transport according to certain criteria such as visualization, software category (commercial or open-source), the public transport infrastructure, the characteristics of public transport vehicles as well as passengers and pedestrians. The authors concluded that the simulators, AIMSUN, VISSIM and Paramics, offer the ability to import aerial images that can be used as background and as a guide in the development of road networks. 3D views offered in VISSIM and AIMSUN are better than Paramics. In Paramics, vehicles appear as 3D blocks without distinct characteristics. However, in VISSIM and AIMSUN, vehicles appear with distinct characteristics. In some simulation software, special attention was paid to the driver settings. For example, VISSIM offers different car-following or lane-changing models. Although VISSIM, AIMSUN and Paramics allow the simulation of bus circuit, only VISSIM simulates disembarkation and embarkation of passengers from the left side of roads. Therefore, only VISSIM allows bus stops on the left side or the right side of the road. The occupancy rate of the bus and the vehicle's capacity can also be defined in VISSIM and Paramics, while the number of doors in buses is only recognized in VISSIM.

In [14] the authors studied the two simulators VISSIM and CORSIM based on an urban network in the North Bund area, Hongkou District, Shanghai, China. They concluded that the use of the software, publishing and network configuration in CORSIM are easier. Due to the existence of different simulation mechanisms, CORISM simulation results can be transmitted quickly, while VISSIM configuration provides an easy interface with a separate output file to the simulation results. While for the calibration effort, the software provides multiple calibration parameters to allow the simulated network to reproduce the real situation. VISSIM is more appropriate for large intersections with broadband traffic, while CORSIM is good for modeling unsaturated intersections. VISSIM's simulation results are closer to the real situation.

The article [15] presents a comparison of seven simulators based on criteria such as visualization, expansion of the simulator and it is oriented agent. The classification of simulators based on their ability to simulate cars, buses, trucks, trains, bicycles and pedestrians, as the management of the sensors (such as traffic lights, electromagnetic loops, cameras, etc). The article found that MITSIM and SUMO do not have 3D visualization. VISSIM, Paramics and AIMSUN have a realistic 3D visualization which is very useful for presenting real-world scenarios. MITSIM does not provide expansion capabilities, but since the simulator is open-source researchers can modify and extend its core. Each simulator has advanced features, for example, VISSIM has parameters allowing the flexibility of its functions, Paramics is suitable for use of resources on distributed machines, AIMSUN provides various options for its extension, SUMO has a flexible architecture, and MAS -T2er Lab. and ITSUMO are oriented agent.

The aim of the research conducted in the article [16] was to provide a comparison of three selected platforms (SUMO, VISSIM and TRANSIMS) on a fragment of a real urban network (in Poland). In general, it seems that the model SUMO had too low capacity compared to the capacity of the actual network, but on the other hand, it is difficult to assess the capabilities of VISSIM and TRANSIMS were consistent with, or better than the actual network. However, despite some differences in quantitative measurements, effects similar to the spread of the traffic flow (appearance of network bottlenecks and traffic jam effects) were observed in all three models.

The article [17] offers a comparative analysis of fourteen simulators focusing on the simulation of traffic at highways, congested urban networks and adaptation of simulators to ITS (Intelligent Transport Systems) applications. This article revealed that AIMSUN, CORSIM and VISSIM are suitable for clogged arteries and highways, but AIMSUN is less suited compared to others. The AIMSUN characteristics are favorable to the creation of large urban and regional networks, but it requires difficult coding. The models AIMSUN, Paramics, INTEGRATION and CORSIM seem most suitable to ITS applications.

4 Comparison of simulators and discussion

Our comparative study is based on a set of criteria chosen so as to draw conclusions about some features of the compared simulators, as well as understanding of their functioning.

4.1 Comparison criteria

4.1.1 Simulation Model

In the field of simulation of road traffic, there are three different models depending on the detail of the simulation.

Microscopic models: These models simulate the characteristics and interactions between individual vehicles. They produce essentially the trajectories of vehicles moving across the network. The processing logic includes algorithms and rules that describe how vehicles move and interact. It also includes acceleration, deceleration, lane changes and overtaking maneuvers.

Mesoscopic models: These models simulate individual vehicles, to this end, traffic is represented by small groups of traffic entities, whose interactions are described in a medium level of detail.

Macroscopic models: These models simulate the flow of traffic. They consider traffic characteristics (speed, flow and density) and their relationships. These models are making on the conservation equations of flows and traffic disturbances that spread in the traffic system. Consequently, they can be used to predict the spatial and temporal congestion that is caused by the traffic demand or incidents in a road network.

4.1.2 Software Category

We can classify traffic simulation software into two categories: open-source or commercial. The open-source designation or "open source code" applies to software whose license specifically meets the criteria established by the Open Source Initiative; that is to say, the possibilities of free redistribution, access to source code and create derivative works. While a commercial software or proprietary means software that does not allow legally or technically, or by any other means whatsoever, to simultaneously perform four software freedoms which are running the software for any type of use, to study its source code (and therefore access to the source code), the distribution of copies, as well as modification and thus the source code improvement.

4.1.3 System

There are two types of systems: discrete (variable changing at regular intervals of time) and continuous (variables change continuously).

4.1.4 Visualization

Visualization can be two-dimensional (2D), three-dimensional (3D) or both. 3D visualization allows to be closer to the real world. Also it gives details of the simulation (eg the vehicle height, better visibility of traffic lights).

4.1.5 Infrastructure

The road network is the basis of any simulation. It includes roads, intersections (intersection, roundabout, etc). In this criterion we will evaluate the difficulty or ease of coding of the road network to be studied (complex junctions, etc) according to three criteria: easy coding, coding with an average difficulty and difficult or hard coding. As well as the coding flexibility of various elements of the infrastructure that may encode (expressways, multivalued panels, etc) according to three criteria: flexible, limited or very limited.

4.1.6 Vehicles and pedestrians

The evaluation of the vehicles will be done on several levels, firstly the type (car, truck, motorcycle, etc), secondly the dimensions (length, width, height), thirdly the priority (priority cars: police car, firefighters, ambulances, etc). As for pedestrians, we will assess whether the simulator supports pedestrian or not. Each simulator will also be evaluated on the ability to simulate the public transport vehicles such as buses and trams.

4.1.7 Scope area

The scope area is the maximum area that the simulator can simulate. In this criterion we will compare simulators for their ability to simulate traffic in a city, a region or an entire country.

4.1.8 Detectors

The simulators use sensors such as electromagnetic loops and cameras, for example to reckon the number of vehicles in the queue or manage traffic speeds. In this criterion we will classify the simulators supported by the type of sensor, wired such as the electromagnetic loop formed of 3 or 4 turns of copper wire embedded in the floor, or wireless sensor such as MICA2 [18], using waves to communicate. Wireless sensors have several advantages such as simplicity of implementation, low cost of installation and maintaining.

4.1.9 GIS (Geographic Information System)

In this criterion we will assess whether the simulator offers the possibility to import maps from geographic information systems to encode road network, using the following three options: yes, partially or not.

4.2 Comparison and discussion

Based on the criteria mentioned above, we have classified the traffic simulator as shown in Table 1 below.

The comparative table shows that only AIMSUN and TransModeler simulate three models at the same time, while others are only microscopic simulators.

VISSIM, SUMO, MATSim and AIMSUN simulate traffic continuously, by opposition to ARCHISIM, CORSIM, Paramics and TRANSIMS which use a discrete system. For MITSIMLab, SimTraffic and TransModeler we could not specify the system used owing to the lack of this information in the literature and in the user guides.

VISSIM and SimTraffic offer an easy coding of the road network, unlike AIMSUN, ARCHISIM and SUMO require difficult or heavy coding. But concerning the coding flexibility in the various infrastructure elements, AIMSUN, Paramics and VISSIM are more flexible than the other simulators.

The most commercial simulators support the type and size of the vehicle, as well as taking into considerations the pedestrians and emergency vehicles such as ambulances and police cars. They have the opportunity to simulate the public transport vehicles such as buses and trams, in contrast to open-source simulators.

All simulators use wired sensors. Additionally AIMSUN, Paramics and VISSIM use wireless sensors which are more efficient and cheaper.

AIMSUN, MATSim, TransModeler and VISSIM support GIS, while other simulators support them partially or not at all. MATSim is the only open-source simulator that uses GIS.

Table 1. Comparative table of traffic simulators.

Simulators	Model			Category		System		Visualization		Infrastructure						Vehicles and pedestrians					Scope Area			Detectors		GIS		
										Difficulty			*Flexibility*															
	I	*E*	*A*	*O*	*C*	*D*	*C*	*2D*	*3D*	*E*	*M*	*D*	*F*	*L*	*VL*	*T*	*D*	*R*	*E*	*O*	*I*	*R*	*O*	*WD*	*WL*	*Y*	*P*	*N*
AIMSUN	√	√	√		√		√	√	√			√	√					√	√	√		√	√	√	√	√		
ARCHISIM	√				√	√		√				√			√						√			√				√
CORSIM	√				√	√		√	√						√			√	√		√	√		√				√
MATSim	√			√			√	√						√							√	√		√		√		
MITSIMLab	√			√				√						√							√			√				√
Paramics	√				√	√		√	√				√					√	√	√	√	√		√	√			
SimTraffic	√				√			√	√	√				√				√	√		√	√		√				√
SUMO	√			√			√	√				√			√						√	√		√				
TRANSIMS	√	√		√		√		√							√							√	√	√				
TransModeler	√	√	√		√			√	√					√					√			√	√	√		√		
VISSIM	√				√		√	√	√	√			√					√	√	√	√	√		√	√	√		

Table 2. Legend.

Criteria		Abbreviation	Signification
Model		I	Microscopic
		E	Mesoscopic
		A	Macroscopic
Category		O	Open-source
		C	Commercial
System		D	Discrete
		C	Continuous
Visualization		2D	Two-dimensional
		3D	Three-dimensional
Infra-structure	*Difficulty*	E	Easy
		M	Medium
		D	Difficult
	Flexibility	F	Flexible
		L	Limited
		VL	Very Limited
Vehicles and pedestrians		T	Type
		D	Dimension
		R	Priority
		E	Pedestrian
		O	Other vehicles (Bus and Tram)
Scope Area		I	City
		R	Region
		O	Country
Detectors		WD	Wired sensor
		WL	Wireless sensor
GIS		Y	Yes
		P	Partially
		N	No

5 Conclusion

The urban road traffic simulators are an vital tool to simulate and evaluate any planned change road network, in order to improve its functioning.

This paper provides an overview of eleven different traffic simulation platforms most used, and a comparison based on their different characteristics to provide a map that can be used as a decision support tool to select the simulator more suited to a need or as a reference for the study and development of new traffic simulators.

We can conclude that the open-source simulators do not simulate wireless sensors. It will be interesting to develop an open-source simulator integrating wireless sensors or integrate this feature in one of the existing open-source simulators.

The same thing for supporting GIS, type of vehicles and pedestrians, we can integrate these features to open-source simulators, or develop a new simulator integrating these features.

References

1. Casas, J., Ferrer, J. L., Garcia, D., Perarnau, J., & Torday, A. (2010). "Traffic simulation with Aimsun". In Fundamentals of traffic simulation (pp. 173-232). Springer New York.
2. Bonte, L., Espié, S., & Mathieu, P. (2006). "Modélisation et simulation des usagers deux-roues motorisés dans ARCHISIM." *JFSMA*, **6**, 17.
3. Halati, A., Lieu, H., & Walker, S. (1997). "CORSIM-corridor traffic simulation model." *In Traffic congestion and traffic safety in the 21st century: Challenges, innovations, and opportunities.*
4. "MATSim | Multi-Agent Transport Simulation", Matsim.org, 2016. [Online]. Available: http://www.matsim.org/. [Accessed: 29- May- 2016].
5. Yang, Q., & Koutsopoulos, H. N. (1996). "A microscopic traffic simulator for evaluation of dynamic traffic management systems." Transportation Research Part C: Emerging Technologies, **4**(3), (pp. 113-129).
6. Cameron, G. D., & Duncan, G. I. (1996). "PARAMICS—Parallel microscopic simulation of road traffic." The Journal of Supercomputing, **10**(1), (pp. 25-53).
7. Husch, D., & Albeck, J. (2000). *SIMTRAFFIC 5.0 User Guide for Windows.*
8. Behrisch, M., Bieker, L., Erdmann, J., & Krajzewicz, D. (2011, October). "SUMO–Simulation of Urban MObility." *In The Third International Conference on Advances in System Simulation (SIMUL 2011), Barcelona, Spain.*
9. Smith, L., Beckman, R., & Baggerly, K. (1995). *TRANSIMS: Transportation analysis and simulation system (No. LA-UR--95-1641). Los Alamos National Lab., NM (United States).*
10. "TransModeler Traffic Simulation Software", Caliper.com, 2016. [Online]. Available: http://www.caliper.com/transmodeler/. [Accessed: 29- May- 2016].
11. Fellendorf, M., & Vortisch, P. (2010). "Microscopic traffic flow simulator VISSIM." *In Fundamentals of Traffic Simulation* (pp. 63-93). Springer New York.
12. *VISSIM 5.10 COM Interface Manual*, PTV Germany, (2009).
13. Ghariani, N., Elkosantini, S., Darmoul, S., & Ben Said, L. (2014, May). "A survey of simulation platforms for the assessment of public transport control systems." *In International Conference on Advanced Logistics and Transport (ICALT)*, (pp. 85-90). IEEE.
14. Sun, D. J., Zhang, L., & Chen, F. (2013). "Comparative study on simulation performances of CORSIM and VISSIM for urban street network." Simulation Modelling Practice and Theory, **37**, (pp. 18-29).
15. Kokkinogenis, Z., Passos, L. S., Rossetti, R., & Gabriel, J. (2011). "Towards the next-generation traffic simulation tools: a first evaluation." *In 6th Iberian Conference on Information Systems and Technologies* (pp. 15-18).
16. Maciejewski, M. (2010). "A comparison of microscopic traffic flow simulation systems for an urban area." Transport Problems, **5**(4), (pp. 27-38).
17. Ratrout, N. T., & Rahman, S. M. (2009). "A comparative analysis of currently used microscopic and macroscopic traffic simulation software." The Arabian Journal for Science and Engineering, **34**(1B), (pp. 121-133).
18. "Wireless Measurement System, MICA2", Crossbow Technology, Inc.41 Daggett Dr.San Jose, CA 95134.

1-D Modelling Comparative Study to Evaluate Performance and Emissions of a Spark Ignition Engine Fuelled with Gasoline and LNG

Ahmet Alper Yontar and Yahya Doğu

Kırıkkale University, Faculty of Engineering, Department of Mechanical Engineering, 71450 Kırıkkale, Turkey

Abstract. In this study, a spark-ignition engine fuelled with gasoline and LNG was modelled in 1-D at wide open throttle by using Ricardo-Wave software. Different engine speeds ranging from 1500rpm to 4500rpm with an increment of 500rpm were studied to evaluate the effects of gasoline and LNG on engine performance and exhaust emissions. It is determined that LNG decreases engine performance and emissions as well, at especially high speeds.

1 Introduction

With developments in engine modelling software, it became possible to make predictions for many crucial engine characteristics such as usage of alternative fuels, combustion visualization in cylinder, examination of combustion chamber geometry, effects of all related geometric and operating conditions in terms of engine performance and emissions.

There has been an ever increasing demand on usage of alternative fuels in internal combustion engines. The most widely available and used alternative fuels are LNG (Liquefied Natural Gas), CNG (Compressed Natural Gas), and LPG (Liquefied Petroleum Gas). LNG and CNG fuels are the most popular fuel varieties and approximately 23 million natural gas vehicles are available by 2015 in the world, led by China (4.44 million), Iran (4.00 million), Pakistan (3.70 million), Argentina (2.48 million), India (1.80 million), and Brazil (1.78 million) [1].

Use of alternative fuels in many countries, especially in CNG (Compressed Natural Gas) and LNG (Liquid Natural Gas) is increasing the use of fuel in internal combustion engines. LNG and CNG fuel are the most popular fuel varieties and approximately 23 million natural gas vehicles by 2015 in the world. The natural gas (NG) consists of methane (90%), ethane, propane, other heavier hydrocarbons, fewer nitrogen, oxygen, carbon dioxide, pollutants such as sulphur compounds and water. LNG is cleaner than NG since hazardous contaminants are removed during liquefaction. LNG has higher density than CNG. Volumetric energy density of LNG is 2.4 times greater than CNG, and 60% more than diesel fuel [2]. Designed usage of LNG engines with higher compression ratios and usage of LNG in engines provide improved performance, fuel economy and avoid engine knock due to high octane number rating (~130)

While it is very important to determine the effects of usage of these alternative fuels especially LNG on engine performance and exhaust emissions, there are limited number of published works in literature.

Ha, J. Y. et al. [3] used a direct-injection engine fuelled with CNG. They observed that combustion characteristics at wide open throttle greatly improved with early injection timing. On the contrary, the late injection timing gave better results at partial throttle openings.

Venugopal, T. and Ramesh A. [4] evaluated effects of throttle opening for butanol–gasoline mixture fuel in a spark ignition engine with dual injection system. They measured engine performance, engine detonation and emissions. Lower hydrocarbon emission was reported for increasing throttle opening. Lower NO_x emission was observed with increasing fuel ratio among 35%-100% throttle openings.

Aslam, M. U. et al. [5] used a retrofitted spark ignition engine with CNG fuel. Tests were carried out for different mixing ratios of gasoline and CNG. They determined optimum engine operating parameters according to fuel consumption, exhaust emissions, engine performance.

Verma, G. et al. [6] tested hydrogen enriched natural gas for various H/C ratios. They found that thermal efficiency increased with increasing H/C ratio. Exhaust gas temperature increased with decreasing H/C ratio. NO_x emission at 55Nm torque became highest for H/C ratio of 4.22, however, NO_x emission at full load had the highest for H/C ratio of 4.5. The mixed fuel with H/C ratio of 4.5 showed best overall performance.

Hooper, P. R. et al. [7] used Ricardo-Wave software to build a 1-D engine model for multi-fuel (indolene, kerosene JET A-1) operation. At 3000 rpm, they calculated minimum specific fuel consumption (SFC) of 273 g/kWh and 310 g/kWh for indolene and JET A-1, respectively.

Jahirul, M. I. et al. [8] tested a spark ignition engine with gasoline and CNG fuels. They measured engine performance parameters and exhaust emission. They investigated effects of throttle opening for both gasoline and CNG fuels. CNG was reported to have less engine performance and emissions for all throttle openings.

Usage of LNG in internal combustion engine provides more flexibility in the design of an internal combustion engine because LNG has an auto-ignition temperature of ~853 K whereas gasoline and diesel auto-ignite at approximately 523 K and 483 K, respectively.

In the current study, a complete 1-D engine model including intake and exhaust manifolds was built in Ricardo-Wave software. The engine is spark ignited and fuelled with gasoline and LNG. The engine performance and exhaust emissions were investigated for wide opening throttle at several engine speeds for both gasoline and LNG.

2 1-D Engine modelling

This work specifically focuses on numerical simulation of a commercially engine with Ricardo-Wave software. Technical specifications of the modelled spark ignition engine are given in Table 1.

The engine 1-D model is built including the entire engine from the beginning of the intake line to the end of the exhaust line as plotted in Figure 1.

At the beginning of the engine modelling phase, each engine components (pistons, cylinders, valves, ports, engine blocks, intake and exhaust manifolds, fuel line, exhaust line and throttle) are separately formed by defining their properties in Ricardo-Wave. Then, they are connected to each other by defining the relevant relations and boundary conditions.

Multi-Component Wiebe model is used as the combustion model. Woschni is used as the heat transfer model. In order to read engine data on the 1-D engine model, many sensors, actuators and signal processor are located at certain points on the model. For each engine speed, the analyses were run for 250 engine cycles in order to ensure fully developed steady conditions before reading the data.

Table 1. Technical specifications of test engine.

Number of cylinders	4
Displacement volume (lt)	1.6
Cylinder diameter (mm)	76.5
Stroke length (mm)	86.9
Compression ratio	11:1
Connecting rod length (mm)	141
Wrist pin offset (mm)	4.2
Number of valves	16

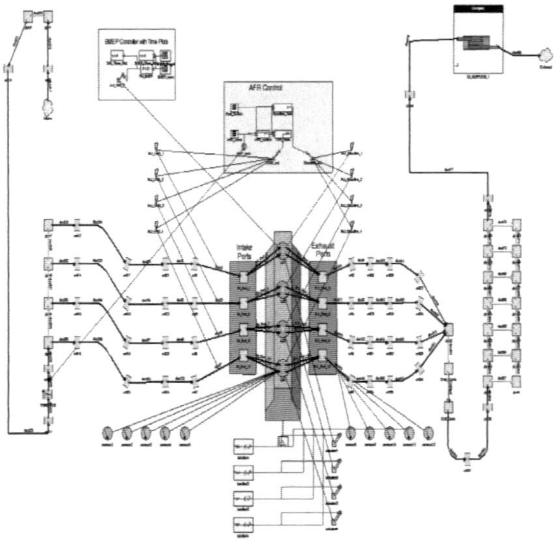

Figure 1. 1-D Engine model.

3 Results and discussion

The 1-D engine simulation model was carried out for seven different engine speeds from 1500 rpm to 4500 rpm at 500 rpm intervals. Parameters of engine performance and exhaust emissions were calculated for all seven engine speeds at wide open throttle.

The results with respect to speeds and different fuels (gasoline and LNG) are presented below in comparative manner by evaluating torque (Fig. 2), power (Fig. 2), fuel consumption (Fig. 3), specific fuel consumption (Fig. 3), indicated mean effective pressure (Fig. 4), volumetric efficiency (Fig. 4), effective efficiency (Fig. 4), NO_x (Fig. 5), exhaust temperature (Fig. 5), CO_2 (Fig. 6) and CO (Fig. 6).

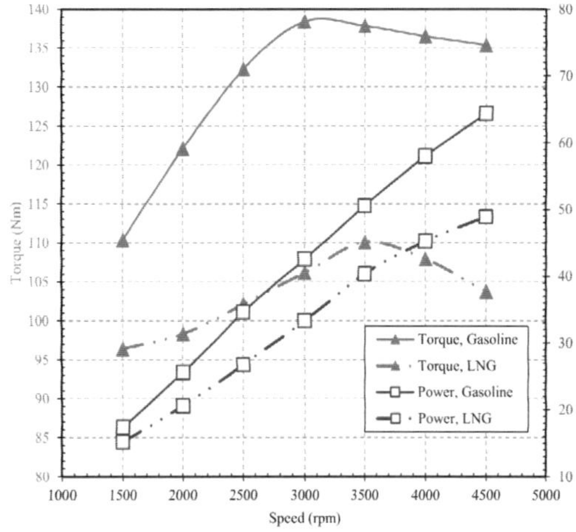

Figure 2. Torque / power – speed.

The torque curve has a nearly parabolic shape with respect to the engine speed for both fuels. These torque curves are consistent with the known engine behaviour which is the torque falls down from its peak due to all kinds of increased losses. The maximum torque values of 138 Nm and 110 Nm are obtained at 3000 rpm and 3500 rpm for gasoline and LNG, respectively. The torque for LNG is about 80% (that varies with speed) less than the gasoline since the lower heating value of LNG for unit fuel mass is 85% less. As a result of this, the indicated mean effective pressure (IMEP) is also low for LNG as plotted in Fig. 3.

As also plotted at the right vertical axis on Fig. 2, the power increases almost linearly due to more fuel intake into cylinder. As a result of less lower heating value of LNG, the power curve for LNG falls under the gasoline curve with increment rates of 1.2 and 1.6 kW/rpm for LNG and gasoline, respectively. The different power slopes are entirely due to the fuel's chemical content.

Figure 3 shows that the volumetric efficiency for LNG is following same trend but lower than gasoline. The calculated volumetric efficiencies are 75-79% and 78-94% for LNG and gasoline, respectively. These values are consistent with reported approximately volumetric efficiency of 75-90% [9]. As also plotted in Fig. 3, effective efficiencies are about thirty percent as expected for typical spark ignition engines.

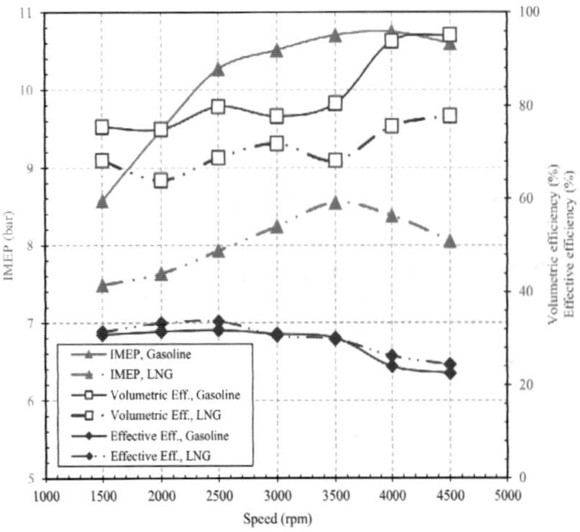

Figure 3. IMEP / volumetric efficiency / effective efficiency – speed.

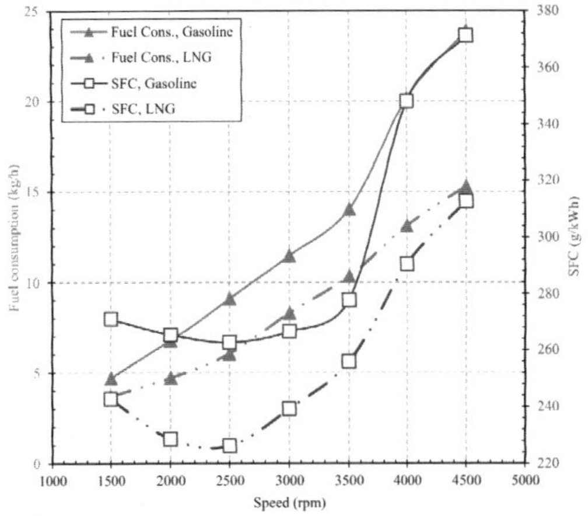

Figure 4. Fuel consumption / SFC - speed.

As seen in Figure 4, the fuel consumption with speed is linearly increasing in parallel to power increase. Gasoline fuel consumption is increasing more than LNG at high engine speeds. The specific fuel consumption (SFC) has roughly a parabolic shape making its minimum at about 2500 rpm (265 g/kWh) and 2300 rpm (225 g/kWh) for gasoline and LNG, respectively. LNG specific fuel consumption curve has very similar trend with gasoline specific fuel consumption curve. These behaviours are consistent with the torque graphs. It is known that the lowest specific fuel consumption is about 270 g/kWh for spark ignition engines [10].

Exhaust temperature and NO_x are plotted in Fig. 5. NO_x with speed increases by making some fluctuations. For the considered range of speed, the average NO_x level for gasoline is slightly less than LNG since the flame temperature of gasoline is lower than LNG. NO_x formation depends on the air-fuel ratio and in-cylinder temperature because nitrogen and oxygen enter the reaction at high temperatures and NO_x are formed. Exhaust temperatures support NO_x trends.

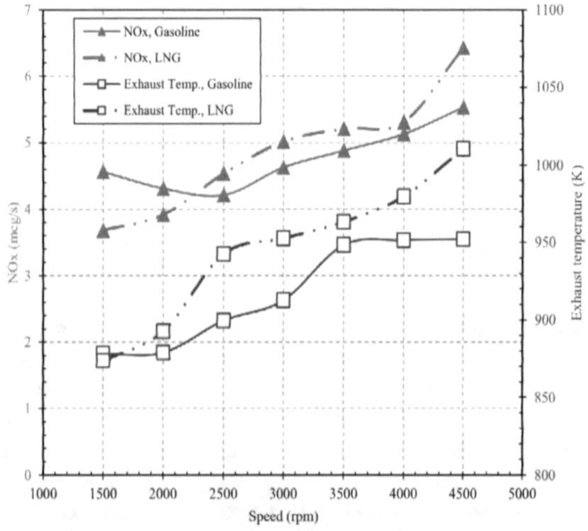

Figure 5. NO_x / exhaust temperature - speed.

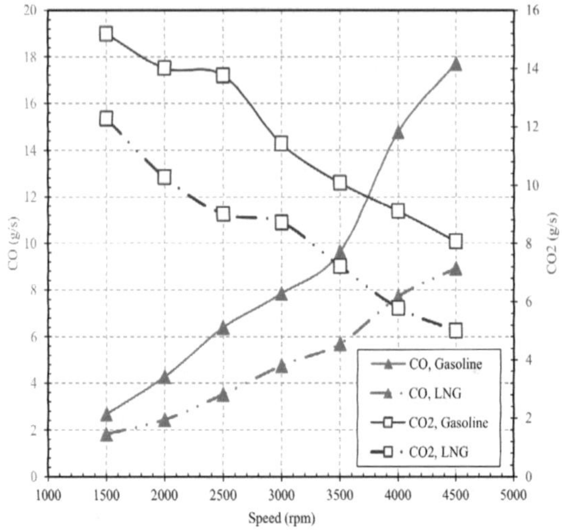

Figure 6. CO_2 / CO - speed.

As seen in Figure 6, CO emission with speed increases faster for gasoline. The emission of CO_2 decreases with increasing speed. LNG has the highest energy/carbon ratio of any fossil fuel and produces less CO_2 and CO per unit of energy. Combustion efficiency decreases at high speed and the partial products of incomplete combustion increases as a result of this CO formation increase. The emissions chart shows that LNG gives lower emissions.

4 Conclusions

In this study, a spark ignition engine performance and exhaust emissions were determined for gasoline and LNG fuels at wide open throttle by using a 1-D model. Findings are listed below.

- The torque and power for LNG is less than gasoline for all speeds since the lower heating value of LNG for unit fuel mass is 85% less.
- Similarly, the IMEP and volumetric efficiency is also lower for LNG.
- Gasoline fuel consumption is increasing more than LNG at high engine speeds. The specific fuel consumption (SFC) is calculated as 2500 rpm (265 g/kWh) and 2300 rpm (225 g/kWh) for gasoline and LNG, respectively.
- LNG has lower emissions for engine operating parameters.

References

1. International Association for Natural Gas Vehicles, *Natural Gas Vehicles Statistics;Summary Data* (2010)
2. James J. Eberhardt, *Fuels of the future for cars and trucks*, U.S. Department of Energy, DEER Workshop, (2002)
3. J.Y. Ha, J.S. Park, and J.H. Kang, *Effects of the throttle opening ratio and the injection timing of CNG on the combustion characteristics of a DI engine,* Int. J. of Auto. Tech., **11**(1): p. 11-17 (2010)
4. T. Venugopal and A. Ramesh, *Effective utilisation of butanol along with gasoline in a spark ignition engine through a dual injection system,* Appl. Therm. Eng., **59**(1): p. 550-558, (2013)
5. M.U. Aslam, H.H. Masjuki, M.A. Kalam, H. Abdesselam, T.M.I. Mahlia, M.A. Amalina, *An experimental investigation of CNG as an alternative fuel for a retrofitted gasoline vehicle,* Fuel, **85**(5): p. 717-724 (2006)
6. G. Verma, R.K. Prasad, R.A. Agarwal, S. Jain, A.K. Agarwal, *Experimental investigations of combustion, performance and emission characteristics of a hydrogen enriched natural gas fuelled prototype spark ignition engine*, Fuel, **178**: p. 209-217 (2016)
7. P.R. Hooper, T. Al-Shemmeri and M.J. Goodwin, *An experimental and analytical investigation of a multi-fuel stepped piston engine,* Appl. Therm. Eng., **48**: p. 32-40 (2012)
8. M.I. Jahirul, H.H. Masjuki, R. Saidur, M.A. Kalam, M.H. Jayed, M.A. Wazed, *Comparative engine performance and emission analysis of CNG and gasoline in a retrofitted car engine*, Appl. Therm. Eng., **30**(14): p. 2219-2226 (2010)
9. Willard W. Pulkrabek, *Engineering fundamentals of the internal combustion engine*, **478** (1997)
10. J.B. Heywood, *Internal combustion engine fundamentals*, **930** (1988)

MATEC Web of Conferences **81**, 05004 (2016) DOI: 10.1051/matecconf/20168105004
ICTTE 2016

The Analysis of Forming Forces in Single Point Incremental Forming

Kyung Hee Koh[1], Jae-Gwan Kang[1] and Jong-Yun Jung[2]

[1]*Division of Mechanical Engineering, Kyungnam University, Changwon, R. Korea*
[2]*School of Industrial Engineering and Naval Architecture, Changwon National University, Changwon, R. Korea*

Abstract. Incremental forming is a process to produce sheet metal parts in quick. Because there is no need for dedicated dies and molds, this process is less cost and time spent. The purpose of this study is to investigate forming forces in single point incremental forming. Producing a cone frustum of aluminum is tested for forming forces. A dynamometer is used to collect forming forces and analyze them. These forces are compared with cutting forces upon producing same geometrical shapes of experimental parts. The forming forces in Z direction are 40 times larger than the machining forces. A spindle and its axis of a forming machine should be designed enough to withstand the forming forces.

1 Introduction

Modern manufacturing requires higher product quality with shorter production time in order to get more profits in competitive markets. Metal forming produces parts in short time upon equipment setup. However, the process needs dies and moulds which are equipped with press machines. The preparation for the setup takes a lot of time for producing dies and moulds.

However, incremental forming does not need the dies and moulds because it forms a shape by gradually deforming materials. Incremental sheet metal forming without dies and moulds is economical because less-equipment is required and shorter production time is spent on. However, the forming forces are great when they are compared with the cutting forces for producing same geometrical shapes.

Research on incremental forming forces required for forming without any defects is not fully studied yet. This paper finds and analyses forming forces in single point incremental forming (SPIF). The pressure to a forming tool is forming force which is studied in this research.

Forming forces in sheet metal forming are important because excessive forming force causes flaws of tears to raw material because it is very thin. However, less force does not make designed shapes of geometries in metal forming. The purpose of this research is to measure forming forces which is expended to a forming tool in SPIF in order to find legitimate forming forces in avoiding flaws or incompleteness. In addition, this paper compares forming forces with cutting forces of same shapes which are required for machining. The results of this research can be used to design a forming machine in structural analysis.

2 Incremental forming

SPIF is a process that produces sheet metal part which is deformed along with gradual changes. This deformation is caused by strong compressional pressure applied to a forming tool of rigid body. Geometrical shapes are formed by a moving tool in layer by layer [1]. Fig. 1 shows a schematic diagram and conceptual description of incremental forming. Raw material, being a sheet metal, is deformed by great pressure of a tool on a forming machine in SPIF.

Movement of a forming tool on a forming machine is controlled by CNC (Computerized Numerical Control), which accurately produces mechanical parts as well as creative arts. Fig. 2 shows typical parts which are produced by incremental forming. Also, auto bodies such as door fenders or engine hoods in Fig. 3 are examples which can be produced by the process.

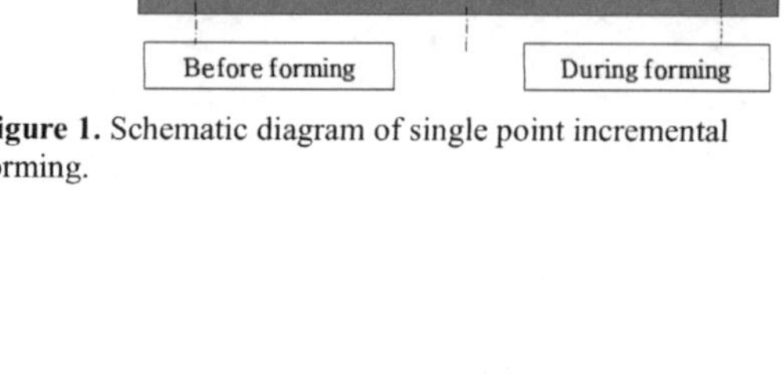

Figure 1. Schematic diagram of single point incremental forming.

DOI: 10.1051/matecconf/20168105004

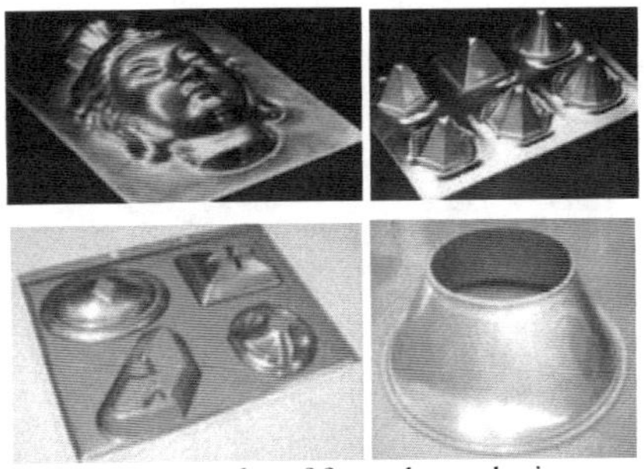

Figure 2. Various examples of formed parts by incremental forming.

Figure 3. Fenders of an auto body.

Figure 4. Door panel part produced by incremental forming.

The main advantage of incremental forming is that it does not need dedicated full pressing dies. In negative incremental forming, any die or supporter is not necessary. For particular cases of positive incremental forming, only a partial die or a supporter is needed under a sheet material in a forming machine.

A photo in Fig. 4 is a part produced by SPIF and is taken from inside an automobile door panel. A schematic diagram of a negative forming is shown in Fig. 5. In the figure, a tool pressing a blank raw material is moving X-Y direction on a plane.

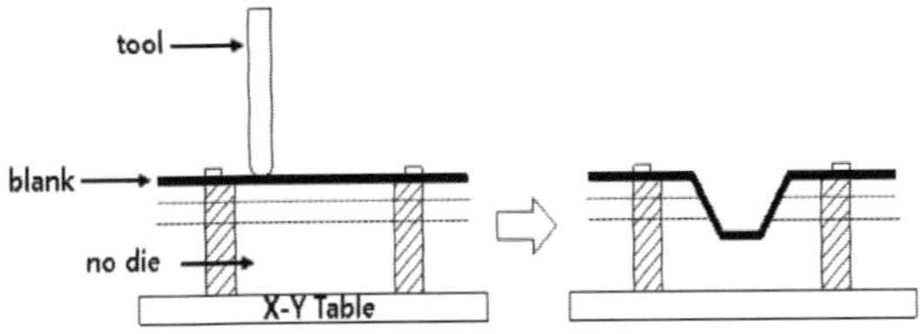

Figure 5. A schematic diagram of negative forming.

As stated in the previous research [1], [7], forming forces and accuracy greatly depend on process parameters. Tool path type for incremental forming is also an effective process parameter. There are two path types. One is constant Z or contoured type, the other is spiral or continuous type. Fig. 6 and 7 show two types of tool path which commonly utilized in SPIF.

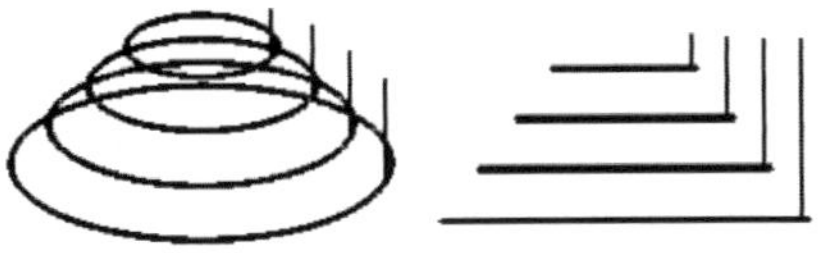

Figure 6. Contoured tool path.

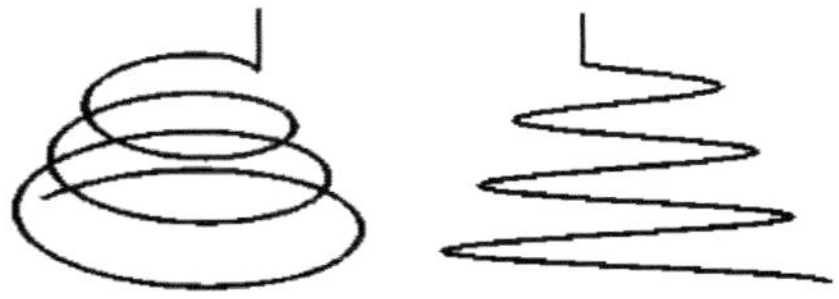

Figure 7. Spiral tool path.

Incremental forming research on mathematical model for forming forces was presented [2]. It suggested a mathematical model for prediction of forming forces in SPIF. It deduced equations through several experiments. Their results were compared with the ones resulted from Finite Element Method (FEM) software. Force prediction using FEM was published [3]. This study examined five different materials and had experimental results. From the result, they obtained regression formulae in order to compute axial and tangential forces in incremental forming. Other research on force analysis of SPIF was studied [4]. The research studied forming forces and formability, which was based on the incremental forming angle. They compared their results with previous work of negative tool path. Another research [5] presented a comparison of forming forces in SPIF with cutting forces. They concluded that forming forces in Z axis was greater than X, Y axis. However, cutting forces in Z axis was the smallest one. Normally, forming forces were greater than machining forces. Research on forces analysis was published in [6]. It found that the forming forces were small when they were compared with deep drawing. The forming force was not depend on a part size, however, it was dependant to a wall angle or a forming angle. On-line control of incremental forming through punch force monitoring was presented in [7]. There were various parameters in incremental forming such as sheet thickness, forming depth, tool diameter, and wall angle. They drew a force equation from regression analysis through various experiments. Forming force was strongly dependant to the process parameters adopted.

Various literatures on forming forces in single point incremental forming have been presented, however, a comparison between forming forces and machining forces are not presented yet. Thus, this research focuses on analysing forces arisen from forming and machining.

3 Experiments

This research forms CAD models for experiments of producing them applying both processes of SPIF and metal cutting. A raw material used for SPIF is aluminium 1050 of 0.8mm thickness. The used forming machine is DLNC-PA manufactured by A Company. Strokes in X, Y, Z axis of the machine are 1000x900x300 mm.

DOI: 10.1051/matecconf/20168105004

The geometrical shape for the experiments used for this research is a cone frustum shown in Fig. 8 with the detail dimensions. Table 1 shows conditions of incremental forming for these experiments. A contoured method is applied to generate tool paths for incremental forming.

Table 1. Forming conditions.

Incremantal Forming	Condition
Forming Tool	Ball type 6mm diameter
Forming Feed	1,000mm/min
Step distance	0.5mm/path
Material thickness	AL 1050 0.8mm

This research adopts negative incremental forming, shown in Fig. 5, which does not need any dedicated dies. A pressing punch tool is moved and controlled by contoured tool path which is shown in Fig. 6.

While the raw material of aluminium is being deformed, forming forces are collected and measured through using a dynamometer. The schematic diagram for the measurement of datum collection is shown in Fig. 9.

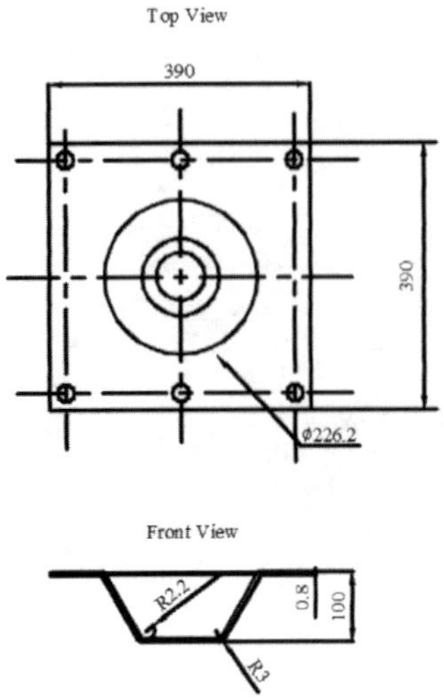

Figure 8. Cone frustum for experiments.

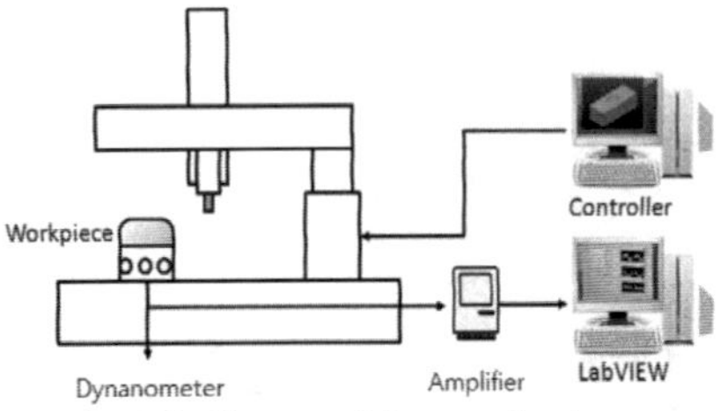

Figure 9. Schematic diagram of datum collection.

The machining forces are also collected while a high speed machining centre cuts an aluminium 1050 block. The machining conditions are shown in Table 2. The cutting forces are also collected and this paper compares the forming forces with the cutting forces.

Table 2. Machining conditions.

Machining	Condition
Cutting tool	Ball end-mill 6mm diameter
Cutting feed	1,000mm/min
Cutting depth	0.5mm
Cutting speed	10,000 rpm
Cutting material	AL1050 block

4 Experimental results

A shot of formed part by incremental forming is shown in Fig. 10. The collected data for forming forces are plotted in Fig. 11. Z direction in the figure is an axis of a forming tool. From the figure, the forces in X, Y direction form sine waves in a distinct time. The curves shape in both positive and negative region while Z directional curve is plotted only in positive. The cutting forces of the experimental part are plotted in Fig. 12. The frequencies of the cutting forces are much higher than those of the forming forces. This is because that a cutting tool revolves around Z axis while a forming tool does not.

Figure 10. Experimental part formed by incremental forming.

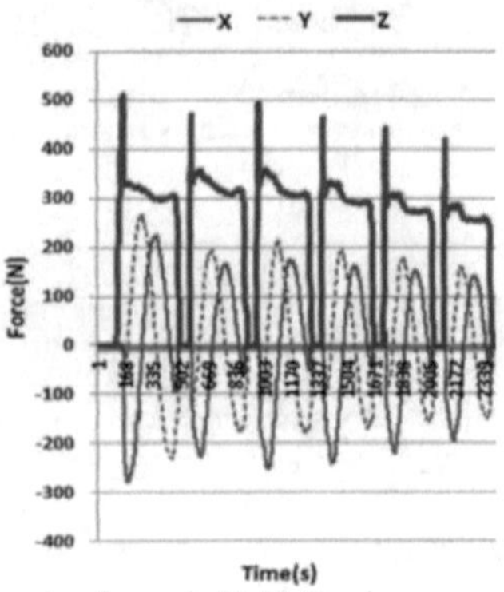

Figure 11. Forming forces in X, Y, Z axis.

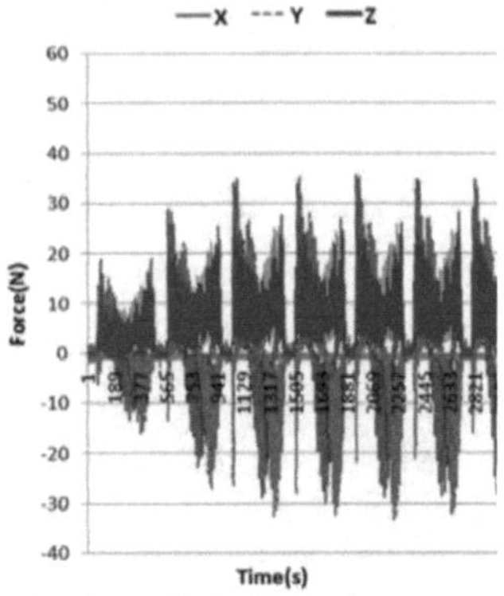

Figure 12. Cutting forces in X, Y, Z axis.

5 Conclusions and future research

This paper presents forming forces in SPIF. The forming forces in Z direction for producing a cone frustum are larger than 300 Newtons in positive. The magnitudes of forces in Y and Z direction are similar to each other but their phases are different. The shapes of both forces trace sine waves. Since the tool path type in the experiments is contour, the forces are distinctly divided in cycle by cycle. The magnitudes of the forming forces in X and Y direction are about 10 times larger than those of the machining forces. However, the forming forces in Z direction require 40 times greater than those in machining. The forming forces in X, Y, Z direction have trends to be declined in the figure 11. This is supposed that deformed area of raw material becomes smaller. However, machining forces are almost constant in all directions of axis.

These results give information that forming requires a lot of forces and energies. When a forming machine is designed, structures of the machine should be strong enough to sustain forming forces. A spindle and its axis, frames, holding devices, and other major elements should be robust to withstand forming forces. In addition, joints in forming machine must be strong enough to hold a punch tool to deform raw material. This is true that the thicker a raw material is, the greater forming forces are.

Forming forces for other different geometrical shapes and different materials are studied for future research. The possible geometrical shape is a frustum of quadrangular pyramid. These forming forces can be compared with cutting forces for same geometrical shapes of experiments. Forming forces and machining forces resulted from this research are analysed and compared with ones from a future research.

References

1. J. Kang, J. Jung, "Effective process parameters on shape dimensional accuracy in incremental sheet metal forming," Journal of Society Korea Industrial and Systems Engineering, vol. **38**, no. 4 (Dec. 2015), pp. 177-183, doi:10.11627/ jkise.2015.38.4.177
2. B. Adrian, B. Octavian, O. Valentin, R. Gabriel, "Mathematical model for determining the forces in SPIF", Proc. International Conference on Manufacturing Science and Education, (2011)
3. R. Aerens, P. Eyckens, A. Van Bael, J.R. Duflou, "Force prediction for SPIF deuced from experimental and FEM observations", International Journal of Advanced Manufacturing Technology, doi: 10.1007/s00170-009-2160-2.
4. A. Fiorentino, E. Ceretti, A. Attanasio, L. Mazzoni, C. Giardini, "Ayalysis of forces, accuracy and formability in positive die sheet incremental forming", International Journal of Material Forum, vol. **2**, (2009), pp. 805-808, doi: 10.1007/s12289-009-0467-z
5. K. Koh, "Comparison between forming and cutting forces in incremental sheet metal forming and metal cutting processes", Unpublished thesis, Kyungnam University, (Jun. 2014)
6. A. Petek, K. Kuzman, J. Kopac, "Deformations and forces analysis of single point incremental sheet metal forming", Archives of Materials Science and Engineering, vol. **2**, (2009), pp. 107-116.
7. L. Filice, G. Ambrogio, F. Micari, "CIRP Annals - Manufacturing Technology, vol. **55**, Issue 1, (2006), pp. 245-248.

MATEC Web of Conferences **81**, 06001 (2016) DOI: 10.1051/matecconf/20168106001
ICTTE 2016

Development of Production Control in Small Batch Production

Péter Németh[1] and Thomas Bernhard Ladinig[2]

[1]*Széchenyi István University, Department of Logistics and Forwarding, 9030 Győr, Hungary*
[2]*Széchenyi István University, Doctoral School of Regional Economics and Business Administration, 9030 Győr, Hungary*

Abstract. Our aim with this paper is to develop a new performance measurement and control system for small batch production in the automotive industry. For this reason, we present our previous research results for warehouse performance measurement and adopt its methodology to production control. The proposed method is based on artificial intelligence (neural networks).

1 Introduction

Stable and predictable production systems are the foundation of profitable and sustainable operations in manufacturing. The external and internal factors of production are constantly hindering these factors. Our aim with this paper is to show the possibility of a new production control and performance management system based on artificial intelligence. First, we examined some earlier research of the Department than a possible solution will be presented for managing production control.

2 Small batch production overview

The small volume production business unit of the automotive OEM has a diverse product mix and many machine set-ups for different production orders that take only 4-12 hours on average to complete. Quality issues arise caused by the many machine set-ups and various products, which results in unstable processes and high variability, as well as complexity. The business unit is an extended workbench and tier-1 supplier, which is focusing on increasing efficiency and productivity within a functional job-shop manufacturing layout with time-phased MRP planning and a multi-level dependent demand system.

The production system consists of three departments that contain all production processes from metal disc to final assembly of all the main products like doors, side panels, roofs, bonnets and hatches for premium small volume sports cars. The first department is the component production where metal components are pressed and laser cut out of aluminum or stainless steel discs that enter the facility pre-cut by a supplier. The second department contains all the assembly processes where the final assemblies are put together out of two to four main components and several smaller purchased parts like reinforcement parts to increase the stability of the products. The third and final department is called "finish" and it is responsible to ensure proper quality in terms of surface quality (which can be impacted by difficult pressing processes) and dimension (which is mostly influenced by the different mating technologies used at the assembly department). At the end the final products are ready to be shipped to the main body shop and subsequently the paint shop of several internal customers of the production system. Furthermore, there is one more department within the small volume job-shop segment which is the project and process management department, responsible for launching new customer projects on time and to ensure continuous smooth operations of all running projects throughout their life cycle.

In addition to the small-series job shop the business unit contains four other, more or less separated segments which are: quality department, logistics, fixture construction and robotics (development and construction of the fixtures, manufacturing cells and robot programs for the assembly department and other customers), tool making (development and construction of pressing tools for the component production and other customers). The high degree of vertical integration makes it difficult to create an adequate value stream and material flow because processes with very different batch sizes and processing times must be tied together efficiently. Compared to normal production series the production system would be divided into two separate segments, one being the pressing plant and the other would be the body shop where relatively high amounts of WIP inventory are quite common. Therefore, it is necessary to have large amounts of inventory to balance the differences in processing time and batch size. Also asset utilization for integration of new projects into existing production assets is quite common, so not all the available time can be used continuously for series production, but also periodical time frames must be planned as implementation time.

3 Proposed method for performance measurement

Based on the previous research activity of the Department the most viable solution for designing a performance measurement system was the artificial neural network.

The main reason behind that is that the tool to be chosen should be able to predict the strategic level result based on operational level inputs. The operational level inputs can be "manipulated" in short term (for example not allowing overtime for the workers) but strategic level outcomes must be accepted and altering them in short term is not possible.

3.1 Artificial neural network

An artificial neural network (ANN) is a mathematical model that is inspired by the structure and functional aspects of biological neural networks. A neural network consists of an interconnected group of artificial neurons, and it processes information using a connectionist approach to computation. In most cases an ANN is an adaptive system that changes its structure based on external or internal information that flows through the network during the learning phase. This learning phase is important for the use of the model. With enough data from the past the ANN is able to learn the connection between the input and output data. Parallel with the learning phase a test phase is taking place where the network is testing itself with other data from the same problem. As a result, after the learning the ANN can be used to predict outputs data from input data.

Lots of different types of artificial neural networks are proposed in the literature [7], [14]. In our application we use Multi-layer Perceptron (MLP) which is one of the most widely known types of ANNs. The topological structure of the MLP type neural network is illustrated in Fig. 1.

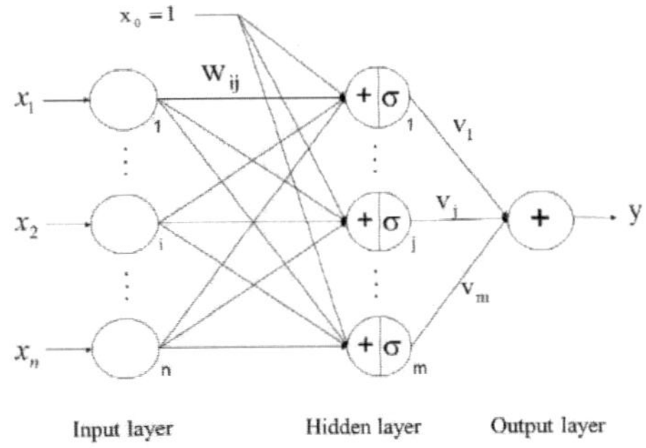

Figure 1. Topological structure of artificial neural network (MLP).

The network contains three layers. The input layer contains the inputs xi. In the hidden layer m hidden neurons are applied. Between the input and hidden layers there are connection weights wij connecting the i-th input and j-th hidden neuron.

There is also a bias input x0 = 1 with weight w0j to the j-th hidden neuron. There is an output layer with output y. Between the hidden and output layers there are weights vj connecting the j-th hidden neuron with the output.

The output of the neural network can be computed as:

$$y = \sum_{j=1}^{m} v_j \cdot \sigma\left(\sum_{i=0}^{n} x_i \cdot w_{ij}\right) \qquad (1)$$

In Equation (1) *s* is the sigmoid function:

$$\sigma(s) = \frac{1}{1 + e^{-K \cdot s}} \qquad (2)$$

where K is the slope parameter of the sigmoid function.

3.2 Training algorithm

Training or learning is the method of modifying the parameters (e.g. the weights) of the neural network in order to reach a desired goal. We can somehow classify learning with respect to the learning mechanism, with respect to when this modification takes place, and according to the manner how the adjustment takes place. In this paper the artificial neural network is trained in supervised, off-line manner by bacterial memetic algorithm [3].

Nature inspired evolutionary optimization algorithms are often suitable for global optimization of even non-linear, high-dimensional, multi-modal, and discontinuous problems. Bacterial Evolutionary Algorithm (BEA) [9] is one of these techniques. BEA uses two operators, the bacterial mutation and the gene transfer operation. These operators are based on the microbial evolution phenomenon. The bacterial mutation operation optimizes the chromosome of one bacterium, the gene transfer operation allows the transfer of information between the bacteria in the population.

Evolutionary algorithms are global searchers, however in most cases they give only a quasi-optimal solution to the problem, because their convergence speed is low. Local search approaches can give a more accurate solution, however they are searching for the solution only in the neighborhood of the search space. Local search approaches might be useful in improving the performance of the basic evolutionary algorithm, which may find the global optimum with sufficient precision in this combined way. Combinations of evolutionary and local-search methods are usually referred to as memetic algorithms [8].

In the following an already completed research is presented. This particular research consisted of the prediction of warehouse performance however we could use its methodology for production control also.

3.3 Use of ANN for warehouse performance measurement

The participant of this project was a central warehouse of a global company acting also as a distribution center for Central Europe and as a finished goods warehouse for the Hungarian factory. An additional task of the warehouse is to serve other warehouses with full truck loads. This makes the warehouse a true logistic service provider despite that it is situated beside a factory. This status comes with additional challenges and problems like the following:

- Different performance factors at outbound flows regarding directions
- Different requirements for the different outgoing flows.

These facts had serious effects on performance measurement that is to be discussed later.

The warehouse itself has 25.000 m2 surface. It has 25 ramps 53 gates. 7 gates are used for outgoing traffic.

The storing technology includes two major storing technologies:

- Shelves with 4 levels
- Block storage

The block storage is used mainly for the factory products.

Other important task of the warehouse is the distribution and collection of pallets within the whole company supply chain.

At the warehouse the strategic level performance measurement is based on various Balanced Scorecard (BSC) based results such as damaged goods, customer complaints, inventory differences among others. Our goal was to develop a tool that is able to create a direct connection between those indicators and various operational level indicators such as incoming goods, delivered goods, transhipment, and placement within the warehouse, capacity usage of the warehouse and overtime of the workers.

As described before the connection between the above mentioned indicators is very complex and inter-related. Simple calculations and regression analysis was not able to model the inter-relations and complexity of the various input parameters and so the prediction of strategic level performance indicators based on operational level actions was inaccurate.

The major problem with this performance measurement system was that it relied on the pre-defined norms. If the performance is high for a given amount of time the management will raise the norm values. However, if the complexity of the tasks is raising the norm values will not be lowered. This one-direction inelasticity can make the operative feedback about performance inaccurate.

The main idea behind the use of the proposed model was that we needed a link between complex Balance Scorecard indicators and easy-to-measure and easy-to-obtain indicators. After visiting the warehouse, we found that the operating SAP-based business software is frequently used to obtain operational level data and is capable to obtain other data not used yet. Based on the above information our first task was to obtain a logical link between some of the already available data and a BSC indicator.

For the pilot test we have chosen the warehouse efficiency Key Performance Indicator (KPI). In the next step we identified operational level data that is most likely being in logical connection with this indicator.

We have chosen the following operational level data and with the help of the warehouse staff we received the required number of them:

- Percentage of not full pallets among outgoing goods
- Work load (actual/norm)
- Warehouse storage capacity utilization (in m^3 and in percentage)

With the exact logical connection described and with the use of hypothetical or forecasted operational level data the approximate value of the KPI could be calculated. For this purpose, we used the aforementioned artificial neural network. After the learning and testing phase the program was able to calculate the number of damaged goods used the aforementioned operational level indicators as input.

3.4 Results and evaluation of the example

In the experiments we performed 10 simulations using 7 hidden neurons and the mean relative error (MRE) of the train and test set was investigated which is defined as:

$$MRE = \frac{1}{p}\sum_{i=1}^{p}\left|\frac{y_i - t_i}{y_i}\right| \qquad (3)$$

where y_i is the output of the network for the i-th pattern, t_i is the desired output for the i-th pattern, and p is the number of patterns.

Table 1 shows the best, the average, and the worst results based on the 10 simulations.

Table 1. Experimental results.

	Train	Test
Average	15.82%	17.42 %
Best	15.56%	17.17 %
Worse	16.01%	17.61 %

In our previous papers [12], [13] we presented one possible use of this model. In that case the strategic level indicator was the percentage of damaged goods at the warehouse. This model was further developed with the evaluation feature.

For further testing and to promote the further possibilities of the ANN we prepared five scenarios based on the test (Fig. 2).

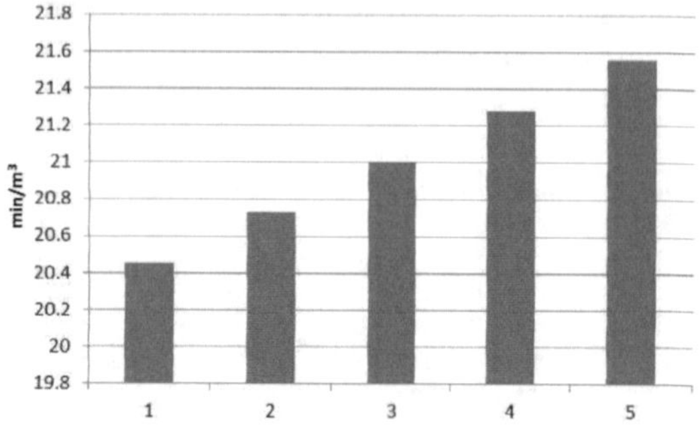

Figure 2. Evaluation of the scenarios.

The first scenario describes a best-case scenario. The warehouse storage capacity utilization is at a low level;

with lowered percentage for the not full pallets and low workload.

The fifth scenario is a worst-case scenario, the warehouse storage capacity utilization at a high level, high work load and large percentage for not full pallets.

For further evaluation of the model we created more scenarios leaving all but one input unchanged. The results showed that "ceteris paribus" – leaving all other inputs un- changed – a rise in the workload level affect the warehouse productivity the most (Fig. 3). By this way we can eliminate all inputs that are not relevant enough for the model. After that – besides the aforementioned parameter – we can insert new inputs for a repeated learning of the ANN. That will further improve the model and help to find all operational level data that matters the most for the warehouse efficiency.

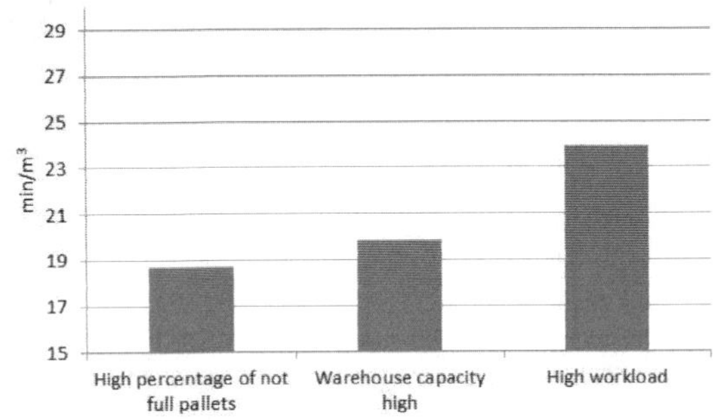

Figure 3. Evaluation of results

4 Adapting research method to production control

The aforementioned method can be used for the production control in small batch production. Our aim is to use it first for the prediction of the production planning accuracy. This accuracy is one of the most important factors in small batch manufacturing. If this accuracy is high enough the production can fulfil the orders in the given time window and can achieve competitive advantage in the long run.

As an input the following data will be used:

- Production quantity
- Production complexity (number of assembly steps)
- Capacity utilization

With enough data of previous production shifts – together with the measured accuracy of production planning – we will be able to simulate this output value with changing the input data according to the production plan. This could be used as a decision making tool: By given number of orders (production quantity and complexity) the adjusting of capacity utilization – by adjusting maintenance time for example – can increase the accuracy of production planning. Or, the other way around, by given maintenance time or pre-defined capacity utilization the number of orders will give the predicted accuracy thus helping the production control with possible output of the current shift - even before production started.

We are currently in the stage of collecting enough data. The results will be presented in later papers.

5 Conclusions

We can state that the proposed method can be adapted to production control. With enough data collected we can create a decision making tool for production logistics decision makers. As stated before, the collection of data started already and first test of the method is planned for the second half of 2016.

Acknowledgement

This work was supported by OTKA Fund (K-105529) of the Government of Hungary.

References

1. Ko M, Tiwari A, Mehnen J, "A review of soft computing applications in supply chain management" *Applied Soft Computing*, **10**, (2010), pp. 661–674
2. Botzheim J, Földesi P, "Fuzzy neural network with novel computation of fuzzy exponent in the sigmoid functions." In*: Proceedings of the 8th International Symposium on Management Engineering*, ISME 2011, Taipei, Taiwan, pp. 285–291
3. Botzheim J, Cabrita C, Kóczy LT, Ruano AE, "Fuzzy rule extraction by bacterial memetic algorithms." In: *Proceedings of the 11th World Congress of International Fuzzy Systems Association, IFSA 2005*, Beijing, China, pp. 1563–1568
4. Bozarth CC, Warsing DP, Flynn BB, Flynn EJ, "The impact of supply chain complexity on manufacturing plant performance." *Journal of Operations Management* **27**(1) (2009), pp. 78–93
5. Dubois A, Hulthén K, Pedersen AC "Supply chains and interdependence: a theoretical analysis." *Journal of Purchasing and Supply Management* **10**(1) (2004), pp. 3–9
6. Gál L, Botzheim J, Kóczy LT, Ruano AE, "Applying bacterial memetic algorithm for training feedforward and fuzzy flip-flop based neural networks." In: *Proceedings of the 2009 IFSA World Congress and 2009 EUSFLAT Conference, IFSA-EUSFLAT 2009*, Lisbon, Portugal, pp. 1833–1838
7. Hecht-Nielsen R, Neurocomputing. Addison-Wesley (1990)
8. Moscato P, "On evolution, search, optimization, genetic algorithms and martial arts: Towards memetic algorithms." Tech. Rep. Caltech Concurrent Computation Program, Report. 826, (1989) California Institute of Technology, Pasadena, California, USA
9. Nawa NE, Furuhashi T: "Fuzzy system parameters discovery by bacterial evolutionary algorithm." *IEEE Transactions on Fuzzy Systems* **7**(5) (1999), pp. 608–616
10. Németh P, "Flexibility in supply chains" *Acta Technica Jaurinensis Series Logistica,* **1**(2) (2008) pp.371–379.
11. Németh P, "Ellátási láncok hatékony irányítása multi kritériumos teljesítmény méréssel" (in Hungarian).

Ph.D. thesis, 2009, Széchenyi István University, Győr, Hungary
12. Németh P, Földesi P, Botzheim J, "Enhancing warehouse performance at a global company" In: *Proceedings of Knowledge Globalization Conference*, Boston, Massachusetts, (October 2011), pp. 7-19.
13. Németh P, Földesi P, Csík Á, "The concept of logistic space in the modelling of supply chain performance" In: *Proceedings of the 22nd Annual Production and Operations Management Society Conference,* Reno, Nevada, (May 2011)
14. Zurada JM, *Introduction to Artificial Neural Systems*, West Publishing Co., St. Paul (1992)
15. Bhagwat R, Sharma MK, "Performance measurement of supply chain management: A balanced scorecard approach", *Computers & Industrial Engineering* **53**(1) (2007) pp. 43–62

MATEC Web of Conferences **81**, 06002 (2016) DOI: 10.1051/matecconf/20168106002
ICTTE 2016

Lean Supply Chain Planning: A Performance Evaluation through Simulation

Matteo Rossini and Alberto Portioli Staudacher

Politecnico di Milano, Management Engineering Department, Milano, Italy

Abstract. Nowadays companies look more and more for improving their efficiency to excel in the market. At the same time, the competition has moved from firm level to whole supply chain level. Supply chain are very complex systems and lacks of coordination among their members leads to inefficiency. Supply chain planning task is to improve coordination among supply chain members. Which is the best planning solution to improve efficiency is an open issue. On the other hand, Lean approach is becoming more and more popular among managers. Lean approach is recognize as efficiency engine for production systems, but effects of Lean implementation out of single firm boundaries is not clear. This paper aims at providing a theoretical and practical starting point for Lean implementation in supply chain planning issue. To reach it, a DES simulation model of a three-echelon and multi-product supply chain has been set. Lean management is a very broad topic and this paper focuses on two principles of "pull" and "create the flow". Kanban system and setup-time and batch-size reductions are implemented in the lean-configured supply chain to apply "pull" and "create the flow" respectively. Lean principles implementations have been analyzed and compared with other supply chain planning policies: EOQ and information sharing (Visibility). Supported by the simulation study, this paper points Lean supply chain planning is a competitive planning policies to increase efficiency.

1 Introduction

In In recent years, the competition is moved from firm vs firm to supply chain vs supply chain [1]. Competition continuously increases and pressures on the performances of the supply chain's members: a great emphasis is given to efficiency and managers are looking for cost reduction everywhere in companies [2]. There are many opportunities to reduce inefficiencies. Main inefficiencies lead to excess of inventory, backlog orders, supply and transportation extra-costing along the chain [3]. Supply chain planning is one of the major tasks for optimizing production, inventory and transportation processes [4].

Lean Management is becoming more and more popular among supply chain managers. The main feature of lean approach is to reduce every type of waste present in the production flow whether it means material waste, time waste or activity waste. It is largely demonstrated that lean approach implementation leads to cost reductions, higher quality level and shorter delivery time in a company [5].

In what direction and how much Lean approach influences the whole supply chain is not well defined [6]. This paper investigates the impact of Lean implementation along a supply chain. The purpose of this research work is to study which impacts has lean approach when implemented at a supply chain level. To achieve this, a simulation model of a multi-product and three-echelon supply chain has been implemented. This study deals with the analysis of supply chain's performance comparing Lean supply chain with two other supply chain planning policies: EOQ and Visibility. In Section 2, a brief literature review of supply chain planning and lean management themes is proposed. Section 3 defines the experimental design, where simulation model and planning policies descriptions and plan of experiments are presented. The outcomes of simulation study are discussed in the Section 4.

2 Literature review

Supply chain planning (SCP) is a complex and difficult task particularly because it involves different actors in different companies. SCP objective is to improve coordination and integration among members of the supply chain [4], [7]. In the next paragraph Visibility that is one of the most famous SCP policy is discussed. The second paragraph briefly discusses the literature of Lean management.

2.1 Visibility

Visibility leads to SC coordination improvement by information sharing practice among SC members [8]. Information sharing about inventory position of SC warehouses and about the members' orders policy helps actors of supply chain to plan their production satisfying customer demand and increasing efficiency [9], [10]. The literature of visibility is very broad.

Lee, in [9], and Cachon, in [10], studied the impact of visibility implementation on bullwhip effect in a SC and both showed that visibility reduce bullwhip effect and inventory level in the system.

Gavirneni, in [11], [12], explored different degree of visibility in a two-echelon SC. He demonstrated that the higher the visibility degree is the greater the SC benefits are in terms of logistic costs and inventories. Bottani, in [13], made a similar simulation study analysing the impact of different visibility degrees on different SC configurations. Different studies simulated the effects that visibility could have on demand forecasts accuracy of SC players [14]. In another work, Gavirneni simulated a two-echelon SC and demonstrated that most profitable order policy change whether or not Visibility is implemented: under no-visibility SC the most efficient order policy resulted (s;S) while under visibility SC the most efficient order policy is (r;Q).

2.2 Lean management

The lean management approach is famous for the struggle against the waste [15]. What is lean is difficult to define, it is possible to state that is a philosophy of work that focuses on what is relevant for the customer [16]. The purpose of lean approach is to improve overall levels of productivity and product quality, waste reduction, integration and interaction across functional departments, and improved work force autonomy. The traditional field of application of lean approach is the manufacturing firm operations, but recently lean academics and lean specialists are moving their attention to other fields that could be positively affected by lean. Supply chain is one of this new fields for Lean [6]. The term Lean supply refers to the extension of lean principles throughout the supply chain, both downstream and upstream [17] Lean supply chain is not properly a new field for Lean. In the past, Lean supply chain appeared in literature under different names: lean logistics, lean distribution, lean supply and lean enterprise. Seminal works have studied which are the practices that "considered Lean enterprises" adopt with their supplier and their customer [18]–[20]. There is a lack in literature of papers that discuss Lean supply chain with quantitative analysis focusing on planning issue.

Up to now little is known about the possible implications of Lean principles implementation for supply chain planning issue, in particular on inventory control.

In the next section it is described the experimental design set to fill this lack of knowledge.

3 Experimental design

Trough simulation tool is possible to reproduce models that are close to the reality and to analyse complex systems with a quantitative viewpoint [21]–[24].

A simulation study has been set up to improve our understanding of lean supply chain. To evaluate lean SC planning performance, a comparison with other SC planning policies is performed: Lean is compared with traditional EOQ policy and with Visibility policy.

The model used in the simulation study has been develop through a Discrete Event Simulation software (Rockwell Arena) and it is kept as basic as possible to avoid any noise that might cloud the sight on causes and effects.

3.1 Supply chain model

The simulation model represents a three echelons supply chain composed by four suppliers, a manufacturer and a retailer.

An input warehouse, a production phase and an output warehouse compose suppliers and manufacturer stages. Retailer stage is a distributor, so it does not process items. An infinite stock at suppliers' input buffers is assumed. The production systems stages a single machine where the items are processed. Many authors have used similar settings to understand the effect of SC planning policies on inventories in single product supply chains [10], [25], [26].

There are 24 different products from 4 different product families in the supply chain. One product family consists of 6 different products. Each of the suppliers is responsible to produce one product family. However, the suppliers are not fully dedicated to this supply chain; they produce other products for other supply chains as well. The manufacturer works all the 24 products and its operations are fully dedicated to this supply chain.

Multi-product SC and three-stage supply chain can provide valuable insights into managing complex systems efficiently.

An infinite number of trucks are available for the transport of finished and semi-finished products and the lead-time to transport an item to the next stage is deterministic and equal to two days.

Trucks leave the stage at the end of the day and shipped items are available to the next stage two days later, at the beginning of the day.

Everyday final customer demands finished product to the retailer and she has to satisfy the demand in Make-To-Stock logic. The retailer has to provide before the shipment time (basically before the end of the day) the demanded pieces. If she does not satisfy the demand there is the stock-out and so the back-log of the order. The way how supply chain members place orders depends on the specific supply chain policy. In the supply chain all the components use the same planning policy. The planning policies studied in this paper are described below.

3.2 SC planning policies

This research work aims to analyse the SC performance while SC has implemented lean approach. Two different SC planning policies are compared with Lean SC planning: Economic Order Quantity (EOQ) and Visibility (VIS). EOQ policy is the basic level of the planning policy task, it is recognize as the first attempt of inventory control optimization. In the major part of the planning simulation works it is used as benchmark. The Visibility approach concerns information sharing practice and it is one of the most known approach to improve the planning function in a system.

3.2.1 EOQ

The EOQ approach is that all the SC members follow the (r,Q) policy. Every warehouse is exposed to a continue check: when the inventory-position goes down a certain level (reorder point), an order is placed upstream to the previous warehouse. If it is an internal order, the production of a batch starts, if it is an inter-stages order, the shipping. The order size and the reorder point could be different for each warehouse and they are fixed during the single replication.

The inventory position is calculated as:
Inventory position = inventory level + ordered items but none arrived – backlog orders [27].

3.2.2 Visibility

Logic of the visibility approach exploits the information sharing among SC members. While visibility is operating, every stages know the inventory position of the downstream stage and can take advantage of that [11], [12], [26]–[28].

By means of this information, the manufacturer is able to postpone the production (optimizing production timing) and so reduces the average WIP.

In this research work the visibility policy bases on the policy developed by Datta and Christopher [29].

3.2.3 Lean

Referring to the lean principles explained by Womack and Jones [15], this paper tests "pull strategy" and "create the flow". "Pull" principle is implemented along the supply chain through a kanban system implementation in all SC stages: the production starts only if there is a consumption of material at the downstream stage. "Create the flow" aims at a flow as levelled as possible, without any kind of interruption. The batch-size and setup time reduction represents this principle in the simulation model.

Table 1. Supply chain planning policies

	EOQ	Visibility	Lean
Batch-size and setup time	100% (EOQ)	100% (VIS)	100% (Lean_100%) 80% (Lean_80%) 60% (Lean_60%) 40% (Lean_40%) 20% (Lean_20%)

Lean_100% is the Lean SC planning policy with the implementation of pull principle. Lean_80%, Lean 60%, Lean_40% and Lean 20% are Lean SC planning policies with the implementation of "pull" principle and implementation of "create the flow" principle (respectively 20%,40%,60% and 80% batch size and setup time reductions). A brief resume of SC planning policies adopted in this simulation study is presented in the Table 1.

3.3 Experimental parameters

Supplier and manufacturer stages have finite production capacity. Moreover supplier is not dedicated only to this supply chain so every day it dedicates a variable amount of time to produce pieces for the manufacturer (65% on average). Process times and demand rates have been defined in order that Manufacturer and Supplier are 80% saturated. This saturation capacity value is a good compromise between what there is in literature, i.e. [26], and in the real world.

Like in the most part of SC simulation works the daily demand follows a Normal distribution.

Many papers stated that demand variability is a relevant parameter on SC planning policy performance. Due to this three different level of demand variability are studied. The coefficient of variation of the daily demand can be low (c.v.=0,4), medium (c.v.=0,6) and high (c.v.=0,8).

The service level of the SC is measured by the mean of the service levels of the single warehouses. The single warehouse service level is the ratio between the number of days of stockout and the overall number of days in the simulated period. The warehouse is in stockout whether it has not handled all the orders at the end of the day. This research work tested the supply chain performance of different service levels (from 92% to 99%).

The inventory level performance considered measures the amount of WIP along the entire SC.

The simulation runs for a period of 2050 days or 410 weeks with the first 50 days as the initialisation period, the statistics from which were not used in the results.

3.4 Plan of experiments

The simulation study considers all the possible scenarios that derive from the combination of the experimental parameters such demand c.v. (3) and service level (15) and the values of experimental variables (7). This means that 315 scenarios are studied.

4 Simulations results

This section discusses the results of simulation study for each of the parameters investigated. Particular attention is paid to use the SC holistic point of view for service level and inventory level.

4.1 Inventory level and service level

It is noteworthy that service level affects inventory levels: the higher the required service level the higher the necessary inventory level is valid for all the SC planning policies. What is interesting to observe is the difference between policies. The Fig. 1 shows all inventory level curves that are related to different SC policies. EOQ SC curve starts at 6553 and arrives to 10024; Visibility SC starts at 5885 and arrives to 9262; Lean_100% starts at 4473 and arrives to 7191. Increasing service level from 93% to 99% EOQ SC inventory has increased by 53%, Visibility SC inventory by 57% and Lean SCs 64%. Service level affect similarly the inventory increase of EOQ and of Visibility: inventory difference at 99% service level and 93% for Visibility is 2,7% lower than for EOQ. It appears that service level affects much more Lean inventory level than Visibility inventory level.

Comparing the inventory values of extreme tested service levels in the graph (93% and 99%), Lean SC absolute inventory increase is at least 20% lower than inventory increase for Visibility. According to this, it is possible to state that, increasing service level, Lean benefits against Visibility benefits increase and so Lean advantages increase in high service level contexts.

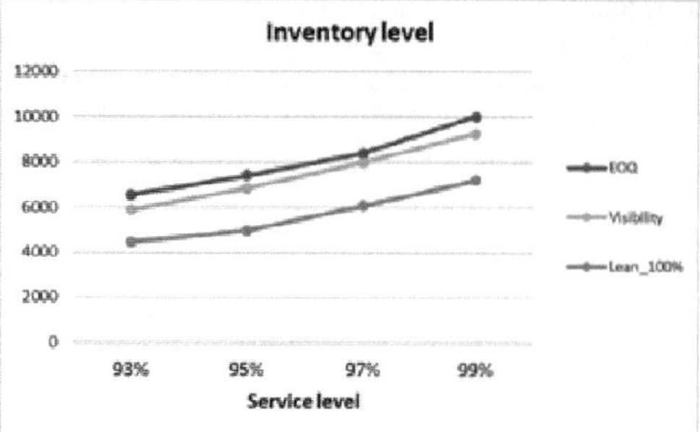

Figure 1. Inventory level curves for different service level.

4.2 Inventory level and variability

Coefficient of variation is relevant for inventory levels. As could been seen in the Figure 2, increasing coefficient of variation of final customer demand there is a general increase of inventory levels. EOQ curve starts at 7815 and increases to 8356, Visibility curve starts at 7220 and increases to 7756, Lean_100% curve starts at 5361 and increases to 6013. Comparing inventory levels of 0,4 c.v. context and 0,8 c.v. context: EOQ SC inventory increase is 6,9%, Visibility SC inventory increase is 7,4% Lean_100% SC inventory increase is 12%. Increasing the coefficient of variation, the relative increase of inventory is greater for Lean than for Visibility. However Lean is not only PULL but also set-up and lot size reduction. Therefore it is necessary to analyse the impact of such element.

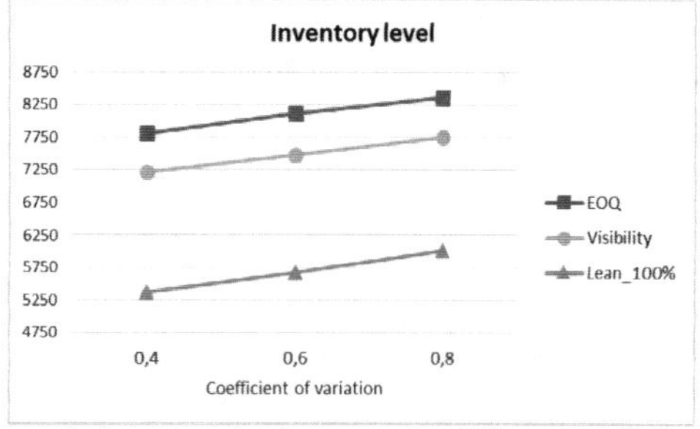

Figure 2. Inventory level curves for different coefficient of variation.

4.3 Inventory level and setup time and batch-size reduction

The average impact of batch-size reduction is given in the Figure 3. Implementing batch-size reduction, inventory level decrease is performed. 20% batch size reduction have been tested in this simulation work. Reducing batch-size a first time (or in other words changing from Lean_100% to Lean_80%) decreases inventory level by 17%. The effect of second batch-size is 12% inventory reduction, the effect of third is 7% and the effect of fourth is 10%. The results show that setup-time reduction benefit on inventory levels is smooth: first batch-size reduction are more effective than the last ones. Moreover, it has been noted a positive effect of batch size and setup time reduction that smooths negative effects of variability on supply chain planning performances.

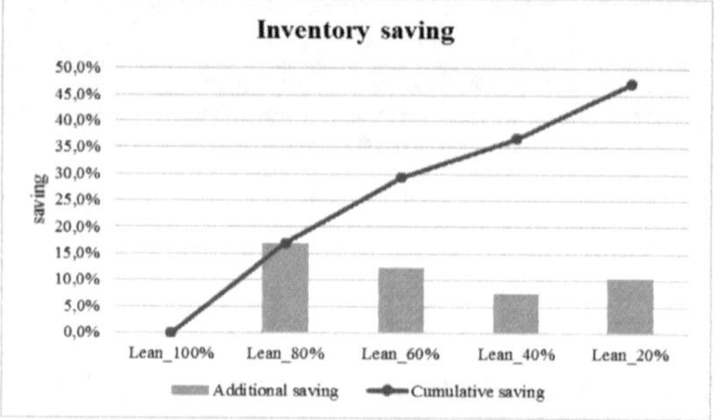

Figure 3. Inventory savings given by setup and batch reductions.

4.4 Inventory savings

An overview of inventory level saving for all the SC planning policies compared with EOQ is presented in fig 4. Visibility leads to an average 7,7% inventory saving. Lean_100%, Lean_80%, Lean_60%, Lean_40% and Lean_20% respectively lead to 30,1%,41,5%, 51,6%, 56% and 63,5% inventory savings. Lean has always greater savings than visibility about inventories. The greater batch-size reduction is the greater the benefits of Lean are. However the effect of the batch size reduction is smooth, as previously discussed. According to the values of this simulation work, the Lean inventory savings are at least three times Visibility inventory savings. Thus, it is possible to conclude that Lean implementation impact on SC planning performances is very relevant and could lead to significant logistic cost improvements.

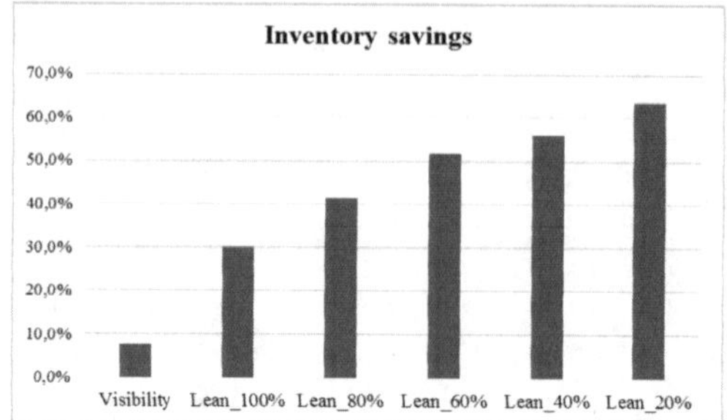

Figure 4. Inventory savings for different supply chain planning policies.

5 Conclusions

This paper aims at improving foundations for lean supply chain implementation. A simulation study has been performed to support our considerations and to analyse impact of lean principles implementation under different contexts. The simulation study shows that Lean planning policy leads to high inventory savings. With the best performing Lean_20%, the average inventory saving reaches to 63%. The simulation also shows that Lean supply chain planning performs better than visibility and EOQ SCs in high service level contexts and high variability contexts.

The results have important managerial implications. When implementing a SC planning policy, the focus should be on specific parameters: there is a trade-off analysis to do because of the different impacts that Lean and Visibility have on the SC inventory carrying cost and transportation costs. Investments on setup-time and batch-size reduction deserve particular attention, and should be based on SC holistic point view. Therefore, this work could be an initial support to implement lean approach in SC's operations systems. The research has scientific implications with respect to the development of supply chain planning task. It provides insight into which sensitivities can be avoided. This simulation model is useful for further researches because it allows to replicate different supply chain planning policies and differently from the most part of the simulation models present in literature that study a single-product and two echelon supply chain, it simulates a multi-product and three echelon supply chain. The embedding of different transportation rules, different production capacity saturations should be added in future research.

References

1. M. W. McCarter and G. B. Northcraft, J. Oper. Manag., vol. **25**, no. 2, pp. 498–511, (2007)
2. K. L. Sim and J. W. Rogers, Manag. Res. News, vol. **32**, no. 1, pp. 37–49, (2009)
3. H. L. Lee, V. Padmanabhan, and S. Whang, "The Bullwhip Effect In Supply Chains 1 Abstract ," vol. **38**, no. 3, pp. 93–102, (1997)
4. J. Olhager, Int. J. Prod. Res., vol. **51**, no. 23–24, pp. 6836–6843, (2013)
5. T. Melton, Chem. Eng. Res. Des., vol. **83**, no. 6, pp. 662–673, (2005)
6. N. V. K. Jasti and R. Kodali, Int. J. Prod. Res., no. August, pp. 1–19, (2014)
7. M. Caridi, L. Crippa, A. Perego, A. Sianesi, and A. Tumino, Int. J. Prod. Econ., vol. **127**, no. 2, pp. 372–383, (2010)
8. P. A. Bartlett, D. M. Julien, and T. S. Baines, Int. J. Logist. Manag., (2008)
9. H. L. Lee, V. Padmanabhan, and S. Whang, Manage. Sci., vol. **43**, no. 4, pp. 546–558, (1997)
10. G. P. Cachon and M. Fisher, Manage. Sci., vol. **46**, no. 8, pp. 1032–1048, (2000)
11. S. Gavirneni, Manage. Sci., vol. **48**, no. 5, pp. 644–651, (2002)
12. S. Gavirneni, R. Kapuscinski, and S. Tayur, Manage. Sci., vol. **45**, no. 1, pp. 16–24, (1999)
13. E. Bottani and R. Montanari, Int. J. Prod. Res., vol. **48**, no. 10, pp. 2859–2886, (2010)
14. A. Angulo, H. Nachtmann, and M. A. Waller, J. Bus. Logist., vol. **25**, no. 1, pp. 101–120, (2004)
15. J. P. Womack, D. T. Jones, and D. Roos, Bus. Horiz., vol. **35**, no. 3, pp. 81–82, (1992)
16. S. Bhasin and P. G. Burcher, "Lean viewed as a philosophy," no. 2002, (2006)
17. A. Reichhart and M. Holweg, Int. J. Prod. Res., vol. **45**, no. 16, pp. 3699–3722, (2007)
18. R. Lamming, Int. J. Oper. Prod. Manag., vol. **16**, no. 2, pp. 183–196, (1996)
19. J. K. Liker and T. Y. Choi, Harvard Business Review, vol. **82**, no. 12. pp. 104–113+149, (2004)
20. J. K. Liker and Y. Wu, Sloan Manage. Rev., no. Fall, pp. 81–93, (2000)
21. H. K. Chan and F. T. S. Chan, Int. J. Prod. Res., vol. **48**, no. 10, pp. 2793–2819, (2010)
22. S. Terzi and S. Cavalieri, Comput. Ind., vol. **53**, no. 1, pp. 3–16, (2004)
23. J. Košturiak and M. Gregor, Comput. Ind., vol. **38**, no. 2, pp. 159–172, (1999)
24. E. Akçalı and S. Çetinkaya, Int. J. Prod. Res., vol. **49**, no. 8, pp. 2373–2407, (2011)
25. F. Chen, Z. Drezner, J. K. Ryan, and D. Simchi-Levi, Manage. Sci., vol. **46**, no. 3, pp. 436–443, (2000)
26. J. K. Kwak and S. Gavirneni, Int. J. Prod. Econ., vol. **132**, no. 2, pp. 271–278, (2011)
27. F. Chen, Manage. Sci., vol. **44**, no. 12-Part-2, pp. S221–S234, (1998)
28. W. Zhu, S. Gavirneni, and R. Kapuscinski, IIE Trans., vol. **42**, no. 3, pp. 173–187, (2009)
29. P. P. Datta and M. G. Christopher, Int. J. Prod. Res., vol. **49**, no. 3, pp. 765–803, (2011)

DOI: 10.1051/matecconf/20168106003

Investigation of Effectiveness of Order Review and Release Models in Make to Order Supply Chain

Kaustav Kundu and Alberto Portioli Staudacher

Politecnico di Milano, Management, Economics and Industrial Engineering Department, 20133, Milano, Italy

Abstract. Nowadays customisation becomes more common due to vast requirement from the customers for which industries are trying to use make-to-order (MTO) strategy. Due to high variation in the process, workload control models are extensively used for jobshop companies which usually adapt MTO strategy. Some authors tried to implement workload control models, order review and release systems, in non-repetitive manufacturing companies, where there is a dominant flow in production. Those models are better in shop floor but their performances are never been investigated in high variation situations like MTO supply chain. This paper starts with the introduction of particular issues in MTO companies and a general overview of order review and release systems widely used in the industries. Two order review and release systems, the Limited and Balanced models, particularly suitable for flow shop system are applied to MTO supply chain, where the processing times are difficult to estimate due to high variation. Simulation results show that the Balanced model performs much better than the Limited model if the processing times can be estimated preciously.

1 Introduction

Customization is one of the major challenges faced by most of the high variety-low volume industries [1]. Usually strategies like make-to-order (MTO), engineer-to-order (ETO), etc. are incorporated to tackle customization in these companies [2]. In MTO companies, there is lot of variation in the process as well as uncertainty in demand [3]. As a result, the companies sometimes struggle to meet the due dates promised to the customers. In order to tackle uncertainty in demand, many authors suggested Lean techniques like Kanban containers, etc. in MTO companies. Although these Lean techniques are efficient in repetitive companies, they are not proved to be effective in MTO companies where there is high variation in product types as well. Later on some authors tried to adopt Workload control (WLC) techniques like order review and release (ORR) systems particularly for job shop environment. These systems try to absorb the variation in demand and reduces the overall throughput time. As a result, the due dates of the orders are met. These systems are adopted successfully in many jobshop companies. The main focus is on production aspects and on the shop floor activities, particularly how to implement WLC techniques in MTO companies' shop floors. But in MTO companies, orders' value stream consists of the tendering phase, the design phase, the purchasing phase and the production phase and activities are not restricted only in the shop floor. Dealing only in the shop floor activities will ignore the other parts of the whole flow and also important issues like departments' integration and coordination [4, 5, 6]. Therefore, WLC in MTO companies should not be confined only in the shop floor but should also address the other parts of the whole supply chain, from the tendering phase to the assembly stage [7,8], similar to a pure flow shop environment.

There are ORR systems which are specifically designed for jobshop companies. These systems try to hold the orders in the pre-shop pool for some time before releasing them to the shop floor. In this way, the system is able to reduce WIP, any wastages due to cancellation of demand or changes in specifications. There are different kinds of load limiting models developed and implemented in jobshop companies. Few authors tried to apply some of these techniques in pure flow shop environment. Limiting (Upper bound limit) is one such Load limiting model which is proved to work best in the flow shop environment. There are some arguments that these systems should have workload balancing capability which is required for good performance in pure flow shop environment [9]. Therefore, a lean based Balanced release (BLR) ORR system is developed by [7] to control the shop floor process in such kind of environment. This BLR system, capable of balancing workload, shows much better performance as compared to Upper bound limited system in the shop floor of non-repetitive companies. But nowadays as the MTO companies are concerned more about the supply chain, the focus of the paper is on internal supply chain of MTO companies. The pure flow shop can be exactly comparable to the internal supply chain. It will be wise to apply those WLC models in supply chain environment where there is high variation

 DOI: 10.1051/matecconf/20168106003

in the process. Therefore, the first research question of this paper is: *How does the MTO supply chain perform under different ORR systems?* To address this question, different simulation scenarios are run for both Limited and BLR ORR systems. Moreover, in MTO supply chain, it is really difficult to predict the expected processing time for each order at each workstation. But the preciseness of processing times can influence the load oriented ORR systems. Therefore, in order to analyze the impact of the preciseness of processing times on the overall performance of ORR systems in MTO supply chain, the second research question is: *Does the prediction of processing times precisely influence the performance of ORR techniques in MTO supply chain?* To answer this question, the expected processing times are grouped into simple classes [10] and simulation runs are performed for both the models.

The remainder of the paper is organized as follows. The issues in MTO companies and the different ORR systems are described in section 2. The simulation model is described in Section 3. The discussion of the results is then presented in Section 4. Finally, in Section 5, conclusions are drawn along with future recommendations.

2 MTO companies and ORR systems

2.1 MTO companies and their internal supply chain

The term internal supply chain is used to address the whole set of stages that, from the reception of the customer order to the delivery [11]. In MTO companies' internal supply chains, there are three main classes of business processes: physical, nonphysical and support processes [5]. Physical processes deal with the physical activities on products. Nonphysical processes particularly include the engineering and procurement activities. Support processes are distinguished by being linked to all the nonphysical and physical processes. When a customer orders a product, both physical and nonphysical activities are carried out. All these processes within the supply chain should be controlled using some suitable techniques otherwise the overall performance of the system will be degraded, particularly when there is high variation in the process.

Another key issue that is highlighted in the literature for MTO companies is the due date (DD) setting at the tendering stage [12]. The high variability and incomplete historical data on lead times in MTO companies make the assessment of due dates very difficult [6]. The authors in [13] tried to show the benefits of WLC along with web functionality in supply chain integration. It is proposed in [14] three different procedures for DD quotation depending on the DD type and orders type of the MTO companies: Negotiable DD and fast order, Negotiable DD and slow order, Fixed DD. But DD determination largely depends on the whole flow of the supply chain within the company. Therefore, the main focus should be on the whole flow which handles the output performances. But there is still less attention in the area how to implement suitable techniques to handle high variation in process in MTO companies, particularly in the supply chain [8].

2.2 Limiting and lean based balanced ORR systems

Most ORR systems are developed and used mostly in job shops [7], [15]. This system implies avoidance of the immediate release of customer orders, and instead creates a pre-shop pool of orders [16]. The benefits of using ORR systems also include easier control over WIP in a shop floor, better workload balance between the stations, reduces throughput times, increases the reliability of due dates [16]-[18]. ORR systems are in line with lean methodology, as they help to streamline the flow [7]. ORR systems consist of 3 main steps [19]: 1. Order entry phase; 2. Pre-shop pool; 3. Order release phase. Order release phase is most important and there should be an effective selection mechanism at this phase [20]. The existing procedures of order release can be divided into two main categories based on the mechanism of release used [19]: load limited methods; time phased methods. Workload limiting and workload balancing are two order release systems based on load limiting methods These systems have been used the most as they offer an easy implementation and a simple control of the WIP.

The workload norm of order release mechanism limits the load present in the shop floor and the release method sequences and releases the orders based on the parameters decided at the management level in the pre-shop pool. The workload norm helps to achieve an implicit balancing of productive resources. This makes it possible to load up all the machines with a workload as close as possible to their limit [15], [13], [21], [22]. There are three main techniques for the workload control within limited workload order release system: upper bound only [15], [23]; lower bound only [11]; upper and lower bound method [17]. In [24], the authors found out that the upper bound limiting order release system performs the best.

An alternative to the limited workload methods is workload balancing method. This order release method balances the load between the different workstations using extended forecast horizon as well as keeping the overall load inside of certain limits [19]. This method prevents the centers to remain inactive and improves the predictability of the crossing times in the system [15], [25]. The lean-based BLR method developed by [7] and applied to non-repetitive manufacturing proves to be more effective than limited workload method. According to this method, the focus is on balancing the release workload, disregarding the current disbalance between the working stations. The choice of orders from the pre-shop pool is done by finding the similar amounts of workloads for each workcentre, taking into account the current release period as well as the future ones (extended schedule visibility). A mathematical model is developed and the objective of the mathematical model is to balance the overloading and underloading of loads at each job station over an extended period of time. The details of the mathematical model are available in the article by [7].

3 Simulation model

3.1 The structure of the supply chain model

A supply chain model is constructed from the point of view of a customer request, which follows a route from an order entry phase to the delivery of an order. In this case, a supply chain resembles a flow shop, where orders follow the predefined sequence of departments. The simulation focuses on one of the main specifics of MTO supply chain – uncertainty in prior evaluation of processing times at a high level of variability. The estimation of expected processing times at release can be simplified by grouping the sample of orders into simple classes [10]. The assumptions made are that all the orders are of the same type, so they follow the same route. Python 3.4 Simpy module [26] is used for simulation and CPLEX 12.6 is used to solve the optimization problem for the BLR model.

The ORR systems are tested in the model in order to discover their applicability to MTO supply chain's performance. Before executing the algorithm of ORR system, a current aggregated workload present in a MTO supply chain and at each department is measured. The workload includes 2 main components: a direct workload of a department, which is found in the queues of each department; and indirect workload, which is the sum of processing times of all the orders in the system that sooner or later have to pass through this department (often referred as an upstream load). Two different ORR systems, Upper bound Limited Workload and BLR Workload, as described in [7] are used for order release in this paper. For simulation, the basic assumptions and the parameters for both the models are almost the same as that given in the article by [7] with little bit modification. According to the classifications by [19], the characteristics of both the ORR systems applied to the SSC simulation model are the same except for two factors. **Aggregation of workload** for Limited Workload is load at each department whereas for BLR Workload is total load in the supply chain; **Schedule visibility** for Limited Workload is limited whereas for BLR Workload is extended.

3.2 Input parameters

In order to simplify the model, the randomness in orders is ignored and they are assumed to be constant. Based on previous studies, the arrival of orders follow a Poisson distribution on average every 16 minutes. Thus on average there are 30 orders per day and all 30 orders arrive to the system just at the beginning of each release period (day in this case). It is decided to use a lognormal distribution of processing times, which is quite common in MTO supply chain [27]. The processing times at each department are assumed to be 29.7 minutes long on an average. As the objective of this paper is related to high variation situation, coefficient of variation of 0.8 is assumed here. Considering the assumptions, the input data is generated with the use of Minitab 17 software. The generated values represent the actual processing times, occurred at each stage of a supply chain. But they cannot be used for planning purposes, as they are not known in advance, before the request is processed. Thus the expected processing times, which are estimated when the request arrives, should be also considered. To do so, each sample is divided into equal parts – classes, and the mean of each class is found. Each class' mean is used for planning purposes as an expected processing time, which is estimated on the requests' arrival time. To answer the second research question, different number of classes are tested. Thus, for a low precision level there are 3 classes of expected processing times: short, medium and long. And for a high precision level in estimating the expected processing times, 7 classes are considered.

3.3 Conduction of experiments

As there is huge variation in processing times of different orders, the system workload cannot be calculated in terms of number of orders. Instead it is assumed to be the total capacity of the entire supply chain and it is defined in terms of minutes. To determine the optimal system workload, some preliminary simulation runs are performed. Simulation is started with very high system workload and then slowly the workload is decreased to lower values. Around 18000 minutes, the performance measures of the system are found to be optimal. Therefore, 18000 minutes is considered to be the system workload for the rest of the experiments. Warm-up period corresponds to the time interval, after which the system becomes stable. The most used method for the calculation of the warm-up period length is the graphical method described by [28]. To determine a warm-up period and length of the simulation runs, an experiment is conducted at 18000 minutes system workload. During the first 300 days the MTO supply chain performance is not stabilized. As the simulation is of a steady-state type and its initial state doesn't represent the real behavior of the system, a warm-up period is set to 300 days, until the output stabilizes. In order to avoid any possible errors, the length of the run is set to 500 days. These values are used for all the experiments. The performances [22] that are measured are: Average Gross throughput time (GTT) - total time from arrival of order till its delivery including waiting time in the preshop pool; Average Supply chain time (SCT) - total time from release of order for actual processing in the supply chain till its delivery; Average Output workload - the load of orders that are released per period.

The experiments are planned in accordance to the research questions of the paper. First set of experiments are conducted with the actual processing times for both the models. The second set of experiments are conducted in order to simulate the MTO supply chain with different precision levels in processing times estimation. With the obtained results from the first set of simulation runs, it will be possible to compare the performance of both models in MTO supply chain. With the obtained results from the second set of experiments, it will possible to analyze the impact of uncertainty in processing times on the performance of upper-bound Limited ORR system and BLR ORR system. In MTO supply chain, the workload does not remain stable. Therefore, it will be wise to test the models under all kinds of situation. Different target workload norms (defined in minutes), which range from very short queues in a MTO supply chain to very long ones (960 to 3500 minutes) are defined to limit the load for each job station in the supply chain. The actual output performances are obtained in minutes but for better understanding, they are converted into hours.

4 Results and discussion

The results of the simulation runs that are carried out to answer the research questions are shown in this section. Figure 1 shows the variation of GTT with respect to SCT. As the workload norms increase, less orders remain in the pre-shop pool and more orders flow into the actual system, thereby decreasing the throughput time of the orders upto certain extent. Therefore, GTT first decreases with SCT and then becomes more or less stable. It can be also seen from the figure that at same SCT, the GTT for BLR model is much lower than that of the Limited model. The difference in performance increases as the workload norms become tighter, i.e., situation when there is short queue within the supply chain. It convinces that the BLR model is better than the Limited model, especially at tighter workload norms. Therefore, it answers our first research question.

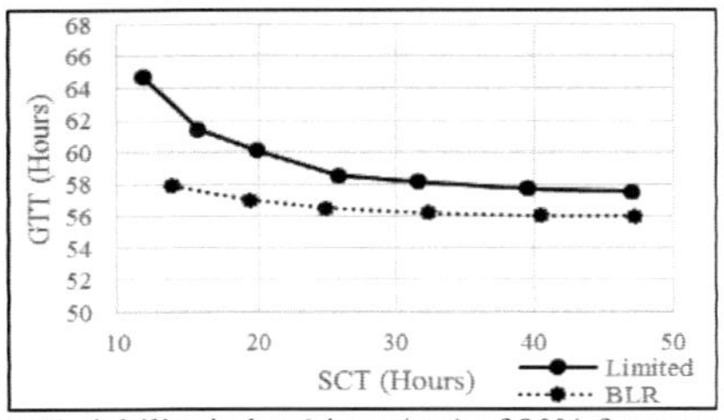

Figure 1. Variability index (sigma/mu) of 80% for actual processing time.

Figure 2 shows that the difference in performance between the Limited and BLR models at low precision level (Class 3). The GTT obtained from both the models at different workload norms are much higher than that obtained using actual processing time information. At some SCT, there is little bit difference between the models but the overall difference in performance is not so significant at low precision level. Therefore, the BLR model does not perform well when it is difficult to predict the processing times precisely.

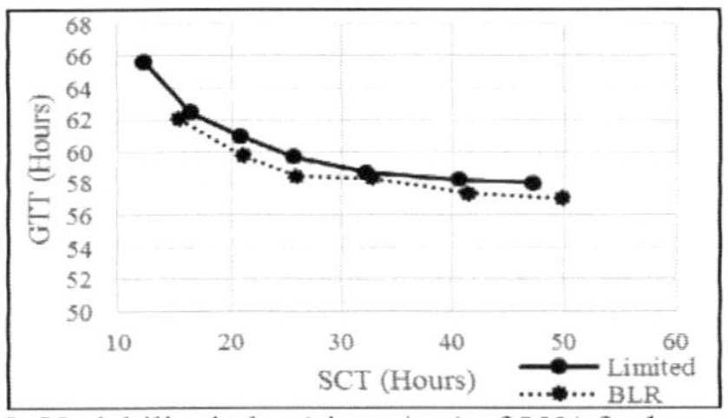

Figure 2. Variability index (sigma/mu) of 80% for low precision (Class 3) processing time.

Figure 3 shows the difference in performance between the two models at high precision level (Class 7). The performance of Limited model is similar to that obtained with actual processing times. But for BLR model, the performance is little bit different at tighter workload norms with that at the actual processing times. The trend of the curves are more or less similar to that in Figures 1 and 2. Both the models perform much better at high precision level than at low precision level. Moreover, there is not significant differences in throughput times for both the models with actual processing times as well as with high precision level. From Figures 2 and 3, it is revealed that BLR model performs better than Limited model when processing times can be estimated accurately at the planning stage. This answers our second research question.

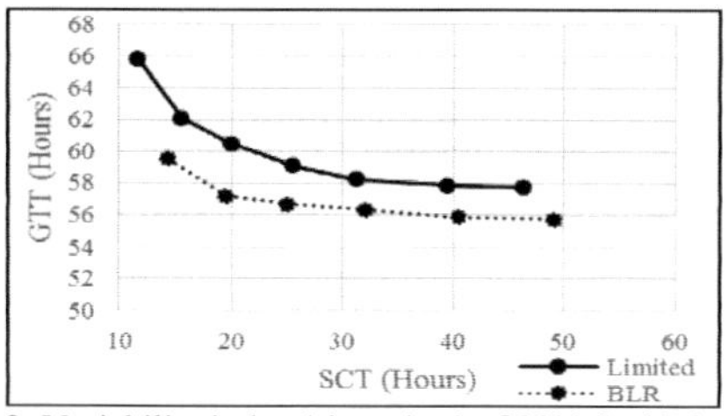

Figure 3. Variability index (sigma/mu) of 80% for high precision (Class 7) processing time.

Order release decision solely depends on the processing times [29]. Both BLR and Limited models depend on the total load, in terms of processing times. Thus any wrong prediction in processing times of different orders will degrade the performance of BLR model. Therefore, BLR model performs better than Limited model only when the processing times can be estimated correctly at the planning stage.

5 Conclusion

This paper looks at the processes of the MTO companies from the point of the internal supply chain as a whole. This gives an opportunity to look at the flow along these processes and eliminate waste, thus leaning the supply chain, and not only its separate stages. Due to variation in processing times and uncertainty in demand, it is really difficult to manage the internal supply chain which directly impacts the customer requirements. In jobshop environment, similar kind of situation is tackled by using workload ORR techniques. But few authors [7] stated the importance of workload ORR techniques in a pure flow shop environment which is suitable for high customization. MTO supply chain is similar to a pure flow shop environment. Continuing the studies of [7] on ORR systems, which proved to be effective in leaning the processes in non-repetitive manufacturing, two types of ORR systems are applied to the MTO supply chain. The Limited and BLR models are run at different workload norms in order to capture all possible situations at 18000 minutes system workload. It is seen form the simulation runs that GTT decreases with increase in SCT, which is quite logical. At actual planning situation, the BLR model performs much better than the Limited model, particularly at tighter workload norms. But in reality it is very difficult to predict the processing times of different orders beforehand. Therefore, processing times are estimated at low and high precision levels and both the models are run at different workload norms. It is seen that the BLR model performs better than the Limited model only when the processing times can be estimated more closely to the actual data. As both BLR and Limited models depend on the load (processing times in this case), the accuracy of the load plays an important role in the performance of ORR systems in MTO supply chains. Therefore, the important contribution of this paper is that it is quite reasonable for the MTO companies to use the workload ORR systems to improve the performance of their internal supply chain but there should be some way to predict the data as close as possible to the original situation in order to obtain the maximum benefits.

This paper is purely theoretical and is based on statistical distributions and generalized model. Though the results are statistically significant, the processing times in the companies might have different distributions and thus have different effects on the performance. But the findings of this paper create a strong base for further research on the applicability of ORR systems to different aspects of MTO supply chain. Here the study is concerned only with the pure flow along the internal supply chain with simple configuration but future studies can concentrate on other factors like the parallel processing of orders, multiple resources, transportation, etc. It will be quite interesting to see how the simulation model could be adapted to the real case. Therefore, another suggestion for further research is to collect data from a company or companies and confirm these theoretical findings empirically.

References

1. P. R. Dean, Y. L. Tu, and D. Xue, Int. J. Prod. Res., **47**, 4, pp. 1071-1087 (2009)
2. E. R. Melchert, M. A .de Mesquita, and P. G. Francischini, 12th Int. Conf. Indus. Eng. Opers. Man. (ICIEOM), Fortaleza, CE, Brasile (2006)
3. W. J. Hopp, and M. L. Spearman, Manufac. Serv. Opers. Man., **6**, 2, pp. 133-148 (2004)
4. B. G. Kingsman, and A. Mercer, Inter. Trans. Oper. Res., **4**, 1, pp. 251-257 (1997)
5. C. Hicks, C. F. Earl, and T. McGovern, IEEE Trans. Eng. Man., **47**, 4, pp. 414-423 (2000)
6. C. Hicks, T. McGovern, and C. F. Earl, Inter. Int. J. Prod. Econ., **65**, pp. 179-190 (2000)
7. A. Portioli, and M. Tantardini, Int. J. Prod. Res., **50**, 12, pp. 3257–3273 (2012)
8. M. Thurer, M. Stevenson, C. Silva, M. J. Land, L. D. Fredendall, and S. A. Melnyk, Prod. Oper. Man., **23**, 3, pp. 463-476 (2014)
9. R. Germs, and J. Riezebos, Int. J. Prod. Res., **48**, 8, pp. 2345–2360 (2010)
10. M. Thurer, M. J. Land, and M. Stevenson, Inter. Int. J. Prod. Econ., **147**, pp. 180-188 (2014)
11. S. Chopra, and P. Meindl, "*Supply Chain Management: Strategy, Planning and Operations*", second ed. Prentice Hall, Upper Saddle River, NJ (2007)
12. D. Corti, A. Pozzetti, and M. Zorzini, Inter. Int. J. Prod. Econ., **104**, pp. 536-554 (2006)
13. M. Stevenson, and L. C. Hendry, Inter. Int. J. Prod. Econ., **104**, 2, pp. 676-693 (2006)
14. M. Zorzini, D. Corti, and A. Pozzetti, Inter. Int. J. Prod. Econ., **112**, pp. 919-933 (2008)
15. M. J. Land, and G. J. C. Gaalman, Inter. Int. J. Prod. Econ., **56-57**, pp. 347-364 (1998)
16. S. A. Melnyk, and G. L. Ragatz, Int. J. Prod. Res., **27**, 7, pp. 1081-1096 (1989)
17. L. Hendry, and B. Kingsman, Int. J. Oper. Prod. Man., **11**, 6, pp. 6-16 (1991)
18. D. Bergamaschi, R. Cigolini, M. Perona, and A. Portioli, Int. J. Prod. Res., **35**, 2, pp. 399-420 (1997).
19. I. P. Tatsiopoulos, Comp. Indus., 1997, **33**, pp. 335-344 (1997)
20. I. Sabuncuoglu, and H. Y. Karapinar, Inter. Int. J. Prod. Econ., **62**, 3, pp. 259–279 (1999)
21. P. Henrich, M. Land, and G. Gaalman, Inter. Int. J. Prod. Econ., **90**, pp. 187-198 (2004)
22. M. Land, Inter. Int. J. Prod. Econ., **40**, pp. 196-209 (2006)
23. W. Bechte, Int. J. Prod. Res., **26**, 3, pp.375–395 (1988)
24. R. Cigolini, and A. Portioli-Staudacher, Prod. Plan. Contr., **13**, 7, pp. 602–613 (2002)
25. H. P. G. Van Ooijen, Inter. Int. J. Prod. Econ., **56–57**, pp. 661–675 (1998)
26. Land, M. J., M. Stevenson, M. Thurer, and G. J. C. Gaalman, Inter. Int. J. Prod. Econ., **168**, pp. 257-266 (2015)
27. T. M. Choi, and T. C. E. Cheng, "*Supply Chain Coordination under Uncertainty*", Springer Science & Business Media, 14 Aug (2011)
28. P. D. Welch, "*The statistical analysis of simulation results*", in The Computer Performance Modeling Handbook, ed. S. Lavenberg, pp. 268-328. New York: Academic Press (1983)
29. M. Thürer, and M. Stevenson, Int. J. Prod. Res., DOI: 10.1080/00207543.2016.1177672 (2016)

Summary of Ammunition Support Studies under Information Condition

Guan Jicheng[1] and Li Liangchun[2]
[1] *Ordnance Engineering College, Shijiazhuang, China*
[2] *Ordnance Technology Institute, Shijiazhuang, China*

Abstract. The informatization demand of military ammunition support was analysed. The research status of ammunition containerization, RFID technology and asset visualization were introduced. The existent problems were pointed out and the future development was predicted.

1 Introduction

The essence of ammunition support is to send the right amount of ammunition to the designated locations in a timely manner by the operation of moving ammunition to complete. In the process of ammunition support, related departments need to know in real-time that what arrived, what has been sent, where is the need, where do the things delivered arrive and what do a point have. The information transfer is slow and lagging seriously rely on traditional methods of information processing. Army ammunition support is experiencing the paradigm shift from the "rough estimates" to "accurately support". It is necessary to use modern information technology to complete the work such as munitions identification, checking, statistics, monitoring, tracking and management quickly and accurately, achieving asset visualization and process controllable of ammunition support, improving rapid response capability, meeting the demand for military general ammunition supply guarantee in local wars under the information condition [1], [2]. Related research relates to three areas, ammunition containerization, RFID technology and asset visualization.

2 Research status

From the mid-90s of the last century, mainly by Ordnance Technology Institute under the guidance of the headquarters to carry out the overall planning and top-level design on ammunition support system of army forces , divisions, groups and other troops at all levels. Explicitly stating the guiding principle that base on reality and face the future, combine the peacetime and wartime and wartime based, overall optimizing and system matching, focus on quality and strive for efficiency, overall planning and step by step. Proposing the system construction goal of quick and rapid reaction, flexible and reliable operation, safe and efficient operations, timely and accurate supply, sustained and effective guarantee.

2.1 Formatting ammunition containerization areas

Ammunition containerization means based on the principles of modern military logistics, forming a certain number of bulk ammunition into a larger transport guarantee unit with advanced container technology to adapt our military mechanization and informatization leapfrog development and meet the demand for ammunition fast supply guarantee in integrated joint operations.

Lujun et al proposed the implementation of infantry ammunition matching packaging and carried out the feasibility analysis by analysing the effect of packaging on ammunition supply support and demand characteristics of infantry ammunition packaging. They described the current application and future development of ammunition packaging in US military supply support. Then they discussed the significance of the army infantry using ammunition matching packaging from three aspects. Studying at infantry ammunition matching packaging, which is in accord with the new military strategic policy, will help to improve the ability of troops to accomplish diverse military tasks. Li liangchun [3] et al analysed the positive effects of containerization on ammunition Support from three aspects and introduced three packaging method. According to the characteristics of army ammunition support, they put forward three packaging proposals, developing pallet container, paying high attention to bracket container and moderately achieving container bags. Gao Xinbao [4] et al analysed the impact of ammunition support on army combat effectiveness in terms of ammunition quality, timeliness and accuracy of the supply. And they analysed the impact of packaging structure and format on support capabilities of ammunition from aspects which are security, transport security mode and information support. In order to improve the ability of ammunition support, they proposed to improve and perfect the ammunition packaging architecture, building information security platform and expanding ammunition packaging security function. Li Tianpeng [5] et al analysed modern warfare demand for ammunition support in two aspects, ammunition storage environment effects (atmospheric environment, mechanical environment, electromagnetic environment) and ammunition supply support efficacy (transport links, information networks). They introduced the status of existing military ammunition packaging and designed and analysed ammunition container model on the protection function, supply function and information function of the ammunition container. Zhao Fanggeng [6] et al analysed the status of military supplies containerization and discussed the strategies of strengthening military supplies containerization, they suggested formulating an overall plan of supplies containerization, establishing a scientific management mode of container, developing advanced container and handing equipment, improving the information level of supplies containerization. Chen Jingsheng et al analyzed the current situation of military ammunition packaging, then pointing out that there are still some problems such as poor protective performance, inconvenient to use and low degree of standardization. Finally they proposed several countermeasures and suggestions for how to carry out ammunition packaging serialization.

2.2 RFID technology areas

Radio Frequency Identification is one of the multitudinous automatic identification technologies. It can work normally without any physical contact. RFID technology uses coupling or propagation of space electromagnetic waves for communications and it can work on the movement of objects or the immobile for data collection [7]. RFID technology started late in China which entered China for the first time in 1994. Whether RFID chip or RFID technology the research is far behind the developed countries and the level of logistics informatization is relatively low. In recent years, with economic development, China has made great progress in RFID standards development and application promotion [8]-[11].

The army also has a very clear demand for applications of RFID technology, the introduction of RFID technology in military logistics management can adapt to military logistics requirements in wartime and peacetime. The General Logistics Department and General Armament Department set up a large number of research projects in RFID and the RFID are also gradually popularized in military logistics. Yan Fengbin [12] et al introduced the basic concepts of packaging and RFID and put forward three kinds of ways of packaging. Aiming at the task-supporting way, they described the writing and revision process of identification data in four stages from supplies packaging to materials consumed. Through the application of RFID technology, it will be solved that military supplies packed identify is inferior in automaticity and difficult to modify. The automated identification of various types of packaging information will be achieved and with modification of the write terminal the precise and timely support will be realized. Cai Junfeng [13] et al elaborated the application of RFID in ammunition supply support from two aspects, ammunition storage and supply security. They explained the issues that the application of RFID faced, such as technology, standardization, confidentiality, safe use, environmental compatibility and storage. Aiming at the application of RFID in ammunition supply support, they proposed three countermeasures, strengthen the research development, unified technical standards and steadily promote pilot applications. They pointed out that informatization is the trends of ammunition support. With the development of military technology and improvement of ammunition security system the army will eventually solve the challenge that the application of RFID faced. RFID technology will play an important role in achieving the army supply support informatization. Xie Guanyou [14] et al introduced the application status of RFID technology in the military field, highlighting the related applications in the US military. By analyzing the feasibility of RFID technology application, they pointed out that hardware and software conditions are ready for extending RFID technology application on ammunition container unit and present the detailed workflow of RFID-based ammunition containerization security. Wang Wei [15] et al introduced the functional requirements of automatic identification system from the environment, life and recognition angles. Then they studied the RFID system solution from three aspects, the electronic tag choice, reader choice, software structure and requirements. Li Hongwei, et al carried out the requirement analysis on ammunition containerized security management system based on RFID technology, expounding the realization of system design from four aspects of the system block diagram, electronic tag selection, the data read and write acquisition and data sheet design. Problems the system faced with and the key technology it needs have also been analysed. They proposed the system has a good promotion prospects and application value, and has been successfully applied in an army ammunition fine management.

2.3 Asset visualization areas

Lin Congguang [16] et al discussed how to advance the construction of logistics assets visualization (LAV) in seven aspects, such as improving and perfecting the strategic planning of LAV, creating the architecture of LAV, promoting logistics system, accelerating integration development, strengthening integrated management and building the core competitiveness. Liu Yanhua [17] et al pointed out that the army total asset visibility system require achieving three visualization, supplies in use and inventory visualization, assets in transport visualization and assets in dealing with visualization. They described the related visualization system to be built. They also elaborated the development approach of Total Asset Visibility, such as adhering to the unified leadership and unified planning, insisting on unified development and unified standards, insisting on open architecture and coordinating stable structure and the balance of environmental change. Yang Zhengshu [18] et al analyzed the army current problems of military supplies support and introduced the advantages of Total Asset Visibility. They said there are four problems of Total Asset Visibility construction need to solve: construction of military logistics network information platform, construction of materials information source data acquisition system, construction of military material transport information tracking and control system and development of common materials management software system. They also described the techniques of each problem and pointed out several considerations of Total Asset Visibility. Liu Hua'an [19] et al discussed the guiding ideology of the army JTAV construction, the construction target (resource visualization, security integration, rapid support, accurately control) and JTAV construction principles. They designed JAVA technology system reference model and JAVA system architecture reference model. Xuan Zhaolong et al introduced ammunition packaging informatization in brief and pointed out the key technology involved are automatic identification and location tracking. They analysed the composition and operation of ammunition packaging information system, explaining how to achieve better ammunition packaging information from four aspects of information standardization, electronic tags, reading and writing equipment and management systems, putting forward that ammunition packaging design should be grounded in becoming the carrier of information network and the realization of ammunition support whole process visualization should base on establishing visual system in each theater.

3 Problems and trends

Overall, the military ammunition support still has the following issues under the new period of informatization warfare conditions:

- The lesser extent of army ammunition containerization is greatly reducing the efficiency of ammunition guarantee.
- RFID technology has just started and standardization has not been unified.
- The information security technology study of ammunition support process is not deep enough.

The military ammunition support should seize the opportunity of informatization, strengthen the construction of ammunition containerization and strengthen the research and application of RFID technology and related information systems, forming the standards and specifications in line with the actual needs of the army in order to achieve the information automatic identification and rapid transfer of ammunition support as soon as possibly [20].

4 Conclusions

Modern wars under conditions of informatization require efficient ammunition storage and supply security. The achievement of informatization is the development trend of ammunition support through information technology. RFID technology with its unique characteristics of non-contact identification has broad prospects for development in ammunition storage management and visible support of ammunition supply. Although currently there are still some problems and challenges in the application of RFID technology in military ammunition storage security, with the development of technology and the improvement of military ammunition storage security system, the problem of the application of RFID in ammunition storage security will definitely be satisfactorily resolved. And RFID technology will play an important role in the army ammunition support.

References

1. Li Liangchun, Xie Guanyou, Pu Zhigang, Tang Bo. Research on Visible Support of General Ammunition Packaging Unit Based on RFID [J]. Logistics Sci-Tech, (2009), **02**:92-95
2. Tian Runliang, Yin Peng, Ren Jie, Zhai Junwei. Application of Container Ro-Ro Pallets in Military Material Containerization Support [J]. Containerization, (2010), **04**:24-26
3. Li Liangchun, Yu Chengguo, Jiao Yang, Xie Guanyou. Development of Ammunition Container Packaging for the New Military Reform [J]. Packaging Engineering, (2007), **11**:8-10
4. Gao Xinbao,Li Tianpeng. Analysis of Influences and Countermeasures of Ammunition Packaging on Army Support Capability [J]. Packaging Engineering, (2011), **23**:154-156
5. Li Tianpeng,Fu Xiaozhong,Xu Haitao Study on Ammunition Integrated Packaging Mode of Storage and Transportation Integration [J]. Packaging Engineering, (2011), **23**:23-25
6. Zhao Fanggeng, Shi Xiaoyan, Strategy Analysis of Enhancing Military Material Containerization Construction [J]. Packaging Engineering, (2013), **09**:127-130
7. Zhang Jun,Mei Zhonghao.Logistics packaging and application Based on the Internet of Things Technology [J]. Packaging Engineering, (2014), **17**:135-139
8. Luo Yanping.Research on Military Logistics Supports Technology RFID Based [D]. Chongqing University, (2014)
9. Zhang Xuechun, Chen Yu, Zhou Dingru, Li Jialiang. Discussion on Application of RFID in Military Product Package [J]. Packaging Engineering, (2007), **12**:66-67
10. Pang Ming. Internet of Things Bar-code Technology and RFID Technology [M]. Beijing, China Supplies, (2011)
11. Zhang Meng. Discuss the Intelligent packaging and RFID technology [J]. Guangdong print, (2011), **05**:47-49
12. Yan Fengbin, Liu Zhenyu, Zhang Baohua. Application of RFID Technology in the Identification and Recognition of Military Material Packaging [J]. Value Engineering, (2015), **03**:160-161

13. Cai Junfeng, Gao Xinbao, Fu Xiaozhong. Research on Key Problems and Countermeasures of RFID Application in Ammunition Reserve and Supply Support [J]. Logistics Engineering and Management, (2012), **12**:63-65
14. Xie Guanyou, Li Liangchun, Xing Yang. The Application of RFID Technology in Ammunition Containerization Support [J]. Journal of Sichuan Ordnance, (2009), **02**:101-103
15. Wang Wei, Xuan Zhaolong, Cheng Ze, Li Jingming, Zhang Xiaoyi. Automatic Recognition System Lised in Ammunition Storage and Transportation Shelter Based on RFID [J]. Packaging Engineering, (2014), **11**:91-95
16. Lin Congguang, Chu Wei. Actively Promoting the Construction of Military Logistical Assets Visualization [J]. Journal of Logistical Engineering University, (2006), **04**:74-79
17. Liu Yanhua, Sun Shaofang, Zhang Lianfeng. The Development Conception and Approach of Army Total Asset Visualization System [J]. Military Economic Research, (2012), **06**:40-41
18. Yang Zhengshu, Yan Hongchao. Study on Military Logistics Total Asset Visualization [J]. Journal of Chongqing University of Technology, (2010), **11**:40-45
19. Liu Hua'an, Yi Mingyi. The Preliminary Exploration of Army Joint Total Asset Visibility System Construction [J]. Automated Command and Computer, (2007), **02**:40-44
20. Zhao Jimin, Chen Wenge, Guo Baohua, Liu Zhenhua. Requirement Analysis of Military packaging Based on Modern Military Logistics [A]. The Thirteenth National Packaging Engineering Conference Proceedings, (2010):**4**

MATEC Web of Conferences **81**, 06005 (2016) DOI: 10.1051/matecconf/20168106005
ICTTE 2016

Efficient Messaging through Cluster Coordinators in Decentralized Controlled Material Flow Systems

Christian Lieberoth-Leden[1], Daniel Regulin[2] and W. A. Günthner[1]

[1] *Technical University of Munich, Institute for Materials Handling, Material Flow and Logistics, Munich, Germany*
[2] *Technical University of Munich, Institute of Automation and Information Systems, Munich, Germany*

Abstract. The modularization of the hard- and software is one approach handling the demand for increasing flexibility and changeability of automated material flow systems. A control that is distributed across several different hardware controllers leads to a great demand for coordination between the modules while planning for example transports, especially if there is a mutual dependency between the modules on the executing tasks. Short-term changes in planning often initiate a rescheduling chain reaction, which causes a high communication load in the system. In the presented approach, module clusters with a centralized coordinator are automatically formed out of multiple modules and substitutional take over the surrounding communication for the modules. As a result, they minimize exchanged messages by focusing on the essential information.

1 Introduction

Present day material flow systems are mostly operated by an individual central control. Flexibility is only facilitated within predefined limits. This is for example the case when a designated alternative route is temporarily used for a higher throughput. New demands on an existing material flow system which require flexibility not originally considered (e.g., an extension of the system) require a manual adaption of the individual control. The manual adjustment of an individual material flow control is performed by skilled employees who know the system functionality and have competencies in programming. A material flow system that provides flexibility beyond predefined limits is characterized as being convertible. Convertible material flow systems can be realized through reusable Plug-and-Play modules that configure autonomously the material flow control. Plug-and-Play material flow systems are able to be modified instantaneously to match the current demands and reduce setup costs, particularly for systems with a high demand for flexibility [1].

In this paper, we define a module as an encapsulated unit that performs predefined functions, such as conveying, buffering or identifying a transport unit (TU). A module possesses all the necessary software functions to control its hardware, to communicate with other modules or superior systems and for the purpose of autonomous self-configuration. Standardized software interfaces and property descriptions enable collaboration between heterogeneous modules, for example the combination of a conveyor with a crane. Consequently, autonomous self-controlled modules enable convertible material flow systems that can be changed during runtime.

In a convertible system that is changed during runtime not every possible material flow situation can be simulated beforehand. Flaws in the material flow like deadlocks must be avoided by planning the execution of transportation in advance. Usually, not all the transport assignments for a specific time slot and module are known in advance, as TUs can, for example, be manually placed in the system. Consequently, TUs are added to the system without a specific sequence, however execution of their transportation is planned instantly. Therefor for each transport specific time slots are exclusively reserved at the modules of the TU path. Later, when new transports are added, they could cause an inefficient planning scenario for a module.

Modules that share resources like a rail, are examples that depend on each other. Only if all the module transports for a specific time slot are initially known, can the utilization of the common resources be optimally planned. Since this is not the case here, the planned schedule must be changed afterwards to achieve optimization. But if the module schedule has to be changed, the changes with all the other modules can potentially lead to a chain reaction with a significant increase in communication traffic.

The communication capacity and computational power of a modular material flow system are technically limited. Therefore, the number of modules in a system is limited, as well as the minimum latency. Regarding a material flow system consisting of self-controlled modules, we aim for a control that plans transports in advance by reserving modules for a designated time slot in a communication efficient manner, even for modules which depend on each other.

To implement a communication efficient control, we can state the following requirements: in particular, we want to reduce the communication load for rearranging existing reservations; by means of frequent and easy rescheduling of firmed reservations we want to increase the throughput; additionally, the new control concept should not cause more traffic for regular reservations, even though they are never rescheduled; the existing individual module control for processing reservation requests should also not be changed to enable a simple implementation. In conclusion, we can only add control functions for more efficient messaging or change the path of communication between different modules.

A solution for the problems for the above-stated requirements is module clusters. The modules build up clusters with a centralized coordinator, which represents the module features to external modules. The coordinator takes over communication with the system and decreases the amount of outgoing messages by focusing on the essential information with respect to the actual planning status.

After the introduction, we provide an overview of agent systems, agent communication concepts and decentralized routing strategies. Based on the routing strategies, we describe in the following section a decentralized material flow control for convertible systems that avoids deadlocks. Afterwards, we introduce module clusters and describe how they affect the messaging between modules for regular reservations and rescheduling by using the transversal carriage system as an example. Subsequently, we evaluate the cluster concept by calculating the message load analytically and conducting a parameter study. In the last section, we summarize and discuss the utilization of cluster coordinators for decentralized controlled material flow systems.

2 State of the art

Autonomous entities that are able to execute predefined tasks such as the modules mentioned in this paper, can be realized by means of agents that control a module and communicate with other modules or systems. Vogel-Heuser stated that the utilization of such agents enables Cyber-Physical Production Systems for Industrie 4.0 applications [2]. In this paper, we use agents to realize a decentralized controlled material flow system. Therefore, we implement a specific routing and messaging strategy for the module agents.

Systems with several cooperating agents are called multi-agent systems. A multi-agent system might involve tasks that can not be performed by a single agent. In this case, multiple agents form a group in order to solve the task together; this process is called coalition forming. Another reason for a coalition is more efficient task handling, since in a group the agents aim for the absolute optimum [3]. In this paper, agents group together to facilitate task handling but we focus on their messaging strategy.

Hellenschmidt et al. have already stated, with respect to an agent system with distributed service providers, that all relevant information has to be communicated to one entity in order to make an optimal decision. Therefore, they developed a model to build a temporary communication group. The group consists of several entities, in which an entity can gather information and broadcast decisions [4]. The focus of this work is not to reduce the communication load, but to broadcast information to all relevant entities.

The focus of the present paper is the planning of transportation in an automatic material flow system. The process is divided into routing and reservation. For the routing, a path from a start to a destination is determined. Various criteria can be considered such as the transport time or the traffic frequency [5]. Then, the reservation process blocks the selected route for other TUs, so that they do not interfere with the transport. Modules only have local knowledge about the actual position of the TUs in the system. Consequently, the modules require knowledge, when to pass on a TU. This knowledge is generated through the reservation process, where for each TU the entire route is reserved in advance for the physical transport [6]. The reservation of a route leads to a lower system throughput. Other TUs are not allowed to use a reserved path till the designated transport has been completed. Other transports have to wait even though they want to use the same path in the same direction and do not interfere with the reserved transport. In this case, a time slot based reservation algorithm increases the throughput. It is not the entire route which is reserved for the whole transportation period, but only parts of the path. In our case module, for example, it refers to the actual required time for transportation [5]. For decentralized controlled modules the reservation process with its communication has already been described. The difficulty with the communication load for rescheduling module reservations that depend on each other has not yet been considered.

If modules depend on each other, the reservation process has to be coordinated with other dependent modules in order to decide, for example, on the utilization of a common resource. Chisu et al. examined different decentralized control strategies based on their throughput for a system with several transversal carriages (TC) travelling on a single rail. The best results are achieved by reserving for each transport the entire path on the rail for the expected process time [7]. Transport rescheduling is not considered, since the control only reserves the rail for actual TUs waiting to be transported and not for ones in the future.

3 Planning transports in decentralized controlled material flow systems

In this section, we focus on the functionality of a time slot based reservation algorithm in a decentralized controlled material flow system. First, we describe the process of finding the shortest path to a destination called the routing. If multiple TUs are in a system at the same time, the transports can interfere with each other causing deadlocks. Therefore, we discuss different strategies to handle deadlocks. Determined by the routing, the path time slots at the corresponding modules are reserved in order to avoid deadlocks.

Prior to the physical transport of a TU, the route has to be determined. Since routing is not a core theme in this paper, we assume that TUs always want to travel on the shortest path in terms of time. Only the modules possess knowledge about their workload at any given time, so their data is needed to determine the shortest path. Sending the total knowledge of each module in the system to a TU with a route request leads to a greater communication load. More efficient communication is possible when only the relevant modules for a route request send messages to each other. For executing transportation, a module only needs to know which neighbouring module a TU has to be transferred to. Therefore, each module possesses a table, structured as a matrix, where, for every possible destination module in the system and time, exactly one neighbouring module is listed that provides the shortest transportation time in this situation. If a module's routing table changes, the updated table is sent to all neighbours. The neighbouring modules examine for each destination module whether a shorter path is available and update their routing table with the new neighbouring module and time for a specific relation [6], [8].

When several TUs are transported at the same time in a system, they can block each others' paths, resulting in a deadlock. Strategies for handling deadlocks are prevention, detection and avoidance [9]. In material flow systems deadlocks are prevented if no deadlocks can occur due to the layout. Since the material flow system is convertible, a layout that prevents deadlocks can not be warranted. The detection and recovery of deadlocks requires a lot of communication between the modules to gather all the necessary information. Since we aim for efficient communication, this strategy is not considered any further. In conclusion, we use an avoidance strategy, where we reserve exclusive time slots for a specific TU at the modules, as explained in the following section.

The reservation process is based on the works of Mors and Mayer [5], [6]. Once a TU transport assignment has been received by a module, the reservation process starts. Each module passes the reservation request to the neighbouring module, which is listed in the module's routing table for the shortest path to the destination module. If the next module is the destination module, the reservation process is successful and a confirmation is returned. The confirmation travels the reversed route and the modules confirm the reservation.

When a module receives a reservation request, the entrance and exit time has to be determined by calculating the process time for the TU at the module. For a continuous conveyor, the processing time is the travelled length divided by the continuous speed. The present paper focuses on a TC system. A TC system is an example of several modules that depend on each other. TCs travel on a rail and can reach several entrance and exit modules. The time required to travel to the entrance prior to the transport has to be considered as well for the process time. If more than one TC uses the same rail, the reservations for the rail from the other TCs have to be considered with respect to the process time. The reason is that those reservations may require a waiting time for an open reservation slot. The earliest possible and latest allowed exit time is then passed on with the reservation request.

4 Application areas and communication of module clusters

Module clusters especially qualify for modules, which depend on each other. Here, they can simplify and reduce communication load, particularly if changes in the reservation planning process are required. In the following, we use a TC system to describe module - clusters as shown in Fig. 1. Every TC can reach every entrance and exit. For every possible destination module in the system, the TC with the path with the shortest time is listed in the routing table as neighbouring module for an entrance.

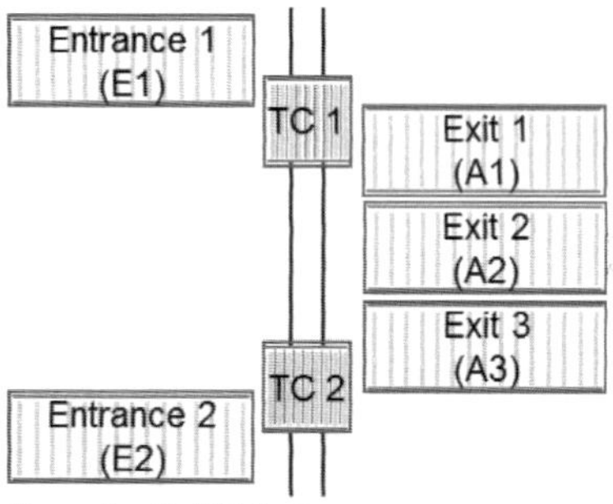

Figure 1. Example of a TC System.

The communication in a decentralized controlled TC system using a time slot reservation algorithm is shown in Fig. 2. When a reservation request is received by an entrance module it is processed and forwarded to the TC as listed in the routing table. The TC also processes and forwards the request to the corresponding exit module as stated in the TC's routing table. After the TC receives the confirmation of the reservation, the required rail segment is firmly reserved for the time slot and the confirmation is passed along to the entrance module.

Each time a TC rail segment reservation for a time slot is confirmed, this information has to be communicated to all other TCs that use the same rail. In conclusion, every TC is aware of the current reservation schedule of the rail. TCs offer transport times to the entrances for every possible entrance exit relation. When a TC offers transportation to an entrance, it calculates the travel time to the entrance, the handling time for loading a TU and the travel time to the exit module. For all those calculations the reservation schedule of the rail has to be taken into account. Each time the reservation schedule is changed, every TC has to recalculate all of their entrance-exit relations to check whether the change affects their offered transports. If transport times for entrance-exit relations change, the new transport times are communicated to the entrance of this relation. The entrance then decides whether the TC still offers the timely shortest path or has to be replaced by another TC.

 DOI: 10.1051/matecconf/20168106005

Module clusters consist of several modules which correlate in one way or another with each other. In the case of limited common resources, the modules need to follow a conflict management strategy to share the resources. In order to collaborate, the modules need a cooperation strategy to coordinate each module for offering enhanced functions.

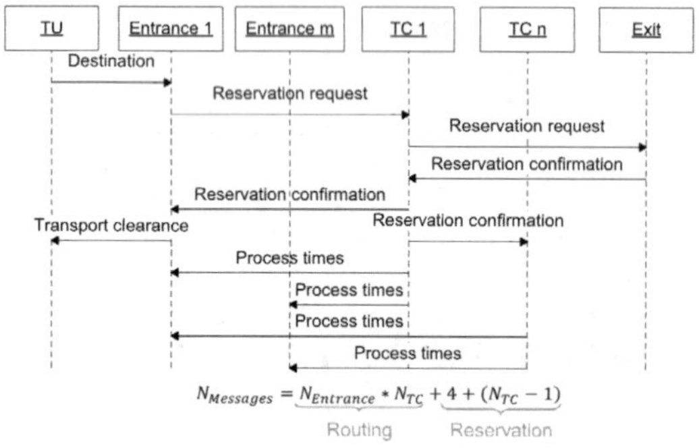

Figure 2. Communication process for a regular reservation without module clusters.

In this paper, we discuss a cluster whose modules depend on each other. We aim to simplify the communication for routing and reservations between the internal modules of a cluster and with respect to other external modules in the material flow system. A cluster consists of one central coordinator, which is responsible for the internal and external communication. The coordinator represents the cluster in the system and behaves like a regular module. The properties are derived from the sum of the represented modules in the cluster, but the coordinator brings its own control agent along. The control agent, or any other properties of the modules in the cluster, are not modified with the exception that all communications now go through the coordinator. Thus, we fulfil the requirement that no modification in a module's control is required. The communication with a coordinator is shown in Fig. 3. For each entrance module, the coordinator is now listed as neighbouring module for all destination modules.

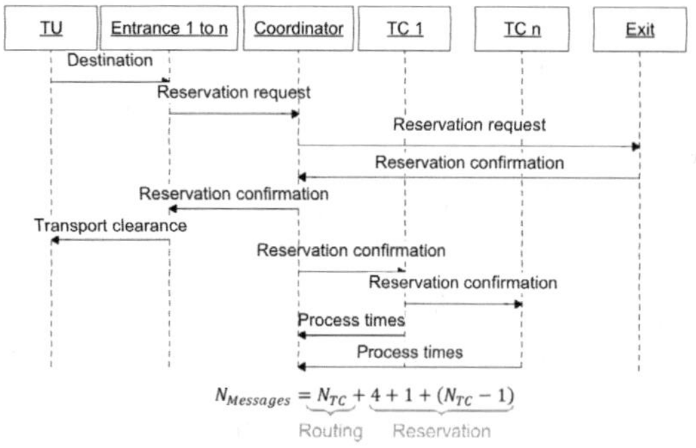

Figure 3. Communication process for a regular reservation with a cluster coordinator.

Hence, all the reservation requests for the TCs are processed by the coordinator. The coordinator can determine the transport times that are based on the expected loads and at this point does not have to assign and reserve a specific TC to a transport. The actual assignment of a transport to a specific TC is done as late as possible to minimize the risks of delays that would cause rescheduling the reservations. Shortly before arrival, the coordinator selects the TC from its routing table with the shortest transport time and forwards a reservation request to the corresponding TC. After a confirmed rail segment reservation, every TC updates its transport times for all entrance exit relations and transmits them to the coordinator, which again updates the routing table.

In the following we describe the rescheduling process with the TC system introduced in Fig. 1. Fig. 4 shows the reservation schedule for the rail over time for both TCs with respect to confirmed transport assignments. The shaded area represents a new reservation request that was added at short notice. Unfortunately, the request slightly overlaps a confirmed reservation from TC 1, so it has to be shifted to the next free time slot in the future. However, in terms of system capacity, a more efficient solution is to slightly shift the predecessor transports to an earlier time slot.

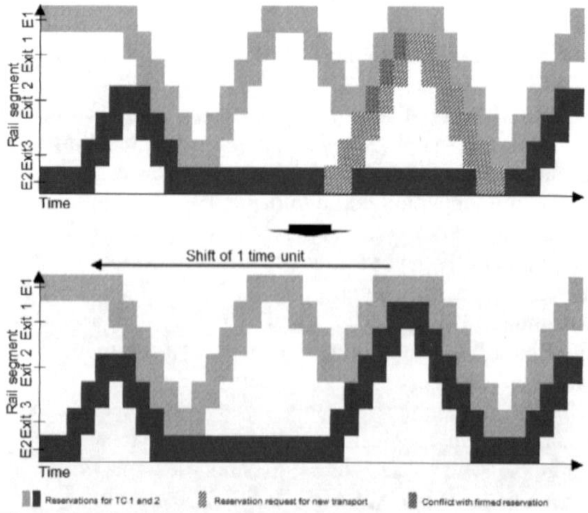

Figure 4. Schedule of confirmed rail segment reservations over time for a TC system and a new reservation request.

In the case of rescheduling, a request has to be sent to all affected modules, as shown in Fig. 5. The notification procedure works as follows: TC 2 requests TC 1 to reschedule one transport. TC 1 detects that, in order to shift the requested transport, two different TC 1 transports and one TC 2 transport have to be moved as well. If the TC reschedules a transport, the corresponding entrance has to be queried whether the transport can start earlier or later. Subsequently a rescheduling request is sent to the affected entrances of TC 2 and TC 1 which then also sends a request to the entrance.

The entrances determine the available buffer for the time slots and inform the requesting modules. The requesting modules also determine the buffer and pass the information along to the module which first sent a request. In the present case, TC 2 decides whether rescheduling is possible and then sends a rescheduling confirmation to all affected modules. In conclusion, rescheduling of firmed reservations can easily trigger a chain reaction of changes that cause a large amount of message traffic in the system.

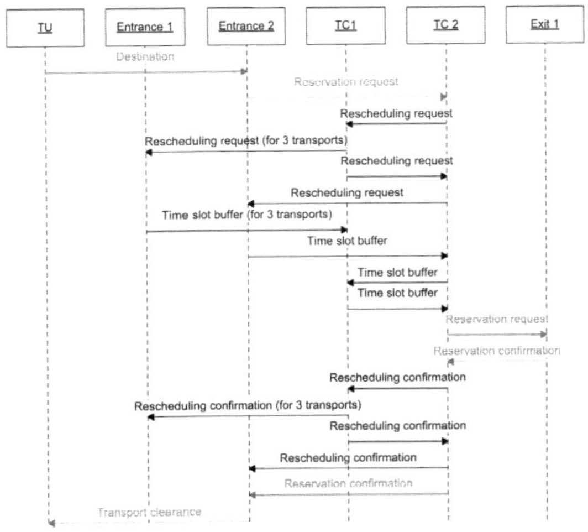

Figure 5. Communication process for rescheduling reservations in the TC system.

An informed coordinator that plans rescheduling and subsequently provides only the affected modules with necessary information is able to reduce the communication load. Since the cluster coordinator assigns transports to a TC just before the arrival of the TU, it only needs to know whether sufficient transport capacity is available. On this basis, new transports are planned and rescheduled with no need to communicate with TCs at this point. The cluster coordinator fulfils the requirement to cut the communication load for rescheduling. Since the coordinator does not use confirmed reservations, but rather estimates, the planning accuracy decreases. For our system, we always expect a high demand for rescheduling, since the modules randomly receive transport assignments. Furthermore, we can now efficiently utilize optimization potential, since transports can easily be rearranged to minimize empty running of TCs. For this purpose, we accept a reduced planning accuracy, since the probability that transportation is conducted as initially planned is minor. However, we still use confirmed reservations for the physical transport to avoid collisions, but we reserve the arrival of the TU at the TC system just beforehand.

5 Analytical evaluation of the communication load of module clusters

The communication load for a reservation and rescheduling process in a TC system with or without a cluster can be analytically calculated, which is shown in the following sections. Disturbances such as an incorrect broadcast of a message are not considered.

The number of broadcasted messages can be analytically calculated in correlation to the number of entrances and TCs. The formulas shown below in Fig. 2 and 3 apply to a successful routing and reservation process. We use both equations to derive another equation that we can use to calculate the difference in the communication load if a cluster coordinator is present, as shown in (1).

$$\Delta N = N_{TC} * (N_{Entrances} - 1) - 1 \qquad (1)$$

By using only one entrance, the coordinator has no effect, and the number of messages stays the same since it acts exactly like a single entrance. With an increasing number of entrances, the coordinator reduces the communication load, because it represents communication for more and more modules. Consequently, the cluster coordinator fulfils the requirement not to cause any more communication, even in the case there is no need.

The number of broadcasted messages can be calculated with the formula shown in Fig. 6. In the first line of the formula, the number of messages that are sent by the TC are calculated, which itself originally sent a change request. In the second line, the number of messages broadcasted by other TCs will be calculated. Those TCs have to reschedule their transports, because of a change request from another TC.

$$N_{messages} = 3 \cdot \underbrace{\sum_{m=1}^{N_{Entrances}} N_{T,Res,m,m}}_{\text{Own entrances}} + \underbrace{3 \cdot N_{TC,Res}}_{\text{Other TCs}} + 3 \cdot \underbrace{\sum_{n=1}^{N_{TC}} \left[\underbrace{\sum_{m=1}^{N_{Entrances}} N_{T,Res,n,m}}_{\text{Own entrances}} + \underbrace{N_{TC,Res,n}}_{\text{Other TCs}} + \underbrace{\sum_{k=1}^{N_{TC}} \sum_{l=1}^{N_{Entrances}} N_{T,Res,n,k,l}}_{\text{Entrances of other TCs}} \right]}_{\text{Other TCs}}$$

Symbols:

$N_{T,Res(m,n,k,l)}$: *Number of affected transports from rescheduling (for TC n,k / Entrance m,l)*

$N_{TC,(Res)}$: *Number of TCs (affected from reschudeling)*

Figure 6. Formula for the calculation of the communication load with respect to rescheduling in the TC System.

We simplify the complex formula with the assumption, that each TC only sends a change request for a single transport. Furthermore, we evaluate the module cluster with two representative scenarios, for which we can simplify the formula further. In both scenarios, a TC sends a rescheduling request to all other TCs.

- In scenario 1, the required new schedule of each TC does not interfere with the transports of other TCs.
- In scenario 2 the required new schedule of each TC interferes with the transports of other TCs. Therefore, each TC additionally sends a rescheduling request to all other TCs.

Finally, all of the TCs send a change request to the corresponding entrances of their affected transports. Starting from the entrances, a message with the time slot buffer is passed back to the module that originally requested rescheduling. After deciding on a rescheduling option, the confirmation is finally distributed to the affected modules in the end. For scenario 1, we can simplify the formula and replace the formula shown in Fig. 6 with 2, and for scenario 2, we can use 3 accordingly.

In a parameter study, we can vary the parameter number of entrances and TCs to observe the effects on the number of broadcasted messages for a change request. The results are shown in Fig. 7. If we double the number of entrances and TCs, the number of messages nearly quadruples for scenario 1 and increases eightfold for scenario 2. In the case of a TC system with a cluster coordinator, the number of messages stays the same. In conclusion, the cluster communicator fulfills the requirement to cut down on the communication for modules that depend on each other.

$$M_{Messages}=3 * N_{Entrances}+3 * (N_{TC} - 1) * (1 + N_{Entrances}) \quad (2)$$

$$M_{Messages}=3 * N_{Entrances} * (1 - N_{TC}) - 3 * N_{TC} + 3 * N^2_{TC} * (1 + N_{Entrances}) \quad (3)$$

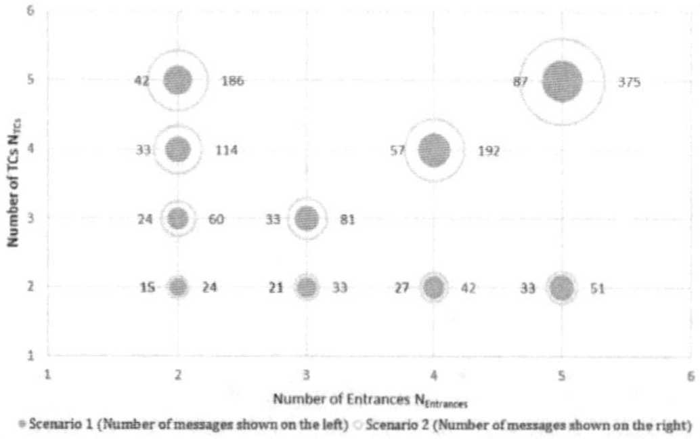

Figure 7. Results of the parameter study to observe the correlation between the number of entrances and TCs on the communication load.

6 Conclusion

In this paper we introduced convertible heterogeneous material flow systems that are realized by making use of autonomous software and hardware modules. We discussed different options to ensure deadlock free transportation in a decentralized controlled system. A routing concept with a time slot based reservation algorithm, as used in previous works for similar systems, appeared to be the best option to prevent deadlocks. In heterogeneous material flow systems, modules might depend on each other, because they use the same resources. Delays, early arrivals and new transport assignments lead to a demand for rescheduling confirmed reservations. We discovered that, in this case, a time-slot-based reservation may cause a very high communication load, especially for modules that depend on each other.

For reducing the communication load we introduced the module cluster with a centralized coordinator. The coordinator takes over the communication from the modules in the cluster. We compared the conventional communication process for each regular reservation and rescheduling process to the process with a cluster coordinator. We used a transversal carriage system as an example system, which is generally representative for modules that depend on each other. We analytically evaluated the number of broadcasted messages in a transversal carriage system. We then stated formulae to calculate the number of messages for each a regular reservation process and rescheduling, each with or without a cluster. In a parameter study, we showed the effect of module clusters depending on the complexity of the cluster. We assessed that module clusters have the greatest effect on cutting message traffic in the case of rescheduling firmed reservations.

Module clusters represent a hybrid solution to an either completely decentralized or centralized material flow control. The concept of a decentralized control for convertible material flow systems has been combined with selective advantages of a central control and provides a greater knowledgebase and reduced demand on coordination. Further studies will analyse automatic formations and additional application areas of module clusters.

Acknowledgment

This research was supported by the German Research Foundation (DFG) as part of the project "Gesteigerte Flexibilität in heterogen aufgebauten Materialflusssystemen auf Basis intelligenter Softwareagenten in selbstkonfigurierender Fördertechnik - iSiKon (Increased flexibility in heterogeneous material flow systems with intelligent software agents for self configuration)".

References

1. K. Furmans, F. Schönung, K. R. Gue, International Material Handling Research Colloquium, *Plug-and-Work Material Handling Systems* (2010)
2. B. Vogel-Heuser, J. Lee, P. Leitão, at - Automatisierungstechnik **63**, *Agents enabling cyber-physical production systems*, 777-789 (2015)
3. O. Shehory, S. Kraus, Artificial Intelligence **101**, *Methods for task allocation via agent coalition formation*, 165–200 (1998)
4. M. Hellenschmidt, R. Wichert, *Self-organization: Things with autonomous cooperation – a system analysis - Selbstorganisation: Dinge in eigenverantwortlicher Kooperation - eine Systemanalyse*, Internet der Dinge, Berlin, (2007)
5. A. W. t. Mors, *The world according to MARP - Multi-Agent Route Planning*, Technische Universiteit Delft, (2010)
6. S. H. Mayer, *Development of a completely decentralized control system for modular continuous conveyor systems*, Karlsruher Institut für Technologie, (2009)
7. R. Chisu, F. Kuzmany, W. A. Günthner, Fachkolloquium der Wissenschaftlichen Gesellschaft für Technische Logistik **5**, *Decentral coordination and cooperation in the internet of things - Dezentrale Koordination und Kooperation im Internet der Dinge*, 35-37 (2009)
8. W. A. Günthner, *Internet of things in the intralogistics - Internet der Dinge in der Intralogistik*. Heidelberg, (2010)
9. E. G. Coffman, M. Elphick, A. Shoshani, ACM Computing Surveys **3**, *System Deadlocks*, 67-78 (1971)

MATEC Web of Conferences **81**, 06006 (2016) DOI: 10.1051/matecconf/20168106006
ICTTE 2016

Business Process Reengineering–A Case Study on Computer Center Service

Soontorn Phiphopsuthipaiboon and Somjai Boonsiri

Department of Mathematics and Computer Science, Faculty of Science, Chulalongkorn University, Bangkok, Thailand

Abstract. This paper used Business process reengineering (BPR) for improve operation of computer service processes in the computer center. Change of service processes were conducted by implementing four-phase: 1) Finding of current process 2) Analysis current process 3) Redesign Process 4) Applying new process and realization framework to support redesign decision making. The framework based on policies and rules of organization. Finally, the new process was implemented at the computer center. The results showed that the new process was better than the current process, time of service decrease and reduce workload.

1 Introduction

Business Process Reengineering (BPR) is a powerful approach to bring in extraordinary improvements in the output of an organization [1]. This approach focus on business process optimization by improvement performance such as speed, cost, shareholder value, customer orientation, quality and service [2][3][4]. BPR were used in every organization for increase satisfaction of customer to get advantages over competition company. By measure quality of service, it has many type of quality and this paper focus on cycle time and performance of service. BPR cycle can be illustrated in the following way (Fig. 1).

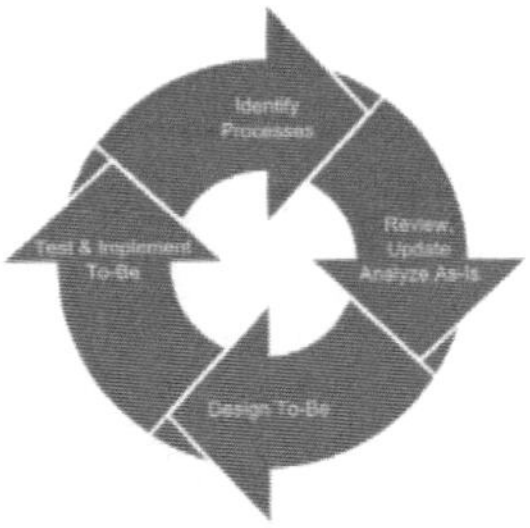

Figure 1. Business process reengineering cycle.

Process is a collection of activities across time and space from start until finish that identifies one or more kinds of input and creates an output that is of value to the customer [5], [6].

In recent years, growth in computer center service has increased not only the number of center but also the size, breadth and complexity of their operations because customer can access to complexity application online or system more than previous time. The computer center service may service many organization or may be devoted to one organization. Computer center service can become a part of every organization such as bank, hospital, restaurant or academy [7].

This paper found problems about request documentation and waiting time for computer service. Data analysis found the process of request documentation had many unnecessary steps and took a long time for assignment task process. When the problems could not solve in time. It had transformed into complicate problems like virus computer that damages on computer and infected to other computers. It affects the performance of the organization. With these reasons it necessary to created new process to solve all problems.

2 Problems statement

The problems found on this paper, show as following;
1) Request documentation process took a long time.
2) Current process has many steps.
3) Increased workload because problem is more complicate due to prolonged waiting time of service.

3 Framework

This paper created framework (Fig. 2) for support redesign decision making [8], as following;
1) Duration time on each step recommended by leader of computer service center, who responsible for effective and efficient functioning of that particular process. He should understand the step in entire process and be able to predict how any proposed changes might affect both the process and organization. By Data analysis from the current process found each period waste time at least 1 hours. So the step took a time over 1 hours called suspected unnecessary step.

2) Identify suspected unnecessary step which considering fact and impact in computer service process.
3) Verify unnecessary step with policy and rules of organization. The step covered by policy or rules of organization then changed status to necessary step.
4) Verify all necessary steps by 3 executives of computer center. When the step could be changed then called unnecessary step.

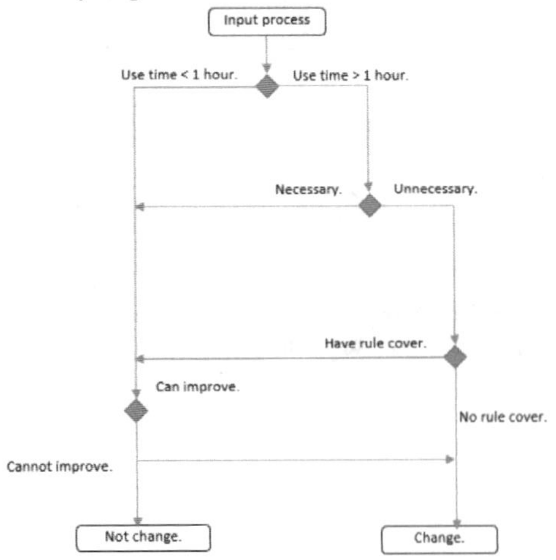

Figure 2. Framework for support redesign decision making.

4 Method

4.1 Finding current process.

Identification of current process and basic information which necessary for create the new process. This paper the researcher had interviewed leader, reviewed old worksheet and observed the behaviour of customers and officers in current process.

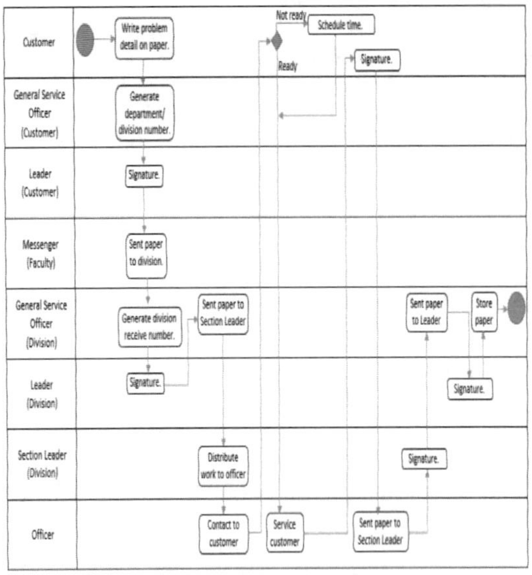

Figure 3. The current business process model.

From Fig. 3, it was divided in 4 phase. The first was documentation stage to request service. The second was assignment stage for distribute work to officer. The third was work stage to resolve problem and the fourth stage was verification stage for verify the job was done correctly. The current process had 8 related persons and each person had role in process as follows:

- Customer – the person who sent request documentation and verified the process performance.
- General Service officer (customer) – the person who sent request documentation to leader.
- Leader (customer) – the person who approved request documentation.
- Messenger – the person who sent request documentation to general service officer (computer center division).
- General Service officer (computer center division) – the person who received request document and coordinator between customer, leader, section leader and officer.
- Leader (computer center division) – the person who is assign task to section leader and verify when the task success.
- Section leader (computer center division) – the person who distributed task to officer and guarantee the task of officer were properly.
- Officer – IT support.

4.2 Current process analysis.

The fishbone (Ishikawa) diagram identifies many possible causes and problem [9] about delay service.

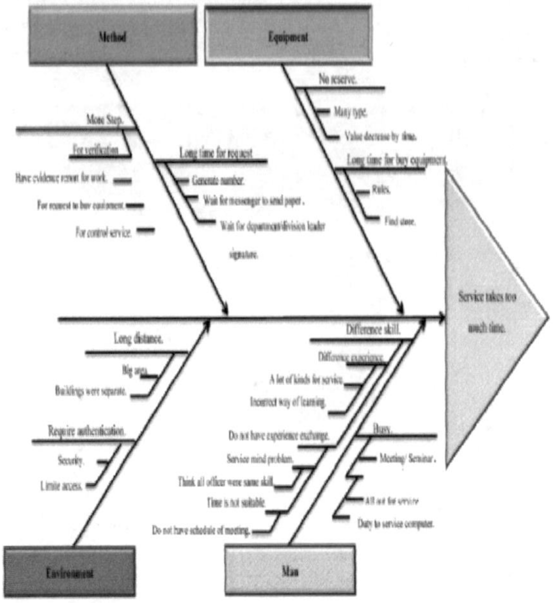

Figure 4. Fishbone diagram.

From Fig. 4 the fishbone diagram define any possible cause specific problems of computer service, which took a long time as following;

Method are divided into 2 cause that are the method has a lot of step and waiting time in request documentation.

Equipment could not reserve such as hard disk, power supply, mouse, keyboard etc. And procurement process has many steps.

Environment, this problem were restrictions on the area such as distance, security and limit access for each department

Man, officer had difference skill and they could not job rotation.

The information analysis can be summarized as follows: 1) Unnecessary processes were several steps 2) Certain regulatory inducing waiting time. Solution: 1) Unnecessary processes elimination 2) Modify processes for reduced waiting time 3) Modify the regulatory restrictions on current processes.

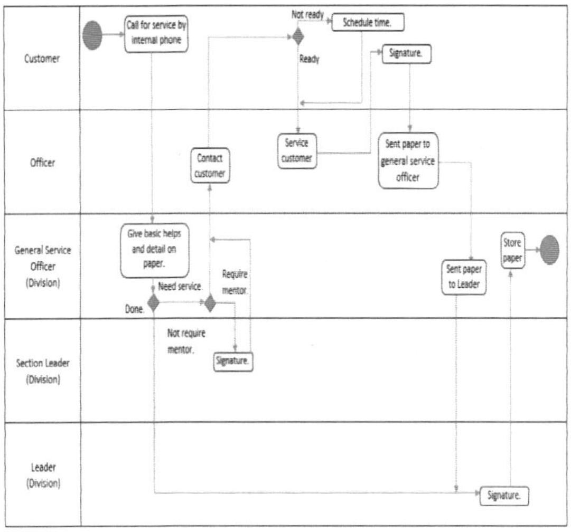

Figure 5. New business process model

4.3 Redesign process

This paper uses a framework for decision making process at each step in the current process. The results have led to creation of the new process. (Fig. 5)

From Table 1 showed status of the step on the framework analysis, each step used framework for decision making, as showed in Fig. 2. The result showed 10 unnecessary steps; 9 steps could be change but only step could not change because this step cover by policy. A necessary could be change for reduce waiting time. Finalize the new process was create with many method like eliminate, create, separate, combination and parallel process, as show in Fig. 5

The new process could reduce related person from 8 to 5 peoples and reduce the step in process from 17 to 11 steps. The new business can describe by person role as following;

- Customer – the person who were call by internal phone to request service.
- General Service officer (computer center division) – the person who were call center, wrote request documentation and sent documentation to officer.
- Leader (computer center division) – the person who verified the final step.
- Section leader (computer center division) – the person who mention officer for complexity job.
- Officer – IT support.

Table 1. The framework of current process

No.	Process tasks	Time using over 1 Hr. (Y/N)	Necessary (Y/N)	Rules cover (Y/N)	Improve (Y/N)	Result	Process owner
1	Write problem detail on paper.	No	-	-	No	Not change	Customer
2	Generate department/ division number.	Yes	No	No	-	Change	General Service Officer (Customer)
3	Signature.	Yes	No	No	-	Change	Leader(Customer)
4	Sent paper to division.	Yes	No	No	-	Change	Messenger(Faculty)
5	Generate division receive number.	Yes	No	No	-	Change	General Service Officer (Division)
6	Signature.	Yes	No	No	-	Change	Leader(Division)
7	Sent paper to Section Leader.	Yes	No	No	-	Change	General Service Officer (Division)
8	Distribute work to officer.	Yes	No	No	-	Change	Section Leader(Division)
9	Contact to customer.	No	-	-	Yes	Change	Officer
10	Schedule time.	No	-	-	No	Not change	Customer
11	Service customer.	Yes	Yes	-	Yes	Change	Officer
12	Signature.	No	-	-	No	Not change	Customer
13	Sent paper to Section Leader.	Yes	No	No	-	Change	Officer
14	Signature.	Yes	No	No	-	Change	Section Leader(Division)
15	Sent paper to Leader.	Yes	Yes	-	No	Not change	General Service Officer (Division)
16	Signature.	Yes	Yes	-	No	Not change	Leader(Division)
17	Store paper.	Yes	No	Yes	No	Not change	General Service Officer (Division)

4.4 Applying new process

The new process was implemented. This paper focus on the service processes and uncontrolled individual factors.

5 Result

The quantity of computer service between the current process and the new process showed 172 working times and 166 working times respectively. The numbers of service in both scenarios were similar, as showed in Table 2

Table 2. Quantity of computer service.

Type of service	Current Process (working times)	New Process (working times)
Software service.	119	93
Network service.	31	42
Hardware service.	22	31
Total	172	166

Table 3. The waiting time for request documentation (minute/working times) between the current process and the new process

Type of service	Current process (minutes)	New process (minutes)
Software service.	1234	30
Network service.	1254	30
Hardware service.	1003	30
Average	1163	30

From data analysis found the process of request documentation had many unnecessary steps and took a long time for assignment task process. After implementation, it reduced waiting time for request documentation 1133 min/working times, as showed in Table 3

Table 4. The Cycle time for service processes (day/working times) between the current process and the new process

Service types	Current process (days)	New process (days)
Software service.	2.92	1.51
Network service.	2.68	1.17
Hardware service.	2.10	1.41
Total	2.57	1.36

From Table 4 presented cycle time for service processes was 2.57 day/sample for current process and 1.36 day/sample for new process. The new process reduced cycle time for service processes 1.21 day/working times

Table 5 presented an impact of time lags was that the effect of the operation did not work as planned, which had rework. The number of rework was 18 times (10.47%) in current process and 2 times in new process (1.20%).

Table 5. Quantity of rework between the current process and the new process

Service types	Current process (sample)	New process (sample)
Software service.	14	2
Network service.	4	0
Hardware service.	0	0
Total	18	2

6 Conclusion

This paper used BPR cycle to improve the process and presented framework for support redesign decision making, which created new process that suitable to environment of organization.

The new process of computer service center is better than the current process by resulting in 3 items of the performance improvement. Because the new process reduced 6 unnecessary steps especially in request documentation that took a long time (Table2). The request documentation was changed to call center system that reduced time 1133 minutes/sample (Table 3) compliance with Hammer and Champy study. Stewart [10] claimed that Bell Atlantic reduced the time of new telecommunication circuits installation from 16 days to just hours. Finally, we could optimized the task assignment due to reducing the rework.

Summary, after applying BPR. It was successful to decreased cycle time and waiting time as Sung J. Shim study [11] and Doomun study [12]. When implement on organization the result going on the same way according to Chou, Ghannouchi and Zhang study [13] [14] [15].

References

1. S. Mohapatra, Business Process Reengineering Automation Decision Points in Process Reengineering Springer, (2013).
2. U. G. Seebacher, Cyber Commerce Reframing: The End of Business Process Reengineering? Springer, (2002).
3. D. J. Elzinga, T. R. Gulledge and Chung-Yee Lee., Business process engineering: advancing the state of the art Springer, (1999).
4. M. Hammer and J. Champy, Reengineering the Corporation: A Manifesto for Business Revolution(Nicholas Brealey Publishing), (2001).
5. M. Hammer and J. Champy, Reengineering the Corporation: A Manifesto for Business Revolution (Harperbusiness), (1993).
6. T.H. Davenport, Process Innovation: Reengineering Work Through Information Technology (Harvard Business School Press), (1993).
7. R.S. Roussey, THIRD-PARTY REVIEW OF THE COMPUTER SERVICE CENTER., Journal of Accountancy, **146,** (1978).
8. M. Abdous, Towards a Framework for Business Process Reengineering in Higher Education, Journal of Higher Education Policy and Management, **33** (2011).
9. D. E Lighter, D. C. Fair, Principles and Methods of Quality Management in Health Care, Aspen Publishers, (2000).
10. T.A. Stewart, Re-Reengineering. The Hot New Management Tool (Fortune), (1993).
11. R.M. Doomun and N.V. Jungum, Business process modelling, simulation and reengineering:call centres, Business Process Management Journal, **14,** (2008).
12. S. J. Shim and A. Kumar, Simulation for emergency care process reengineering in hospitals, Business Process Management Journal, **16,** (2010).
13. Y.C. Chou, B.Y. Chen and others, Prescription-filling process reengineering of an outpatient pharmacy, Journal of Medical Systems, **36,** (2012).
14. S.A. Ghannouchi, K. Mabrouk and S. Ghannouchi, Proposal of data warehouse in the context of healthcare process reengineering, Business Process Management Journal, **16,** (2010).
15. Zhang, R. J. Jiao, Q. Ma, Accountability-based order fulfillment process reengineering towards supply chain management, Journal of Manufacturing Technology Management, **21**, (2010).

Projects Delay Factors of Saudi Arabia Construction Industry Using PLS-SEM Path Modelling Approach

Ismail Abdul Rahman[1], Nashwan Al-Emad[2] and Sasitharan Nagapan[1]

[1]*Faculty of Civil and Environmental Engineering, Universiti Tun Hussein Onn Malaysia, 86400 Parit Raja – Batu Pahat, Johor, Malaysia*
[2]*Sariya Contracting Company Ltd, Novotel Business Park, Tower 2, 3rd Floor, Dammam,-Khobar Highway, Post Box: 4486, Al Khobar 31952, Kingdom of Saudi*

Abstract. This paper presents the development of PLS-SEM Path Model of delay factors of Saudi Arabia construction industry focussing on Mecca City. The model was developed and assessed using SmartPLS v3.0 software and it consists of 37 factors/manifests in 7 groups/independent variables and one dependent variable which is delay of the construction projects. The model was rigorously assessed at measurement and structural components and the outcomes found that the model has achieved the required threshold values. At structural level of the model, among the seven groups, the client and consultant group has the highest impact on construction delay with path coefficient β-value of 0.452 and the project management and contract administration group is having the least impact to the construction delay with β-value of 0.016. The overall model has moderate explaining power ability with R^2 value of 0.197 for Saudi Arabia construction industry representation. This model will able to assist practitioners in Mecca city to pay more attention in risk analysis for potential construction delay.

1 Introduction

Structural Equation Model (SEM) using Partial Least Square (PLS) technique is also known as PLS-SEM path modelling. It is a second generation of multivariate statistical analysis which simultaneously evaluates multiple variables. SEM comprises of two types that are covariance based SEM known as CB-SEM and PLS path modelling known as PLS-SEM. Principally, PLS-SEM is used for developing a structural relationship model while CB-SEM is for confirming the structural relationship of the developed model [1]. PLS-SEM path model consists of two components. First component is structural model also known as inner model and the second component is measurement models called outer models. Outer model displays the relationships between the independent variables and manifest variables while the inner model displays the relationships between the independent and dependent variables. Variables also called constructs are not directly measured while manifests also called indicators or items are the directly measured proxy variables that contain the raw data [2].

2 PLS-SEM model

A hypothetical model of delay factors for Saudi Arabia construction industry was constructed where it consisted of 37 delay factors (manifests/indicators/items), 7 groups of the factors (independent constructs/variables) generated through factor analysis technique and one dependent variable which is the construction delay. These delay factors were identified through rigorous literature view and verify by experts in pilot study. The model displays the concept/theory with its key elements (i.e. constructs) and cause-effect relationships (i.e. paths) [2, 3].

2.1 Model's data

Input data for the model were derived from the questionnaire survey involved 100 respondents who are working in Mecca city construction industry. They rated each of the 37 factors using 5-point likert scale on degree of significance toward construction delay. These yielded 3700 data used to develop the PLS structural model for causative factors of construction delay in Mecca city. These data are recorded in *sav* format of excel software and then converted into the .csv format by using (*save As*) function in the software. Once the data is converted, the following step is to create a new project sheet in SmartPLS software for developing a model based on the hypothetical model and the converted (.csv) data file is imported into the model.

2.2 Model's construction

Construction of the model in SmartPLS software involved several steps and amongst the important steps are constructing the model according to the hypothetical model, assigned name of constructs (groups), connecting the independent variables/groups with dependent variable (construction delay) and assigning the manifests(factors) to the respective independent variables. Once all the manifests are assigned to the respective constructs then the colour of the constructs will change from red to blue and the manifest displayed in yellow colour.

Constructed model comprised of 7 groups (independent variables) for the 37 delay factors (manifests) where the collected data were assigned to them. These 7 groups are converged to one dependent variable. The dependent variable needs input data with single item that is either each of the factors are causing delay or otherwise (1 or 0 respectively). Prior to this modelling, it was determined that all the factors are causing delay which means value of 1 for each factor was applied to the model. Based on this information, the model was constructed as Fig. 1.

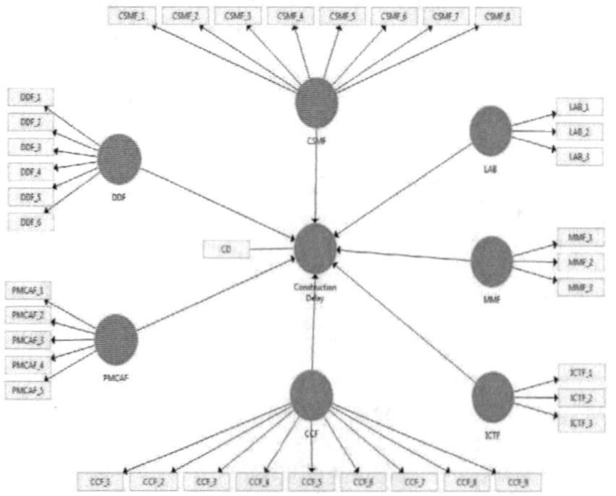

Figure 1. The developed PLS model.

Based on Fig. 1, each of the 7 independent variables consists of several factors/manifests which are as clarified below:

1) CSMF is contractor's site management related factor group which consisted: CSM1 *for Poor qualification of staff assigned to the project,* CSM2 *for Shortage of professionals in organization*, CSM3 *for Inadequate Contractor's experience*, CSM4 *for Inaccurate estimation of project duration during the bidding stage*, CSM5 *for Incompetent subcontractors*, CSM6 *for Delay in preparation of shop drawings*, CSM7 *for Delay in preparation and submission of materials samples* and CSM8 *Difficulties in financing project by contractor*.
2) CCF is client and consultant related factor group which consisted: CCF1 *is Delay in approving of change orders,* CCF2 *is Frauds practice among the parties involved,* CCF3 *is Inadequate consultant experience,* CCF4 *is Lack of coordination with authorities,* CCF5 *is Slow decision-making process,* CCF6 *is Difficulty in accessing the site,* CCF7 *is Delay in obtaining permits from Municipality,* CCF8 *is Delay in approving shop drawings* and CCF9 *is Delay in progress payment.*
3) DDF is design and documentation related factor group which comprised: DDF1 *is Mistakes in design documents*, DDF2 *is Changes in design documents*, DDF3 *is Late in approving of design documents*, DDF4 is *Inadequate details provided in drawings*, DDF5 is *Delays in producing design documents*, and DDF6 *is Insufficient data collection and survey before design.*
4) PMCAF is project management and contract Administration related factors group which consisted: PMCAF1 *is Poor contract management*, PMCAF2 *is Unrealistic contract duration*, PMCAF3 Poor *site management and supervision*, PMCAF4 *is Improper planning and scheduling of the project* and PMCAF5 is *lack of teamwork.*
5) MMF is material and machinery related factor group which comprised: MMF1 *for late procurement of materials*, MMF2 *for Delay in materials delivery* and MMF3 *for Delay in equipment delivery.*
6) LAB is labour related factors croup which consisted: LAB1 *for Low productivity level of labour*, LAB2 *for Unqualified workforce* and LAB3 *for Shortage of manpower.*
7) ICTF is the information and communication related factor group which comprised: ICTF1 is *Poor communication between parties*; ICTF2 *is Poor coordination between parties* and ICTF3 *is Ineffective monitoring and controlling of the project progress.*

After the model has developed, simulations on the model are carried out using PLS algorithm function for calculating the loadings on each manifest as well as estimating the structural model's path coefficients. The software generates various required parameters for model assessment as shown in Fig. 2.

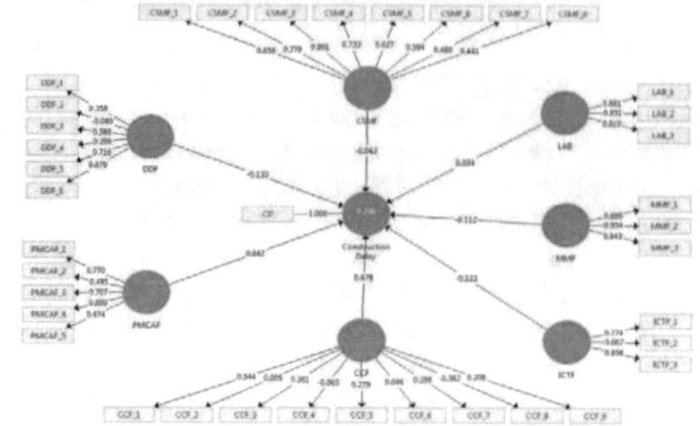

Figure 2. Results from simulation process.

Fig. 2 shows the generated values for assessment at two stages which are at measurement level for internal consistency to ensure the adequacy of the relationships between independent and manifest variables and also at the structural model for examining the relationship between dependent and independent variables. Required assessments for the measurement and structural model were carried out on accordance to [4], [5] suggestions.

3 Assessments on measurement model

Assessment of measurement model (outer model) is to check internal consistency of the model whether relationships between independent and manifest variables are adequate. It is carried out in two stages; first stage is checking the model performance after each iteration of computation (*Individual item reliabilities and convergent validities*) and the second stage is the analysis on the model's final iteration (*discriminant validities*).

3.1 Model's performance

After each simulation process, the measurement model needs to be assessed for its convergent validity and individual item reliability values which were generated concurrently. Three parameters used to determine convergent validity are Average Variance Extracted (AVE), Convergent Validity (CR) and Cronbach's alpha. Required threshold values for the parameters are Average Variance Extracted (AVE) $\geq$ 0.5, Convergent Validity (CR) $\geq$ 0.7 and Cronbach's alpha $\geq$ 0.7 [6, 7, 8]. For individual item reliability, each manifest is considered significant if its loading value is more than 0.5 [1, 4]. However, if any of manifests having loading value less than 0.5, it should be omitted and dropped. Then the computational process and model performance are reassessed in several iterations until it achieved the stated values for convergent validity and individual item reliability. For this study 6 iterative processes were carried out before reaching reliability for all the factors. A total of 11 weak factors/manifests were omitted and left 26 significant manifests which are reliable for the final outputs. After iteration 6, the required threshold values for the average convergent validity are Average Variance Extracted (AVE) =0.658, Composite Reliability (CR) =0.870 and Cronbach's Alpha (Alpha) = 0.820 [6, 8]. This means that the model has achieved the required validity process. The generated parameters values for the 7 groups generated from the final iteration as shown in Table 1.

Table 1. Convergent validity of measurement model.

Group/ Construct	Convergent validity parameters		
	AVE ($\geq$ 0.5)	CR ($\geq$ 0.7)	Alpha ($\geq$ 0.7)
Material and Machinery Related Factors	0.867	0.951	0.929
Contractor's Site Management Related Factor	0.553	0.879	0.842
Design and Documentation Related Factor	0.569	0.839	0.784
Information and Communication Technology Related Factors	0.654	0.847	0.827
Labour Management Related Factors	0.778	0.913	0.861
Client and Consultant Related Factors	0.543	0.822	0.757
Project Management and Contract Administration Related Factors	0.640	0.841	0.739
Average	**0.658**	**0.870**	**0.820**

Table1 presents the convergent validity values for all the groups/independent variables. All the values are above the threshold and these indicate that the measurement model has achieved convergent validity. Since all the manifests attained loading factor equal or more than 0.5, then the measurement model has achieved individual item reliability. Thus, the measurement model had achieved the consistency of all the manifests toward the construction delay.

3.2 Discriminant validity assessment

Next step of measurement assessment for PLS-SEM model is to assess discriminant validity which reflects the extent to which a construct is truly distinct from other constructs by empirical standards [4]. It involves Cross-Loading and Average Variance Extracted analysis [4, 6, 8] which are as follows:

3.2.1 Cross loading analysis

It is to examine that each manifest variable should have higher loading correlation with their relative independent variable. Specifically, an indicator's outer loading on the associated construct should be greater than all of its loadings on other constructs [4]. Generated values of this model found that each manifest variable is higher in their relative independent variable than other independent variables as in Table 2. This indicates that the model has achieved its discriminant validity.

Table 2. Analysis of cross-loading factors.

Factor	Independent variables/construct						
	CCF	CSMF	DDF	ICTF	LAB	MMF	PMCAF
CCF1	0.926	0.337	0.383	0.115	0.213	0.345	0.126
CCF3	0.754	0.375	0.397	0.344	0.345	0.307	0.358
CCF5	0.584	0.394	0.430	0.226	0.297	0.298	0.316
CCF7	0.637	0.516	0.318	0.158	0.288	0.602	0.234
CSMF1	0.351	0.863	0.265	0.466	0.502	0.552	0.457
CSMF2	0.374	0.795	0.227	0.484	0.442	0.402	0.355
CSMF3	0.316	0.814	0.219	0.452	0.431	0.428	0.337
CSMF4	0.504	0.740	0.358	0.276	0.429	0.434	0.275
CSMF5	0.438	0.629	0.411	0.333	0.442	0.506	0.358
CSMF6	0.216	0.577	0.384	0.354	0.325	0.303	0.414
DDF1	0.517	0.351	0.599	0.192	0.275	0.201	0.387
DDF3	0.455	0.224	0.777	0.164	0.348	0.287	0.454
DDF5	0.333	0.320	0.822	0.252	0.373	0.283	0.446
DDF6	0.349	0.371	0.797	0.123	0.388	0.259	0.259
ICTF1	0.295	0.378	0.274	0.774	0.364	0.296	0.608
ICTF2	0.305	0.299	0.240	0.667	0.316	0.165	0.597
ICTF3	0.170	0.553	0.166	0.958	0.543	0.318	0.574
LAB1	0.245	0.450	0.411	0.463	0.881	0.266	0.436
LAB2	0.295	0.586	0.459	0.552	0.951	0.473	0.568
LAB3	0.342	0.441	0.346	0.380	0.810	0.576	0.368
MMF1	0.378	0.474	0.293	0.332	0.438	0.895	0.363
MMF2	0.425	0.596	0.329	0.325	0.439	0.954	0.367
MMF3	0.404	0.506	0.323	0.298	0.476	0.943	0.342
PMCAF1	0.211	0.379	0.399	0.511	0.427	0.235	0.771
PMCAF3	0.375	0.469	0.399	0.445	0.403	0.258	0.712
PMCAF4	0.137	0.410	0.365	0.607	0.469	0.371	0.905

3.2.2 Average Variance Extracted (AVE) analysis

It is to check the correlation amongst the independent variables for adequate discriminant validity. It is also known as the Fornell-Larcker criterion and it is the second and more conservative approach to assess discriminant validity of PLS-SEM model. It compares the square root of the AVE values with the independent variable correlations. Specifically, the square root of each construct's AVE should be greater than its highest correlation with any other construct [4]. The correlations amongst independent variables for this model generated from this analysis are shown in Table 3. This indicates that the diagonal correlation values of the independent variables are higher than non-diagonal values and this satisfies the discriminant validity of the tested model.

Table 3. Analysis of Average Variance Extracted (AVE).

	Independent variables / Construct						
	CSMF	**DDF**	**LAB**	**MMF**	**PMCAF**	**CCF**	**ICTF**
CSMF	0.743						
DDF	0.401	0.754					
LAB	0.569	0.466	0.882				
MMF	0.576	0.342	0.482	0.931			
PMCAF	0.511	0.464	0.537	0.379	0.800		
CCF	0.451	0.466	0.324	0.435	0.267	0.737	
ICTF	0.550	0.222	0.541	0.335	0.658	0.238	0.808

4 Structural model assessments

Once the groups' manifests are reliable and valid, the next step is to assess the results of structural model. This involves examining relationship between dependent variable with independent variables. It is carried out by checking the strength of impact path of independent variable to the dependent variable and also the explanatory power (R^2).

4.1 Impact strength

Impact strength of variables toward to the dependent variable is assessed using path coefficients or β-value [1, 9, 10]. A structural model can be considered acceptable if all the β-value is above 0.1 [4]. According to [8], the higher β-value shows the stronger impact of a predictor independent variable on the dependent variable.[4] mentioned that the path coefficients have standardized values between -1 and + 1, where the estimated path coefficients close to + 1 represent strong positive relationships (and vice versa for negative values) that are almost always statistically significant. The closer the estimated coefficients are to 0, the weaker the relationships. And the very low values close to 0 are usually non-significant. According to [4], the signage of β-value whether it is positive or negative is not a concern because the impact of path is the absolute value of the β-value. The results of structural model obtained from SmartPLS are presented in Fig. 3.

Fig. 3 presents the path co-efficient (β values) of each path which was found as -0.223 for CSMF, -0.157 for DDF, -0.016 for PMCAF, 0.452 for CCF, -0.116 for ICTF, -0.168 for MMF and 0.060 for LAB. These path co-efficient values of the model substantiates that CCF (Client and consultant group) has the highest co-efficient value of 0.452. However the PMCAF (Project Management and Contract Administration group) is having the smallest co-efficient value of -0.016 which indicates that PMCAF group is giving the least impact to the construction delay. Thus, it can be concluded that Client and consultant group of factors has the strongest impact toward the construction delay.

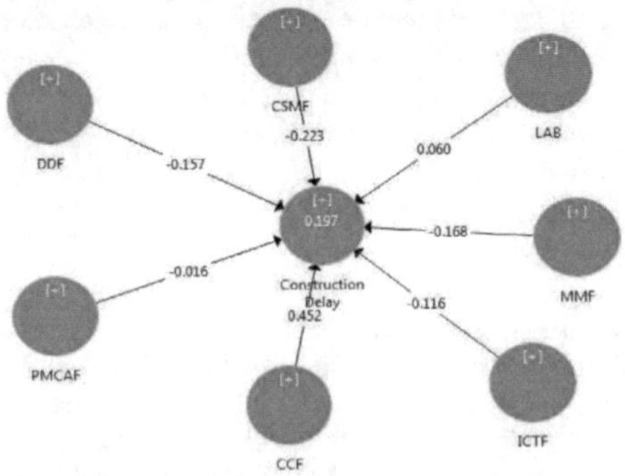

Figure 3. Path coefficients values (β) of structural model.

4.2 Explanatory POWER

Explanatory Power is to examine the overall ability of the model in representing the impact of independent variables toward the dependent variable. [4] stated that the most commonly used measure to evaluate the structural model is the coefficient of determination (R^2 value). The indicator used to describe the explanatory power is the model's coefficient of determination which is (R^2) value [1, 8, 6]. According to [11], a model can be assessed as substantial if (R^2) equal or more than 0.26, moderate if (R^2) equal or more than 0.13 and weak if (R^2) equal or less than 0.02. In this study the result of developed model is as illustrated in Fig. 3 which shows that (R^2) is 0.197. This means that the developed model has moderate explaining power in representing the impact of the 7 groups of factors toward the construction delay (CD).

5 Conclusions

This study has examined 37 factors that contribute to construction delay in Saudi Arabia construction industry using PLS-SEM modelling in SmartPLS software. The developed PLS-SEM path model comprised of these factors in 7 groups which contribute to construction delay is to show graphical representation of relationships of construction delay factors. Assessment on the model found that in the outer model all the manifests in the model are reliable and valid and for the inner model, it was found that the client and consultant group is the most dominant path with β-value 0.452. In contrast, Project Management and Contract Administration group is having the least impact to the construction delay with β-value of 0.016. Finally, the overall model has moderate explaining power ability to generalize the model for Saudi Arabia construction industry representation. This model will be helpful to persons who are involved in Mecca construction industry in analysing risk for delay and also researchers in the field of construction.

Acknowledgment

The authors are grateful to University Tun Hussein Onn Malaysia for supporting this research. Furthermore, the authors are thankful to construction practitioners who are working in Mecca city for their contributions and providing comprehensive and important information.

References

1. Memon, A.H., Rahman, I.A., Akram, M., Ali, N.M., Significant factors causing time overrun in construction projects of Peninsular Malaysia, Modern Applied Science, Canadian Center of Science and Education, **8**,(4), pp. 16-28, (2014).
2. C. M. Ringle, M. Sarstedt, and E. A. Mooi , *Response Based Segmentation Using Finite Mixture Partial Least Squares Theoretical Foundations and an Application to American Customer Satisfaction Index Data*. In R. Stahlbock et al. (eds.), Data Mining, Annals of Information Systems, pp. 19-49, (2010).
3. R. Fellows, and A. Liu, *Research Methods for Construction*. 3rd ed. Wiley-Blackwell: Utopia Press, (2008).
4. J.F.Hair, G.T.M.Hult, C.M.Ringle, and M.Sarstedt, *A Primer on Partial Least Squares Structural Equation Modeling (PlS-SEM).* United State of America: SAGE Publications, pp. 95-198, (2014).
5. J. Henseler, C. M. Ringle, and R. R. Sinkovics, "The Use of partial least Squares Path modeling in International Marketing", *Advances in International Marketing*, **20**, pp. 277–319, (2009).
6. S. Akter, J. D'Ambra, and P. Ray, "Trustworthiness in mHealth Information Services: An Assessment of a Hierarchical Model with Mediating and Moderating Effects using Partial Least Squares (PLS)", *Journal of the American Society for Information Science and Technology*, **62**, pp. 100-116, (2011).
7. J. F. Hair, C. M. Ringle, and M. Sarstedt, "PLS-SEM: Indeed a Silver Bullet", *The Journal of Marketing Theory and Practice*, **19**(2), pp. 139–152, (2011).
8. A. A. Aibinu, and A. M. Al-Lawati, "Using PLS-SEM technique to model construction organizations' willingness to participate in e-bidding", *Automation in Construction*, **19**, pp. 714–724, Feb. (2010).
9. [9] W. W. Chin, *The partial least squares approach for structural equation modeling*, In: Marcoulides, G.A. (Ed.), Modern Methods for Business Research. Lawrence Erlbaum Associates, London, **295**(2), pp. 295–336, (1998).
10. N. Urbach, and F. Ahlemann, "Structural Equation Modeling in Information Systems Research Using Partial Least Squares", *Journal of Information Technology Theory and Application*, **11**(2), pp. 5–40, (2010).
11. J. Cohen, *Statistical power analysis for the behavioral sciences* (2nd ed.), Hillsdale, Lawrence Erlbaum Associates, NJ, (1988).

MATEC Web of Conferences **81**, 07002 (2016)
ICTTE 2016
DOI: 10.1051/matecconf/20168107002

Comparison between ARIMA and DES Methods of Forecasting Population for Housing Demand in Johor

Ahmad Rizal Alias, Noor Yasmin Zainun and Ismail Abdul Rahman

Universiti Tun Hussein Onn Malaysia, Faculty of Civil and Environmental Engineering.

Abstract. Forecasting accuracy is a primary criterion in selecting appropriate method of prediction. Even though there are various methods of forecasting however not all of these methods are able to predict with good accuracy. This paper presents an evaluation of two methods of population forecasting for housing demand. These methods are Autoregressive Integrated Moving Average (ARIMA) and Double Exponential Smoothing (DES). Both of the methods are principally adopting univariate time series analysis which uses past and present data for forecasting. Secondary data obtained from Department of Statistics, Malaysia was used to forecast population for housing demand in Johor. Forecasting processes had generated 14 models to each of the methods and these models where evaluated using Mean Absolute Percentage Error (MAPE). It was found that 14 of Double Exponential Smoothing models and also 14 of ARIMA models had resulted to 1.674% and 5.524% of average MAPE values respectively. Hence, the Double Exponential Smoothing method outperformed the ARIMA method by reducing 4.00 % in forecasting model population for Johor state. These findings help researchers and government agency in selecting appropriate forecasting model for housing demand.

1 Introduction

Housing sector is very significant and acts as a catalyst to a country development including Malaysia. This is because housing sector is one part of construction industry which plays vital role in economic and social growth [1]. The balance of housing supply and demand will ensure economic stability which contributes to better Gross Domestic Product (GDP) for a country. According to [2], there is significant relationship between population and housing demand where the number of housings in an area will approximately reflect the number of household's formation. This means that when there is a population growth, it will definitely affect the increasing number of households and leads to a growth in housing demand. [3] pointed out that the important thing need to be considered before forecasting housing demand is the appraisal of population growth. Hence, the population projection is important in producing a good forecasting housing demand model.

Rapid advancement of computer has contributed to the development of predication methods which able to forecast many things including population trend. However, not all these methods can forecast accurately due to their limitations. In this study, two methods which are identified suitable for predicting the population are Autoregressive Integrated Moving Average (ARIMA) and Double Exponential Smoothing. Both methods ignore independent variable in making a prediction and principally adopting univariate time series analysis for forecasting. ARIMA method has been widely used in various fields such as socio-economic, agricultural and construction. This method is most popular in forecasting univariate time series data [4, 5]. While in contrast, Exponential Smoothing method is more convenient and practical to use because the behaviour of model is easy to understand and has a low level of complexity compare with ARIMA method [6]. Hence, Exponential Smoothing had been used widely with almost 13% by industry [7]. Therefore, this paper presents a comparison between two models which are ARIMA and DES models for forecasting yearly population of Johor state for the period of 10 years.

2 Methodologies

In the first method, ARIMA model is used for the non-stationary time series data. Type of non-stationary model is discussed by Box and Jenkins, [4] and mathematically the ARIMA (p,d,q) model can be expressed as in eq.(1):

$$\phi_p(B)(1-B)^d Y_t = \theta_q(B)at \quad (1)$$

where;
ϕ_p = Autoregressive parameters.
θ_q = Moving average parameters.
d = Order of differencing
(1-B) = Yt−Yt−1
Y_t = Actual value at time period t
a_t = Unknown random errors (or residuals)

In Box-Jenkins method, it does not assume any particular pattern in the historical data of the series to be forecasted. Rather, it uses three main steps in modelling process which are; (1) identification of data; (2) parameter estimation and; (3) diagnostic checking to determine the best model to be applied for the final model. The Box-Jenkins forecast method is schematically shown in Fig. 1:

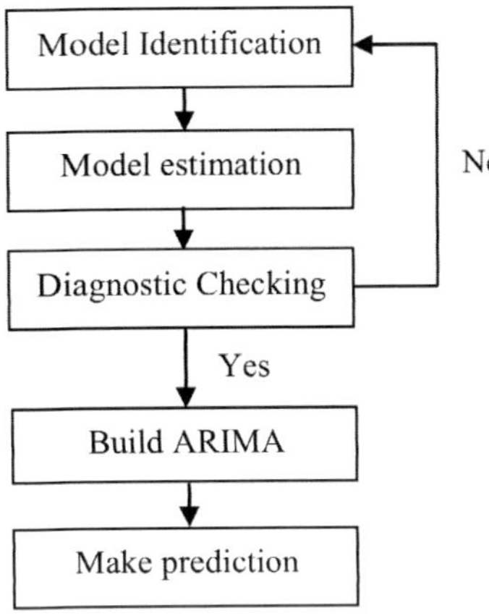

Figure 1. Box-Jenkins method for model selection [4].

Important step to make an appropriate ARIMA model is to determine optimal model parameters (p,d,q) using autocorrelation function (ACF) and partial autocorrelation function (PACF). At the initial stage, the parameters values of ACF and PACF are determined using three types of non-seasonal theoretical Box-Jenkins models as summarized in Table 1.

Table 1. Non-seasonal theoretical Box-Jenkins Models [8].

Model	ACF	PACF
Moving average (MA) of order q	Cuts off after lag q	Dies down
Autoregressive (AR) of order p	Dies down	Cuts off after lag p
Mixed autoregressive-Moving average (p,q)	Dies down	Dies down

In a second method, Double Exponential Smoothing which is a part of Exponential Smoothing method use moving average of time series data. Principally, this method use weight past technique within the data for smoothing purposes to generate smoothed data set. The smoothed data is then use for forecasting application. The following eq. (2) and eq. (3) are used in double exponential smoothing method with separate for level (Lt) and trend (Tt):

$$L_t = \alpha X_t + (1 - \alpha)(L_{t-1} + T_{t-1}) \quad (2)$$

$$T_t = \beta(L_t - L_t\text{-}1) + (1 - \beta)T_{t-1} \quad (3)$$

where;
α = Smoothing constants for level
β = Smoothing constants for trend
X_t = Current actual value
L_t = Level
T_t = Trend adjustment

In determine the best double exponential smoothing model, there are four steps to be followed which are; (1) obtain initial estimates for Lt and Tt; (2) update the estimates Lt and Tt by using some predetermined values of smoothing constants; (3) find the best combination of α and β that minimizes error (MAPE) and (4) make prediction using the followed eq. (4) :

$$Ft(k) = Lt + k.Tt \quad (4)$$

where;
Ft = Forecasting equation
k (1,2,3….n) = Value for next level of forecasting

Following stage is to apply both models with population time series data for Johor state from the Department of Statistic, Malaysia. This population data comprises of age ranged from 15 to 80+ years with an interval of five years from the period of 1970 up to 2015. However for the model 13 & 14 which used age of 75-79 and 80+, there are 10 sets of missing data from period 1970 until 1979. This data is divided into two sets which is for Training (Tn) and Testing (Ts) purposes.

Data from 1970-2010 contributed to 41 sets are used for constructing the models. While, 5 sets of data from 2011 to 2015 are used for testing process. These data sets are also used for validation of the models. In view of these, various performance measures are proposed in literature to estimate forecast accuracy and to compare different models namely Mean Absolute Percentage Error (MAPE), Mean Square Error (MSE), Mean Absolute Error (MAE), Root Mean Square (RMS), Mean Percentage Error (MPE) and sum of squared error (SSE) [9]. According to [10], the two most commonly used statistical methods for validations are RMS and MAPE. Therefore, for this study, MAPE is used for validation. MAPE is defined according to the following eq. (5):

$$MAPE = |\,PE\,| / n \quad (5)$$

where;
|PE | = actual value –forecasted value
n = number of forecast

Lewis classified models as "best" if the MAPE less than 10%, "good" with MAPE range between 10% - 20%, "acceptable" for MAPE 20% - 50% and "false" if MAPE more than 50% [11].

3 Materials and study area

Data of population for Johor state obtained from the Department of Statistics, Malaysia was used as a data input for the models. The data was separated according to age category with the range of five years. Based on the selected methods which are Autoregressive Integrated Moving Average (ARIMA) and Double Exponential Smoothing, 14 forecasting models were developed to each of the methods. Using the developed models of forecasting designed, the input data are then analysed using Software Package for Social Science (SPSS) for

Mean Absolute Percentage Error (MAPE) which was used to measure the accuracy of the models

4 Results and discussion

Comparison between two methods, ARIMA and DES aims to identify which methods are suitable to be use as a population model in Johor. Forecasting methods is chosen based on the low value of MAPE in the models where it shows that the DES has a high accuracy rate. Fig. 2-15 shows the comparison between ARIMA and Double Exponential Smoothing (DES) against actual data in term of accuracy of Training (Tn) and Testing (Ts) process.

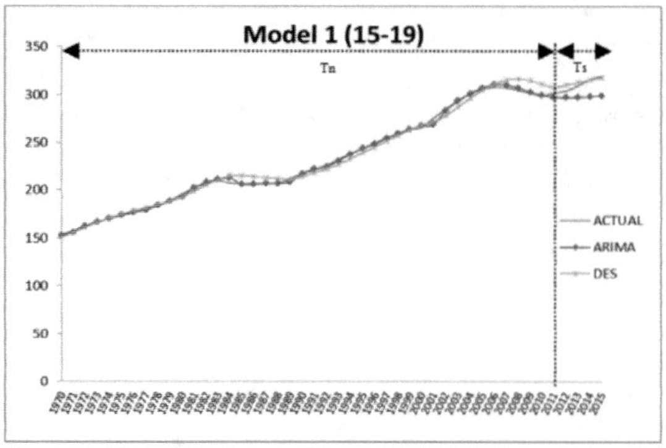

Figure 2. Tn and Ts data set for model 1 (15-19).

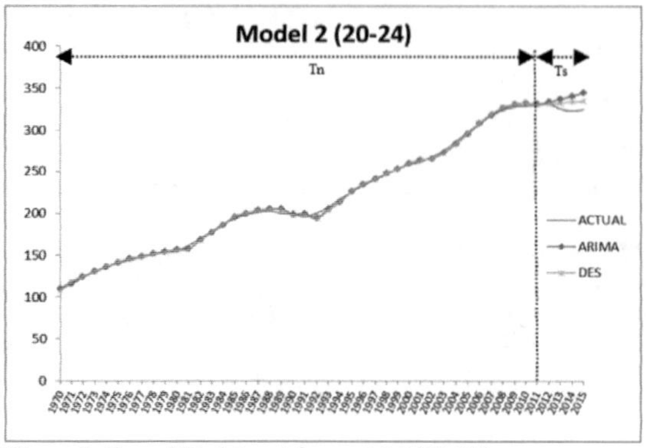

Figure 3. Tn and Ts data set for model 2 (20-24).

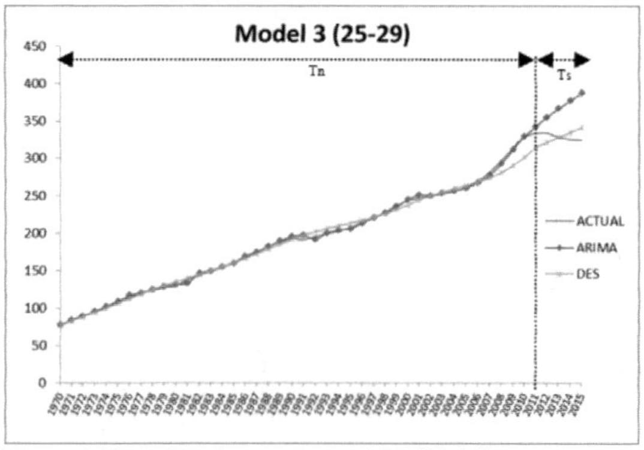

Figure 4. Tn and Ts data set for model 3 (25-29).

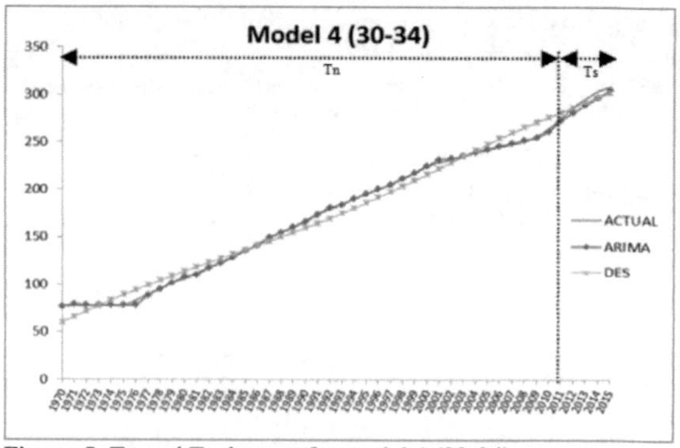

Figure 5. Tn and Ts data set for model 4 (30-34).

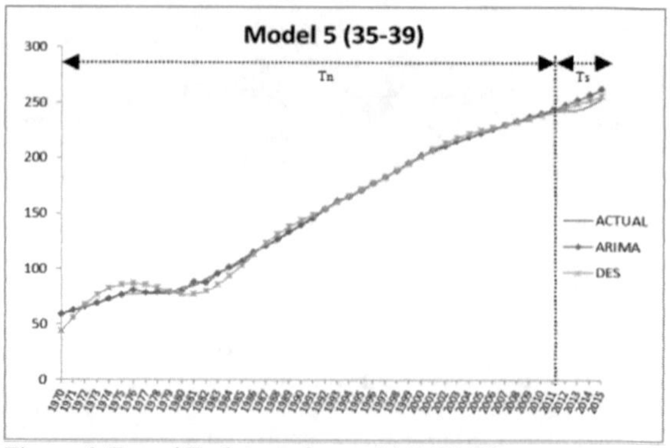

Figure 6. Tn and Ts data set for model 5 (35-39).

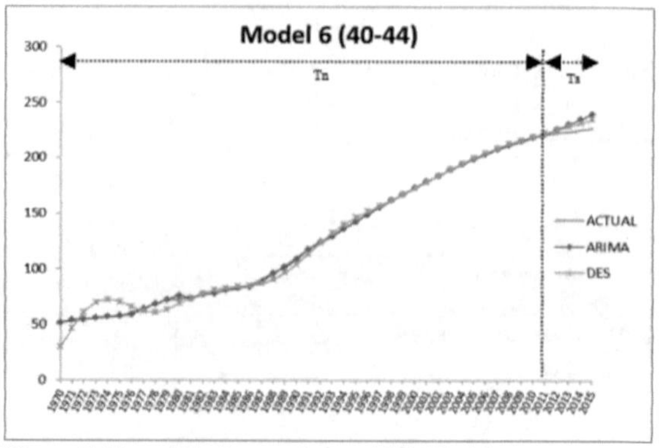

Figure 7. Tn and Ts data set for model 6 (40-44).

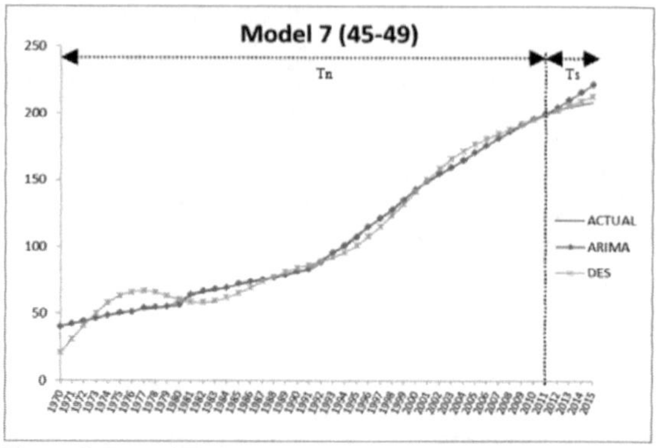

Figure 8. Tn and Ts data set for model 7 (45-49).

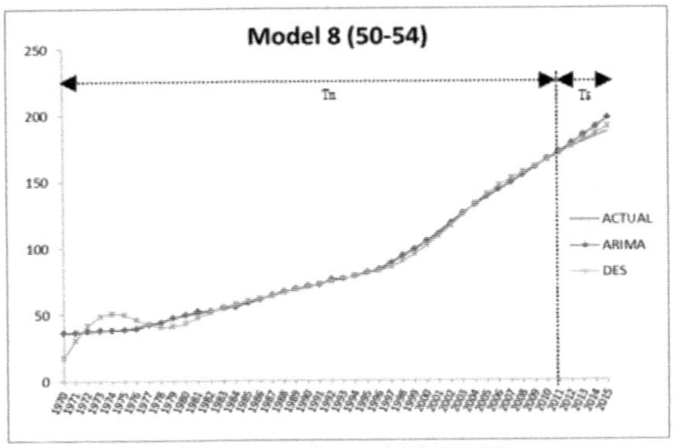

Figure 9. Tn and Ts data set for model 8 (50-54).

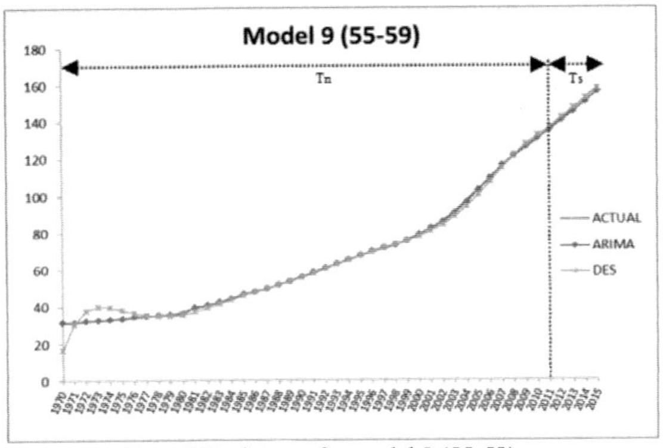

Figure 10. Tn and Ts data set for model 9 (55-59).

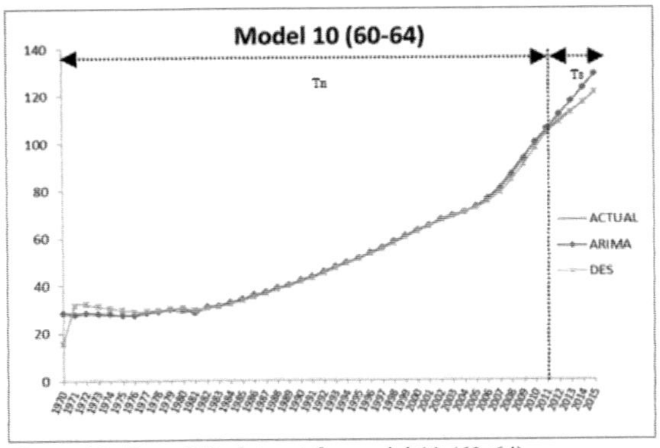

Figure 11. Tn and Ts data set for model 10 (60-64).

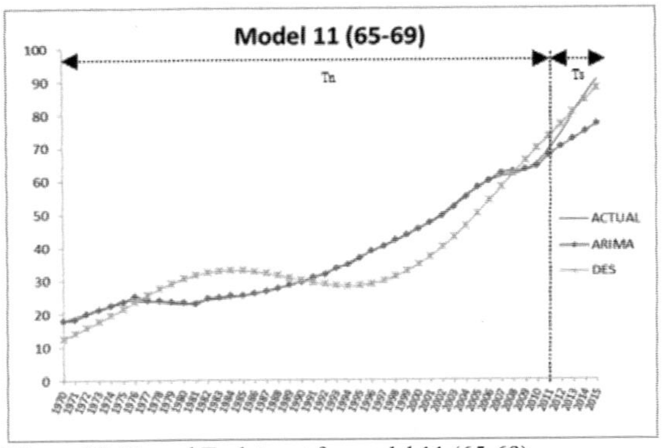

Figure 12. Tn and Ts data set for model 11 (65-69).

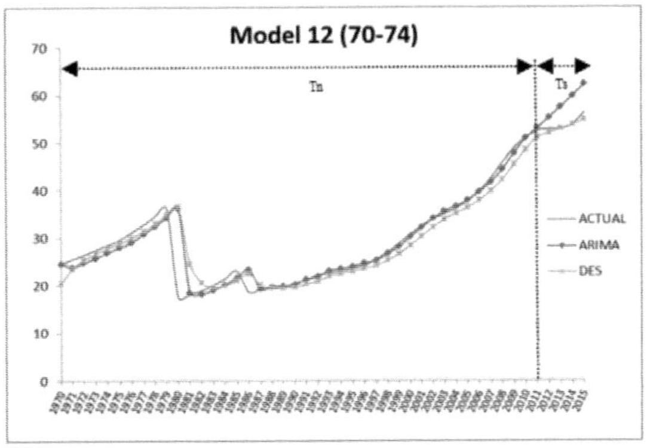

Figure 13. Tn and Ts data set for model 12 (70-74).

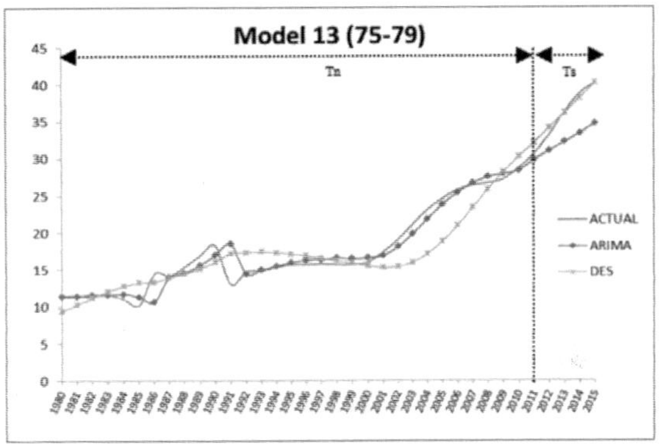

Figure 14. Tn and Ts data set for model 13 (75-79).

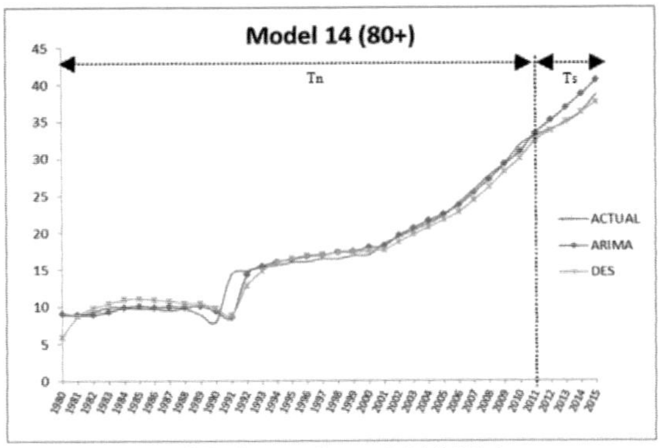

Figure 15. Tn and Ts data set for model 14 (80+).

Based on the Fig. 2-Fig. 15, obviously it shows that the DES model is more accurate for validation process rather than ARIMA model using the data set 2011-2015. In term of testing, ARIMA model is seems good for that particular part but it is not enough for the purpose of model development without accuracy in future prediction (testing). According to Lewis [11], best model development if the MAPE less than 10%, and MAPE range between 10% - 20%, is consider acceptable. In view of this, both of model is less than 20% error, but DES is superior with every single model is less than 5% MAPE comparing with ARIMA models. Table 2 shows the comparison between ARIMA and DES model in term of model parameters and error (MAPE).

Table 2. Comparison between ARIMA and DES Models.

Model No	Age group	MODEL			MAPE	
		ARIMA (p,d,q)	DOUBLE EXPONENTIAL SMOOTHING (DES)			
			Alpha (α)	Beta (β)	ARIMA	DES
1	15-19	(1,1,0)	0.5123	0.1416	3.930	1.17
2	20-24	(1,1,0)	0.9612	0.9569	3.240	1.72
3	25-29	(1,1,0)	0.2123	0.1121	11.16	3.64
4	30-34	(1,1,0)	0.0157	0.9571	1.720	1.64
5	35-39	(1,1,0)	0.2351	0.9836	2.162	1.20
6	40-44	(1,1,0)	0.2863	0.9462	2.951	1.97
7	45-49	(1,1,0)	0.2565	0.5121	2.979	0.90
8	50-54	(1,1,0)	0.2863	0.8562	2.738	0.79
9	55-59	(1,1,0)	0.5234	0.4651	0.800	0.42
10	60-64	(1,1,0)	0.9325	0.1813	3.706	0.45
11	65-69	(1,1,0)	0.0611	0.5086	9.034	2.86
12	70-74	(1,2,0)	0.6420	0.0391	11.294	1.46
13	75-79	(1,2,0)	0.1051	0.7596	9.410	1.93
14	80+	(1,2,0)	0.6092	0.1331	12.210	1.16
AVERAGE					**5.524**	**1.522**

Referring to table 2 in ARIMA method, the 14 models generated an average MAPE value of 5.524 with the lowest MAPE of 0.8 for model no. 9 (Age group 55-59) and the highest MAPE of 12.210 for the model no. 14 (age group 80+). While, in DES method, values of MAPE recorded for all models are below to 3 except for model no 3 (age group 25-29) having MAPE 3.640. In addition, every single model in DES generated lowest MAPE compare with ARIMA model. Therefore, the DES is considered a better method for modelling population for Johor state for the next 10 years. Hence, DES method performs better than ARIMA with these data.

5 Conclusion

This study has shown comparison between ARIMA and DES methods which are popularly used for linear data trend analysis. The comparison applied Johor state yearly population data sets. It was found that the most appropriate forecasting method for population of Johor state is DES method with lower MAPE for all 14 models. DES method for all models based on age group have the lower MAPE can be assumed better accurate models for forecasting population in Johor for the next 10 years.

Acknowledgement

This research is fully supported by MyBrain 15 and MOSTI. The authors fully acknowledged Ministry of Higher Education (MOHE) and Ministry of Science, Technology and Innovation for the approved fund for which makes this important research viable and effective.

References

1. I.A. Rahman, A.H. Memon, A. Azis and N. Abdullah. Modeling causes of cost overrun in large construction projects with partial least square-SEM approach: Contractor's perspective. Engineering and Technology **5**, 1963–1972, (2013).
2. C.H. Mulder. Population and housing: A two sided relationship. Demographic Research **15**, 401-412. (2006).
3. W.S Thompson. Population growth and housing demand. The ANNALS of the American Academy of Political and Social Science **190**, 131-137, (1937).
4. G.E.P Box & G. Jenkins. *Time series analysis, forecasting and control,* Holden-Day, San Francisco, California. (1970).
5. G.P. Zhang. Time series forecasting using a hybrid ARIMA and neural network model, Neurocomputing **50**, 159–175, (2003).
6. K.K. Lai, L. Yu, S.Y. Wang and W. Huang. Hybridizing exponential smoothing and neural network for financial time series predication. Computational Science–ICCS. Springer Berlin Heidelberg **3994**, 493-500, (2006).
7. G.L. Lilien, P. Kotler. Marketing decision making: A model-building approach. Marketing research **21**, 339-341, (1983).
8. B.L Bowerman, R.T. Connell, and A.B. Koehler. *Forecasting, Time Series, and Regression*, South-Western College Pub; 4th edition, (2005).
9. R.Adhikari, & R.K. Agrawal. *An Introductory Study on Time Series Modeling and Forecasting*. 21-45, (2013).
10. N.Y. Zainun. Computerized Model to Forecast Low-Cost Housing Demand in Urban Area in Malaysia Using Artificial Neural Networks (ANN). Loughborough University, (2011).
11. C.D. Lewis. *Industrial and Business Forecasting Methods*. Butterworths. **2**,194-196, (1982).

MATEC Web of Conferences **81**, 08001 (2016) DOI: 10.1051/matecconf/20168108001
ICTTE 2016

An Analysis of Environmental Dimensions Affected in Adoption of Hydrogen Fuel Cell Vehicles: A Study in Shah ALAM Industrial AREA, Selangor

Rusinah Siron[1], Tan Kwe Lu[2] and Mohd Amin Tasripan[2]

[1] *Graduate Business School, College of Graduate Studies, Universiti Tenaga Nasional, 43000 Kajang, Malaysia*
[2] *Institute of Energy Policy and Research, Universiti Tenaga Nasional, 43000 Kajang, Malaysia, College of Foundation and General Studies, Universiti Tenaga Nasional, 43000 Kajang, Malaysia*

Abstract. The aim of the study is to identify the perceptions of respondents on environmental dimensions hat affected in adoption of hydrogen fuel cell vehicles. The study was conducted at Shah Alam industrial areas of Selangor, Malaysia, with the number of respondents are 120 respondents with various job positions that related with engineering and automobiles industry. The findings of the research shows that the dimensions of HFCV Internal Environmental total score of the items statement is 3.40 with the percentage of agreement in implementation is 3.72 percent, HFCV Environmental Information Systems shows that the total score of the items statement is 3.63 with the percentage of agreement on use to great extend is 42.5 percent, HFCV Cooperation with Customers shows that the total score of the items statement is 3.81 with the percentage of agreement on implementation is 44.2 percent. The findings on HFCV Eco Design shows that the total score of items statement is 4.02 with the percentage of agreement on implementation is 42.3 percent, HFCV Environmental Organizational Culture shows that the total score of the items statement is 3.37 with the percentage of agreement is 34.2 percent, HFCV Environmental Leadership shows that the total score of the items statement is 3.34 with the percentage of agreement is 48.2 percent. HFCV Proactive Green Innovation shows that the total score of items statement is 4.10 ahead of automobile got the highest mean score of 4.32 with the percentage of agreement is 41 percent. HFCV Environmental performance shows that the total score of the items statement is 3.87 with the percentage of agreement is 39 percent and the last environmental dimensions was HFCV Environmental Risks shows that the total score of the item statement is 4.00 with the percentage of agreement is 40 percent

1 Introduction

A hydrogen fuel cell vehicle is a type of electric vehicle that uses a fuel cell instead of a battery to power the car. Hydrogen cars are basically electric cars. They have electric motors that drive the wheels just like an electric cars but the difference lies in the battery. A hydrogen fuel cell creates a chemical reaction with the hydrogen that generates electricity, water and heat. Just like an electric car, there are no harmful exhaust emissions that can affect environment. But the continuous improvement on the management and behavioural dimensions of the HFCV manufacturers and users are needed to safe and secure the environment for the long term. The initiatives of HFCV management and behavioural dimensions that consists of HFCV internal environmental management, HFCV proactive in green management, HFCV environmental organizational culture, HFCV environmental leadership and HFCV environmental performance is in need to continuously care and cautions on the impact of environment in an adoption of HFCV.

2 Literature review

In the perspective of environmental impact, hydrogen fuel cells are a clean, reliable, quiet, and efficient sources of high-quality electric power. The hydrogen use as a fuel to drive an electrochemical process that produces electricity, with water and heat as the only by-products.

In the perspectives of environmental impact of hydrogen fuel cell vehicles, there is a few advantages of using hydrogen fuel cell as follows:

1) Converting hydrogen gas into electricity produces only water and heat as a product, meaning fuel cell vehicles did not create tailpipe pollution when they are driven
2) Hydrogen fuel cell vehicles combines the range and refuelling of conventional cars with the recreation and environmental benefits of driving on electricity.

3 Research methodology

3.1 The research objective

DOI: 10.1051/matecconf/20168108001

The research objective of the study is to identify the perceptions of the respondents on environment dimensions in adoption of hydrogen fuel cell vehicles in Malaysia.

3.2 The Respondents

The study was conducted at Shah Alam Industrial Area, Selangor. Shah Alam is the state capital of Selangor, Malaysia and situated within the Petaling District and a small portion of the neighbouring Klang District. It is located about 25 kilometres west of the country's capital, Kuala Lumpur. The respondents of the study was consists of employees at managerial and operation level in related industry as follows:

Industry Category	No of industry	Respondents
▪ Transportation equipment manufacturing	7	38
▪ Automobile Industries	2	43
▪ Car sales and service centre	15	22
▪ Metal product manufacturing	3	10
▪ Machinery manufacturing	4	7

3.3 The research instrument and measurement

The instrument and measurement of the questionnaire items was by use of "five-point scale from 1 to 5". The rating constructs in this study were in following:

Measurement	Scales
1. **HFCV Internal Environmental Management** Adopted (Source: Zhu et al; Kenneth W. Green Jr and Pamela J. Zelbst, 2012, and revised the items to suit with the study requirement.	Five-point scale (1 –not considering it; 2 – planning to consider it; 3 – considering it currently; 4 – initiating implementation; 5 – implementing successfully)
2. **HFCV Proactive Green Innovation** Adopted (Source: O'Connor et al., 2006; Chen YS et al. (2012) and revised the items to suit the study requirement.	Five point Likert scale (1 – strongly disagree to 5 – Strongly agree)
3. **HFCV Environment Information Systems** Adopted (Source: Esty and Winston, 2006, Kenneth W. Green Jr and Pamela J. Zelbst 2012, and revised the items to suit with the study requirement.	Five-point scale (1 – not used at all; 5 – used to a great extent)
4. **HFCV Cooperation with customers** Adopted (Source: Zhu et al; Kenneth W. Green Jr and Pamela J. Zelbst, 2012, and revised the items to suit with the study requirement.	Five-point scale (1 –not considering it; 2 – planning to consider it; 3 – considering it currently; 4 – initiating implementation; 5 – implementing successfully)
5. **HFCV Eco Design** Adopted (Source: Zhu et al; Kenneth W. Green Jr and Pamela J. Zelbst, 2012, and revised the items to suit with the study requirement.	Five-point scale (1 –not considering it; 2 – planning to consider it; 3 – considering it currently; 4 – initiating implementation; 5 – implementing successfully)
6. **HFCV Environmental Organizational Culture** Adopted (Source: Chen, 2011, Chen et al. 2012)	Five point Likert scale (1 – strongly disagree to 5 – Strongly agree)
7. **HFCV Environmental Leadership** Adopted (Source: Chen, 2011, Chen et al. 2012)	Five point Likert scale (1 – strongly disagree to 5 – Strongly agree)
8. **HFCV Environmental Performance** Adopted (Source: Zhu et al; Kenneth W. Green Jr and Pamela J. Zelbst, 2012, and revised the items to suit with the study requirement.	Five point scale (1 – not at all; 2 – a little bit; 3 – to some degree; 4 – relatively significant; 5 – significant
9. **HFCV Environmental Risks** Adopted (Source: Sonya Graci and Jaqueline Kuehnel; downloading @ http://green.hotelscombined.com)	Five point Likert scale (1 – strongly disagree to 5 – Strongly agree)

3.4 Data analysis

Data analysis used in this study was descriptive statistics with mean score and the percentage of the agreement for the assessment of respondent perception on items measurement.

4 Research findings

The result of the findings were presented accordingly based on the following:

Section A: Respondent Demographic Profile

Job Title	Number
▪ Plant Manager	5
▪ Operation Manager	6
▪ Purchasing Manager	12
▪ Logistics Manager	8
▪ Sales Manager	15
▪ Engineering Manager	18
▪ Industrial Waste Manager	4
▪ Supply Chain Manager	6
▪ Information Systems Manager	7
▪ Executives and Administrative Officer	18
▪ Technician	21
TOTAL	**120**
Industry Category	**Number**
▪ Transportation equipment manufacturing	7
▪ Automobile Industries	2
▪ Car sales and service centre	15
▪ Metal product manufacturing	3
▪ Machinery manufacturing	4

The number of respondents involved in the study was 120 respondents. Demographic profile for respondents in Shah Alam Industrial Area, Selangor was summarized in majority of respondents as per shown in Table 1.

Table 1. Summary of demographic profile of respondents

No	Demographic Variable	Majority	Percentage (%)
1	Gender	Male	68
2	Citizen	Citizen	100.0
3	Age	31- 45 Years	65
4	Race	Malay	75
5	Religion	Muslim	80
6	Education Level	Diploma/Degree	58
7	Job Category	Administration	31.8
8	Monthly Income	RM 1,001 – RM 5,000	81.2

Section B: The Findings of Variables

1) HFCV Internal Environmental Management

The findings on HFCV Internal Environmental Management is shown in Table 2.

Table 2. The mean score of hfcv internal environmental management

HFCV Internal Environmental Management	Mean Score	% of Agreement on Implementation
▪ Commitment and support of HFCV environmental practices from top management towards	3.58	45%
▪ HFCV Environment Total Quality Management	3.45	38%
▪ HFCV Environmental compliance and auditing program	3.33	35%
▪ HFCV Environmental Program Systems	3.25	30%
▪ **Total Score**	**3.40**	**37.2%**

The findings on HFCV Internal Environmental Management shows that the total score of the items statement is 3.40 with the percentage of agreement in implementation is 37.2 percent.

2) HFCV Environment Information Systems

The finding on HFCV Internal Environmental Management is shown in Table 3.

Table 3. The mean score of HFCV environmental information systems

HFCV Environment Information Systems	Mean Score	% of Agreement on use to great extend
▪ HFCV can reduce transportation costs	4.01	52%
▪ Tracking HFCV environmental information (such as toxicity, energy used, water used and air pollution)	3.85	50%
▪ Monitoring emissions and waste production	3.77	48%
▪ Providing information to encourage choices of car vehicles by consumers	3.69	45%
▪ Improving decision making by executives by highlighting sustainability issues	3.55	40%
▪ Reducing energy consumption	3.45	38%
▪ Supporting the generation and distribution of renewable energy	3.40	35%
▪ Limiting carbon and other emissions	3.35	32%
▪ **Total Score**	**3.63**	**42.5%**

The findings on HFCV Environmental Information Systems shows that the total score of the items statement is 3.63 with the percentage of agreement on use to great extend is 42.5 percent.

3) HFCV Cooperation with Customers

The findings on HFCV Internal Environmental Management is shown in Table 4.

Table 4. The mean score of HFCV cooperation with customers

HFCV Cooperation with Customers	Mean Score	% of Agreement on Implementation
▪ Cooperation with customers for eco design of HFCV	4.11	54%
▪ Cooperation with customers for cleaner HFCV production	3.95	45%
▪ Cooperation with customers for HFCV green manufacturing	3.65	40%
▪ Cooperation with customers for using less energy during HFCV product transportation	3.55	38%
▪ **Total Score**	**3.81**	**44.2%**

The findings on HFCV Cooperation with Customers shows that the total score of the items statement is 3.81 with the percentage of agreement on implementation is 44.2 percent.

4) HFCV Eco Design

The findings on HFCV Eco Design is shown in Table 5.

Table 5. The MEAN Score of HFCV eco design

HFCV Eco Design	Mean Score	% of Agreement on Implementation
▪ Design of HFCV for reduced consumption of material/energy	4.21	48%
▪ Design of HFCV for reuse, recycle, recovery of material and/or component parts	4.01	41%
▪ Design of HFCV to avoid or reduce use of hazardous and /or their manufacturing process	3.85	38%
▪ **Total Score**	**4.02**	**42.3%**

The findings on HFCV Eco Design shows that the total score of items statement is 4.02 with the percentage of agreement on implementation is 42.3 percent.

5) HFCV Environmental Organizational Culture

The findings on HFCV Environmental Organizational Culture is shown in Table 6.

Table 6. The mean score of HFCV environmental organizational culture

HFCV Environmental Organizational Culture	Mean Score	% of Agreement
▪ The company concerns the knowledge of environmental management and protection of HFCV	3.55	45%
▪ The company concerns the collaboration of environmental management and protection of HFCV	3.40	34%
▪ The company concerns of environmental agreements of HFCV	3.30	30%
▪ The company concerns of the responsiveness of environmental management and protection of HFCV	3.25	28%
▪ **Total Score**	**3.37**	**34.2%**

The findings on HFCV Environmental Organizational Culture shows that the total score of the items statement is 3.37 with the percentage of agreement is 34.2 percent.

6) HFCV Environmental Leadership

The findings on HFCV Environmental Leadership is shown in Table 7.

Table 7. The mean score of HFCV environmental leadership

HFCV Environmental Leadership	Mean Score	% of Agreement
▪ The leaders within the company inspire a shared vision of the organization as environmentally sustainable, creating or maintaining green values throughout the company	3.55	58%
▪ The leaders within the company utilize well-developed approaches to environmental management which generally centre around a program customized to the company's specific business and market	3.40	50%
▪ The leaders within the company create partnerships with the company's stakeholders to solve environmental problems and to accomplish environmental goals	3.33	45%
▪ The leaders within the company can take on the responsibility of environmental education with the intent of engaging employees in environmental management initiatives	3.10	40%
▪ **Total Score**	**3.34**	**48.2%**

The findings on HFCV Environmental Leadership shows that total score of the items statement is 3.34 with the percentage of agreement is 48.2 percent.

7) HFCV Proactive Green Innovation

The findings on HFCV Proactive Green Innovation is shown in Table 8.

Table 8. The mean score of HFCV proactive green innovation

HFCV Proactive Green Innovation	Mean Score	% of Agreement
▪ The company often undertakes active environment-related innovation in order to take initiatives new practices or products ahead of automobiles industry	4.32	45%
▪ The company actively commit to continuously invest resources in green innovations to successfully seize opportunities and to lead in the market	4.05	40%
▪ The company actively improve manufacturing processes to reuse, recycle, and reduce materials to decrease cost	3.95	38%
▪ **Total Score**	**4.10**	**41%**

The findings on HFCV Proactive Green Innovation shows that the total score of items statement is 4.10 ahead of automobile got the highest mean score of 4.32 with the percentage of agreement is 41 percent.

8) HFCV Environmental Performance

The findings on HFCV Environmental Performance is shown in Table 9.

Table 9. The mean score of HFCV environmental performance

HFCV Environmental Performance	Mean Score	% of Agreement on Significant
▪ Reduction of air emission	4.11	48%
▪ Reduction of effluent waste	4.05	45%
▪ Reduction of solid waste	3.85	38%
▪ Decrease in consumption for hazardous/harmful/toxic materials		
▪ Decrease in frequency for environmental accidents	3.70	35%
▪ Improvement in and environmental situation	3.65	28%
▪ **Total Score**	**3.87**	**39%**

The findings on HFCV Environmental Performance shows that the total score of the items statement is 3.87 with the percentage of agreement is 39 percent.

9) HFCV Environmental Risks

The findings on HFCV Environmental Risks is shown in Table 10.

Table 10. The mean score of HFCV environmental risks

HFCV Environmental Risks	Mean score	% of Agreement
▪ Water and land contamination	4.21	45%
▪ Air and noise pollution	4.00	42%
▪ Waste management	3.98	38%
▪ Supply chain environmental practices	3.84	35%
▪ **Total Score**	**4.00**	**40%**

The findings on HFCV Environmental Risks shows that the total score of the items statement is 4.00 with the percentage of agreement is 40 percent.

5 Recommendation and conclusions

5.1. Recommendation

Based on the study findings, the following recommendations to strengthen the environmental impact positively towards the adoption of HFCV are as follows:

1. **HFCV Internal Environmental Management**
   - The top management should give full commitment on internal environmental management that can encourage proactive action and productive output on internal assessment and management of HFCV.
2. **HFCV Environment Information Systems**
   - Provide information effectively and efficiently in HFCV
3. **HFCV Cooperation with Customers**
   - The HFCV management should cooperate with customers in eco design, cleaner production, green product design and using less energy.
4. **HFCV Eco Design**
   - The initiatives of designing the products that can reduced consumption of material and energy for reuse, recovery of material and recycle the component or material in HFCV
5. **HFCV Environmental Organizational Culture**
   - The company that involve with HFCV must be concern on knowledge of environmental management, environmental agreements, responsiveness and the vision of environmental issues of HFCV.
6. **HFCV Environmental Leadership**
   - The leaders in the organization that involves in HFCV adoption should be motivated to inspire a shared vision on environmental issues.
7. **HFCV Proactive Green Innovation**
   - The management should be active in environment-related innovation in order to take initiatives for HFCV to be ahead than other vehicles.
8. **HFCV Environmental Performance**
   - The initiatives to reduce of air emissions, effluent waste, solid waste and the decrease in consumption for hazardous and harmful materials should be efficient.
9. **HFCV Environmental Risks**
   - Risk minimization is now viewed as increasingly intertwined with good corporate social responsibility and governance.

6 Conclusion

Based on the findings, it indicates that the environmental dimensions in adoption of HFCV could generate many constraint and benefits of the HFCV success in future. This study verifies the issues of environment that need to be in precaution on action of HFCV implementation.

References

1. Chen, Y.S. *Green Organizational Identity: Sources and Consequences*. Management Decision, **49**, 3, pp. 384-404, (2011).
2. Chen, Y.S, Chang, C.H, Wu,F.S. *Origin of Green Innovations: The Differences between Proactive and Reactive Green Innovation.* Management Decision, **50**, 3, pp. 368-398, (2012).
3. Esty, D and Winston, A. *Green to Gold: How Smart Companies Use Environmental Strategy to Innovate, Create Value, and Build Competitive Advantage,* Yale University Press, New Haven, CT, (2006).
4. Green Jr. K.W., Zelbst, P.J. Meacham, J. and Bhadauria V.S. *Green Supply Chain Management Practies: Impact on Performance.* Supply Chain Management: An International Journal, **17/3**, pp. 290-305, (2012).
5. Graci, S. and Kuehnel, Jaqueline. *How to increase your bottom line by going green.* Downloading @ http://green.hotelscombined.com
6. Zhu, Q.,Sarkis, J. and Lai, K. *Confirmation of a Measurement model for green supply chain management practices implementation*. International Journal of Production Economics, **111**, 2, pp. 261-73. (2008).

MATEC Web of Conferences **81**, 08002 (2016)
ICTTE 2016
DOI: 10.1051/matecconf/20168108002

Torsional Oscillation Analysis of a Hybrid Electric Bus's Parking Condition

Xiaoguang Guo and Hongwen He

National Engineering Laboratory for Electric Vehicles, Department of Mechanical and Vehicular Engineering, Beijing Institute of Technology, Beijing100081, China

Abstract. The aim of this study is to solve the problem of torsional oscillation in the parking condition of a hybrid electric bus. The dynamic amplification factor formula is derived, and its main optimization variables is found by the method of sensitivity analysis. The maximum torque equation of the hybrid electric bus is constructed and optimized, and the optimization results is verified by experiment finally.

1 Introduction

Compared with the internal combustion engine bus, the structure of hybrid electric bus's transmission system is more complex. Both of the transmission system constitution's changing and the rapid vary of the motor torque put forward new challenges in the hybrid electric bus's shafting systems [1], so the torsional oscillation is the main problem that the transmission system needs to face.

In the parking condition, a kind of hybrid electric bus's engine is killed by the torque from the ISG motor, midway through spline shaft and torsion shock absorber. When the torsional frequency of the ISG motor and the natural frequency of the transmission system is the same or similar, the resonance phenomenon will be produced [2]-[5]. When this starts to happen, the torque value of the transmission system should be much greater than the torque provided by the ISG motor, and the shaft may be broken off [6].

2 Dynamic amplification factor

The structure of a hybrid electric bus transmission system is shown in Fig. 1. The transmission system is composed of the engine, the torsion shock absorber, the spline shaft, the ISG motor.

In the parking condition, torque provided by ISG motor passes through the spline shaft and torsional shock absorber to the engine starts. In this process, the ISG motor is the driving source, and the spline shaft, the torsion shock absorber, the engine is the vibration's receiving terminal.

In order to express convenience, the spline shaft, torsional shock absorber, engine components as part one, and the ISG motor is treated as part two. When resonances occurred in transmission systems, the natural frequency of part one is the same or similar as part two.

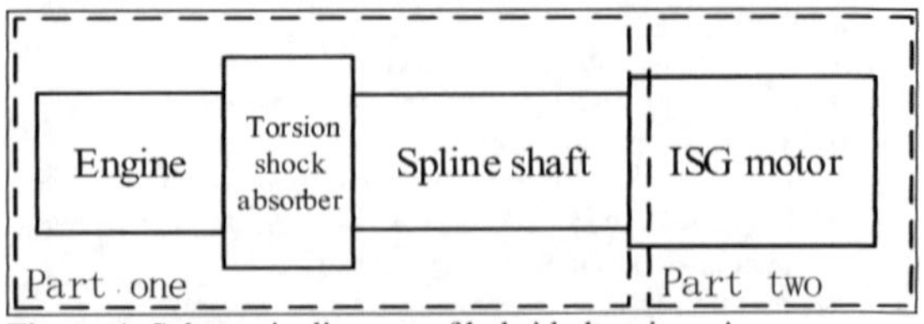

Figure 1. Schematic diagram of hybrid electric car's transmission system

2.1 Nature frequency

The mechanical model of Fig. 1 is shown in Fig. 2.

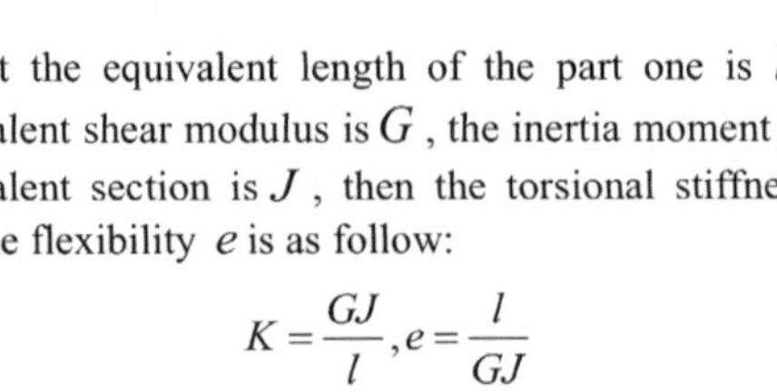

Figure 2. Mechanical model of hybrid electric car's transmission system

Set the equivalent length of the part one is l , the equivalent shear modulus is G , the inertia moment of the equivalent section is J , then the torsional stiffness K and the flexibility e is as follow:

$$K=\frac{GJ}{l},e=\frac{l}{GJ} \tag{1}$$

Set the angular displacement of any instantaneous part one is ϕ ,then the elastic moment is:

$$U = -K\phi \quad (2)$$

The negative sign in this equation means the direction of elastic moment is always opposite to the angular displacement ϕ .

According to D'Alembert principle, the differential equations of motion is constructed as follow [7]:

$$S = -I\ddot{\phi} \quad (3)$$

The inertia moment and elastic moment are balanced, so we may get the equation:

$$S + U = 0 \text{, thus : } I\ddot{\phi} + \frac{1}{e}\phi = 0 \quad (4)$$

The solution of the differential equation is the equation of part one:

$$\phi = A\sin(pt + \varepsilon) \quad (5)$$

$$\text{which : } p = \sqrt{\frac{1}{Ie}} = \sqrt{\frac{GJ_p}{Il}} \quad (6)$$

In formula (6) p is part one's natural frequency expression.

2.2 Dynamic amplification factor

The dynamic amplification factor here means the ratio of toque on part one and toque issued by ISG motor.

In formula (4), the torque of ISG motor's expression is as follow:

$$T = M\sin\omega t \quad (7)$$

Set damping moment as:

$$R = -c\dot{\phi} \quad (8)$$

thus,

$$I\ddot{\phi} + c\dot{\phi} + \frac{1}{e}\phi = M\sin\omega t \quad (9)$$

Deforming formula (9) as:

$$\ddot{\phi} + 2n\dot{\phi} + p^2\phi = h\sin\omega t \quad (10)$$

$$\text{which: } n = \frac{c}{2I} \quad (11)$$

The general solution for differential formula (10) is:

$$\phi = Ae^{-nt}\sin(p_d t + \varepsilon) + A_1\sin(\omega t - \psi) \quad (12)$$

Solving differential equations (10), and the dynamic amplification factor will be obtained as:

$$\beta = \frac{1}{\sqrt{\left[1 - (\frac{\omega^2}{p^2})\right]^2 + 4(\frac{n^2}{p^2})(\frac{\omega^2}{p^2})}} \quad (13)$$

Getting formula (6) and (11) into formula (13), we may safely get the following results:

$$\beta = \frac{1}{\sqrt{\left(1 - \omega^2\frac{Il}{GJ_p}\right)^2 + \left(\frac{c}{I}\right)^2 \omega^2 \left(\frac{Il}{GJ_p}\right)^2}} \quad (14)$$

3 Sensitivity analysis

3.1The design of sensitivity analysis

The independent variable is used to scatter sample points in design space, so as to constructing the mathematical expression of the independent variable and the objective function, and then the construction of pareto chart between the variable and the objective function will be completed, thus the sensitivity analysis will be finished.

As the upper and lower limit of the independent variable is very wide, and in order to cover the design space with the limited number of sampling points, the construction of sample points is completed by Latin hypercube method. Latin hypercube method is a statistical method for generating a sample of plausible collections of parameter values from a multidimensional distribution.

In the context of statistical sampling, a square grid containing sample positions is a Latin square if (and only if) there is only one sample in each row and each column [8] . A Latin hypercube is the generalization of this concept to an arbitrary number of dimensions, whereby each sample is the only one in each axis-aligned hyperplane containing it. Fig. 3 direct comparison the difference between Latin hypercube method and full factorial method.

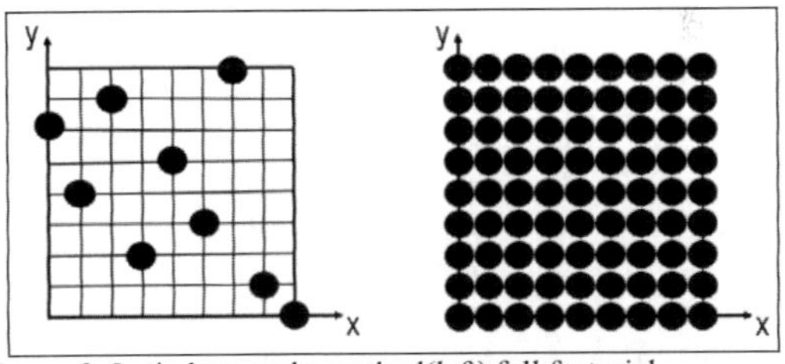

Figure 3. Latin hypercube method(left) full factorial method(right)

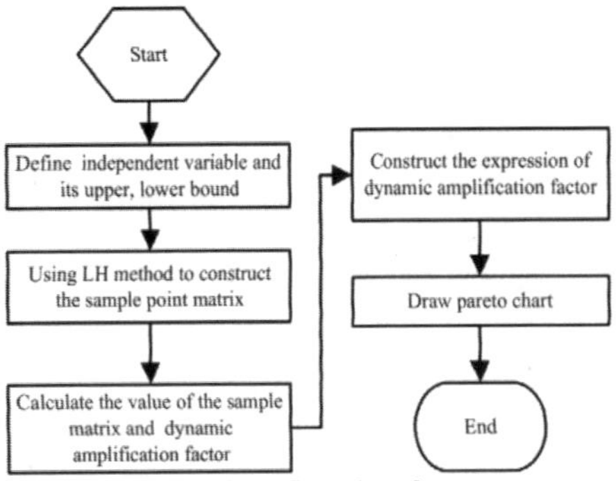

Figure 4. Sensitivity analysis flow chart for

transmission system

Fig. 4 shows the specific processes of the sensitivity analysis for transmission system.

3.2 Sensitivity analysis results

According to Fig. 4, formula (14) should be expressed as:

$$\begin{aligned}\beta &= 10.62\omega - 10.58GJ + 8.08GI \\ &+ 6.68G\omega - 6.31L\omega + 6.24IJ + 5.54GL \\ &+ 4.68J^2 - 4.57cG - 4.53I^2 + ...\end{aligned} \quad (15)$$

Using formula (15) to draw the pareto chart of each independent variable's dynamic magnification factor, and the results shows in Fig. 5.

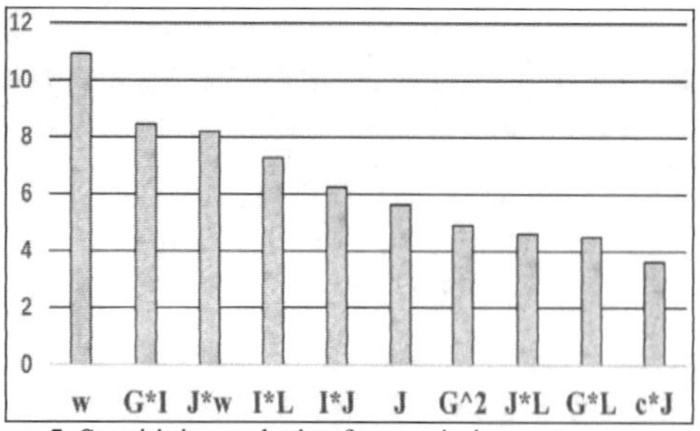

Figure 5. Sensitivity analysis of transmission system - Pareto chart

From Fig. 5, we may safely find out that the independent variable of ω is the highest sensitivity to dynamic amplification factor. The second highest sensitivity to dynamic amplification factor is the product of the shear modulus ***G*** and the moment of inertia ***I***. The third one is the inertia moment of the equivalent section J and ω.

4 Torque optimization

4.1 Independent variables

Generally speaking, the design variables are the greatest impact on the sensitivity of the target function. Considering the sensitivity analysis's result, the actual engineering structure design, parts processing and other factors, the shear modulus of elasticity ***G*** is chosen as the independent variable.

4.2 Constraint condition

In order to make the transmission system work normally in all the working conditions, the maximum torque of the shaft section is less than the allowable value, that is:

$$|T| \leq T^* \quad (16)$$

$|T|$ is the representative of the actual absolute torque, and T^* is the representative of the allowable stress.

4.3 Objective function

Minimization of each axis section's torque T is chosen as the objective function, which $T=T_{ISG}\beta$ .In order to ensure the hybrid electric bus in the parking condition does not occur the torsion form of resonance, it is need to iterates over the bus's frequency in the parking condition. It is need to traverse frequency 0~50Hz, and the step size is 0.1.

The torque optimization equation of the transmission system can be described as:

$$\begin{cases} loop\ i = 0, 0.1, 50 \\ find\ G \\ \min\ T_i(G) = T_{ISGi}\,\beta_i \\ s.t.\ \max\{T_i(G)\} \leq T^* \\ G^L \leq G \leq G^U \end{cases} \quad (17)$$

4.4 Selection algorithm

To prevent the optimization into local optimization and couldn't find global optimal value, here the algorithm of genetic algorithm(GA) is chosen.

4.5 The flow chart of optimization

The flow chart of the torque optimization is shown in Fig. 6.

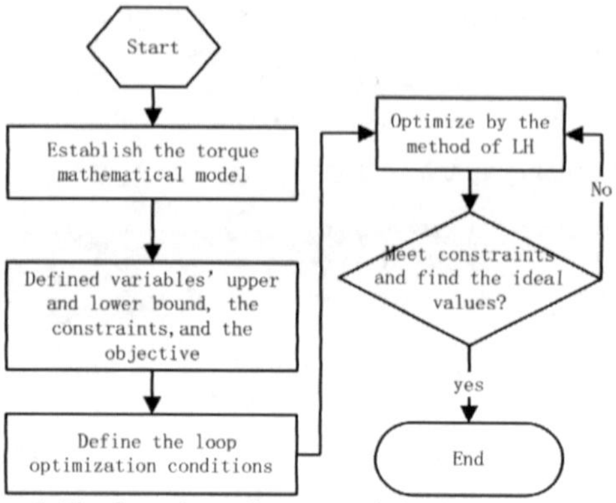

Figure 6. The flow chart of torque optimization for transmission system

4.6 Optimization result analysis

Using the GA algorithm, the optimal value is obtained after 224 iteration calculation, and the maximum torque history curve of the transmission system is shown in Fig. 7.

Before optimization, in the hybrid electric bus's parking condition, the ISG motor and transmission system has occurred the torsional resonance, and the maximum torque of the transmission system is 1520Nm, which beyond the allowable stress 1200Nm and it is in the danger stage.

DOI: 10.1051/matecconf/20168108002

After optimization, the maximum torque of transmission system is 1100Nm, which has been down by 27.6%, and the result is also in the range of safety torque.

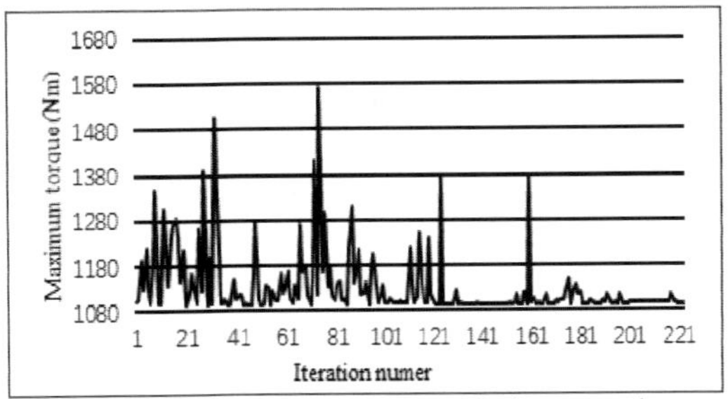

Figure 7. The history curve of the transmission system's maximum torque

5 Experimental verification of the parking condition

The experimental platform of the hybrid electric bus transmission system is built, as shown in Fig. 8.

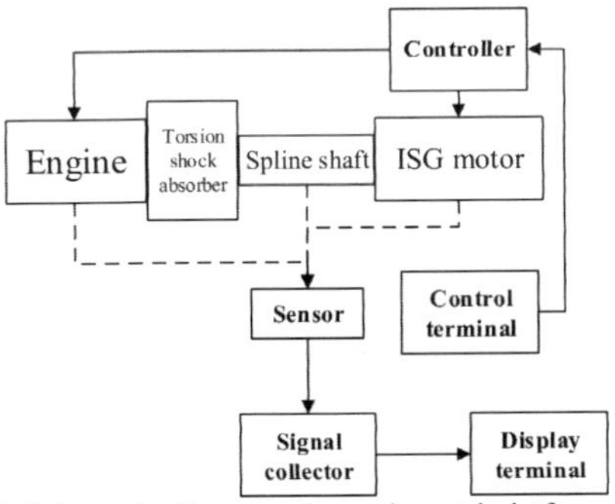

Figure 8. Schematic diagram of experimental platform of transmission system

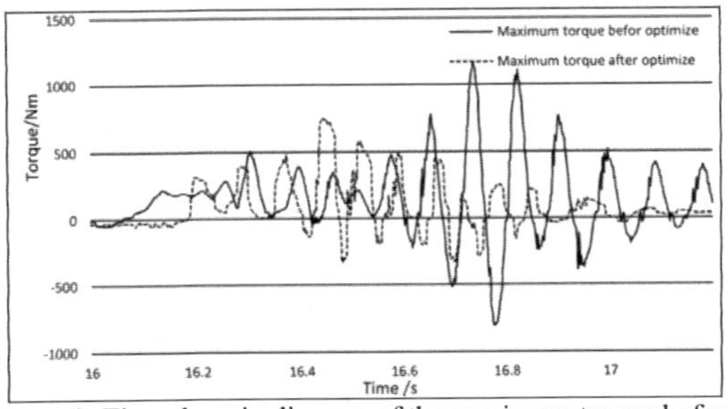

Figure 9. Time domain diagram of the maximum torque before and after the optimization

The transmission system's shear modulus of elasticity G is changed according to the optimization results, and Fig. 9 shows the maximum torque of the drive system. We can obtain that, after optimization, the maximum torque of the drive system is reduced to 749.01Nm.

The maximum torque of the hybrid electric bus transmission system is reduced by the way of shifting frequency, and this means the hybrid electric bus is also provided a more reliable security.

6 Conclusions

This paper takes hybrid electric bus's transmission system as the research object. The expression of the dynamic amplification factor for the transmission system is deduced, and the sensitivity analysis is carried out. The maximum torque equations of the parking condition for the hybrid electric bus is constructed and optimized.

The main conclusions is as follows:

1) By the sensitivity analysis of the dynamic amplification factor expression, the transmission system of shear modulus of elasticity G, moment of inertia J, rotational inertia $\boldsymbol{I}$ are coupling.

2) In order to solve the problem of the torsional oscillation in hybrid electric bus, using the method of sensitivity analysis to find out the maximum sensitivity of the parameter in the expression of dynamic amplification factor, and making this parameter as independent variable to reduce the torque of the optimization process is feasible. These work will provide a set of guidelines for the issue of the torsional oscillation in the future.

References

1. Kim H, Kim J, Lee H, et al. Mode Transition Control Using Disturbance Compensation for a Parallel Hybrid Electric Vehicle[J]. Proceedings of the Institution of Mechanical Engineers Part D Journal of Automobile Engineering, **225**(225):150-166, (2011).
2. Walker P D, Zhang N. Modelling of dual clutch transmission equipped powertrains for shift transient simulations[J]. Mechanism & Machine Theory, **60**(1):47–59, (2013).
3. Zhang J, Tang X, Yu H, et al. Multi-body dynamics and noise analysis for the torsional vibration of a power-split hybrid driveline[J]. Proceedings of the Institution of Mechanical Engineers, Part K: Journal of Multi-body Dynamics, **228**(4):366-379, (2014).
4. Iman, R.L.; Helton, J.C.; and Campbell, J.E.. "An approach to sensitivity analysis of computer models, Part 1. Introduction, input variable selection and preliminary variable assessment". Journal of Quality Technology **13**(3): 174–183, (1981).
5. Iman, R.L.; Davenport, J.M.; Zeigler, D.K.. Latin hypercube sampling (program user's guide), (1980).
6. Mckay M D, Conover R J B W J. A Comparison of Three Methods for Selecting Values of Input Variables in the Analysis of Output From a Computer Code[J]. Technometrics, **42**(1):55-61, (2012).
7. Eglajs, V.; Audze P.. "New approach to the design of multifactor experiments".Problems of Dynamics and Strengths. **35** (in Russian) (Riga: Zinatne Publishing House):104–107, (1977).
8. Iman, R.L.; Helton, J.C.; Campbell, J.E.. "An approach to sensitivity analysis of computer models, Part 1. Introduction, input variable selection and preliminary variable assessment". Journal of Quality Technology **13** (3): 174–183, (1981).

MATEC Web of Conferences **81**, 08003 (2016)
ICTTE 2016 DOI: 10.1051/matecconf/20168108003

LPV Control for a Semi-Active Suspension Quarter of Car-One Parameter Case

Jorge de Jesús Lozoya Santos[1], Juan Carlos Tudon-Martinez[1] and Ricardo A. Ramirez-Mendoza[2]

[1]*Dirección de Investigación, DIECI, Universidad de Monterrey, 66238 San Pedro Garza Garcia, Nuevo Leon, Mexico*
[2]*Dirección de Investigación, Escuela de Ingeniería, Tecnológico de Monterrey 64810 Monterrey, Nuevo Leon, Mexico*

Abstract. The actual semi-active suspension control systems with a balance between comfort and road holding goals are not optimal because in these solutions one goal or the other always dominates in the suspension performance. This paper is centered in a new proposal to control an automotive semi-active suspension to achieve the comfort and maintain the road holding.The output in the control strategy is the electric current. A nonlinear quarter of vehicle model simulation compares and validates the proposal versus different controllers. The controller is designed with the H∞ criteria and the Linear Varying Parameter (LPV) considering the saturation and sigmoid shape of the F-V characteristic diagram. Unlike the solutions in literature, which use at least two scheduling parameters, the proposed LPV controller scheme for a semi-active suspension uses only one scheduling parameter.

1 Introduction

A control strategy for semi-active suspension in a Quarter of Vehicle (*QoV*) consists in giving the control goals, and in developing the controller, and the algorithm to map the controller output to the electric current. Most of the semi-active suspension controllers are devoted to a specific goal, while in control theory the controllers are independent of the goals. Hence, in such cases, the control strategy consists in the selection of the controller and the mapping algorithm. Seminal results show that a semi-active suspension can decrease up to 52 % of vertical sprung mass acceleration and up to 20 % of vertical sprung mass displacement, [1].

Two general classifications of semi-active control exist. The first type is the *Continuously variable Semi-Active (CSA) control*. The type of manipulation generates a continuous manipulation over an interval in the semi-active interface [2]. The second type is the *On-off Semi-Active (OSA) control*. The manipulation has only two values in the semi-active interface, according to the applied damping coefficient: hard for road holding or soft for ride comfort, [3].

The controllers can be categorized according to the number of goals to optimize: comfort, road holding and deflection. Some controllers optimize one objective, while the actual goal is the compromise between comfort and road holding. Some recent works are focused on the decrease in the compromise between comfort and road holding [4]. An interesting strategy is the LPV/H ∞ approach which solves the compromise adequately and with low on-line computation, [1], [5]. In this paper, a controller for an automotive semi-active suspension adaptable to the controllable shock absorber nonlinearities in order to ensure the full exploiting of the damping variability will be developed. The design will specify the same units in the controller output and the manipulation input of the controllable damper and it will seek to reach the damping steady state to fully explore the semi-activeness. The proposed controller must ensure good performances in both comfort and road holding. The pseudo-Bode and transient response plots will be the qualitative criteria in the frequency and time domain respectively. The quantitative criteria are the *Power Spectral Density (PSD)* and the *RMS*.

2 Controller design

The actual semi-active suspension control systems with a balance between comfort and road holding goals are not optimal because in these solutions one goal or the other always dominates in the suspension performance. The control strategy considers the piston velocity and displacement as the control system inputs. These inputs ensure the mapping of nonlinear characteristics of the suspension in control laws. The output in the control strategy, in both controllers, is the electric current. A nonlinear *QoV* model simulation compares and validates the proposal versus different controllers.

The control system consists in a gain-scheduling approach. A simple *MR* damper model is used allowing the scheduling of the damping coefficient nonlinearity in the control law. It is inspired in the work of [1] and [6]. The new contribution is the inclusion of a simple *MR* damper model and the use of only one scheduling parameter. The controller is designed with the H∞ criteria and the Linear Varying Parameter (*LPV*) considering the saturation and sigmoid shape of the *F-V* characteristic diagram.

The electric current is the input signal that allows exploring in a full way the characteristics of the *MR* damper. Using this signal as the controller output and the chosen inputs, a complete exploration of the semi-active zone in the damper characteristic is ensured. However, the use of the electric current as a controller output is not common. Most of the actual research works use the force or damping coefficient, and then a conversion algorithm for the electric current. This can be one of the causes of the actual compromise between comfort and road holding in semi-active suspensions.

2.1 LPV/H∞ controller design

The representation of state space of a QoV model in the LPV framework by including an MR damper in the suspension considering straight direction and D = 0 can be defined as:

$$\begin{bmatrix} \dot{z}_s \\ \ddot{z}_s \\ \dot{z}_{us} \\ \ddot{z}_{us} \end{bmatrix} = A_s \begin{bmatrix} z_s \\ \dot{z}_s \\ z_{us} \\ \dot{z}_{us} \end{bmatrix} + [\, B_s \; B_{s1}] \begin{bmatrix} u \\ z_r \end{bmatrix}, y = C_s \begin{bmatrix} z_s \\ \dot{z}_s \\ z_{us} \\ \dot{z}_{us} \end{bmatrix} \quad (1)$$

where

$$A_s = \begin{bmatrix} 0 & 1 & 0 & 0 \\ -\frac{k_s+k_p}{m_s} & -\frac{c_p}{m_s} & \frac{k_s+k_p}{m_s} & \frac{c_p}{m_s} \\ 0 & 0 & 0 & 1 \\ \frac{k_s+k_p}{m_{us}} & \frac{c_p}{m_{us}} & \frac{-k_s-k_p-k_t}{m_{us}} & -\frac{c_p}{m_s} \end{bmatrix} \quad (2)$$

$$B_s = \begin{bmatrix} 0 \\ \frac{c_{MR}\frac{\dot{z}}{\epsilon+|||\dot{z}|||_{i-k}^{k}}\left[\frac{\tanh\left(\frac{u}{I_{sat}}\right)}{\frac{u}{I_{max}}}\right]}{m_s} \\ 0 \\ \frac{c_{MR}\frac{\dot{z}}{\epsilon+|||\dot{z}|||_{i-k}^{k}}\left[\frac{\tanh\left(\frac{u}{I_{sat}}\right)}{\frac{u}{I_{max}}}\right]}{m_{us}} \end{bmatrix}, B_{s1} = \begin{bmatrix} 0 \\ 0 \\ 0 \\ \frac{k_t}{m_{us}} \end{bmatrix}, C_s = \begin{bmatrix} 0 \\ 1 \\ 0 \\ -1 \end{bmatrix}^T \quad (3)$$

where u is the LPV QoV model exogenous input. The MR damper is represented by the Vmax model [5]:

$$F_D = c_p\dot{z} + k_p z + Ic_{MR}\rho \quad (4)$$

$$\rho = \dot{z}\frac{1}{|||\dot{z}|||_{i-k}^{i}+\epsilon} \quad (5)$$

where ρ describes the damping force behavior due to the pre-yield and post-yield regimes of the MR fluid. In this case, a very narrow pre-yield zone is assumed in order to emphasize the independence of the MR effect of the velocity.

The equations (1-3) were obtained by inserting equations (4-5) in the space state representation of a QoV model that lacks a damper. Thus, the passive damping force component is inserted in the A matrix, and the semi-active component force is inserted in the B matrix. The electric current, input of the VMax model, is represented as variable u. The *tanh* function saturates the model input. According to the system, equation (1), the input constraints on the LPV QoV model are:

1) Semiactiveness. The input of the QoV LPV system, equation (1) must be positive, [7]:

$$u \geq 0, \forall t \geq 0 \quad (6)$$

where u is the electric current to be applied. In order to take into account this input constraint, the absolute value of the LPV-based controller filtered output, u_f, is defined as equal to the electric current magnitude to be applied, $u = |u_f|$.

2) Saturation. The control input provided by the semi-active damper must be bounded to the finite interval between $[0, I_{max}]$ A. First, the MR damper model included in the LPV QoV model is:

$$F_D = c_p\dot{z} + k_p z + c_{MR}\frac{\dot{z}}{\epsilon+|||\dot{z}|||_{i-k}^{k}}\left[\frac{\tanh\left(\frac{u}{I_{sat}}\right)}{\frac{u}{I_{max}}}\right]u \quad (7)$$

Hence, the proposed solution to saturation can be seen by simplifying equation (7), deriving on:

$$F_D = c_p\dot{z} + k_p z + c_{MR}I_{max}\tanh\left(\frac{u}{I_{sat}}\right)\frac{\dot{z}}{\epsilon+|||\dot{z}|||_{i-k}^{k}} \quad (8)$$

equation (8) is a version of the V_{Max} model where the electric current input is saturated to I_{max}. The term $c_{MR}I_{max}$ defines the maximum *MR* force F_{MR} to be applied in the whole velocity span. The term:

$$\frac{\dot{z}}{\epsilon+|||\dot{z}|||_{i-k}^{k}} \quad (9)$$

represents the sigmoid form of the *MR* force component oscillating between -1 and 1; it assigns the sign to the force component and the maximum *MR* force for the maximum velocity in the last k samples. The term:

$$\tanh\left(\frac{u}{I_{sat}}\right) \quad (10)$$

oscillates between 0 and 1, where I_{sat} defines the saturation slope of the electric current to be applied, as well as limits the exogenous input, u_c. High values allow I_{sat} a fast saturation, and low values get slow saturation. Therefore, this parameter limits the maximum electric current to be applied. Higher values of I_{sat} will obtain a fast response of the change between I_{min} and I_{max} (i. e. a road holding oriented *LPV* control), while lower values will obtain a slow response (i. e. a comfort oriented *LPV* control). The value of I_{sat} is obtained from simulations of the frequency responses of the vertical acceleration and tire deflection to observe which values of I_{sat} are better for comfort, road holding, or both.

According to equation (7), the measurable and scheduling parameter is defined. The parameter is:

$$\rho^* = \frac{\dot{z}}{\epsilon+|||\dot{z}|||_{i-k}^{k}}\left[\frac{\tanh\left(\frac{u}{I_{sat}}\right)}{\frac{u}{I_{max}}}\right], \rho^* \in [-1, 1] \quad (11)$$

where u represents the electric current magnitude which is proportional to the maximum damping force to be applied by the *MR* damper and the piston velocity modifies the damping coefficient. Equation (7) derives on:

$$F_D = c_p\dot{z} + k_p z + c_{MR}\rho^* u \quad (12)$$

3) Dissipativity. The damping coefficient must always be positive. This is shown using the *MR* force component, F_{MR}:

$$F_{MR} = c_{MR} u \frac{\dot{z}}{\epsilon + |||\dot{z}|||_{i-k}^{k}} \left[\frac{\tanh\left(\frac{u}{I_{sat}}\right)}{\frac{u}{I_{max}}} \right] \quad (13)$$

and dividing equation (13) by z derivative, the variable damping coefficient (c_u) is obtained:

$$c_u = \frac{F_{MR}}{\dot{z}} = \frac{I_{max} c_{MR}}{\epsilon + |||\dot{z}|||_{i-k}^{k}} \tanh\left(\frac{u}{I_{sat}}\right) \quad (14)$$

where c_u is always positive since I_{max} and c_{MR} are constant, and the terms $\epsilon + |||\dot{z}|||_{i-k}^{k}$ and $\tanh\left(\frac{u}{I_{sat}}\right)$ are always positive.

4) Gain of the MR mechanism. The maximum damping coefficient generated by the *MR* damper must be as close as possible to the critical damping coefficient of the mass of interest. If the *MR* damping is not enough, a gain of damping force *GI* is needed, allowing the *MR* damper to equal the required critical damping coefficient of the sprung mass or the unsprung mass in order to achieve the control goals. A continuously variable semiactive suspension should have an off-state damping ratio of 0.1 to 0.2 and an on-state damping ratio of 1.0 or greater, [8]. These design values will improve ride comfort and suspension deflection while approximately maintaining the same road holding as the passive suspension. If the semi-active suspension control goals include the comfort and the road holding, the critical damping of the unsprung mass for an ideal damping ratio, $\xi = 1.0$ must be equal to the maximum damping coefficient of the *MR* damper:

$$2\sqrt{(k_s + k_t) m_{us}}\xi = \frac{\overline{F_D(\dot{z})}}{\dot{z}} = GI\, I_{max} c_{MR} + GI\, c_p \quad (15)$$

where $\frac{\overline{F_D(\dot{z})}}{\dot{z}}$ is the maximum damping force that the *MR* damper can dissipate. If the goal is only the comfort, the critical damping of the sprung mass for an ideal damping ratio, $\xi = 1.0$, must be equal to the maximum damping coefficient of the *MR* damper:

$$2\sqrt{k_r m_s}\xi = \frac{\overline{F_D(\dot{z})}}{\dot{z}} = GI\, I_{max} c_{MR} + GI\, c_p \quad (16)$$

where k_r is the effective stiffness of the suspension and tire springs called ride rate, $k_r = \frac{\eta^2 k_s + k_t}{\eta^2 k_s k_t}$. The computed gain *GI* will be included in the *QoV* model used in the *LPV* controller synthesis. This gain ensures the controller output considers the critical damping ratio of the *QoV* model. The parameters c_p and c_{MR} in matrices A_s and B_s will be multiplied by the gain *GI*:

$$A^*_s = \begin{bmatrix} 0 & 1 & 0 & 0 \\ -\frac{k_s + k_p}{m_s} & -\frac{GI\, c_p}{m_s} & \frac{k_s + k_p}{m_s} & \frac{GI\, c_p}{m_s} \\ 0 & 0 & 0 & 1 \\ \frac{k_s + k_p}{m_{us}} & \frac{GI\, c_p}{m_{us}} & \frac{-k_s - k_p - k_t}{m_{us}} & -\frac{GI\, c_p}{m_s} \end{bmatrix}, B^*_s = \begin{bmatrix} 0 \\ -\frac{GI\, c_{MR}\, \rho^*}{m_s} \\ 0 \\ \frac{GI\, c_{MR}\, \rho^*}{m_{us}} \end{bmatrix} \quad (17)$$

2.2 Controller synthesis

The generalized system for the H∞/LPV controller synthesis for one scheduling parameter is not proper for the *LPV*-based controller synthesis. According to the definition presented in [2] for an ideal linear design of a *LPV* system for the controller synthesis, a proper filter, equation (18) is added to the input of the *LPV* system, equation (1):

$$F_1: \begin{pmatrix} \dot{x}_f \\ u_f \end{pmatrix} = \begin{bmatrix} A_f & B_f \\ C_{f\,\rho} & 0 \end{bmatrix} \begin{pmatrix} x_f \\ u_o \end{pmatrix} \quad (18)$$

In this way, ρ^* will be in the states transition matrix A^*_s. Fig. 1 shows the obtained *QoV* structure by using the *MR* damper model with saturated input, equation (8) and the ideal linear design.

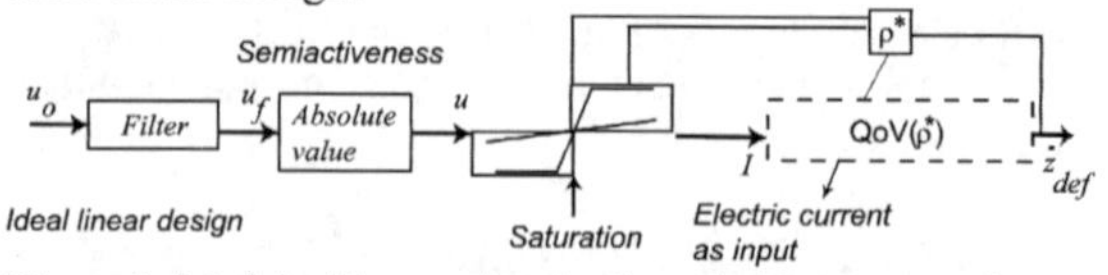

Figure 1. Model with a semi-active bounded input saturation.

Hence, the new proposed *LPV* system, equation (19), to perform the *LPV* controller synthesis is defined by the scheduling variable ρ^*:

$$\begin{bmatrix} \dot{z}_s \\ \ddot{z}_s \\ \dot{z}_{us} \\ \ddot{z}_{us} \\ \dot{x}_f \end{bmatrix} = A(\rho^*) \begin{bmatrix} z_s \\ \dot{z}_s \\ z_{us} \\ \dot{z}_{us} \\ x_f \end{bmatrix} + Bu + B_1 w, y = C \begin{bmatrix} z_s \\ \dot{z}_s \\ z_{us} \\ \dot{z}_{us} \\ \dot{x}_f \end{bmatrix} \quad (19)$$

where

$$A(\rho^*) = \begin{bmatrix} A^*_s & \rho^* B^*_s C_f \\ 0 & A_f \end{bmatrix}, B = \begin{pmatrix} 0 \\ B_f \end{pmatrix} \quad (20)$$

$$B_1 = \begin{pmatrix} B_{s1} \\ 0 \end{pmatrix}, \; C = \begin{pmatrix} C_s \\ 0 \end{pmatrix}^T \quad (21)$$

and the electric current to be applied is a function of u:

$$I(u) = I_{max} \tanh\left(\frac{u}{I_{sat}}\right) \quad (22)$$

In order to meet the control specifications, two H∞ weighting functions, $W_{\ddot{z}_s}$ and $W_{z_{us}-z_r}$, obtained in case 1in [9], were used according to the comfort performance without affecting the road holding, Fig. 2.

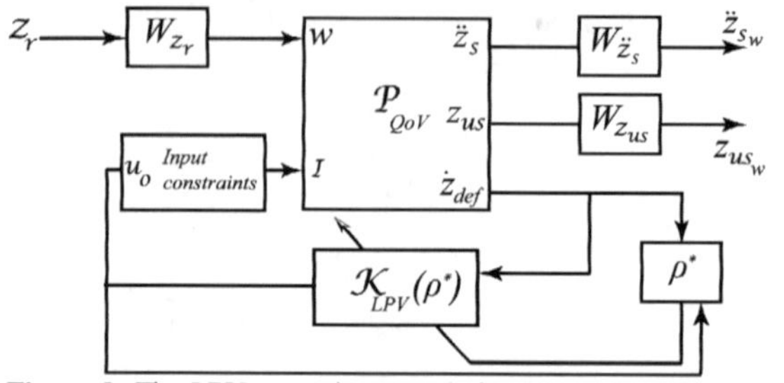

Figure 2. The LPV control approach for the QoV model with semiactive suspensión.

The *LPV*-based controller was obtained through the solution of a *Linear Matrix Inequalities* problem, [2]. The controlled output vector is:

$$z = (\ddot{z}_s, z_{us})^T \quad (23)$$

Three *LPV*-based controllers were synthesized according to:

$$2\sqrt{k_s m_s}\xi = 7{,}945\ N/m \qquad (24)$$

that corresponds to a damping ratio $\xi = 1.1$ in the resonant frequency of the unsprung mass of the *QoV* model. By solving equation (16), one gets the desired gain of the damping value GI which will allow to apply the appropriate value of electric current:

$$GI = \frac{2\sqrt{(k_s+k_t)m_{us}\xi}}{I_{max}c_{MR}+c_p} = \frac{7912}{3192} = 2.72 \qquad (25)$$

where the values of equation (25) were obtained from an experimental lumped QoV lumped parameter model for k_s, k_t, and m_{us}; as well as the values of c_p and c_{MR}. A maximum electric current $I_{max} = 5$ A was defined according to the industrial practice for this type of MR damper. The values of I_{sat} were 0.8, 1.6, and 2.8 for road holding, tradeoff and comfort oriented performances whose LPV-based controllers are named LPV-Road holding, LPV-tradeoff, and LPV-Comfort, respectively.

2.3 Control schemes validation

The nomenclature, validation and the set of the benchmark controllers are done and specified according to [5].

3 Results and discussion

The nonlinear QoV model was simulated with the controllers specified as specified in [10] using the test based on pseudobode test [2], Fig. 3.

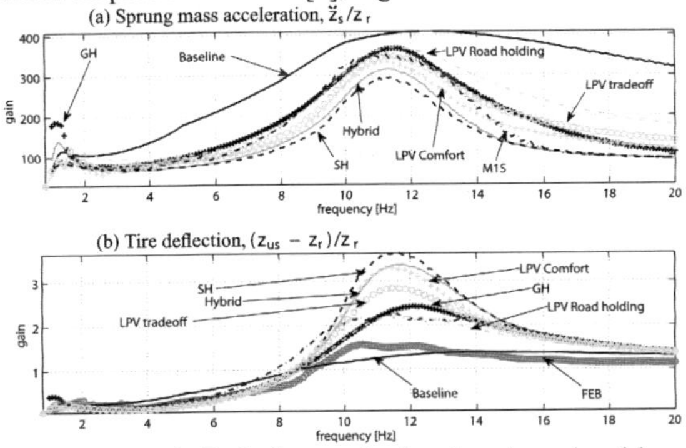

Figure 3. Pseudo-Bode for transfer functions in a closed loop simulation for a nonlinear QoV model: (a) sprung mass acceleration, (b) tire deflection.

The baseline suspension was also simulated. It must be noted that all the controllers are better than the baseline suspension in comfort, Fig. 3(a). The proposed controller (LPV-based) have a better performance below 1.6 Hz than the Hybrid controller. In the secondary ride frequencies, the LPV-Co, LPV-Tradeoff, Hybrid controllers offer good gains. Qualitatively, it can be said that the proposed approaches are better than the Hybrid controller in the primary ride, and they have similar performances in the tirehop resonant frequency. When dealing with the road holding goal, the best performance corresponds to the GH controller. The LPVComfort, SH and Hybrid controller are not well suited for this performance, see Figure 3c. The LPV Road holding controllers behave better than the GH controller. The Hybrid controller does not achieve a good tradeoff performance for comfort and road holding at the same time. Also, the baseline suspension is optimized for the suspension deflection and road holding with a sensitive payload in comfort. The frequency domain analysis validates the LPV-based proposed controllers as the best options in comfort and road holding performances.

The implementation of this LPV controller design based on one parameter case must be explored using a synthesis for LPV controllers with low implementation complexity, [10], in order to validate it using an experimental test rig and in an in-vehicle tests.

4 Conclusion

In summary, numerical results show *LPV*-based controllers as the best options to control comfort and road holding performances simultaneously. The jerk in the sprung mass in controllers as *SH*, *GH* and *Hybrid* is present in numerical results, while the proposed controllers do not have this problem. If the GI parameter reacts to a road estimation then GI can be proposed as an adjustment of the semi-active suspension according to the road.

References

1. P. Barak, No. 922140, SAE Technical Paper, (1992)
2. C. Poussot-Vassal, O. Sename, L. Dugard, P. Gaspar, Z. Szabo, J. Bokor, CEP **16**, 17 (2008)
3. S. M. Savaresi, C. Spelta, JDSMC **129**, 11 (2007)
4. F. D. Goncalves, M. Ahmadian, SHOCK VIB **10**, 11 (2003)
5. J. de-J. Lozoya-Santos, R. Morales-Menendez, R. A. Ramírez-Mendoza, MATH PROB ENG **2012**, (2012)
6. J. Mohammadpour, C. W. Scherer, Springer Science & Business Media, 2012
7. A. L. Do, O. Sename, L. Dugard, *American Control Conference (ACC)*, 2010.
8. L. R. Miller, *Proceedings of the 27th IEEE Conference on In Decision and Control*, (1988)
9. A. L. Do, S. Boussaad, J de-J. Lozoya-Santos, O. Sename, L. Dugard, R. A. Ramirez-Mendoza, *12th Mini conference on vehicle system dynamics, identification and anomalies (VSDIA)* (2010)
10. C. Hoffmann, S. M. Hashemi, H. S. Abbas and H. Werner, IEEE T CONTR SYST T, **22**, 6 (2014)

MATEC Web of Conferences **81**, 08004 (2016) DOI: 10.1051/matecconf/20168108004
ICTTE 2016

Comprehensive Evaluation of Car-Body Light-Weighting Scheme Based on LCC Theory

Qing-lan Han and Lan Gao

Accounting Department, Business School, Central South University, China

Abstract. In this paper, a comprehensive evaluation model of light-weighting scheme is established, which is based on three dimensions, including the life cycle costs of the resource consumed by the designed objects (LCC), willingness to pay for the environmental effect of resource consumption (WTP) and performance (P). Firstly, cost of each stage is determined. Then, based on the resource classification, which is based on cost elements, determine the material list needed, and apply WTP weight coefficient to monetize life cycle environmental impact and obtain the life cycle comprehensive cost of designed scheme (TCC). In the next step Performance (P) index is calculated to measure the value of the life cycle costs by applying AHP and SAW method, integrated (TCC) and (P) to achieve comprehensive evaluation of light-weighting scheme. Finally, the effectiveness of the evaluation model is verified by the example of car engine hood.

1 Introduction

On the background of sustainable development, automobile light-weighting has become a focus of academic research in the current society in which, resources and environment problems have become increasingly prominent. Light-weighting is achieved mainly through the structure improvement, material replacement and advanced manufacturing technology selection, of which the effect of material replacement is the most significant. In the designing process of light-weighting, the choice of the raw material, the corresponding production process and the connection process, have important influence on the life cycle cost, the environmental impact and the performance of the lightweight scheme. The usage of a large number of cars has caused a lot of problems such as energy shortage and environmental pollution to human society. Therefore, from the perspective of LCC, research on how to reduce the life cycle cost and environmental impact of automotive products, and improve the product life cycle performance, so as to guide the selection of light-weighting designation program is of great significance.

Study on the comprehensive evaluation of light quantization schemes abroad: Ineˆs Ribeiro, Paulo Pec, as etc. [1], integrated LCA, LCC considering the material properties, to analyze the life cycle costs and environment impact of low carbon steel, high strength steel, and Aluminum Alloy, in order to provide selection concerning car fender material replacement and material selection. Robert A. Witik, Jérôme Payet etc. [2] analyzed the life cycle costs and environmental impact of 6 different materials used for replacing car boot partition, based on the theory of LCC and LCA, to obtain the comprehensive evaluation value. While Ahmad T. Mayyas etc. [3], proposed designation oriented for life cycle assessment, as ecological designation decision support tools used for optimization design and lightweight material selection for white car bodies. Domestic literature Lv Yining, Lv Zhenhua [4], by setting the thin wall plate structure rigidity for objective, proposed that body sheet bending strain energy and membrane strain energy can be considered as performance evaluation, so as to guide the light-weighting scheme evaluation and selection, but they did not consider the cost and environmental impact of each scheme. Li Yanping, Liu Haijiang etc. [5] set the goal of reducing costs and weight and improving performance, put forward that light-weighting scheme evaluation should be based on multi-objective decision satisfaction mechanism, by verifying car engine hood. Han Qinglan, Zhang Yang etc. [6] applied light-weighting material as the main input variables, using AHP method to establish the LCC and LCA integrated evaluation model, at the same time, put the model into the car door material selection case.

Though scanning domestic and foreign literature, it is not difficult to find that foreign literature focuses on the life cycle assessment of light-weighting materials, domestic literatures focus on light-weighting designation method research, however, few are based on LCC perspective for comprehensive quantitative evaluation of light-weighting design. Taking this as the breakthrough point, this paper builds a comprehensive evaluation model of light-weighting program from three dimensions of economy, environment and performance from the perspective of LCC.

2 Comprehensive evaluation model of lightweight scheme

In this paper, the light-weighting designation is mainly considered upon the selection of light-weighting materials and the use of advanced technology. The life cycle cost, environmental impact and performance of the scheme are different because of different materials and processes. Therefore, in order to evaluate the scheme comprehensively, above factors must be fully considered from the perspective of LCC. Comprehensive evaluation model is shown in Fig. 1:

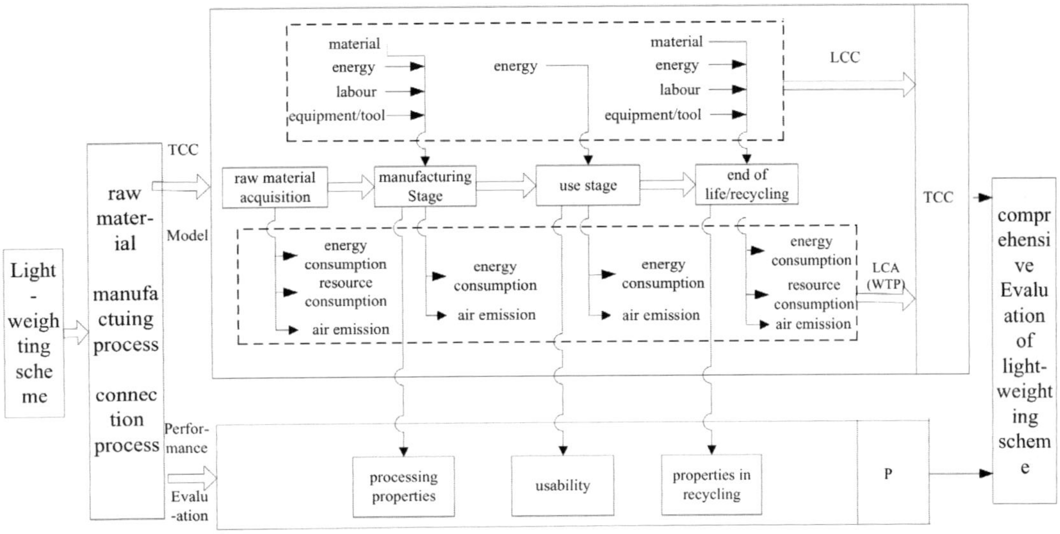

Figure 1. Comprehensive evaluation model of light weight scheme.

Evaluation scope includes raw material acquisition stage, manufacturing stage, usage stage and recovery stage. The input-output model based on process, the elements put into each stage of the life cycle such as materials, energy, labour, equipment and other costs are main data to compute LCC while environmental factors generated by the consumption of resources at each stage of the life cycle are main data used for the LCA detailed list analysis. Monetizing LCA to achieve the integration of LCC and LCA so that a comprehensive cost of light-weighting program can be achieved. Then, evaluate performance within manufacturing process, usage process and recovery phase using AHP and SAW methods to get the performance value. Finally, integrate comprehensive cost and performance value to obtain the comprehensive quantitative evaluation of the light-weighting scheme

2.1 Life cycle cost model

The life cycle cost relates to different life cycle stages, so establishing cost model based on process ac-cording to input-output model is needed. Input factors are divided into material cost C_M labor cost C_L energy cost C_E equipment and tools $C_{T/E}$ calculate the cost of each process, as shown in the formula (1):

$$C_P = C_M + C_L + C_E + C_{T/E} \qquad (1)$$

The calculation formula of each input element in the process model is (2)-(5):

$$C_M = \sum_{i=1}^{I} w_i.P_i.(1+r_i) \qquad (2)$$

w_i , P_i , r_i represent material quantity, unit price and scrap rate

$$C_L = \sum_{k=1}^{n} P_k.t_k.n_k \qquad (3)$$

P_k, t_k, n_k represent reward per unit time, working hours and the number of workers in the k process.

$$C_E = \sum_{n=1}^{N} E_n.P_n \qquad (4)$$

E_n, P_n represent of energy consumed in the nth energy and the price per unit of energy. represent material quantity, unit price and scrap rate.

$$C_{T/E} = \sum_{m=1}^{M} I_m.r.(1+r)^{mn} \Big/ ((1+r)-1)T/UT \qquad (5)$$

I_m , mn , T , UT represent The cost of investment, investment return rate, the payback period of the mth investment, the running time of the year, the total service life of them.

$$LCC = w_p.C_{PP} + w_U.C_{PU} + w_R.C_{RR} \quad (6)$$

C_{PP}, C_{PU}, C_{RR} represent the cost of manufacturing phase, usage phase and recovery phase, w_P, w_U, w_R represent the weight of manufacturing phase, usage phase and recovery phase.

2.2 Assessment model of life cycle environmental impact

Currently, carbon footprint and LCA method are most frequently applied to value the environmental impact, of which LCA applies more EI99 to quantify. In this paper, based on the LCA method, environmental impact is monetized by applying WTP weighted index and then incorporated into the comprehensive life cycle cost. Specific steps are as follows:

Classification: analyze detailed data to classify various factors as types.

Characterization: convert factors Q_i into influencing outcomes with unified units. Then, through totalling influencing value is achieved which can reflect the environmental impact. The formula is as follows:

$$S_k = \sum_{i=1}^{I} Q_i.EP_i \quad (7)$$

Q_i, stands for the analysis measurement of each unit per function releasing i, EP_i is the equivalent factor.

Weight processing: S_k characterized and the coefficient of willingness to pay, V_k are processed together, to get WTP_k to achieve the goal to monetize WTP, the formula is:

$$WTP = \sum_{k=1}^{K} S_K.V_k \quad (8)$$

2.3 Life cycle performance evaluation model

Firstly, determine the performance evaluation index of each stage. Secondly, collect data to get the attribute value of each index and make the qualitative index quantitative. Then, normalize the index value by mathematical transformation. Finally, performance values are obtained by using AHP method and SAW method. The calculation formula is as follows:

$$P_i = \sum_{j=1}^{J} w_j.mn_{ij} \quad (9)$$

P_i, w_j represent performance value, weight of scheme i (i=1,2,…,N), m_{ij} represent the attribute value of j(j=1,2,3…,J) of scheme

$$P_i = \sum_{j=1}^{J} = w_j.m_{ijnorm} \quad (10)$$

Efficiency index standardization

$$m_{ijnorm} = \frac{m_{ij}}{\max[m_{ij}]_1^N} \quad (11)$$

Cost index standardization

$$m_{ijnorm} = \frac{\min[m_{ij}]_1^N}{m_{ij}} \quad (12)$$

2.4 Comprehensive evaluation model

The output of TCC that integrated economy and environment is cost, however, the result of life cycle performance evaluation is non-dimensional value. To realize the integration of the two, the standardization of TCC is needed. The formulation is as follows:

$$TCC'_{knorm} = \frac{\min[TCC]_1^N}{TCC_k} \quad (13)$$

Formula (14) is used to get the comprehensive evaluation value of total life cycle cost, environment and performance.

$$Value_k = P_k / TCC'_{knorm} \quad (14)$$

3 Model application

In this paper, engine hood is used as example to evaluate light-weighting. According to literature [7], 3 types of designation plan can be seen as Table 1: includes inner p.

Table 1. Light-weighting scheme of engine hoods.

Scheme	Scheme 1	Scheme 2	Scheme 3
Outer plate Inner plate	0.75 mm strengthened steel 0.60 mm strengthened steel	1.50 mm aluminum alloy 0.75 mm strengthened steel	1.5 mm aluminum alloy 1.5 mm aluminum alloy
Connection type	Flanging, rolling depression, splice	Flanging, rolling depression, splice	Flanging rolling depression, splice
Manufacturing process	hydraumatic	Aluminum/stamping, strengthened steel/hydraumatic	stamping

Table 2. Cost of each phase.

Scheme	manufacturing phase/[\$] (w_P=0.64)	Use phase/[\$] (w_U=0.28)	Recovery phase/[\$] (w_R=0.07)	Total cost/[\$]
Scheme 1	50.39	133.42	-0.33	69.58
Scheme 2	60.01	128.18	-5.45	73.92
Scheme 3	74.32	107.26	-9.68	76.92

3.1 Life cycle cost estimation

3.1.1 Cost estimation in manufacturing phase

Material cost is obtained based on data collected and formula (2) based on manufacturing, connection characteristics of each scheme, equipment costs mainly consist of molds costs and fashioned equipment costs. Therefore, labor costs, energy costs, equipment costs and total costs can be calculated based on data collected and formula (3) (4) (5) (1).

3.1.2 Use phase cost estimate

The main cost is the cost of fuel consumption and maintenance in this stage, According to the literature [6], automobile fuel consumption and weight are highly correlated. Consumption of gasoline will in-crease 0.004L when weight of each car traveling 100 kilometers increases by 1 kilogram. Therefore, this paper only considers the cost of fuel consumption, i.e. energy cost, which is directly related to the de-sign scheme. Assuming that the mileage during life cycle of a car is 25 million kilometers, 93 gasoline price was 133.42 $/ L

3.1.3 Cost estimation within recovery phase

This stage mainly considers the equipment cost, energy cost and labor cost involved in dismantling and recycling activities. Cost of raw materials is the in-come of recycled materials. Assume that the recovery rates of highly strengthened steel and aluminum alloy were 85% and 95%, respectively. The price of aluminum scrap and steel scrap are is 1.1 $ /kg and 0.12 $ /kg. Therefore, cost within the recovery stage is calculated through formula (2) - (5).

As costs incurred during each stage of the life cycle belong to different subjects, designers consider the manufacturing costs as well as costs incurred within usage stage and recycle stage from the perspective of company interest. As a result, weight is needed to value different stages in the life cycle, this paper adopts weight data from literature [7] which sets manufacturing as 0.64, usage as 0.28 and recycling as 0.07, to get the LCC of different scheme through formula (6) The cost of different phase show in Table 2.

3.2 Life cycle environmental impact

By using the collected data, and considering the influence of producing 1kg steel and aluminum on the environment, the detailed list of raw material data in acquisition stage. [8], [9] can be calculated, as well as the detailed list data of manufacturing phase through consumption and unit emission of oil, coal and electricity. Index of environmental impact in using phase is determined by data of use stage and environmental impact of unit oil consumption amount. In recovery phase, improvement of environment brought by resource and energy conservations of material recycling is mainly considered. Finally, on the basis of materials and processes in all schemes, environmental impact data in all phases of the life cycle are determined

The characteristic factor is obtained through the research of existing references. According to domes-tic air emission, resource and energy consumption, the resource tax rate as well as the reference document [10], [11], a list of social-willingness-to-pay is obtained CO_2 is 3.4E-02, NOx is1.39E-01, SO2 is 9.74 E-02, petroleum is 0.98E-03, coal is 8.40E-03, Ironore is 1.71E-02.According to the composition of electricity in China, electric energy can be converted into the consumption of coal, and 1KWh electricity =0.424kg Coal [12], assuming that gasoline consumption is equivalent to petroleum consumption. According to the formula (7) and (8), the life cycle environment impact in each scheme is obtained, as shown in Table 3:

Table 3. Life cycle environment impact in each scheme.

Scheme	Raw materials acquisition phase	Manufacturing phase	Using phase	Recovery phase	WTP/ [$]
1	1.52	0.77	10.79	-0.86	12.22
2	3.17	1.07	10.36	-1.94	12.66
3	4.37	1.23	8.67	-2.74	11.53

Table 4. Life cycle performance values of each scheme.

		Index							
		The manufacturing phase			The using phase			The recovery phase	
Scheme	B_j	Feasibility	Complexity	Productivity	Rigidity	Modal	Weight	Recovery rate	P_i
	w_j		0.08	0.05	0.36	0.21	0.08	0.07	
1		0.42	1	0.33	1	0.87	0.80	0.89	0.82
2		0.71	0.6	0.55	0.74	0.78	0.84	0.94	0.74
3		1	0.42	0.78	0.33	1	1	1	0.69

Table 5. The comprehensive evaluation values of each scheme in life cycle.

Scheme	LCC/[$]	WTP/[$]	TCC/[$]	Standardization	P	P/TCC (Value)
1	69.58	12.22	81.80	1	0.82	0.82
2	73.92	12.66	86.58	0.94	0.74	0.79
3	76.92	11.53	88.45	0.92	0.69	0.75

3.3 Life cycle performance evaluation

As a closure member of car-body, engine hood is designed to protect the engine, so that the rigidity and strength affect crashworthiness directly. The weight directly influences fuel costs in using process. The modal reflects vibration reducing performance of cars. Therefore, rigidity, first-order modal and weight are selected as performance evaluation indexes in using process.

While light-weighting scheme is being designed performance of manufacture process and recovery process, as well as performance of products using phase are all considered. As a consequence, in this paper, the feasibility, complexity and productivity of the process are selected as the evaluation indexes of the process performance in the manufacturing stage, and the recovery rate as the evaluation index of the ending stage. Based on the existing references, expert advice and the 5 Component [13], the manufacturing index values are determined.

It is different of the importance of each stage in the product life cycle. It is also different of the importance each performance index counted in each stage. In this paper, the AHP method is applied to determine the weight of each stage of the life cycle and the weight of each performance index counted in each stage. So as to determine the weights of manufacturing stage, use stage and recovery stage were 0.27, 0.65 and 0.07.

Using the same method, the weight of feasibility, the complexity and the productivity in the manufacturing stage are determined as 0.53, 0.3 and 0.17. The weights of rigidity, modal and weight in using stage are 0.56, 0.32 and 0.12. Since only one index is chosen in recovery phase, the weight is 1. After standardization of the performance values in all phases by using formula (11) and (12), and calculation of weight of life cycle indexes by using formula (10), performance values of each scheme are as shown in Table 4.

3.4 Comprehensive evaluation

TCC of each scheme is obtained by totaling LCC and WTP, and standardized TCC is got by formula (13). Then by integrating P with formula (14), life cycle comprehensive evaluation value of all schemes have been obtained, as shown in Table 5.

In summary, with consideration of material and process, and based on the perspective of LCC, 3 light-weighting schemes are evaluated comprehensively from three dimensions -economics, environment and performance. The best one is program 1, which can be seen in Table 5.

4 Conclusions

This paper, based on LCC theory, and from the perspective of the resource-based view, analyzes resource consumptions of the designed object in each stage of the life cycle, and the environmental impacts caused by resource consumptions. Meanwhile, this paper takes into account the performance factor of the designed object, and constructs light-weighting design synthetic evaluation model. On the basis of this model, this paper establishes the life cycle cost model, the environmental impact assessment model and the product performance evaluation model. According to the consumption of each stage, the calculation formula, the calculation process and the solving method of each model are provided

Then, taking three design schemes of the automobile engine hood as examples, the material cost, manufacturing cost, consumption cost and recovery cost are first calculated by using the cost model. Then, the BOM that impacts environment is determined according to the cost of each stage. Besides, by using environmental impacts evaluation model, environmental impacts in each stage are monetarily measured, so as to realize the unification of two measurement standards. In the end, by applying the performance evaluation model, the performance indexes are calculated in 3 stages — hood manufacturing, usage and recovery stages. By means of AHP method, the weights of life cycle stages and performances indexes are determined, which comes to the only value (P) of metric properties, and realization of quantitative unification of the three dimensions of the comprehensive evaluation model. Then, the fuzziness of the qualitative evaluation is solved.

Finally, the validity and feasibility of the model are proved by an example, and proved to be applicable for wider use.

Acknowledgement

This paper was financially supported by the National Natural Science Foundation(71172101).The project is Building and Application Knowledge base with Theory of Life Cycle Cost.

References

1. I. Ribeiro, Paulo Pec etc., Journal of Cleaner Production, **16**, 1887-1899 (2008)

2. R. A. Witik, J. Payet, Composites Part A: Applied Science and Manufacturing. **42**, 1694-1709 (2011)
3. A. T. Mayyas, A. Qattawi, Energy, **39**, 412-425 (2012)
4. L. Ying-ning, L. Zhen-hua, Journal of Mechanical Engineering, no. 12, 289-294 (2009)
5. L. Yan-ping, L. Hai-jiang T. Rong-hui, Computer Integrated Manufacturing Systems, no. 01, 37-44 (2011)
6. H. Qing-lan, Z. Yang, Science and Technology Management Research. **7**. 180-184 (2015)
7. W. Hong-yan, G. Wei-ming, Journal of Tong Ji U (Nature Science), vol. 43, no. 8, 1098-1103 (2006)
8. L. Zhen, Research on Environmental Consciousness Design and Evaluation Method of Electromechanical Products, (2013).
9. W. Shuai, Reserch on the EIA of Read-mixed concrete during the life cycle, (2009)
10. C. Wei-qiang, W Hong-yan, W. Juan-ni etc., (2009). Light Metal, **5**. 2-10 (2009)
11. L. Zhi-chao, Life Cycle Assessment Meth-odology of Original Manufacturing and Remanu-facturing of an Engine, (2013)
12. Liu S N, Lin Z S, Zhang X W, China Environmental Science, **18**, 328-332
13. Chen SJ, Hwang, Lecture notes in economics and Mathe-matical systems, Springer, (1992)
14. L. Shi-hui, H. Qing-lan, Accounting Research, **7**, 35-42 (2013)

MATEC Web of Conferences **81**, 08005 (2016) DOI: 10.1051/matecconf/20168108005
ICTTE 2016

Fatigue Analysis of Automobile Control Arm Based on Ncode

Ren Huanmei[1] and Chen Lihui[2]

[1] *No. 1111, Hucheng Ring Road, Pudong New District Shanghai City, PRC*
[2] *No. 145, Tiantai Road, Hantan District, Hanzhong, Shaanxi, PRC*

Abstract. In order to improve the vehicle chassis structure durability, the fatigue analysis and optimization design of the low control arm (LCA) was taken. A finite element model was established. By using this model, the stress distribution and lowest point of lifetime of the control arm under fatigue load was calculated. Based on the results of analysis, the optimization scheme according to the structure characteristics of components was presented, and a solution to improve the control arm fatigue life was given out. The research provided reference for engineering application of calculation and optimization of chassis components fatigue life.

1 Introduction

Suspension system plays an important role in the chassis structure, it transfers force and torque between the wheels and the body, a guiding and controlling the relative movement of the wheel and body. It mitigates body shock transferred from the road and attenuates vibration of system and so on. Therefore, the suspension design is very important in automobile chassis design and the development of automobiles. McPherson suspension system has the advantages of compact structure, little change of wheel alignment parameters with beating of wheels, stability, and due to the cancellation of the upper arm, the arrangement of engine and steering system is convenient [1]. It is a durable independent suspension of with strong ability to adapt to the road. It has been used most widely as the independent suspension in the car.

In early stage of product development, it is important to focus on investigation of the structure strength and fatigue life of the key pieces. Suspension is an important force transferring system, its strength and fatigue life is required to be very strict. For McPherson independent suspension, the lower arm is the affordability of the subject [2]. Therefore the strength and service life of lower arm play an important role in suspension work [3]. Reliable prediction of fatigue life, analysis of influence factors of fatigue life , are of great significance to guide the design.

2 Fatigue load

For calculation, there are two categories of fatigue load named classic fatigue load and fatigue load spectrum [4]. Classic fatigue condition is defined as conditions in certain standards and specifications based on experience and accumulation, which is combinations of vertical beating, lateral impact, longitudinal impact, steering (left and right), braking and acceleration (left and right). Classic fatigue condition depends on long time accumulation and maturity standards. The standard load spectrum is measurements of the direct or indirect time-varying load at hard points of components.

The load spectrum are from the following: 1) Large vehicle factory's load spectrum accumulation of vehicles on the same platform. These measured spectrum, multiplied by a coefficient, can be used as a initial assessment of fatigue life in the early stage of the development of new models. 2) Directly measuring load spectrum of concerned parts. Due to measurement on a single component of single model, this method has a disadvantage of poor. 3) To measure spectrum of the wheel core with Six Axis Force method, then through load decomposition to obtain the load spectrum of concerned parts is obtained[7]. As this paper is for a new vehicle model development, the classic fatigue condition method was used to calculate fatigue life of the control arm.

3 Control arm fatigue life prediction

There are two categories of fatigue life calculation method commonly used. 1) Based on S-N curve, the fatigue life was calculated through application of linear fatigue cumulative damage rule. This method is also called nominal stress approach, or SN approach. 2) Get life of crack with stress-strain estimation, then added with crack propagation life from fracture mechanics to get total life[5,6]. This is called EN method. SN method is applied in high cycle fatigue, EN method is applied in low cycle fatigue. For automobile chassis component, most of the damage is low cycle fatigue failure, so we used EN method to calculate the fatigue life[7].

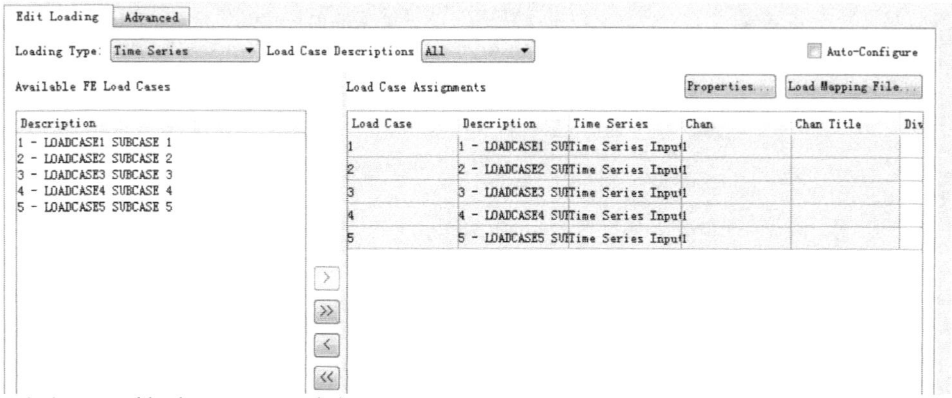

Figure 1. Analysis step and load spectrum correlation.

3.1 Setting fatigue load

The stress results were calculated for control arm under fatigue condition by using Natran linear solver. The results file was put into Ncode software as the input file of fatigue calculation. The half sine wave is set simulating bench test loading process. The waveform curve loading and the analysis step are correlated to simulating a drive cycle [8], [9]. The vertical beating, lateral impact, longitudinal impact, steering (left and right), brake and acceleration were respectively cyclic loaded once as a big cycle. The correlation results are shown in Figure 1.

3.2 Material parameters

Upon experience, SAPH440 was used for the control arm with reference the same vehicle classification. Material parameters of SAPH440 was input by using material editor in Ncode. The stress-strain curve was generated, which is shown in Figure 2. The result was used to solve the fatigue life of the control arm.

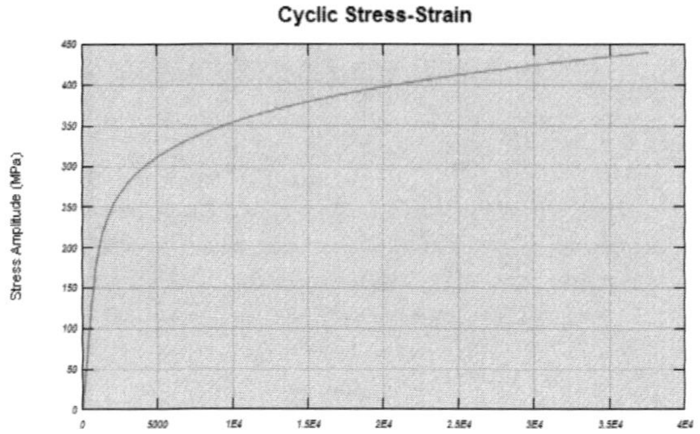

Figure 2. Stress-strain curves of SAPH440.

The output was set up to complete the fatigue process. The complete solving process was shown in Figure 3.

According to the calculation result, the control arm had a minimum life cycle of 1.5E+5, which did not meet the life target of 1E+6 cycle. The minimum life presents in the intensive board. Therefore, it is necessary to optimize the design of reinforcing plate in order to meet the minimum life target.

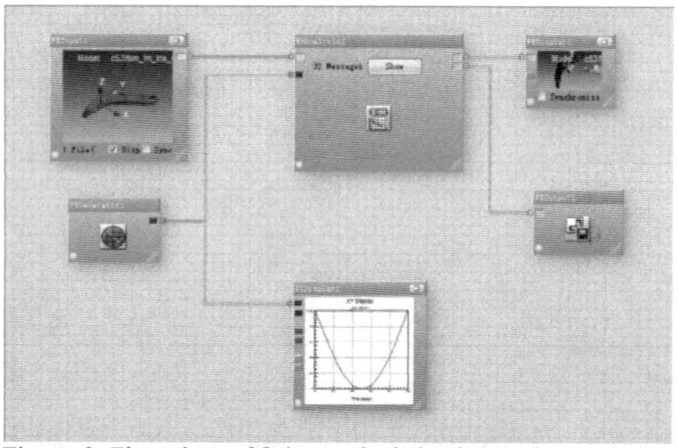

Figure 3. Flow chart of fatigue calculation in Ncode.

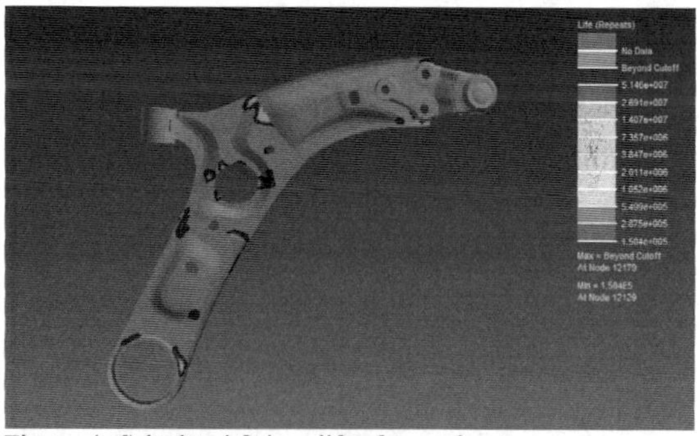

Figure 4. Calculated fatigue life of control arm.

As results of analysis, life of the control arm body should be more than 1E+7 cycle, only local area of lower reinforcement did not meet the minimum life requirements. Lower life in this area was maily due to unreasonable structure. The bevel was too close to the weld bead, causing stress concentration in local area, which led to low fatigue life.

The original control arm body was retained, meanwhile the reinforcing plate was optimized [10]. The bevel of the reinforcing plate was shifted away from end of the weld bead. This migrated the stress concentration in the area. Resulted structure is shown in Figure 5.

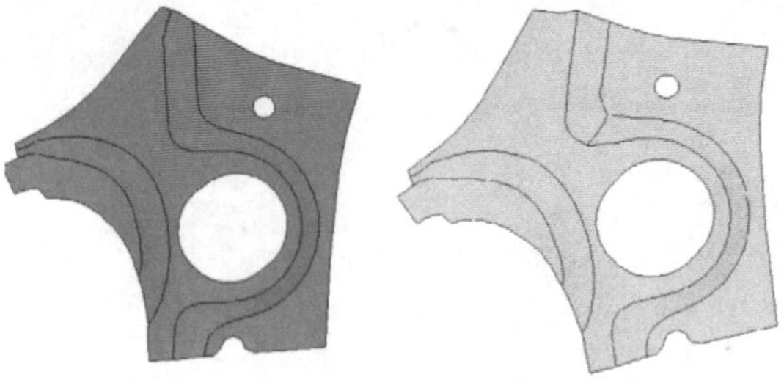

Before optimization After optimization

Figure 5. Comparison of the reinforcing plate before and after optimization.

3.3 Optimization and evaluation

Optimization of the bevel position in reinforcing plate effectively migrated stress concentration with very little effect on other parts of the control arm. Fatigue life of this area was increased to 1.1E+6 cycle. After the optimization, control arm had lowest life of 1.1E+6 cycles, and fatigue life of the vast majority of the arm were higher than 7E+7 cycle. After optimization, control arm fatigue life met the target. The lifetime results of optimization is shown in Figure 6.

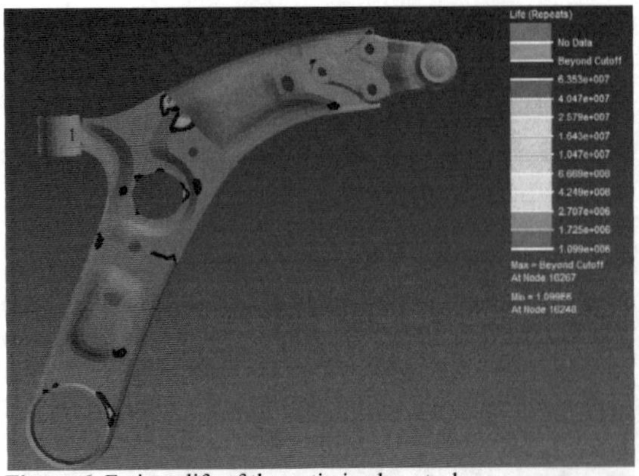

Figure 6. Fatigue life of the optimized control arm.

4 Conclusions

In this paper, a basic method of fatigue analysis and fatigue life prediction of control arm in early stage of vehicle development is presented, the minimum fatigue life was determined through calculation of control arm's fatigue life. The optimization scheme was presented on basis of this. With this scheme, making of control arm meeting the application requirements will be assured. The method allows fatigue life of control arm to be predicted and design to be optimized in early stage of the product development, the design of product also can be optimized upon it. This greatly reduced the product development cycle [11]. It is of significant importance to direct the engineering design.

References

1. Liu Yongchen, Wang Guolin, Sun Li. Fatigue damage analysis and life prediction for vehicle control arm. *Journal of agricultural engineering*. Vol. **35**, 57-59 (2013)
2. Liu Yongchen, Wang Guolin, Sun Li. Reserch of the Vehicle Front Suspension Assembly Durability Simulation. *The Fifth China CAE Annual Conference*. 173-177 (2009)
3. Nagendra Iranna Jamadar and K K DhandeTransient. Analysis of Vehicle Passive Suspension System. *IJMERR*. Volume **3**, No. 3, July, 451-464 (2014)
4. Huang Minfeng, Jiang Chun. Analyzing fatigue life of automobile components based on FEM. *Mechanical Research & Application*. Vol. **21**, 57-60 (2008)
5. Zhang Haibo, Zhang Ruijun. A Finite Element Analysis of Automobile Suspension Control Arm Base on ANSYS. *Modular Machine Tool & Automatic Manufacturing Technique*. Vol. **45**, 42-45 (2015)
6. Thanh-Tung, Tran, Hirosh. Advanced Passive Suspension with Inerter Devices and Optimization Design for Vehicle Oscillation. *IJMERR*. Volume **4**, No. 4, October, 354-360 (2015)
7. HB Zhang, RJ Zhang, Y Chang. Finite Element Analysis of Automobile Suspension Control Arm. *Applied Mechanics & Materials*. Vol. 752-753, 859-863 (2015)
8. BC Song, HS Park, YM Kwon. YC Park: Structural Design of an Upper Control Arm, Considering Static Strength. *Transactions of the Korean Society of Automotive Engineers*. Vol. 17, 190-196 (2009)
9. B Ashtekar and A G Thakur. Simulink Model of Suspension System and its Validation on Suspension Test Rig. *IJMERR*. Volume **3**, No. 3, July, 811-818 (2014)
10. Maotao Zhu, Jiahuan Guo, Tao Liu. Finite Element Analysis and Fatigue Analysis of Control Arm. *The Series Lecture Notes in Electrical Engineering*. Vol. 364, 457-468 (2015)
11. Raghav Goyal, Anitej Tripathy, Mandeep Singh Gujral. Producing Suspension. *IJMERR*. Volume **3**, No. 2, April, 239-244 (2014)

MATEC Web of Conferences **81**, 08006 (2016) DOI: 10.1051/matecconf/20168108006
ICTTE 2016

Investigating reliability centered maintenance (RCM) for public road mass transportation vehicles

James Hopkinson , Noel Perera , Evran Kiazim

Department of Mechanical and Construction Engineering, Faculty of Engineering and Environment, Northumbria University, Wynne Jones Building, Newcastle Upon Tyne, NE1 8ST,England

Abstract. Presently public road mass transportation organisations maintain their vehicles using scheduled corrective maintenance which is carried out at the discretion of the organisation. Operating a corrective maintenance policy means that equipment are allowed to operate to failure and are only repaired or replaced when it fails. Corrective maintenance is generally recognised as an inefficient approach to maintaining equipment as the costs in terms of finance and time can be excessive. In many industries, from aerospace to oil production, a system of reliability centered maintenance (RCM) is utilised. RCM estimates when equipment failure can be expected to occur and ensures that maintenance is performed before it does. Using the RCM approach usually results in assets being more reliable, with low equipment failure rates within its mechanical systems. This often results in significant financial benefit to the organisation as operating equipment become more reliable allowing the vehicles to be in service for longer and achieving higher safety standards. This investigation has predicted that potential equipment failure could take place between the vehicle mileage of 6000 -12000 miles. Using the consecutive reading /sampling method together with a set oil spray diameter size alert value of 130 mm, it was possible to further narrow down the potential equipment failure to a vehicle mileage of 10000 miles.

1 Introduction

In recent times the public road mass transportation industry has faced challenging demands with respect to its environmental effects, emissions, reliability, safety and commercial competition. In meeting these challenges its important that its workforce and equipment performed daily at its optimum level. One potential option of addressing these challenges was the adoption of the reliability centered maintenance (RCM) approach. A key concept of RCM is reliability which is important for improving safety. Researchers [1,2] have described reliability as "probability that an item will perform a required function without failure under stated conditions for a specified period of time". RCM has existed in industry for over 30 years presenting optimisation of preventive maintenance techniques [3]. RCM was first developed in the airline industry in the late 1960s [4] to address concerns with the introduction of the wide body jets. As maintenance cost started to increase sharply, the airline industry at that time started to become concern that it would be financially unfeasible to use conventional maintenance techniques to maintain aircrafts that were larger and more complex. As conventional maintenance techniques were interval based this further contributed to the increase in maintenance cost relative to the other operating cost This contributed to the growth of RCM which was based on equipment performance data and criticality. The RCM cost remained approximately the same as the cost of the conventional maintenance techniques however aircraft availability and reliability improved. This was because more attention was devoted to maintaining the equipment that was most likely to fail. Thus the objectives of RCM [5,6] are to reduce maintenance and support cost as well as increase equipment reliability and safety. Developing a maintenance strategy and plan is key to the implementation of RCM within an organisation/ industry. The two possible RCM techniques that can be adopted are preventative and predictive maintenance techniques.

Preventive maintenance is carried out on a working piece of equipment to avoid unanticipated failure. Therefore it is maintenance which is performed frequently to reduce the possibility of equipment failure. The equipment time to failure and trend of failure rate can be monitored to increase the efficiency and cost effectiveness of employing the preventive maintenance strategy. Published work [7] has shown that replacing equipment which has a diminishing failure rate might increase its possibility of failing. However replacing equipment which has a constant failure rate has shown to have no effect on its possibility of failing. On the other hand replacing equipment which has a rising failure rate

DOI: 10.1051/matecconf/20168108006

has shown that it can reduce its possibility of failing. This is illustrated in figure 1 showing the bath tub curve for an equipment standard life cycle. Following this preventive maintenance should be carried out on equipment that is in the 3rd section of the bath tub curve i.e. equipment that is in its life cycle period of experiencing increasing failure.

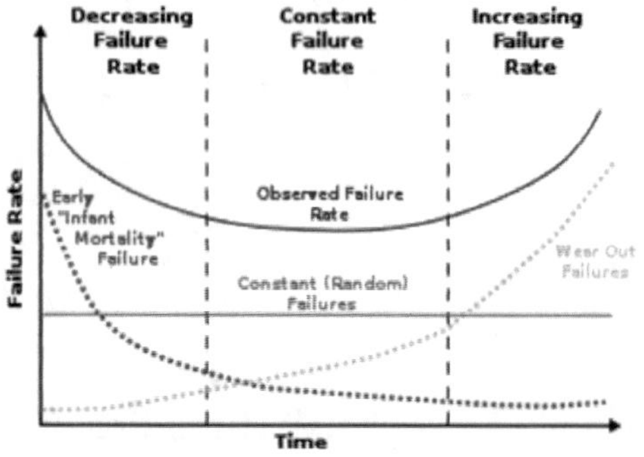

Figure 1. Bath tub curve [5]

Predictive maintenance is carried out to ascertain the working condition of the equipment whilst in operation to predict when maintenance should be performed. Most equipment failure modes are not time dependent but do produce indications that failure is imminent or is occurring. In this case if these indications can be detected it would be possible to identify the failing equipment and take remedial action before it experiences total failure.

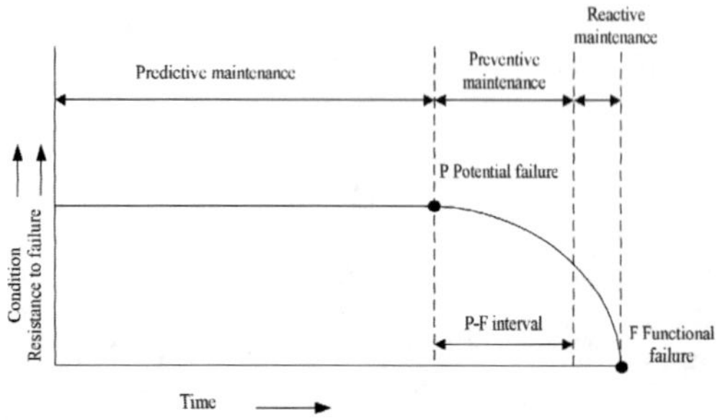

Figure 2. The maintenance potential failure (P-F) curve [8]

Figure 2 above shows the potential failure (P-F) curve which illustrates the point P where failure can first be detected followed by point F when failure begins. The point in the failure process where it would be possible to detect whether failure is occurring or about to occur is known as the potential failure, point P [8]. If failure is not detected and addressed at point P it continues to deteriorate until it reaches point F, functional failure.

Another important aspect of the reliability centered maintenance strategy is the drafting of the reliability programme which identifies all the proactive tasks that have to be completed. Reliability programmes do not react to single events but is utilised to detect system problems and trends that may trigger many other subsequent events [9]. A reliability program must be a written program approved by an Authority. It has to specify responsibilities and procedures within an operator's organisation which will ensure successful implementation of a reliability program. In the reliability programme identified tasks should clearly be recorded as completed by named operator(s), date and time of completion and description of the level of completion. In industry, an operator can vary a number of parameters used. For example pilot complaints per 100 flights, technical delays above 15 minutes per 1000 flights, component replacements per 1000 component flight hours, unscheduled component replacements per 1000 flight hours, engine in flight shut downs and unscheduled changes per 1000 engine hours, repetitive pilot complaints, long lasting technical issues and significant findings during maintenance activities. For all of the measured parameters, an upper control limit or alert value has to be determined. An alert value is used to recognise and react to significant deviations from statistically acceptable limits in reliability. An alert value is a pre-determined threshold that triggers an alert. Upper control level or alert value is a statistical value which shows the limit below which deviations are considered statistically acceptable. In the case of an exceeded alert limit value, system reliability is considered to be unstable. An appropriate corrective action must be applied to stop a negative trend due to an alert value being exceeded repeatedly.

2 Experimental test

For the purpose of this paper an investigation was carried into the oil contamination of the air braking system of a public road mass transportation vehicle [10]. The braking system on these vehicles operated as a compressed air system. Air was taken in to the system, filtered and compressed and stored in the air reservoirs at a pressure of around 9 bar. The compressed air was then delivered via a control system. The air compressor mechanism was powered via the engine timing gears or a belt running off the crankshaft pulley. This mechanism had to be cooled and lubricated to avoid overheating or seizure. This was achieved through the engines cooling and lubrication system. As the air can contain contaminants such as oil and water, normally due to worn piston rings in the compressor, it had to be funnelled through an air dryer to remove the impurities in the air. This reduced the potential for corrosion and subsequent failure in the air braking system. The pressurised air was then held in a vessel known as the air reservoir tank until it was distributed throughout the system via a 4-way protection valve. This valve channelled the air to the front and rear circuit brake air tanks, parking brake air tank and an air tank for the auxiliary equipment. This air braking system also included safety air release valves to prevent the air lines from becoming over-pressurised and potentially rupturing.

An experimental test was initiated to investigate the effectiveness of the proposed modifications to the air dryers within the air braking system to address the issue of oil contamination. The purpose of this investigation was to identify the most effective modification to the air dryer system to generate the greatest reduction of oil contamination within the air braking system. The most effective modification was then implemented on the fleet

DOI: 10.1051/matecconf/20168108006

of this public road mass transportation vehicle. The proposed modifications were:

Modification 1: The air dryer was fitted with an Oil Separator Cartridge (O.S.C.).

Modification 2: Air dryer was overhauled by using Fluorocarbon O-rings where possible.

Modification 3: Air dryer cartridge renewed with standard fitment.

Each modification was applied to one vehicle. For example, modification 1 was applied to Vehicle 1 and modification 2 was applied to Vehicle 2 etc.

2.1 Test procedure

In the air braking system the compressed air was stored within the air reservoir which was fabricated with a drain plug. The purpose of this plug was to allow air to be manually purged from the system. It was possible to visually gauge the level of oil contamination in the air system when purging it by measuring the resulting oil spray output from this reservoir. The oil was sprayed onto a piece of paper and the larger the diameter of the oil spray over a fixed distance, the higher the level of oil contamination within the air system. This was known as the blot test. A blot test was carried out every 1000 miles of vehicle operation and the experimental detail is shown in table 1. The results were then tabulated and plotted as a graph.

Table 1. Experimental detail

Experiment 1	Modification 1:	The air dryer was fitted with an Oil Separator Cartridge (O.S.C.).
Experiment 2	Modification 2:	Air dryer was overhauled by using Fluorocarbon O-rings where possible.
Experiment 3	Modification 3:	Air dryer cartridge renewed with standard fitment.

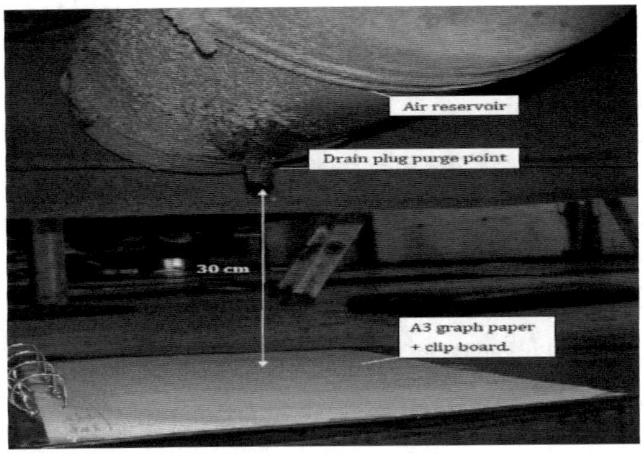

Figure 3. Blot test set up [10]

3 Results

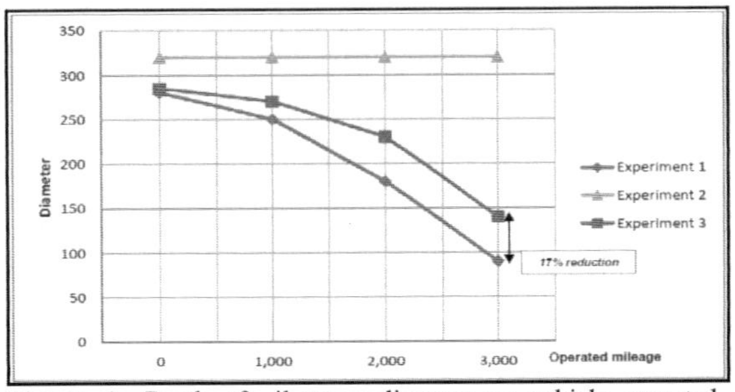

Figure 4. Graph of oil spray diameter vs vehicle operated mileage [10]

It can be deduced from the graph shown in figure 4 that experiment 2 had no effect on the level of oil contamination within the air brake system. However experiments 1 and 3 have had a positive effect on the oil contamination levels. Experiment 1 achieved a 17% reduction in oil contamination when compared with experiment 3. Experiment 1 and 3 continued to remove the oil contaminants from the air braking system up to an operated mileage of 3000 miles. Hence it can be assumed that if the experiments were allowed to continue beyond 3000 miles, the oil contamination would continue to reduce at a similar rate. However it is anticipated that this trend would continue until the applied modifications could no longer remove any further oil contaminants from the system, causing it to plateaux.

Similar to other equipment within a public road mass transportation vehicle's mechanical system, the oil separator cartridge (O.S.C.) has to be replaced several times within the vehicle's operation period due to degradation and inevitable corrosion. Constant intake of oil contaminants contributed to a build-up of residue in the filter which resulted in performance degradation or equipment failure. The loss of performance in the O.S.C meant that significantly more oil contaminants were able to enter the air braking system potentially damaging it. For that reason, it's essential that the compromised O.S.C cartridge be replaced before significant damage can be caused to the air braking system.

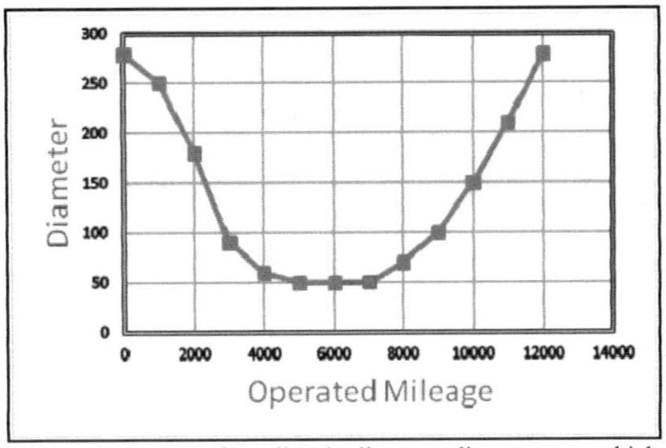

Figure 5. Graph of predicted oil spray diameter vs vehicle operated mileage

The graph showed in figure 5 highlights for experiment 1 the predicted oil spray diameter with respect to the vehicle operation mileage over a higher operating mileage. Utilising the bath tub curve methodology, the authors have predicted the possible O.S.C performance beyond the vehicle operated 3000 mileage by extrapolating the experimental data. The extrapolated experimental data highlighted a steady reduction of the oil spray output diameter as it reached a minimum output diameter of 50 mm at a mileage of 6000 miles. This reduction in the level of the oil contamination can be attributed to the O.S.C performing to its optimum capability. However beyond the 6000 mile point the size of oil spray diameter then began to slowly increase at an accelerated rate till it reached 12000 miles. At this point the extrapolated experimental data showed that the O.S.C would have the same performance level as at the beginning of the experiment.

4 Discussion of results

The gradual accelerated increase in the size of the oil spray diameter can be attributed to the impending failure of the O.S.C. The predicted performance of the O.S.C as it began to experience potential failure from the 6000 mile point to its functional failure at the 12000 mile point can provide maintenance engineers with an estimated timeline for scheduling maintenance activities. However it is important to note that the graph shown in figure 5 was based on extrapolated experimental data beyond the vehicle's experimental operating mileage of 3000 miles. Hence the accuracy of the predicted performance of the O.S.C as shown in figure 5 is debateable without a proper validation of the experimental results. The O.S.C should be replaced at any point when it experiences potential failure to functional failure as shown in figure 2. However it not cost effective to replace the O.S.C without first determining whether it's at the end of its service life. Despite experiencing potential failure the O.S.C could still continue to perform at an acceptable level for an indefinite period of time. Also incorrect blot test readings could be recorded leading to the replacement of a potentially perfect performing O.S.C. Therefore to address these issues the outcomes of including the alert value and consecutive reading/sampling techniques were investigated.

For this investigation an alert value was set at the oil spray diameter size of 130 mm which is shown as a straight horizontal line in the graph in figure 6. During the service life of the O.S.C, the size of the oil spray diameter may start above the alert value but when performing to its optimum capability the size of the oil spray diameter would remain below the alert value. However, as the performance levels of the O.S.C. decreased, the size of the oil spray diameter increased. For a brief period after this occurrence, no maintenance activity was required as the O.S.C. continued to perform at an acceptable level. However, the blot test circled in figure 6 showed that when the size of the oil spray diameter exceeded 130mm, it triggered the immediate replacement of the O.S.C. Although the alert value could be considered as an effective method, a potential drawback was the possibility that one anomalous result above the alert value could cause the premature replacement of the O.S.C before its failure point.

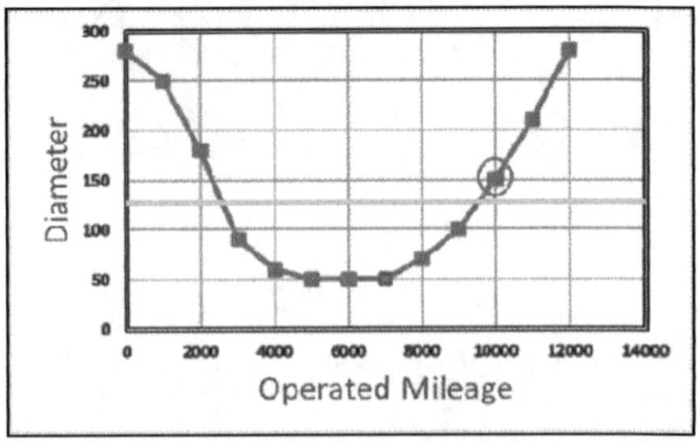

Figure 6. Graph of predicted oil spray diameter vs vehicle operated mileage with indicated alert value

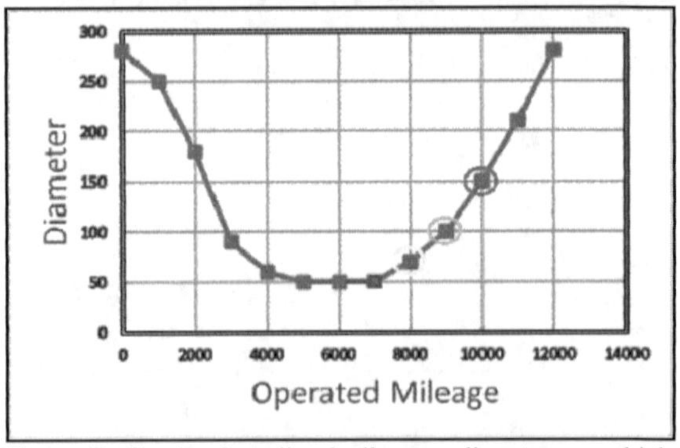

Figure 7. Graph of predicted oil spray diameter vs vehicle operated mileage with highlighted significant blot tests

Using the consecutive reading /sampling method the graph in figure 7 showed the first circled indication of an O.S.C. failure at the 8000 mile point. This is the first indication that the size of the oil spray diameter had increased after the O.S.C. had been performing at its optimum capability. The circled second indication of the O.S.C failure was at the 9000 mile point with an increase in the size of the oil spray. Subsequently the circled third indication of the O.S.C failure was at the 10000 mile point with a further increase in the size of the oil spray diameter. It was at the third indication highlighting 3 consecutive increases in size of the oil spray diameter that triggered the replacement of the failing O.S.C. This method could be considered to be more reliable than the alert value method, as a single anomalous result would not result in the immediate replacement of an O.S.C. performing at an acceptable level. However, this method also has its weakness. Potentially an O.S.C could be replaced if it generated 3 consecutive blot readings showing increasing oil spray diameter sizes. This is despite the increases being comparatively small indicating that the O.S.C was still performing at an acceptable level.

5 Conclusion

The authors have shown the potential effectiveness of implementing a RCM approach in maintaining equipment reliability for the public road mass transportation industry. However the challenges to its implementation are the time and resource implications which equates to additional capital expenditure. With regards to the blot tests, it would take considerable amount of resources to set up, test and record the result of every vehicle at intervals of 1000 miles. Then there is also the added financial implication to the organisation when these vehicles have to be removed from service every 1000 miles. A positive option would be to incorporate the blot test into the interim vehicle maintenance service interval. This would negate the requirement for these vehicles to be removed from service specifically for the blot test.

References

1. Marquez Garcia F. P., Schmid F. and Collado J. C., A reliability centered approach to remote condition monitoring, Reliability Engineering and System Safety, 80 (1) 33-40 (2003).
2. Dummer, G., Tooley, M., Winton, R., *An elementary guide to reliability*, Butterworth – Heinemann, Oxford, UK, Chap. 2, (1997)
3. Desphande VS, Modak JP, Application of RCM to a medium scale industry. Reliability Engineering and System Safety, 77, pp 31-43, (2002).
4. International Atomic Energy Agency (IAEA), *Application of Reliability Centred Maintenance to Optimize Operation and Maintenance in Nuclear Power Plants*, (May 2007).
5. Bae Chulho, Kim Hyunjun, Son Youngtak, Lee Hoyong, Han Seokyoun, Suh Myungwon, Development of a Web based RCM system for the driverless Rubber-Tired K-AGT system, Journal of Mechanical Science and Technology 23, pp 1142-1156 (2009)
6. Selvik, JT, Aven, T, A framework for reliability and risk centered maintenance, Reliability Engineering and System Safety, 96, pp 324-331, (2011)
7. O'Connor, P., *Practical Reliability Engineering* (third edition revised), John Wiley and sons, Chichester, Chap 14 (1997)
8. Moubray, John, Reliability-centered Maintenance, 2nd Edition, Butterworth Heinemann (1997)
9. Zeljko Mausic, Borivoj Galovic, Omer Pita, Optimizing maintenance reliability program for small fleets, Transport, 22:3, 174-177, (2007).
10. Kiazim, Evran, *MSc Dissertation*, Northumbria University (2014)

AUTHOR INDEX

AUTHOR INDEX